DEVELOPMENTAL INSTABILITY

Contemporary Issues in Genetics and Evolution

VOLUME 2

The titles published in this series are listed at the end of this volume.

Developmental Instability: Its Origins and Evolutionary Implications

Proceedings of the International Conference on Developmental Instability:
Its Origins and Evolutionary Implications,
Tempe, Arizona, 14–15 June 1993

Edited by
THERESE ANN MARKOW

Contributions with an asterisk in the table of contents were first published in *Genetica*, Volume 89 (1993)

Springer Science+Business Media, B.V.

Library of Congress Cataloging-in-Publication Data

Developmental instability : its origins and evolutionary implications
/ edited by Therese Ann Markow.
p. cm. -- (Contemporary issues in genetics and evolution ; v.
2)
Some of "these papers appeared in a special issue of Genetica and
served as the framework for a conference held in Tempe, Arizona in
June 1993"--Editorial overview.
ISBN 978-94-010-4357-1 ISBN 978-94-011-0830-0 (eBook)
DOI 10.1007/978-94-011-0830-0
1. Developmental genetics. I. Markow, Therese Ann. II. Series.
QH453.D49 1994
575.2--dc20 94-1934

ISBN 978-94-010-4357-1

Printed on acid-free paper

Cover illustration by **Scott Pitnick and Michal Polak**

Contents

* Contributions indicated with an asterisk were first published in *Genetica*, Volume 89 (1993).

T. A. Markow (ed.), Developmental Instability: Its Origins and Evolutionary Implications, 1-3, 1994.

Editorial overview

This volume is composed of two parts. The first is a collection of prepared contributions. These papers appeared in a special issue of Genetica and served as the framework for a conference held in Tempe, Arizona in June 1993. Transcripts from that conference, along with a primer, form the second part of this book.

Prior to the conference, participants received copies of the manuscripts so that issues of interest could be identified for discussion. Thus the conference did not consist of oral renditions of the manuscrips, but instead, discussions, often lively, of the their contents. Papers and subsequent discussions at the conference focussed upon a number of themes, all related to the questions of origins and evolutionary implications of developmental instability, or DI.

Fluctuating asymmetry, or FA, is a term often used interchangeably with developmental instability. This is unfortunate, because, as brought out during the conference, it confuses pattern with process, and in the case of DI, the process is unknown. Another problem with using FA to refer to developmental instability is that it is not the only meaningful or useful measure. Several chapters in the book deal with lesser known and sometimes novel approaches for thinking about and measuring developmental instability. The primer by A. R. Palmer represents the latest methodological synthesis for treating departures from symmetry, while chapters by W. C. Freeman, J. Graham and J. Emlen, and by E. Hoyme, present alternative approaches for studies in plants and humans, respectively. J. Emlen, W. C. Freeman, and J. Graham make an interesting case for the use of fractal analysis in the evaluation of morphogenetic patterns. These same authors, in a different chapter (Graham, Freeman and Emlen) present their rationale for considering AS (antisymmetry) and DA (directional asymmetry), in addition to FA, as potential indicators of developmental instability.

The role of genetic factors in developmental homeostasis is addressed in a number of contributions. Assessing the influence of relative levels of heterozygosity compared to homozygosity on development is the theme of papers by F. R. Leary, F. Allendorf and K. Knudsen, by J. Mitton, by C. Woolf, and by G. Livshits and P. Smouse. However, these studies are divided as to whether heterozygosity increases developmental stability. It is clear that the need to identify reasons underlying their contrasting findings will fuel additional work. However, this work must be properly designed in order to descriminate among potential explanations for correlations between genotypes and phenotypes. E. Zouros clarifies the basis for statistical tests of the associative overdominance hypothesis for correlations observed between heterozygosity and phenotypic traits. Chapters by G. Clarke and by J. McKenzie and K. O'Farrell provide support for the alternative view that coadaptation rather than overdominance underlies developmental homeostasis.

The developmental processes leading to what we perceive as unstable phenotypes are not well understood. J. Emlen, W. C. Freeman and J. Graham discuss mechanisms by which random perturbations in biosynthetic pathways can produce growth disturbances which result in morphogenetic abnormalities. Their thesis is that because enzyme kinetics are typically nonlinear, feedback mechanisms can spiral out of control and lead to abnormal developmental phenotypes. Whatever the mechanism, it is clear from the paper by A. Chippindale and A. R. Palmer, that in some organisms such as branchyuran crabs, animals are not always able to compensate for early disturbances in development, as symmetry departures persist through successive molts. A curious observation, made by a number of investigators is the lack of correlations between FA estimates for different characters in the same individual. This situation is nicely exemplified in the chapter by L. Leamy, in which very large samples were required to detect an individual asymmetry parameter (IAP) for a series of characters of the mouse mandibile.

Several of the contributions deal with the evolutionary potential of developmentally unstable morphological variants. P. Parsons reminds us that bizarre phenotypes arising in more central, less extreme habitats, are more likely to survive than those arising in harsh, marginal environments. P. Parsons, and also J. Kieser, suggest that environmental shifts could then result in punctuated change over time. The assimilation of phenodeviants is the subject of the paper of V. Zakharov. However, A. R. Palmer, A. Chippindale and C. Strobeck make the point that with respect to asymmetries, evolutionary potential will be a function of genetic architecture underlying the traits. These ideas are explored in the paper by J. Collins, M. Sredl and K. Zerba, in their effort to understand the processes leading to the evolution of new character states in tiger salamanders.

Somewhat problematic is the question of how to envision the action of selection on these phenotypes. T. Markow and I. Gottesman advance the concept of behavioral phenodeviance as the level at which selection, either natural or sexual, acts on these phenotypes. Behavioral phenodeviance is described as the behavioral manifestation of developmental instability, translated into subtle behavioral differences which render individuals less fit. Several contributors discuss specific characters and model systems in this context. R. Yeo and S. Gangested test the hypothesis that deviations from expected hand preference in man are a function of developmental instability. Evidence is reviewed by A. Moller and A. Pomiankowski that wing and tail asymmetries are important predictors of mating success in barn swallows. Another and very useful model system for directly studying the relationship between the causes of developmental instability and the evolutionary outcome is described in the chapter by M. Polak. The cactophilic *Drosophila nigrospiracula* has an endoparasite (nematode) and an ectoparasite (mite), both of which can be manipulated, along with the fly genotype to test a variety of hypotheses.

Discussion at the conference revealed a number of issues requiring clarification in future studies. First priority should be given to the identification of the mechanisms underlying developmental instability, especially when manifest as FA. We need to elucidate the responsible mechanisms in order to construct viable hypotheses about factors which can increase its expression and about its evolutionary implications. While much of the contention has focussed upon whether heterozygosity or coadaptation is important for stability, my own view is that it is not productive to think about mechanisms in mututally exclusive terms. The possibility exists that there are multiple mechanisms by which FA may be generated and that it may, in some cases, even be heritable. While studies are usually done to descriminate among mechanisms, these studies should be done with an open mind. Because of implications for human health, some of the studies should target human populations. For example, as F. C. Fraser points out, the evidence supporting a relationship between FA and inbreeding in man is rather weak.

It is also clear that traditional approaches of examining genetic contributions by looking at polymorphisms in structural genes may be missing an important role of regulatory genes. Thom Kaufman, in his presentation, provided a developmental genetic framework within which future investigations could address the question of mechanisms. There was also the feeling by participants that much of the data in the classical and widely cited literature should be reevaluated in light of recent statistical methodology. Furthermore, to address much of the criticism of earlier work, as well as to advance a complete picture of when developmental instability can be expected to occur, negative data should be reported. New empirical studies should ideally look at the influence of choice of trait (ie. merisitic or morphometric) on the outcome of the study. In order for this field to advance to the next level of understanding, new investigators should carefully read and think about the discussions presented in the transcripts of this conference.

Acknowledgement

This volume would never be in existence if it were not for the generous advice, assistance, and participation of a number of people. John McDonald provided tremendously helpful direction in organizing the volume as well as the conference. Without the tireless efforts of Patricia Sanchez, neither the volume nor the conference would ever have even gotten off the ground. Scott Pitnick was responsible for audiorecordings of the conference sessions and Jennifer Ashley for their transcription. I would also like to thank people who have been important in developing and shaping my interest in the topic of developmental instability. These include my advisor, Charles Woolf, who told me about crooked toed chickens when I was still a graduate student, and my own wonderful students, Chris Armstrong, Chris Breitmeyer, Scott Pitnick, Mike Polak, Rhonda Snook, Jennie Wakefield, and Albert Yuan Yue, who are all constant sources of stimulation and excitement to me. I thank all of these students as well as Ruby Jo Sheffer and Philip Miller for their assistance with preparations for the conference. The conference was supported by the National Science Foundation, The National Institutes of Health, Genetrix, Inc., Department of Zoology, Arizona State University, and Kluwer Academic Publishers. Views expressed by contributors to this volume should not be taken as representing the views of the National Science Foundation or National Institutes of Health.

Therese Ann Markow
Tempe, Arizona

T. A. Markow (ed.), Developmental Instability: Its Origins and Evolutionary Implications, 5–15, 1994.

Null alleles at two lactate dehydrogenase loci in rainbow trout are associated with decreased developmental stability

Robb F. Leary, Fred W. Allendorf & Kathy L. Knudsen
Division of Biological Sciences, University of Montana, Missoula, MT 59812, USA

Received 15 April 1993 Accepted 5 May 1993

Key words: bilateral asymmetry, null allele, gene duplication, polyploidy, salmonid

Abstract

Previous studies with rainbow trout (*Oncorhynchus mykiss*) have shown that allozymic heterozygotes have increased developmental stability, as measured by reduced fluctuating bilateral asymmetry. In this paper, we examine the phenotypic effects of null alleles at two lactate dehydrogenase (LDH) loci. If the association between allozymic heterozygosity and developmental stability is due largely to linked chromosomal segments, then we would expect null allele heterozygotes to have increased developmental stability. In contrast, heterozygotes for LDH null alleles in three populations have reduced developmental stability. This suggests that the reduction in enzyme activity at these loci is having a deleterious effect on development that is strong enough to mask any beneficial effects that may be associated with heterozygosity for these chromosomal segments. The LDH loci examined in this study are members of two different paralogous pairs of duplicate genes produced by the polyploidization of the ancestral salmonid genome. The apparent deleterious effects of these null alleles in heterozygotes could retard the possible loss of duplicate gene expression.

Introduction

Associations between allozymic variation and developmental stability, as measured by reduced fluctuating asymmetry of bilateral characters, have been described in many organisms (reviewed by Mitton & Grant, 1984; Allendorf & Leary, 1986; Palmer & Strobeck, 1986; Zouros & Foltz, 1987). In most cases, it is unknown whether variation at the loci examined influences developmental stability or variation at loci in linkage disequilibrium with those examined affects developmental stability.

We have previously described significant positive associations between allozymic heterozygosity and developmental stability in hatchery and natural populations of rainbow trout, *Oncorhynchus mykiss* (Leary, Allendorf & Knudsen, 1984a; 1992). Here we examine the potential effects of inactive or null alleles at two lactate dehydrogenase (LDH; EC 1.1.1.27) loci on developmental stability of rainbow trout. LDH plays an important role in the regulation of energy metabolism during rainbow trout development (Boulekbache, 1981) which is dependent upon the degradation of endogenous compounds (glycogen and lipids) stored in the egg yolk. We expected individuals heterozygous or homozygous for the null alleles to have reduced LDH concentration. This in turn could disrupt the regulation of energy production during development resulting in reduced developmental stability of both genotypes.

There is some indirect evidence that suggests that LDH null alleles may be associated with reduced developmental stability. Ferguson *et al.* (1988) have provided evidence for reduced developmental rate of null allele heterozygotes and homozygotes at one of the loci we studied. Other data have indicated that rainbow trout with increased developmental rate tend to have increased developmental stability (Leary, Allendorf & Knudsen, 1984b; Danzmann *et al.*, 1986; Leary, R. F., unpublished results).

Although we predicted the LDH null alleles to be

Table 1. Amount of tissue-specific LDH isozyme activity in rainbow trout with the common genotype (*100/100*) at all five loci. ++ = strong activity; + = weak activity; 0 = no detectable activity.

Locus	Tissues						
	Muscle	Liver	Eye	Heart	Brain	Stomach	Kidney
LDH-A1	++	0	0	0	0	0	0
LDH-A2	++	0	0	0	0	0	0
LDH-B1	+	0	++	++	++	++	+
LDH-B2	+	++	++	+	++	++	++
LDH-C	0	0	++	0	0	0	0

associated with reduced developmental stability, it is possible that this deleterious effect could be masked by other factors. For example, if the null alleles are in linkage disequilibrium with other loci having a heterotic effect on developmental stability, then null allele heterozygotes could be developmentally as stable as active allele homozygotes and only null allele homozygotes would display reduced developmental stability. This would occur when the advantageous effects of chromosomal heterozygosity counter-balance the deleterious effects of null alleles in heterozygotes. Another possible outcome is that null allele heterozygotes are developmentally more stable than either homozygote. This would occur when advantageous effects of chromosomal heterozygosity outweigh the potential deleterious effects of null alleles. We considered this outcome to be the least likely, since previous studies had generally failed to demonstrate strong heterotic effects associated with individual allozyme loci (Leary, Allendorf & Knudsen, 1984a).

Fishes of the family Salmonidae (trout, Pacific salmon, char, whitefish, and grayling) are derived from an ancient autotetraploid ancestor (Ohno, 1970). This is supported by the increased amounts of DNA, greater number of chromosome arms, and the number of isozyme loci coding for particular enzymes in these fishes (Allendorf & Thorgaard, 1984). Lactate dehydrogenase is encoded by at least five loci in rainbow trout (Allendorf & Thorgaard, 1984; Table 1). *LDH-A1* and *LDH-A2* are expressed only in skeletal muscle and are paralogous genes that are orthologous to the *LDH-A* locus of other vertebrates (Bailey & Wilson, 1968; Markert, Shaklee & Whitt, 1975). Similarly, the *LDH-B1* and *LDH-B2* loci are paralogous duplicates arising from the orthologous *LDH-B* locus found in many groups of fish (Wright, Heckman & Atherton, 1975; Markert, Shaklee & Whitt, 1975). *LDH-C* is the only duplicate of the *LDH-C* locus (Markert, Shaklee & Whitt, 1975) that can still be detected. We have previously published (Allendorf, Stahl & Ryman, 1984) an electropherogram showing this pattern of tissue-specific expression.

Materials and methods

Putative null alles were detected from electropherograms at *LDH-B1* in the Erwin strain, and at *LDH-A1* in the Spokane strain and in a naturally reproducing population from the Jewel Basin Lakes, Montana (Table 2). Electrophoresis was conducted using the buffer system and stain recommended by Allendorf *et al.* (1977); nomenclature of loci and alleles follows Shaklee *et al.* (1990), with the exception that null alleles are designated by *n* rather than by *Q0*.

The Erwin strain is a hatchery population maintained by the United States Fish and Wildlife Service (USFWS) at the Ennis National Fish Hatchery, Ennis, Montana. In 1983, the fertilized eggs from 252 full-sib families were labeled and placed in separate compartments of Heath incubators. The genotype of each parent at *LDH-B1* was determined subsequently by horizontal starch gel electrophoresis of vitreous humor homogenates. We chose 14 of these families for study: 12 were active allele homozygotes, *LDH-B1*100/100*, crossed with null allele heterozygotes, *LDH-B1*100/n*; one was a cross between two heterozygotes; and the final one was a cross between a heterozygote and null allele homozygote, *LDH-B1*n/n*.

These families were raised in a partitioned trough at the Creston National Fish Hatchery, USFWS, Creston, Montana. We had planned to compare developmental stability among full sibs with different *LDH-B1* genotypes. Analysis of several isozyme loci, however, revealed that progeny phenotypic ratios at *LDH-B1* and other loci did not conform to expected Mendelian ratios in most families. Family data from this and other strains of rainbow trout indicate that the electromorphs examined in the Erwin strain segregate in Mendelian

Table 2. Observed (and expected) genotypes at two LDH loci in three populations of rainbow trout.

Locus	Population	Genotype			Freq (*n*)
		100/100	*100/n*	*n/n*	
LDH-A1	Jewel Basin	39 (36.8)	39 (43.4)	15 (12.8)	0.371
	Spokane-A	48 (47.2)	4 (5.7)	1 (0.2)	0.057
	Spokane-B	172 (173.2)	32 (29.5)	0 (1.2)	0.078
LDH-B1	Erwin	390 (380.4)	87 (106.2)	17 (7.4)	0.122***

*** = $P < 0.001$; test for binomial (Hardy-Weinberg) genotypic proportions.

ratios (Ferguson *et al.*, 1988). In all cases, the deviation from Mendelian ratios in the Erwin families was due to an excess of phenotypes that were most frequent in the strain. This suggests that inadvertent mixing of eggs had occurred. We therefore sampled individuals from each compartment and combined them into a single sample as described below.

Individuals (50 of each genotype) were collected when they were one year old. We have previously shown that the counts of the meristic characters used in this study are determined by this age in rainbow trout (Leary, Allendorf & Knudsen, 1984b). In order to reduce potential effects due to environmental and genetic differences between compartments on meristic counts, the first three or four individuals identified as being *LDH-B1*100* homozygotes and null allele heterozygotes were selected for analysis from each compartment. Whether three or four individuals of each genotype were analyzed from a compartment was determined using random numbers. As expected based upon parental genotypes, individuals homozygous for *LDH-B1*n* were detected in only two compartments and the first 25 identified in each were chosen for meristic analysis.

The Spokane strain is a hatchery population of rainbow trout maintained by the Washington State Department of Game at the Spokane State Trout Hatchery, Spokane, Washington. A sample (Spokane-A) of 53 fish was taken from this strain on 5 July 1981 for electrophoretic analysis. On 3 October 1983, 204 one-year-old fish were sampled from the population for LDH activity measurements and morphological analysis (Spokane-B). *LDH-A1* genotypes of these individuals were determined by electrophoresis of skeletal muscle homogenates. Meristic counts were taken in Spokane-B from the first 45 individuals identified as being *LDH-A1*100* homozygotes and the 32 heterozygous individuals in the sample. This sample contained no individuals homozygous for *LDH-A1*n*. Meristic counts were not taken from Spokane-A fish because the presence of only four heterozygous individuals precluded meaningful comparisons.

The Jewel Basin Lakes are four small, interconnected lakes in northwestern Montana south of Glacier National Park. Rainbow trout are not native to this region (Allendorf *et al.*, 1980). These lakes were probably originally barren of fish and this population is believed to have been established in the late 1920s by human introductions. On 4 September 1986, 93 fish were collected from the lakes and the genotype of each at *LDH-A1* was determined. Meristic counts were taken from all of the fish.

Developmental stability

The counts of the following bilateral meristic chatacters, all of which demonstrate fluctuating asymmetry in rainbow trout (Leary, Allendorf & Knudsen, 1984a), were taken on the left and right side of each fish: rays in the pectoral fins, rays in the pelvic fins, gillrakers on the upper first branchial arches, gillrakers on the lower first branchial arches, and mandibular pores. Developmental stability was measured in two ways: the number of asymmetric characters per individual and the magnitude of asymmetry, computed as the absolute value of the left minus the right count summed over all the characters on an individual. In the analysis of the data, both parameters yield qualitatively the same result, so we restrict our attention to the former measure.

Enzyme activity
The amount of LDH activity in various tissues of individuals with different genotypes was estimated with a Beckman DU-8 spectrophotometer with the assay procedure described by Shaklee *et al.* (1977). The amount of LDH activity in brain, heart, liver, and skeletal muscle was measured in ten individuals homozygous for *LDH-B1*100*, ten heterozygous individuals, and three *LDH-B1*n* homozygotes. These fish were identified in a random sample from the Erwin population the year before the fish from this strain used for meristic analysis were created. The amount of LDH activity in skeletal muscle was determined similarly in 15 individuals of each *LDH-A1* genotype detected in the Spokane-B and Jewel Basin Lakes samples. Comparisons are based on the relative amount of LDH activity per unit weight of tissue with the mean of the *100/100* genotype standardized to 1.00 in each tissue.

Results

Null alleles
There are two possible interpretations of the electrophoretic variation at *LDH-A1* (Fig. 1). The variant allele may encode a polypeptide with the same electrophoretic mobility as that encoded by the common allele at *LDH-A2*, or it could be a null (i.e., enzymatically inactive) allele. An analogous situation occurs at *LDH-B1*; this polymorphism was first described and shown to have a Mendelian basis in rainbow trout by Wright, Heckman and Atherton (1975).

We measured LDH activities of fish with different genotypes at both loci to distinguish between these hypotheses. Under the comigration hypothesis, we expected all genotypes to have the same amount of LDH activity in all tissues where the locus is strongly expressed. Under the null allele hypothesis, we expected the genotypes with the null allele to have less LDH activity in such tissues. The measurements of LDH activity indicate that the variation detected at *LDH-A1* and *LDH-B1* is caused by null alleles (Table 3).

The amount of LDH in brain and heart, where *LDH-B1* is strongly expressed, varies significantly among the *LDH-B1* genotypes (analysis of vari-

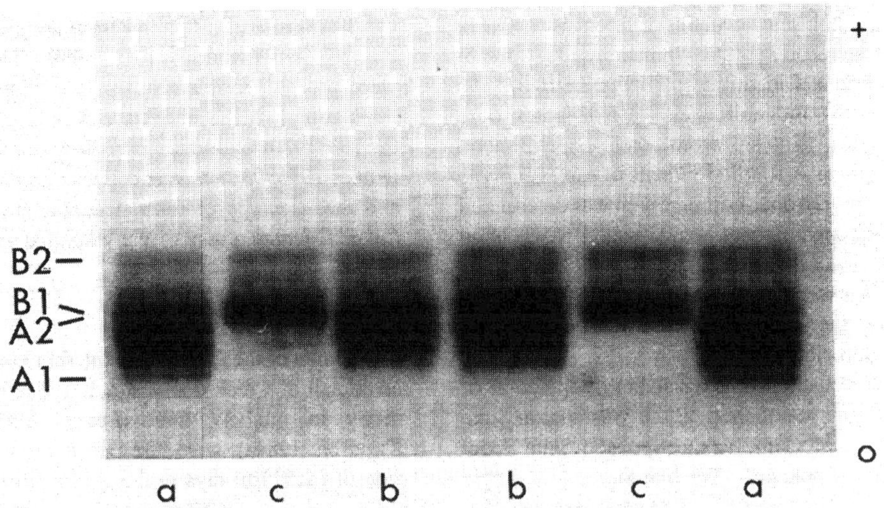

Fig. 1. Electrophoretic variation at *LDH-A1*: a = *100/100*, b = *100/n*, c = *n/n*. The lines on the left designate the location of the homotetramers produced by the common alleles of the four LDH loci expressed in muscle.

Table 3. Relative lactate dehydrogenase activities, given as mean (SE), in tissues of individuals with and without null alleles.

		Genotype			
Locus/Sample	Tissue	*100/100*	*100/n*	*n/n*	Prob.
LDH-A1/Jewel Basin	Muscle	100.0 (11.6)	95.9 (6.8)	64.0 (6.6)	**
LDH-A1/Spokane-B	Muscle	100.0 (7.2)	78.7 (4.9)	-	*
LDH-B1/Erwin	Muscle	100.0 (10.0)	109.6 (12.0)	119.4 (11.8)	NS
	Liver	100.0 (5.1)	95.3 (5.7)	84.6 (18.5)	NS
	Heart	100.0 (4.8)	63.1 (3.0)	30.1 (2.8)	**
	Brain	100.0 (4.7)	85.1 (4.0)	69.5 (5.4)	**

* $P < 0.05$; ** $P < 0.01$.

ance; $P < 0.01$; Table 3). Individuals homozygous for the null allele have only 30% of the amount of LDH activity in heart and 70% of the LDH activity in brain as individuals homozygous for *LDH-B1*100*. Heterozygous individuals have intermediate levels of LDH in these tissues. The difference in the percent reduction of LDH between brain and heart reflects the relatively greater predominance of *LDH-B1* in heart compared to brain. Further support for the contention that this is a null allele is provided by the observation that LDH activities among *LDH-B1* genotypes does not differ in liver and muscle, where the locus is only weakly or not expressed.

The amount of LDH in skeletal muscle significantly varies in the direction predicted of a null allele among the *LDH-A1* genotypes in the Jewel Basin Lakes sample ($P < 0.01$) and between the genotypes in the Spokane sample ($P < 0.05$; Table 3). Individuals homozygous for *LDH-A1*n* have 64% of the LDH as *LDH-A1*100* homozygotes in the Jewel Basin Lakes sample. Heterozygotes have 96% of the LDH present in *LDH-A1*100* homozygotes in this sample and 79% of the LDH present in active allele homozygotes in the Spokane sample.

Morphology

There is litte indication of differences in the mean total counts (left + right) of the five meristic characters among genotypes in the populations (Table 4). Moreover, there is no trend for individuals with null alleles to have greater or lesser total counts.

The only statistically significant difference (Kruskall-Wallis test, $P < 0.001$) out of 15 comparisons is for pectoral fin rays in the Erwin strain among *LDH-B1* genotypes. This difference is caused by lower counts of the *n/n* homozygotes; the mean counts of the *100/100* and *100/n* genotypes are not significantly different. We believe that this difference results from genetic differences among families, rather than being an effect of *LDH-B1* genotypes. All *n/n* homozygotes came from only two of the 14 compartments contributing to this sample. This trait has a high heritability (Leary, Allendorf & Knudsen, 1985a), and we therefore expect large genetically determined differences in this trait among families. We counted pectoral rays from 25 heterozygotes in one of the two compartments containing *n/n* homozygotes. There is no difference between the counts of heterozygotes (27.84) and *n/n* homozygotes (27.92) sampled from this compartment. This result is in agreement with the hypothesis that the difference in pectoral ray counts results from genetic differences among families rather than being an effect of *LDH-B1* genotypes.

Developmental stability

Bilateral characters of an organism are often asymmetric: that is, they differ in size, shape, or number. Fluctuating asymmetry occurs when symmetry is the normal state of expression and there is no ten-

Table 4. Mean total meristic counts of five traits for LDH genotypes in three populations of rainbow trout.

Locus/Sample Trait	Genotype			Prob.
	100/100	*100/n*	*n/n*	
LDH-A1/Jewel Basin				
Lower gillrakers	22.03	22.33	21.93	NS
Mandibular pores	15.49	16.35	16.07	NS
Pectoral fin rays	29.35	29.54	29.33	NS
Pelvic fin rays	19.43	19.30	19.40	NS
Upper gillrakers	16.38	16.79	16.33	NS
LDH-A1/Spokane-B				
Lower gillrakers	23.80	23.25	–	NS
Mandibular pores	16.51	16.38	–	NS
Pectoral fin rays	29.62	29.59	–	NS
Pelvic fin rays	20.49	20.66	–	NS
Upper gillrakers	18.44	18.28	–	NS
LDH-B1/Erwin				
Lower gillrakers	21.42	21.46	21.24	NS
Mandibular pores	13.12	13.10	13.16	NS
Pectoral fin rays	29.06	28.68	27.88	***
Pelvic fin rays	20.12	19.90	19.78	NS
Upper gillrakers	17.30	17.36	17.46	NS

*** $P < 0.001$.

dency for the character on one side of individuals to have a larger value than on the other side (Van Valen, 1962). Fluctuating asymmetry reflects a lack of precision in the control of development between the two sides of an individual and, therefore, is considered to provide a convenient measure of developmental stability (Van Valen, 1962; Soulé, 1979; Leary, Allendorf & Knudsen, 1984a; 1985b). Increased fluctuating asymmetry indicates reduced developmental stability.

The levels of fluctuating asymmetry exhibited by the five bilateral meristic characters indicate that individuals with *LDH-B1*n* alleles have reduced developmental stability. Fish heterozygous or homozygous for *LDH-B1*n* have a significantly higher average number of asymmetric characters per individual than *LDH-B1*100* homozygotes (Table 5, Fig. 2). There is no significant difference, however, in the amount of fluctuating asymmetry exhibited by the two null allele genotypes. These and the spectrophotometric results suggest *LDH-B1*n* is additive on effects of LDH activity but dominant for slightly deleterious effects on developmental stability, as revealed by increased fluctuating asymmetry.

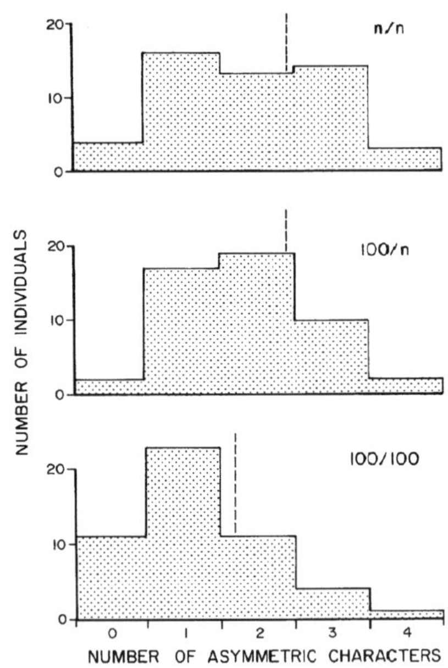

Fig. 2. Distribution of fluctuating asymmetry in rainbow trout with different *LDH-B1* genotypes. The dotted line shows the mean.

Table 5. Fluctuating asymmetry, given as mean (SE), at five bilateral meristic characters in rainbow trout with and without LDH null alleles. The average amount of asymmetry does not significantly differ (P < 0.05) between the genotypes in parentheses.

Locus/Sample	Genotype *100/100*	*100/n*	*n/n*	Wilcoxon two-sample test
LDH-A1/Jewel Basin	1.44 (0.15)	1.90 (0.16)	1.60 (0.36)	100/100 < 100/n (100/100 = n/n) (100/n = n/n)
LDH-A1/Spokane-B	1.73 (0.07)	2.03 (0.16)	-	(100/100 = 100/n)
LDH-B1/Erwin	1.22 (0.13)	1.86 (0.13)	1.92 (0.15)	100/100 < 100/n 100/100 < n/n (100/n = n/n)

The data from the Jewel Basin Lakes fish suggest that *LDH-A1*n* may also be dominant with respect to decreased developmental stability. The null allele heterozygotes have a significantly higher level of asymmetry than the *LDH-A1*100* homozygotes ($P < 0.05$; Table 5). The null allele homozygotes have an intermediate level of asymmetry that does not significantly differ from the amount possessed by the other genotypes (Table 5).

In the Spokane fish, the amount of asymmetry does not significantly differ between the heterozygotes and *LDH-A1*100* homozygotes (Table 5). The difference between the means, however, is in the same direction observed in the previous data sets.

Discussion

Phenotypic effects of null alleles

The association between developmental stability and the presence of LDH null alleles could be either the result of reduced LDH concentration or of the action of loci in linkage disequilibrium with the LDH loci. Previous results have shown that heterozygosity at isozyme loci in salmonid fishes is usually associated with increased developmental stability (Leary, Allendorf & Knudsen, 1984a; 1985a, b, c). This suggests that the reduced LDH concentration is responsible for the reduced developmental stability of null allele heterozygotes.

We cannot conclude that linked loci have no influence on developmental stability. Nevertheless, any increase in developmental stability caused by chromosomal heterozygosity apparently is being overshadowed by the deleterious effects of the null alleles. Thus, single loci can have detectable effects on developmental stability. We do not know how general this may be, but we do not believe that it is restricted to null alleles. We have shown that differences in tissue-specific expression of a phosphoglucomutase (EC 5.4.2.2) locus in rainbow trout are associated with differences in developmental rate and developmental stability (Allendorf, Knudsen & Leary, 1983). In addition, DiMichele, Powers and DiMichele (1986) have shown that allozymes of the *LDH-B* locus in the mummichog (*Fundulus heteroclitus*) have kinetic differences that are associated with differences in oxygen consumption and developmental rate during embryogenesis. Finally, an allele responsible for diazinon resistance in the Australian sheep blowfly, *Lucilia cuprina*, disrupts developmental stability when it is first introduced to a population (McKenzie & Clarke, 1988; Clarke & McKenzie, 1992).

The counts of meristic characters in fishes appear to be determined during certain 'critical periods' of development (Tanning, 1950; Lindsey, 1954; MacCrimmon & Kwain, 1969; Fowler, 1970; Lindsey & Harrington, 1972; Ali & Lindsey, 1974). We have not been able to ascertain from the literature the critical periods of the meristic characters we used to quantify developmental stability. Consideration of the ontogenetic expression of LDH loci in rainbow trout, however, suggests that the counts of some of the characters are not determined until after hatching. The isozymes of *LDH-B1* and *LDH-B2* are the predominant forms of LDH during embryonic development and the isozyme of *LDH-A1* is not electrophoretically detectable until shortly before hatching (Ferguson, Danzmann & Allendorf, 1985). Thus, if the counts of the meristic characters used to estimate developmental stability were determined before hatching we view it unlikely that *LDH-A1*n* alleles would disrupt the determination of the counts.

The finding that null allele homozygotes are not developmentally less stable than heterozygotes was unexpected. Our estimator of developmental stability may not be sensitive enough to detect differences between these genotypes. Fluctuating asymmetry is bounded between zero and five, which limits its ability to detect differences as development becomes less stable. In addition, the average asymmetry of *LDH-B1*n* homozygotes is based upon fish from only two families, and, as with the total meristic counts, we have found this average to be strongly influenced by genetic differences between families (Leary, Allendorf & Knudsen, 1985a).

Null alleles can be produced by a variety of genetic mechanisms (e.g., deletions and point mutations). The similar phenotypes of the heterozygotes and *n/n* homozygotes suggest that these null alleles result from point mutations. If the null alleles result from deletions larger than the chromosomal region responsible for producing LDH, then we would expect large deleterious effects in null allele homozygotes. Ferguson *et al.* (1988) have shown that there are no departures from Mendelian expectations in full-sib families segregating at the *LDH-B1* locus in the Erwin strain. This indicates that homozygosity for the null allele at this locus has no detectable effect on early survival under hatchery conditions

Frequency of null alleles

Enzymatically inactive alleles are rare in populations of diploids. Null alleles have been found to occur at a mean frequency of 0.0024 at 20 loci in two populations of *Drosophila melanogaster* (Voelker *et al.*, 1980; Langley *et al.*, 1981). Similar mean frequencies of null alleles have been found (Allendorf, Knudsen & Blake, 1982) in populations of ponderosa and red pine (0.0031 at 29 loci in *Pinus ponderosa* and 0.0028 at 27 loci in *P. resinosa*).

Null alleles are likely to be rare in diploids because of harmful phenotypic effects in homozygotes. For example, induced null mutations at *LDH-A* in the mouse (*Mus musculus*) have been found to be lethal when homozygous because of a block of glycolysis causing early postimplantation death (Merkle *et al.*, 1992). However, these authors found no evidence of harmful effects of null alleles in heterozygotes.

Haldane (1933) first suggested that one of the two loci of a paralogous pair may become nonfunctional through the drift and eventual fixation of mutations at one locus, while the other locus continued to perform the original function. As expected, null alleles are more common at the many duplicated loci in salmonids (Stoneking, May & Wright, 1981; Allendorf, Stahl & Ryman, 1984; Fujio, Tsuyuki & Sasaki, 1985).

Fluctuating asymmetry appears to be associated with reduced fitness in salmonid fishes (Leary, Allendorf & Knudsen, 1984a; 1985a, b, c). The existence of these null alleles at relatively high frequencies in these populations is somewhat enigmatic. We have previously described an *LDH-A1* null allele that is frequent and widespread in populations of brown trout (*Salmo trutta*) in Swedish lakes (Allendorf, Stahl & Ryman, 1984). We suggested then that the effects on fitness of such null alleles would vary with environmental conditions. For example, reduced amounts of LDH activity may have little metabolic effect on trout living in a hatchery or a lake. Such reductions, however, may have significant effects in trout living or making spawning migrations through fast-flowing streams and rivers that are much more energetically demanding.

Rainbow trout in hatcheries exist in a benign environment relative to natural populations. Under hatchery conditions, these null alleles may not have substantial selection differential and may behave as effectively neutral with regard to fitness. Furthermore, hatchery populations sometimes experience temporary reductions in effective population size due to founding events or management practices (Allendorf & Ryman, 1987). Thus, the high frequency of the null alleles in the Erwin and Spokane strains may be attributable to genetic drift possibly in conjunction with a founder event or subsequent bottlenecks.

The Jewel Basin Lakes population is believed to have been established by stocking hatchery fish in the late 1920s or early 1930s. These fish have a generation time of three to four years. This population, therefore, has probably existed for less than 20 generations. A founder event could account for the high frequency of *LDH-A1*n* in this population. Moreover, most natural populations of rainbow trout exist, at least during part of their life cycle, in a fluvial environment. Such an environment is expected to be more energetically taxing than the lacustrine environment of the Jewel Basin Lakes. Thus the *LDH-A1*n* allele may not have as strong a selection differential in the latter environment as it may in the former.

Li (1980) has reviewed the theoretical models that have considered the expected frequency of null alleles at duplicate loci and rate of loss of duplicate gene expression. The loss of duplicate gene expression in polyploid fishes, such as salmonids, has been a common occurrence (Allendorf, Utter & May, 1975; May, Stoneking & Wright, 1980; Li, 1980; Fujio, Tsuyuki & Sasaki, 1985). However, in a number of instances duplicate gene expression has been retained in salmonids. In these circumstances, the duplicate loci often have either not attained disomic inheritance (Allendorf & Thorgaard, 1984) or exhibit a bidirectional mode (Ferris & Whitt, 1979) of tissue-specific regulatory divergence: that is, one locus of the pair predominates or is the sole locus expressed in some tissues, while the other locus is predominately expressed in other tissues. Natural selection is expected to hinder the increase in frequency of null alleles in both of these situations (Li, 1980).

Our extensive surveys of natural populations of rainbow trout, cutthroat trout, *Oncorhynchus clarki*, and brown trout and a search of the literature supports the premise that the high frequency of the null alleles detected in the populations reported in this paper is unusual and is probably due to near-selective neutrality in certain environments. We are not aware of any reports of null alleles at the two LDH loci that are unique to a particular tissue (*LDH-B2* in liver and *LDH-C* in eye; Table 1). Furthermore, we are not aware of any non-hatchery populations in which a *LDH-B1* null allele occurs at a frequency greater than 0.05. The predominance of the *LDH-B1* enzyme in heart (Table 1) thus appears to result in selection against null alleles in natural populations. These results are in agreement with the expectation that the establishment of bidirectional regulatory divergence may preclude the loss of duplicate gene expression.

In contrast, the *LDH-A1* and *LDH-A2* loci do not exhibit tissue-specific (Table 1) or ontogenetic (Ferguson, Danzmann & Allendorf, 1985) regulatory divergence. This type of duplicate gene expression is expected to provide the greatest opportunity for the fixation of a null allele at one locus (Li, 1980). This assumes that a null allele at only one locus will not be selected against. However, the effects of *LDH-A1* null alleles on developmental stability and the exiguous state of these alleles outside lacustrine or hatchery environments suggests that they are generally selected against. This supports the hypothesis of Takahata and Maruyama (1979) that the slow rate of gene silencing in salmonids results at least partially from some selection against all genotypes containing a null allele.

Acknowledgements

The senior author is grateful to the Society for the Study of Evolution for their partial support of this work through the 1984 Dobzhansky Prize. This work was also supported by National Science Foundation grants BSR-8300039 and BSR-8600108 awarded to FWA. We are indebted to Mike Alberts of the Washington Department of Wildlife; B. Joe Huston and Kevin Sage of the Montana Department of Fish, Wildlife and Parks; Wes Orr and Tom Pruitt of the United States Fish and Wildlife Service; and H. McPherson for making samples available to us.

References

Ali, M. Y. & C. C. Lindsey, 1974. Heritable and temperature-induced meristic variation in the medaka, *Oryzias latipes*. Canad. J. Zool. 52: 959-976.

Allendorf, F. W., D. M. Espeland, D. T. Scow & S. Phelps, 1980. Coexistence of native and introduced rainbow trout in the Kootenai River drainage. Proc. Montana Acad. Sci. 39: 28-36.

Allendorf, F. W., K. L. Knudsen & G. M. Blake, 1982. Frequencies of null alleles at enzyme loci in natural populations of ponderosa and red pine. Genetics 100: 497-504.

Allendorf, F. W., K. L. Knudsen & R. F. Leary, 1983. Adaptive significance of differences in the tissue-specific expression of a phosphoglucomutase gene in rainbow trout. Proc. Nat. Acad. Sci. USA 80: 1397-1400.

Allendorf, F. W. & R. F. Leary, 1986. Heterozygosity and fitness in natural populations of animals, pp. 57-76 in Conservation Biology: The Science of Scarcity and Diversity, edited by M. Soulé. Sinauer Assoc., Sunderland, MA.

Allendorf, F. W., N. J. Mitchell, N. Ryman & G. Stahl, 1977. Isozyme loci in brown trout (*Salmo trutta*): Detection and interpretation from population data. Hereditas 86: 179-190.

Allendorf, F. W. & N. Ryman, 1987. Genetic management of hatchery stocks. pp. 141-159 in Population Genetics and Fisheries Management, edited by N. Ryman & F. M. Utter. Univ. Washington Press, Seattle, WA.

Allendorf, F. W., G. Stahl & N. Ryman, 1984. Silencing of duplicate genes: a null allele polymorphism for lactate dehydrogenase in brown trout (*Salmo trutta*). Mol. Biol. Evolution 1: 238-248.

Allendorf, F. W. & G. H. Thorgaard, 1984. Polyploidy and the evolution of salmonid fishes, pp. 1-53 in Evolutionary Genetics of Fishes, edited by B. J. Turner. Plenum, New York.

Allendorf, F. W., F. M. Utter & B. P. May, 1975. Gene duplication in the family Salmonidae: II. Detection and determination of the genetic control of populations, pp. 415-432 in Isozymes IV: Genetics and Evolution, edited by C. L. Markert. Academic Press, New York.

Bailey, G. S. & A. C. Wilson, 1968. Homologies between isoenzymes of fish and those of higher vertebrates. J. Biol. Chem. 243: 5843-5853.

Boulekbache, H., 1981. Energy metabolism in fish development. Amer. Zool. 21: 377-389.

Clarke, G. M. & J. A. McKenzie, 1992. Coadaptation, developmental stability, and fitness of insecticide resistance genotypes in the Australian sheep blowfly, *Lucilia cuprina*: a review. Acta Zoologica Fennica 191: 107-110.

Danzmann, R. G., M. M. Ferguson, F. W. Allendorf & K. L. Knudsen, 1986. Heterozygosity and developmental rate in a strain of rainbow trout (*Salmo gairdneri*). Evolution 40: 86-93.

DiMichele, L., D. A. Powers & J. A. DiMichele, 1986. Developmental and physiological consequences of genetic variation at enzyme synthesizing loci in *Fundulus heteroclitus*. Amer. Zool. 26: 201-208.

Ferguson, M. M., R. G. Danzmann & F. W. Allendorf, 1985. Developmental divergence among hatchery strains of rainbow trout (*Salmo gairdneri*). I. Pure strains. Canad. J. Gen. Cytol. 27: 289-297.

Ferguson, M. M., K. L. Knudsen, R. G. Danzmann & F. W. Allendorf, 1988. Developmental rate and viability of rainbow trout with a null allele at a lactate dehydrogenase locus. Biochemical Genetics 26: 177-189.

Ferris, S. D. & G. S. Whitt, 1979. Evolution of the differential regulation of duplicate genes after polyploidization. J. Mol. Evol. 12: 267-317.

Fowler, J. A., 1970. Control of vertebral number in teleosts - an embryological problem. Quart. Rev. Biol. 45: 148-167.

Fujio, Y., H. Tsuyuki & N. Sasaki, 1985. Loss of duplicated gene expression in Japanese char, *Salvelinus pluvius*. Tohoku J. Agricultural Research 36: 35-47.

Haldane, J. B. S., 1933. The part played by recurrent mutation in evolution. Amer. Natur. 67: 5-19.

Langley, C. H., R. A. Voelker, A. J. Leigh Brown, S. Ohnishi, B. Dickson & E. Montgomery, 1981. Null allele frequencies at allozyme loci in natural populations of *Drosophila melanogaster*. Genetics 99: 151-156.

Leary, R. F., F. W. Allendorf & K. L. Knudsen, 1984a. Superior developmental stability of enzyme heterozygotes in salmonid fishes. Amer. Natur. 124: 540-551.

Leary, R. F., F. W. Allendorf & K. L. Knudsen, 1984b. Major morphological effects of a regulatory gene: Pgml-t in rainbow trout. Mol. Biol. Evol. 1: 183-194.

Leary, R. F., F. W. Allendorf & K. L. Knudsen, 1985a. Inheritance of meristic variation and the evolution of developmental stability in rainbow trout. Evolution 39: 308-314.

Leary, R. F., F. W. Allendorf & K. L. Knudsen, 1985b. Developmental instability and high meristic counts in interspecific hybrids of salmonid fishes. Evolution 38: 1318-1326.

Leary, R. F., F. W. Allendorf & K. L. Knudsen, 1992. Genetic, environmental, and developmental causes of meristic variation in rainbow trout. Acta Zoologica Fennica 191: 79-95.

Leary, R. F., F. W. Allendorf, K. L. Knudsen & G. H. Thorgaard, 1985c. Heterozygosity and developmental stability in gynogenetic diploid and triploid rainbow trout. Heredity 54: 219-225.

Li, W.-H., 1980. Rate of gene silencing at duplicate loci: a theoretical study and interpretation of data from tetraploid fishes. Genetics 95: 237-258.

Lindsey, C. C., 1954. Temperature-controlled meristic variation in the paradise fish *Macropodus opercularis*. Canad. J. Zool. 30: 87-98.

Lindsey, C. C. & R. W. Harrington, Jr., 1972. Extreme vertebral variation induced by temperature in a homozygous clone of the self-fertilizing fish *Rivulus marmoratus*. Canad. J. Zool. 50: 733-744.

MacCrimmon, H. R. & W. Kwain, 1969. Influence of light on early development and meristic characters in rainbow trout, *Salmo gairdneri* Richardson. Canad. J. Zool. 47: 631-637.

Markert, C. L., J. B. Shaklee & G. S. Whitt, 1975. Evolution of a gene. Science 189: 102-114.

May, B., M. Stoneking & J. E. Wright, 1980. Joint segregation of biochemical loci in Salmonidae. II. Linkage associations from a hybridized *Salvelinus* genome. Genetics 95: 707-726.

McKenzie, J. A. & G. M. Clarke, 1988. Diazinon resistance, fluctuating asymmetry, and fitness in the Australian sheep blowfly, *Lucilia cuprina*. Genetics 120: 213-220.

Merkle, S., J. Favor, J. Graw, S. Hornhardt & W. Pretsch, 1992. Hereditary lactate dehydrogenase A-subunit deficiency as cause of early postimplantation death of homozygotes in *Mus musculus*. Genetics 131: 413-421.

Mitton, J. B. & M. C. Grant, 1984. Associations among protein heterozygosity, growth rate, and developmental homeostasis. Ann. Rev. Ecol. Syst. 15: 479-499.

Ohno, S., 1970. Evolution by gene duplication. Springer, New York.

Palmer, A. R. & C. Strobeck, 1986. Fluctuating asymmetry: measurement, analysis, patterns. Ann. Rev. Ecol. Syst. 17: 391-421.

Shaklee, J. B., F. W. Allendorf, D. C. Morizot & G. S. Whitt, 1990. Gene nomenclature for protein-coding loci in fish. Transactions of the American Fisheries Society 119: 2-15.

Shaklee, J. B., J. A. Christiansen, B. D. Sidell, C. L. Prosser & G. S. Whitt, 1977. Molecular aspects of temperature acclimation in fish: contributions of changes in enzyme activities and isozyme patterns to metabolic reorganization in the green sunfish. Jour. Exper. Zool. 210: 1-20.

Soulé, M. E., 1979. Heterozygosity and developmental stability: another look. Evolution 33: 396-401.

Stoneking, M., B. May & J. E. Wright, 1981. Loss of duplicate gene expression in salmonids: evidence for a null allele polymorphism at the duplicate aspartate aminotransferase loci in brook trout (*Salvelinus fontinalis*). Biochem. Genet. 19: 1063-1077.

Takahata, N. & T. Maruyama, 1979. Polymorphism and loss of duplicate gene expression: a theoretical study with application to tetraploid fish. Proc. Nat. Acad. Sci. USA 76: 4521-4525.

Tanning, V. A., 1950. Influence of the environment on number of vertebrae in teleostean fishes. Nature 165: 28.

Van Valen, L., 1962. A study of fluctuating asymmetry. Evolution 16: 125-142.

Voelker, R. A., C. H. Langley, A. J. Leigh Brown, S. Ohnishi, B. Dickson & B. Montgomery, 1980. Enzyme null alleles in natural populations of *Drosophila melanogaster*: frequencies in a North Carolina population. Proc. Nat. Acad. Sci. USA 77: 1091-1095.

Wright, J. E., J. R. Heckman & L. A. Atherton, 1975. Genetic and developmental analyses of LDH isozymes in trout, pp. 375-401 in Isozymes III: Developmental Biology, edited by C. L. Markert. Academic Press, New York.

Zouros, E. & D. W. Foltz, 1987. The use of allelic isozyme variation for the study of heterosis. Isozymes: Current Topics in Biological and Medical Research 13: 1-59.

T. A. Markow (ed.), Developmental Instability: Its Origins and Evolutionary Implications, 17–25, 1994.

The genetic basis of developmental stability. I. Relationships between stability, heterozygosity and genomic coadaptation

Geoffrey M. Clarke
CSIRO Division of Entomology, GPO Box 1700, Canberra, ACT, 2601, Australia

Received 5 April 1993 Accepted 27 April 1993

'Read not to contradict nor to believe but to weigh and consider' BACON

Many readers will be familiar with this quotation from Bacon not only because of its relevance when reading scientific literature but also as it was used as the frontispage to I. Michael Lerner's 1954 treatise entitled *Genetic Homeostasis*. It was really the publication of this book that fuelled subsequent investigations and efforts to elaborate the genetic basis of developmental stability, so in a way this quotation is doubly relevant to what follows.

Abstract

The genetic basis of developmental stability has been much debated over the last four decades. Two major hypotheses have been developed over this period, one that argues that the level of stability is a reflection of the underlying level of genomic heterozygosity and the other that stability reflects the general level of genomic coadaptation, or genic balance. In this paper I have attempted to critically review the evidence for these hypotheses from both practical and theoretical perspectives. On balance there is little convincing evidence to suggest that heterozygosity plays a significant role in the determination and maintenance of developmental stability, whereas there is considerable support for the genomic coadaptation hypothesis.

Introduction

Developmental homeostasis can be defined as the ability of an organism to withstand genetic and environmental disturbances encountered during development so as to produce a predetermined optimum phenotype (Waddington, 1942; Lerner, 1954). Such homeostasis can be viewed as having two principal components, *viz.* canalisation and developmental stability (reviewed by Zakharov, 1989). Canalisation refers to the developmental processes by which consistent phenotypes are produced under a range of genetic and environmental conditions, whereas developmental stability refers to those processes that ensure common developmental outcomes under specified conditions. Canalisation processes act to reduce the phenotypic variation associated with a particular trait that might otherwise result from genetic and environmental variability, whereas developmental stability relates to processes that reduce the phenotypic variation resulting from developmental accidents.

It is undeniable that the level of developmental stability of an organism is influenced by both genotype and environment, as evidenced by different strains displaying different levels of stability under the same environmental conditions and identical strains showing different levels of stability under different environmental conditions (see Zakharov, 1989, 1992 for reviews). The genetic basis of developmental stability has been the subject of much debate over the last fifty years among evolutionary and developmental geneticists. There are those who argue that developmental stability is affected by heterozygosity, the *heterozygosity theory*, with heterozygous individuals being developmentally more stable than their homozygous counterparts due to some inherent superior biochemical efficiency present in heterozygotes (Lerner, 1954; Soulé, 1979). Indeed, there are a number of studies that have shown a positive relationship between heterozygosity and stability (Soulé, 1979; Kat, 1982; Vrijenhoek & Lerman, 1982; Biémont, 1983; Leary, Allendorf & Knudsen, 1983, 1984). Others argue that developmental stability is a result of genomic coadaptation, or genic balance established over the evolutionary history of the organism via natural selection, the *genomic coadaptation theory* (Dobzhansky, 1950, 1970; Mather, 1953, 1973; Thoday, 1955). Studies have shown that the level of stability decreases upon disruption of coadapted gene complexes (Zakharov, 1981; Graham & Felley, 1985; Leary, Allendorf & Knudsen, 1985a, 1985b; Clarke & McKenzie, 1987, 1992; McKenzie & Clarke, 1988). However, for both the heterozygosity and genomic coadaptation theories there are examples that fail to show any such relationships with developmental stability (see Palmer & Strobeck, 1986, for review).

The ability of an organism to efficiently regulate its development to produce an optimal predetermined phenotype can be considered as selectively advantageous; thus developmental stability can be viewed as an integral component of individual fitness. Evidence for this has been demonstrated in a limited number of cases (Leary, Allendorf & Knudsen, 1984, 1985a, 1992; McKenzie & Clarke, 1988; Clarke & McKenzie, 1992). Developmental stability has been shown to have relatively low heritability, indicating low levels of additive genetic variance but appreciable levels of dominance genetic variation, thus indicating its relationship with fitness.

Measurement of developmental stability

A variety of measures have been used to assess the level of developmental stability. These include assessment of the frequency of phenodeviants, i.e. minor morphological variants from the normally expressed phenotype of a given character (Lerner, 1954; Rasmuson, 1960) and assessment of the level of within-individual variability. Such intraindividual variability can be assessed either as differences in phenotypic value between similar characters on repeated structures or as fluctuating asymmetry, i.e. non-directional differences between left and right sides for paired bilateral characters (Thoday, 1955, 1958).

The most commonly used measure is that of fluctuating asymmetry (FA). The underlying assumption of FA analysis is that the development of the two sides of a bilaterally symmetrical organism are presumably under the influence of an identical set of genes and therefore any differences between the sides must be environmental in origin and result from accidents occurring during development, a reflection of Waddington's (1942) so-called developmental 'noise'. Thus the level of FA will be a reflection of the level of efficiency of such developmental stability mechanisms. The biochemical and physiological processes underlying these homeostatic mechanisms are poorly understood, although they are thought to operate via a form of negative feedback control (Lerner, 1954; Zakharov, 1989).

The extent to which FA reflects the general level of developmental stability of an organism depends on the characters chosen for analysis. Where symmetry of a character is essential for normal functioning of an organism, such that even minor decreases in symmetry impose significant costs to fitness, such characters can be expected to be so well buffered, or canalised, that asymmetry is extremely rare. At the other extreme some extremely labile characters, that do not rely on strict symmetry for functioning, may exhibit high levels of asymmetry even under optimal conditions. Analysis of the level of FA of either of these types of characters in an effort to characterise the general level of developmental stability may be misleading. Ideally, characters should have a functional basis for symmetry and be able to withstand minor fluctuations in symmetry without concomitant significant losses in individual fitness.

Heterozygosity

It was Lerner's 1954 essay, *Genetic Homeostasis*, that first put forward comprehensively the hypothesis that developmental stability, or homeostasis, was a result of biological feedback mechanisms that are controlled by the genotype, in particular by the level of genomic heterozygosity. Put simply, the hypothesis argued that heterozygous individuals are developmentally more stable than their homozygous counterparts as they are better able to buffer development against perturbation due to some innate biochemical and physiological superiority of being heterozygous. Furthermore, under a system of stabilising selection (Schmalhausen, 1949) whereby the modal or optimum phenotypes are selected for at the expense of the extremes, populations could be expected to maintain relatively high levels of genomic heterozygosity. Much of the evidence for Lerner's hypothesis comes from artificial selection programs aimed primarily at animal improvement and assessment of the level of phenotypic variability of heterozygotes and homozygotes.

I believe there are a number of problems with the heterozygosity hypothesis on both theoretical and practical grounds. Since the publication of *Genetic Homeostasis* there have been numerous reports in the literature of studies designed to investigate any relationship between developmental stability and heterozygosity (e.g. Soulé, 1979; Vrijenhoek & Lerman, 1982; and see Mitton & Grant, 1984, and Palmer & Strobeck, 1986 for reviews). Despite the common belief by many workers in the field that there exists an abundance of evidence that shows a clear positive correlation between stability and heterozygosity, in reality conclusive examples are quite rare and are greatly outnumbered by studies that reveal no such relationship. Interestingly, there are those who believe that the evidence for such a relationship is so strong that it is possible to infer the level of heterozygosity within a population by assessing the level of fluctuating asymmetry in that population (Palmer, 1986).

In general, the superiority of heterozygotes has been determined by comparing the level of developmental stability as assessed by fluctuating asymmetry between populations displaying different levels of such heterozygosity. In nearly all cases the level of heterozygosity has been calculated as the percentage of a limited number of electrophoretically mobile enzyme systems that display heterozygosity.

The most compelling (and most cited) evidence comes from a series of studies conducted on rainbow trout, *Salmo gairdneri* (Leary, Allendorf & Knudsen, 1983; 1984, 1992; Leary *et al.*, 1985). In this case there is a significant positive correlation between heterozygosity (based on 13 polymorphic loci) and stability as measured by fluctuating asymmetry and the number of asymmetric characters per individual. In addition, this correlation was valid for both specific-locus heterozygosity (for two loci) and the general level of genomic heterozygosity. While it is virtually impossible to put forward any reasonable alternative hypothesis to explain these data, the same cannot be said for many of the other 'classic' examples.

Many of these experiments were undertaken in such a manner that it is impossible to categorically state that the level of heterozygosity *per se* was responsible for the observed level of developmental stability. For such statements to be made, experiments need to be conducted such that other changes in the genetic organisation of the populations under examination, such as disruption of genic balance and the unmasking of deleterious recessive alleles, which may also influence stability, are minimised. Unfortunately, this has rarely been the case for many of the studies that claim to show a relationship between stability and heterozygosity.

It is worth examining a number of these cases in more detail to highlight the problems in attempting to ascribe definitive relationships between stability and heterozygosity. Vrijenhoek and Lerman (1982) examined developmental stability, as measured by fluctuating asymmetry, in *Poeciliopsis monarcha*, a normally sexual diploid fish species. A cline in heterozygosity was observed from highly homozygous upstream populations to heterozygous downstream populations. This cline was matched by a cline in developmental instability with the upstream homozygous populations being more unstable than their heterozygous downstream counterparts. The high levels of homozygosity in the upstream populations were a result of founder events, as these populations lived in small pools that occasionally dried up and were later recolonised, whereas the downstream populations lived in permanent environments that could be considered closer to the optimum for survival. Thus there is considerable potential for fixation of deleterious recessive alleles in the upstream populations that may contribute to the observed levels of instability. It is also likely that the marginal habitat experienced by these populations may be stressful to normal developmental processes, thus also contributing to the high levels of asymmetry (see Zakharov, 1987, 1989, 1992; Parsons, 1990). The use of a triploid, highly heterozygous, clonal population that inhabits both habitats as the control population is fraught with danger, as the genetic constitution of such a species is likely to be considerably different to that of the diploid species and thus may not constitute a valid comparison.

Soulé's (1979) paper, in which he reanalyses morphometric data collected for island and mainland populations of the lizard *Uta stansburiana* in light of information gained on the levels of heterozygosity of these populations, led to a change in the hypothesis as to the cause of the observed patterns of asymmetry. Island populations tended to display greater levels of asymmetry than mainland populations, the level of asymmetry being correlated with the size of the island and time since colonisation. This pattern correlates well with the observed levels of heterozygosity of the populations, with the small island populations being less heterozygous than the larger island and mainland populations. The interesting point to this example is that Soulé himself first argued (1967) that the observed pattern of asymmetry could be adequately explained by founder events and breakdown of coadapted gene complexes. Similar conclusions were drawn by Sarre and Dearn (1991) in a virtually identical study (in which Soulé's original work was not cited) of island and mainland populations of another lizard species, *Trachydosaurus rugosus*.

Similarly, Kat (1982) examined the relationship between heterozygosity and stability in peripheral and central populations of the unionid bivalve, *Elliptio complanata*. Peripheral populations displayed significantly lower levels of heterozygosity and higher levels of asymmetry compared with populations located in the centre of the species range. Again although there is a clear correlation between the level of heterozygosity and developmental stability, there is no evidence to suggest that heterozygosity is the responsible agent for the lower asymmetry values in these populations. It is not possible to eliminate potential founder effects, in the form of fixation of deleterious alleles, and again environmental factors may be important. While the author acknowledges both these possibilities he chooses to ignore them in summarising the data. The potential for inbreeding, directional selection and genetic drift to influence the level of developmental stability in peripheral populations due to changes in the genetic constitution of such populations has been discussed by Levin (1970).

Wayne, Modi and O'Brien (1986) showed that the level of asymmetry was significantly greater in the genetically monomorphic cheetah, *Acinonyx jubatus*, compared with three other feline species that display significantly greater levels of allozyme heterozygosity. These results were later confirmed by Modi, Wayne and O'Brien (1987). Willig and Owen (1987), however, criticised these data on the basis of inappropriate statistical analyses and the failure of the analyses to differentiate between the different types of asymmetry, i.e. fluctuating, directional and antisymmetry (see Van Valen, 1976; Palmer & Strobeck, 1986 for definition and discussion). Subsequent analyses taking into consideration the concerns of Willig and Owen, conducted by Kieser and Groeneveld (1991), have shown that overall *A. jubatus* does not show significantly higher levels of asymmetry, and therefore reduced developmental stability, than other feline species used for comparison.

Many studies have used inbreeding programs to generate homozygous individuals for comparison with outbred heterozygotes (e.g. Robertson & Reeve, 1952; Beardmore, 1960; Angus & Shultz, 1983; Leamy, 1984, 1992; Shackell & Doyle, 1991). Inbreeding is known to cause a breakdown in coadapted gene complexes and also results in the unmasking of deleterious recessive alleles (Mather, 1973) which are known to influence the level of developmental stability in inbred populations (see further). Brückner (1976) attempted to overcome the problems associated with deleterious recessives by inbreeding the haplo-diploid honeybee, *Apis mellifera*. Due to haploidy of the males in this system, deleterious recessives should not be able to accumulate in this system, as they will be constantly exposed in the haploid part of the genome and thus be eliminated by selection. Results of this study showed that inbreeding resulted in increased asymmetry in the diploid females, hypothesised to be a result of increased homozygosity. Unfortunately, inappropriate statistical methods employed in this study and a critical arithmetical error in the analysis question the validity of these results. Attempts to repeat this study have shown no relationship between asymmetry and heterozygosity in this system (Clarke, Brand & Whitten, 1976; Clarke, Oldroyd & Hunt, 1992).

Other studies have found a strong negative correlation between heterozygosity and intrapopulation or intra-individual morphological variance as measured across a number of characters; i.e. the higher the level of heterozygosity, the lower the overall variance (Eanes, 1978; Mitton, 1978; Handford, 1980; Fleischer, Johnson & Klitz, 1983). In these studies morphological variation is taken to reflect the level of developmental stability. However, as Eanes correctly states, 'Problems certainly arise when assuming morphological variability to be a reasonable measure of developmental homeostasis.' Eanes then goes on to correctly argue that such an association between heterozygosity and morphological variance could reflect underlying patterns of genetic variation.

Perhaps the most damning of all arguments against the heterozygosity hypothesis is the number of examples which clearly demonstrate no relationship between stability and heterozygosity (e.g. Tebb & Thoday, 1954; Jinks & Mather, 1955; Thoday, 1958; Bader, 1965; McAndrew, Ward & Beardmore, 1982; Beacham & Withler, 1985; Ferguson, 1986; Wooten & Smith, 1986; Strauss, 1987; Patterson & Patton, 1990).

Theoretically, it seems incongruous that something as fundamental to the normal functioning of all organisms as developmentally stability should be controlled and maintained by a genetic system than is not fundamental to all types of organisms. For example, in haplo-diploid systems whereby one sex (usually the male) is haploid and therefore effectively 100% homozygous (the other sex being diploid and thereby able to maintain levels of heterozygosity), it would be foolish to assume that developmental stability is only possible for the diploid sex. Similar arguments apply for many parthenogenetic organisms and organisms that reproduce asexually, where the level of individual heterozygosity can be expected to be low.

While it is often argued that many populations maintain high levels of heterozygosity (see Lewontin, 1974), in reality the average values are low, with the highest values not much greater than 20% (Selander, 1976), meaning of course that 80% of loci are in fact homozygous. If the level of heterozygosity was the major factor contributing to developmental stability, would not the level of heterozygosity be expected to be considerably higher than that observed? In fact, due to the nature of techniques used to assess heterozygosity, i.e. electrophoresis, the level of heterozygosity is determined using what may be a biased subset of all loci. Such estimates may not reflect the general level of genomic heterozygosity. Also, in nearly all studies to date, the correlation between heterozygosity and stability has only been shown at the population level, as both heterozygosity and developmental stability estimates are based on population averages. Little if any attempt has been made to determine if the relationship holds at the individual level, which is critical. In addition, while the theoretical basis of selection favouring heterozygotes is well understood, in reality there are very few examples in the literature that have conclusively shown heterozygote advantage to be other than a rare occurrence. In many cases where hybrid products produced by heterozygotes have been well studied, it is often unclear what, if any, advantage the product has over a homozygous product when compared within the normally narrow operating limits experienced within most organisms (Johnson, 1976). Kimura (1983) has argued quite convincingly that it is possible to achieve and maintain the observed levels of heterozygosity in populations under a neutralist theory without the need for the heterozygotes to have any selective advantage. Furthermore many of the loci which are found to be heterozygous are not active during embryogenesis and early development and thus could play no role in buffering against developmental perturbation. As Thoday (1955) states 'We thus return to segregation as the only essential function of the heterozygous state. All other integrations of heterozygous systems must ultimately be a consequence of this.'

Genomic coadaptation

Coadaptation refers to the selection process by which harmoniously collaborating genes become accumulated in the gene pool of a population (Dobzhansky, 1951). The results of such coadaptation, so called coadapted gene complexes, involve both a relational balance of alleles between chromosomes and an internal balance of loci within chromosomes (Mather, 1973). Under a system of stabilising selection, the integrity of these complexes is maintained by a tendency to retain previously selected, well-integrated combinations of genes. Any change in the genetic constitution of the organism, either as a result of mutation or gene flow, may be at a selective disadvantage due to a breakdown of such complexes. Thus it will be eliminated from the population. Such a selection regime, after many generations, will result in uniform phenotype across a range of genotypes within populations.

Much of the evidence for the relationship between developmental stability and genomic coadaptation has come from the examination of levels of asymmetry in hybrid populations between species. In fitness terms, hybridisation has been shown to result in both increased fitness of the hybrids, so-called heterosis, and decreased fitness, or outbreeding depression. In general the explanations for such dichotomous results argue that the resulting fitness of hybrid populations is dependent on the time since divergence of the two parental populations. Artificial hybrids of species that have diverged recently in evolutionary terms may be expected to show elevated levels of fitness compared with parental strains, whereas hybrids between very distantly related species typically show outbreeding depression. Specifically this is thought to be a reflection of the potential for differentiation in patterns of genomic coadaptation since time of divergence, i.e. species recently derived from a common ancestor will still share much of their genomic organisation compared with species that diverged much earlier, thus allowing time for reorganisation and reassociation of genes and genic combinations in response to differing patterns of selection (see Hewitt, 1988). In addition, natural hybrids derived from a stable long-standing hybrid zone may display few differences compared with parental species given the potential for selection of newly coadapted gene complexes within the hybrid zone (Graham, 1992).

This pattern is consistent with that observed for developmental stability (see Graham, 1992, for review). Natural and laboratory hybrids of salmonid fishes have been observed to have significantly lower levels of developmental stability than their parental species that is matched by reduced fitness in such hybrids (Leary, Allendorf & Knudsen, 1985b). The evolutionary divergence between these species has been shown to be high (Leary, Allendorf & Knudsen, 1983). Similarly, hybrids between hatchery strains of rainbow trout also display lower developmental stability and fitness compared with parental strains (Ferguson, 1986). Graham and Felley (1985) have shown similar results in natural hybrids of sunfish, *Enneacanthus gloriosus* and *E. obesus* derived from recent regions of hybridisation. Hybrid populations displayed greater levels of asymmetry than either parental species. Similar results have been shown in *Drosophila* (Markow & Ricker, 1991), fire ants, *Solenopsis* spp. (Ross & Robertson, 1990) and mice (Bader, 1965).

In contrast Jackson (1973) and Felley (1980) using hybrid populations of lizards, *Sceloporus* spp. and sunfish, *Lepomis* spp. respectively, failed to show any differences in the level of asymmetry of the hybrids compared with parentals. Graham and Felley (1985) argue that these results reflect the fact that the hybrid zones examined in each of these cases were long-standing stable populations and thus there was considerable potential for selection of newly coadapted gene complexes in the area of hybridisation. Lamb, Novak and Mahoney (1990) found no differences in the level of developmental stability in hybrid populations of treefrogs, *Hyla* spp. although the hybrid zone was recent in origin and the parental populations showed relatively high levels of genetic divergence. They argue that these results reflect the unique nature of amphibian hybridisation potential as a result of conservation of chromosomal and/or regulatory evolution despite increasing genetic divergence.

Besides providing much support for the genomic coadaptation hypothesis, these hybridisation studies have indirectly cast doubt on the heterozygosity theory. Hybrid formation not only results in the breakdown of coadapted gene complexes but also leads to increased levels of heterozygosity in the hybrid progeny. If heterozygosity was the major responsible agent for developmental stability, then it might be expected that hybrids would show elevated levels of stability compared with the more homozygous parentals. To date there are no examples in the literature where this has been shown to be true (Graham, 1992).

Inbreeding in a normally outbreeding species also reduces the level of genomic coadaptation in inbred individuals. However, because inbreeding also increases the level of homozygosity and unmasks deleterious recessive alleles, inbreeding studies have little to offer in trying to delineate between genomic coadaptation and heterozygosity theories. In an attempt to address these problems Clarke, Oldroyd and Hunt (1992) have used a haplo-diploid system. In theory it is possible to increase the level of homozygosity in the diploid part of such a system by inbreeding without affecting the level of genomic coadaptation, which is predominantly an internal balance of genes within rather than between chromosomes, or unmasking deleterious recessive alleles. Results of this study clearly demonstrated that the level of heterozygosity is not important for the determination or maintenance of developmental stability in this system and added considerable support for the genomic coadaptation theory.

Genomic coadaptation can be disrupted in other ways besides hybridisation. Changes in the genetic constitution of an organism, either by mutation or gene flow, can be expected to cause changes in the level of genic balance. Gene flow between genetically isolated populations of sticklebacks, *Gasterosteus aculeatus*, has been shown to result in increased levels of asymmetry (Zakharov, 1981). Jokela and Portin (1991) have shown that stocks of *Drosophila* that carry an extra Y chromosome are developmentally less stable than normal strains. Sakai and Shimamoto (1965) have argued that there are specific mutations responsible for developmental instability in tobacco, *Nicotiana tabacum*.

Arguably, the most convincing evidence for the relationship between genomic coadaptation and stability comes from a series of experiments investigating the evolution of insecticide resistance in the blowfly, *Lucilia cuprina* (Clarke & McKenzie, 1987, 1992; McKenzie & Clarke, 1988; McKenzie, Batterham & Baker, 1990). The initial introduction of the resistance allele into the genome is detrimental to normal development in this species due to disruption of the genic balance, as evidenced by an increase in asymmetry and lowered fitness of resistance genotypes. Subsequent selection pressure by the insecticide led to the selection for modification of the genetic background to restore the balance and ameliorate the effects of the resistance allele, resulting in a concomitant reduction in asymmetry and enhancement of the fitness of resistance genotypes to pre-perturbation levels. In this case the background modification is due to a single gene.

Concluding remarks

Although this has not been an exhaustive review of the subject of the genetic basis of developmental stability, I believe it is an unbiased representation of the available literature. I have deliberately excluded the abundance of literature on developmental stability in human populations due to the inherent difficulties in determining the underlying genetic structure of these populations. I have tried to critically analyse the available evidence in terms of both practical and theoretical aspects of developmental stability theory.

Theoretically it might be argued that genomic coadaptation is a better candidate for providing a genetic basis for developmental stability than heterozygosity. Mather (1973) has argued the importance of genomic coadaptation and the evolution of genic balance as a necessary component of evolutionary success common for all organisms. In some cases homozygous balance will be advantageous for evolutionary survival while heterozygous balance, permitting the necessary genetic variation necessary for evolution, carries the burden of the production of relatively ill-adapted segregants (Darlington, 1939; Thoday, 1955). This is obviously true for organisms that are obligate inbreeders and have a haploid phase to their life cycle. In addition, the genomic coadaptation theory does not rely on any inherent biochemical or physiological superiority of well-adapted gene combinations, but rather on structural and epistatic interactions between genes.

The available evidence would tend to support these theoretical conjectures. While there is little conclusive evidence to show a clear relationship between heterozygosity and stability, there are numerous examples that fail to show any relationship. Of critical importance is the need to examine any such relationship at the individual level rather than at the population level. On the other hand, there is a growing body of evidence to support the genomic coadaptation theory and, as importantly, little evidence that contradicts this hypothesis.

One of the major problems in trying to delineate between these two theories, if it is possible at all, is that for the majority of experimental designs used to date it has been virtually impossible to manipulate either the level of heterozygosity or genic balance without concomitantly affecting the other. In fact, there are very few model systems where this is even possible. In addition, the use of natural populations is fraught with danger unless there is a good understanding of the genetic structure and evolutionary history of the populations under examination.

In reality, it is likely that the level of developmental stability in natural populations will reflect a balance between heterozygosity and genomic coadaptation, with that balance depending on the breeding system and genetic structure of the system under examination. These, of course, will reflect past and present evolutionary processes, thus emphasising the significance of developmental stability as an integral part of the evolutionary process.

References

Angus, R. A. & R. J. Schultz, 1983. Meristic variation in homozygous and heterozygous fish. Copeia 1983: 287-299.

Bader, R. S. 1965. Fluctuating asymmetry in the dentition of the house mouse. Growth 29: 291-300.

Beacham, T. D. & R. Withler, 1985. Heterozygosity and morphological variability of chum salmon *(Oncorhynchus keta)* in southern British Columbia. Heredity 54: 313-322.

Beardmore, J. A., 1960. Developmental stability in constant and fluctuating temperatures. Heredity 14: 411-422.

Biémont, C., 1983. Homeostasis, enzymatic heterozygosity and inbreeding depression in natural populations of *Drosophila melanogaster*. Genetica 61: 179-189.

Brückner, D., 1976. The influence of genetic variability on wing symmetry in honeybees *(Apis mellifera)*. Evolution 30: 100-108.

Clarke, G. M. & J. A. McKenzie, 1987. Developmental stability of insecticide resistant phenotypes in blowfly; a result of canalizing natural selection. Nature 325: 345-346.

Clarke, G. M. & J. A. McKenzie, 1992. Coadaptation, developmental stability and fitness of insecticide resistance genotypes in the Australian sheep blowfly, *Lucilia cuprina*: A review. Acta Zoologica Fennica 191: 107-110.

Clarke, G. M., G. W. Brand & M. J. Whitten, 1986. Fluctuating asymmetry: a technique for measuring developmental stress caused by inbreeding. Australian Journal of Biological Sciences 39: 145-153.

Clarke, G. M., B. P. Oldroyd & P. Hunt, 1992. The genetic basis of developmental stability in *Apis mellifera*: Heterozygosity versus genic balance. Evolution 46: 753-762.

Darlington, C. D., 1939. The Evolution of Genetic Systems. Cambridge University Press, Cambridge.

Dobzhansky, T., 1950. Genetics of natural populations. XIX. Origin of heterosis through natural selection in populations of *Drosophila pseudoobscura*. Genetics 35: 288-302.

Dobzhansky, T., 1951. Genetics and the Origin of Species 3rd ed. Columbia University Press, New York.

Dobzhansky, T., 1970. Genetics of the Evolutionary Process. Columbia University Press, New York.

Eanes, W. F., 1978. Morphological variance and enzyme heterozygosity in the monarch butterfly. Nature 276: 263-264.

Felley, J., 1980. Analysis of morphology and asymmetry in bluegill sunfish *(Lepomis macrochirus)* in the southeastern United States. Copeia 1980: 18-29.

Ferguson, M. M., 1986. Developmental stability of rainbow trout hybrids: genomic coadaptation or heterozygosity? Evolution 40: 323-330.

Fleischer, R. C., R. F. Johnston & W. J. Klitz, 1983. Allozymic heterozygosity and morphological variation in house sparrows. Nature 304: 628-629.

Graham, J. H. & J. D. Felley, 1985. Genomic coadaptation and developmental stability within introgressed populations of *Enneacanthus gloriosus* and *E. obesus* (Pisces, Centrarchidae). Evolution 39: 104-114.

Graham, J. H., 1992. Genomic coadaptation and developmental stability in hybrid zones. Acta Zoologica Fennica 191: 121-132.

Handford, P., 1980. Heterozygosity at enzyme loci and morphological variation. Nature 286: 261-262.

Hewitt, G. M., 1988. Hybrid zones - natural laboratories for evolutionary studies. Trends in Ecology and Evolution 3: 158-167.

Jackson, J. F., 1973. A search for the population asymmetry parameter. Systematic Zoology 22: 166-170.

Jinks, J. L. & K. Mather, 1955. Stability in development of heterozygotes and homozygotes. Proceeding of the Royal Society of London Series B 143: 561-578.

Johnson, G. B., 1976. Genetic polymorphism and enzyme function. In F. J. Ayala (ed.) Molecular Evolution pp. 46-59. Sinauer, Sunderland, MA.

Jokela, P. & P. Portin, 1991. Effect of extra Y chromosome on number and fluctuating asymmetry of sternopleural bristles in *Drosophila melanogaster*. Hereditas 114: 177-187.

Kat, P. W., 1982. The relationship between heterozygosity for enzyme loci and developmental homeostasis in peripheral populations of aquatic bivalves (Unionidae). American Naturalist 119: 824-832.

Kieser, J. A. & H. T. Groeneveld, 1991. Fluctuating odontometric asymmetry, morphological variability, and genetic monomorphism in the cheetah *Acinonyx jubatus*. Evolution 45: 1175-1183.

Kimura, M., 1983. The Neutral Theory of Molecular Evolution. Cambridge University Press, Cambridge.

Lamb, T., M. Novak & D. L. Mahoney, 1990. Morphological asymmetry and interspecific hybridisation: a case study using hylid frogs. Journal of Evolutionary Biology 3: 295-309.

Leamy, L., 1984. Morphometric studies in inbred and hybrid house mice. V. Directional and fluctuating asymmetry. American Naturalist 123: 579-593.

Leamy, L., 1992. Morphometric studies in inbred and hybred house mice. VII. Heterosis in fluctuating asymmetry at different ages. Acta Zoologica Fennica 191: 111-120.

Leary, R. F., F. W. Allendorf & K. L. Knudsen, 1983. Developmental stability and enzyme heterozygosity in rainbow trout. Nature 301: 71-72.

Leary, R. F., F. W. Allendorf & K. L. Knudsen, 1984. Superior developmental stability of heterozygotes at enzyme loci in salmonid fishes. American Naturalist 124: 540-551.

Leary, R. F., F. W. Allendorf & K. L. Knudsen, 1985a. Inheritance of meristic variation and the evolution of developmental stability in rainbow trout. Evolution 39: 308-314.

Leary, R. F., F. W. Allendorf & K. L. Knudsen, 1985b. Developmental instability and high meristic counts in interspecific hybrids of salmonid fishes. Evolution 39: 1318-1326.

Leary, R. F., F. W. Allendorf & K. L. Knudsen, 1992. Genetic, environmental and developmental causes of meristic variation in rainbow trout. Acta Zoologica Fennica 191: 79-96.

Leary, R. F., F. W. Allendorf, K. L. Knudsen & G. H. Thorgaard, 1985. Heterozygosity and developmental stability in gynogenetic diploid and triploid rainbouw trout. Heredity 54: 219-225.

Lerner, I. M., 1954. Genetic Homeostasis. Wiley, New York.

Levin, D. A., 1970. Developmental instability and evolution in peripheral populations. American Naturalist 104: 343-353.

Lewontin, R. C., 1974. The Genetic Basis of Evolutionary Change. Columbia University Press, New York.

MacAndrew, B. J., R. D. Ward & J. A. Beardmore, 1982. Lack of a relationship between morphological variance and enzyme heterozygosity in the plaice, *Pleuronectes platessa.* Heredity 48: 117-125.

Markow, T. A. & J. P. Ricker, 1991. Developmental stability in hybrids between the sibling species pair, *Drosophila melanogaster* and *Drosophila simulans.* Genetica 84: 115-121.

Mather, K., 1953. Genetical control of stability in development. Heredity 7: 297-336.

Mather, K., 1973. Genetical Structure of Populations. Chapman and Hall, London.

McKenzie, J. A. & G. M. Clarke, 1988. Diazinon resistance, fluctuating asymmetry and fitness in the Australian sheep blowfly, *Lucilia cuprina.* Genetics 120: 213-220.

McKenzie, J. A., P. Batterham & L. Baker, 1990. Fitness and asymmetry modification as an evolutionary process. A study in the Australian sheep blowfly, *Lucilia cuprina* and *Drosophila melanogaster.* In J. S. F. Barker, W. T. Starmer and R. J. MacIntyre (eds.) Ecological and Evolutionary Genetics of Drosophila pp. 57-73. Plenum, New York.

Mitton, J. B., 1978. Relationship between heterozygosity for enzyme loci and variation of morphological characters in natural populations. Nature 273: 661-662.

Mitton, J. B. & M. C. Grant, 1984. Association among protein heterozygosity, growth rate, and developmental homeostasis. Annual Review of Ecology and Systematics 15: 479-499.

Modi, W. S., R. K. Wayne & S. J. O'Brien, 1987. Analysis of fluctuating asymmetry in cheetahs. Evolution 41: 227-228.

Palmer, A. R., 1986. Inferring relative levels of genetic variability in fossils: the link between heterozygosity and fluctuating asymmetry. Palaeobiology 12: 1-5.

Palmer. A. R. & C. Strobeck, 1986. Fluctuating asymmetry: measurement, analysis, patterns. Annual Review of Ecology and Systematics 17: 392-421.

Parsons, P. A., 1990. Fluctuating asymmetry: an epigenetic measure of stress. Biological Reviews 65: 131-145.

Patterson, B. D. & J. L. Patton, 1990. Fluctuating asymmetry and allozymic heterozygosity among natural populations of pocket gophers *(Thomomys bottae).* Biological Journal of the Linnean Society 40: 21-36.

Rasmusson, M., 1960. Frequency of morphological deviations as a criterion of developmental stability. Hereditas 46: 511-536.

Robertson, F. W. & E. C. R. Reeve, 1952. Heterozygosity, environmental variation and heterosis. Nature 170: 286.

Ross, K. G. & J. L. Robertson, 1990. Developmental stability, heterozygosity, and fitness in two introduced fire ants *(Solenopsis invicta* and *S. richteri)* and their hybrid. Heredity 64: 93-103.

Sakai, K-I. & Y. Shimamoto, 1965. Developmental instability in leaves and flowers of *Nicotiana tabacum.* Genetics 51: 801-813.

Sarre, S. & J. M. Dearn, 1991. Morphological variation and fluctuating asymmetry among insular populations of the sleepy lizard, *Trachydosaurus rugosus* Gray (Squamata: Scincidae). Australian Journal of Zoology 39: 91-104.

Schmalhausen, I. I., 1949. Factors of Evolution. Blakiston, Philadelphia.

Selander, R. K., 1976. Genic variation in natural populations. In F. J. Ayala (ed.) Molecular Evolution pp. 21-45. Sinauer, Sunderland, MA.

Shackell, N. L. & R. W. Doyle, 1991. Scale morphology as an index of developmental stability and stress resistance of tilapia *(Oreochromis niloticus).* Canadian Journal of Fisheries and Aquatic Science 48: 1662-1669.

Soulé, M. E., 1967. Phenetics of natural populations. III. Asymmetry and evolution in a lizard. American Naturalist 101: 141-159.

Soulé, M. E., 1979. Heterozygosity and developmental stability: another look. Evolution 33: 396-401.

Strauss, S. H., 1987. Heterozygosity and developmental stability under inbreeding and crossbreeding in *Pinus attenuata.* Evolution 41: 331-339.

Tebb, G. & J. M. Thoday, 1954. Stability in development and relational balance of X-chromosomes in *Drosophila melanogaster.* Nature 174: 1109-1110.

Thoday, J. M., 1955. Balance, heterozygosity and developmental stability. Cold Spring Harbor Symposium on Quantitative Biology 20: 318-326.

Thoday, J. M., 1958. Homeostasis in a selection experiment. Heredity 12: 401-415.

Van Valen, L., 1962. A study of fluctuating asymmetry. Evolution 16: 125-142.

Vrijenhoek, R. C. & S. Lerman, 1982. Heterozygosity and developmental stability under sexual and asexual breeding systems. Evolution 36: 768-776.

Waddington, C. H., 1942. Canalization of development and the inheritance of acquired characters. Nature 150: 563-565.

Wayne, R. K., W. S. Modi & S. J. O'Brien, 1986. Morphological variability and asymmetry in the cheetah *(Acinonyx jubatus),* a genetically uniform species. Evolution 40: 78-85.

Willig, M. R. & R. D. Owen, 1987. Fluctuating asymmetry in the cheetah: methodological and interpretive concerns. Evolution 41: 225-227.

Wooten, M. C. & M. H. Smith, 1986. Fluctuating asymmetry and genetic variability in a natural population of *Mus musculus.* Journal of Mammology 67: 725-732.

Zakharov, V. M., 1981. Fluctuating asymmetry as an index of developmental homeostasis. Genetika 13: 241-256.

Zakharov, V. M., 1987. Animal asymmetry: a population-phenogenetic approach. (In Russian). Nauka, Moscow.

Zakharov, V. M., 1989. Future prospects for population phenogenetics. Soviet Scientific Reviews F Physiology and General Biology 4: 1-79.

Zakharov, V. M., 1992. Population phenogenetics: Analysis of developmental stability in natural populations. Acta Zoologica Fennica 191: 7-30.

T. A. Markow (ed.), Developmental Instability: Its Origins and Evolutionary Implications, 27–35, 1994.

Does homozygosity contribute to the asymmetry of common white leg markings in the Arabian horse?

Charles M. Woolf
Department of Zoology, Arizona State University, Tempe, AZ 85287, USA

Received and accepted 26 April 1993

Key words: asymmetry, homozygosity, Arabian horse, white leg markings

Abstract

Common white and facial markings have a multifactorial mode of inheritance in *Equus caballus* and result from the absence of melanocytes in the unpigmented areas. Directional asymmetry and fluctuating asymmetry apparently account for the total asymmetry of common white leg markings. Using computerized records obtained from the Arabian Horse Registry of America, Inc., and the International Arabian Horse Association, studies were carried out to determine if homozygosity increases the total asymmetry in common white leg markings by presumably promoting fluctuating asymmetry. The results were as follows: (1) Arabian horses that are symmetrical and asymmetrical for common white leg markings have similar distributions of inbreeding coefficients; (2) Arabian and half-Arabian horses have similar concordance values, in general, for specific white markings in both their forelegs and hind legs. It is concluded that homozygosity does not contribute to the total asymmetry of common white leg markings in the Arabian horse.

Introduction

Common white facial and leg markings in *Equus caballus* have a multifactorial mode of inheritance (Nebe, 1984; Woolf, 1989, 1990) and dam-foal regression analyses suggest that about two-thirds of the phenotypic variance in facial and total leg markings in the Arabian horse (Woolf, 1989, 1990) is attributable to genetic differences. Common white markings may occur when a threshold number of genes is reached, and the extent of the markings is influenced by the number of genes that is present above the threshold number (Woolf, 1989, 1990). Horses below the threshold number, and devoid of white markings, can vary as to the number of genes they possess, as shown by the contributions they make to their offspring (Woolf, 1989). Black (a/a; E/–), bay (A/–; E/–) and chestnut (A/–; e/e or a/a; e/e) Arabian horses vary for the quantitative expression of common white markings, with black horses being the least heavily marked and chestnut horses being the most heavily marked. Although polygenes (perhaps with equal and unequal additive effects) play an important role, the observed differences among black, bay and chestnut horses suggest that specific genes with major effects also influence the variation of these markings, unless the differences are due to linkage disequilibrium (See Woolf, 1991, 1992 for discussion).

Common white facial and leg markings occur at a relatively high frequency in many breeds of *Equus caballus* because of artificial selection practices by past and present breeders. Common white facial and leg markings were not present, or were very rare, in the primeval wild horse, as evidenced by their absence in Przhevalski's horse (*Equus przewalskii*) of central Asia and related equidae (Geurts, 1977). The Arabian horse is recognized as the oldest living breed handed down to modern times and the gene pools of many other modern breeds in the United States, such as the Thoroughbred horse, the Quarter horse, the American Saddle horse, and the Morgan horse, have contributions from the gene pool of the Arabian horse (Crowell, 1951). White markings have been present in the Arabian horse since ancient times, and were selected for by early Bedouins living in the deserts of Arabia, Egypt, and Persia (Forbis, 1976). The relatively high heritability value of white markings (Woolf, 1989, 1990) indicates that the early Bedouins would have realized a good response to their efforts to increase the frequency and the extent of these markings. Arabian horses with white markings are popular with many current breeders, and chestnut horses with 'flashy' white facial and leg markings are a special favorite of some fanciers.

Common white markings in *Equus caballus* are characterized by a lack of pigment in hairs and in the skin beneath the hairs, resulting from the lack of melanocytes in the hair follicles and skin. In mammals, melanocytes in skin and hair follicles are clonally derived from a fixed, small number of primordial melanoblasts in the neural crest region (Mintz, 1967, 1974). Although the failure of certain melanoblasts to survive in peripheral locations may be the explanation for common white markings in *Equus caballus*, these markings may also result from the failure of melanoblasts to migrate and proliferate fully in presumptive facial and limb tissues (Seidel & Elsden, 1987). A sharp mid-dorsal separation exists for the primordial melanoblasts in the neural crest region so that migration and clonal proliferation occur autonomously on each side of the embryo. If genes are present for common white markings in *Equus caballus* and the survival or distribution pattern of the melanoblasts is similar on each side of an embryo, symmetry will be present for these markings; the boundaries of a white facial marking will be the same on each side of the face and the leg markings on each side will be similar, even though the markings in the forelegs may be different from those in the hind legs. Even though the development of both sides of a horse is assumed to be under the influence of similar genetic and intra-uterine environmental factors, accidents or random events can influence the survival or the migration and clonal proliferation of melanoblasts in presumptive limb and facial tissues, resulting in asymmetry. These random developmental accidents resulting in phenotypic variation have been termed developmental noise (Mather, 1953; Waddington, 1942, 1957; Thoday, 1955; Mintz, 1974).

Asymmetry for common white facial and leg markings occurs frequently in *Equus caballus* and studies have shown that a directional asymmetry is present for white leg markings, with the left legs being more heavily marked than the right legs (Dreux, 1966; Woolf, 1990). This directional asymmetry, which has a genetic basis, contributes only minimally to the total asymmetry that is present. A study has not been carried out to determine if directional asymmetry also occurs for white facial markings.

It has been shown in various bilateral organisms that a positive correlation exists between homozygosity and the asymmetry of certain traits. For recent studies and discussions see Livshits and Kobyliansky (1991); Markow and Ricker (1991, 1992); Clarke, Oldroyd and Hunt (1992); Leary, Allendorf and Knudsen (1992); Palmer and Strobeck (1992); and Zakharov (1992). In these organisms homozygosity leads to the lack of developmental stability, and in the presence of environmental stress, homozygous individuals are more apt to show phenotypic variation from the norm. One commonly used measure of developmental stability is fluctuating asymmetry, defined as random departures from symmetry in either direction. Thus, highly homozygous individuals show more fluctuating asymmetry, on the average, for given traits than random individuals from large Mendelian populations. It is evident however, that developmental instability, and consequently increased fluctuating asymmetry, can occur for reasons other than homozygosity, including adverse environmental and genetic effects.

Although the correlation between developmental instability and homozygosity has been studied and discussed for many years, the biological explanation for this relationship is still unknown (Clarke, Oldroyd & Hunt, 1992, Lerner, 1954; Leamy, 1984; Leary, Allendorf & Knudsen, 1984; Thoday, 1955; Van Valen, 1962; Waddington, 1942). Waddington (1942) and Lerner (1954) proposed that Mendelian populations evolve highly integrated genetic systems that increase the likelihood that a zygote will realize its genetic potential when exposed to adverse environmental and genetic factors. Lerner (1954) championed the model that heterozygosity at specific polygenic loci is a major component of some of these genetic systems, and individuals who are multiply heterozygous are better buffered during development in the presence of perturbations. Buffering leads to an increased uniformity of phenotypic expression. In these Mendelian populations, extreme homozygous segregants may show adverse phenotypes described by Lerner as phenodeviants. Lerner emphasized that it is not homozygosity as such that is disadvantageous, but homozygosity at particular loci, and importantly, that the buffering property of heterozygotes is the result of previous selection in a given genetic background. Hence, the capacity for buffering can be reduced by decreasing the amount of heterozygosity or altering the genetic background. This concept of coadapted genetic systems conforms to the views of Dobzhansky (1950), who concluded that inversion heterozygotes in *Drosophila pseudoobscura* may show heterosis by the action of polygenic complexes which have been coadapted by natural selection. It can be conjectured, therefore, that coadapted gene complexes consist of (1) obligate levels of heterozygosity at specific polygenic loci and (2) a coadapted genetic background, with all the alleles of both components working in an integrated manner to help insure a balance of regulatory and enzyme systems needed for developmental stability. The loss of certain alleles by inbreeding or selection leading to homozygosity at certain loci and adjoining regions would reduce the buffering potential, resulting in reduced developmental stability and therefore increased fluctuating asymmetry. The presence of major mutant genes (including deleterious recessive genes uncovered by inbreeding) and chromosomal aberrations could have a similar disruptive influence on coadapted gene complexes, resulting in increased fluctuating asymmetry. The developmental instability resulting from inbreeding or selection might be corrected by outbreeding to other strains or subspecies, thus restoring essential regulatory and enzyme systems and the proper levels of these systems. It follows, however, that outbreeding *per se* may not be effective, especially if it involves individuals from far removed geographic regions (Dobzhansky, 1950). It also follows that interspecies crosses may disrupt coadapted gene complexes resulting in acute developmental instability.

A worthy question is whether homozygosity promotes fluctuating asymmetry of common white markings in the Arabian horse on top of directional asymmetry. Utilizing records obtained from the Arabian Horse Registry of America, Inc., and the International Arabian Horse Association, studies were carried out to answer this question. The role of homozygosity was studied by (1) comparing the inbreeding coefficients of Arabian horses that are symmetrical and symmetrical for common white leg markings, and (2) comparing the symmetry of common white leg markings in Arabian and half-Arabian horses. The asymmetry of white facial markings was not considered in this study because the computerized records do not yield any information on this subject.

Colors and common white leg markings in the Arabian horse. The Registry recognizes only five colors: bay, black, chestnut, gray, and roan. The genetics of these colors in the Arabian horse has been discussed elsewhere (Woolf & Swafford, 1988; Woolf, 1992). Computerized records obtained from the Arabian Horse Registry of America, Inc., Westminster, Colorado (known hereafter as the Registry) have been used to test various hypotheses concerning the genetics of white markings in the Arabian horse (Woolf, 1989, 1990, 1991, 1992). Although these records must be used cautiously in genetic studies because of their changing nature with time and the presence of certain errors, they can be used to good advantage when their characteristics are understood.

For the purposes of registration by the Registry, white leg markings are defined as follows: 1) *coronet*-white marking extending no more than an inch above the coronet band; 2) *pastern*-white marking anywhere between the coronet marking and the bottom of the fetlock joint; 3) *fetlock*-white marking extending above the pastern but below the top of the fetlock joint; 4) *sock*-white marking extending above the top of the fetlock but below the midpoint of the cannon; 5) *stocking*-white marking extending above the midpoint of the cannon. Using these descriptions of white leg markings, the amount of whiteness in each leg can be scored from 0 to 5 as follows (Woolf, 1990): 0, no white marking; 1, coronet or whiteness on the hoof; 2, pastern; 3, fetlock; 4, sock; 5, stocking. Adjectives such as partial or high are ignored; thus a leg with a partial sock is scored as if it has a sock. A hoof is always white, or partially white, if there is whiteness above it, but a hoof may be partially white, completely white, or have a heel (white marking on the heel of the hoof), without having any whiteness above it. Any whiteness on the hoof with no whiteness above it is grouped with coronet, and that leg is given a score of 1. Although these scores are artificial constructs, they have shown to be useful for testing various hypotheses of biological interest (Woolf, 1990).

Methods

Inbreeding coefficients. Sixteen sire families were used by Woolf (1990) in a study of white leg markings in the Arabian horse. A sire family consists of all the foals produced by one sire and the dam of each foal. For the present study the computerized records of the 16 sire families were scrutinized, and ignoring color, foals were selected that had a sock or stocking in at least one of the legs. Each leg was then scored for white markings. The foal was classified as symmetrical if the differences between the foreleg scores and the hind leg scores were both zero. The foal was classified as asymmetrical if the absolute difference between either the foreleg scores or the hind leg scores was at least 4. Using microfiche stud records obtained from the Registry the pedigree of each foal was traced backwards for five generations. The procedure was terminated when the pedigrees of n = 198 symmetrical and n = 198 asymmetrical foals were available, with an equal number coming from each sire family. Inbreeding coefficients were then determined for each foal and the distributions of the inbreeding coefficients were compared for central tendency.

White leg markings in Arabian and half-Arabian bay and chestnut horses. The Registry was requested to screen its registration records and to send computerized information about large samples of bay and chestnut horses with descriptions of white leg markings (Woolf, 1990). In order to facilitate the scoring procedure, a request was also made for the information to appear on computer tapes, rather than on printouts. Included on the tapes were the names, registration numbers, colors, and the descriptions of white facial and leg markings in the bay and chestnut horses.

Half-Arabian horses are registered by the International Arabian Horse Association, Denver, Colorado. Tapes were provided by this registry containing information about white facial and leg markings in large samples of bay and chestnut half-Arabian horses. One parent of the half-Arabian horses was full Arabian and the other was some other breed. Information was not available on the color or white markings present in the parents of these half-Arabian horses.

The tapes received from the two different registries were read by a program and four files were prepared giving the color and white leg markings present in the Arabian and half-Arabian horses: File 1, Arabian bay horses; File 2, Arabian chestnut horses; File 3, half-Arabian bay horses; File 4, half-Arabian chestnut horses.

The optimal situation would be to have Arabian and half-Arabian samples with genetically similar types of white leg markings. A negating factor, however, is the genetic heterogeneity of the non-Arabian parents of half-Arabian horses, with some of these parents possessing dominant genes or large numbers of polygenes producing a collection of pied patterns (such as tobiano, overo, and sabino), often called 'pinto' or 'paint' in the United States if the white areas are large enough (Geurts, 1977; Sponenberg & Beaver, 1983). From the descriptions of the white markings present in some of the half-Arabian horses, such as white legs and whiteness in the body and face, it became evident that these pied patterns were represented at a relatively high frequency in the non-Arabian parents. Certain of these pied types (sabino and overo) occur at a low frequency in Arabians but if the white areas in the body are too large, the horse cannot be registered as an Arabian. To maintain the 'purity' of the Arabian gene pool, there is often selection against these types by breeders, especially if the objective is to produce show horses.

Because the genetic mechanisms giving rise to these pied patterns are different from the polygenic system responsible for common white markings, all horses were omitted from the files described as having one or more white legs and/or whiteness in the body. Using this screening procedure, a relatively large number of half-Arabians and a very small number of Arabians were removed from the files. Because of the nature of the program used to execute this procedure, it is unknown exactly how many half-Arabians and Arabians were removed. It was realized that the screening procedure would not remove all horses with one of these pied patterns because of the phenotypic variation that is present among them. Pied horses showing a minimal expression may have whiteness in all four legs, or in a lesser number, but little or no whiteness in the body. Although the whiteness in a leg may be described in the computerized records as 'white leg', it may also be described as 'high stocking' or merely as 'stocking'. Thus, some horses with a minimal expression of a pied pattern remain in the files, but the proportion is much higher in the half-Arabian files than in the Arabian files. It is also expected that the half-Arabians possess other genotypes influencing white markings that are not present in the Arabians.

Each file was then read by a program that converted the description of the white marking (i.e., none, hoof/coronet, pastern, fetlock, sock, and stocking) in each leg into one of the scores ranging from 0 to 5, and mean scores for each leg (right foreleg, left foreleg, right hind leg, left hind leg) were calculated.

Although the means or variances of absolute differences are commonly used to study asymmetry in different populations, these statistics are not appropriate for comparing the degrees of asymmetry of white leg markings in a lightly marked horse population and a heavily marked horse population. If leg scores vary from 0 to 5 in both populations, and the leg scores have different means in the two populations, the means and variances of the absolute differences between the paired leg scores would have lower values in the lightly marked population than in the heavily marked population, even though the propensity for symmetry was similar in both populations.

With the evidence that half-Arabian horses are less heavily marked than Arabian horses (see Results), even though the half-Arabians have a much higher proportion showing a minimal expression of one of the pied markings, a method of analysis was needed that did not use absolute differences. The method chosen was to screen the forelegs and hind legs of Arabian and half-Arabian horses for a specific score and then to determine the frequency of concordance of that score in the paired legs. The concordance and discordance values for forelegs and hind legs were then compared in Arabian and half-Arabian horses by color using 2 × 2 contingency tables.

Only three of the markings, sock (score = 4), fetlock (score = 3) and pastern (score = 2), were used in this concordance study. Stocking was omitted because a higher proportion of the half-Arabians in the files have a minimal expression of one of the pied phenotypes. Both forelegs and/or both hind legs tend to be white when these patterns are present, and the inclusion of stocking in this study would introduce a statistical bias favoring concordance for stockings in half-Arabian horses. Coronet/hoof was also omitted because of the concern that reporting is incomplete for white hoof markings in the records of the Arabian horses used in this study.

Results

Inbreeding coefficients. The mean inbreeding coefficients for the symmetrical and asymmetrical Arabian foals, each based on n = 198 values, were 0.0348 and 0.0295, respectively. Little difference exists between these two means, and the mean for the asymmetrical foals is slightly lower. Comparing the frequency distributions of the inbreeding coefficients by using a 2 × 4 contingency table yields a chi square value of 1.67 with 3 degrees of freedom. It is concluded that the two distributions are similar (0.70 P > 0.50).

White white leg markings in Arabian and half-Arabian horses. The mean leg scores for the Arabian

Table 1. Frequency distributions of inbreeding coefficients in Arabian horses that are symmetrical and asymmetrical for white leg markings. The symmetrical horses have a sock or a stocking in at least one leg and are concordant for the markings they possess. The asymmetrical horses have a sock or stocking in at least one of the legs, and an symmetry difference of at least 4 in either the forelegs or hind legs.

	Arabian horses	
Inbreeding coefficients	Symmetrical	Asymmetrical
0.000 - 0.005	108	106
0.005 - 0.050	51	59
0.050 - 0.020	29	27
0.020 - 0.500	10	6
Total	198	198

Chi square value with 3 degrees of freedom:
1.672 (0.70 < P < 0.50)

Table 2. Leg scores in Arabian and half-Arabian horses.

Breed	Color	n	Means Foreleg Right	Foreleg Left	Hind leg Right	Hind leg Left
Arabian	Bay	17,786	0.99	1.12	1.79	1.83
Half-Arabian	Bay	17,223	0.83	0.94	1.68	1.85
Arabian	Chestnut	12,251	1.65	1.79	2.58	2.54
Half-Arabian	Chestnut	19,208	1.35	1.47	2.34	2.54

and half-Arabian horses in the four files are given in Table 2. Supporting the results of previous studies (Dreux, 1966; Woolf, 1990), (1) chestnut horses are more heavily marked in each leg than bay horses, (2) directional asymmetry tends to be present, with the left side being more heavily marked than the right side, and (3) the hind legs are more heavily marked than the forelegs. It is also noteworthy that with the exception of the left hind legs, the half-Arabians are less heavily marked than the Arabians even though a higher proportion of the half-Arabians possess minimal expressions of a pied marking.

The concordance values obtained in the comparative study of the forelegs and hind legs of Arabian and half-Arabian bay horses are given in Table 3. In each case the difference between the concordance values for socks, fetlocks and pasterns is small and not significant, as shown by the chi square values. It can be concluded that the probability of concordance for socks, fetlocks and pasterns in either pair of legs is similar in Arabian and half-Arabian bay horses.

The comparative study of Arabian and half-Arabian chestnut horses gave similar results with one exception (Table 4). All differences in concordance percent values are small and not significant except for the case of pasterns in the forelegs. In this case the concordance percent was 7.8% (n = 1,289) in Arabian chestnut horses and 12.8% (n = 1,870) in half-Arabian chestnut horses. The difference is highly significant ($P < 0.001$). However, since the other three differences involving pasterns (fore- and hind legs of bay horses, and hind legs of chestnut horses) are not significant, the one significant difference is likely spurious, possibly resulting from sampling error.

Table 3. Percent of concordance for socks, fetlocks, and pasterns in the forelegs and hind legs of Arabian and half-Arabian bay horses.

	Score	Arabian n	Arabian Percent	Half-Arabian n	Half-Arabian Percent	Chi square*	P
Foreleg *white marking*							
Sock	4	1,664	23.8	1,652	21.9	1.894	0.20 < P < 0.10
Fetlock	3	1,884	19.4	1,864	19.1	0.088	0.80 < P < 0.70
Pastern	2	1,380	10.7	1,641	12.2	1.856	0.20 < P < 0.10
Hind leg *white marking*							
Sock	4	3,226	37.8	3,852	37.0	0.438	0.70 < P < 0.50
Fetlock	3	2,715	25.3	3,231	25.4	0.023	0.90 < P < 0.80
Pastern	2	1,462	13.6	2,139	12.0	2.276	0.10 < P < 0.05

* Chi square value with Yates' correction.

Table 4. Percent of concordance for socks, fetlocks, and pasterns in the forelegs and hind legs of Arabian and half-Arabian chestnut horses.

		Arabian		Half-Arabian			
	Score	n	Percent	n	Percent	Chi square*	P
Foreleg *white marking*							
Sock	4	1,868	18.3	2,314	19.0	0.387	070 < P < 0.50
Fetlock	3	1,398	12.4	1,819	14.1	1.951	0.20 < P < 0.10
Pastern	2	1,289	7.8	1,870	12.8	19.947	< 0.001
Hind leg *white marking*							
Sock	4	2,721	26.2	3,841	27.4	1.127	0.30 < P < 0.20
Fetlock	3	1,787	14.8	2,386	13.8	0.993	0.50 < P < 0.30
Pastern	2	1,323	10.8	2,002	10.2	0.031	0.90 < P < 0.80

* Chi square value with Yates' correction.

Discussion

If homozygosity were a major contributing factor to the total asymmetry of common white markings, it might be expected that Arabian foals symmetrical for white leg markings would have lower inbreeding coefficients, on the average, than Arabian foals asymmetrical for these markings, resulting in distributions with different central tendencies. Because of the varying degrees of inbreeding that have gone on in Arabian horses throughout the history of the breed and the amount of homozygosity that could have been produced at times by the founder effect and small breeding populations (Crowell, 1951; Forbis, 1976), it might also be expected that half-Arabian horses would show a higher frequency of concordance for specific white leg markings than Arabian horses. These two expectations were not realized in this study. All statistical tests except one led to the acceptance of the null hypothesis of no differences between Arabian and half-Arabian horses, and the one disparate test may have resulted from sampling error. These overall results lead to the conclusion that homozygosity does not contribute to the total asymmetry of white leg markings in the Arabian horse.

This conclusion is not unexpected considering the trait being studied. Studies are now under way to determine the influence of genetic background on the symmetry of common white leg markings, but data obtained in the present study suggest that chance plays a major role in determining whether paired legs will be symmetrical or asymmetrical for these markings. Table 5 shows the white markings present in the hind legs of 3,226 Arabian horses selected because they have a sock in one or both of these legs. Although horses concordant for socks (37.8%) occur the most frequently, the majority (62.2%) of the horses are discordant for their markings and horses with no white markings in one of the legs occur at a relatively high frequency (37.5%). Similar results occur for other markings in the paired legs of Arabian and half-Arabian bay and chestnut horses, with concordance and the absence of a marking in one leg tending to be the two most frequent phenotypes. This observation suggests

Table 5. White leg markings in the hind legs of 3,226 Arabian bay horses selected for a sock in at least one of these legs.

Leg markings	Percent
Sock-stocking	5.6
Sock-sock	37.8
Sock-fetlock	11.8
Sock-pastern	4.8
Sock-coronet/hoof	2.5
Sock-none	37.5

that even though the total amount of common white markings in a horse is influenced by polygenic inheritance (Woolf, 1989, 1990), developmental noise influences significantly the migration or survival of melanoblasts in a given limb bud. An adaptive advantage of symmetry for a given marking and an evolutionary history would undoubtedly bring order to the migration or survival of melanoblasts in presumptive leg tissues, but at present developmental noise plays a major role in determining whether symmetry or asymmetry will be present. It is expected that this chance effect would obfuscate other effects, if any exist, influencing the symmetry or asymmetry of common white leg markings.

Even in the absence of a major influence of developmental noise, the hypothesis can be presented that the levels of asymmetry of common white leg markings should be similar in inbred and outbred strains of *Equus caballus*. Although examples of positive correlation between fluctuating asymmetry and homozygosity are abundant in the literature, in all of these cases the symmetry of the trait (e.g., veins in the wings of insects, teeth in mammals, bones in vertebrates, and bristles in insects) being measured or counted is genetically programmed, and the actions of the genes responsible for the trait are undoubtedly integrated with the actions of the genes comprising coadapted gene complexes. If the capacity for buffering requires a previous meshing by natural selection of the genotypes responsible for traits with coadapted gene complexes, buffering should be absent, or perhaps coincidental, for traits caused by recent mutations, or rare traits with a polygenic basis, which lack a strong genetic mechanism for symmetry and which have been increased in frequency in a population solely by artificial selection. This hypothesis could be tested by studying appropriate traits caused by mutant genes or polygenic systems in inbred and outbred strains of *Drosophila* and mice. According to the hypothesis, the levels of asymmetry should be similar in these different types of strains.

Acknowledgements

I wish to thank Ron M. Ballantine, Director of Data Administration, Arabian Horse Registry of America, 12000 Zuni Street, Westminster, Colorado 80234, and Charles Nininger, Computer Project Leader, International Arabian Horse Association, P.O. Box 33695, Denver Colorado 80233-0696 for supplying the computerized records used in these studies. I am greatly indebted to Kreg W. Kunz and Douglas E. Berry, Support Systems Analysts, Arizona State University Computing and Network Consulting Services, for providing the procedures for transforming the descriptions of white leg markings on tapes into numerical scores that could be used in the comparative study of Arabian and half-Arabian horses. I also wish to thank Carla M. Weber for her able assistance in compiling pedigrees and determining inbreeding coefficients.

References

Clarke, G. M., B. P. Oldroyd & P. Hunt, 1992. The genetic basis of developmental stability in *Apis mellifera*: heterozygosity versus genic balance. Evolution 46: 753-762.

Crowell, P., 1951. Cavalcade of American Horses. McGraw-Hill, New York.

Dobzhansky, T., 1950. Genetics of natural populations. XIX. Origin of heterosis through natural selection in populations of *Drosophila pseudoobscura*. Genetics 35: 288-302.

Dreux, P., 1966. Introduction statistique a la genetique des marques blanches limitees chez le cheval domestique. Ann. Genet. 9: 66-72.

Forbis, J., 1976. The Classic Arabian Horse. Liveright, New York.

Geurts, R., 1977. Hair Colour in the Horse. J. A. Allen, London.

Leamy, L., 1984. Morphometric studies in inbred and hybrid house mice. V. Directional and fluctuating asymmetry. Am. Nat. 123: 579-593.

Leary, R. F., F. W. Allendorf & K. L. Knudsen, 1984. Superior developmental stability of heterozygous enzyme loci in salmonid fishes. Am. Nat. 124: 540-551.

Leary, R. F., F. W. Allendorf & K. L. Knudsen, 1992. Genetic, environmental and developmental causes of meristic variation in rainbow trout. Acta Zoologica Fennica 191: 79-96.

Lerner, I. M., 1954. Genetic Homeostasis. Wiley, New York.

Livshits, G. & E. Kobyliansky, 1991. Fluctuating asymmetry as a possible measure of developmental homeostasis in humans: A review. Hum. Biol. 63: 441-466.

Markow, T. A. & J. P. Ricker, 1991. Developmental stability in hybrids between the sibling species pair, *Drosophila melanogaster* and *Drosophila simulans*. Genetica 84: 115-121.

Markow, T. A. & J. P. Ricker, 1992. Male size, developmental stability, and mating success in natural populations of three *Drosophila* species. Heredity 69: 122-127.

Mather, K., Genetical control of stability in development. Heredity 7: 297-336.

Mintz, B., 1967. Gene control of mammalian pigmentary differentiation. I. Clonal origin of melanocytes. Proc. Nat. Acad. Sci. 58: 344-351.

Mintz, B., 1974. Gene control of mammalian differentiation. Annu. Rev. Genet. 8: 411-470.

Nebe, H. D., 1984. Die Farbevererbung beim Pferd unter besonder Berucksichtigung der Vererbung weisser Abseichen (Inaugural dissertation). Fachbereich Veterinarmedizin und Tierzucht, Justus-Liebig-Universitat, Giessen, Germany.

Palmer, A. R. & C. Strobeck, 1992. Fluctuating asymmetry as a measure of developmental stability: Implications of nonnormal distributions and power of statistical tests. Acta Zool. Fennica 19157-72.

Seidel, G. E., Jr. & R. P. Elsden, 1987. Why identical twins may be different. Hoard's Dairyman 132: 740.

Sponenberg, D. P. & B. V. Beaver, 1983. Horse Color. Texas A & M University Press, College Station.

Thoday, J. M., 1955. Balance, heterozygosity and developmental stability. Cold Spring Harbor Symp. Quant. Biol. 20: 318-326.

Van Valen, L., 1962. A study of fluctuating asymmetry. Evolution 16: 125-142.

Waddington, C. H., 1942. Canalization of development and the inheritance of acquired characters. Nature 150: 563-565.

Waddington, C. H., 1957. The Strategy of Genes. McMillan, New York.

Woolf, C. M., 1989. Multifactorial inheritance of white facial markings in the Arabian horse. J. Hered. 80: 173-178.

Woolf, C. M., 1990. Multifactorial inheritance of common white markings in the Arabian horse. J. Hered. 81: 250-256.

Woolf, C. M., 1991. Common white facial markings in bay and chestnut Arabian horses and their hybrids. J. Hered. 82: 167-169.

Woolf, C. M., 1992. Common white facial markings in Arabian horses that are homozygous and heterozygous for alleles at the A and E loci. J. Hered. 83: 73-77.

Woolf, C. M. & J. R. Swafford, 1988. Evidence of eumelanin and pheomelanin producing genotypes in the Arabian horse. J. Hered. 79: 100-106.

Zakharov, V. M., 1992. Population phenogenesis: analysis of developmental stability in natural populations. Acta Zool. Fennica 191: 7-30.

T. A. Markow (ed.), Developmental Instability: Its Origins and Evolutionary Implications, 37–48, 1994.

Associative overdominance: evaluating the effects of inbreeding and linkage disequilibrium

E. Zouros
Department of Biology, Dalhousie University, Halifax, N.S., Canada, B3H 4J1 and Department of Biology, University of Crete, Iraklion, Crete, Greece

Received and accepted 27 April 1993

Key words: heterozygosity, associative overdominance, inbreeding, linkage disequilibrium

Abstract

Expressions are obtained for the expected phenotypic values of homozygous and heterozygous genotypes for a neutral marker locus linked to a locus segregating for a recessive deleterious gene. The phenotypic values are functions of the allele frequencies at the marker locus, the inbreeding coefficient and the degree of association of the deleterious gene with the marker alleles. The analysis is extended to more than two alleles at the marker locus. Either linkage disequilibrium or inbreeding alone can produce an apparent superiority of heterozygotes for the marker locus (unless specified otherwise, the terms 'homozygote' and 'heterozygote' will refer to the marker locus). The effect of linkage disequilibrium on the difference between the heterozygote and homozygote values can be positive (associative overdominance) or negative (associative underdominance), depending on the frequencies of the marker alleles and the degree of their association with the deleterious gene. Inbreeding has always a positive effect. In the presence of both, the possibility of associative overdominance due to linkage disequilibrium is higher. In general, the expected value of a homozygote is a positive function of its allele frequency. When the various homozygous genotypes are combined into one class and the various heterozygous genotypes into another, the phenotypic difference of the two classes is a function of the evenness of the allelic frequency distribution. Inbreeding is a more likely explanation of associative overdominance if the frequency of the deleterious gene is low, but its effect on the character high. Conversely, linkage disequilibrium is more likely if the frequency is high and the effect low. The degrees of association between marker alleles and the deleterious gene can, in principle, be estimated from the observed phenotypic scores and used to calculate expected multi-locus genotype scores. This could provide the basis for statistical tests of the associative overdominance hypothesis as an explanation of observed correlations between multi-locus heterozygosity and phenotypic traits.

Introduction

The correlation of multi-locus heterozygosity with developmental and other fitness-related characters (see Zouros & Foltz, 1987; Houle, 1989; and Zouros & Pogson, 1993 for reviews) poses several theoretical problems of interpretation that remain poorly understood. At the most basic level, there is the question of whether heterozygosity *per se* is responsible for the correlation, or the correlation is the indirect result of a negative association that may exist between heterozygosity for the scored loci (traditionally allozymes and more recently single locus DNA variants) and homozygosity for linked deleterious genes. I will refer to these competing explanations as the 'direct effect' and the 'associative overdominance' hypotheses, respectively. The term 'direct effect' is used to refer, collectively, to the various mechanisms through which the scored loci themselves produce a correlation between multi-locus heterozygosity and a phenotypic trait (see Zouros & Mallet, 1989, for a discussion of these mechanisms). The term 'associative overdominance', originally proposed by Frydenberg

(1963), is maintained for historical reasons and, also, because it conveys the main element of the hypothesis, i.e. that the correlation is due to associated rather than to scored genes. The 'direct effect' versus 'associative overdominance' dichotomy has been at the center of discussions concerning the explanation of the multi-locus heterozygosity association with fitness related traits (see, for example, Koehn, 1990, and Zouros, 1990).

At a second, more advanced level the multi-locus correlations must be understood in terms of how genotypes affect phenotypes. If the homozygosity/heterozygosity state of the scored loci affects directly the character under consideration, this has to be understood in terms of enzyme properties and controls of the biochemical and developmental pathways in which the products of these loci are involved. If the correlation is caused by linked deleterious genes, these genes must be ultimately identified and their function studied. Even though progress at one level of explanation would certainly facilitate progress at the other, the two levels may progress independently. An indication of this is the demonstration in a number of cases of a negative correlation between catabolic energy expenditure and multi-locus heterozygosity (Koehn & Shumway, 1982; Hawkins *et al.*, 1986).

This communication is about problems at the first level, and in particular about the hypothesis of associative overdominance. It is well known that genotypic correlations between neutral marker loci and loci under selection can result from inbreeding (Weir & Cockerham, 1969), linkage (gametic-phase) disequilibrium (Ohta, 1971) or strong directional selection (Maynard Smith & Haigh, 1974). The search for inbreeding or linkage disequilibrium as explanations of the correlation between multi-locus heterozygosity and fitness-related characters has produced mixed results. The most compelling evidence for such a correlation comes from studies of growth rate in natural populations of marine bivalves (Mitton & Grant, 1984; Zouros & Foltz, 1987). In a review of these studies, Zouros (1987) observed that the correlation was more common among populations in which the numbers of homozygotes were much higher than Hardy-Weinberg ratios. In a more recent study, employing the largest number of loci examined for this purpose, Gaffney *et al.* (1990) reported a positive correlation between locus-specific heterozygote deficiencies and locus-specific contributions to the overall correlation of multi-locus heterozygosity with growth. By drawing a connection between the homozygosity excess and the heterozygosity/growth correlation, these observations appear to suggest that associative overdominance caused by inbreeding is the explanation for the correlation. In contrast, linkage disequilibria do not appear to be common in populations in which the correlation was observed (Ahmad & Hedrick, 1985; Zouros & Mallet, 1987). The interpretation of these results is severely hampered from the lack of an appreciation of the relative importance of inbreeding and linkage disequilibrium in causing an apparent correlation between heterozygosity and phenotype.

Here I formulate an explicit model of associative overdominance and produce quantitative expressions for the joint and separate effects of inbreeding and linkage disequilibrium on the mean phenotypic values of homozygous and heterozygous genotypes. The results can be useful in evaluating the possibility of associative overdominance and deciding whether inbreeding or linkage disequilibrium is the most likely cause of a correlation between degree of heterozygosity and a phenotypic score.

The model

I consider a quantitative character C (which can be either continuous or metric) and two loci A and B. I refer to locus A as the *marker* locus and to locus B as the *agent* locus. Two alleles segregate at each locus, A_1 and A_2 at locus A with frequencies p and $1 - p$, and B_1 and B_2 at locus B with frequencies q and $1 - q$. All three genotypes at locus A are scorable and 'neutral' with respect to the phenotypic character under consideration. This does not mean that locus A does not affect the character in some way (e.g. it may code for an enzyme involved in a pathway that produces energy needed for the development of the character); it simply means that there is no difference in the effects of the three genotypes on the character. Locus B differs from locus A in two respects: (a) its three genotypes cannot be scored with the assay in use, and (b) the three genotypes have a differential effect on the character. In particular, if we assign the value 1 to a reference state of the character, the three genotypic values for locus B are defined as $1 - s$ for B_1B_1, $1 - hs$ for B_1B_2 and 1 for B_2B_2 ($0 < s < 1$). For characters that are directly related to fitness, such as growth, viability or fertility rates, the highest of the three genotype means is usually given the value 1 and the parameters s and h (known, respectively, as the selection and dominance coefficient) are estimated from the mean phenotypic values of the two other genotypes. For a character like bilateral asymmetry or a continuous morphological character, the value 1 is assigned to the state of perfect symmetry or to the average phenotype in the sample, and s and h are measured as deviations from that phenotype.

The purpose of the model is to account for differences in the phenotypic values of the three genotypes of the marker locus at the time of the sampling, using only the information provided by the sample. This has been done using two assumptions. First, the individuals scored are assumed to be the products of the union of gametes drawn from the same pool and to represent an age cohort grown under the same environment. Second, the phenotypic character measured is assumed to have no effect on the probability of survival from the time of fertilization to the time of scoring. This second assumption may be interpreted to mean that the genotypic differences on survival to the age of scoring are negligible compared to differences on fertility or post-scoring viability. These conditions appear to be true for most studies that have reported strong and persistent correlations between multilocus heterozygosity and phenotypic scores. In most marine species spawning is seasonal, communal and external, and age cohorts are formed. Characters usually measured include body size, bilateral asymmetry or other morphological scores, whose effect on pre-scoring viability can be considered to be small.

Table 1. Two-locus gametic frequencies. Locus A is the marker-locus and locus B is the agent-locus.

	B_1	B_2	Total
A_1	g_{11} = pa	g_{12} = (1-a)p	p
A_2	g_{21} = b(1-p)	g_{22} = (1-b) (1-p)	1-p
Total	q	1-q	1

Besides mean phenotypic values for the three genotypes at locus A, the sample provides estimates for the two allele frequencies p and $1 - p$ and the inbreeding coefficient f. Let g_{11}, g_{12}, g_{21} and g_{22} be the four sample gametic frequencies (for A_1B_1, A_1B_2, A_2B_1 and A_2B_2, respectively). Given that locus B cannot be assayed, these frequencies are unknown. Assume that a fraction a of A_1-carrying gametes also carry allele B_1 and that a fraction b of A_2-carrying gametes also carry allele B_1. I will refer to a and b as *degrees of association*. Then the four gamete frequencies can be written as in Table 1.

The system can be described either in terms of three independent parameters, p, a and b, in which case the allele frequency at the agent locus B is obtained from

$$q = ap + b(1-p) \tag{1}$$

or in terms of p, q and a, in which case b is obtained from

$$b = (q - ap)/(1-p) \tag{2}$$

Traditionally (Lewontin & Kojima, 1960), a two-locus-two-allele system is formulated in terms of p, q and D, where D, the disequilibrium parameter, is defined as $g_{11} - D$ and estimated as $D = g_{11}g_{22} - g_{12}g_{21}$. The parameterization of Table 1 introduces no additional parameters, but leads to simpler expressions that can be readily extended to loci with any number of alleles. The fact that variation at locus B cannot be observed directly also argues against using q as a defining parameter in models of associative overdominance. In fact, this locus is detected only through the association of its alleles with the alleles at the scored locus, which is the definition of equation (1). The two parameterization systems are related as follows:

$$D = p(a-q) = (1-p)(q-b) = p(1-p)(a-b) \tag{3}$$

At linkage equilibrium, $a = b = q$.

With the four gametic frequencies and the inbreeding coefficient defined, it is easy to obtain expressions for the nine two-locus genotypes treating the gametic pool as a gene pool with four alleles (see, for example, Hedrick, 1983, pp. 84-87). The first draw will produce an A_1B_1 gamete with fre-

Table 2. Two-locus genotype frequencies. f is the inbreeding coefficient; other symbols are defined in Table 1.

	B_1B_1	B_1B_2	B_2B_2	Total
A_1A_1	ap[f + (1-f) ap]	2 (1-f) a (1-a) p^2	(1-a) p [f + (1-f) (1-a) p]	fp + (1-f)p^2
A_1A_2	2 (1-f) abp (1-p)	2 (1-f) (a + b-2 ab) p (1-p)	2 (1-f) (1-a) (1-b) p (1-p)	2(1-f)p(1-p)
A_2A_2	b (1-p) [f + (1-f) b (1-p)]	2 (1-f) b (1-b) $(1-p)^2$	(1-b) (1-p) [f + (1-f) (1-b) (1-p)]	f(1-p)+(1-f)$(1-p)^2$
Total	fq + (1-f) q^2	2 (1-f) q (1-q)	f (1-q) + (1-f) $(1-q)^2$	1

quency ap. The second draw may also produce an A_1B_1 gamete, which will be identical to the first by descent with probability f and identical in state with probability $(1 - f)ap$. Thus, the frequency of genotype $A_1A_1B_1B_1$ will be $ap[f + (1 - f)ap]$. Continuing for the other three possible gametes (A_1B_2, A_2B_1, A_2B_2) for the second draw and then repeating the exercise for the other three possibilities for the first draw produces the frequencies of all genotypes as shown in Table 2.

The next step is to obtain expressions for the mean phenotypic values of the three genotypes for the marker locus. For simplicity of treatment (but also because deleterious genes are normally considered to be recessive to their wild alleles), we will assume $h = 0$. It follows from Table 2 that the mean phenotypic value of A_1A_1, $P(A_1A_1)$, is:

$$P(A_1A_1) = 1 - s\left\{\frac{ap[f+(1-f)ap]}{fp+(1-f)p^2}\right\} = 1 - s\left\{\frac{a[f+(1-f)ap]}{f+(1-f)p}\right\} \quad (4)$$

Likewise, the mean phenotypic value of the heterozygote is:

$$P(A_1A_2) = 1 - s\left\{\frac{2(1-f)abp(1-p)}{2(1-f)p(1-p)}\right\} = 1 - sab \quad (5)$$

Expressions (4) and (5) can be used to investigate the effects of linkage disequilibrium and inbreeding on the phenotypic difference of various genotypes.

Difference between the two homozygotes

Let $p > 1 - p$. Then the phenotypic difference of the less common homozygote A_2A_2 from the more common A_1A_1 is:

$$T_1 = P(A_1A_1) - P(A_2A_2) = s\left\{\frac{b[f+(1-f)b(1-p)]}{f+(1-f)(1-p)} - \frac{a[f+(1-f)ap]}{f+(1-f)p}\right\} =$$
$$s\left\{\frac{f^2(b-a)+(1-f)^2(b^2-a^2)p(1-p)}{[f+(1-f)(1-p)][f+(1-f)p]}\right\} + s\left\{\frac{f(1-f)[(1-p)(b^2-a)+p(b-a^2)]}{[f+(1-f)(1-p)][f+(1-f)p]}\right\} \quad (6)$$

The difference is zero only when both inbreeding and linkage disequilibrium are absent from the population ($f = 0$, $a = b = q$). In the presence of inbreeding, but absence of disequilibrium, the numerator becomes $f(1 - f)\, q(1 - q)\,(2p - 1)$ which for $p > 1/2$ is positive and increasing with p. In the presence of disequilibrium and absence of inbreeding $T_1 = s\,(b^2 - a^2)$, i.e. the difference can be positive or negative, depending on the difference of the association of A_1 and A_2 with the deleterious gene B_1. Its maximum value, at $a = 0$, is $sq^2/(1 - p)^2$ and its minimum, at $b = 0$, is $-sq^2/p^2$.

Difference between the heterozygote and a homozygote

Following the above procedure, the phenotypic difference between A_1A_2 and A_1A_1 can be written as follows:

$$T_2 = P(A_1A_2) - P(A_1A_1) = s\left\{\frac{a[f+(1-f)ap]}{f+(1-p)p} - ab\right\} = s\left\{\frac{af(1-b)+a(a-b)(1-f)p}{f+(1-f)p}\right\} \quad (7)$$

This is again zero only when both $f = 0$ and $a = b$. For $a = b = q$ and $f > 0$,

$$T_2 = s\left\{\frac{fq(1-q)}{f+(1-f)p}\right\} \qquad (8)$$

which is always positive and has a maximum at $q = 0.5$. For $f = 0$ and $a \neq b$, $T_2 = s\,a(a - b)$, which can be positive (for $a > q$) or negative (for $a < q$). Its maximum value, at $b = 0$, is sq^2/p^2 and its minimum, at $a = q/2$, is $-s\,q^2/4(1 - p)$.

Difference between the heterozygote and the weighted mean of the two homozygotes

In most studies of multilocus heterozygosity the various types of homozygotes at a locus are pooled together to form the 'homozygote' class, and the various types of heterozygotes are also pooled together to form the 'heterozygote' class. It is, therefore, important to compare the mean phenotypic scores of these combined genotypic classes. The mean score of the homozygous class is:

$$P(A_1A_1 + A_2A_2) =$$
$$1 - s\left\{\frac{ap[f+(1-f)ap] + b(1-p)[f+(1-f)b(1-p)]}{[fp+(1-f)p^2] + [f(1-p)+(1-f)(1-p)^2]}\right\} = \qquad (9)$$
$$1 - s\left\{\frac{fq+(1-f)[q^2 - 2abp(1-p)]}{1-2(1-f)p(1-p)}\right\}$$

Thus the difference between the heterozygote and homozygote classes is:

$$T_3 = P(A_1A_2) - P(A_1A_1 + A_2A_2) =$$
$$\{\;-sab\} - \left\{1 - s\frac{fq+(1-f)[q^2 - 2abp(1-p)]}{1-2(1-f)p(1-p)}\right\} =$$
$$s\left\{\frac{fq+(1-f)q^2 - ab}{1-2(1-f)p(1-p)}\right\} \qquad (10)$$

For $a = b = q$, the numerator becomes $sfq(1 - q)$, which is always positive and has a maximum at $q = 0.5$. Thus in the lack of linkage disequilibrium, the heterozygote score always exceeds the weighted mean of homozygotes, the excess being a monotonically increasing function of f. In the presence of linkage disequilibrium, T_3 will be negative if $fq + (1 - f)\,q^2 < ab$. Substituting b for $(q - ap)/(1 - p)$ (from 1), this condition defines the range of a for which T_3 takes negative values:

$$\frac{q - R^{1/2}}{2p} < a < \frac{q + R^{1/2}}{2p} \qquad (11)$$

where

$$R = q^2 - 4p(1-p)q[f + (1-f)q] \quad (\text{for } R > 0). \qquad (12)$$

Because R decreases with f, the range of disequilibrium values for which T_3 is negative becomes narrower as the amount of inbreeding increases. Thus inbreeding promotes associative overdominance both in the presence or absence of linkage disequilibrium. It is obvious from (10) that associative overdominance will become more pronounced as (i) either a or b approach zero, that is as most deleterious alleles at the agent locus tend to co-occur with only one or the other allele of the marker locus (the situation of complete linkage disequilibrium), (ii) q becomes larger and (iii) $p(1 - p)$ approaches its maximum limit of 0.25 ($p = 0.5$). T_3 has a minimum at $a = q/(2\,p)$ [or $b = q/2\,(1 - p)$]. At $p = 0.5$ the value of a and b at which T_3 is minimized is the same. Since T_3 is always positive for $a = b$, it follows that at $p = 0.5$ the heterozygote score is always larger than that of the combined score of homozygotes.

From (3), $a = (pq + D)/p$ and $b = [(1 - p)q - D]/(1 - p)$. Substituting into (10) we obtain

$$T_3 = \left\{\frac{s}{[1-2(1-f)p(1-p)]p(1-p)}\right\} \times \qquad (13)$$
$$\{fq(1-q)p(1-p) + D[D - q(1-2p)]\}$$

which allows for the partitioning of the associative overdominance effect into two components, one due to inbreeding and one due to linkage disequilibrium.

Extension to multiple alleles

The above treatment can be easily extended to the case of more than two alleles at the marker locus. We assume that n alleles ($A_1, A_2, \ldots, A_n$) segregate at locus A with frequencies $p_1, p_2, \ldots, p_n$, respectively. We also assume that a fraction $a_1, a_2, \ldots, a_n$ of chromosomes $A_1, A_2, \ldots, A_n$ carry allele B_1 of locus B. The frequency of B_1 is, then, $q = \Sigma\, a_i p_i$, where the summation is for $i = 1$ to n.

The frequency of genotype $A_iA_iB_1B_1$ is (from Table 2) $a_ip_i[f + (1 - f)\, a_ip_i]$ and that of the genotype $A_iA_jB_1B_1$ is $2(1 - f)a_ia_jp_ip_j$. The phenotypic scores of the homozygote A_iA_i and the heterozygote A_iA_j are given by (4) and (5) by replacing a and b by a_i and a_j, and p and $(1 - p)$ by p_i and p_j. The new expressions for T_1 and T_2 are obtained likewise from (6) and (7). It is worth noting that for $a_i = a_j$ and $p_i > p_j$ relation (4) gives $P(A_iA_i) > P(A_jA_j)$, i.e. for the same degree of association with the deleterious gene, homozygotes for a more frequent allele have a higher expected phenotypic value than homozygotes for a less frequent allele.

The expressions for the expected phenotypic value of the homozygote and heterozygote classes and their difference are

$$P(\text{Hom}) = 1 - s\left\{\frac{fq + (1-f)[q^2 - 2\Sigma a_ip_ia_jp_j]}{1 - 2(1-f)\Sigma p_ip_j}\right\} \quad (14)$$

$$P(\text{Het}) = 1 - s\left\{\frac{\Sigma a_ip_ia_jp_j}{\Sigma p_ip_j}\right\} \quad (15)$$

$$T_4 = s\left\{\frac{q[f + (1-f)q]\Sigma p_ip_j - \Sigma a_ip_ia_jp_j}{[1 - 2(1-f)\Sigma p_ip_j]\Sigma p_ip_j}\right\} \quad (16)$$

where the summation is for $i \neq j$.

Note that $\Sigma p_ip_j = (1 - \Sigma p_i^2)/2 = H_e/2$, where H_e is the expected heterozygosity at the marker locus. If $H_o = 1 - H_e$ is the expected homozygosity at the marker locus, an alternative expression for T_4 is

$$T_4 = s\left\{\frac{q[f + (1-f)q]}{H_o + fH_e} - \frac{2\Sigma a_ip_ia_jp_j}{H_e[H_o + fH_e]}\right\} \quad (17)$$

From (16), the requirement for $T_4 > 0$ is $q[f + (1 - f)q]\,\Sigma p_ip_j > \Sigma a_ip_ia_jp_j$. For $a_1 = a_2 = \ldots = a_n = q$, this is always true and T_4 becomes

$$T_4 = \left\{\frac{sfq(1-q)}{H_o + fH_e}\right\} \quad (18)$$

For $f = 0$

$$T_4 = s\left\{\frac{q^2}{H_o} - \frac{2\Sigma a_ip_ia_jp_j}{H_oH_e}\right\} \quad (19)$$

and the requirement for $T_4 > 0$ is $q^2 > (\Sigma a_ip_ia_jp_j)/\Sigma p_ip_j$. For $p_i = p_j = \ldots = p_n = 1/n$, $H_e = (n - 1)/n$ and $H_o = 1/n$ (the maximum and minimum values, respectively), and T_4 becomes

$$T_4 = s\left\{\frac{n(n-1)q[f + (1-f)q] - 2\Sigma a_ia_j}{(n-1)[1 + f(n-1)]}\right\} \quad (20)$$

which is an increasing function of n. All expressions in this section reduce to the corresponding ones in the previous sections by setting $a_1 = a$, $a_2 = b$, $p_1 = p$, $p_2 = 1 - p$, and $a_i = 0$, $p_i = 0$ for $i > 2$.

To partition the associative overdominance effect into the inbreeding and linkage disequilibrium components, we can write $a_i = q - d_i$, subject to $\Sigma p_id_i = 0$. Then

$$\Sigma a_ip_ia_jp_j = q^2\Sigma p_ip_j + \Sigma d_ip_id_jp_j - q\Sigma(d_i + d_j)p_ip_j \quad (21)$$

Inserting in (16) we obtain

$$T_4 = \frac{s}{[1 - 2(1-f)\Sigma p_ip_j]\Sigma p_ip_j} \times \{fq(1-q)\Sigma p_ip_j + [q\Sigma(d_i + d_j)p_ip_j - \Sigma d_ip_id_jp_j]\} \quad (22)$$

which is analogous to (13). In a simpler form, it can be written as

$$T_4 = \left\{\frac{2s}{H_e[1 - (1-f)H_e]}\right\}(C_I + C_{LD}) \quad (23)$$

where

$$C_I = (1/2)fq(1-q)H_e$$
$$C_{LD} = q\Sigma(d_i + d_j)p_ip_j - \Sigma d_ip_id_jp_j = (1/2)q^2H_e - \Sigma a_ip_ia_jp_j \quad (24)$$

From this follows that fraction $C_I/(C_I + C_{LD})$ of the homozygote depression is due to inbreeding and fraction $C_{LD}/(C_I + C_{LD})$ is due to linkage disequilibrium. From (14) and (15) it can be seen that the quantity $\Sigma a_ip_ia_jp_j$ can be expressed in terms of the observed homozygous and heterozygous scores. As a result both components of T_4 can be expressed as functions of known genetic parameters (H_e and f), the known character scores of the animals in the sample and the unknown frequency of the deleterious gene. This can lead to some insights about the associative overdominance hypothesis, as will be shown in a following section.

Estimation and testing

In empirical studies of multi-locus phenotypic correlations, allele frequencies for the marker loci are directly estimated as $p_i = D_{ii} + (\Sigma_j H_{ij})/2$, where D_{ii} is the frequency of the homozygote and H_{ij} the frequencies of heterozygotes for the ith allele at the marker locus. The inbreeding coefficient can also be estimated from the genotypic frequencies [e.g. through the method of maximum likelihood; see Zouros *et al.*, 1980, for an example]. The individual phenotypic scores can be used to estimate the expected phenotypic value of each genotype. Take one marker locus at a time, say A, with n scorable alleles. Let L_{ij} be the mean score (e.g. body size) of genotype A_iA_j in the sample (where $i = j$ for homozygotes and $i \neq j$ for heterozygotes) and L_R be the *reference* score. L_R can be, for example, the highest individual score in the sample. Then we may define $P(A_iA_j)$ as

$$P(A_iA_j) = (L_R - L_{ij}) / L_R = 1 - L_{ij} / L_R = 1 - K_{ij} \quad (25)$$

K_{ij} is, then, an estimate of the term $s_A\{.\}$ in expressions (14) and (15), and s_A measures the effect of homozygosity for the deleterious gene in the neighborhood of locus A. We expect s to vary among marker loci, as deleterious alleles at different parts of the genome may affect the character in a different way. For a locus with n alleles, there are $n(n+1)/2$ genotypes, but not all of these may be present in the sample. Employing (14) and (15) generates $n(n+1)/2$ equations with $n+1$ unknowns (the vector of degrees of association, a_i, and the selection coefficient, s). For a two-allele locus, there will be three equations in three unknowns yielding the solution (for $f \neq 0$):

$$a_1 = \frac{K_{12} f[K_{11}c_1 + K_{12}d_2]}{K_{11}K_{22}c_1c_2 - K_{12}^2 d_1 d_2},$$
$$a_2 = \frac{K_{12} f[K_{22}c_2 + K_{12}d_1]}{K_{11}K_{22}c_1c_2 - K_{12}^2 d_1 d_2}, \quad s = \frac{K_{12}}{a_1 a_2} \quad (26)$$

where $c_i = f + (1-f)p_i$ and $d_i = (1-f)p_i$. For loci with more than two alleles there will be more genotypes than there are parameters to estimate, so a method of estimation, like maximum likelihood, must be employed.

Once the selection coefficient s and the vector a_i are estimated for each locus, the expected value K for each genotype at that locus can be obtained from:

$$K_{ii} = s\left\{\frac{a_i[f + (1-f)a_i p_i]}{f + (1-f)p_i}\right\}, \quad K_{ij} = s a_i a_j \quad (27)$$

A comparison with the observed K necessitates knowledge of the expected variance of K, which contains the standard errors of the allele frequencies and the inbreeding coefficient and the variances of L_{ij}, both genetic and non-genetic. Instead of calculating an expected value for each genotype and comparing it with the observed value, a more appealing approach would be to calculate the expected phenotypic score for each individual animal in the sample and compare it with the observed score. This will give a single test for the entire data set. An animal's expected score can be obtained from the multiplicative-fitness model as $\Pi(1-K)$ or from the additive model as $1 - \Sigma K$ (where the product or summation is over all loci). A problem with this test is that an erroneous choice for the multi-locus fitness function will complicate the interpretation of a poor fit between observed and expected values.

An application

Zouros *et al.* (1988) measured shell length and scored the genotypes at seven enzyme loci in a sample of 376 individuals from a two-year-old cohort of blue mussels. The mean fixation index across loci (here treated as inbreeding coefficient) was $f = 0.053$. The individuals were grouped according to degree of heterozygosity. The mean shell lengths of the 'zero' heterozygosity class and of the 'six' heterozygosity class were 50.55 and 54.20 mm, respectively (no animal heterozygous for all seven loci was observed). Let s_m be the mean length reduction associated with homozygosity at a marker locus. This is not equivalent to s, the reduction due to homozygosity for a deleterious gene, as a collection of homozygous animals for the marker locus will contain homozygous as well as heterozy-

Table 3. Frequency of the deleterious gene (q), the relative reduction of shell length because of homozygosity for the deleterious gene (s), and the relative contribution of inbreeding (C_I) and linkage disequilibrium (C_{LD}) to the difference between heterozygotes and homozygotes at six allozyme loci. Data from Zouros *et al.* (1988). The inbreeding coefficient is f = 0.053 and mean shell length of non-homozygotes for the deleterious gene at vicinity of any given allozyme locus is L_m = 51.61. H is the observed heterozygosity, and L_o and L_e are the observed mean shell lengths of homozygotes and heterozygotes, respectively. One locus, Pgm, is not examined because at this locus homozygotes have a higher mean shell than heterozygotes. See text (An Application) for details.

	q	s	C_I	C_{LD}
Locus:	Pep - 1,	H = 0.634,	L_o = 50.70,	L_e = 51.30
	0.081	1	0.862	0.138
	0.10	0.721	0.741	0.259
	0.15	0.364	0.530	0.470
	0.20	0.220	0.401	0.599
	0.25	0.147	0.314	0.686
Locus:	Pep - 2,	H = 0.501,	L_o = 50.60,	L_e = 51.60
	0.079	1	0.386	0.614
	0.10	0.703	0.329	0.671
	0.15	0.355	0.235	0.765
	0.20	0.214	0.178	0.822
	0.25	0.143	0.140	0.860
Locus:	Pgi,	H = 0.678,	L_o = 50.97,	L_e = 51.11
	0.346	0.081	1	0
	0.35	0.079	0.984	0.016
	0.45	0.049	0.668	0.332
	0.40	0.062	0.808	0.192
	0.50	0.041	0.552	0.448
Locus:	Lap - 1,	H = 0.136,	L_o = 51.00,	L_e = 51.53
	0.083	1	0.664	0.336
	0.10	0.739	0.580	0.420
	0.15	0.373	0.415	0.585
	0.20	0.225	0.314	0.686
	0.25	0.137	0.225	0.775
Locus:	Lap - 2,	H = 0.480,	L_o = 50.89,	L_e = 51.27
	0.081	1	0.991	0.009
	0.10	0.718	0.853	0.147
	0.15	0.362	0.610	0.390
	0.20	0.219	0.462	0.536
	0.25	0.146	0.362	0.638
Locus:	Odh,	H = 0.221,	L_o = 50.94,	L_e = 51.42
	0.084	1	0.550	0.450
	0.10	0.747	0.485	0.515
	0.15	0.378	0.347	0.653
	0.20	0.228	0.263	0.737
	0.25	0.152	0.206	0.794

gous animals for the associated deleterious gene. In fact, there should always be $s > s_m$. Assigning the highest relative phenotypic value of 1 to the mean of the class with the highest degree of heterozygosity (54.20), the relative value of the zero heterozygosity class becomes $(1 - s_m)^6$, from which we obtain $(1 - s_m)^6 = 50.55/54.20$, or $s_m = 0.0116$.

Out of seven loci scored, the observed average number of heterozygous loci per individual was 2.8, implying that the average expected length is $L_m = 54.20 \times 1^{2.8} \times (1 - s_m)^{7-2.8} = 51.61$. For any particular locus, we may consider this to be the expected shell length for individuals non-homozygous for a deleterious gene at the vicinity of that locus. Then P(Hom) = L_o/L_m and P(Het) = L_e/L_m, where L_o and L_e are the mean shell lengths of homozygotes and heterozygotes for the marker locus. L_o and L_e are obtained from the data. Having estimates for L_m, L_o and L_e allows for the expression of the components of inbreeding and linkage disequilibrium in terms of only one unknown parameter, the frequency of the deleterious gene, q, as follows.

In (14) and (15), P(Hom) and P(Het) can be replaced with L_o/L_m and L_e/L_m, respectively. Dividing (14) by (15) gives

$$\Sigma a_i p_i a_j p_j = \frac{1}{2} \times \frac{H_e q(L_m - L_e)[f + (1-f)q]}{(L_m - L_o)[1-(1-f)H_e] + (L_m - L_e)(1-f)H_e} \quad (28)$$

which can be substituted in (24) to give an expression of the linkage disequilibrium components in terms of q. It will be also noted that from (23) the depression, s, due to homozygosity for the deleterious gene can be obtained as

$$s = \frac{H_e[1-(1-f)H_e](L_e - L_o)}{2L_m(C_I + C_{LD})} \quad (29)$$

The results shown in Table 3 were obtained as follows. Using the observed heterozygosity, H_e, the average length of homozygotes and heterozygotes for each locus (Table 2 of Zouros *et al.*, 1988), the observed f (0.053) and the estimated L_m (51.61), a value for $\Sigma a_i p_i a_j p_j$ was obtained from (28) starting with an arbitrarily low frequency of q. This was then used to obtain s, C_I and C_{LD} from (29) and (24), respectively. If any of these three values were outside the permissible range of zero to one, a new (higher) value of q was chosen. The first q value given in the table is the minimum for which s and C_I (and, therefore C_{LD}) are between zero and one. The value of q was then increased at intervals of 0.05, starting with the round-value closest to the minimum q and proceeding for four steps.

The first observation is that, for the inbreeding coefficient observed, fairly high frequencies of the deleterious gene are required to explain the data on the basis of the associative overdominance hypothesis. For the majority of allozyme loci the associated deleterious gene frequency must be at least 0.1, implying that the effect on shell length of these recessive deleterious genes must be high ($s = 0.7$ or more). For lower effects (e.g. $s = 0.15$), the frequency of the gene must be around 0.25.

These calculations provide an explicit illustration of the most serious problem of the associative overdominance explanation: How do deleterious genes achieve high frequencies in the population? There appears to be no escape from this difficulty, except to assume that the phenotypic trait in question is only weakly related to fitness. This might be a difficult assumption for characters found to be most strongly correlated with heterozygosity (growth rate, bilateral asymmetry and morphometric deviations from the population mean; see Zouros & Foltz, 1983). Even if the character is assumed to be completely neutral, it will be difficult to see how the frequency of the 'score-reducing' gene will drift to the required level, particularly in large stable populations like those of marine bivalves. A possible escape will be to assume that in these species enzyme markers are tightly flanked with strings of loci, each of which affects the character in question and mutates at a high rate. But even this hypothesis will require some degree of non-complementarity or synergistic interaction among these mutants.

A second observation is that, because of the inverse relationship between q and s, the 'genetic load' for the character at each locus remains relatively small. In the above example, for $q = 0.1$, s is on average 0.725 for five out of six loci, giving a load of $0.725 \times (0.1)^2 = 0.007$. For $q = 0.25$, the average s is 0.145, giving a load of 0.009. The load is comparable for the other locus, Pgi, where the minimum q is 0.35. This argument of a relatively small load depends critically on the observed difference between the two extreme heterozygosity classes. However, in most bivalve studies that reported a correlation between heterozygosity and body size, the degree of heterozygosity accounts for about 4% of the variance in shell length, suggesting that a genetic load of 0.01 per allozyme locus might be a good and general approximation for this character in this group of organisms.

The third point to be made from Table 3 is the dependence of the relative contributions of inbreeding and linkage disequilibrium on the frequency of the deleterious gene. For the same genetic load (sq^2), the contribution of linkage disequilibrium is expected to be low for low frequencies of the deleterious gene. Thus, inbreeding will be a more appealing explanation of associative overdominance in the face of difficulties to account for high frequencies of deleterious genes. Conversely, linkage disequilibrium will be the preferable explanation if depression coefficients are suspected to be low.

Discussion

Even though the underlying causes of associative overdominance have been well recognized, there has been no attempt to obtain a quantitative expression for the expected phenotypic difference between heterozygous and homozygous genotypes under this model of heterosis. I develop here a model that does this, considering the two most important forces of associative overdominance, inbreeding and linkage disequilibrium. The model produces explicit expressions for the expected phenotypic values of single locus genotypes (and, therefore, of the expected differences between heterozygotes and homozygotes) in terms of allelic frequencies at the marker locus, the inbreeding coefficient of the population and the degrees of association between marker alleles and the deleterious gene. The model is extended to loci with multiple alleles, an almost absolute requirement for present-day studies of heterosis that employ hypervariable loci. In these studies (as indeed in most allozyme studies), it is not possible to single out a major allele and pool all the other alleles into one heterogeneous class for the purpose of reducing the polymorphism into a two-allele system.

The analysis produced several qualitative results, some more intuitively obvious than others.

They can be summarized as follows:

1. The associative overdominance effect is stronger the higher the frequency of the deleterious gene.
2. For associative overdominance to occur, there is no need that both inbreeding and linkage disequilibrium be present in the population. Conversely, associative overdominance is absent when both inbreeding and linkage disequilibrium are absent.
3. The higher the amount of inbreeding, the stronger the effect of the associative overdominance.
4. The effect of linkage disequilibrium depends mainly on the summation of the pair-wise products of the degrees of association. The smaller this product is (i.e. as one allele at the marker locus tends to become completely disassociated with the deleterious gene), the stronger the effect of the associative overdominance.
5. In the absence of linkage disequilibrium (but presence of inbreeding) there will always be associative overdominance, the magnitude of which is an increasing function of the inbreeding coefficient. The converse is not true. In the absence of inbreeding there may or may not be associative overdominance.
6. For a given amount of homozygote depression, the relative contribution of inbreeding and linkage disequilibrium depends critically on whether the depression is caused by highly deleterious genes segregating in relatively low frequency (in which case the contribution of inbreeding is high) or by moderately deleterious genes segregating at high frequencies (in which case the contribution of linkage disequilibrium is high).
7. Considering one locus at a time, the fitness depression of a homozygous genotype due to associative overdominance is larger the lower the frequency of the allele. This 'allele frequency effect' is independent of the number of alleles per locus.
8. For the simplest case of two alleles at the marker locus, it is shown that associative overdominance is an increasing function of the allele frequency product. With multiple alleles and for the case of linkage equilibrium, associative overdominance is shown to be an increasing function of the summation of the cross-products of allele frequencies. This suggests a general 'allele frequency evenness effect' on associative overdominance.

The last two points may provide the basis for tests of the hypothesis of associative overdominance versus the competing hypothesis of direct allozyme effects. In most empirical studies of phenotype correlations with heterozygosity, the raw data consist of the phenotype score and the multilocus genotype of a set of animals. In large studies employing many multi-allele loci, one may assume that the effects of linkage associations are sufficiently randomized across loci. Under this assumption, the 'allele frequency effect' suggests an 'intra-locus' test: taking one locus at a time, a positive correlation between $P(A_iA_i)$ and p_i would favor the hypothesis of associative overdominance. Likewise, the 'allele frequency evenness effect' suggests an 'inter-locus' test: a positive correlation between P(Het) - P(Hom) and an index of allele frequency evenness (such as the expected heterozygosity $H_e = 1 - \Sigma p_i$, or the effective number of alleles ($n_e = 1/\Sigma p_i^2$) will favor the associative overdominance hypothesis. Unfortunately, these predictions are not unique to associative overdominance, and in fact are shared by most balanced selection models with either fixed or frequency-dependent selection coefficients (see, for example, the equilibrium adaptive distance model of Smouse, 1986). These tests would appear to be most informative when negative, because they will then render the associative overdominance explanation less likely.

The theory allows for the estimation of the degrees of association of scored alleles with deleterious genes at neighboring loci and these estimates may, in principle, be used to test the null hypothesis of associative overdominance by comparing observed and expected phenotypic values for each individual in the sample. Unfortunately, the power of these tests depends on a number of factors about which we may have little information. These include the standard errors of the gene frequencies and the inbreeding coefficient, the genetic and non-genetic variances of the various genotypic values, and the epistatic interation among loci. At present it is difficult to choose between tests based on the comparison of expected and observed individual scores and tests based on 'trends' between mean genotypic scores and gene frequencies.

An interesting result of the analysis is that the components of inbreeding and linkage disequilibrium can be separated. If the frequencies of the deleterious genes can be estimated independently from their phenotypic effect (as can be done in *Drosophila melanogaster*, with the use of the balanced chromosome technique), then the relative contributions of inbreeding and linkage disequilibrium can be estimated. This, in turn, will be useful in our attempts to obtain an indication not only about the amount of the genetic load in natural populations, but also about its distribution over the genome of a species. These questions will become more and more tempting as more and more single-locus hypervariable polymorphisms become available.

The theory developed here considers the population at the moment of sampling and provides expressions for the expected phenotypic values of the various genotypes on the basis of allele frequencies, inbreeding, linkage associations and selection coefficients at that particular moment. These parameters are, of course, not static in time unless the population is at equilibrium. But the usefulness of the analysis does not depend on the assumption of equilibrium. If we could evaluate the degree to which associative overdominance may or may not account for the observations at the time the observations were made, we would have made considerable progress in our effort to explain the correlation between heterozygosity and phenotype, even if we are not clear about the origin and the fate of the forces that cause the associative overdominance. The simplifying assumptions mentioned above, as well as the assumption of no viability differences between genotypes, will have no serious effect on the ability of the theory to help decide between direct and associated marker effects and evaluate the contributions of inbreeding and linkage disequilibrium, as long as these assumptions do not affect seriously the qualitative predictions of the theory. Both the 'allele frequency effect' and the 'allele frequency evenness effect' appear to be fairly robust. Equally robust appears to be the relationship between the frequency of the deleterious gene and its effect on the character, and the relative contributions of inbreeding and linkage disequilibrium. If neither a high selection coefficient nor a high frequency for the deleterioius gene can be entertained as viable options, we might have to abandon the hypothesis of associated overdominance. If a high selection coefficient appears more likely than a high gene frequency, then most of the effect should be attributed to homozygosity for deleterious genes. The opposite should be true if a high frequency is more likely than a high selection coefficient.

There are indications in the literature that inbreeding may be the explanation of heterozygosity/fitness correlations in plants (Strauss, 1986, but see Bush *et al.*, 1987, for a difference in opinion). In marine mollusks, where the correlation is apparently more prominent, heterozygote deficiencies are common but highly uneven among loci. Contrary to inbreeding, there is no evidence for linkage disequilibrium in these studies, but this is to be expected given that most marker loci are unlinked or separated by large recombination distances (Foltz, 1986). The employment of DNA markers for the study of heterosis (e.g. Pogson & Zouros, in prep.) may be most useful in this respect. In such studies it is possible to design a protocol that will allow the scoring of variability at pairs of closely linked marker loci. In such pairs the fitness depression of the double homozygote due to associative overdominance is expected to be less than the depression of double homozygotes for non-linked marker loci (Zouros & Mallet, 1989). Thus it is possible that the associative overdominance hypothesis will soon be subject to a more rigorous testing than has been the case so far.

Acknowledgements

I benefitted from discussions with Drs G. Pogson and M. Ball and Mr. C. Batargias. Dr. Phil Hedrick provide insightful comments on an earlier draft of the paper. This research is supported by a grant from NSERC and a grant from the European Communities.

References

Ahmad, M. & P. W. Hedrick, 1985. Electrophoretic variation in the common mussel, *Mytilus edulis*: random association of alleles at different loci. Heredity 55: 47-51.

Foltz, D. W., 1986. Segregation and linkage studies of allozyme loci in pair crosses of the oyser *Crassostrea virginica*. Biochem. Genet. 24: 941-956.

Frydenberg, O., 1963. Population studies of a lethal mutant in *Drosophila melanogaster*. Behaviour in populations with discrete generations. Hereditas 50: 89-116.

Gaffney, P. M., T. M. Scott, R. K. Koehn & W. J. Diehl, 1990. Interrelationships of heterozygosity, growth rate, and heterozygote deficiencies in the coot clam, *Mulinia lateralis*. Genetics 124: 687-699.

Hawkins, A. J. S., B. L. Bayne & A. J. Day, 1986. Protein turnover, physiological energetics and heterozygosity in the blue mussel, *Mytilus edulis*: The basis of variable age-specific growth. Proc. Roy. Soc. London B. 229: 161-176.

Hedrick, P. W., 1983. Genetics of Populations. Science Books International. Boston.

Houle, D., 1989. Allozyme-associated heterosis in *Drosophila melanogaster*. Genetics 123: 789-801.

Koehn, R. K., 1990. Heterozygosity and growth in marine bivalves: comments on the paper by Zouros, Romero-Dorey, and Mallet (1988). Evolution 44: 213-216.

Koehn, R. K. & S. E. Shumway, 1982. A genetic/physiological explanation for differential growth rate among individuals of the American oyster, *Crassostea virginica* (Gmelin). Marine Biology Letters 3: 35-42.

Lewontin, R. C. & K. Kojima, 1960. The evolutionary dynamics of complex polymorphisms. Evolution 14: 450-472.

Maynard Smith, J. & J. Haigh, 1974. The hitchhiking effect of a favorable gene. Genet. Res. 23: 23-35.

Mitton, J. B. & M. C. Grant, 1984. Associations among protein heterozygosity, growth rate, and developmental homeostasis. Ann. Rev. Ecol. & Syst. 15: 479-499.

Ohta, T., 1971. Associative overdominance caused by linked detrimental mutations. Genet. Res. 18: 277-286.

Smouse, P. E., 1986. The fitness consequences of multiple-locus heterozygosity under the multiplicative overdominance and inbreeding depression models. Evolution 40: 946-957.

Strauss, S. H., 1986. Heterosis and allozyme loci under inbreeding and crossbreeding in *Pinus attenuata*. Genetics 113: 115-134.

Weir, B. S. & C. C. Cockerham, 1969. Pedigree mating with two linked loci. Genetics 61: 923-940.

Zouros, E., 1987. On the relation between heterozygosity and heterosis: An evaluation of the evidence from marine mollusks. Isozymes: Current Topics in Biological and Medical Research 15: 255-279.

Zouros, E., 1990. Heterozygosity and growth in marine bivalves: response to Koehn's comments. Evolution 44: 216-218.

Zouros, E., S. M. Singh & H. E. Miles, 1980. Growth rate in oysters: An overdominant phenotype and its possible explanations. Evolution 34: 856-867.

Zouros E. & D. W. Foltz, 1987. The use of allelic isozyme variation for the study of heterosis. Isozymes: Current Topics in Biological and Medical Research 13: 1-59.

Zouros, E. & A. L. Mallet, 1987. Contingency chi-square tests with small expectations. Heredity 59: 117-118.

Zouros, E., M. Romero-Dorey & A. L. Mallet, 1988. Heterozygosity and growth in marine bivalves: Further data and possible explanations. Evolution 42: 1332-1341.

Zouros, E. & A. L. Mallet, 1989. Genetic explanations of the growth/heterozygosity correlation in marine mollusks, pp. 317-324 in Reproduction, Genetics and Distributions of Marine Organisms: XXIII European Marine Biology Symposium, edited by J. S. Ryland and P. A. Tyler. Olsen and Olsen, Fredensborg, Denmark.

Zouros, E. & G. H. Pogson, 1993. The present status of the relationship between heterozygosity and heterosis, in Genetics and Evolution of Aquatic Organisms, edited by A. Beaumont. Chapman and Hall (in press).

T. A. Markow (ed.), Developmental Instability: Its Origins and Evolutionary Implications, 49–67, 1994.

Enzyme heterozygosity, metabolism, and developmental stability

Jeffry B. Mitton
Department of Environmental, Population, and Organismic Biology
The University of Colorado Boulder, CO 80309, USA

Received 20 April 1993 Accepted 24 May 1993

Key words: developmental homeostasis, enzyme heterozygosity, physiological efficiency, routine metabolic cost

Abstract

Developmental homeostasis, measured as either fluctuating asymmetry or variance of morphological characters, increases with enzyme heterozygosity in many, but not all, natural populations. These results have been reported for *Drosophila*, monarch butterflies, honeybees, blue mussels, side-blotched lizards, killifish, salmonid fishes, guppies, Sonoran topminnows, herring, rufous-collared sparrows, house sparrows, brown hares, white-tailed deer, and humans. Because heterozygosity at a few loci can not predict heterozygosity of the entiry genome, these loci must be detecting localized zones that influence the developmental environment.

Studies of malate dehydrogenase in honeybees, *Apis mellifera*, and lactate dehydrogenase in killifish, *Fundulus heteroclitus*, revealed that developmental homeostasis varied with heterozygosity of individual loci. Heterozygotes differed from homozygotes in fluctuating asymmetry, morphological variance, and in correlations between morphological characters.

The protein loci in these studies code for enzymes, and therefore do not directly influence morphological characters. However, some enzymatic loci substantially influence metabolism, and contribute to variation in the amount of energy available for development and growth. This argument can be made most convincingly for the LDH polymorphism in killifish. LDH genotypes differ in enzyme kinetic properties that measure differences in physiological efficiency, and these differences produce measurable and predictable differences in physiology and development. Under environmental conditions which impose a stress upon development, genotypes at these loci may have different amounts of energy available for development, and consequently exhibit different levels of developmental homeostasis.

Introduction

Developmental homeostasis or developmental stability is the buffering of developmental processes against environmental and physiological variability. This phenomenon is often traced back to I. Michael Lerner (1954), who compared the variability of phenotypes in pure breeding lines with the variability in the progeny of crosses between lines. In many cases, the variability in the progeny of crosses between lines is lower than the variability in the pure breeding lines. Because the F_1 progeny are heterozygous for all loci differing between the pure breeding lines, Lerner concluded that developmental homeostasis increased with heterozygosity.

The work of Lerner, which involved crosses between strains, has led us to think in terms of developmental homeostasis reflecting the heterozygosity of the entire genome. However, the relationship between heterozygosity and developmental stability is still apparent when smaller sets of genes are used to define the groups. For example, Robertson and Reeve (1952) demonstrated that morphological variation in *Drosophila melanogaster* decreased regularly with the number of chromosomes made heterozygous by controlled breeding with balancer

Table 1. The squared coefficient of variation of wing length in *Drosophila melanogaster* as a function of chromosomal heterozygosity.

	# Heterozygous chromosomes			
Sex	0	1	2	3
female	2.54	2.23	1.44	1.20
male	1.64	1.31	1.19	

Note: from Robertson and Reeve (1952)

chromosomes. They estimated morphological variation with the squared coefficient of variation of wing length, and compared groups carrying zero to three chromosomes in heterozygous condition. Variability decreased regularly with chromosomal heterozygosity; the most homozygous females were more than twice as variable as the most heterozygous females (Table 1).

The viabilities of individuals made homozygous or heterozygous for chromosomes from natural and experimental populations were estimated in *Drosophila pseudoobscura*, *D. persimilis*, *D. prosaltans*, and *D. melanogaster* (Dobzhansky & Wallace, 1953). Each chromosomal genotype was replicated across environments differing randomly in density, amount of food and quality of food. In general, the viabilities of homozygous genotypes differed significantly among replicates, while the viabilities of heterozygotes were homogeneous. Dobzhansky and Wallace concluded that heterozygotes were better able than homozygotes to tolerate a range of environmental conditions.

The work by Robertson and Reeve and by Dobzhansky and Wallace demonstrated that levels of developmental homeostasis differed among groups of flies defined by heterozygosity or homozygosity for a particular chromosome. More recently, empirical studies have demonstrated that enzyme variability is associated with developmental homeostasis (reviewed by Mitton & Grant, 1984; Zouros & Foltz, 1987). These studies demonstrate that differences in developmental homeostasis appear when groups from a population are defined with a few loci, and in a few cases, with a single locus. This treatment will focus upon those studies, and the questions that they raise. The objective here is to relate developmental stability to patterns of variation and patterns of selection in natural populations.

Measures of developmental stability

Developmental homeostasis has been measured with both the degree of morphological variation (Mitton, 1978; Eanes, 1978; Handford, 1980; Mitton & Koehn, 1985; Govindaraju & Dancik, 1987; Yezerinac, Lougheed & Handford, 1992; Hartl *et al.*, 1993) and with fluctuating asymmetry (Leary, Allendorf & Knudsen, 1983, 1984, 1985; Clarke, Brand & Whitten, 1986; Hartl *et al.*, 1993). Fluctuating asymmetry is generally conceded to be the optimal measure of developmental homeostasis, for the genetic environment is held constant by comparing the sizes, shapes, or numbers of elements between the opposing elements of bilateral characters. This measure is then compared among groups differing by treatment or level of genetic variation. In contrast, when morphological variation is used to estimate developmental homeostasis, variance is calculated among the individuals within a group, and the levels of variability are compared among groups.

Chakraborty and Ryman (1983) and Leary, Allendorf and Knudsen (1984, 1985) faulted the use of phenotypic variance as an estimate of developmental stability. Chakraborty and Ryman (1983) asserted that heterozygotes should have intermediate phenotypic values and lower phenotypic variance than homozygotes. Their assertion, however, seems to apply only to the genes directly determining the development of a character. Leary, Allendorf and Knudsen (1984, 1985) make the same point; they discount the use of morphological variance to estimate developmental homeostasis because the differences between heterozygotes and homozygotes can be accounted for by simple additive genetic variation controlling these characters. Once again, this caveat only applies to the genes directly determining the trait, or genes in strong linkage disequilibrium with those genes. While these concerns are valid, direct involvement is not possible in the examples considered here, for the loci used to define the degree of heterozygosity are enzyme polymorphisms; these genes code for proteins, and do not directly influence morphological characters. Linkage disequilibrium is common in inbreeding species (Allard, 1975). In outbreeding species, however, linkage disequilibrium is evident between inversions and loci immediately adjacent to them (Prakash & Lewontin, 1968, 1971; Prakash

& Merritt, 1972; Prakash, 1974; Policansky & Zouros, 1977), but otherwise, linkage disequilibrium does not appear to characterize the genomes of outbreeding species (Mukai, Mettler & Chigusa, 1971; Mukai & Yamaguchi, 1974).

Morphological variance can be estimated with a single character or multiple characters. When multiple characters are used, the variances or coefficients of variation can be averaged across characters, or, if there are differences among the means, comparisons can be made with the coefficient of variation, the standard deviation divided by the mean. Alternatively, the variability can be quantified with the generalized variance (Van Valen, 1978), which provides the advantage of utilizing both the variances and the correlations among characters to describe the variation within a group. The generalized variance is proportional to the volume occupied by the data points in multivariate hyperspace. It is estimated as the determinant of the variance-covariance matrix, or equivalently, the determinant of the correlation matrix multiplied by all of the univariate variances. Thus, there are alternate advantages to recommend either fluctuating asymmetry or morphological variances for estimating developmental stability.

Expected relationships with allozyme loci

Morphological variance increases with heterozygosity among populations

Among populations, morphological variation should increase with genetic diversity. That is, if mutation, genetic drift and selection influence all portions of the genome in similar ways, then populations with high levels of protein polymorphism should also have high levels of genetic variation for morphological characters. This result was first reported from a survey of variation in the side-blotched lizard, *Uta stansburiana* (Soulé, 1967, 1971), and has been extend to freshwater sculpins (Strauss, 1989, 1991). This relationship is probably general, but it has nothing to do with developmental homeostasis.

Developmental homeostasis increases with heterozygosity among populations

Measures of phenotypic variability and measures of variability of production are generally lower in the progeny of crosses between pure breeding lines than in the pure breeding lines. Because pure breeding lines are expected to have lost some genetic variability, and because the progeny of a cross between lines will be heterozygous for any genes fixed for different alleles in those lines, crosses between lines should produce higher levels of genetic variability than is maintained within pure breeding lines. Thus, the progeny of crosses between lines, which generally have enhanced developmental homeostasis, also have enhanced heterozygosity. Furthermore, developmental homeostasis generally declines as inbreeding accumulates in populations. These general relationships lead us to associate developmental stability with heterozygosity. Among populations, fluctuating asymmetry is expected to decrease with heterozygosity. But a prediction for morphological variation is difficult, for variation in heterozygosity among populations is often associated with variation in the numbers of polymorphic loci and the numbers of alleles. An increase in heterozygosity might produce an increase in developmental homeostasis and a decrease in morphological variation. But if the increase in heterozygosity is accompanied with an increase in the numbers of alleles and polymorphic loci, then morphological variation would probably increase, as explained above.

Developmental homeostasis increases with heterozygosity within populations

Developmental homeostasis is expected to increase with heterozygosity within a population and conceivably within families. This relationship should be detectable with both fluctuating asymmetry and morphological variance. Comparisons within populations are preferable to comparisons among populations, for environmental variation and variation in levels of genetic variation among populations could obscure relationships between genetic variation and developmental homeostasis.

Observations among populations

Fluctuating asymmetry has a biogeographic pattern and is associated with genetic variation in the side-blotched lizard, *Uta stansburiana* (Soulé, 1967, 1979). Fluctuating asymmetry was measured with four bilateral characters in two populations on the mainland and 18 populations on islands in the Sea of Cortez. Two geographic patterns emerged; asymmetry increased with distance from the mainland and decreased with island size. There did not seem to be an organism-wide level of buffering, for asymmetries were distributed independently among individuals within a population. However, if a character was highly asymmetric in a population, it was likely that asymmetry was relatively high at other characters; this pattern of variation prompted Soulé to propose the population variation parameter, or PVP, the avarage asymmetry for a population. Genetic variation for some of these population samples was estimated with population surveys of 18 allozyme loci (Soulé & Yang, 1974). The correlation between the population variation parameter and average heterozygosity for 15 island populations was $r = -0.55$ ($P < 0.05$).

A study employing a triploid parthenogenetic fish and one of its diploid progenitor species permitted a unique dissection of a pattern of geographic variation in developmental stability (Vrijenhoek & Lerman, 1982). *Poeciliopsis 2 monacha-lucida* is a triploid, gynogenetic species that uses the sperm of one of its progenitor species, *P. monacha*, to initiate development. The genes carried by the sperm are not incorporated into the egg. These sympatric taxa occupy a drainage system in Sonora, Mexico, that varies from ephemeral pools at the upper end to a river at the lower end. Genetic variation at 25 allozyme loci in the sexual species varied across this gradient, from average heterozygosity of 0.1% at the upper end to 4.4% at an intermediate site to 8.1% at the lower end. Genetic variation was probably lost in recurrent founder events in the ephemeral pools at the upper extreme of the drainage system. The parthenogenetic species, of course, did not vary from site to site, for individuals collected at different sites were all members of an extended clone. Developmental homeostasis of both species was estimated with fluctuating asymmetry at eight bilateral characters. In the sexual species, developmental homeostasis increased with average heterozygosity. In contrast, developmental homeostasis did not vary across the same environmental gradient in the asexual species. The parthenogenetic species served as a baseline for comparison in this study, and it showed that the range of environmental variation sampled did not have a demonstrable effect upon developmental homeostasis. In this particular case, variation in developmental homeostasis among populations was clearly associated with differences in genetic variation.

The genetic variation that influences developmental homeostasis may also be associated with other components of fitness. For example, a study of geographic variation in the Sonoran topminnow revealed that allozyme variation was associated with both survival and developmental homeostasis (Quattro & Vrijenhoek, 1989). Three populations of the endangered Sonoran topminnow, *Poeciliopsis occidentalis occidentalis*, were compared for allozyme heterozygosity and components of fitness. Populations from Monkey Springs, Tule Springs, and Sharp Springs had average heterozygosities of 0.0%, 1.5%, and 3.7%, respectively. Ten gravid females were captured from these remnant populations, and their offspring were raised in the laboratory. The viability of fish from Monkey Springs was significantly lower than the viabilities in the other populations, which did not differ from one another. Growth rates differed among localities, and increased regularly with heterozygosity. Fecundities also differed among populations, and again increased with heterozygosity. Fluctuating asymmetry at seven morphological characters was used to estimate developmental stability; fluctuating asymmetry differed among populations, and decreased with heterozygosity.

Observations within populations

Many studies have reported associations between individual heterozygosity and components of fitness, including developmental homeostasis. Of these, comparisons within populations are most convincing, for they avoid the environmental heterogeneity among populations. Some of these studies are briefly summarized here.

Poikilotherms

Morphological variation decreases with allozyme heterozygosity in the monarch butterfly, *Danaus plexippus* (Eanes, 1978). Morphological variation was described with two characters, the length of the forewing and the diameter of the major forewing spot divided by the length of the adjacent wing vein. The means and variances of these characters were compared among the genotypes at six allozyme polymorphisms. Analyses of variance did not reveal any differences among the means of the genotypes segregating at a locus. However, when the variances were compared between heterozygotes and homozygotes at a locus, heterozygotes had lower variances in 19 of 24 tests, and five of these reached statisctical significance.

Morphological variation in the blue mussel, *Mytilus edulis*, decreases with allozyme heterozygosity (Mitton & Koehn, 1985). Morphological variation was measured with 12 shell characters, and enzyme genotypes at four polymorphic loci were obtained for 103 mussels collected from a single site in the Cape Cod Canal. Morphological variation was summarized with principal components analysis; the first axis was clearly associated with size, while principal axes 2, 3, and 4 summarized various aspects of shape. The projections of individuals heterozygous for 0, 1, 2, 3, and 4 loci onto these principals axes were compared to determine whether these groups differed in size, in shape, or in the variability of shape. The heterozygosity classes did not differ in mean size, or in shape, but they differed in the variability of shape, with highly heterozygous individuals exhibiting the lowest variance in shape. Imagine a population of mussels represented by points in multivariate morphological space; the most heterozygous individuals were concentrated at the nucleus of this cluster, while the most homozygous individuals tended to be randomly distributed around the periphery.

Developmental homeostasis increases with allozyme heterozygosity in a variety of salmonid fishes. This relationship was first revealed in a laboratory study of five bilateral characters and 42 polymorphic loci in rainbow trout, *Salmo gairdneri* (Leary, Allendorf & Knudsen, 1983). In a subsequent study, the heritabilities of meristic characters and fluctuating asymmetry were estimated in 14 full sib families. Five bilateral characters and three additional meristic characters were used to estimate developmental homeostasis, and 13 protein polymorphisms were used to estimate genetic variation. The heritability of meristic characters was high (mean $h^2 = 0.65$, range 0.18-0.90), but the heritability of fluctuating asymmetry was only $h^2 = 0.02$. The coefficient of variation is negatively correlated with the heritability of the character ($r = -0.40$). The average number of asymmetric traits per individual declined with the expected allozyme heterozygosity for the full-sib family.

Developmental homeostasis also increases with protein heterozygosity in natural populations of rainbow trout, in cutthroat trout *Salmo clarki lewisi*, and brook trout *Salvelinus fontinalis* (Leary, Allendorf & Knudsen, 1985). Fourteen tests of the relationship between fluctuating asymmetry and heterozygosity were tabulated, and the relationship was negative in 13 of the 14 tests (sign test, $P < 0.01$), and five of the 13 correlation coefficients were significantly negative. Analysis of these data suggests that the strength of the association with homeostasis varies among loci. Fluctating asymmetry was also higher in individuals with serious deformities (truncated upper jaw, extra pelvic fins, missing pectoral fins), indicating that fluctating asymmetry revealed biologically relevant differences in developmental stability. A separate study involving a null LDH allele revealed that null heterozygotes had significantly higher, not lower, fluctuating asymmetry. This result suggested a role for the individual enzyme and its activity, rather than associated heterozygosity.

Allozyme heterozygosity was associated with developmental homeostasis and other components of fitness in the guppy, *Poecilliopsis* (Beardmore & Shami, 1979). This study examined genetic variation and components of fitness in a population maintained in a greenhouse. Both growth rate and viability increased with individual heterozygosity. Because fecundity increases with size, and size was correlated with individual heterozygosity, fecundity increased with individual heterozygosity. Mortality in young and adults produced normalizing selection upon the number of caudal fin rays. In addition, morphological variability decreased with individual heterozygosity; that is, the most heterozygous individuals were concentrated around the mean number of caudal fin rays, while predominantly homozygous individuals more frequently had high or low values.

Enzyme heterozygosity is associated with both growth and developmental homeostasis in the herring, *Clupea harengus* (King, 1985). Mean length of fish increased with heterozygosity at five loci, and the asymptotic length of heterozygotes (324 mm) exceeded that of homozygotes (308 mm). In

70 comparisons of variances, homozygotes were more variable than heterozygotes in 49 comparisons, and 13 of these were statistically significant. Just as in killifish (Mitton, 1978), contrasts were most consistent for the LDH polymorphism.

Homeotherms

Because the majority of the observations linking developmental stability with allozyme variation have been reported in poikilotherms, the importance of this phenomenon in homeotherms has been suspect. Development is expected to occur under a wider range of temperatures in poikilotherms than in homeotherms, conceivably reducing the associations between heterozygosity and developmental homeostasis in homeotherms. Handford (1980) was the first to test this association in a homeotherm; he found no relationship between protein heterozygosity and morphological variation in the rufous-collared sparrow, *Zonotricchia capensis*. However, that study was criticized for its small sample sizes and lack of statistical power (Eanes, 1981). A subsequent study employing the same species but with large samples (N = 479) yielded very different results (Yezerinac, Lougheed & Handford, 1992). Approximately 20 males were captured at each of 25 collection localities. Heterozygosity was monitored at 12 allozyme polymorphisms, and morphological variation was measured with eight external and twelve skeletal measurements. The dimensionality of the morphological data was then reduced and summarized with principal components. Morphological variation increased with heterozygosity among populations, although this relationship was not strong. In 95 of 100 single locus tests, the variability of heterozygotes was less than the variation in homozygotes (sign test, $P < 0.001$). Results were heterogeneous among loci, for comparisons with 6-phosphogluconate dehydrogenase did not reveal the differences revealed by the other loci. The low level of genetic variation, which is typical for birds, restricted multilocus comparisons to groups with 0, 1 and >1 heterozygous loci. Sixteen of twenty tests of homogeneity of variance were heterogeneous among the heterozygosity classes. Morphological variation did not decrease monotonically with heterozygosity, but the most heterozygous group consistently had the lowest level of morphological variability.

Two to four polymorphic loci, depending on locality, and 14 skeletal measurements were employed for a study of developmental homeostasis in the house sparrow, *Passer domesticus* (Fleischer, Johnston & Klitz, 1983). Comparisons were generally made between individuals heterozygous for 0, 1, and 2 loci. Size did not differ among the genotypes at any single locus. However, size increased with individual heterozygosity in females, and in two of four analyses of males. Variances did not differ among the genotypes of individual loci, but variances generally decreased with individual heterozygosity. Thus, the most heterozygous birds tended to be both the largest and the least variable.

Heterozygosity at nine polymorphic loci is related to body size and antler size and form in white-tailed deer, *Odocoileus virginianus* (Scribner, Smith & Johns, 1989; Scribner & Smith, 1990; Smith *et al.*, 1991). In a study of 558 bucks hunted at the Savannah River Ecology Lab in 1978 and 1979, body mass, kidney fat, antler beam length and antler beam diameter increased with heterozygosity. Highly heterozygous individuals had slower growth during the early years, but had greater asymptotic body mass, fat levels, and antler size (Scribner, Smith & Johns, 1989). Boone and Crockett scores, which quantify the size and symmetry of deer horns, were 27% higher in highly heterozygous bucks in 1978 and 1979, and were 17% higher in highly heterozygous bucks in 1985 and 1986.

A study of 13 polymorphic loci and fluctuating asymmetry in 27 meristic and nine metric bilateral skull and mandible traits in the brown hare, *Lepus europaeus*, revealed a novel pattern of association between heterozygosity and developmental stability (Hartl *et al.*, in prep.). While no relationship was found between heterozygosity and metric traits, fluctuating asymmetry of meristic characters decreased with protein heterozygosity.

Developmental homeostasis is also associated with protein polymorphisms and blood group loci in humans (Livshits & Kobyliansky, 1985; Kobyliansky & Livshits, 1985). Within a sample of 200 Jewish men of similar age and ancestry, there was a remarkable correspondence between morphological variation and heterozygosity at blood group loci (Livshits & Kobyliansky, 1984). Men were placed in five heterozygosity classes on the basis of their genotypes at the MN, Ss, and Duffy systems and

Table 2. Analyses of variance and fluctuating asymmetry in honeybees, comparing dimensions of wings between homozygotes and heterozygotes for the Mdh locus.

Trait	Hom	Het	F	P	FA
L11	27.69	27.96	7.23	**	NS
L12	11.60	11.63	0.25	NS	Hom > Het*
L13	10.26	10.20	0.48	NS	NS
L14	4.93	5.09	9.77	**	NS
L15	5.34	5.47	3.35	NS	NS

Note: * = P < 0.05, ** = P < 0.01, NS = non significant. Values under 'Homozygotes' and 'Heterozygotes' are mean values. 'FA' uses analysis of variance to test differences in fluctuating asymmetry, employing a one-tailed test.

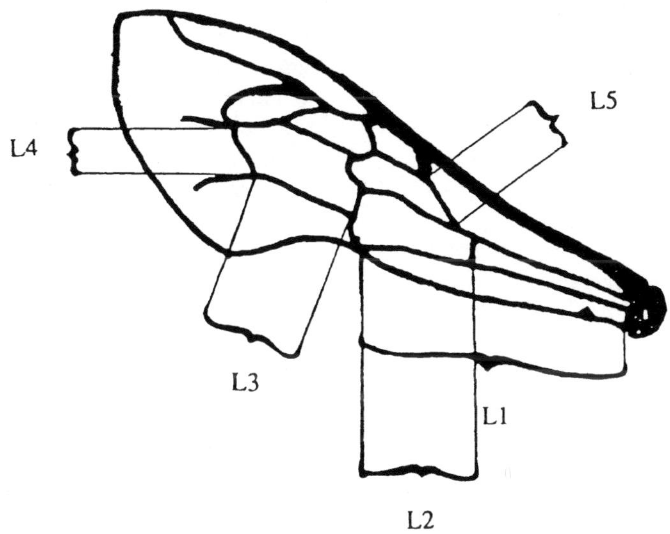

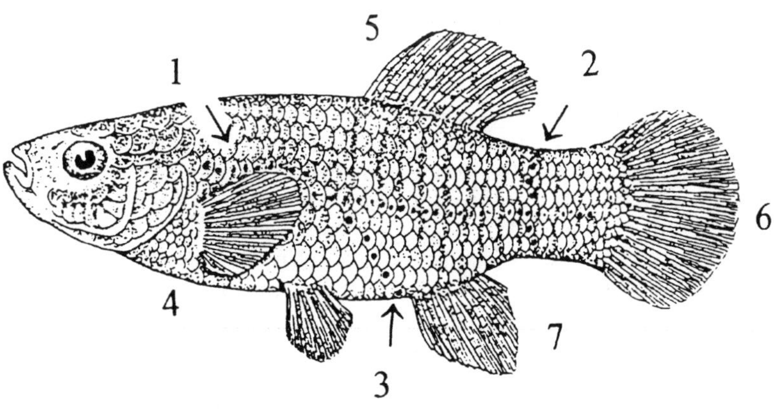

Fig. 1. Morphological characters measured form the fore-wings of honeybees, *Apis mellifera*, and from killifish, *Fundulus heteroclitus*. Characters from killifish are (1) scales along the lateral line; (2) circumpeduncular scales; (3) scales below the lateral line; (4) pectoral fin rays; (5) dorsal fin rays; (6) caudal fin rays, and (7) anal fin rays.

the C locus of the Rhesus system. Morphological variation was measured with ten morphological characters. For each of the ten morphological characters, men were assigned to the lower extreme (25%), the upper extreme (25%) or the modal group (± 0.67 standard deviations of the mean, 50%). For nine of the ten characters, the frequency of individuals in the modal group increased with heterozygosity and seven of these correlations were statistically significant (mean of nine r values = 0.85). Similarly, the coefficients of variation were negatively correlated with heterozygosity at nine of the ten characters, and four of these were statistically significant (mean of nine correlations = 0.82). Only one character deviated from this general pattern; the modal frequency at this character decreased with heterozygosity ($r = -0.975$, $P < 0.01$) and the coefficient of variation increased with heterozygosity ($r = 0.959$, $P = 0.01$). A multivariate analysis of morphological and genetic characters revealed that heterozygotes were closer than homozygotes to the multivariate mean in multidimensional morphological space (Kobyliansky & Livshits, 1985).

MDH in honeybees

The haplodiploidy of honeybees (*Apis mellifera*) and the control that can be exercised over their breeding system have made them a fine subject for the study of the genetic mechanisms underlying developmental homeostasis (Bruckner, 1976; Clarke, Brand & Whitten, 1986; Clarke, Oldroyd & Hunt, 1992). The haplodiploidy of honeybees allows comparisons between haploid drones and diploid workers. Furthermore, the breeding system can be controlled to alter degrees of inbreeding. Measurements of distances on wings provide estimates of fluctuating asymmetry. Females are consistently more symmetric than drones from the same hive, an observation consistent with both a hypothesis based upon heterozygosity and a hypothesis based upon genic balance (Clarke, Oldroyd & Hunt, 1992). Results of studies of developmental stability associated with the loss of heterozygosity resulting from inbreeding have been inconsistent; Bruckner (1976) reported a decrease in developmental stability with inbreeding, but subsequent studies (Clarke, Brand & Whitten, 1986; Clarke, Oldroyd & Hunt, 1992) found no consistent differences associated with inbreeding.

The haplodiploidy of social insects reduces the probability of balanced polymorphism, and therefore it is not surprising that honey bees have low levels of protein polymorphism. However, malate dehydrogenase (MDH) is polymorphic in the honeybee throughout both its natural and introduced ranges. The genotypes of MDH differ in respiration rates during hovering, and the sexes exhibit different patterns of variation (Coelho & Mitton, 1988). Oxygen consumption differs among the diploid genotypes of the workers, with the highest value in FF, an intermediate value in FS, and the lowest value in SS. In haploid drones, which are larger and therefore hover with a higher internal temperature for any given ambient temperature, the oxygen consumption of the F haplotype is significantly lower than the oxygen consumption of the S haplotype. Selection favoring one homozygous genotype in workers and the alternate haplotype in males could balance this genetic variation, explaining the occurrence of a widespread, common polymorphism in a species with relatively little genetic variation.

Heterogeneous rates of respiration among MDH genotypes suggest that the MDH polymorphism (or a locus in linkage disequilibrium with it) has an important impact upon metabolism. To test whether this locus influences developmental homeostasis, measurements of five distances were taken from both wings of workers (Fig. 1) from a single hive, and their MDH genotypes were identified with starch gel electrophoresis (Messier & Mitton, in prep., Table 2). Character two had greater fluctuating asymmetry in homozygotes than in heterozygotes. Analysis of variance was used to compare the means of the characters measured from the right wings. Characters one and four were significantly larger in heterozygotes than in homozygotes; this trend was not seen in all characters, indicating that heterozygotes and homozygotes differed in wing shape. The characters L3 and L4 were positively correlated in heterozygotes ($r = 0.47$) and negatively correlated in homozygotes ($r = -0.20$), and these correlations were significantly different ($P < 0.01$). A discriminant function analysis of the five wing characters revealed significant differences between the wings of heterozygotes and homozygotes ($P = 0.01$); this discriminant function correctly assigned 73% of the bees to their genotypic groups.

LDH in killifish

When killifish collected from a natural population were divided into two groups, heterozygous and homozygous for a single allozyme polymorphism, the heterozygous group had lower phenotypic variation at seven meristic variables (sign test, $P < 0.01$; Mitton, 1978). This was the first observation of differences in phenotypic variability associated with a single locus, and it was an unexpected result. Vertebrates have genomes containing between 50,000 and 100,000 genes, and for this reason, groups defined by heterozygosity at a single locus were not expected to reveal detectable differences in developmental stability. Comparisons between heterozygotes and homozygotes were most striking for the lactate dehydrogenase (LDH) polymorphism, encouraging further analyses, presented below.

To gain further insights into the associations between heterozygosity and developmental stability, additional analyses were conducted upon the genetic and morphological data gathered from the killifish collected from Flax Pond, New York, in 1971 (Mitton & Koehn, 1975, 1976; Mitton, 1978, 1993). From twelve polymorphic loci, lactate dehydrogenase (LDH), transferrin, serum esterase, phosphohexose isomerase, glucose-6-phosphate dehydrogenase and phosphoglucomutase were chosen for this analysis for their moderate to high frequencies of heterozygotes. The following meristic characters (Fig. 1) were used to estimate morphological variance: scales along the lateral line, circumpeduncular scales, scales below the lateral line, and fin rays in the pectoral, dorsal, caudal and anal fins. Only fish with complete data for all genotypes and morphological characters were included in the analyses, yielding a total sample size of 219 (Table 3).

Fish were divided into groups defined by individual heterozygosity at six loci. To maintain group sizes of at least 20 for unbiased estimates of variance, fish with zero and one heterozygous loci were combined into one group, and fish with four or more heterozygous loci were joined into a single group. This yielded sample sizes of 61, 65, 69, and 24 fish in the groups with 0-1, 2, 3, and 4-6 heterozygous loci, respectively. Univariate variances were compared between the most homozygous and the most heterozygous groups with a one tailed test of the ratio of variances. The mean values did not differ between these groups, but one of the seven tests of the ratio of variances reached statistical significance ($P < 0.05$). The variance of the number of scales along the lateral line was significantly lower in the most heterozygous group.

Table 3. Measures of morphological variation for a sample of *Fundulus heteroclitus* divided into heterozygosity classes.

	Heterozygosity			
	0,1	2	3	4-6
Σ	0.055	0.035	0.060	0.017
N	61	65	69	24
det	0.663	0.639	0.540	0.232

Note: Fish were collected from Flax Pond, NY in 1971. Heterozygosity is based upon 6 allozyme polymorphisms, morphological variation is based upon 7 meristic characters. Σ is the generalized variance, N is the sample size, and det is the determinant of the correlation matrix.

The generalized variance also suggested that the highly heterozygous group had lower morphological variation (Table 3). For fish with 0-1, 2, 3, and 4-6 loci heterozygous the generalized variances were 0.055, 0.035, 0.060, and 0.017, respectively. The determinants of the correlation matrices for these groups progress monotonically from 0.663 to 0.232, indicating that the correlations in these matrices tend to increase in absolute value with increasing heterozygosity. Higher correlations among morphological characters suggest more regular development. I have chosen to focus upon LDH because of the wealth of enzyme kinetic data and physiological data suggesting that this locus has a major impact upon whole animal physiology and development (elaborated under Interpretations of the data).

The 219 fish from Flax Pond were divided into two groups, those heterozygous and those homozygous for LDH (Table 4). Although the differences are not statistically significant for any of the characters, the variances of the seven morphological characters tend to be smaller in heterozygotes than in the homozygotes. This trend was also apparent for the generalized variance, which was 0.041 for heterozygotes, and 0.072 for homozygotes. The generalized variances suggest that the volume occupied in morphological hyperspace by LDH heterozygotes is only 57% of the volume occupied by LDH homozygotes.

Table 4. Measures of morphological variance and the correlation between scales below the lateral line and dorsal fin rays for *Fundulus heteroclitus* grouped by Ldh genotype.

	Flax Pond		West Meadow Creek	
	het	hom	het	hom
Σ	0.041	0.072	0.006	0.024
N	69	150	18	34
det	0.590	0.702	0.098	0.349
r	-0.34**	0.21**	-0.41†	0.40*

Note: Σ is the generalized variance, N is the sample size, det is the determinant of the correlation matrix, and r is the correlation between the number of scales below the lateral line and the number of dorsal fin rays. Het and hom refer to heterozygous and homozygous at the Ldh locus. † = $P < 0.10$, * = $P < 0.05$, ** = $P < 0.01$.

The same genetic and morphological data were available for fish captured from West Meadow Creek, and these data revealed similar patterns when fish are grouped by genotype at the LDH locus (Table 4). The generalized variances for heterozygotes and homozygotes were 0.006 and 0.024, respectively. In this data set, heterozygotes occupy 25% of the volume of hyperspace occupied by homozygotes.

The volume of hyperspace occupied by different population samples seems to correspond closely to the sample size (Tables 3, 4). Is this trend a statistical artifact? Variances estimated by samples of less than 20 are biased downwards, but only one of the sample size in this analysis is less than 20, and it is 18. The trends seen in the generalized variances are paralleled by the determinants of the correlation matrices (Tables 3, 4). Correlation coefficients are unbiased in the range of sample sizes utilized here, so these trends in multivariate morphological variance appear to be real.

Inspection of the correlation matrices of LDH heterozygotes and homozygotes revealed a striking difference. At Flax Pond, the number of dorsal fin rays and scales below the lateral line (figure 1) were negatively correlated in heterozygotes ($r = -0.34$, $P < 0.01$) but positively correlated in homozygotes ($r = 0.21$, $P < 0.01$). These correlation coefficients were significantly different ($P < 0.001$). The same pattern was evident at West Meadow Creek. The correlation between dorsal fin rays and scales below the lateral line was negative in heterozygotes ($r = -.41$ $P < 0.10$) and positive in homozygotes ($r = 0.40$, $P < 0.05$). Once again, the correlation coefficients are different ($P < 0.02$) and the trend is the same in these population samples.

These analyses revealed several consistent patterns. When populations are divided into groups heterozygous or homozygous for a single locus, univariate variances of morphological characters tend to to be lower in heterozygotes than in homnozygotes. These comparisons are occasionally statistically significant with a test of the ratio of variances, but the general trend is apparent when a sign test is used to summarize many comparisons (Mitton, 1978; Eanes, 1978; Yezerinac, Lougheed & Handford, 1992). Consistent with this pattern is the decrease in generalized variance with individual heterozygosity. While this pattern is partially attributable to the pattern of univariate variances, there appears also to be an effect of the correlations among morphological characters, measured by the determinant of the correlation matrix. The determinants decrease with individual heterozygosity (Table 3), and exhibit consistent differences between groups defined solely by the genotype at the LDH locus (Table 4).

Null observations

Certainly not all of the studies testing relationships between allozyme variation and morphological variation have revealed significant results. An extensive study of morphological and genetic variation revealed no association in the plaice, *Pleuronectes platessa* (McAndrew, Ward & Beardmore, 1982). No relationship was found between esterase variation and morphological variation in the mosquito *Aedes aegypti* (Kasule & Cook, 1988), and allozyme variation is independent of shell shape variation in the land snail, *Cerion* (Booth, Woodruff & Gould, 1990). A series of experiments have consistently revealed no relationship between allozyme variability and developmental homeostasis in chum salmon *Oncorhynchus keta*, and pink salmon, *Oncorhynchus gorbuscha* (Beacham & Withler, 1985, 1987; Beacham, 1991). Hartl *et al.* (1991) found no relationship between protein variability and the developmental homeostasis measured by antler shape in red deer, *Cervus elaphus*. Wooten and Smith (1986) found no relationship between heterozygosity at 16 polymorphic loci and asymmetry in kidney weight and size of the testicles in

Mus musculus. Handford (1980) initially reported no relationship between allozyme variation and morphological variation in the rufous-collared sparrow, but a later study with larger sample sizes revealed a general decrease in morphological variability with heterozygosity (Yezerinac, Lougheed & Handford, 1992). It is not expected that relationships between genetic variation and developmental stability would be constant; these relationships are expected to vary with environmental conditions, reaching their maximal strength under some intermediate level of environmental stress (developed more fully below).

Interpretations of the data

Genomic heterozygosity

It is important to appreciate what a handful of polymorphic loci might do, and what they cannot do. Most allozyme surveys examine 15-30 loci, and because the percentage of polymorphic loci is typically in the range of 33% to 50%, the number of polymorphic loci might be five to perhaps 15. The number of genes in *Drosophila melanogaster* is on the order of 5,000, and the number of genes in *Homo sapiens* is 50,000 to 100,000. If we assume that protein polymorphisms are selevtively neutral, then these polymorphisms could be used to estimate of the level of genetic variability within a species or the relative amounts of genetic variation in various populations within a species. A sample of protein polymorphisms may also be used to rank individuals in a sample for their individual heterozygosity, or the number of loci heterozygous in an individual for the set of sampled loci (Mitton & Pierce, 1980). While some biologists have speculated that individual heterozygosity of a few polymorphisms might predict the individual heterozygosity of the entire genome (Zouros & Foltz, 1987), this is clearly not possible. This speculation concerns the correlation of a part with a whole, and the number of the loci in a sample (part) is simply too limited to predict the heterozygosity of the entire genome (Mitton & Pierce, 1980; Chakraborty, 1981, 1987). The associations between allozyme heterozygosity and developmental homeostasis are not attributable to heterozygosity of the entire genome.

Chromosomal heterozygosity

The work of Robertson and Reeve (1952) and Dobzhansky and Wallace (1953) clearly demonstrated that the genetic characters influencing developmental homeostasis of *Drosophila melanogaster* were located on all of the major chromosomes. Modern maps of *D. melanogaster* indicate that all of the large chromosomes bear enzymes. Are the associations between enzyme heterozygosity and developmental homeostasis attributable to a correlation between the heterozygosity of one or a few enzymes and the heterozygosity of the genes linked to these markers? This question is similar to the one above, but differs slightly in scope. It is conceivable that the heterozygosity of an enzyme polymorphism is correlated with the heterozygosity of a few tightly linked genes. Note here that the critical issue is not linkage, but linkage disequilibrium. It is certainly true that each gene is linked to hundreds or thousands of other genes, but how far along a chromosome does linkage disequilibrium extend? Linkage disequilibrium does not characterize the genome of obligately outcrossing species, such as the ones mentioned in this review (Mukai, Mettler & Chigusa, 1971; Mukai & Yamaguchi, 1974). Thus, the individual heterozygosity estimated with a few marker loci cannot predict the individual heterozygosity of the entire genome, and it may have very limited capability for estimating the heterozygosity of tightly linked genes. This is not to say that this would never happen, only that it is not likely to be the explanation for a general phenomenon.

Enzymes directly influence developmental homeostasis

Associations between enzyme genotypes and developmental homeostasis suggest the possibility that it is the enzymatic variation that produces measurable differences among genotypes. After all, enzymes catalyze metabolic reactions, and thereby modulate metabolism and other physiological processes. Indeed, most of the kinetic analyses of protein polymorphisms have revealed biochemical differences among genotypes (Koehn, Zeba & Hall, 1983). But if this hypothesis is to be taken seriously, there should be evidence that enzyme polymorphisms influence physiological phenotypes and are evident as differences in growth and physiological efficiency. Are these data available?

The first report of an association between protein heterozygosity and growth rate was from a study of growth in the American oyster, *Crassotrea virginica* (Singh & Zouros, 1978). A second experiment with the oyster produced very similar results; the mean weight of the most homozygous and the most heterozygous oysters were 4.28 grams and 7.33 grams, respectively. A similar study was conducted with the blue mussel, and once again growth rate increased with heterozygosity (Koehn & Gaffney, 1984). Growth rate has also been reported to increase with heterozygosity in the coot clam, *Mulinia lateralis* (Koehn, Diehl & Scott, 1988), the tiger salamander *Ambystoma tigrinun* (Pierce & Mitton, 1982), white-tailed deer, *Odocoileus virginianus* (Cothran *et al.*, 1983), sheep (Baker & Manwell, 1977), pigs (Makaveev, Vener & Baulov, 1978) and humans (Bottini *et al.*, 1979). However, there have been studies that detected no relationship between heterozygosity and growth, with perhaps the most exhaustive study reporting no relationship in humans (Ward *et al.*, 1985).

Numerous empirical studies suggest that physiological efficiency increases with protein heterozygosity. These studies employ enzyme polymorphisms to estimate individual heterozygosity and oxygen consumption at rest to estimate basal metabolic cost. Basal metabolic cost is the energetic cost of keeping the metabolic machinery running. Weight corrected oxygen consumption rates decreased with heterozygosity at 5 enzyme polymorphisms in the American oyster, *Crassostrea virginica* (Koehn & Shumway, 1982). Basal oxygen consumption also decreased with heterozygosity in the gastropod, *Thais hemostomata* (Garton, 1984), the coot clam, *Mulinia lateralis* (Garton, Koehn & Scott, 1984), the blue mussel (Diehl, 1985; Diehl, Gaffney & Koehn, 1986), rainbow trout (Danzmann, Ferguson & Allendorf, 1987, 1988) and the tiger salamander (Mitton, Carey & Kocher, 1986). Weight loss in starving oysters was least severe in highly heterozygous individuals (Rodhouse & Gaffney, 1984).

If heterozygous individuals are, indeed, more physiologically efficient than homozygous individuals, then heterozygous individuals may have more energy to put into reproduction, and may be more fecund than homozygotes. Data consistent with this hypothesis come from a study of patterns of mating and fecundity of *Drosophila melanogaster* (Serradilla & Ayala, 1983). Heterozygotes for the α-glycerophosphate locus, the alcohol dehydrogenase locus, and the acid phosphatase locus produced approximately 50% more eggs than homozygotes. Data consistent with these were obtained in a study of components of fitness associated with the alcohol dehydrogenase polymorphism in *D. melanogaster* (Bijlsma-Meeles & Bijlsma, 1988). Relative fecundities estimated for slow homozygotes (SS), heterozygotes (SF), and fast homozygotes (FF) were 0.88, 1.00, and 0.61, respectively. The production of overwintering cysts per female increases from 61 to 527 as individual heterozygosity ranges from zero to three in the brine shrimp, *Artemia franciscana* (Gajardo & Beardmore, 1989). Fecundity also increases with heterozygosity in both the blue mussel (Rodhouse *et al.*, 1986) and the guppy (Beardmore & Shami, 1979). Fecundity differs among the Pgi genotypes of the butterfly, *Colias philodice eriphyle* (Watt, 1992). The number of eggs laid per day by the 34, 33, and 44 genotypes were 16, 12, and 6, respectively ($P < 0.001$).

Although the mechanism providing greater physiological efficiency of heterozygotes is not yet identified, greater insight into the problem has been revealed by studies of protein turnover rate (Hawkins, Bayne & Day, 1986; Hawkins *et al.*, 1989). The half-life of soluble enzymes is on the order of 24 h (Koehn, 1991), so that many of the thousands of enzymes needed to control metabolism must be replaced each day. Consequently, a substantial portion of basal metabolic cost, perhaps 20% to 40%, is spent to break down damaged proteins and synthesize new proteins. The protein turnover rate in the blue mussel decreases with enzyme heterozygosity, with the turnover rate of the most homozygous class exceeding the rate of the heterozygous class by 25%. The energy saved in heterozygous individuals can then be invested in growth or reproduction.

The studies relating allozyme heterozygosity to developmental homeostasis have employed a few genetic polymorphisms to define the relative heterozygosities of populations or of individuals within a population. Some biologists may interpret these groups as reflecting genomic heterozygosity, or, alternatively, the degree of inbreeding. It is difucult to reject these hypotheses convincingly in most studies. But the genotypes at a single locus can not predict either the individual heterozygosity of the entire genome or the degree of inbreeding. The studies of MDH in honeybees and LDH in killifish indicate that differences in developmental homeostasis can be related to genotypes at a single locus. The study of developmental homeostasis in honeybees was prompted by our impression that the MDH variation was balanced by natural selection. Remember that MDH is variable in all populations, despite a general paucity of genetic variation in honeybees, and that MDH genotypes differ in oxygen consumption while hovering (Coelho & Mitton, 1988). These observations suggest physiological differences attributable to MDH genotypes but there are, as yet, no kinetic studies of honeybee MDH to provide the critical test of this hypothesis.

The study of MDH needs to be replicated because all of the bees were sampled from a single hive, and therefore were probably a mixture of half-sibs and full-sibs. This raises the possibility of a substantial degree of linkage disequilibrium in this particular study. But the study of killifish does not share this problem. The killifish were sampled from a natural population of many thousands of individuals. Furthermore, there is abundant and convincing evidence demonstrating that the LDH polymorphism in killifish has a profound impact upon physiology, fitness and development (Powers *et al.*, 1993).

Two alleles segregate at the LDH-B locus (Mitton & Koehn, 1975), and allelic frequencies vary from nearly fixed for the *a* allele in northern Florida to nearly fixed for the *b* allele in Nova Scotia (Powers & Place, 1978). The clinal variation at this locus is consistent with differences in biochemical properties among the enzyme genotypes (Place & Powers, 1979). Catalytic efficiencies, estimated with the ratio of the rate constant, k_{cat}, to the Michaelis constant, K_m (Hall & Koehn, 1983; Watt, 1985, 1986) are heterogeneous among genotypes, and the pattern of differences varies with temperature. At temperatures near 40 °C, the catalytic efficiency is highest in the *aa* homozygote, but at temperatures near 10 °C, the *bb* homozygotes has the highest catalytic efficiency. The genotype with the highest catalytic efficiency at high temperatures is most abundant in warm environments, and the genotype with the highest catalytic efficiency at low temperatures is most abundant in relatively cold environments.

The LDH genotype is tightly correlated with the amount of ATP in red blood cells (Powers, Greany & Place, 1979). ATP is the major allosteric modifier of the hemoglobin of *F. heteroclitus*, decreasing the affinity for hemoglobin. The higher ATP levels in red blood cells of *bb* homozygotes depress the affinity of hemoglobin for oxygen, allowing this genotype to deliver more oxygen to tissues active in exercise. The amount of oxygen delivered to the muscles of the *bb* homozygote at the maximum swimming speed is estimated to be 18% to 40% greater than that of the *aa* homozygote.

The enzyme kinetic and physiological differences mentioned above suggested specific predictions concerning the time of hatching and the maximium sustained swimming speeds of the LDH genotypes (DiMichele & Powers, 1982a, b; Powers, DiMichele & Place, 1983; Powers *et al.*, 1993). At 10 °C the sustained swimming speeds of the LDH homozygotes differed by 20%. In all cases, differences in whole animal physiology were predictable solely from the kinetics of the LDH genotypes. These results, therefore, are directly attributable to the LDH locus.

In addition to differences in hatching times and sustained swimming speeds, LDH kinetics also predict differential survival of the LDH homozygotes at high (30 °C) temperature (DiMichele & Powers, 1991). The relative fitnesses for the *aa*, *ab*, and *bb* genotypes were 0.66, 1.0, and 0.25, respectively.

The relationship between LDH genotype and hemoglobin-oxygen affinity and the role of low oxygen concentration in hatching suggested the hypothesis that LDH genotypes would hatch at different times. The *aa* homozygote, having the highest affinity for oxygen, was predicted to produce the lowest oxygen tensions, and therefore to hatch before the other genotypes (DiMichele & Powers, 1982a). LDH genotypes and hatching times of more than 20,000 eggs were examined in a series of mass crosses and pair crosses. The hatching times of the different genotypes were significantly heterogeneous, and as predicted, eggs bearing the fast homozygote hatched before the other genotypes. The mean hatching times for the *aa* homozygote, the *ab* heterozygote, and the *bb* homozygote were, respectively, 12.2 days, 12.5 days, and 13.6 days.

Further studies of developing eggs revealed the mechanism by which LDH-B genotypes influence hatching times of eggs. In both air and in water, the oxygen consumption of the *aa* homozygotes exceeded that of *bb* homozygotes at 20 °C (DiMichele & Powers, 1984). These differences in oxygen consumption revealed differences in rates of development. Within 24 h of fertilization, *aa* homozygotes have higher rates of oxygen consumption and levels of lactate than *bb* homozygotes (Paynter *et al.*, 1991). The *aa* homozygotes increase metabolic rate dramatically between 10 and 15 hours after fertilization, while *bb* do not. Metabolic rate of very young eggs appears to depend upon lactate accumulation and utilization, both controlled by LDH-B. Direct involvement of LDH enzymes in development was demonstrated by injecting LDHs of known genotypes into developing eggs (DiMichele, Paynter & Powers, 1991). Injection of the *aa* enzyme into an egg with the LDH-*bb* accelerated metabolism, while injection of the *bb* enzyme into an egg with LDH-*aa* slowed metabolism. If the LDH locus directly influences the physiological environment of the egg and its developmental rate, then it is possible that this locus also has a direct influence upon developmental homeostasis.

Developmental homeostasis reflects both genetics and environment

Developmental homeostasis is not a constant or fixed attribute of a population or strain, but a variable that reflects both the genotype of the individual and the environment in which that genotype develops. Developmental homeostasis declines with virtually any environmental change that imposes a drain upon upon the energy budget and thus a stress upon development. Insufficient food, minerals or vitamins, high or low temperature, and disease can all challenge a developing embryo and disturb normal development (Soulé, 1982, and references therein). These stresses may disrupt cellular structures, such as the endoplasmic reticulum, but for the purposes of this presentation, the currency of stress will be the net energy balance that an individual has to work with (Mitton & Koehn, 1985). The capability of an organism to meet the energetic demands of development and environmental stress will be related to its energy reserves or net energy balance. Developing individuals with critically low energy reserves are at risk to any stresses that cost still more energy to maintain homeostasis, and if available energy is not sufficient to buffer against the insult, the event will be recorded in the development of an unusual character or an asymmetry.

While many studies have reported associations between protein heterozygosity and developmental homeostasis, numerous thorough studies have found no correspondence between heterozygosity and either morphological variance or fluctuating asymmetry. This makes good sense. Stressful environments accentuate differences among genotypes, and increase the likelihood of detecting overdominance (Parsons, 1971, 1973, 1987; Mitton & Grant, 1984). For example, Koehn and Shumway (1982) found that oxygen consumption decreased with protein heterozygosity in oysters under normal conditions, but also showed that the relationship was much stronger when oysters were kept at a combination of temperature and salinity which imposed a stress. Similarly, allozyme genotypes are correlated with growth under stress conditions (Diehl, 1989). The feeding efficiency of *Peromyscus polionotus* is independent of enzyme genotype when food quality is high, but feed efficiency increases with heterozygosity when food quality is low (Teska, Smith & Novak, 1990). The relationship between allozyme heterozygosity and growth in the coot clam, *Mulinia lateralis*, is critically dependent upon the environment (Scott & Koehn, 1990). Growth was independent of genotype in both control and highly stressful environments, but growth increased with heterozygosity in two environments with moderate degrees of stress. Relationships between enzyme genotypes and developmental homeostasis are not constant because animals develop in heterogeneous environments, and their energy budgets vary with both genotype and environmental conditions.

Enzymes may reflect energy balance

Insight into the role of protein polymorphisms in developmental homeostasis can be gained by considering the balanced energy equation as presented by Bayne and Newell (1983):

$$Pg + Pr = C^*AE - (R_m + R_r)$$

where Pr and Pg are production of gametes and somatic tissues, C is the consumption of energy, AE is the efficiency of absorption of consumed energy, R_m is the metabolic cost of maintenance, and R_r includes all the costs of activity, seeking food and mates, defending territories, etc. Thus, the terms on the left side of the equation represent growth and reproduction. For developing animals, we may assume that all energy is apportioned to development and growth, and therefore Pr = 0. On the right side of the equation, the first term represents the absorbed energy (A), while the second term represents costs. If stress is defined as any change in the environment that reduces the fitness of an organism, then any change in the right hand side of the equation that reduces the terms on the left hand side of the equation is a stress. Developmental homeostasis declines when energy is insufficient to sustain normal growth and development. Differences among individuals in any of the variables on the right hand side of the equation could contribute to variation among individuals for Pg.

R_m, which is typically estimated as oxygen consumption at rest and in the absence of food, is related to enzyme heterozygosity. R_m decreases with allozyme heterozygosity in a wide range of invertebrates and vertebrates (summarized above). Although variation in R_m is typically used to explain differences in growth rate (Koehn & Shumway, 1982; Mitton, Carey & Kocher, 1986; Hawkins, Bayne & Day, 1986, Hawkins *et al.*, 1989; Koehn, Diehl & Scott, 1988; Koehn & Bayne, 1988), the same reasoning reveals how enzyme polymorphisms can contribute to variation in developmental homeostasis. Just as energy is needed for growth, energy is needed for normal development.

Consideration of the balanced energy equation helps to explain why some studies report associations between protein heterozygosity and developmental homeostasis, while other studies report no associations, and still other studies report associations with some loci, but not with others. For example, when animals are developing under optimal conditions with abundant food, subtle differences in R_m do not produce important variation in Pg, so all individuals appear to have high levels of developmental homeostasis. It is when Pg is lowered by some form of stress (insufficient food, disease, high or low temperature, etc.) that differences in R_m become important and measurable as variation in morphological variance or fluctuating asymmetry.

One of the critically important issues in this line of reasoning is whether the products of a single gene can have a sufficient impact upon physiology to produce the observed differences in developmental stability between heterozygotes and homozygotes at a single locus. Two observations suggest that this is possible. Koehn calculated the metabolic cost of protein turnover for a range of proteins, and demonstrated that enzymes with moderate turnover rates, maintained in high concentration in muscle, could cost up to several percent of the mass-specific energy demands for maintenance metabolism (R_m). Because minor changes in the amino acid sequence can dramatically change the half-life of a protein (Rogers, Wells & Rechsteiner, 1986), alternate genotypes at a locus could differ in their energetic costs. But the most convincing evidence comes from the LDH locus in *Fundulus heteroclitus*. At hour 20 of development, the respiration rate of the LDH-*aa* genotype is four times as high as the respiration rate of the LDH-*bb* homozygote (Paynter *et al.*, 1991). There are substantial differences in lactate accumulation, lactate utilization, glucose utilization, and heat dissipation as well. Injections of purified enzymes into *Fundulus* eggs (DiMichele, Paynter & Powers, 1991) clearly demonstrated that these differences are due to the lactate dehydrogenase enzymes, not to genomic heterozygosity, not to heterozygosity of particular chromosomes, not to some mystical gene in linkage disequilibrium with the LDH locus. The physiological differences produced by the LDH enzymes are profound, and are probably responsible for the differences in phenotypes measured in adult fish (Tables 3, 4).

Conclusion

For the better part of a century, geneticists have been examining the phenomena of inbreeding depression and heterosis. The genetic basis for the continuum between inbreeding depression and heterosis received extensive study for decades, but it was never determined decisively whether this continuum was more directly attributable to overdominance or to recessive deleterious alleles. The controversy went stale before it was ever resolved, but I suspect that the majority of contemporary geneticists would favor the hypothesis relying upon deleterious recessive alleles. Developmental homeostasis was most formally recognized by Lerner (1954), a scant 50 years ago, as a phenomenon related to inbreeding depression and heterosis. Throughout these studies, the individual genes underlying developmental homeostasis have been detected but unseen and unidentified, invisible in the conceptual black box. I believe that we have now found some of those genes. Enzyme polymorphisms, which were introduced to evolutionary biologists as the quintessential neutral markers, are typified by biochemical differences in performance among genotypes, and at least some of them produce physiological differences that contribute to differences in growth, viability, fecundity, mating success, and development.

In the early years, developmental homeostasis was measured almost exclusively in laboratory and synthetic field populations. It is now clearly evident that developmental stability can be measured in natural populations, with the genetic markers that have been used by population geneticists for decades. Developmental homeostasis can be estimated with either fluctuating asymmetry or the variance of morphological characters. Studies of the MDH polymorphism in honeybees and the LDH polymorphism in killifish suggest that some differences in the genetic and developmental environments will influence correlations among morphological characters. These influences may be detected with the generalized variance and the determinant of the correlation matrix.

References

Allard, R. W., 1975. The mating system and microevolution. Genetics 79s: 115-126.

Baker, C. M. A. & C. Manwell, 19077. Heterozygosity of the sheep: Polymorphism of 'malic enzyme', isocitrate dehydrogenase (Nadp), catalase and esterase. Aust. J. Biol. Sci. 30: 127-140.

Bayne, B. L. & R. C. Newell, 1983. Physiological energetics in marine molluscs. In A. S. M. Saleudden and K. M. Wilburg (eds.) The Mollusca 4: 407-515. Academic Press, New York.

Beacham, T. D. & R. E. Withler, 1985. Heterozygosity and morphological variability of chum salmon (*Oncorhynchus keta*) in southern British Columbia. Heredity 54: 313-322.

Beacham, T. D. & R. E. Withler, 1987. Developmental stability and heterozygosity in chum (*Oncorhynchus keta*) and pink (*Oncorhynchus gorbuscha*) salmon. Can. J. Zool 65: 1823-1826.

Beacham, T. D., 1991. Developmental stability, heterozygosity, and genetic analysis of morphological variation in pink salmon '*Oncorhynchus gorbuscha*'. Can. J. Zool. 69: 274-278.

Beardmore, J. A. & S. A. Shami, 1979. Heterozygosity and the optimum phenotype under stabilising selection. Aquilo. Ser. Zool. 20: 100-110.

Bijlsma-Meeles, E. & R. Bijlsma, 1988. The alcohol dehydrogenase polymorphism in *Drosophila melanogaster*. Fitness measurements and predictions under conditions with no alcohol stress. Genetics 120: 743-753.

Booth, C. L., P. S. Woodruff & S. J. Gould, 1990. Lack of significant associations between allozyme heterozygosity and phenotypic traits in the land snail *Cerion*. Evolution 44: 210-213.

Bottini, E., F. Gloria-Bottini, P. Lucarelli, Al Polzonetti, F. Santoro & A. Ververi, 1979. Genetic polymorphisms and intrauterine development: Evidence of decreased heterozygosity in light for dates human newborn babies. Experientia 35: 1565-1567.

Bruckner, D., 1976. The influence of genetic variability on wing symmetry in honeybees. (*Apis mellifera*). Evolution 30: 100-108.

Chakraborty, R., 1981. The distribution of the number of heterozygous loci in an individual in natural populations. Genetics 98: 461-466.

Chakraborty, R., 1987. Biochemical heterozygosity and phenotypic variability of polygenic traits. Heredity 59: 19-28.

Chakraborty, R. & N. Ryman, 1983. Relationship of mean and variance of genotypic values with heterozygosity per individual in a natural population. Genetics. 103: 149-152.

Clarke, G. M., G. W. Brand & M. J. Whitten, 1986. Fluctuating asymmetry: A technique for measuring developmental stress caused by inbreeding. Aust. J. Biol. Sci. 39: 145-153.

Clarke, G. M., B. P. Oldroyd & P. L. Hunt, 1992. The genetic basic of developmental stability in *Apis mellifera*: heterozygosity versus genic balance. Evolution 46: 753-762.

Clarke, G. M. & J. A. McKenzie, 1987. Developmental stability of insecticide resistant phenotypes in the blowfly: A result of canalizing natural selection. Nature 325: 345-346.

Coelho, J. R. & J. B. Mitton, 1988. Oxygen consumption during hovering is associated with genetic variation of enzymes in honey-bees. Functional Ecology 2: 141-146.

Cothran, E. G., R. Chesser, M. H. Smith & P. E. Johns, 1983. Influences of genetic variability and maternal factors on fetal growth in white-tailed deer. Evolution 37: 282-291.

Danzmann, R. G., M. M. Ferguson & F. W. Allendorf, 1987. Heterozygosity and oxygen-consumption rate as predictors of growth and developmental rate in rainbow trout. Physiol. Zool. 60: 211-220.

Danzmann, R. G., M. M. Ferguson & F. W. Allendorf, 1988. Heterozygosity and components of fitness in a strain of rainbow trout. Biol. J. Linn. Soc. 39: 285-304.

Diehl, W. J., 1989. Genetics of carbohydrate metabolism and growth in *Eisemia foetida* (Oligochataea: Lumbricidae). Heredity 61: 379-387.

Diehl, W. J., P. M. Gaffney & R. K. Koehn, 1986. Physiological and genetic aspects of growth in the mussel *Mytilus edulis*. I. Oxygen consumption, growth, and weight loss. Physiol. Zool. 59: 201-211.

Diehl, W. P., P. M. Gaffney, J. H. McDonald & R. K. Koehn, 1985. Relationship between weight standardized oxygen consumption and multiple-locus heterozygosity in the marine mussel *Mytilus edulis* L. (Mollusca), pp. 531-536 in Proceedings of the 19th European Marine Biology Symposium, edited by P. Gibbs. Cambridge University Press, Cambridge.

DiMichele, L., K. Paynter & D. A. Powers, 1991. Lactate dehydrogenase-B allozymes directly affect development of *Fundulus heteroclitus*. Science 253: 898-900.

DiMichele, L. & D. A. Powers, 1991. Developmental heterochrony and differential mortality in the model teleost, *Fundulus heteroclitus*. Physiological Zoology 64: 1426-1443.

DiMichele, L. & D. A. Powers, 1982a. Physiological basis for swimming endurance differences between LDH-B genotypes of *Fundulus heteroclitus*. Science 216: 1014-1016.

DiMichele, L. & D. A. Powers, 1982b. LDH-B genotype specific hatching times of *Fundulus heteroclitus* embryos. Nature 296: 560-563.

DiMichele, L. & D. A. Powers, 1984. Developmental and oxygen consumption differences between LDH-B genotypes of *Fundulus heteroclitus* and their effect on hatching times. Physiol.. Zool. 57: 52-56.

Dobzhansky, Th. & B. Wallace, 1953. The genetics of homeostasis in *Drosophila*. Proc. Natl. Acad. Sci. 39: 162-171.

Eanes, W. F., 1978. Morphological variance and enzyme heterozygosity in the monarch butterfly. Nature 276: 263-264.

Eanes, W. F., 1981. Enzyme heterozygosity and morphological variance. Nature 290: 609-610.

Fleischer, R. C., R. F. Johnston & W. J. Klitz, 1983. Allozymic heterozygosity and morphological variation in house sparrows. Nature 304: 628-630.

Garton, D. W., 1984. Relationship between multiple locus heterozygosity and physiological energetics of growth in the estuarine gastropod *Thais haemastoma*. Physiol. Zool. 57: 530-543.

Garton, D. W., R. K. Koehn & T. M. Scott, 1984. Multiple-locus heterozygosity and the physiological energetics of growth in the coot clam, *Mulinia lateralis*, from a natural population. Genetics: 445-455.

Gajardo, G. M. & J. A. Beardmore, 1989. Ability to switch reproductive mode in *Artemia* is related to maternal heterozygosity. Mar. Ecol. Prog. Ser. 55: 191-195.

Govindaraju, D. R. & B. P. Dancik, 1987. Allozyme heterozygosity and homeostasis in germinating seeds of jack pine. Heredity 59: 279-283.

Hall, J. G. & R. K. Koehn, 1983. The evolution of enzyme catalytic efficiency and adaptive inference from steady-state kinetic data. Evol. Biol. 16: 53-96.

Handford, P., 1980. Heterozygosity at enzyme loci and morphological variation. Nature 286: 261-262.

Hartl, G. B., G. Lang, F. Klein & R. Willing, 1991. Relationships between allozymes, heterozygosity and morphological characters in red deer (*Cervus elaphus*), and the influence of selective hunting on allele frequency distributions. Heredity 66: 343-350.

Hartl, G. B., F. Suchentrunk, R. Willing & R. Petznek, 1993. Allozyme heterozygosity and fluctuating asymmetry in the brown hare (*Lepus europaeus*): a test of the developmental homeostasis hypothesis. (in preparation).

Hawkins, A. J. S., B. L. Bayne & A. J. Day, 1986. Protein turnover, physiological energetics and heterozygosity in the blue mussel *Mytilus edulis*: the basis of variable age-specific growth. Proc. R. Soc. Lond. B 229: 161-176.

Hawkins, A. J. S., B. L. Bayne, A. J. Day, J. Rusing & C. M. Worrall, 1989. Genotype-dependent interrelations between energy metabolism, protein metabolism and fitness, pp. 283-292 in Reproduction, Genetics and Distributions of Marine Organisms, edited by J. S. Ryland and P. A. Tyler. Olsen and Olsen, Fredensborg, Denmark.

Kasule, F. K. & L. M. Cook, 1988. Phenotypic variability and heterozygosity at an esterase locus in the mosquito *Aedus aegypti*. Heredity 61: 427-431.

King, D. P. F., 1985. Enzyme heterozygosity associated with anatomical character variance and growth in the herring (*Clupea harengus* L.). Heredity 54: 289-296.

Kobyliansky, E. & G. Livshits, 1985. Differential fertility and morphological constitution of spouses. Z. Morph. Anthrop. 76: 95-105.

Koehn, R. K., 1991. The cost of enzyme synthesis in the genetics of energy balance and physiological performance. Biol. J. Linn. Soc. 44: 231-247.

Koehn, R. K. & B. L. Bayne, 1988. Towards a physiological and genetical understanding of the energetics of the stress response. Biol. J. Linn. Soc. 37: 157-171.

Koehn, R. K., W. J. Diehl & T. M. Scott, 1988. The differential contribution by individual enzymes of glycolysis and protein catabolism to the relationship between heterozygosity and growth rate in the coot clam, *Mulinia lateralis*. Genetics 118: 121-130.

Koehn, R. K. & P. M. Gaffney, 1984. Genetic heterozygosity and growth rate in *Mytilus edulis*. Mar. Biol. 82: 1-7.

Koehn, R. K. & S. E. Shumway, 1982. A genetic/physiological explanation for differential growth rate among individuals of the American oyster, *Crassostrea virginica* (Gmelin). Mar. Biol. Letters 3: 35-42.

Koehn, R. K., A. J. Zera & J. G. Hall, 1983. Enzyme polymorphism and natural selection, pp. 115-136 in Evolution of Genes and Proteins, edited by M. Nei and R. K. Koehn. Sinauer Associates Inc. Sunderland, Mass.

Leary, R. F., F. W. Allendorf & K. L. Knudsen, 1983. Developmental stability and enzyme heterozygosity in rainbow trout. Nature 301: 71-72.

Leary, R. F., F. W. Allendorf & K. L. Knudsen, 1984. Superior developmental stability of heterozygotes at enzyme loci in salmonid fishes. Amer. Natur. 124: 540-551.

Leary, R. F., F. W. Allendorf & K. L. Knudsen, 1985. Inheritance of meristic variation and the evolution of developmental stability in rainbow trout. Evolution 39: 308-314.

Lerner, I. M., 1954. Genetic homeostasis. Oliver and Boyd, Edingburgh, 154 pp.

Livshits, G. & E. Kobyliansky, 1984. Biochemical heterozygosity as a predictor of developmental homeostasis in man. Ann. Hum. Genet. 48: 173-184.

Livshits, G. & E. Kobyliansky, 1985. Lerner's concept of developmental homeostasis and the problem of heterozygosity level in natural populations. Heredity 55: 341-353.

Makaveev, T., I. Venev & M. Baulov, 1978. Investigations on activity level and polymorphisms of some blood enzyme in farm animals with different growth energy. II. Correlations between homo- and heterozygosity of some protein and enzyme phenotypes and fattening ability and slaughter indices in various breeds of fattened pigs. Genet. Sel. 10: 229-236.

McAndrew, B. J., R. D. Ward & J. A. Beardmore, 1982. Lack of relationship between morphological variance and enzyme heterozygosity in the plaice, *Pleuronectes platessa*. Heredity 48: 117-125.

Messier, S. & J. B. Mitton. Heterozygosity at the malate dehydrogenase locus and fluctating asymmetry in *Apis mellifera*. Evolution, in review.

Mitton, J. B., 1978. Relationship between heterozygosity for enzyme loci and variation of morphological characters in natural populations. Nature 273: 661-662.

Mitton, J. B., 1993. Theory and data pertinent to the relationship between heterozygosity and fitness, in The Natural History of Inbreeding and Outbreeding, edited by W. Shields and N. Thornhill. University of Chicago Press.

Mitton, J. B., C. Carey & T. D. Kocher, 1986. The relation of enzyme heterozygosity to standard and active oxygen consumption and body size of tiger salamanders, *Ambystoma tigrinum*. Physiol. Zool. 59: 574-582.

Mitton, J. B. & M. C. Grant, 1984. Associations among protein heterozygosity, growth rate, and developmental homeostasis. Ann. Rev. Ecol. Syst. 15: 479-499.

Mitton, J. B. & R. K. Koehn, 1975. Genetic organization and adaptive response of allozymes to ecological variables in *Fundulus heteroclitus*. Genetics 79: 97-111.

Mitton, J. B. & R. K. Koehn, 1985. Shell shape variation in the blue mussel, *Mytilus edulis* L., and its association with enzyme heterozygosity. J. Exp. Mar. Biol. Ecol. 90: 73-80.

Mitton, J. B. & B. A. Pierce, 1980. The distribution of individual heterozygosity in naural populations. Genetics 95: 1043-1054.

Mukai, T., L. E. Mettler & S. Chigusa, 1971. Linkage disequilibrium in a local population of *Drosophila melanogaster*. Proc. Natl. Acad. Sci. USA 68: 1065-1069.

Mukai, T. & O. Yamaguchi, 1974. The genetic structure of natural populations of *Drosphila melanogaster*. XI. Genetic variability in a local population. Genetics 76: 339-366.

Parsons, P. A., 1971. Extreme environment heterosis and genetic loads. Heredity 26: 479-483.

Parsons, P. A., 1973. Genetics of resistance to environmental stresses in *Drosophila* populations. Ann. Rev. Genet. 7: 239-265.

Parsons, P. A., 1987. Evolutionary rates under environmental stress. Evolutionary Biology 21: 311-347.

Paynter, K. T., I. DiMichele, S. C. Hand & D. A. Powers, 1991. Metabolic implications of LDH-B genotype during early development in *Fundulus heteroclitus*. J. Exp. Zool. 257: 24-33.

Pierce, B. A. & J. B. Mitton, 1982. Allozyme heterozygosity and growth in the tiger salamander, *Ambystoma tigrinum*. J. Hered. 73: 250-253.

Place, A. R. & D. A. Powers, 1979. Genetic variation and relative catalytic efficiencies: LDH-B allozymes of *Fundulus heteroclitus*. Proc. Natl. Acad. Sci. U.S.A. 76: 2354-2358.

Powers, D. A., L. DiMichele & A. R. Place, 1983. The use of enzyme kinetics to predict differences in cellular metabolism, developmental rate, and swimming performance between LDH-B genotypes of the fish, *Fundulus heteroclitus*, in Isozymes: Current Topics in Biological and Medical Research. Volume 10, edited by G. Whitt and G. C. Markert. Academic Press, New York.

Powers, D. A., G. S. Greaney & A. R. Place, 1979. Physiological correlation between lactate dehydrogenase genotype and haemoglobin function in killifish. Nature 277: 240-241.

Powers, D. A. & A. R. Place, 1978. Biochemical genetics of *Fundulus heteroclitus* (L.). I. Temporal and spatial variation in gene frequencies of Ldh-B, Mdh-A, Gpi-B, and Pgm-A. Biochem. Genet. 16: 593-607.

Prakash, S., 1974. Gene differences between the Sex Ratio and Standard gene arrangements of the X chromosome and linkage disequilibrium between the Standard gene arrangements and the X chromosome in *Drosophila pseudoobscura*. Genetics 77: 795-804.

Prakash, S. & R. C. Lewontin, 1968. A molecular approach to the study of genic heterozygosity in natural populations. III. Direct evidence of coadaptation in gene arrangements of *Drosophila*. Proc. Natl. Acad. Sci. U. S. A. 59: 398-450?

Prakash, S. & R. C. Lewontin, 1971. A molecular approach to the study of genic heterozygosity in natural populations. V. Further direct evidence of coadaptations in inversions of *Drosophila*. Genetics 69: 405-408.

Prakash, S. & R. B. Merritt, 1972. Direct evidence of genic differentiation between Sex Ratio and Standard gene arrangements of X chromosome in *Drosophila pseudoobscura*. Genetics 72: 169-175.

Policansky, D. & E. Zouros, 1977. Gene differences between the sex ratio and standard gene arrangements on the X chromosome in *Drosophila persimilis*. Genetics 85: 507-511.

Powers, D. A., M. Smith, I. Gonzalez-Villasenor, L. DiMichele, D. Crawford, G. Bernardi & T. Lauerman, 1993. A multidisciplinary approach to the selectionist/neutralist controversy using the model teleost *Fundulus heteroclitus*. Oxford Surveys in Evolutionary Biology, Oxford University Press.

Quattro, J. M. & R. C. Vrijenhoek, 1989. Fitness differences among remnant populations of the endangered sonoran topminnow. Science 245: 976-978.

Robertson, F. W. & E. C. Reeve, 1952. Heterozygosity, environmental variation and heterosis. Nature 170: 286-287.

Rodhouse, P. G. & P. M. Gaffney, 1984. Effect of heterozygosity on metabolism during starvation in the American oyster, *Crassostrea virginica*. Mar. Biol. 80: 179-188.

Rodhouse, P. G., J. H. McDonald, R. I. E. Newell & R. K. Koehn, 1986. Gamete production, somatic growth and multiple locus heterozygosity in *Mytilus edulis* L. Mar. Biol. 90': 209-214.

Rogers, S., R. Wells & M. Rechsteiner, 1986. Amino acid sequences common to rapidly degraded proteins: the PEST hypothesis. Science 234: 364-368.

Scott, T. M. & R. K. Koehn, 1990. The effect of environmental stress on the relationship of heterozygosity to growth rate in the coot clam *Mulinia lateralis* (Say). J. Exp. Mar. Biol. Ecol., 135: 109-116.

Scribner, K. T., M. H. Smith & P. E. Johns, 1989. Environmental and genetic components of antler growth in white-tailed deer. J. Mamm. 70: 284-291.

Scribner, K. T. & M. H. Smith, 1990. Genetic variability and antler development, pp 460-473 in Horns, Pronghorns, and Antlers, edited by G. A. Bubenik and A. B. Bubenik. Springer, New York.

Serradilla, J. M. & F. J. Ayala, 1983. Alloprocoptic selection: A mode of natural selection promoting polymorphism. Proc. Natl. Acad. Sci. USA 80: 2022-2025.

Singh, S. M. & E. Zouros, 1978. Genetic variation associated with growth rate in the American oyster (*Crassostrea virginica*). Evolution 32: 342-353.

Smith, M. H., K. T. Scribner, P. E. Johns & O. E. Rhodes, Jr., 1991. Genetics and Antler development. Proc. 18th Congr. Internat. Union Game Biol., Krakow, Poland.

Soulé, M. E., 1967. Phenetics of natural populations. II. Asymmetry and evolution in a lizard. Amer. Natur. 101: 141-160.

Soulé, M. E., 1971. The variation problem: the gene-flow-variation hypothesis. Taxon 20: 37-50.

Soulé, M. E., 1979. Heterozygosity and developmental stability: another look. Evolution 33: 396-401.

Soulé, M. E., 1982. Allomeric variation. 1. The theory and some consequences. Amer. Natur. 120: 751-764.

Soulé, M. E. & S. Y. Yang, 1974. Genetic variation side-blotched lizards on islands in the Gulf of California. Evolution 27: 593-600.

Strauss, R. E., 1989. Associations between genetic heterozygosity and morphological variability in freshwater sculpins, genus *Cottus* (Teleostei: Cottidae). Biochem. Syst. Ecol. 17: 333-340.

Strauss, R. E., 1991. Correlations between heterozygosity and phenotypic variability in *Cottus* (Teleostei: Cottidae): character components. Evolution 45: 1950-1956.

Teska, W. R. M. H. Smith & J. M. Novak, 1990. Food quality, heterozygosity, and fitness correlates in *Peromyscus polionotus*. Evolution 44: 1318-1325.

Van Valen, L., 1978. The statistics of variation. Evol. Theory 4: 33-43.

Vrijenhoek, R. C. & S. Lerman, 1982. Heterozygosity and developmental stability under sexual and asexual breeding systems. Evolution 36: 768-767.

Ward, R. D., M. Sarfarazi, C.Azimi-Garakani & J. A. Beardmore, 1985. Population genetics of polymorphisms in Cardiff newborn: Relationship betwene blood group and allozyme heterozygosity and birth weight. Hum. Hered. 35: 171-177.

Watt, W. B., 1985. Bieonergetics and evolutionary genetics: opportunities for new synthesis. Amer. Natur. 125: 118-143.

Watt, W. B., 1986. Power and efficiency as indexes of fitness in matabolic organization. Amer. Natur. 127: 629-653.

Watt, W. B., 1992. Eggs, enzymes, and evolution-natural genetic variants change insect fecundity. Proc. Natl. Acad. Sci. USA 89: 10608-10612.

Wooten, M. C. & M. H. Smith, 1986. Fluctuating asymmetry and genetic variability in a natural population of *Mus musculus*. J. Mammalogy 67: 725-732.

Yezerinac, S. M., S. C. Lougheed & P. Handford, 1992. Morphological variability and enzyme heterozygosity: individual and population level correlations. Evolution 46: 1959-1964.

Zouros, E. & D. W. Foltz, 1987. The use of allelic isozyme variation for the study of heterosis, pp 2-59 in Isozymes: Current Topics in Biological and Medical research, Volume 13, edited by M. C. Rattazzi, J. G. Scandalios and G. S. Whitt. Alan R. Liss, Inc., New York.

T. A. Markow (ed.), Developmental Instability: Its Origins and Evolutionary Implications, 69–78, 1994.

Modification of developmental instability and fitness: malathion-resistance in the Australian sheep blowfly, *Lucilia cuprina*

J. A. McKenzie & K. O'Farrell
Department of Genetics, University of Melbourne, Parkville, Victoria 3052, Australia

Received and accepted 23 April 1993

Key words: *Lucilia cuprina*, malathion-resistance, fluctuating asymmetry, fitness

Abstract

The evolution of resistance to malathion by *Lucilia cuprina* initially results in an increase in fluctuating asymmetry. Resistant flies are at a selective disadvantage, relative to susceptibles, in the absence of the insecticide. A fitness/asymmetry modifier of diazinon-resistant phenotypes ameliorates these effects resulting in malathion-resistant phenotypes of relative fitness and asymmetry similar to susceptibles. For the nine genotypic combinations of the modifier and malathion-resistance alleles, developmental time increases linearly with increasing asymmetry. Percentage egg hatch decreases linearly with increasing asymmetry. The initially disruptive effect of the malathion-resistant allele was partially dominant, the effect of the modifier dominant. The results are discussed in terms of developmental perturbation, asymmetry estimation and relative fitness to consider whether it is adequate to use changes in fluctuating asymmetry alone as measures of developmental instability. It is suggested that in some circumstances antisymmetry may indicate developmental instability and that the diazinon/malathion-resistance systems in *L. cuprina* may allow the relative importance of genetical and/or environmental developmental perturbations to be ascertained.

Introduction

In 1976, in the British newspaper *The Observer*, Clive James the author, critic and television personality commented that: 'It is only when they go wrong that machines remind you how powerful they are.' His comment was technologically based but is no less apt when applied to biochemical 'machines.' The capacity of an organism to follow a developmental norm in the face of perturbing conditions, both genetical and environmental, is impressive. Departures from the norm re-enforce the power of the usual process but also allow observation and experimentation at the interface of development, fitness estimation and evolution.

Measurements at this interface have usually been restricted to the phenotypic level of a multi-tiered hierarchy. Therefore, unifying principles have frequently been conjectural being based on observations hindered by difficulties of definition and appropriate statistical analysis. This has often been the case when departures from symmetry have been used to indicate that the 'machine' has gone wrong (Palmer & Strobeck, 1992). However, the approach still has much to recommend it.

One expectation of the evolution of developmental patterns is that newly evolved phenotypes will be more variable than long established ones when developmental constraints have evolved through canalizing selection (Maynard Smith *et al.*, 1985; Scharloo, 1991; Zakharov, 1992). Departures from developmental stability caused by environmental or genetic stresses may be measured by comparing differences in asymmetry between normally bilaterally symmetrical organisms (Van Valen, 1962; Palmer & Strobeck, 1986, 1992; Leary & Allendorf, 1989; Parsons, 1990, 1992). Interpretation of results can be difficult when the genotype is ill-defined (Graham, 1992; Parsons, 1992). The diazinon-resistance system in the Australian sheep blowfly, *Lucilia cuprina*, is one in which this difficulty has been minimized (Clarke & McKenzie, 1987, 1992; McKenzie & Clarke, 1988; McKenzie, Batterham & Baker, 1990).

Diazinon was the principle agent in the chemical control of *L. cuprina* between 1958 and 1980 when a new insecticide, cyromazine, became widely used. Diazinon is still used on many properties but post-1980 its pre-eminence in the control of the blowfly has declined (Hughes & McKenzie, 1987). Resistance was observed ten years after the introduction of the chemical, resulting primarily from allelic substitution at the *Rop-1* locus on chromosome IV (McKenzie, 1987). Diazinon was used for a number of years after resistance developed because a more efficacious chemical was not available and because, in normal circumstances, it provided adequate protection from blowfly strike in sheep. That is, it provided a protection period that was sufficient for economic control and sheep management (Hughes & McKenzie, 1987). The biochemical basis of resistance involves an altered carboxylesterase, E_3, encoded by the *Rop-1* locus (Hughes & Devonshire, 1982; Parker *et al.*, 1991).

The frequency of resistant flies in natural populations of *L. cuprina* increased rapidly post-1968. Approximately 90% of flies in populations were resistant by the early 1970s and by 1980 many populations were close to fixation for the *Rop-1* allele (McKenzie *et al.*, 1980; Hughes & McKenzie, 1987). The frequency of the *Rop-1* allele has remained relatively constant over the last decade (McKenzie, 1993).

The introduction of the *Rop-1* allele into a genome adapted to susceptible genotypes resulted in phenotypes of lowered relative fitness in the absence of the insecticide, and increased bilateral asymmetry (McKenzie, Whitten & Adena, 1982; Clarke & McKenzie, 1987, 1992; McKenzie & Game, 1987; McKenzie & Clarke, 1988). The use of diazinon for a significant period after resistance evolved selected for a modifier that ameliorated these effects. The modifier is unlinked to the *Rop-1* locus mapping to the *w* region of chromosome III. Data are consistent with the modifier being allelic to *Scalloped wings* (*Scl*) (McKenzie, Batterham & Baker, 1990) which is homologous to the *Notch* gene of *Drosophila melanogaster* (A. Davies *et al.*, unpublished data).

The initially disruptive effect of the *Rop-1* allele (as assessed by an increase in asymmetry of resistant phenotypes) was dominant, as is the influence of the modifier in returning resistant phenotypes to levels of asymmetry observed in susceptibles (McKenzie & Clarke, 1988). A similar relationship is observed for fitness modification. For example, the developmental time of resistant phenotypes increases relative to susceptibles in a genetic background without the modifier, but is comparable in all *Rop-1* genotypes when the modifier allele is heterozygous or homozygous (McKenzie & Game, 1987). There is a significant positive linear relationship between asymmetry score and developmental time (McKenzie, 1993).

It should be noted that all these fitness estimates were made in a diazinon-free environment. Selection for the modifier occurred in the presence of the chemical during the period that the *Rop-1* allele was common in natural populations of *L. cuprina* (McKenzie, 1987). Fitness differences between modifier and non-modifier resistant genotypes are therefore also expected in this case. This expectation is met with modifier genotypes being relatively more fit (McKenzie, 1993).

Fluctuating asymmetry was observed for phenotypes susceptible to diazinon and for resistant flies carrying the modifier. The distribution of phenotypes of resistant genotypes without the modifier was antisymmetric (McKenzie & Clarke, 1988). The evolution of resistance and subsequent selection of the modifier therefore represent a transition from fluctuating asymmetry through antisymmetry back to fluctuating asymmetry. McKenzie and Clark (1988) and McKenzie, Batterham and Baker (1990) have interpreted these trends as a perturbation, followed by the re-establishment, of developmental stability paralleled by a transition in the relative fitness of resistant individuals. Palmer and Strobeck (1992) have questioned this interpretation essentially because they argue that changes in fluctuating asymmetry alone can be used as a measure of developmental stability. Their contribution is important to ongoing debates incorporating firmly held but disparate views (see also Palmer & Strobeck, 1986; McKenzie, Batterham & Baker, 1990; Zakharov & Graham, 1992; Graham, Freeman & Emlen, 1993 for discussion and references). The malathion-resistance system of *L. cuprina* offers another comparative area of study to contribute to these debates.

Malathion, like diazinon, is an organophosphorus pesticide. Beginning in the mid-1950s it was used to control the ectoparasites, lice, itch-mite and ked of sheep (Hughes & McKenzie, 1987). While malathion was not specifically used in the control of sheep blowfly, by 1968 some populations of *L. cuprina* had become resistant to the chemical due to allelic substitution at the *Rmal* locus. Resistant individuals were neither common nor widespread. The *Rmal* locus is closely linked to *Rop-1* and codes for a malathion-specific carboxylesterase known not to be E_3 (Hughes, Green & Reichmann, 1984; Raftos, 1986; Raftos & Hughes, 1986; Russell *et al.*, 1990).

The present study ascertained that, in the absence of the insecticide, substitution of the *Rmal* allele resulted in increased fluctuating asymmetry and lower fitness than in susceptibles. The modifier enhanced the fitness of resistant individuals and returned their levels of asymmetry to those of susceptibles. These results show that the modifier influences the phenotype of genotypes resistant to an organophosphorus chemical other than diazinon. Furthermore, as asymmetry differences are restricted to altering levels of fluctuating asymmetry, the system meets the definition of Palmer and Strobeck (1992) as one that estimates changes in developmental stability. When these results and those observed for diazinon-resistant phenotypes (McKenzie & Clarke, 1988) are compared, they question a definition of developmental stability that is limited only to considerations of fluctuating asymmetry.

Materials and methods

Strains

A strain homozygous for the modifier (*M/M*) and for the susceptible allele (+) at the *Rmal* locus carrying the recessive markers *ck* and *ru* (McKenzie & Game, 1987) was crossed to a non-modifier (+) malathion-resistant strain without the markers. The crossing regime used to generate the pure-breeding modifier resistant strain is shown in Figure 1. Similar crossing regimes, involving progeny testing in the case of susceptible strains (McKenzie & Game, 1987), were used to produce the remaining three genotypic combinations doubly homozygous for modifier/non-modifier and resistant/susceptible alleles in comparable genetic backgrounds. These strains were crossed to generate the nine possible modifier/resistance genotypes used for asymmetry and fitness estimation.

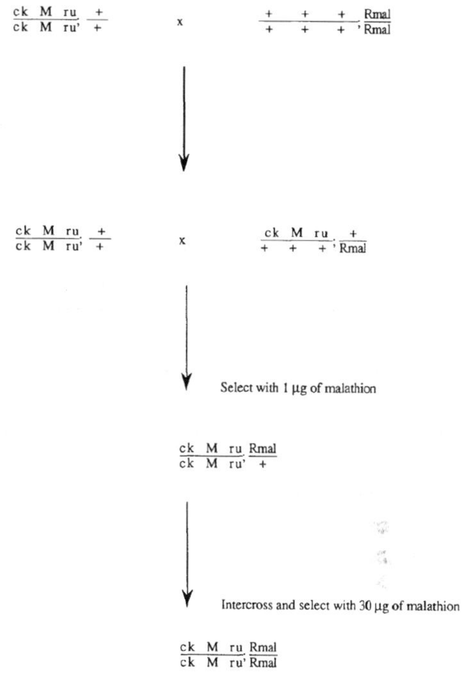

Fig. 1. Crossing regime for generation of a malathion-resistant strain homozygous for the modifier. Malathion concentrations of 1 μg and 30 μg discriminate against susceptibles and heterozygotes respectively. The modifier is flanked by the recessive markers *ck* (crooked bristle) and *ru* (rusty body).

Asymmetry estimation

The methods followed Clarke and McKenzie (1987) and McKenzie and Clarke (1988). Asymmetry of a fly was estimated from the absolute difference in bristle count between the left and right sides of the frontal head stripe, the outer wing margin and the R_{4+5} wing vein. The asymmetry of 50 flies was scored for each genotypic combination.

Fitness estimation

The methods followed those of McKenzie and Game (1987). The nine possible genotypic combinations were scored for percentage egg hatch, the percentage of 100 first instar larvae that developed on standard medium through larval and pupal

Table 1. Analysis of variance of the asymmetry values summarized in Figure 2 for the nine genotypic combinations at the *M* and *Rmal* loci.

Source of variation	d.f.	Mean square	F	P
Between genotypes	8	7.11	5.12	<0.001
Within genotypes	441	1.39		

stages to emerge as adults and the developmental time, in days, from egg to adult. Forty trials were conducted for each comparison at 27 °C.

For egg hatch estimation, females were allowed to lay on a small piece of liver. One hundred eggs were removed from the liver with a brush moistened with water and placed randomly on a moist filter paper in a 7 cm diameter petri dish. The percentage of eggs hatching after 24 h was recorded.

For the percentage of larvae developing to adults, 50 first instar larvae were placed on 110 ml of standard medium in paper cups held on vermiculite in plastic containers. Following pupation in the vermiculite, the number of flies eclosing and the time of their emergence was recorded.

Results

Asymmetry estimation

Significant differences are observed between asymmetry levels of the nine genotypic combinations of modifier and *Rmal* (Fig. 2, Table 1). Least significant difference comparisons (Sokal & Rohlf, 1981) show that *+/+; Rmal/Rmal* has greater asymmetry than *+/+; Rmal/+* ($P < 0.05$) and all other combinations at least the 0.01% level of probability. Phenotypes of *+/+; Rmal/+* are significantly different, at least $P < 0.05$, for all comparisons but those with *M/M; Rmal/Rmal* and *M/+; Rmal/Rmal*. No other comparisons differ significantly. The effect of the *Rmal* allele in increasing asymmetry therefore appears to be partially dominant. The influence of the modifier in returning asymmetry levels of resistant genotypes to those of susceptibles is dominant (Fig. 2).

The increase in asymmetry observed for resistant phenotypes without the modifier is best explained as an increase in fluctuating asymmetry. For each of the three characters measured for each of the genotypes, the signed (L-R) differences are normally distributed about zero. Resistant individuals

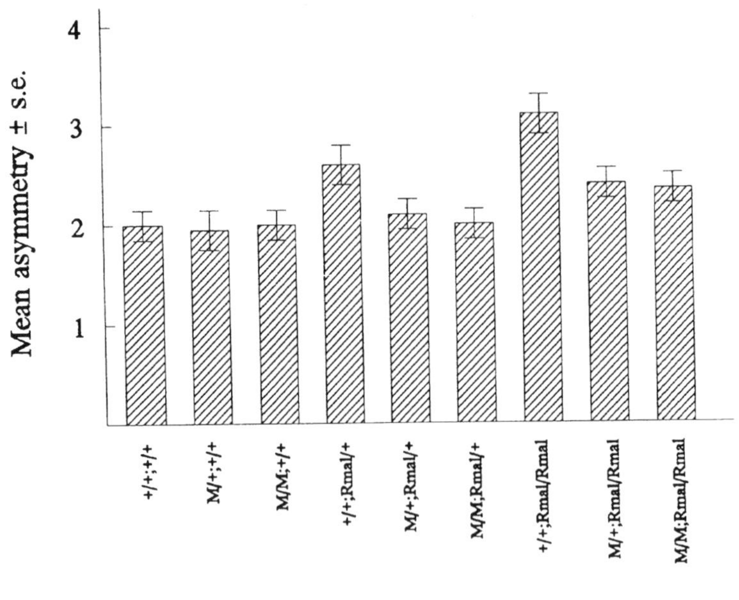

Fig. 2. Mean asymmetry ± s.e. of the nine genotypic combinations of modifier and malathion-resistance genotypes. Fifty flies are measured for each combination.

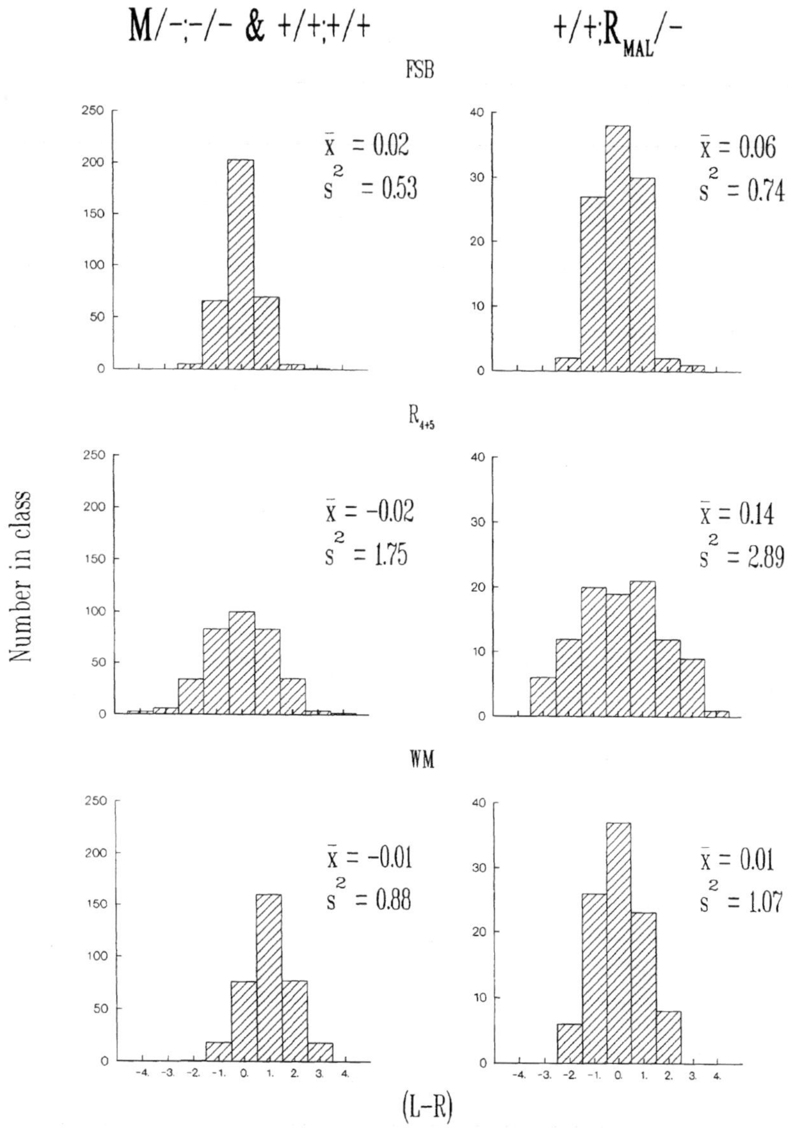

Fig. 3. Distributions, with means and variances, of signed differences between left and right sides of flies measured for frontal stripe bristles (FSB), R_{4+5} wing vein and wing margin (WM). Data are pooled for the susceptible genotype without the modifier (*+/+*; *+/+*) and all modified genotypes (*M/-*; *-/-*) and for nonmodified resistant genotypes (*+/+*; *Rmal/-*).

Table 2. Means ± standard errors for percentage egg hatch, percentage of larvae reaching the adult stage of the life cycle and developmental time for each of the nine modifier/resistance genotypes. Standard errors for percentage egg hatch and percentage to adult are expressed in the angularly transformed scale.

Genotype	Percentage egg hatch	Percentage to adult	Developmental time
+/+; Rmal/Rmal	65.7 ± 2.4	58.5 ± 1.3	15.39 ± 0.08
+/+; Rmal/+	67.6 ± 1.1	65.9 ± 1.6	15.15 ± 0.03
+/+; +/+	70.6 ± 1.7	61.5 ± 1.5	15.06 ± 0.07
M/+; Rmal/Rmal	64.7 ± 1.2	64.5 ± 1.2	15.06 ± 0.04
M/+; Rmal/+	69.5 ± 1.2	64.6 ± 1.3	15.04 ± 0.05
M/+; +/+	73.4 ± 1.2	62.7 ± 1.5	15.04 ± 0.04
M/M; Rmal/Rmal	71.0 ± 1.6	62.3 ± 1.7	14.99 ± 0.06
M/M; Rmal/+	73.1 ± 1.2	63.1 ± 1.1	15.01 ± 0.03
M/M; +/+	75.2 ± 1.0	62.6 ± 1.7	15.01 ± 0.08

without the modifier show a consistent, but nonsignificant, increase in variance relative to the other classes (Fig. 3).

Fitness estimation

The percentage egg hatch, the percentage of larvae reaching the adult stage and the egg to adult developmental time of each of the genotypes are provided in Table 2. There are significant differences observed between genotypes for percentage egg hatch and developmental time (Table 3). Least significant difference analysis shows *+/+; Rmal/Rmal* develops at a significantly (P < 0.01) slower rate than all other genotypes. The developmental time of the *+/+; Rmal/+* genotype is intermediate between the former and all other genotypes, but is not significantly different from any of the latter, although the P < 0.05 level is bordered in several instances.

The patterns for egg hatch percentage are less clear. Where significance (P < 0.05) occurs, the genotypes *+/+; Rmal/Rmal*, *+/+; Rmal/+* and *M/+; Rmal/Rmal* display lower percentage egg hatches than the genotypes to which they are compared.

Significant correlations are observed between asymmetry levels (× variable) and percentage egg hatch (P < 0.05) and developmental time (P < 0.01). The relationships are linear (percentage egg hatch, y = 66.81 − 4.36 ×, $F_{1,7}$ = 8.90, P < 0.05; developmental time, y = 14.44 + 0.28 ×, $F_{1,7}$ = 21.30, P < 0.01) with egg hatch decreasing and developmental time increasing with increasing asymmetry (Fig. 4). There is no significant association between the percentage of larvae reaching the adult stage of the life cycle and asymmetry level, however, the trend is one of decreasing completion of development with increasing asymmetry.

Discussion

Departures from symmetry are potentially powerful tools in addressing applied, developmental or evolutionary questions (Parker & Strobeck, 1986, 1992; Clarke & McKenzie, 1992a; Markow & Ricker, 1991; Clarke, Oldroyd & Hunt, 1992; Møller & Höglund, 1992; Thornhill, 1992; Zakharov & Graham, 1992; Graham, Freeman & Emlen,

Table 3. Analysis of variance of percentage egg hatch, percentage to adult and developmental time data summarized in Table 2. Percentage egg hatch and percentage to adult data were angularly transformed before analysis.

		% to Egg Hatch		% to Adult		Developmental time	
Source of variation	d.f.	Mean square	F	Mean square	F	Mean square	F
Between genotypes	8	200.62	2.35*	63.40	0.76	0.61	4.61***
Within genotypes	351	85.48		83.49		0.13	

* P<0.05; *** P <0.001

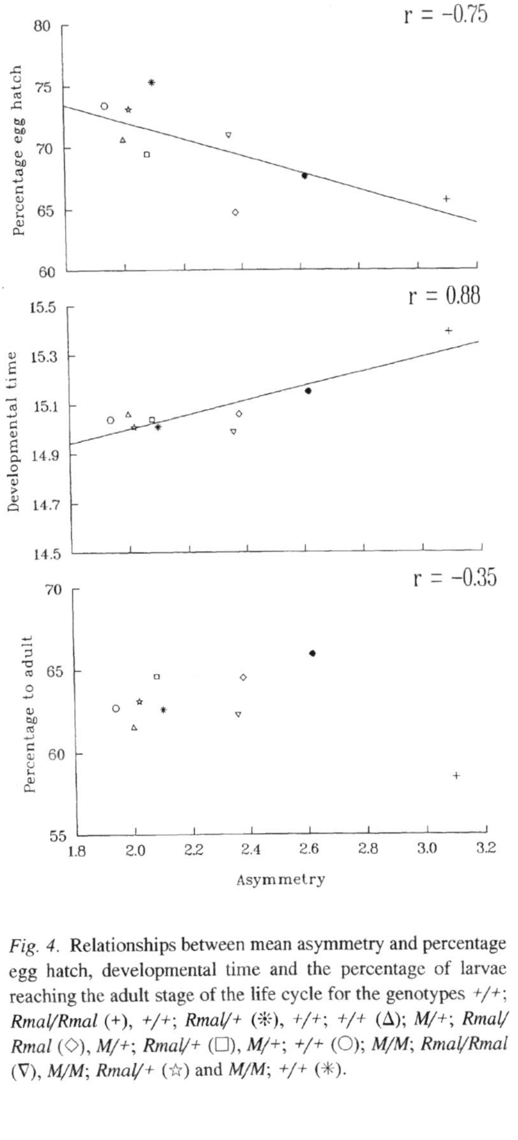

Fig. 4. Relationships between mean asymmetry and percentage egg hatch, developmental time and the percentage of larvae reaching the adult stage of the life cycle for the genotypes *+/+; Rmal/Rmal* (+), *+/+; Rmal/+* (✻), *+/+; +/+* (Δ); *M/+; Rmal/Rmal* (◇), *M/+; Rmal/+* (□), *M/+; +/+* (○); *M/M; Rmal/Rmal* (∇), *M/M; Rmal/+* (☆) and *M/M; +/+* (✳).

1993). Changes to the normal developmental 'route' are accompanied by a physiological 'cost' (Maynard Smith *et al.*, 1985); therefore fitness is inclusive of the developmental process. The cost may be reflected in an increase in the level of asymmetry observed in an individual or a population. Frequently, however, unambiguous interpretation of departures from symmetry is complicated as it is not possible to partition the influence of the genotype and/or the environment, the genotype being ill-defined (Palmer & Strobeck, 1986; 1992; McKenzie & Clarke, 1988; Leary & Allendorf, 1989; Graham, 1992; Leary, Allendorf & Knudsen, 1992). So, in many instances, is the ecological history of evolutionary change. In these circumstances the relationship between developmental stability and patterns of asymmetry comes into question (Palmer & Strobeck, 1992). Previously reported data on diazinon and dieldrin resistance systems of *L. cuprina* (Clarke & McKenzie, 1987; McKenzie & Clarke, 1988; McKenzie, Batterham & Baker, 1990), when coupled with the present data for malathion-resistant phenotypes, are particularly important in this context, as the genotypes and their evolutionary history are defined.

In each of these systems, the introduction of the resistance allele into the genome results in an increase of asymmetry. The effect is dominant for diazinon- and dieldrin-resistant phenotypes (Clarke & McKenzie, 1987; McKenzie & Clarke, 1988) and partially dominant for malathion-resistant individuals (Fig. 2). At the time resistance initially evolves, the relative fitness of resistant phenotypes, in the absence of the insecticide, is lower than that of susceptibles in each instance (Whitten, Dearn & McKenzie, 1980; McKenzie, Whitten & Adena, 1982; Table 2; Fig. 4). For both diazinon- and malathion-resistance systems a dominant modifier has been selected that ameliorates these effects resulting in asymmetry and fitness levels similar to those of susceptibles (McKenzie & Game, 1987; McKenzie & Clarke, 1988; Table 2; Figs. 2 & 4). A fitness/asymmetry modifier was not selected for dieldrin-resistant phenotypes. This is not surprising given the limited time the chemical was used after resistance evolved (Hughes & McKenzie, 1987; Roush & McKenzie, 1987).

There is a clear association between changes in asymmetry and fitness in each of these resistance systems. The association is interpreted as representing a measure of underlying developmental perturbation followed, in the case of resistance to diazinon and malathion, by the re-establishment of a developmental norm. That is, periods of normal developmental stability are interrupted, prior to the selection of the modifier, by a period of developmental instability. The interpretation is potentially constrained by observed patterns of asymmetry and their use as estimators of stability of the developmental process (Palmer & Strobeck, 1992). Susceptible phenotypes show classical fluctuating asymmetry as do modified resistant phenotypes of the diazinon- and malathion-resistance systems (Clarke & McKenzie, 1987; McKenzie & Clarke, 1988; Fig. 3). Dieldrin-resistant and unmodified diazinon-resistant phenotypes are antisymmetric. The trend in unmodified malathion-resistant phenotypes is one of increasing fluctuating asymmetry relative to susceptible and modified resistant phenotypes (Fig. 3).

Palmer and Strobeck (1992) do not regard antisymmetric distributions as reflecting developmental instability. There are clearly instances, such as their example of claw size in fiddler crabs, where normal development will result in an asymmetrical phenotype. This is not indicative of developmental perturbation but neither is it typical of asymmetry measures where it is usual to measure several characters. Furthermore, if an antisymmetric distribution results as a departure from the developmental norm it does not seem reasonable to exclude such systems as indicators of developmental stability changes (Graham, Freeman & Emlen, 1993). This point is highlighted by a consideration of the results for the diazinon- and malathion-resistant systems. In both cases the initial substitution of the resistant allele results in increased asymmetry and lower fitness of resistant phenotypes relative to susceptibles. In both cases selection of a fitness/asymmetry modifier results in resistant phenotypes with asymmetry and fitness values comparable to susceptibles. Asymmetry changes associated with malathion-resistance are fluctuating while those related to diazinon resistance include antisymmetry. The argument that the former reflects an instance of changes in developmental stability while the latter does not seems difficult to sustain.

The modifier was selected through its effect on the fitness of *Rop-1* genotypes. The low frequency of the *Rmal* alleles in natural populations of *L. cuprina* (Hughes & McKenzie, 1987) means that the conditions necessary for the selection of a modifier specific to this chemical were not met. In particular, *Rmal* homozygotes were not in sufficiently high frequencies for a sufficient period to afford this evolutionary opportunity (Roush & McKenzie, 1987). Resistance to diazinon and malathion involves alteration of different esterases (Russell *et al.*, 1990). The actual mode and site of action of these esterases is not known. In the absence of this information the interaction between gene products of the modifier and the resistance loci remains open to speculation. On the basis that the modifier is allelic to *Scl* (McKenzie, Batterham & Baker, 1990), comparisons with the numerous pleiotropic effects of the homologous *Notch* locus offer many possibilities (Artavanis-Tsakonas, 1988). One intriguing possibility is that the esterases show homologies with cell adhesion molecules of the nervous system. Such molecules may be associated with cell/cell interactions of neurogenic transfer (Hortsch & Goodman, 1991). This could indicate a direct interaction between the gene products of *Scl* and *Rop-1* and *Rmal* or reflect the level of cell adhesion which may influence the determination of bristle phenotypes of the *Scl* locus.

Novel environmental and genetical changes can perturb development but the capacity to partition the relative importance of these influences is difficult (Leary, Allendorf & Knudsen, 1992; Graham, 1992; Parsons, 1992). The malathion- and diazinon-resistance systems provide an experimental opportunity to do this because of the similarities and differences of fitness and asymmetry observed. Specific effects can be related to specific genotypes and in this circumstance the influence of environmental stress can be ascertained.

Three key stresses in the ecology of the evolution of insecticide resistance in *L. cuprina* are larval density (Whitten, Dearn & McKenzie, 1980), insecticide concentration (McKenzie, 1987, 1993) and developmental temperature (McKenzie, 1990). There is evidence that these parameters may influence developmental stability (Clarke & Ridsdill-Smith, 1990; Clarke & McKenzie, 1992a). Therefore, by varying environmental components of known ecological and developmental relevance for all combinations of modifier and resistance genotypes, the nature of evolutionary change in stressful and benign environments may be estimated (Hoffmann & Parsons, 1991; Parsons, 1991). This will assist in providing a more robust definition of developmental stability and ultimately enable an evolutionary process to be studied at population, developmental, biochemical and molecular levels (Templeton *et al.*, 1990).

Acknowledgements

Comments on the manuscript by Phil Batterham, Geoff Clarke and Andrew Davies are gratefully acknowledged as is technical assistance by Kate Anderson, Stuart Newstead and Janet Yen. The Australian Research Council and the Research Trust Fund of the Wool Research and Development Corporation provided financial support.

References

Artavanis-Tsakonas, S., 1988. The molecular biology of the Notch locus and the fine tuning of differentiation in Drosophila. Trends in Genetics 4: 95-101.

Clarke, G. M. & J. A. McKenzie, 1987. Developmental stability of insecticide resistant phenotypes in blowfly; a result of canalizing natural selection. Nature 325: 345-346.

Clarke, G. M. & J. A. McKenzie, 1992. Coadaptation, developmental stability, and fitness of insecticide resistance genotypes in the Australian sheep blowfly, Lucilia cuprina: a review. Acta. Zool. Fennica 191: 107-110.

Clarke, G. M. & L. J. McKenzie, 1992a. Fluctuating asymmetry as a quality control indicator for insect mass rearing processes. J. Econ. Entomol. (In Press).

Clarke, G. M. & T. J. Ridsdill-Smith, 1990. The effect of avermectin B_1, on developmental stability in the bush fly, Musca vetustissima, as measured by fluctuating asymmetry. Entomol. exp. appl. 54: 265-269.

Clarke, G. M., B. P. Oldroyd & P. Hunt, 1992. The genetic basis of developmental stability in Apis mellifera: heterozygosity versus genic balance. Evolution 46: 753-762.

Graham, J. H., 1992. Genomic coadaptation and developmental stability in hybrid zones. Acta Zool. Fennica 191: 121-131.

Graham, J. H., D. C. Freeman & J. M. Emlen, 1993. Antisymmetry, directional asymmetry and dynamic morphogenesis. Genetica (This Volume).

Hoffmann, A. A. & P. A. Parsons, 1991. Evolutionary genetics and environmental stress. Oxford University Press, Oxford.

Hortsch, M. & C. S. Goodman, 1991. Cell and substrate adhesion molecules in Drosophila. Ann. Rev. Cell. Biol. 7: 505-557.

Hughes, P. B. & A. L. Devonshire, 1982. The biochemical basis of resistance to organophosphorus insecticides in the sheep blowfly, Lucilia cuprina. Pestic. Biochem. Physiol. 18: 289-297.

Hughes, P. B. & J. A. McKenzie, 1987. Insecticide resistance in the Australian sheep blowfly, Lucilia cuprina: speculation, science and strategies, pp. 162-177 in: Combating resistance to Xenobiotics. Biological and chemical approaches, edited by M. G. Ford, D. W. Hollowman, B. P. S. Khambay and R. M. Sawicki, Ellis Horwood, Chichester.

Hughes, P. B., P. E. Green & K. G. Reichmann, 1984. Specific resistance to malathion in laboratory and field populations of the Australian sheep blowfly Lucilia cuprina (Diptera: Calliphoridae). J. Econ. Entomol. 77: 1400-1404.

Leary, R. F. & F. W. Allendorf, 1989. Fluctuating asymmetry as an indicator of stress: implications for conservation biology. Trends Ecol. Evol. 4: 214-217.

Leary, R. F., F. W. Allendorf & K. L. Knudsen, 1992. Genetic, environmental and developmental causes of meristic variation in rainbow trout. Acta Zool. Fennica 191: 79-95.

Markow, T. A. & J. P. Ricker, 1991. Developmental stability in hybrids between the sibling species pair, Drosophila melanogaster and Drosophila simulans. Genetica 84: 115-121.

Maynard Smith, J., R. Burian, S. Kauffman, P. Alberch, J. Campbell, B. Goodwin, R. Lande, D. Raup & L. Wolpert, 1985. Developmental constraints and evolution Quart. Rev. Biol. 60: 266-287.

Møller, A. P. & J. Höglund, 1992. Patterns of fluctuating asymmetry in avian feather ornaments: implications for models of sexual selection. Proc. Roy. Soc. Lond. B. 245: 1-5.

McKenzie, J. A., 1987. Insecticide resistance in the Australian sheep blowfly - messages for pesticide usage. Chem. Ind. 8: 266-269.

McKenzie, J. A., 1990. Selection at the dieldrin resistance locus in overwintering populations of Lucilia cuprina (Wiedemann). Aust. J. Zool. 38: 493-501.

McKenzie, J. A., 1993. Measuring fitness and intergenic interactions: The evolution of resistance to diazinon in Lucilia cuprina. Genetica (In Press).

McKenzie, J. A. & G. M. Clarke, 1988. Diazinon resistance, fluctuating asymmetry and fitness in the Australian sheep blowfly, Lucilia cuprina. Genetics 120: 213-220.

McKenzie, J. A. & A. Y. Game, 1987. Diazinon resistance in Lucilia cuprina; mapping of a fitness modifier. Heredity 59: 381-391.

McKenzie, J. A., P. Batterham & L. Baker, 1990. Fitness and asymmetry modification as an evolutionary process. A study in the Australian sheep blowfly, Lucilia cuprina and Drosophila melanogaster, pp. 57-73. In: Ecological and evolutionary genetics of Drosophila, edited by J. S. F. Barker, W. T. Starmer and R. J. MacIntyre, Plenum Press, New York.

McKenzie, J. A., J. M. Dearn & M. J. Whitten, 1980. Genetic basis of resistance to diazinon in Victorian populations of the Australian sheep blowfly, Lucilia cuprina. Aust. J. Biol. Sci. 33: 85-95.

McKenzie, J. A., M. J. Whitten & M. A. Adena, 1982. The effect of genetic background on the fitness of diazinon resistance genotypes of the Australian sheep blowfly, Lucilia cuprina. Heredity 49: 1-9.

Palmer, A. R. & C. Strobeck, 1986. Fluctuating asymmetry: measurement, analysis, patterns. Ann. Rev. Ecol. System. 17: 391-421.

Palmer, A. R. & C. Strobeck, 1992. Fluctuating asymmetry as a measure of developmental stability: Implications of non-normal distributions and the power of statistical tests. Acta Zool. Fennica 191: 57-72.

Parker, A. G., R. J. Russell, A. C. Delves & J. G. Oakeshott, 1991. Biochemistry and physiology of esterases in organophosphate-susceptible and -resistant strains of the Australian sheep blowfly, Lucilia cuprina. Pest. Biochem. Physiol. 41: 305-318.

Parsons, P. A., 1990. Fluctuating asymmetry: an epigenetic measure of stress. Biol. Rev. 65: 131-145.

Parsons, P. A., 1991. Evolutionary rates: stress and species boundaries. Ann. Rev. Ecol. System. 22: 1-18.

Parsons, P. A., 1992. Fluctuating asymmetry: a biological monitor of environmental and genomic stress. Heredity 68: 361-364.

Raftos, D. A., 1986. The biochemical basis of malathion resistance in the Australian sheep blowfly, Lucilia cuprina. Pestic. Biochem. Physiol. 26: 302-309.

Raftos, D. A. & P. B. Hughes, 1986. Genetic basis of a specific resistance to malathion in the Australian sheep blowfly, Lucilia cuprina (Diptera: Calliphoridae). J. Econ. Ent. 79: 553-557.

Roush, R. T. & J. A. McKenzie, 1987. Ecological genetics of insecticide and acaricide resistance. Ann. Rev. Ent. 32: 361-380.

Russell, R. J., M. M. Dumancic, G. G. Foster, G. L. Weller, M. J. Healy & J. G. Oakeshott, 1990. Insecticide resistance as a model system for studying molecular evolution, pp. 293-314. In: Ecological and evolutionary genetics of Drosophila, edited by J. S. F. Barker, W. T. Starmer and R. J. MacIntyre, Plenum Press, New York.

Scharloo, W., 1991. Canalization: genetic and developmental aspects. Ann. Rev. Ecol. System. 22: 65-93.

Sokal, R. R. & F. J. Rohlf, 1981. Biometry, 2nd edn., W. H. Freeman, San Francisco.

Templeton, A. R., H. Hollocher, S. Lawler & J. S. Johnston, 1990. The ecological genetics of abnormal abdomen in Drosophila mercatorum, pp. 17-35. In: Ecological and evolutionary genetics of Drosophila, edited by J. S. F. Barker, W. T. Starmer and R. J. MacIntyre, Plenum Press, New York.

Thornhill, R., 1992. Fluctuating asymmetry, interspecific agression and male mating tactics in two species of Japanese scorpion flies. Behav. Ecol. Sociobiol. 30: 357-363.

Van Valen, L., 1962. A study of fluctuating asymmetry. Evolution 16: 125-142.

Whitten, M. J., J. M. Dearn & J. A. McKenzie, 1980. Field studies on insecticide resistance in the Australian sheep blowfly, Lucilia cuprina. Aust. J. Biol. Sci. 33: 725-735.

T. A. Markow (ed.), Developmental Instability: Its Origins and Evolutionary Implications, 79–96, 1994.

Nonlinear growth dynamics and the origin of fluctuating asymmetry

J. M. Emlen,[1] D. C. Freeman[2] & J. H. Graham[3]
[1] *National Fisheries Research Center U.S. Fish and Wildlife Service Naval Station Puget Sound Seattle, WA 98115, USA*
[2] *Dept. Biological Sciences Wayne State University Detroit, MI 48202, USA*
[3] *Department of Biology Berry College Mt. Berry, GA 30149, USA*

Received and accepted 5 May 1993

Key words: developmental instability, fluctuating asymmetry, fractal, growth

Abstract

The nonlinear, complex nature of biosynthesis magnifies the impacts of small, random perturbations on organism growth, leading to distortions in adaptive allometries and, in particular, to fluctuating asymmetry. These distortions can be partly checked by cell-cell and inter-body part feedback during growth and development, though the latter mechanism also may lead to complex patterns in right-left asymmetry. Stress can be expected to increase the degree to which random growth perturbations are magnified and may also result in disruption of the check mechanisms, thus exaggerating fluctuating asymmetry.

The processes described not only provide one explanation for the existence of fluctuating asymmetry and its augmentation under stress, but suggest additional effects of stress as well. Specifically, stress is predicted to lead to decreased fractal dimension of bone sutures and branching structures in animals, and in increased dimension of growth trace patterns such as those found in mollusc shells and fish otoliths and scales.

A basic yet broad primer on fractals and chaos is provided as background for the theoretical development in this manuscript.

Introduction

A large and rapidly growing literature supports the proposition that chronically stressed organisms generally exhibit evidence of developmental instability. Developmental instability has been suggested (Soule, 1967; Valentine, Soule & Samollow, 1973; Ames, Felley & Smith, 1979; Jagoe & Haines, 1985; Zakharov & Rubin, 1985; Clarke, 1992) and is currently being used as a monitoring and early warning tool in the detection and assessment of human-induced disturbances in natural ecosystems. Developmental instability also arises from genetic stress in the form of disruption of coadaptive gene complexes (Graham, 1992; Clarke, Oldroyd & Hunt, 1992) via inbreeding (Mather, 1953; Bader, 1965; DiBennardo & Bailit, 1978; Leary, Allendorf & Knudsen, 1985) and outbreeding depression (Manley & Ledig, 1979; Zakharov & Bakulina, 1984; Graham & Felley, 1985; Graham, 1992), from strong directional selection (Thoday, 1958), and from various chromosomal abnormalities (Miller & Giroux, 1966; Shiono & Kadowaki, 1975; Barden, 1980; Shapiro, 1983) and other genetic defects (Adams & Niswander, 1972; Woolf & Gianas, 1976, 1977; Sofaer, 1979; Markow & Gottesman, 1989). Finally, adverse environmental conditions lead to developmental instability in human populations (Bailit *et al.*, 1970; Harris & Nweeia, 1980; Kieser, Groeneveld & Preston, 1986).

Traditionally, developmental instability has been measured by right-left asymmetry of normally distributed bilateral traits (fluctuating asymmetry, or FA). Unfortunately, while organisms believed to be stressed generally exhibit exaggerated fluctuating asymmetry, this is not always the case (e.g., Siegel & Doyle, 1975; Ames, Felley & Smith, 1979;

Felley, 1980; Jagoe & Haines, 1985). To our knowledge no one has found instances of reduced FA in the presence of stress, but it is not at all uncommon to find no response. Is the lack of response an indication of limited utility for this early warning and monitoring tool (Parsons, 1992)? Are those negative results merely the consequence of choosing inappropriate traits for examination? Is it possible that the putative stress has been ameliorated by a consequent drop in population density and correspondingly lowered population feedback stress? Soule and Cuzin-Roudy (1982) addressed the matter of appropriate choice of traits, suggesting that those with high heritability - those less critical to fitness - should be the most sensitive to stress. The accuracy of this possibly useful prediction must remain speculative until we find a mechanism or mechanisms explaining FA's existence and its usual exaggeration under stress. Discovery of an underlying mechanism also should boost confidence in the use of FA as an indicator of stress. We assert that FA arises from two causes. One involves the nonlinear dynamics of biosynthetic processes in cells, the other the nonlinear feedback occurring among developing body parts. In this paper we address the former. The latter is explored in Graham, Freeman and Emlen (1993).

Basic processes

The growth of organisms depends on the growth and proliferation of cells and the replacement of dead cells, all of which ultimately depend on the synthesis of proteins, lipids, nucleic acids, etc. The pathways by which these substances are synthesized are often complex. For example, construction of phosphatidyl ethanolamine, a lipid common in animal tissues, requires first the synthesis of its fatty acid precursors and then 15 additional enzymatic steps involving 11 enzymes to put these fatty acids together (Lehninger, 1971). Construction of proteins, composing the bulk of non-water cell material by weight, requires first the synthesis of amino acids, and then the complex process of migration to and build-up upon the ribosomes. Substrates, enzymes and an energy source such as ATP must all be present at minimal (threshold) concentrations within the same immediate vicinity for interactions to take place, so the dynamics of synthesis must reflect rate of production, diffusion rates of various substances, and the rate at which precursors are used up and replaced. Contributing to the complex dynamics of synthesis is the fact that enzymes involved in one synthetic pathway do not act independently, but rather function in many multi-enzyme sequences that are interconnected (Lehninger, 1971). Finally, enzyme kinetics are on the whole nonlinear, providing a mechanism whereby feedback can spiral out of control much like an overly sensitive rheostat.

Consider the effect of a random perturbation in biosynthesis within a cell: a shift in dispersion of some substrate involved in synthesis, for example. Such a shift, no matter how minute, will alter the spatial and temporal details of synthesis in the cell, thereby introducing changes in the spatial pattern of diffusion rates, further affecting ATP, substrate and enzyme distribution. Owing to the complexity, both functional and spatial, of biosynthesis, such random perturbations will often become magnified, leading to variations in biosynthesis rates over time. The consequence is that unless checked by cell-cell feedback, the rate of biosynthesis will vary randomly among cells. This will lead, though probably in something much more complicated than a linear fashion, to differential rates of cell division as well. Therefore, if cells grew with no intercommunication, random variation in growth rates of individual cells would compound over all component cells within growing structures, leading to random variation in size and shape of even genetically identical organisms; identical twins would probably appear very different from one another. Such apparent randomness in development is not the norm. Therefore, there must exist mechanisms to check runaway random drift. Since such drift would probably impact fitness adversely by destroying allometric relations, the evolution of check mechanisms is hardly unexpected.

Consider a situation in which two structures (corresponding right and left structures, for example) grow under identical genetic direction and constraints, and in identical physical and chemical milieus. In keeping with the argument above, these structures should drift randomly in relative growth, leading to FA that is naturally held in bounds by evolved check mechanisms.

In addition to placing right and left structures under identical controls and constaints, there are at least two possible ways to check runaway drift:

(1) Negative feedback among growing cells within each structure. Recall that the variance of the sum of two variables is the sum of the variances plus twice the covariance. Negative feedback among cells would lead to negative covariances in growth among cells, thereby damping the cumulative drift over all cells in a structure. A drop in growth variance between right and left paired structures translates into less drift toward FA. Cell-cell feedback acting to suppress biochemical synthesis or cell proliferation is poorly understood, but there seems no doubt it occurs. Terayama *et al.* (1969) discussed evidence of regulatory feedback on cell division via suppression of DNA synthesis by thymidine deoxynucleotides, and suggest these compounds may be important in regulating cell proliferation in general. Turkington (1971) reviewed evidence of negative feedback in epidermal cells, and Koji *et al.* (1988) described density feedback on cell proliferation in cultures of adult rat hepatocytes.

(2) Feedback between right and left structures. Communication between right and left sides can take place via the nervous or circulatory system. Positive feedback, promoting catchup growth on the lagging side, generally should promote equal growth rates, but its effectiveness will vary depending on the communications time lag, the strength of the feedback signal, and the presence or absence of other, systemic feedback systems. For example, if the time lag were sufficiently large and the signal strong, overcompensation could occur, leading to right-left side oscillations and increased asymmetry at any point in time. If combined cumulative growth of the right and left structures, or larger structures of which they were a part, stimulated systemic feedback to slow or halt growth, headstarts of one side over the other could become locked in, particularly if right-left oscillations were occurring, leading to antisymmetry. Feedback between sides, by its almost invariably non-linear nature, should result in growth waves; if the feedback time lags are not too great, it could promote phase-lock in these waves, right and left (Graham, Freeman & Emlen, 1993). Phase-locking would discourage divergence in size between right and left corresponding structures. Indeed, phase-locking is a mechanism by which appropriate relative rates of development of diverse structures can be controlled, thereby preserving adaptive allometries. We should expect feedback mechanisms promoting phase-locking to have evolved. Because different structures exhibit different growth rates and the wave characteristics of feedback lags between these structures will vary with their proximities, the phase-lock cycles of one set of structures can be expected to differ from those of other structure groups. Accordingly, we expect natural selection to have favored a plethora of oscillatory processes acting to appropriately entrain the diverse developmental rhythms of different body parts. The pervasiveness of oscillations, from circadian to those of much shorter duration - oscillations in enzyme synthesis, enzyme concentrations, NADH, light scattering and ion fluxes in suspensions of mitochondria, volume changes in mitochondria, redox potentials, muscle fiber membrane potentials, muscle tonicity (Hess & Boiteux, 1971), cyclicity of stomatal openings, water content of guard and other cells, and plant transpiration (Gumowski, 1987) - speak for the importance of periodicity in physiological processes. One might even argue that if biological rhythms did not exist for other reasons, they would eventually evolve simply as mechanisms to control body allometry.

Efficacy of the phase-lock mechanisms would be greatest if the wave-producing interactions were such as to produce sharp wave peaks, i.e. growth spurts, so we might expect natural selection to have encouraged saltatory growth. This expectation is supported by observations on growth in human infants (Lampl, Veldhuis & Johnson, 1992).

While discouraging right-left divergence, interstructure feedback may also confer some rather interesting features on development, including oscillations in relative growth of the two sides, and complex, perhaps chaotic temporal growth patterns (see Graham, Freeman & Emlen, 1993, for further elaboration).

The effects of stress

Regardless of the presence of the various mechanisms for checking right-left drift in growth, the underlying existence of perturbations, with their potential for cascading effects on differential growth rates and replication among cells, remains. Under normal circumstances the resulting asymmetry in growth should be small, but its elimination can never be assured. The thesis of this paper is

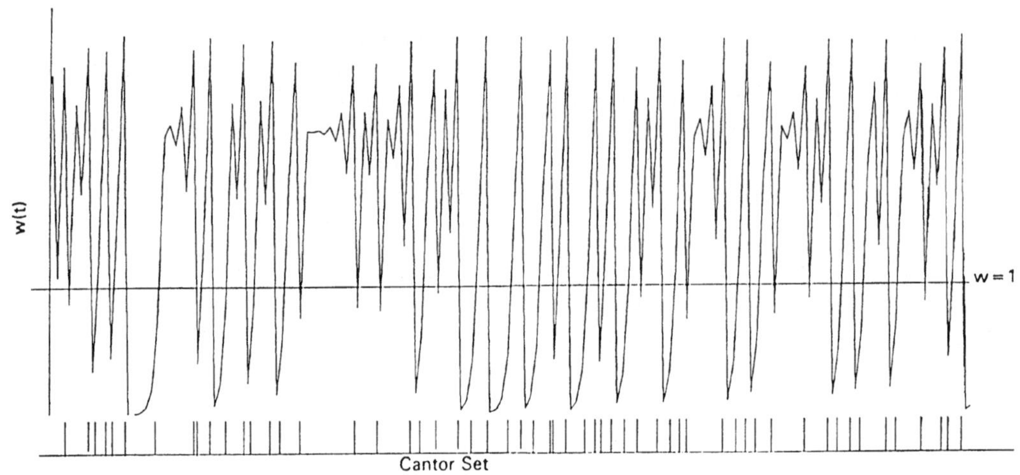

Fig. 1. Hypothetical plot of the concentration of the biosynthesis-limiting factor divided by the threshold level for biosynthesis against time. Synthesis occurs so long as w(t) > 1. The points where w(t) crosses the threshold (w = 1) line form a Cantor set.

that, under stress, the impact of random perturbations on drift will increase. This and the possible stress-induced breakdown or alterations of check mechanisms leads, in turn, to increased FA.

To explore further the role of cell-level perturbations in FA and its exaggeration under stress, it is necessary to examine the process of biosynthesis. To do this requires a foray into fractal geometry and its relation to chaotic dynamics. A primer on these topics is the subject of the appendix. The reader unfamiliar with fractals and chaos is advised to read the appendix before proceeding.

We have asserted that even minute perturbations in the process of biosynthesis within cells can become magnified, leading to among-cell variance in growth. This assertion is based on the fact that functional and spatial complexity of the degree found in biosynthesis leads almost inevitably to chaotic dynamics. That is, growth occurs in waves compounded on waves in such complexity that the precise dynamics at any one point in time are unpredictable from even a complete historical knowledge of the process. It follows that if the oscillatory patterns of either of two initially identical growth processes are even minutely disturbed, they rapidly become uncorrelated, or independent of one another.

Note that the immediately limiting precursor to molecular synthesis (of whatever product) may vary over time. If we plot the course of this precursor, whatever it happens to be, divided by its threshold concentration for synthesis, over time, we obtain a chaotic curve w(t), such as that in Figure 1. This figure describes a statistical fractal; the Cantor set defined by w(t) crossing the value w = 1 (above which synthesis occurs) is also a statistical fractal. Let the parameters of this Cantor set be a and b (a, $b > 0$). That is, the relative expected frequency of spaces between Cantor set points separated by a distance $1/b^n$ is a^n (with some random error in the case of statistical fractals). Consider the case where mean w(t) = 1. Uninterrupted pulses of w(t) > 1, that is, when growth occurs, occur with relative duration 1, 1/b, ... at relative frequencies 1, a, ... Let us examine the consequences of this dynamic.

Consider a time interval t and its various growth waves of duration t/b, t/b^2, ... Then the expected cumulative growth over the interval is given by

$$E[G(t)] = \frac{ag(t)/b + a^2 g(t)/b^2 + \ldots}{a + a^2 + ..} = g(t)\frac{(a/b) + (a/b)^2 + (a/b)^3 + ..}{a + a^2 + a^3 + ..} \equiv \frac{g(t) * X_t}{A}, \tag{1}$$

where g(t) is the growth rate and the denominator normalizes the contributory terms such that the sum of frequencies is 1.0. Similarly, over the interval (0, t/b), we have

$$E[G(t/b)] =$$

$$\frac{ag(t)b^2 + a^2 g(t)/b^3 + \ldots}{a + a^2 + ..} =$$

$$\frac{g(t)}{aA}[(a/b)^2 + (a/b)^3 + ..] = \qquad (2)$$

$$\frac{g(t)}{aA}[X_t - a/b] = \frac{E[G(t)]}{a} - \frac{g(t)}{bA}$$

This equation has two components, a growth rate unrelated to the chaotic nature of growth, *per se*, and another portion specifically relating to chaos via considerations of scale. The latter part of the equation, denoted with a subscript, *s*, is the same as equation A3 in the Appendix, but with E[G(t)] in place of N_x.

$$E[G(t)]_s \propto \frac{1}{1/a} E[G(t/b)]_s \, .$$

Therefore,

$$E[G(t)]_s \propto t^{-\alpha} \equiv t^{\gamma}$$

where (3)

$$\gamma = -\alpha = -\log(1/a)/\log(b) = \log(a)/\log(b).$$

Finally,

$$E[G(t)] = cg(t) + c' t^{\gamma}$$

where *c*, *c′* are constants.

Of more pertinence to this paper is the growth *variance*. The probabilities used in the derivation of equation (1) are a/k, a^2/k ... where $k = a + a^2 + a^3 + \ldots$. Then, because growth at any one scale, $1/b^n$, is statistically independent of that at other scales, and because presence or absence of growth in any randomly chosen interval is binomial, the variance becomes

$$V[G(t)] =$$

$$\left(\frac{a}{k}\right)\left(1-\frac{a}{k}\right)\frac{g(t)}{b} + \left(\frac{a^2}{k}\right)\left(1-\frac{a^2}{k}\right)\frac{g(t)}{b^2} + \ldots = \qquad (4a)$$

$$\frac{g(t)}{k}\left\{\left[\left(\frac{a}{b}\right)+\left(\frac{a}{b}\right)^2 + ..\right] - \left[\frac{a^2}{bk}+\frac{a^2}{b^2 k}+\ldots\right]\right\} \equiv$$

$$[g(t)/k]*(X' - Y').$$

Similarly,

$$V[G(t/b)] =$$

$$\left(\frac{a}{k}\right)\left(1-\frac{a}{k}\right)\frac{g(t)}{b^2} + \left(\frac{a}{k}\right)^2\left(1-\frac{a^2}{k}\right)\frac{g(t)}{b^3} + \ldots =$$

$$\frac{g(t)}{ak}\left[\left(\frac{a}{b}\right)^2 + \left(\frac{a}{b}\right)^3 + ..\right] - \frac{g(t)}{a^2 k}\left[\frac{a^4}{b^2 k}\frac{a^6}{b^3 k} + \ldots\right] = \qquad (4b)$$

$$\frac{g(t)}{ak} * (X' - a/b) - \frac{g(t)}{a^2 k} * (Y' - a^2/bk).$$

From (4b),

$$Y' - a^2/bk = \frac{[g(t)/ak]*(X' - a/b) - V[G(t/b)]}{g(t)/a^2 k} =$$

$$a(X' - a/b) - a^2 k V[G(t/b)]/g(t).$$

Substituting this into (4a), (5)

$$V[G(t)] = \frac{g(t)}{k}\{X' - [a(X' - a/b) -$$

$$a^2 k V[G(t/b)]/g(t) + a^2/bk\} = \ldots =$$

$$\frac{g(t)}{k} * [X'(1 - a + a^2/b - a^2/bk^2] + a^2 V[G(t/b)].$$

Because $a > 1$, $k = a + a^2 + a^3 + \ldots$ becomes very large and the first term on the right in equation (5) vanishes. Hence

$$V[G(t)] \to a^2 V[G(t/b)] = \frac{1}{1/a^2} V[G(t/b)],$$

so that (6)

$$V[G(t)] \propto t^{2\gamma} : \gamma = \log(a)/\log(b) > 0.$$

Equation (6) indicates that because of the chaotic nature of the growth process, the variance in cumulative growth (i.e., the uncertainty in growth among cells) increases over time. Furthermore, as the complexity of the growth process increases, the Cantor set defined by w = 1 becomes more dense implying larger log(a)/log(b) and, therefore, a faster rate in variance increase. The chaotic nature of the growth process acts to *magnify* drift arising from random perturbations.

West (1990) stated that chaotic processes are resistant to disturbance because they already incorporate variation in their strange attractors. Thus perturbations simply move a process to another point on its attractor and have very little effect on random drift of the processes over time. This is reflected in the fact that variance in a chaotic process, w(t), increases with t^{β} (equation A5), where $\beta = 2H$ (equation A9) $= 2(2 - D)$, and D is dimension. The higher the dimension of a chaotic process, the slower the drift. This conclusion seems at odds with the above conclusion that increased dimension leads to *greater* drift. The apparent inconsistency is cleared up when we note that it is not the chaotic process, w(t), whose drift we are concerned with. Growth dynamics are not the same as w(t); they merely arise as a *consequence* of w(t).

To explore the consequences of stress, note that each cell is capable of synthesizing protein at a rate of up to more than 1000 molecules/second (Lehninger, 1971). That is, synthesis may be a nearly continuous process. This, in turn, means that the curve in figure 1 may lie above the w = 1 threshold most of the time. Such a curve exhibits a greater frequency of long waves, and thus a lower γ value, than a curve lying closer to the x-axis, that is, a curve representing a slower growth process. It follows that accelerated growth is associated with lower dimension of the dynamic process, w(t), and consequently a reduced among-cell variance in growth.

Note next that stress results in the drawing off of energy otherwise used in growth and maintenance to repair processes and other stress-related physiological responses (Wedemeyer, Burton & McLeay, 1990; Alekseeva, Zinichev & Zotin, 1992; Ozernyuk *et al.*, 1992). Indeed, stress is so intimately associated with energy drain that it is often measured using the adenylate energy charge,

$$\frac{[ATP]+\frac{1}{2}[ADP]}{[ATP]+[ADP]+[AMP]}$$

(Ivanovics & Wiebe, 1981). But as an energy-dissipative force, stress lowers the growth rate, hence raising the variance in among-cell growth and accelerating the drift process. Stress in the form of ionizing radiation or oxygen-free radicals can be expected to make destructive hits on the biosynthetic pathways at random (aperiodic) times, cutting short what long growth waves occur, thus raising the dimension of w(t) and thereby augmenting among-cell variance in growth.

The cell-cell negative feedback check process, by virtue of its importance in controlling among-cell variance in growth and thus discouraging aberrations in allometry, can be expected to have evolved to perform its function efficiently, that is, to a degree limited only by biochemical and physical shortcomings of the communications channels. As such, any disturbance (due to stress) can be expected to compromise that efficiency, weakening resistance to the increasing among-cell growth variance, and exaggerating FA.

The phase-lock mechanisms produced by feedback among the various body parts still operate under stress. However, because stress is likely to alter growth rates differently in different structures, the normal, and presumably adaptive pattern of wave frequencies arising from feedback is likely to be disrupted to the detriment of homeostasis in allometry. Also, as the rate of drift increases, it may exceed the capacity of these mechanisms to haul diverging structures back into phase-lock and, instead, result in the growth wave cycle of one structure advancing into phase-lock with the *next* growth cycle of the other. This would result in a discrete random (though not necessarily Brownian) walk in relative right-left structure size. This subject is explored more fully in Graham, Freeman and Emlen (1993).

Where these waves describe growth precursors or morphogens triggering development of sequential structures (e.g., pectoral fin rays in fish), drift will result in shifting positions of these structures or, with phase-lock slippage, deviation in the number of such structures right and left.

Discussion

Once fluctuating asymmetry has begun to develop, differential growth distortions right and left will impact development of adjoining tissues. As such, the impacts of FA on one set of paired structures may magnify size differentials elsewhere. For example, right-left asymmetry in vascular growth will cause right-left asymmetry in blood supply, thus promoting differential right-left growth in structures depending on that blood supply. In addition, discordant growth may cause twisting, thereby altering shape as well as size of structure. Finally, shape is affected also by differential drift in different portions or components of a structure. Size, by virtue of its averaging the drift of component parts, carries less information about stress than shape, which therefore can be expected to be the more sensitive indicator (see Soule, 1982; Alados, Escos & Emlen, 1992).

Patterns such as stripes, spots and buds (myomeres, meristic traits, limb buds, leaf buds, etc.) may arise by any of a number of processes. One is the laying down of spatial traces by growth wave events already mentioned. Another is the formation of standing waves of morphogen concentrations (Turing patterns; see for example, Meinhardt, 1982; Eilbeck, 1989; Nagorcka, 1989; Castets *et al.*, 1990; Agladze, Dulos & De Kepper, 1992) and perhaps accompanying cell migrations (Maini *et al.*, 1991). In all cases, the nature of the pattern formed depends on production rates of chemical activitors, and in the latter two, sinks or inhibitors. Diffusion rates, dependent in turn on production rates, are also critical to the pattern. In the case of the Turing model, minute changes in these parameters may have dramatic effects on the patterns formed. Thus, stress can be expected to affect right-left positions *and other orientational aspects* of color patterns and structures on organisms. An illustration of right-left placement arising from one such mechanism and its disturbance under stress is found in the black locust, *Robinia pseudoacacia*. Here the leaves are pinnate compound and in unstressed environments maintain a fairly accurate right-left symmetry in position of leaflet petioles. Under stress from air pollution, corresponding petiole positions drift as expected, their random walk seemingly curtailed by some phase-lock mechanism. Phase-lock slippage was never observed in unstressed areas, but became quite common at highly polluted sites (Freeman, Graham & Emlen, 1993).

Branching structures

By virtue of altering pattern and bud position, stress can be expected to impact biological branching structures such as vascular systems, nerve networks, biliary networks, bronchial trees, modular plant structure, and leaf venation.

Consider any such system and suppose that the size x of branches of any given branching generation is a consequency of growth as described above. More specifically, suppose the size of a branch is proportional to the time spent growing it. Then at any branching generation, x will be distributed as

(7)

Because almost nothing is known of the mechanism underlying branch formation, this is largely conjecture. Nevertheless, West (1990) used this density function to derive a distribution of average branch size over branching generations that fits a wide range of observed branching structures of many animal taxa remarkably well.

Now recall that stress raises the value of γ. Stress, therefore, preferentially diminishes the relative number of finer branches at all generations, lowering the completeness with which the branching structure fills space. That is, we can expect the dimension of the structure to drop under stress. Freeman (pers. comm.) has obtained data suggesting that dimension of the vasculature of human retinas is significantly depressed among samples of hyperstressed, myopic, diabetic, and senile humans.

The above argument should hold generally for animals. In plants, however, an apical bud suppresses growth of collateral buds. Here, stress may interfere with this feedback, permitting development of increased numbers of subsidiary branches. In plants, therefore, stress might *increase* dimension. It is not clear in this instance which trend is most likely.

Bone sutures

Bone sutures develop in a variety of ways (see, for example, Pritchard, Scott & Girgis, 1956), but can be characterized generally as outgrowths of ossification in a cartilagenous matrix. Presumably the same sort of growth dynamics applies to this process as to growth of other tissues. As such, the forward wave of ossification can be expected to vary chaotically in its advance, showing irregular processes that themselves grow out irregularly. The complex shape that develops could easily be tamed by promoting short-range feedback, but in the case of sutures this feedback is disadvantageous. Fractal sutures have increased edge to which bones on either side can adhere, thus increasing strength. Suture strength may be enhanced still further by recurvations generated by the chaotic advance of ossification. Given such considerations, the negative feedback evolved as a normal trait in cells elsewhere probably will have been either suppressed or reversed here.

Because we can only surmise the situation *vis a vis* cell communication in suture formation, we can only make educated guesses as to the impacts of stress. However, phase-lock slippage rates between different sections of the suture probably would be increased and positive feedback or suppression of negative feedback among cells probably would be compromised. The consequences would be still deeper curves at large wavelengths, but a suppression of smaller irregularities. This fits well with the observation that fingering (longer wavelengths) is exaggerated in skull sutures of hydrocephalics (Walker, 1961), and also fits with observations (C. L. Alados, J. Escos and J. M. Emlen, unpubl. data) that sagittal sutures in North African gazelles (*Gazella cuvier* and *G. dorca*) exhibit depressed dimensionality in populations stressed by inbreeding.

Hartwig (1991), using a coastline approach with human sagittal skull sutures, found a curvilinear log-log plot, and interpreted these results as evidence for two superimposed processes of different dimension. It seems more likely that he was seeing evidence of self-affinity (see figure [A9]), but his conclusion may be warranted. In either case, unless correction for self-affinity can be made, it is wise to designate a specific range in scale over which dimensions are to be calculated and compared.

Growth traces

Bouts of rapid proliferation of material can be expected to result in corresponding bouts of increase in tissue volume and tissue density. In some hard tissues, such as mollusc shells, such density variations would become locked in, leaving a trace of past growth dynamics. Cross section luminance profiles reveal chaotic patterns of high and low density, presumably with some characteristic fractal dimension. In such cases, the imposition of stress, by superimposing noise and thus increasing the dimensionality of the growth process, should also raise the dimension of the fractal profile. Emlen (unpubl. data) has examined shells of mussels (*Mytilus edulis*) taken from heavily polluted, slightly polluted (with the same contaminants), and clean sites in Puget Sound and found dimensions to vary with contamination as expected. An R/S analysis (Feder, 1988) permits determination of fractal dimension over spatial scales that filter out random noise due to less than perfect specimen preparation or microscope focus.

Hard structure patterns probably reflect the overlapping of several growth processes. If so, we can expect to find beats in the physical pattern, also with a fractal distribution. Perhaps the growth rings of fish otoliths or clam shells, or the circuli of fish scales reflect either the existence of growth spurts or beats, or both.

What sorts of traits are most likely to display developmental instability in response to stress? Soule and Cuzin-Roudy (1982) argued that traits with high heritability should be sensitive because such traits are least important to fitness and are, therefore, least canalized. The argument can be put on firmer footing. We mentioned previously that physiological mechanisms should exist to shunt energy and substrate out of pathways where growth is too rapid or occurring at the expense of growth elsewhere. Presumably, these mechanisms will have evolved so that processes most critical to fitness are the best regulated. As stress draws off energy, depriving growth processes across the board, these controlling mechanisms should act to ensure continued energy and material to the most critical pathways. This process may, indeed, be the central basis of canalization. The less important pathways thus experience exaggerated loss of energy and material, driving $w(t)$ down relative to the threshold $w = 1$. As noted above, this preferentially eliminates the longer growth waves, increasing γ and, therefore, the rate of right-left drift. Presumably the same argument applies to branching structures, sutures and growth traces.

Certain organisms may be more likely to give evidence of stress than others. Species occupying spatially and temporally constant environments have less need of strong physiological canalizing mechanisms. Thus stress, when it comes to these species, may be more likely to cause developmental instability. On the other hand, if canalization consists at least in part of the adaptive shunting of energy and substrate to properly regulate traits critical to fitness, the less important traits will not be significantly more deprived under stress than the important ones. Thus a wider range of traits may show a DI response to stress in stenotopic species.

Conclusions

We have described a process by which asymmetry might be generated, and by which FA can be expected to increase progressively with stress. The arguments are not dependent on any model form for growth; they follow simply and directly from an assertion that the complexity of the multi-step, interlocking, multiple-chain enzyme pathways involved in molecular synthesis lead almost certainly to chaotic oscillations in growth. The argument predicts also that many, if not all, physiological processes in living organisms will be complex-cyclic, a prediction that seems amply borne out by experimental data. While the conclusion *vis a vis* increased FA under stress does not depend on this prediction, it is certainly strengthened by it.

The arguments go well beyond the predictions *vis a vis* biological rhythms and predicted impacts of stress on FA. They lead also to expectations of structural pattern change in the presence of stress. Specifically, they predict that stress will result in lowered dimension in animal branching structures and bone sutures, and increased dimension in the growth traces characteristic of hard structures such as mollusc shells. All of these predictions are borne out by data, although the breadth of examples is still extremely limited.

References

Alekseeva, T. A., V. V. Zinichev & A. I. Zotin, 1992. Energy criteria of reliability and stability of development. Acta Zool. Fennica 191: 159-165.

Adams, M. S. & J. D. Niswander, 1967. Developmental 'noise' and a congenital malformation. Genet. Res. 10: 313-317.

Agladze, K., E. Dulos & P. De Kepper, 1992. Turing patterns in confined gel and gel-free media J. Phys. Chem. 96: 2400-2403.

Alados, C. L., J. Escos & J. M. Emlen, 1992. Developmental instability as an indicator of stress in the Pacific hake (*Merluccius productus*). Fishery Bulletin (in press).

Ames, L. J., J. D. Felley & M. H. Smith, 1979. Amounts of asymmetry in centrarchid fish inhabiting heated and non-heated reservoirs. Trans. Amer. Fish. Soc. 108: 489-485.

Bader, R. S., 1965. Fluctuating asymmetry in the dentition of the house mouse. Growth 29: 291-300.

Bailit, H. L., P. L. Workman, J. D. Niswander & C. J. MacLean, 1970. Dental asymmetry as an indicator of genetic and environmental conditions in human populations. Hum. Biol. 42: 626-638.

Barden, H. S., 1980. Fluctuating dental asymmetry: a measure of developmental instability in Down syndrome. Amer. J. Phys. Anthro. 52: 169-173.

Castets, V., E. Dulos, J. Boissonade & P. De Kepper, 1990. Experimental evidence of a sustained standing Turing-type nonequilibrium chemical pattern. Phys. Rev. Lett. 64: 2953-2956.

Clarke, G. M., 1992. Fluctuating asymmetry: A technique for measuring developmental stress of genetic and environmental origin. Acta Zool. Fennica 191: 31-36.

Clarke, G. M., B. P. Oldroyd & P. Hunt, 1992. The genetic basis of developmental stability in *Apis mellifera*: Heterozygosity versus genic balance. Evolution 46: 753-762.

DiBennardo, R. & H. L. Bailit, 1978. Stress and dental asymmetry in a population of Japanese children. Amer. J. Phys. Anthro. 48: 89-94.

Eilbeck, J. C., 1989. Pattern formation and pattern selection in reaction-diffusion systems, pp. 31-41 in Theoretical Biology: Epigenetic and Evolutionary Order from Complex Systems, edited by B. Goodwin and P. Saunders. Edinburgh University Press.

Feder, J., 1988. Fractals. Plenum Press, New York, 283 pp.

Felley, J., 1980. Analysis of morphology and asymmetry in bluegill sunfish (*Lepomis machrochirus*) in the southeastern United States. Copeia 1980: 18-29.

Freeman, D. C., J. H. Graham & J. M. Emlen, 1993. (This volume).

Graham, J. H., 1992. Genomic coadaptation and developmental stability in hybrid zones. Acta Zool. Fenn. 192: 121-132.

Graham, J. H. & J. D. Felley, 1985. Genomic coadaptation and developmental stability within introgressed population of *Enneacanthus gloriosus* and *E. obesus* (Pisces, Centrarchidae). Evolution. 39: 104-114.

Graham, J. H., D. C. Freeman & J. M. Emlen, 1993. (This volume).

Gumowski, I., 1987.Oscillatory plant transpiration: Problem of a mathematically correct model, pp. 13-18 in Chaos in Biological Systems, NATO ASI series, edited by H. DFegn, A. V. Holden and L. F. Olson. Plenum Press, New York.

Harris, E. F. & M. T. Nweeia, 1980. Dental asymmetry as a measure of environmental stress in the Ticuna Indians of Columbia. Amer. J. Phys. Anthro. 53: 133-142.

Hartwig, W. C., 1991. Fractal analysis of sagittal suture morphology. J. Morphology 210: 289-298.

Hess, B. & A. Boiteux, 1971. Oscillatory phenomena in biochemistry. Ann. Rev. Biochem. 40: 237-258.

Ivanovics, A. M. & W. J. Wiebe, 1981. Toward a working definition of stress: A review and critique, pp. 13-27 in Stress Effects on Natural Ecosystems, edited by G. W. Barrett and R. Rosenberg. Wiley and Sons, Ltd.

Jagoe, C. H. & T. A. Haines, 1985. Fluctuating asymmetry in fishes inhabiting acidified and unacidified lakes. Can. J. Zool. 63: 130-138.

Kieser, J. A., H. T. Groeneveld & C. B. Preston, 1986. Fluctuating dental asymmetry as a measure of odontogenic canalization in Man. Amer. J. Phys. Anthro. 71: 437-444.

Koji, T., P. K. Nakane, M. Murakoski, K. Watanabe & H. Terayama, 1988. Cell density dependent morphological changes in adult rat hepatocytes during primary culture. Cell Biochem. and Function 6: 237-243.

Lampl, M., J. D. Veldhuis & M. L. Johnson, 1992. Saltation and stasis: A model of human growth. Science 258: 801-803.

Leary, R. F., F. W. Allendorf & K. L. Knudsen, 1985. Developmental instability as an indicator of reduced genetic variation in hatchery trout. Trans. Amer. Fish. Soc. 114: 230-235.

Lehninger, A. L., 1971. Bioenergetics, the Molecular Basis of Biological Energy Transformations. W. A. Benjamin, Inc., 245 pp.

Maini, P. K., M. R. Myerscough, K. H. Winters & J. D. Murray, 1991. Bifurcating spatially heterogeneous solutions in a chemotaxis model for biological pattern generation. J. Math. Biol. 53: 701-719.

Mandelbrot, B., 1982. The fractal Geometry of Nature. W. H. Freeman, New York.

Manley, S. A. M. & F. T. Ledig, 1979. Photosynthesis in black and red spruce and their hybrid derivations: ecological isolation and hybrid adaptive inferiority. Can. J. Bot. 57: 307-314.

Markow, T. A. & I. I. Gottesman, 1989. Fluctuating dermatoglyphic asymmetry in psychotic twins. Psychiatry Res. 29: 37-43.

Mather, K., 1953. Genetical control of stability in development. Heredity 7: 297-336.

Meinhardt, H., 1982. Models of Biological Pattern Formation. Academic Press, 230 pp.

Miller, J. R. & J. Giroux, 1966. Dermatoglyphics in pediatric practice, J. Pediatrics 69: 302-312.

Nagorcka, B. N., 1989. Wavelike isomorphic prepatterns in development. J. Theoret. Biol. 137: 127-162.

Ozernyuk, N. D., V. I. Dyomin, E. A. Prokofyev & I. M. Androsova, 1992. Energy, homeostasis and developmental stability. Acta Zool. Fennica 191: 167-175.

Parsons, P. A., 1992. Fluctuating asymmetry: a biological monitor of environmental and genomic stress. Heredity 68: 361-364.

Pritchard, J. J., J. H. Scott & F. G. Girgis, 1956. The structure and development of cranial and facial sutures. J. Anat. 90: 73-86.

Shapiro, B. L., 1983. Down syndrome. A disruption of homeostasis. Amer. J. Med. Genet. 14: 241-269.

Shiono, H. & J. Kadowaki, 1975. Dermatoglyphics of congenital abnormalities without chromosomal aberrations. Clim. Pediatr. 14: 1003-1013.

Siegel, M. I. & W. J. Doyle, 1975. Stress and fluctuating limb asymmetry in various species of rodents. Growth 39: 363-369.

Sofaer, J. A., 1979. Human tooth-size asymmetry in cleft lip with or without cleft palate. Arch. Oral Biol. 24: 141-146.

Soule, M. E., 1967. Phenetics of natural populations, II. Asymmetry and evolution in a lizard. Amer. Natur. 101: 141-160.

Soule, M. E., 1982. Allomeric variation. 1. The theory and some consequences. Amer. Natur. 120: 751-764.

Soule, M. E. & J. Cuzin-Roudy, 1982. Allomeric variation. 2. Developmental instability of extreme phenotypes. Amer. Natur. 120: 765-786.

Terayama, H., H. Otsuka, K. Sakuma & H. Yamagata, 1969. Molecular aspects of the release of tumor cells from the control of proliferation, pp. 327-336 in Nucleic Acid Metabolism, Cell Differentiation and Cancer Growth, edited by E. V. Cowdry and S. Seno. Pergamon Press, New York.

Thoday, J. M., 1958. Homeostasis in a selection experiment. Heredity 12: 401-415.

Turkington, R. W., 1971. Hormonal regulation of cell proliferation and differentiation, pp. 315-355, in Developmental Aspects of the Cell Cycle, edited by I. L. Cameron, G. M. Padilla and A. M. Zimmerman. Academic Press, London.

Valentine, D. W., M. E. Soule & P. Samollow, 1973. Asymmetry analysis in fishes: A possible statistical indicator of environmental stress. Fish. Bull. 71: 357-370.

Walker, D. G., 1961. Malformations of the Face. Livingstone, London.

Wedemeyer, G. A., B. Burton & D. McLeay, 1990. Stress and Acclimation, pp. 451-490 in Methods for Fish Biology, edited by C. Schreck and P. Moyle. Amer. Fisheries Society, Bethesda, Maryland.

West, B. J., Fractal Physiology and Chaos in Medicine World. Scientific.

Woolf, C. M. & A. D. Gianas, 1976. Congenital cleft lip and fluctuating dermatoglyphic asymmetry. Amer. J. Hum. Genet. 28: 400-403.

Woolf, C. M. & A. D. Gianas, 1977. A study of fluctuating dermatoglyphic asymmetry in the sibs and parents of cleft lip propositi. Amer. J. Hum. Genet. 29: 503-507.

Zakharov, V. M. & E. D. Bakulina, 1984. Disturbance of developmental stability at crossing different strains of *Drosophila virilis* (variation in the number of ovarioles taken as an example), Genetika 20: 1390-1391 (in Russian).

Zakharov, V. M. & G. I. Rubin, 1985. The disturbance of developmental stability as an indicator of anthropogenic influence on animal populations in the Baltic Sea basin pp. 526-536 in Symp. on Ecol. Investig. of the Baltic Sea Environment. Riga, USSR, 1983, Valkion Paintuskeskus, Helsinki.

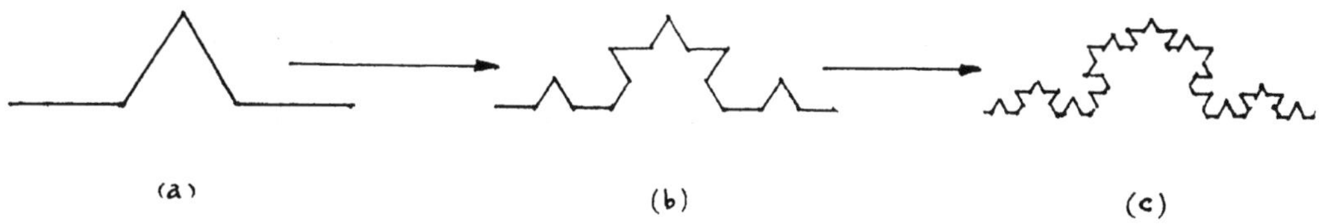

Fig. A1. A Koch curve at three scale levels.

Appendix: A primer in fractals and chaos

Although most developmental instability (DI) work has involved investigation of bilateral symmetry, there are several other types of symmetry available for DI analysis (Graham, Freeman & Emlen, 1992; Freeman, Graham & Emlen, 1993). Of particular relevance to this paper is symmetry of scale. Symmetry of scale occurs when structures are composed of substructures whose appearances, though at a smaller spatial scale, are similar to that of the structure itself. An (non biological) example is to be found in the triangular 'Koch curve' (Fig. A1). To construct this curve, start with a straight line, remove the center third and replace it with two lines each equal to 1/3rd the original line. Then modify each new line segment in a similar manner. Repeat this process an infinite number of times. The result is a curve that has an identical appearance at all spatial scales; it is said to be 'self-similar'.

Fractals

Objects that appear the same at all spatial scales are known as fractals (Mandelbrot, 1982), and they typically possess non-integer dimension. To appreciate this, picture the simple line in Figure A2, and measure its length with a ruler of length 1 unit.

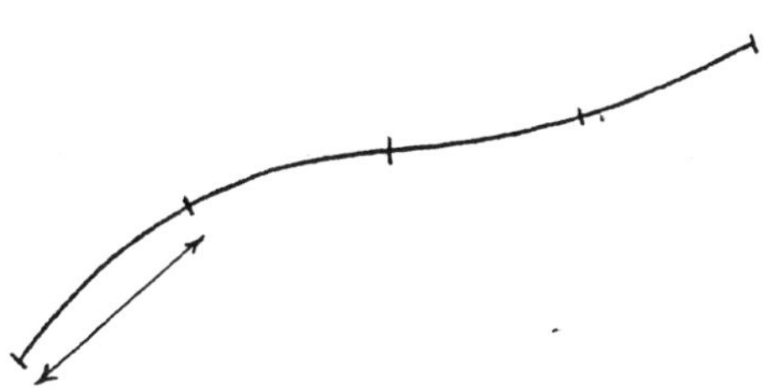

Fig. A2. A line in 2-space.

Suppose, as shown in the figure, that 4/1 = 4 such rulers are required to span the line. Now repeat the excercise using a ruler of length $^1/_2$. This time we will need $4/^1/_2 = 8$ ruler lengths to cover the line. If we were to use a ruler of length 1/3, we would need $4/(1/3) = 12$ ruler lengths. In the general case, the number of rulers required is proportional to $1/x^1$, where x is the length of the ruler used. This relation is precise, but only as x becomes very small, because a long ruler cuts across curves in the line.

Now consider a 2-dimensional object such as an ellipse and ask how many squares of side x would be needed to cover it. Because the area of each square is x^2, the number of squares needed to cover the *area* of the object is proportional to $1/x^2$. Again, this relation becomes precise only as x becomes very small. In the case of the line example, the exponent in the denominator is 1. For the 2-dimensional example it is 2. This exponent, in fact, *defines* the dimension.

Now consider the Koch curve. If we measure its length with a ruler one-third the length of the original straight line used in construction, we find the needed number of rulers to be *not* 3, but rather 4 (see Figure A1-a). If we use a ruler one-ninth the original line we find that the required number of rulers rises to 16. In this case the number of rulers needed to cover the curve varies *not* with $1/x^1$ as with a simple line, or with $1/x^2$, but rather with $1/x^D$, where D = 1.2618. Using the definition of dimension given above, the Koch curve has dimension 1.2618.

Denote, by N_x the number of units of length x needed to cover a line. Then the dimension of any line can be found by writing

(A1)

A plot of the latter relation yields D as its slope.

Note that the length of a curve, L, is given by the product of the number of rulers required to cover it and the length of the ruler. For continuous lines this is proportional to

$$L \propto [(1/x^1)(x) = 1],$$

so that L is constant. For a fractal line, however,

$$L \propto [(1/x^D)(x) = x^{1-D}],$$

so that as x decreases, the length of the line goes to either zero (D < 1) or infinity (D > 1). Fractal lines have no characteristic length!

With knowledge of how a curve is constructed, D can be calculated more directly. Start with a ruler that exactly covers the line (which we arbitrarily scale to one unit). Then, if the number of rulers needed rises from 1 to a to a^2 to . . ., as the scale (ruler length) drops from 1 to 1/b to $1/b^2$ to . . . , we have

$$N_x = a^n,$$

where

$$x = (1/b)^n = b^{-n}.$$

But $N_x \propto x^{-D}$ (equation [A1]). Therefore,

$$a^n = (x)^{-D} = (b^{-n})^{-D} = b^{nD}, \text{ and } n\log(a) = nD\log(b).$$

Solving for D,

$$D = \log(a)/\log(b) \qquad \text{(A2)}$$

One more expression will prove useful. The required number of rulers of length x/b is a times the required number of rulers of length x. That is,

$$N_{x/b} = a\,N_x, \text{ or } N_x = \frac{1}{a}N_{x/b}. \qquad \text{(A3)}$$

Whenever we encounter this relation we know that

$$Nx \propto x^{-D} : D = \log(a)/\log(b).$$

In the Koch curve in Figure A1, the number of rulers needed increases by a factor of a = 4 whenever the scale drops by a factor of b = 3. Therefore,

$$D = \log(4)/\log(3) = 1.2618.$$

Calculation of D by walking rulers of various lengths along an object, as above, is referred to as the coastline method.

Branching structures also have fractional dimension. The object depicted in Figure A3 cannot conveniently be measured with rulers but, as with the 2-dimensional ellipse, we can ask how many squares of side x are needed to cover it. Using the grid superimposed on the object we see that as x drops from one ninth the width of the figure to 1/18, to 1/36, to 1/72, the number rises from 40 to 100 to 220 to 492. Plotting log(N) against log(x), we find a slope of D = 1.20. The dimension determined by using squares of decreasing size is referred to as the box dimension.

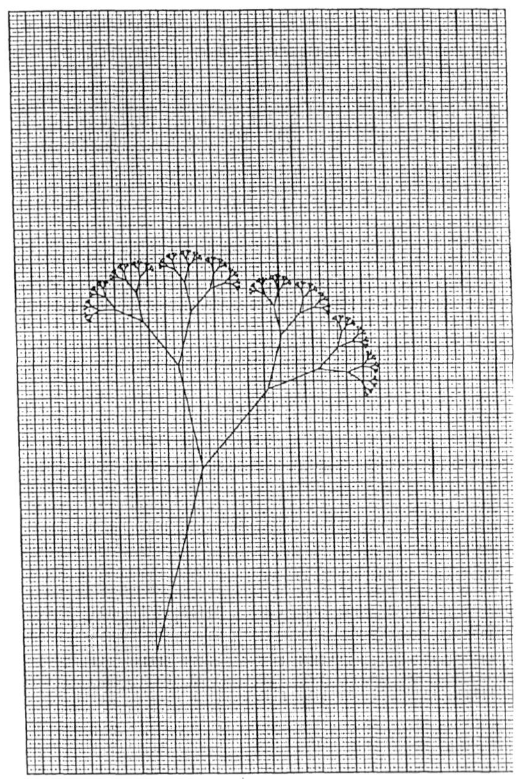

Fig. A3. A fractal branching pattern set against a grid for calculation of fractal dimension.

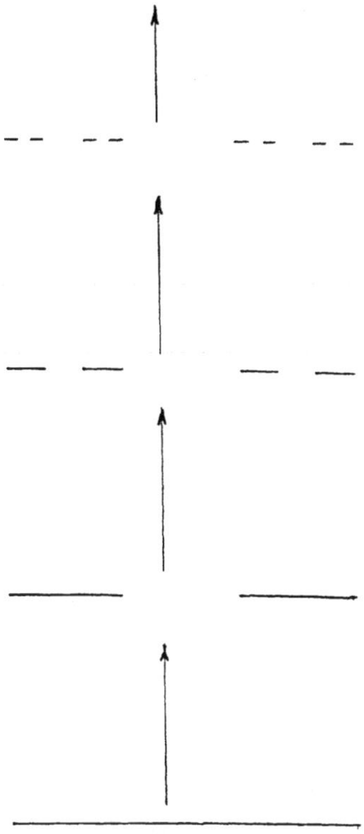

Fig. A4. Construction of a Cantor set.

The values of D for the above examples indicate that both objects fill space more completely than a simple line (1-dimension), but less than a 2-dimensional object. In fact, the fractal dimension, D, can be thought of as the completeness with which a fractal object fills the space in which it is embedded.

There exist different definitions of dimension. Throughout this paper we will be concerned only with one, the 'Haussdorf-Besichovitch' (or simply Haussdorf) dimension. Strictly speaking, the coastline and box dimensions are not Haussdorf dimensions. They are, however, numerically equivalent to the corresponding Haussdorf measures, so distinctions will not be drawn. In the literature on fractal objects, 'dimension' almost always refers to Haussdorf dimension. This is not necessarily true in discussions of chaotic processes and their associated fractal descriptions.

Consider an object constructed by drawing a line, removing the center third of that line, then removing the center of each remaining line segment, and so on (Fig. A4). In this case, it requires only one ruler of length 1 to span the entire set; it requires two rulers of length 1/3, four of length 1/9, and so on. That is, a = 2, b = 3. We write, after equation A3,

$$N_x = \frac{1}{a} N_{x/b} = \tfrac{1}{2} N_{x/3}, \quad \text{so that } N_x \propto x^{-D}:$$

$$D = \log(a)/\log(b) = \log(2)/\log(3) = 0.6309 < 1.0.$$

The set of segments generated using this procedure fills more than 0-dimensional space (a point), but does not entirely fill a 1-dimensional line. Sets of disjoint segments (or points distributed in like fashion) are known as Cantor sets.

Consider a line composed of the sum of sine waves of wavelength and amplitude, respectively,

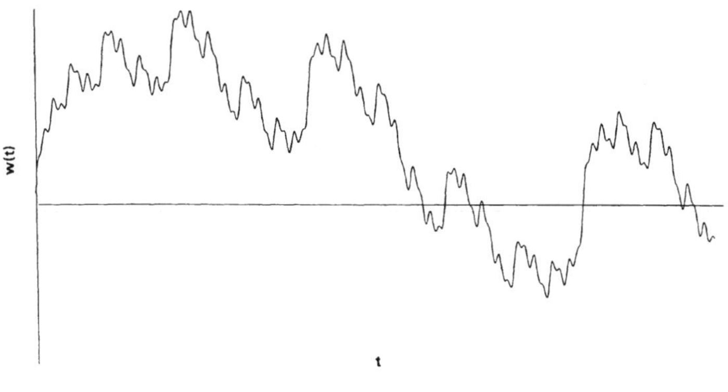

Fig. A5. Compound curve made up of sine waves of periods 1, 1/2, 1/4, etc., with corresponding amplitudes 1, 1.5, 1.5^2, etc.

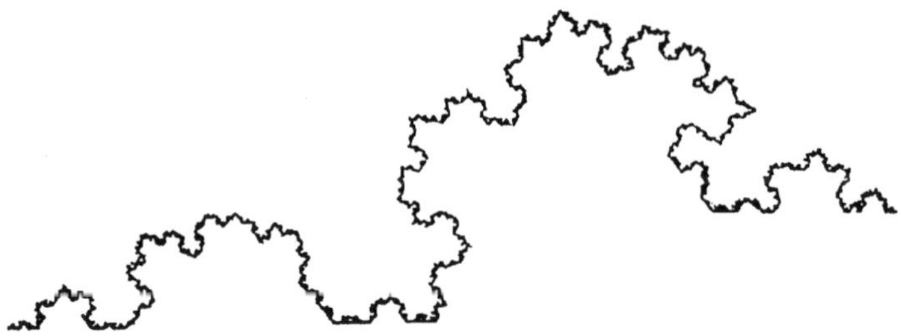

Fig. A6. A statistical Koch curve.

1 and 1, b^{-1} and a, b^{-2} and a^2 .... Figure A5 shows a curve, with a = 1.5, b = 2. For reasons given below, under the section on self-affine objects, there may be difficulties in obtaining consistent fractal dimensions of such objects using the coastline approach. For the moment, then, look at the distribution of points at which the curve intersects the t axis. These points clearly define a Cantor set (the dimension of which is 0.415). The relation of D to a and b in this case is not so simple as in equation A2.

Just as Figure A5 was constructed by compounding a number of different, simple waves, fractal curves can be decomposed into component waves of various amplitudes and wavelengths. In addition, the intersections of such curves with any horizontal line from Cantor sets.

Statistical fractals

Look again at the Koch curve, but this time let the size of the removed center section vary randomly and the length of the replacement line segments also (Fig. A6). Now the object is no longer *identical* at all scales, but is merely *similar* in appearance. So long as the number of rulers required to cover the curve still rises, *on average*, by a factor of a whenever scale is reduced to $1/b$, though, dimension still is given by D = log(a)/log(b). The object is now considered to be *statistically* self-similar.

Figure A7 provides another example of a statistically self-similar object, this time a branching structure.

If the coastline or box approaches are used to ascertain dimension for the objects in Figures A6 or A7, the $\log(N_x)$ vs. log(x) plot will show irregularities because of the statistical nature of the fractal object. To obtain a statistically reliable measure of D, therefore, it is necessary that the sample size, that is number of ruler lengths or boxes, be large. Put differently, more accuracy is obtained over that part of a finite length plot where x is small. Note that branching structures, composite waves and Cantor sets can all be statistical fractals.

Self-affine objects

Consider a set of connected rectangles of identical length-to-width ratio and with relative lengths $1:(1/b):(1/b^2)$ ... This set of rectangles (Figure A8-a) constitutes a self-similar set. Now consider another set of rectangles (figure A8-b) for which length declines as $1:(1/b):(1/b^2)$ ..., as before, but for which width declines in some *other* ratio, $1:(1/b'):(1/b'^2)$ .... As we look at this second structure at finer and finer scales, it looks more and more fore-

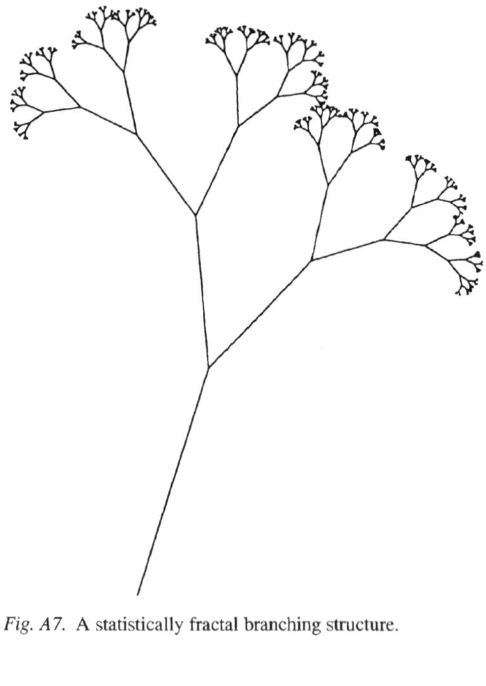

Fig. A7. A statistically fractal branching structure.

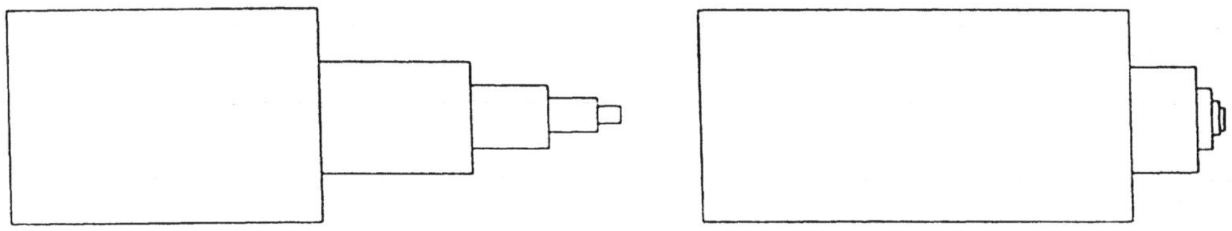

Fig. A8. Examples of self-similar and self-affine sets.

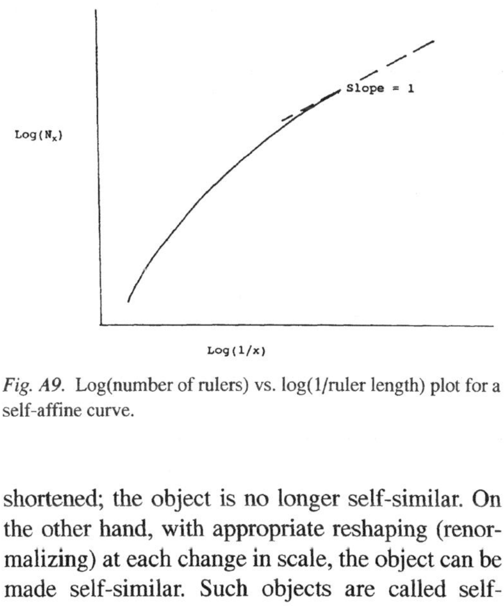

Fig. A9. Log(number of rulers) vs. log(1/ruler length) plot for a self-affine curve.

shortened; the object is no longer self-similar. On the other hand, with appropriate reshaping (renormalizing) at each change in scale, the object can be made self-similar. Such objects are called self-affine.

The dimension of a self-affine object is not uniquely defined. For example, suppose we tried the coastline approach on the graph in figure A5. We would find that the plot of $\log(N_x)$ vs $\log(1/x)$ was *not* a straight line (Fig. A9). In other words, the dimension would appear to change as a function of scale x. It is possible, though, with appropriate reshaping (renormalizing) of the curve, or progressive reshaping with scale of the boxes used to calculate box dimension, to define a unique fractal dimension for compound waves. Let us examine compound waves in more detail.

The structure of wave forms

Any wave can be broken down into component parts. To show how this is accomplished, look at the simple sine wave of wavelength λ in Figure A10. Start at any point, t, on the horizontal axis and relate the position of the wave, w(t), at that point to that at t + λ by calculating the autocovariance,

$$Cov[w(t),\, w(t+\lambda)] =$$

$$\lim_{T\to\infty}\frac{1}{T}\int[w(t)-w][w(t+\lambda)-w]dt.$$

Because w(t) = w(t + λ) for all t, this covariance equals Cov[w(t), w(t)] = Var[w(t)], and thus gives

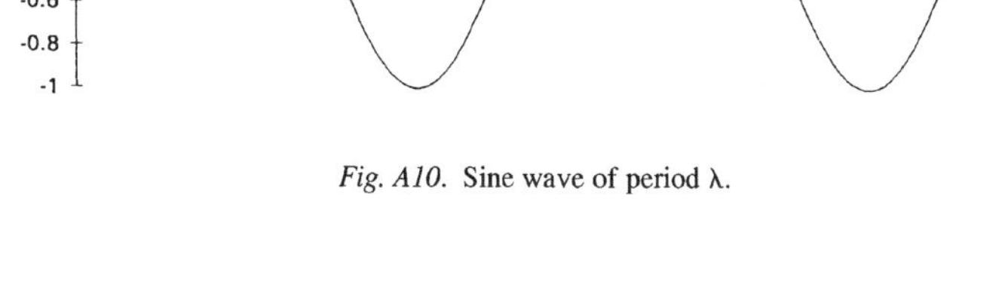

Fig. A10. Sine wave of period λ.

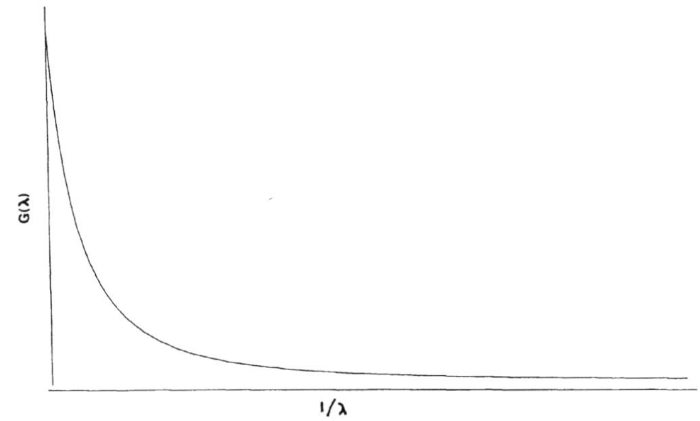

Fig. A11. Power spectrum of a self-affine, complex wave process.

the mean squared deviation in w(t) from w. Suppose now that another wave of another wavelength and amplitude is superimposed on this first wave. The new wave will raise the difference between w(t) and w(t + λ) for some values of t, lower it for others. But over a sufficiently long time interval, because the two waves both vary about the same mean w, the plus and minus effects of the second wave on w(t) - w(t + λ) must exactly cancel. That is, the contributions to Cov[w(t), w(t + λ)] of any other wave, over a sufficiently long time interval, is zero. The covariance still reflects the mean squared deviation of that particular wave.

Call this covariance C(λ). Because C(λ) gives the mean squared deviation from w for a wave of length λ, it seems a good candidate for describing the contribution of such a wave to any compound wave process w(t). Unfortunately, some further manipulation is needed. The fact that w(t) is perfectly correlated with w(t + λ) means it also is perfectly correlated with w(t + 2λ), w(t + 3λ) .... To use C(λ) as an indicator of contribution to w(t) would be misleading, for it would imply contributions from waves of length 2λ, 3λ ..., and in Figure A10 there are no such waves. What is needed is an alternate measure G(λ) that is unique to λ, but is faithful to the measure of mean squared deviation. One such measure is

$$G(\lambda) = \lim_{T\to\infty} \frac{1}{T} \int c(t) \cos(2\pi t / \lambda) dt. \qquad \text{(A4)}$$

For every t an integer multiple of λ, there is a single value of G(λ) specific to λ, that is, G(λ) ≠G(nλ), for any n ≠1.

Consider any time interval δ ≠λ. The value of C(δ), as applied to figure A10, is non zero for all δ ≠$^1/_2$λ, implying contributions from non-existent waves of all wavelengths ≠$^1/_2$λ. But application of equation A4 gives G(δ) = 0 unless δ = λ (see below). Thus equation A4 filters out the spurious contributions indicated by the covariance terms and provides an unambiguous measure of mean squared error (variance) for a wave of length λ. G(λ) is known as the Fourier Transform of C(λ) and can be plotted against λ (or, commonly, 1/λ) to give a 'power' curve (Fig. A11). Often (within some finite range in λ), and always for self-affine and self-similar waves, the mean squared deviation, or power, varies as a power law λ^β.

Let us now attempt to measure the dimension of the compound wave, w(t). Because we already know a coastline approach may lead us astray, we start with a box dimension approach, forewarned that to compensate for changing curve shape we will have to reshape the boxes into appropriately shorter and shorter rectangles as scale increases. From the foregoing discussion we know that the mean squared deviation (i.e. variance) in w(t) contributed by a wave of length τ varies with τ^β. It follows that the variance in w(t) over an interval of length t = τ is the sum of contributed variances by all waves in that interval, or

$$\tau^\beta + (\tau / b)^\beta + (\tau / b^2)^\beta + \ldots \propto \tau^\beta. \qquad \text{(A5)}$$

Note that a plot of log variance against log(t) will give an oscillatory rather than a straight line because equation A5 holds only when t is the wavelength of a contributory wave.

Consider again the interval, τ. The range in w(t) over this interval does *not* scale as the standard deviation, $\propto \tau^{1/2\beta}$, because w(t) may lie on an upward or downward trend of the next longer wave (of length $b\tau$). Therefore range, R_T, varies, on average, with the standard deviation of this next longer wave, $(b\tau)^{1/2\beta}$, times the proportion of that next wave it comprises, 1/b, corrected for the shrinkage in wave height as scale increases from τ to $b\tau$. Suppose wave height varies with scale as τ^H. Then

$$R_T \propto (b\tau)^{\frac{1}{2}\beta} * 1/b * (b\tau)^H \propto \tau^{H+\frac{1}{2}\beta}.$$

Therefore, denoting the standard deviation by S_T, we have

$$R_T / S_T \propto \tau^{H+\frac{1}{2}\beta} / \tau^{\frac{1}{2}\beta} \propto \tau^H, \text{ or}$$
$$\log(R_T / S_T) = c - H\log(\tau) \qquad \text{(A6)}$$

where *c* is some constant. When t is not equal to the wavelength of a contributory wave, this relation is not precise. Hence, a plot of $\log(R_T/S_T)$ vs. log(t) will yield an oscillatory line of average slope H. The value of H can be determined by regressing $\log(R_T/S_T)$ on $\log(\tau)$, a procedure known as rescaled range analysis (Feder, 1988). It remains to show how H is related to D.

The width of an interval of unit length, measured in boxes of side x, is 1/x. The expected average vertical range over an interval x is proportional to x^H. Therefore, the average number of stacked boxes needed to cover an interval x is $x^H/x = x^{H-1}$. Finally, then, the expected number of boxes needed to cover the w(t) curve over the unit interval is

interval width in number of boxes × average number of boxes in any stack of width × across that interval

$$= (1/x) * (x^{H-1}) = x^{H-2}.$$

But, by the definition of D, the expected number of boxes needed to cover the curve is proportional to x^{-D}. Thus

$$D = 2 - H. \qquad \text{(A7)}$$

The various conclusions reached with regard to compound waves, w(t), can be summarized as follows:

(1) Variance in w(t) over an interval, τ, varies with τ^β, and

(2) D = 2-H, where H can be found by regression from

(3) $\log(R/S) = c + H \log(\tau)$

Chaos

Figure A12 shows the relation between population sizes in successive years for several values of the intrinsic growth rate, r, as given by the (discrete) logistic equation,

$$\Delta n(t) = rn(t-1)[1-n(t-1)] = rn(t-1) - rn(t-1)^2. \qquad \text{(A8)}$$

The diagonal is the replacement line, where n(t) = n(t - 1). In Figure A12a an initial value of n(t - 1) = 1.2 leads to a slight downward adjustment to n(t) = 0.84. As shown in Figure A13a, n displays a damped oscillation and eventually comes to rest at $n_{equil.} = 1$. In Figure A12b the population overadjusts to below $n_{equil.}$ and then (Fig. A13b) bounces back and forth in a sustained 'limit cycle' with wavelength 2. As r increases further, the behavior of n(t) becomes more and more complex, with the maximum and minimum extents of the limit cycle bifurcating into four values among which n(t) fluctuates. The path of n(t) now displays two component waves of lengths 2 and 4, with different amplitudes. With still further rise in r, these four values bifurcate. Then the resulting eight bifurcate yet again, and so on until, when r reaches a value just under 3, the number of points among which n fluctuates approaches infinity (Fig. A13d). At this point n(t) describes a curve with component waves of length 2, 4, 8, 16 ....

A wave composed of an infinity of different waves never repeats the same pattern. Therefore, the value of n at any time is unpredictable even with complete information about previous n values; the curve is referred to as chaotic.

The continuous form of the logistic does *not* show chaotic behavior, nor do many systems of continuous equations describing *pairs* of interact-

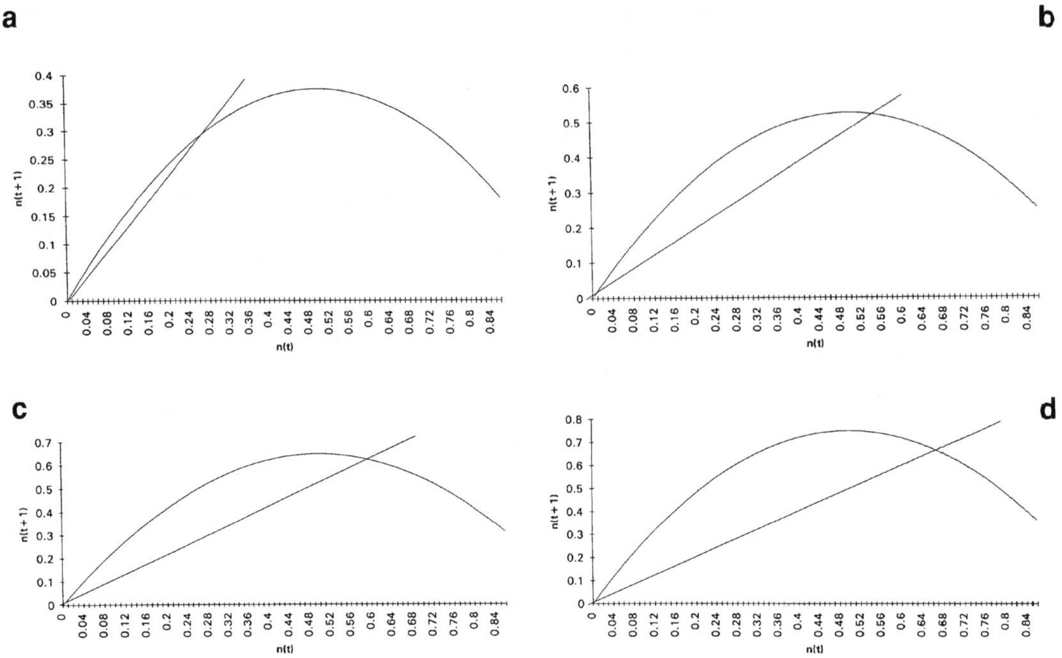

Fig. A12. Plot of population density, n, against n in the previous time interval for a discrete logistic process. (a) r = 1.2; (b) r = 2.1; (c) r = 2.52; (d) r = 3.0.

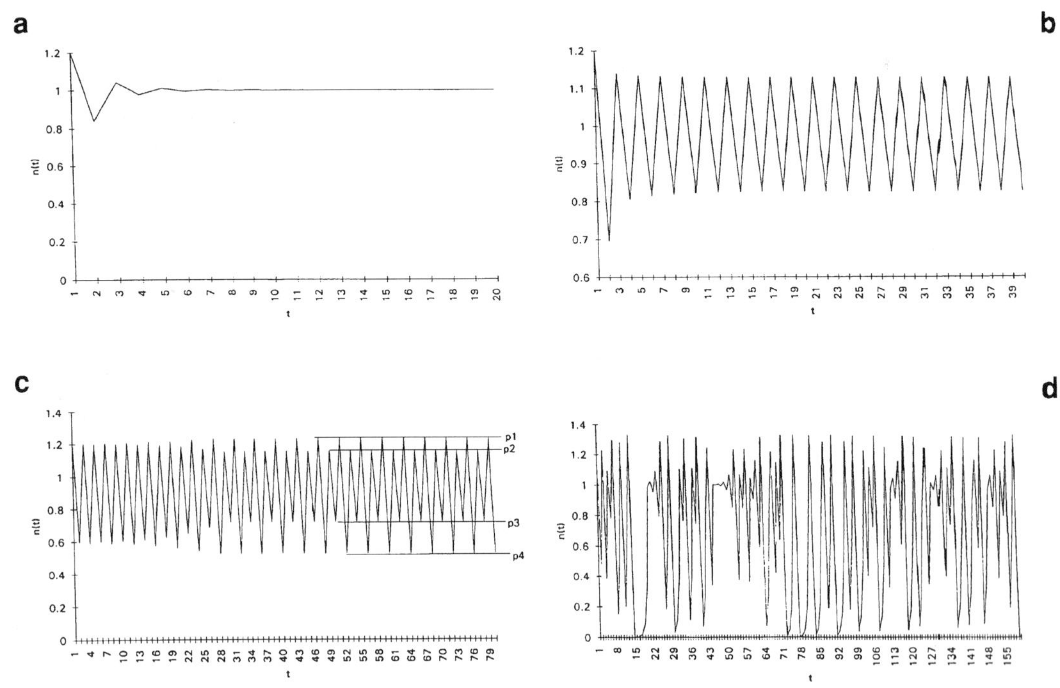

Fig. A13. Time traces of population densities corresponding to logistic plots in Fig. A12.

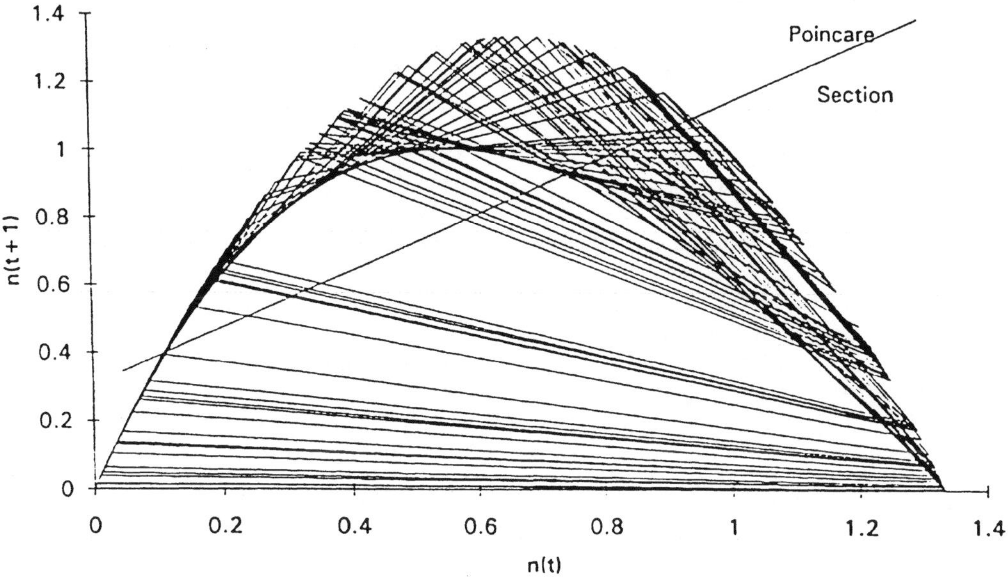

Fig. A14. Phase space plot showing path of the population from time t - 1 to time t.

ing variables. However, if the number of interacting variables reaches 3 or more, and if their behavior is non linear (i.e., squared or higher order terms are involved, as in equation A8), the chances for chaos jump dramatically. Chaos becomes more and more likely as the number of interacting variables and the complexity of each variable's dynamics increases.

As discussed earlier, chaotic curves - curves composed of an infinite number of waves of different length and amplitude - describe a fractal. The dimension of this fractal takes on different values depending on how the process is viewed; it can be calculated

1) using rescaled range analysis (equation A6),
2) using the Cantor set defined by the curve as it crosses some line, w = constant,
3) using the Cantor set described by the distribution of w values among which the process oscillates (points p_1, p_2, p_3, p_4 in figure A13c), and
4) using a plot of successive points, w(t) vs w(t - 1), for all t, in phase space. This describes what is called a strange attractor (Fig. A14). A 'Poincare' section drawn across this figure (a straight line cutting through it) intersects the graph at a series of points that form different Cantor sets depending on where and at what angle the section is drawn. The set described in (2) above is one of these Cantor sets.

One final consideration: we noted earlier that variance in w(t) over the interval t increases with t^{β} : $\beta > 0$. This is equivalent to saying that the variance in w(t) over several parallel processes is proportional to t^{β}. In other words, the uncertainty of w(t) grows with t. The reason is that as time progresses we pick up the influences of longer and longer waves, and the position of w(t) seems to wander randomly up and down farther and farther from its origin. But the influence of longer and longer waves can be characterized also by the progressively increased (affine) scaling. Thus, mean-squared drift should be proportional to $(t^H)^2 = t^{2H}$. Since mean-squared drift also is proportional to t^{β}, we have

$$\beta = 2H. \qquad \text{(A9)}$$

Thus $D = 2 - H = 2 - \beta/2$. To put this in perspective, a Brownian random walk results in the variance in w(t) increasing with t^1. Therefore, the dimension of Brownian drift is

$$D = 2 - \beta / 2 = 2 - ½ = 1.5.$$

We return finally to proving that the Fourier Transform in fact filters out unwanted wave contributions. For a wave of length λ, the autocovariance over an interval δ is given by

$$C(\delta) = \lim_{T\to\infty}\frac{1}{T}\int_o^T [A*\cos(2\pi t/\lambda)]$$
$$[A*\cos(2\pi t+\delta]/\lambda)]t,$$

where A is the amplitude. Applying the rule for cosines of sums gives us

$$\lim_{T\to\infty}\frac{1}{T}A^2\int_o^T \cos(2\pi t/\lambda)*[\cos(2\pi t/\lambda)\cos(2\pi\delta/\lambda)-$$
$$\sin(2\pi t/\lambda)\sin(2\pi\delta/\lambda)]dt =$$

$$\lim_{T\to\infty}\frac{1}{T}A^2\cos(2\pi\delta/\lambda)*\int_o^T \cos^2(2\pi t/\lambda)dt +$$

$$\lim_{T\to\infty}\frac{1}{T}A^2\sin(2\pi\delta/\lambda)\int_o^T \cos(2\pi t/\lambda)*\sin(2\pi t/\lambda)dt.$$

But the integral of sin (2πt/λ)*cos(2πt/λ) over t is zero (the products in first and second, the third and fourth quadrants cancel). Thus

$$C(\delta) =$$

$$\lim_{T\to\infty}\frac{1}{T}A^2\cos(2\pi\delta/\lambda)\int_o^T \cos^2(2\pi t/\lambda)dt =$$

$$A^2\cos(2\pi\delta/\lambda)*2\int_o^\pi \cos^2(2\pi t/\lambda)dt =$$

$$A^2\cos(2\pi\delta/\lambda)*2[(1+x+\sin x*\cos x)/2]^\pi{}_0 =$$

$$2A^2\cos(2\pi\delta/\lambda)*\pi, \text{ so } C(\delta) = 2A^2\pi*\cos(2\pi\delta/\lambda).$$

Then $G(\delta) =$

$$\lim_{T\to\infty}\frac{1}{T}\int_o^T C(t)*\cos(2\pi t/\lambda)dt =$$

$$\lim_{T\to\infty}\frac{1}{T}\int_o^T 2A^2\pi*\cos(2\pi t/\lambda)\cos(2\pi t/\delta)dt.$$

But the integral of cos(x)cos(y) over unlimited time is equal to the integral of sin(x)sin(y). Thus

$$\int_o^\pi \cos(x)\cos(y)dt = \tfrac{1}{2}\int_o^\pi [\cos(x)\cos(y)+$$
$$\sin(x)\sin(y)]dt = \tfrac{1}{2}\int_o^\pi [\cos(x-y)]dt,$$

and the above expression becomes

$$G(\delta) = \lim_{T\to\infty}\frac{1}{T}2A^2\pi\int_o^T \cos[2\pi t\left(\frac{1}{\lambda}-\frac{1}{\delta}\right)]dt.$$

But the integral of cos(x) over unlimited time is zero. Therefore, unles 1/λ = 1/δ, in which case the cosine is 1.0, the above expression vanishes. We are left with

$$G(\delta) = \lim_{T\to\infty}\frac{1}{T}2A^2\pi\int_o^T \cos(0)dt = A^2\pi \propto A^2 \text{ for}$$
$$\delta = \lambda,\ G(\delta) = 0, \text{ for } \delta \neq \lambda.$$

The contributions of all wavelengths δ not equal to λ are filtered out.

T. A. Markow (ed.), Developmental Instability: Its Origins and Evolutionary Implications, 99–121, 1994.

Developmental stability in plants: symmetries, stress and epigenesis

D. Carl Freeman[1], John H. Graham[2] & John M. Emlen[3]
[1] *Department of Biological Sciences, Wayne State University, Detroit MI 48202, USA*
[2] *Department of Biology, Berry College, 430 Mount Berry Station, Mount Berry, GA 30149, USA*
[3] *National Fisheries Research Center, U. S. Fish and Wildlife Service, Seattle, WA 98115, USA*

Received and accepted 5 May 1993

Key words: Plant developmental stability, symmetry, fractals, epigenesis

Abstract

Plant developmental stability has received little attention in the past three or four decades. Here we review differences in plant and animal development, and discuss the advantages of using plants as experimental subjects in exploring developmental stability. We argue that any type of developmental invariant may be used to assess developmental stability and review the use of fluctuating asymmetry in studies of plant developmental stability. We also examine the use of deviations from translatory, radial, and self-symmetry as measures of developmental instability. The role of nonlinear dynamics and epigenesis in the production of the phenotype is also discussed.

Introduction

Developmental stability has been used in environmental monitoring to explore the effects of anthropogenic and natural stressors (Clarke & McKenzie, 1987; Clarke, 1992; Zakharov, 1989; Zakharov, 1990, Zakharov *et al.*, 1991) and in evolutionary genetics to explore the effects of inbreeding (Leary, Allendorf & Knudsen, 1983, 1984, 1985) and hybridization (Markow & Ricker, 1991; Graham, 1992) on natural and artificial populations. Historically, the study of developmental stability has focused largely on animals. Of the hundreds of papers dealing with developmental stability, we are aware of fewer than one dozen dealing with plants. As discussed below, plants have a number of properties that make them ideal experimental systems for resolving some of the issues vexing the field today. For example, Palmer and Strobeck (1986, 1992) have argued that several types of asymmetry should not be used to estimate developmental stability because they may have a genetic origin. But, because plants have modular construction and can be cloned, one can fix both the genetic constitution of the plant and tightly regulate its environment, allowing one to partition the variance in the phenotype among the genetic and environmental components (see Paxman, 1956, for an example).

Below we review the basic notion of developmental stability and the use of symmetry as an indicator of developmental stability. We extend the arguments for using symmetry beyond bilateral symmetry to include other types of symmetries found in (but not unique to) plants. We review some of the basic differences in animal and plant development, and then examine the few studies we are aware of that have dealt with plant developmental stability. We have drawn heavily on our own unpublished work, in large part because the existing literature is so sparse. Finally, we present a brief discussion on the role of nonlinear dynamics and stress in generating developmental instability.

Developmental stability

Developmental stability involves the consistent generation of the same phenotype in a given environment (Leary, Allendorf & Knudsen, 1992). This requires the suppression of potential sources of variation (both genetic and environmental; Waddington, 1957; Leary, Allendorf & Knudsen, 1992). Developmental stability is greatest when organisms are raised under optimal conditions, and decreases when organisms are grown under suboptimal conditions (see Zakharov, 1989, 1992 and Graham, Freeman & Emlen, 1993a, for an introduction to this literature). Similar decreases in developmental stability occur as a result of endogenous (e.g. genetic) stress brought about by the disruption of coadapted gene complexes such as occurs in hybridization between disparate taxa (see Graham, 1992, for a review) or inbreeding (Leary, Allendorf & Knudsen, 1983, 1984, 1985; Bush, Smouse & Ledig, 1987; Strauss, 1987; Clarke, Oldroyd & Hunt, 1992).

While the mechanisms producing developmental stability have not been explored previously, at least by those investigating fluctuating asymmetry (but see Graham, Freeman & Emlen and Emlen, Freeman & Graham, this volume, and Zeeman, 1989, for work by others), the basic assumption is that exogenous stress disrupts the physiology of the organism, and to the extent that the organism is undergoing development, development is altered. Similarly, the disruption of coadapted gene complexes alters the regulation of physiological processes and thus development of the phenotype (Burton, 1990a,b; McKenzie & O'Farrell, this volume). Implicit in this argument is the assertion that the phenotype an organism displays is not merely a linear product of the organism's genotype (See Mitchison, 1980; Meinhardt, 1982; Cocho & Rius, 1989; Eilbeck, 1989; Goodwin, 1989; Huberman, 1989; Savageau, 1989; Saunders & Kubal, 1989; Thom, 1989; Desbiez, Tort & Thellier, 1991). The production of the phenotype involves regulating not only the kinetics of a particular mRNA species, but, also, the complex interconnected metabolic pathways coded for by the coadapted gene complexes. It is these pathways that ultimately regulate the production of the phenotype (see Emlen, Freeman & Graham and Kieser in this volume). We expect organisms exhibiting the greatest degree of developmental stability are those whose physiological processes are best buffered against environmental insults.

Given that stress is expected to cause deviations in developmental patterns, how can we know that the phenotype has been altered? The early workers in developmental stability hit upon a remarkably simple and elegant solution - examine characters that ought not to deviate in development (Mather, 1953).

Symmetry

The notion of symmetry has played a prominent role in the study of developmental stability. 'By symmetry we mean an invariance against change: something stays the same in spite of some potentially consequential alteration' (Schroeder, 1991). Because the vast majority of studies have involved animals, bilateral symmetry has received special attention. But Graham, Freeman & Emlen, 1993a have argued that all symmetries represent potential developmental invariants and may be used in a manner analogous to that of bilateral symmetry. In addition to bilateral symmetry, some plants exhibit radial symmetry in their leaves, flowers, inflorescences, and fruits. Plants grow as modules, and one can legitimately ask how similar one module is to another- i.e. one can examine within-plant variation (Paxman, 1956; Graham, Freeman & Emlen, 1993a). Many plants exhibit acropetal growth, and the size of their leaves, for example, may decrease regularly as one proceeds up the stem. In such a case the plants can be said to exhibit a translatory symmetry with scaling. Finally, branching and venation patterns of many plants exhibit a symmetry with respect to scale, i.e. the same pattern is repeated over many scales. This self-similarity is typical of fractal structures (see Lindenmayer, 1968; Prusinkiewicz & Hanan, 1985; appendix 1 in Emlen, Freeman & Graham, this volume). In Figure 1 we show how iterating a complex subset of Euclidean spaces leads to both self-symmetry across many scales and to structures with non-Euclidean dimensions.

We suggest that each of these symmetries represents an *a priori* idealized state. Deviations away from the ideal are taken as measures of developmental instability- i.e. development was not adequately buffered against environmental insults. Implicit in this idea is the assumption that the organism represents a single genotype without differential gene expression among the parts being compared, i.e. all the leaves on the plant are genetically identical (but see Cullis 1977, 1983, 1986,

1987, for examples where the assumption of genetic uniformity does not hold). Thus, deviations away from the idealized state are attributed to environmental causes (Palmer & Strobeck, 1986). Clearly, where these assumptions are not met the disruption of symmetry will not necessarily reflect developmental instability (Palmer & Strobeck, 1992). However, even where the assumption of genetic uniformity holds, and all the variation in the phenotype is environmental, it does not necessarily follow that the variation in the phenotype represents developmental instability. For example, many plants produce two different types of leaves, one in sun and the other in shade (Fig. 2). This represents normal variability in these species. As Zakharov (1989) has argued, the genetic principle is that developmental instability is associated with departures from normal developmental patterns, regardless of the nature of the norm. Thus, in using symmetries to measure developmental stability, it

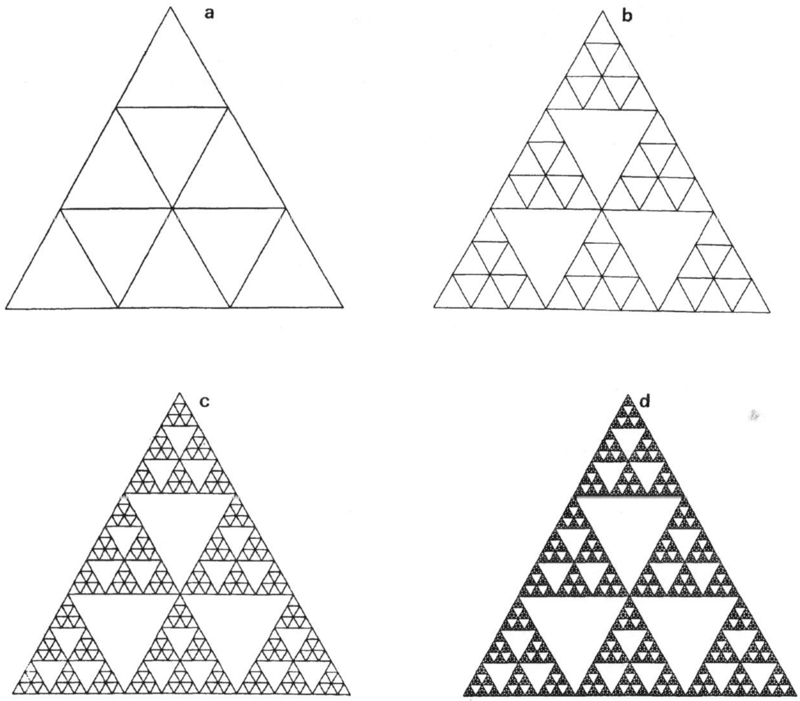

Fig. 1. Fractals. Self-similar fractals are invariant with respect to scale. A magnified portion of the image resembles the whole image. Two conditions are necessary to produce a fractal. First there must be iteration of a pattern across multiple scales, and second the pattern must represent a complex subset of Euclidean space. Botanically, successive branching of structures is analogous to the iteration, and because only a fraction of the three-dimensional embedding space is occupied, the complex subset criteria is also usually satisfied. Self-similar fractals are produced by repeated application of a single scaling rule. In the above example, we begin with a triangle and divide each side into thirds (the scaling rule) forming nine smaller triangles in the interior of the original triangle (a). We then remove the central three triangles from the plane. It is the removal of these triangles which ultimately makes the space a complex subset of Euclidean space. We then divide the remaining six triangles into thirds (b). The central three smaller triangles are removed again (c). Figure d shows the fourth iteration of this process. Further iterations of the above rules result in a structure which no longer fills all of the space occupied by the original triangle. The dimension of such an image is a measure of the degree to which it fills space. The fractal dimension of an image generated using the above rules is given by $\ln 6/\ln 3 = 1.63$. Such fractals are known as Sierpinski gaskets and were produced using Fractint software (Wegner & Peterson, 1991). The reader is referred to Emlen, Freeman and Graham, this volume, for a further discussion of fractal dimension.

Fig. 2. Shade (left) and sun (right) leaves of *Quercus velutin.*

is important to ascertain the normal developmental state; but in general, the symmetrical state appears to approximate that norm. Graham, Freeman and Emlen (this volume) argue that both endogenous deterministic and exogenous stochastic processes may cause deviations away from the symmetrical state. Furthermore, the symmetrical state need not represent a single point solution (point attractor in the terminology of chaos theory) as much as a broad basin of attraction. Thus, some variation, even under ideal conditions, is to be expected.

Bilateral symmetry of morphological structures has clearly received the most attention. In measuring departures from bilateral symmetry it is important to note that one is examining the distributions of the difference in the values of traits measured on the right and left sides of a number of individuals, i.e. these are population studies.

Three basic types of asymmetry occur in regards to right and left corresponding structures. First, handedness (directional asymmetry), where the character on one side of the organism consistently has a higher value than that on the other side. The average difference in the value of the characters between the right and left sides will not have a mean of zero, but will reflect the handedness of the organisms. Second, antisymmetry occurs when there is a bimodal distribution of the difference between the right and left sides (Timofeeff-Ressovsky, 1934), and third, fluctuating asymmetry occurs when there is a unimodal distribution of right-left sides with a mean of zero. Fluctuating asymmetry has been widely used as an indicator of developmental stability (Mather, 1953; Palmer & Strobeck, 1986; Clarke, 1992, for reviews).

Palmer and Strobeck (1986, 1992) have argued

against using directional asymmetry and antisymmetry as indicators of developmental stability because they believe such traits are genetically determined and do not reflect environmental or developmental modification. They have even extended their argument to suggest that some apparent cases of fluctuating asymmetry may result from combining distributions that really exhibit other types of asymmetry and thus may have a genetic basis; to that extent, fluctuating asymmetry may be a poor indicator of developmental stability. These views are not widely supported (Graham, Freeman & Emlen, this volume; McKenzie & O'Farrell, 1993) because, in part, the assumptions need not hold. Handedness, for example, need not have a genetic basis.

Handedness is not normally associated with plants, but it does occur in some plants. Coconut trees display either a right-handed or a left-handed spiral of their leaves. The two forms can be distinguished by examining the position of the spadices with respect to the subtending leaf (Davis, 1962, 1963). What is surprising in this case is that the yield of nuts is associated with handedness (plants with left-handed spirals produce significantly greater yields), but handedness is not heritable. Indeed handedness has been shown to change both within an individual and within a clone (Davis, 1962, 1963). Though he did not consider it a departure from random expectations, Davis did find a significantly larger number of left handed plants among those receiving supplemental iron fertilizer than one would expect by chance.

Similarly, Kojima *et al.* (1955) and Sueoka and Mukai (1956) have shown that the ability of Einkorn wheats to produce alternating left and right handed spikelets is inherited. But since both types of handedness occur on the same individual, it is doubtful that there exists a gene for handedness *per se*. Rather it would appear that physicochemical gradients may have a standing wave pattern which produces alternating spikelets, thus this trait appears to be a property of the physiological system's dynamics. Based upon these studies of plants, we question the *a priori* rejection of the use of structures showing handedness in examining developmental stability.

Differences between animal and plant development

The majority of studies regarding developmental stability involve animals. Animal development involves growth, differentiation, and morphogenesis. In animals, growth involves cell multiplication and expansion. Cell division via cleavage produces independent entities. Differentiation results in groups of cells that alter their physiology in rather specific ways as a result of selective gene expression. The end result is that groups of cells become different from one another, i.e. tissues. Morphogenesis is the establishment of a given shape or arrangement of tissues and often involves the movement of cells within the developing embryo. The overall development of most animals is inextricably linked to the movement and behavior of cells (Gilbert, 1991). Morphogenesis in animals results in the organism assuming a fixed shape and, in the case of most large animals, bilateral symmetry in many of their features.

Several internal structures of animals also exhibit another kind of symmetry. Structures like lungs, kidneys, nerves, and the vascular system are highly branched. In the case of the lung, each trachea branches into two smaller branches, with the smaller branches resembling the larger trachea. Such successive branching produces structures with large surface areas in small volumes. These structures display non-integer dimensions and are fractal (West & Goldberger, 1987; West & Shlesinger, 1989, 1990; West, 1990). The higher the dimension the greater the amount of the three dimension embedding space occupied by the structure. Fractal structures maximize the respiratory and other surface areas within fixed volumes. Plants that have open growth systems lack a fixed volume, and as discussed below, have external features that appear fractal (Prusinkiewicz & Hanan, 1985; Tatsum *et al.*, 1989; Schroeder, 1991). Plants, too, have large surface areas relative to the three dimensional volume occupied. In the case of root or stem branches, the surface is external; with leaf venation, the surface is internal.

In many vertebrates, stem cells may remain undifferentiated, providing a source of regenerative potential (Gilbert, 1991), but there is no permanently distinctive tissue such as the plant cambium, which can regenerate all the tissues of the adult

(Fahn, 1974). Regeneration, however, is possible in flatworms and amphibians. Many tissues of animals exhibit hypertrophy, having a fixed number of cells throughout life. This is known to occur in adipose tissue (Ward & Armitage, 1981) and skeletal musculature (Gilbert, 1991), while liver tissue can form new structures throughout life (Goss, 1966). Developmental stability can only be influenced within the time frame that the particular structure of interest is undergoing development.

Plants do not develop in the same way as animals. Cytokinesis is not needed for plant growth. Plants increase in volume first, and then initiate internal cell wall formation, repartitioning the existing volume of the organism. While cells do elongate after cell division, the same growth takes place even when cross wall formation is suppressed (Kaplan & Hagemann, 1991). Thus, growth can occur without cell division. Cell division in animals is like adding a new sausage to a chain of sausages. Cell division in plants is like taking a meter long sausage and dividing it into link sausages.

Cell division in plants is not by cleavage, but by the formation of a phragmoplast (Fahn, 1974). The phragmoplast forms interior walls, preserving portions of the endoplasmic reticulum and golgi between cells (plasmodesmata). Thus, the cells do not become independent entities, and for the most part they do not move during morphogenesis. Rather, morphogenesis is directed by the formation of various hormone and electric gradients within the embryo. The strength of these gradients diminishes with distance and is subject to nonlinear dynamics. Thus, the study of morphogenesis or pattern formation in plants involves examining the topology of these fields and the nonlinear feedback systems that govern them (Turing, 1952; Jaffe & Nuccitelli, 1977; Weisenseel *et al.*, 1979; Jaffe, 1980; Weisenseel & Kicherer, 1981; Mitchison, 1980; Meinhardt, 1982; Toko *et al.*, 1987, 1990; Souda *et al.*, 1990; Desbiez, Tort & Thellier, 1991; Trewavas, 1991; Hecks, Hejnowicz & Sievers, 1992). Animals also have gradients, but superimposed upon the topology of these gradients are the migration patterns of cells, rendering animal development a much more complex phenomenon (Tornheim, 1986; Minzoni & Alsono- de Florida, 1989).

Growth in most plants is indeterminate and occurs by adding modular units (Watkinson & White, 1986). In woody plants, last year's modular units increase in diameter, but not length, and this year's modular unit is similar in morphology, or at least shows an affinity in shape, to the modular unit produced last year. Branches of mature trees resemble young trees. If the branches were magnified their shape would closely resemble that of the mature tree. This self-similarity or self-affinity with respect to scale is common in plants. Some, like Lindemayer (1968), have used this property of plants to construct computer generated images that are remarkably lifelike (Fig. 3).

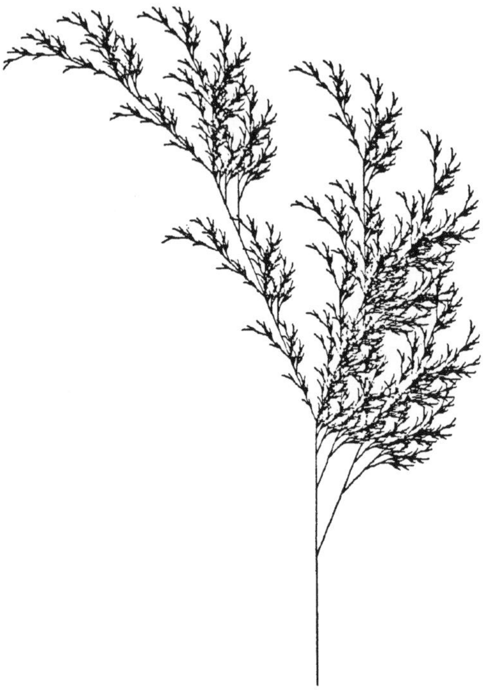

Fig. 3. Lindenmayer Plant Image. Lindenmayer (1968) found that computer generated images very similar to real plants could be generated by successive application of simple rules at a progressively smaller scale. This image was generated using Fractint software (Wegner & Peterson, 1991).

Because animal development involves the movement of independent entities in morphogenesis, the disruption of this movement may lead to major disruptions in the phenotype. In mammals, for example, cranial neural crest cells migrate and contribute to the formation of the first pharyngeal arch which gives rise to jawbones, and contributes to the formation of the palate, nose, ear, and much of the facial cartilage and musculature (Gilbert, 1991). Disrupting these migrations would markedly alter the phenotype, and have large adverse impacts on fitness. Thus, the animals which survive are often those least impacted, particularly in early developmental stages. Analysis of only surviving adults may lead to some bias in the estimates of the effects of various treatments. Because plants don't have cell migration they may be better able to survive stress.

Like teeth, plants grow by accretion, with little if any reformation of structures. Once xylem has been formed it is extremely persistent. While branches may be broken off, they are not reabsorbed. Thus, plants may provide a more permanent record of the environmental quality they have experienced. Since plants do not have significant cell migration, we expect them not to exhibit the catastrophic developmental aberrations engendered by the disruption of cell movement in animals. On the other hand, because growth is by accretion, small differences in development will become magnified over time.

Plants have one peculiar and significant disadvantage for use in studies of developmental instability- they don't move. Because of this, they exhibit a relatively fixed orientation. Since growth is by accretion, slight differences in the environment on one side may become magnified, as can be attested by any one who has raised plants in windows. Thus, it is important in studies with plants to ensure that such microsite variation is not incorporated into the variance component attributed to developmental instability. This can be done with judicious sampling procedures both within and among individuals, and by examining a variety of within-plant measures of variation. The variance component that estimates the among branch variance is much more likely to be subject to this type of error than, say, the term that describes the within-leaf variation.

Below, we examine deviations from perfect bilateral, translatory, radial, and self-symmetry. We also examine within-individual variability in several ways.

Fluctuating asymmetry and plants

Fluctuating asymmetry has been used to measure developmental stability in plants. In their investigation of *Nicotiana* species, Sakai and Shimamoto (1965) examined 11 different varieties of tobacco for deviations from bilateral symmetry. They examined leaf width at the widest part and the distance between two major veins. They found significant differences among the varieties for both traits, and significant differences among plants within a variety for vein deviation. The Ambalema variety of tobacco consistently exhibited the greatest amount of leaf asymmetry, while the Connecticut Broad Leaf variety exhibited the lowest amount of leaf asymmetry. Sakai and Shimamoto also examined the issue of directional versus fluctuating asymmetry to determine if there existed either a varietal specific or genotype specific directional asymmetry. Their data showed that handedness for both leaf traits was normally distributed among leaves of the same plant (regardless of position) and among plants within all varieties. There appears to be no genetic basis for the fluctuating asymmetry they observed.

Fluctuating asymmetry has also been used to investigate the response of plants to stress in the field as well. To illustrate the effects of stress on simple branching patterns, we examined three populations of the brown alga *Fucus furcatus latifrons* growing in the intertidal regions off the coast of Washington (Tracy *et al.* unpublished data). The three environments differed in the degree of pollution, but not other environmental factors such as wave action or salinity. Because *Fucus* has dichotomous branching we can compare the lengths of the members of each pair (Fig. 4). In Figure 5 we have plotted the mean difference in branch length against both the harbor from which the alga came and the branch order. The results show that increased fluctuating asymmetry is associated with higher levels of pollution ($F_{2,72} = 5.13$, $P < 0.01$). regardless of the branch order.

Terrestrial plants also exhibit fluctuating asymmetry in response to anthropogenic stressors. We have examined populations of plants growing at various distances away from chemical production facilities in both Ukraine and the Russian Federation. In Ukraine we measured the length of the right and left posterior leaf lobes of the annual plant *Convolvulus arvensis* (Graham, Freeman & Emlen, 1993a). The absolute value of the difference between the sides of the leaf decreased with the distance away from the facility ($F_{2,27} = 16.69$, $P < 0.001$, Fig. 6). Similarly, we measured the differ-

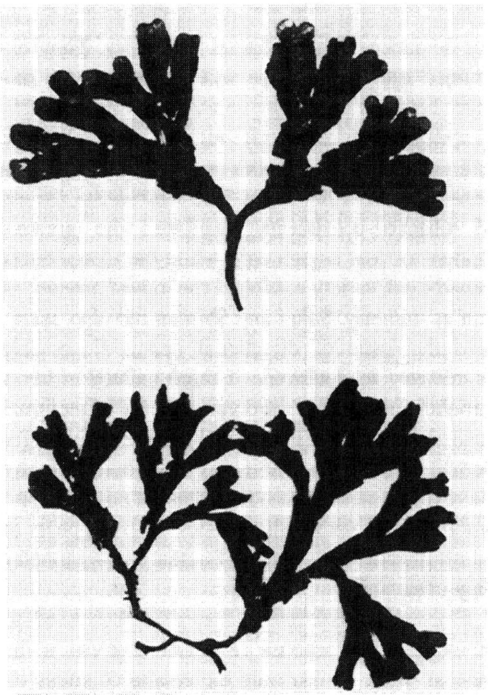

Fig. 4. Fucus furcatus latifrons growing in unpolluted (top) and polluted (bottom) waters near Seattle, Washington. The first branch pair are considered order one, while the smallest branches are fourth order.

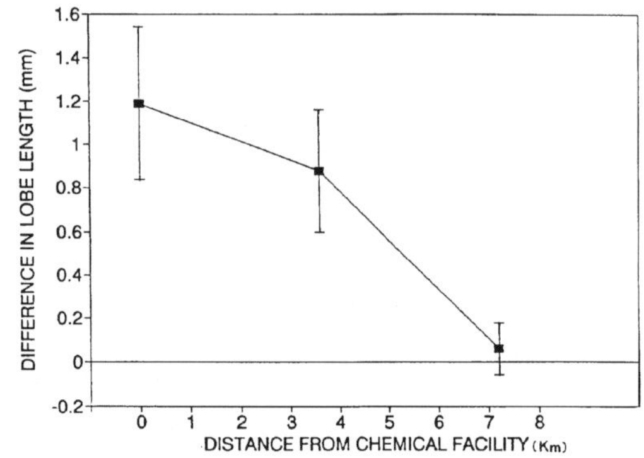

Fig. 6. Fluctuating asymmetry of the posterior leaf lobes of *Convolvulus arvensis L.* growing near a chemical production facility near Odessa, Ukraine.

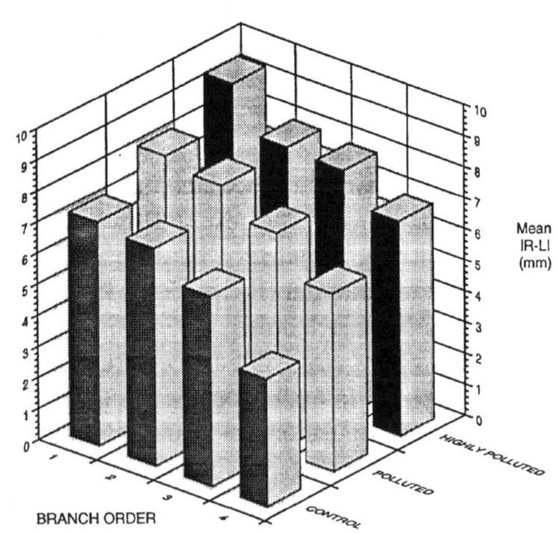

Fig. 5. Fluctuating asymmetry between the branches of *Fucus furcatus latifrons* as a function of both branch order and the pollution. The largest branches at the base are first order branches, while the smallest branches at the tips of the plant are fourth order branches.

Fig. 7. Influence of pollution on leaf symmetry. Leaves of *Robinia pseudoacacia L.* growing on the premises of a chemical production facility (a) and more than seven kilometers away from the facility (b).

ence in the point of origin for right and left leaflets of leaves from black locust trees. Locust trees have compound pinnate leaves with the leaflets usually originating in pairs directly opposite one another. Again we found that the difference in the origin of the right and left leaflet decreased as we moved away from the facility ($F_{2,27}$ = 3.68, P < 0.05, Fig. 7 and Fig. 8, Graham, Freeman & Emlen, 1993a).

In 1992 we examined populations of a small herbaceous umbel, *Aegopodium podagaria* and a short lived perennial herb, *Epilobium angustifolium*, growing at various distances away from a chemical production facility in northern Russia. In the case of *Aegopodium podagaria* we examined populations from four sites, and 15 plants per site. We examined the length of the right and left leaflet blades and the rachises supporting the leaflet blades. The results of a multivariate analysis of variance show that fluctuating asymmetry decreases as one moves away from the facility ($F_{6,118}$ = 3.90, P < 0.001, Fig. 9). For *E. angustifolium*, we examined four leaves per plant, and 20 plants per site at four sites. For each leaf, we measured the length of the leaf and the distance from the midvein of the leaf to the right and left leaf margin. This distance was measured at the mid-length of each leaf. Again, deviations away from perfect bilateral symmetry declined with distance ($F_{3,76}$ = 4.83, P < 0.003, Fig. 10).

The studies described above have dealt with chemical stresses, but other stressors also influence developmental stability. In a preliminary survey, we examined soybeans (*Glycine max*) at varying distances from a high voltage transmission power line in central Ohio (Turner & Freeman, unpublished data). The line carries 765kv of electricity and produces a magnetic field of 0.032-0.046 gauss and an electric field of 3-10 kilovolts/m directly underneath the power lines (Jones, pers. comm.). These fields should decrease as the square of the distance as one moves away from the power line. Thus, when one is 100m away from the power line, the magnetic field should be 1/10,000 of what it was under the power lines.

Soybeans are trifoliate, having three leaflets per leaf. We have contrasted the right and left leaflets. Two leaves were sampled per plant. Twenty plants were examined per site at each of three sites. here we simply report on deviations in rachis length. The data were analyzed using a nested multivariate

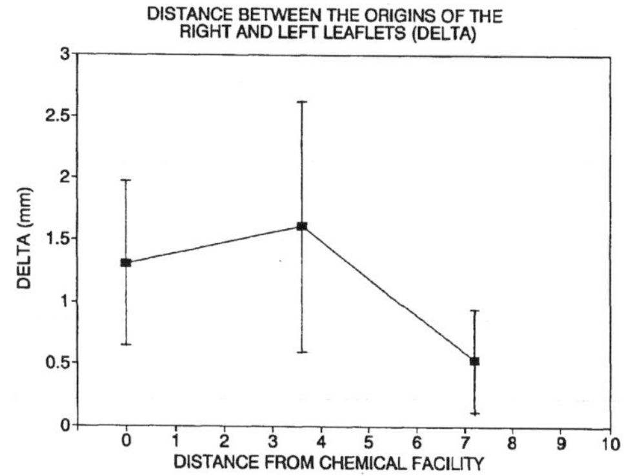

Fig. 8. Fluctuating asymmetry of leaflet positions. The average distance along the rachis between the origin of the right and left leaflets of *Robinia pseudoacacia L.* growing at various distances away from a chemical production facility.

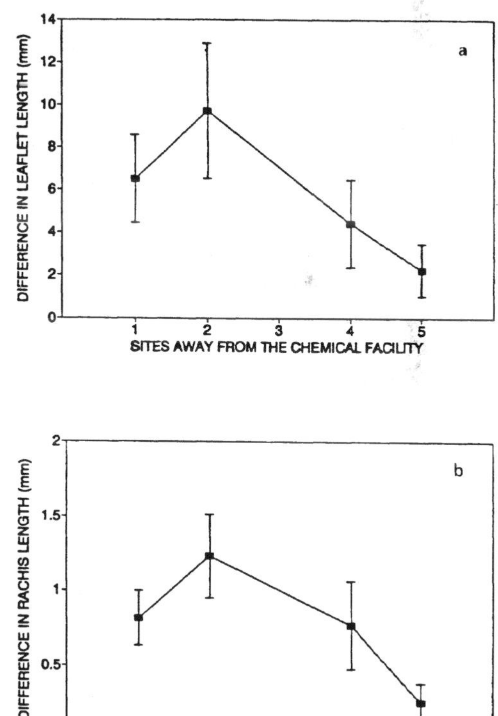

Fig. 9. Fluctuating asymmetry of *Aegopodium* leaves. Fluctuating asymmetry in the blade length (a) and petiole length (b) of *Aegopodium podagaria L.* were examined at various distances away from a chemical production facility in northern Russia. Site 1 was on the facility premises: site 2 was 2 km distant; site 4 was 9 km from the plant, and site 5 was 20 km from the plant.

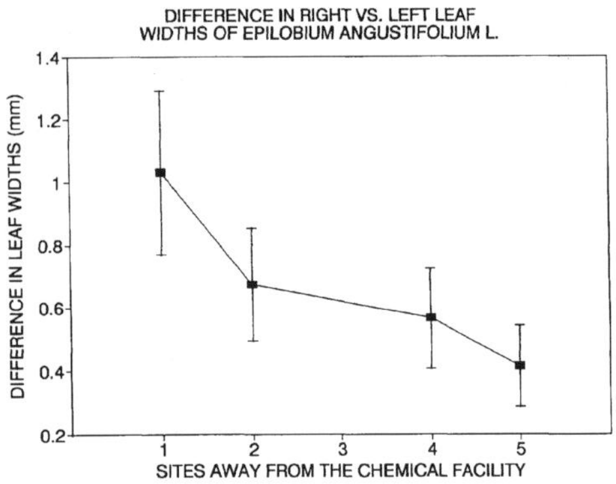

Fig. 10. Fluctuating asymmetry in leaf width of *Epilobium angustifolium L.* as a function of the distance away from a chemical production facility in northern Russia. Site 1 was on the facility premises: site 2 was 2 km distant; site 4 was 9 km from the plant, and site 5 was 20 km from the plant.

analysis of variance. The difference between the right and left sides fit Palmer and Strobeck's (1986) criteria, i.e. the mean was not significantly different from zero and the data were normally distributed about that mean. Handedness was also normally distributed among the plants at each site. We infer that the variation in rachis length among the leaflets is nongenetic fluctuating asymmetry. We have plotted the average absolute difference between the right and left rachis lengths as a function of distance away from the transmission lines ($F_{2,55}$ = 10.25, $P < 0.001$, Fig. 11, means and 95% confidence intervals are shown). Clearly, the most asymmetrical plants are found underneath the power lines.

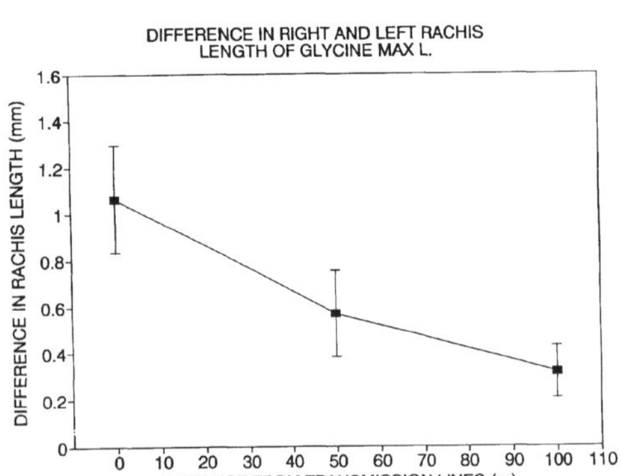

Fig. 11. Fluctuating asymmetry in rachis length of *Glycine max L.* as a function of the distance away from a 765 kv electric transmission line.

Within plant similarities- variances, euclidean distance measures, and stress

Because plants exhibit modular growth, it is legitimate to ask how similar the modules are to one another (i.e. how symmetrical is the plant.) In their examinations of tobacco, Sakai and Shimamoto (1965) examined the lengths of pistils and stamens from multiple flowers of the same individual. They used the variance in pistil or stamen lengths between flowers on the same plant and the within-flower variance as measures of developmental stability. They found significant variation among the varieties for pistil length and the within-flower variance of stamen lengths, but not for the between-flower variance of stamen lengths. They further found that the foliar measures of instability (see above) were significantly correlated with one another, but were not correlated with the variation of floral characters. The variability in pistil and stamen lengths were positively correlated with each other. Thus, there appear to be two independent blocks of genes: one that controls foral traits and the other that controls foliage traits.

Sakai and Shimamoto (1965) again found that developmental stability, using either floral or foliage traits, had a genetic basis since some varieties were more stable than others. This result is similar to that reported by Paxman (1956) for *N. rustica*. Paxman used a set of diallel crosses among five varieties. He found significant variation in pistil and stamen lengths among the varieties, and showed that the variation was both additive and nuclear. Nevertheless, the heritability of the traits he examined did decline during the inclement year of 1954 (as opposed to 1953) as one would expect. What is surprising about Paxman's work is that the inbred parents of *N. rustica* did not differ in stability compared with their F_1 progeny. Such a difference would be expected if, as Lerner (1954) suggested, heterozygosity necessarily conferred stability. Strauss (1987) found similar results using inbred and outbred pines. He too concluded 'that the relationship of heterozygosity to homeostasis for fitness components is neither simple nor monotonic'.

There are other ways of analyzing the similarities among modules of the same plant. Using a narrow hybrid zone between subspecies of basin big sagebrush (*Artemisia tridentata tridentata* × *A. t.*

Table 1. Measure of within plant similarity/dissimilarity using resemblance functions computed with data from the fall 1991.

		Basin	Hybrid	Mountain
Euclidean distance among leaves from the same branch	$\bar{X}$	0.11_a	0.26_b	0.41_c
	σ	0.07	0.20	0.29
Chord distances among leaves from the same branch	$\bar{X}$	1.21_a	0.31_b	1.36_b
	σ	0.13	0.07	0.12
Mean absolute distance among leaves from the same branch	$\bar{X}$	0.24_a	0.59_b	0.91_c
	σ	0.15	0.43	0.64
Jaccard index for leaves from the same branch	$\bar{X}$	0.79_a	0.68_b	0.56_b
	σ	0.14	0.13	0.22

vaseyana), Byrd (1992) examined the similarities in composition, concentration and proportions of terpenes produced by hybrid and parental plants. Terpenes are nonsaponifiable lipids and represent a portion of the volatile compounds produced by sagebrush. These biosynthetic endpoints are believed to function as pesticides, phytotoxins, bacteriostats, and antiherbivore agents (Duke, Paul & Lee, 1988; Kelsey, Stephens & Sharizadeh, 1982; Kelsey, 1984. Normally, to analyze the developmental stability of terpenes produced in common among all taxa examined by Byrd, one could conduct a multivariate analysis of variance, using the within-plant variance in concentration as the measure of developmental stability. However, this approach carries with it all of the assumptions of analysis of variance, which are severely violated by these terpene data. We prefer a nonparametric approach, which also allows one to use data from all the terpenes, and has fewer restrictive assumptions.

Conceptually, we are asking how similar the terpene profiles are between two leaves on the same branch, or between two branches on the same plant. The simplest approach is to use Jaccard's index of similarity (Ludwig & Reynolds, 1988), which is based upon presence/absence data. The index is simply the percent of the compounds produced in common by two or more leaves on the same branch or individual. Alternatively, one can treat the concentration of each terpene as a coordinate in a Cartesian space. If we were interested in only two compounds, the concentration of the first could be treated as the x coordinate, while the concentration of the second could be used as the y coordinate. Each leaf would specify a point in space. The Euclidean distance between the two points is a measure of the dissimilarity between the leaves. For our purposes this dissimilarity is being treated as a measure of developmental instability. Finally, one can determine if two or more leaves differ in the proportion of compounds produced by projecting the Euclidean distances onto a unit circle and measuring the length of the chord separating the two points (Ludwig & Reynolds, 1988; Byrd, 1992). Using these measures, the hybrids showed intermediate stability. The most developmentally unstable sagebrush taxa was *A. t.* ssp. *vaseyana*. This was true regardless of the nonparametric measure used (Table 1). Similarly, Graham *et al.* (in review) have shown that hybrids are, on average, intermediate in seed production, and that individuals of *A. t.* ssp. *vaseyana* produced significantly fewer seeds than either hybrids or the individuals of *A. t.* ssp. *tridentata*. Finally, morphological measures of developmental instability did not differ among the taxa (Graham *et al.*, in review).

While the study of developmental stability has been dominated by morphological and genetic studies, the use of biochemical systems may allow us to better explore the mechanism by which stability is produced and maintained. The regulation of morphology is poorly understood compared to the regulation of some physiological processes (See Zakharov, 1989, 1992 for an introduction to some physiological measures of developmental stability in animals.)

Translational symmetry

Many plants show a regular progression of leaf sizes with the largest leaf being found at the base and the smallest at the tip (acropetal growth). Using a regression analysis, Paxman (1956) assessed developmental stability among the varieties in his diallel crosses (see above). The standard error of the regression was taken as a measure of instability. Again, he found significant variation among the varieties of *N. rustica* and concluded that there was a genetic basis for this measure of stability.

We have used measures of blade length and petiole length regressed against the node number on which the leaf was borne to assess developmental stability in *Convolvulus arvensis* (described above). Data from the plants at each of the three sites was regressed separately. In Figure 12 we have plotted the coefficients of determination for both blade length and petiole length of plants from each site. The results clearly show that as one moves away from the source of pollution, the regression accounts for progressively more of the variance for both measures.

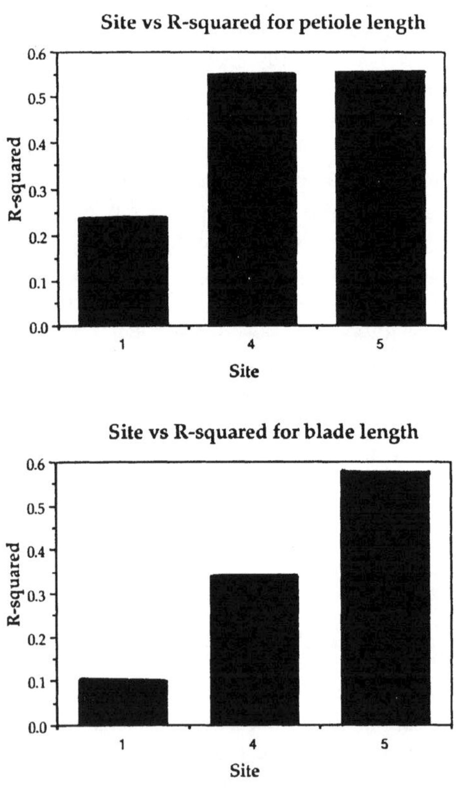

Fig. 12. Influence of pollution on scaling and translation symmetry. The correlation coefficient for petiole length (a) and blade length (b) of *Convolvulus arvensis L.* as a function of node number. Note how the correlation breaks down as one samples plants closer to the chemical production facility. Site 1 was on the facility premises, while site 4 and 5 were 3.6 and 7.7 km from the facility.

Radial and rotational symmetry

One of the characteristic features of many plants is that new leaves or flower parts originate at a constant angle from the last previously formed leaf or flower part. This gives rise to a spiral pattern as one moves up the stem (see Richards, 1951; Mitchison, 1977). We are unaware of studies that examine within-plant variability of this angle in the meristems, where such variation is most easily measured. In our study of *E. angustifolium*, we did examine the number of leaves and the number of rotations passed through in traveling the spiral from one leaf to the leaf directly above it. We found no difference among the sites for this trait. As noted above, we did find a difference in the fluctuating asymmetry of the leaves among the sites. We suggest that phyllotaxis may be more tightly canalized than leaf width. Nevertheless, Richards (1951) does note that wounding plants may cause a change in the angle. In the case he examined, the normal spiral was built around an angle of 137.5°, but after wounding the new spiral was built around a 99.5° angle. We would like to encourage further investigation of this subject.

Many species of plants such as roses or Jasminum exhibit radial symmetry in their flowers. While variation in petal number by flowers of the same plant is known to occur (see Roy, 1963) we are unaware of studies that have examined the genetic or environmental basis of such variation. We have, however, examined the length of leaflets of *Lupinus sulphureus* growing in control sites and sites exposed to sulfanylurea, an herbicide. Lupine has compound palmate leaves with leaflets radiating out from a central point. The within-plant variance in leaflet length and leaflet number were again taken as measures of instability. The result showed that the sites differed significantly ($F_{1,34}$ = 6.21, P < 0.05) with leaves exposed to the herbicide being most variable in terms of leaflet length ($F_{1,34}$ = 13.35, P < 0.001 and number ($F_{1,34}$ = 6.99, P < 0.01). The plants at the control site averaged more leaflets per leaf.

Fig. 13. Influence of pollution on Norway maple. Leaves of *Acer platanoides L.* growing 20 km from the facility (top), and on the facility premises (middle and bottom). Note that the leaves from trees on the facility have both narrower vein angles, and folds of tissue that protrude beyond the plane defined by the leaf veins.

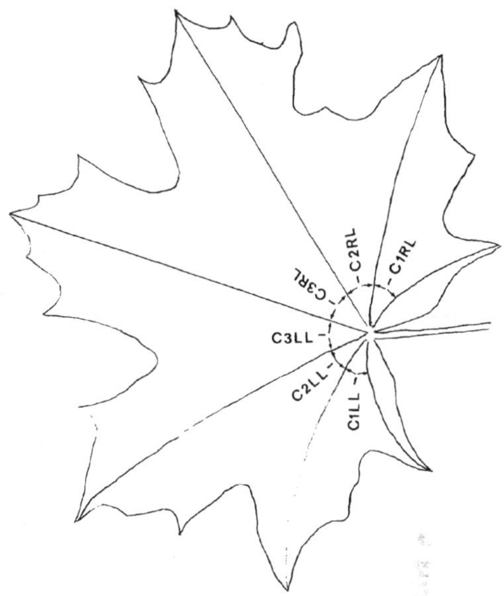

Fig. 14. Vein angles examined in leaves of *Acer platanoides L.*

Similarly, we have examined populations of Norway maple around a chemical production facility in northern Russia. Maples growing on the premises of the facility had more acute angles between the large veins of their leaves than did maples growing at a control site some 20km away ($F_{6,55}$ = 27.75, P < 0.001, Figs. 13-15).

It is interesting to note that not only did the angle decrease, but the fluctuating asymmetry between the right and left angle pairs was also greater for plants growing on the facility premises than those growing some 20km distant ($F_{3,58}$ = 16.15, P < 0.001, Fig. 16). We also found that the veins in the leaves were longer at the stressed site than at the control site. Furthermore, the lamina of the leaves at the stressed site was greater in area than could be accommodated within the narrower angle. This caused the lamina to protrude out of the plane defined by the leaf veins. Leaves from plants at the control site had lamina confined within the plane (Fig. 13).

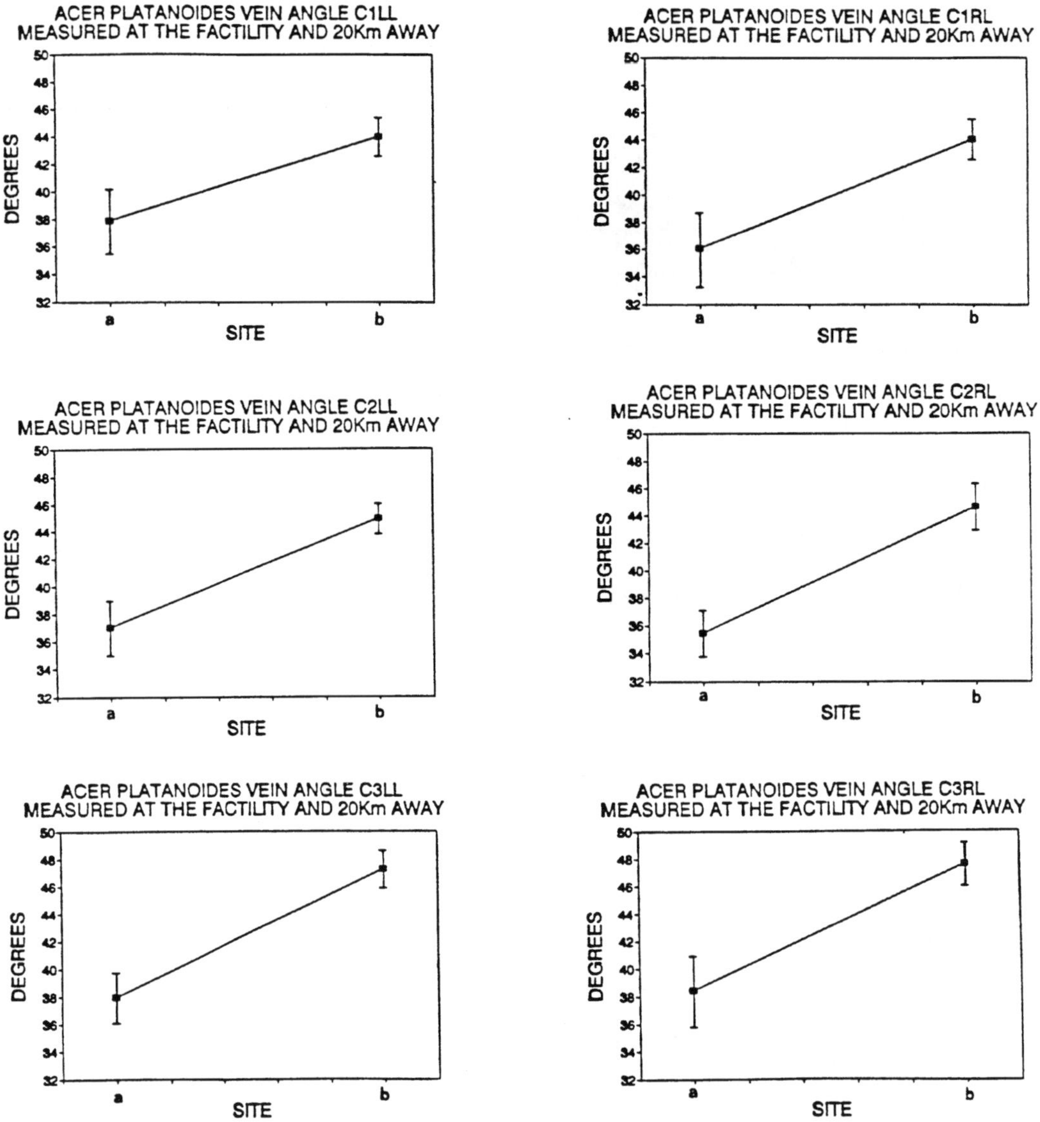

Fig. 15. Influence of pollution on rotational symmetry. Vein angle from *Acer platanoides L.* growing 20 km from the facility (a), and on the facility premises (b). See Figure 14 for vein angle notation.

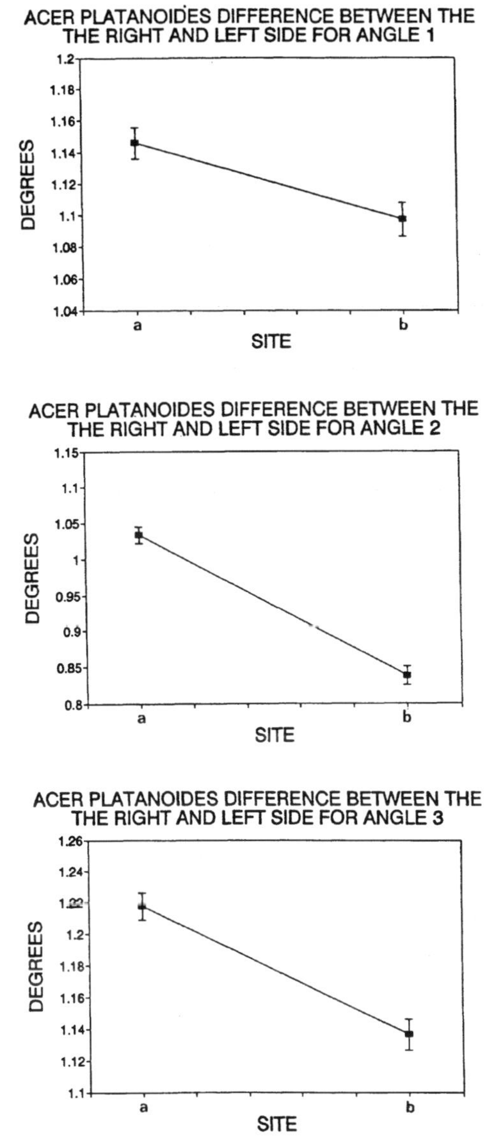

Fig. 16. Fluctuating asymmetry between pairs of vein angles for leaves of *Acer platanoides L.* growing on the premises (a) and 20 km away from a chemical production facility (b).

Self-similarity or self affinity and stress

Lindenmayer (1968) was perhaps the first to notice that plants can be modelled using a series of rewriting rules. In using Lindenmayer or analogous systems (see Aono & Kunii, 1984) to produce computer images of plants, one begins with a simple structure called a generator (see Fig. 17 for example). This generator is then subjected to a rewriting rule that is then repeated at successively smaller scales. The rewriting rules correspond to transformational codes. By using several rewriting rules, it is possible to produce complex images such as those in Figure 3. Because multiple rules are imposed, the final structures are best thought of as self affine rather than self-similar (see Figs. 1 and 17 and appendix 1 in Emlen, Freeman & Graham, this volume, for examples and explanations).

Figure 17 can also serve to illustrate another important point. If structures are the result of several processes, or if more than one variable is involved in a single process, and stress differentially influences the processes or variables, then stress may cause a change in the overall shape of a structure as in the case of the maple leaves (Fig. 16) or the differences seen among Figures 17**c**-**g**. In such cases, linear measures of developmental stability, such as fluctuating asymmetry, may prove inappropriate because they presuppose a change along a single dimension.

Lindenmayer systems are useful in exploring how changes at small scales may affect the overall appearance of the whole structure. Thus, they may serve as a heuristic illustration of how the effects at small scales may influence the morphology at larger scales. In Figure 18 we have used the same generator to produce four different Lindenmayer figures by simply varying one feature of the rewriting rules and the angle at which the figures were plotted. Figures 18**a** and **b** have exactly the same length; they differ from one another only by the rotation of one part of the structure. Yet when the two rules are repeatedly applied at smaller and smaller scales the resulting structures are remarkably different. Figure 18**c** was produced using the same rule as that for Figure 18**a** and similarly, Figure 18**d** was generated using the same rule as Figure 18**b**. Figures 18 **c** and **d** differ from **a** and **d** by being plotted at 60° instead of 90°. Like Lindenmayer figures, plants grow by adding on to previously developed structures, and thus preserve a developmental history that may be used to decipher the quality of the environment the plant experienced during development.

To illustrate the effects of stress on simple branching patterns, we examined three populations of *Fucus furcatus latifrons* growing in the intertidal

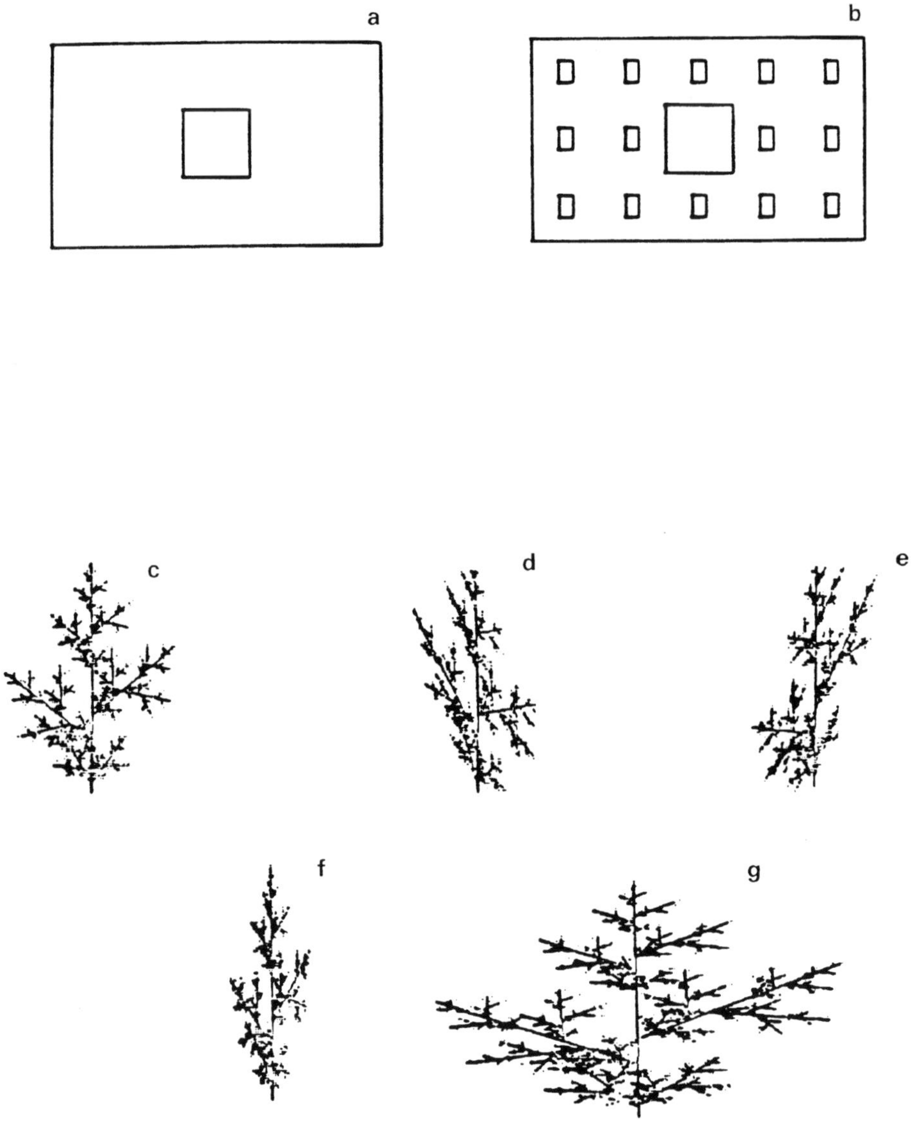

Fig. 17. Self-affine fractals. To generate self affine structures one must apply two or more scaling rules. In figures a and b we began with a rectangle 30 units by 18 units. The width was divided into thirds, and the length into fifths. This resulted in the production of squares. We then removed the central square from the plane (a). Had we used a single scaling rule, rectangles and not squares would be produced. To generate b, we again divided the width of the remaining squares into thirds and the length into fifths. This resulted in the production of rectangles, but the orientation changes from the original rectangle. The length is now smaller than the width. Many biological structures have long been known to have multiple scaling rules, i.e. to show allometric growth. Galileo (1638) showed that the diameter of bones must increase faster than length in order to support an increase in weight. Many self-affine transformations involve rotating, shrinking or stretching a figure. To illustrate this we begin with (c). We rotate the figure in (d), differentially shrink (e) or stretch (f) one axis. The transformations involve changing shape, just as the maple leaves in Figure 13.

regions off the coast of Washington (as described above). *Fucus* exhibits dichotomous branching with each branch pair resembling the larger pair below it (Fig. 4). The fractal dimension was estimated using a box counting method. One simply plots the natural log of the number of occupied boxes against the natural log of the box length. The dimension D is the absolute value of the slope of the line (Schroeder, 1991). The data were analyzed using a Kruskal-Wallis test. The results showed that the dimensionality of the plants differed significantly among three populations (χ^2 = 37.05, $P < 0.001$). Plants at the most polluted sites had the highest dimensions while those at the clean site had the lowest (Fig. 19).

In the examples discussed above, we show that a variety of symmetries may be used to investigate developmental stability in plants. With the exception of phyllotaxis, all of the symmetries became distorted under stress. However, we are unaware of

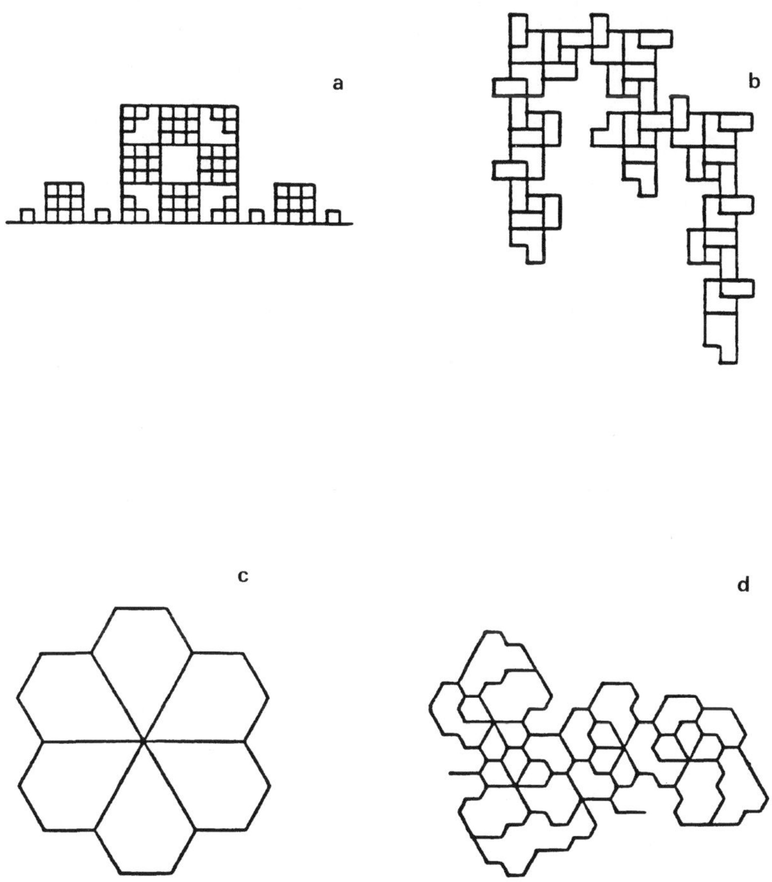

Fig. 18. Influence of small alterations in code on fractal structures. These four images have codes which are closely related. They were produced using Lindenmayer rules. In constructing Lindenmayer figures F means go forward one unit. The rule for Figure a was to replace the original F by FF + F + F + F + FF. Then in each of the succeeding generations each F was replaced by this string. The + indicates a left turn and FF indicates to move forward two units. This procedure can be repeated endlessly. Figure b differs from a by one right turn in the code i.e. the rule was F = FF + F - F + F + FF. Both a and b were plotted at 90°. Figures c and d have exactly the same codes as a and b respectively, but differ from a and b in that they were plotted at 60°.

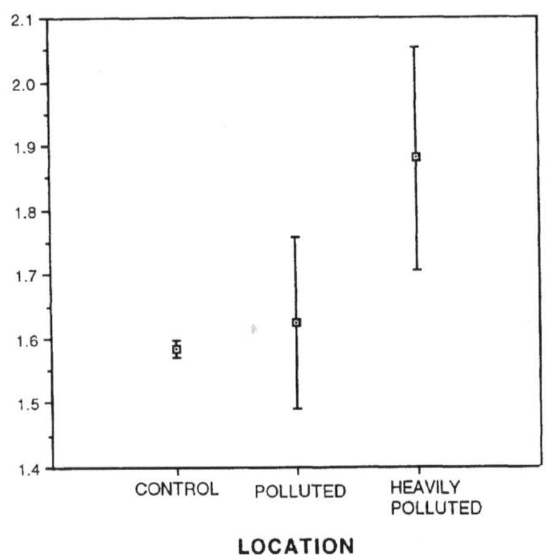

Fig. 19. The fractal dimension of *Fucus furcatus latifrons* as a function of pollution.

any studies that have evaluated several or all of the symmetries simultaneously within the context of a common experiment. Thus, we are unable to evaluate the symmetries with respect to either (a) their differing degrees of sensitivities to one or more stressors or (b) the depth of their canalization. We suspect that some of the symmetries will be more sensitive to stress than others, and that this will vary among taxa. For example, in our experience locust trees are less developmentally stable with respect to the position of their leaflets than are ash trees. Understanding how sensitivity varies among the symmetries and among various taxa would be extremely useful for those using developmental stability in environmental monitoring, conservation biology, or probing epigenesis.

Discussion

Mechanisms- effects of stress on growth

We have shown that a number of symmetries may be used to estimate the degree of developmental instability in plants. However, not all of our results are intuitively satisfying. In both *Fucus* and *Acer*, stressed plants had greater growth than those at controle sites. How is this possible?

In their study of *Phaseolus vulgaris*, Konsens, Ofir and Kigel (1991) examined plants raised under various temperature regimes. They found that branching was enhanced three-fold under high temperature stress, but that pod production was depressed. Thus, our two studies, as well as this one by Konsens show that under moderate stress, growth is enhanced, but developmental stability and fitness decline Konsens, Ofir and Kigel. (1991) and Kigel, Konsens, and Ofir (1991) argue that, for snap beans, this is because of the loss of inhibition. Their suggestion also explains the enhanced growth of *Fucus* and the vein angles of *Acer*.

There are two basic models of vein formation and growth in plants. Mitchison (1980), following the experiments of Sachs (1975), suggested that there is a signal (most likely an auxin) that causes a pathway (vein) to differentiate in such a manner that the capacity of the pathway to transport the signal increases as the flux of the signal transported increases. His model incorporates two-dimensional diffusion patterns which oscillate both in time and space. Finally, run away growth is inhibited because the diffusion rate is a function of the flux and decreases as the flux increases. Sachs' work also shows that the pathway can become saturated and non-responsive. This, too, will serve as a brake on growth. Using Mitchison's model, the spacing between veins is a function of the flow rate. Oscillations in flux of the signal also cause the formation of small pathways connecting existing pathways.

Meinhardt (1982) has proposed a model which differs from that of Mitchison in that Meinhardt's model does not assume that the properties of the pathway are a function of the flux through it, nor is the spacing of the veins directly related to the flux of the signal. Rather, Meinhardt stipulates an autocatalytic activator, and an inhibitor (perhaps ethylene or abscisic acid?) that must diffuse faster than the activator. Recent work, however, has shown that this condition need not be met (Vastano *et al.*, 1987; Pearson & Horsthemke, 1989). Equal diffusion coefficients are sufficient.

The Mitchison and Meinhardt models are not mutually exclusive, and both belong to a general class of models known as Turing models (Turing, 1952). Both models have some form of autocatalysis, nonlinear diffusion rates, and some mechanism that ultimately impedes growth. We suspect that the form of the equations is what is important (Graham, Freeman & Emlen and Emlen, Freeman & Graham, this volume). Both models can be used to explain the increased growth and narrow vein angles produced under stress.

Using Mitchison's model, if we suppose that stress causes a reduction in the production or transport of the signal, then this would tend to cause a relative increase in the concentration of the auxin at the tip versus the base of the leaf. Since veins grow towards the highest auxin concentration, the angle will become more acute. If transport had increased, producing a more even distribution of the signal, the vein angle would approach 90°. This would also explain the greater vein length as autocatalysis was allowed to go on for a greater time before the signal was drained away. Similarly, if transport is inhibited, the signal should remain in the leaf longer and cause additional growth of the maple leaf lamina.

Using Meinhardt's model, suppose that the production and or diffusion of the inhibitor was itself inhibited; this would then cause the veins to be produced at a closer interval, and therefore a more acute angle. Inhibiting the inhibitor would also lead to increased growth as we see with maples, beans, and *Fucus*.

We are suggesting that it is the dynamics of the system that causes symmetry to be either preserved or distorted. No more striking illustration of this can be found than in the work of Desbiez, Tort & Thellier (1991). Using *Biden pilosa L.*, they show that the symmetry breaking process was influenced by the presence of an apical bud, the time of day when the apical bud was removed, and the mode of removal. They also found an underlying temporal oscillation in the ability of the plant to respond. If they pricked a cotyledon at the lowest point in the temporal oscillation, the ability to respond was greatly reduced. If the plant was pricked at the peak of the temporal oscillation, the response was greater. Interestingly, the response was independent of the number of pricks. Their results are easily interpreted in light of Turing models. By pricking just one member of each cotyledon pair, they altered the standing wave pattern in the growth signal. As long as the apical bud was present, the expression of growth by the lateral bud was suppressed because of normal apical dominance. By removing the apical bud, they removed the structure that regulated the auxin waves and maintained the feedback between the right and left sides. Now the inherent asymmetry in the auxin waves induced by the prick could stimulate growth, and because growth is by accretion, the structures end up being asymmetrical. We suspect that phase locking was the only mechanism holding the two sides of the plant in symmetry. Removing the feedback via removing the apical bud (the controller) greatly weakened the phase locking (Graham, Freeman & Emlen, this volume). The puncture itself may also have altered the intercellular geometry which may have altered the wave pattern (see Emlen, Freeman & Graham, this volume). As Desbiez, Tort and Thellier point out, 'At the cellular level, the pricking of one cotyledon caused a number of cells, which were within the meristem of the bud associated with the pricked cotyledon and were in cell-cycle phases S or G2, to undergo cellular division and then be blocked in phase G1, whereas the cells of the opposite bud were practically unchanged'.

If asymmetry was determined by some static factor (e.g. heterozygosity or dominance), then when the bud was removed would be of no consequence. Clearly, in order to understand developmental stability, we must understand the dynamics of the system, and how the dynamics change in response to perturbations.

Evolutionary implications of developmental stability

The data which we have presented indicate that some genetic varieties of plants are better buffered, i.e. more stable, than other varieties. We interpret this in one of two ways. Either the more stable variety is physiologically better adapted to the growing conditions imposed, or coadaptation of the genes is better evolved in that genotype. Either way, as organisms become better adapted to their environments, all measures of developmental instability should decline. Thus, these measures may be used as a tool by conservation biologists and others interested in the well-being of natural populations.

If our argument that harmonious coadapted physiological processes determine both developmental stability and fitness is valid, then selection may operate not only to confer adaptation (via selection on functional structural genes), but also to promote coadapted gene complexes. The results of Eanes (1984, 1987) clearly demonstrate that the effect an allele has on fitness is dependent upon the allelic composition at other loci. Eanes (1984, 1987) has been using *Drosophila melanogaster* to study variants of alleles coding for the first two enzymes in the pentose phospate pathway i.e. glucose-6-phosphate dehydrogenase (G6PD) and 6 phosphogluconate dehydrogenase (6PGD). In the presence of the normal 6PGD allele, the tetrameric form of G6PD has no adverse affect on fitness. But, in the presence of a null or a low activity allele of 6PGD, the tetrameric form of G6PD is semilethal. The dimeric form of G6PD has a lower enzymatic activity than the tetrameric form, and less of an adverse impact on fitness when in the presence of the null or low activity forms of 6PGD. However, in the presence of the normal allele of 6PGD, there is no difference in the fitness of individuals possessing either G6PD variant (see Hughes & Lucchesi, 1977; Eanes 1984, 1987 for an introduction to this literature).

Similarly, Burton's (1987, 1990a,b) work with copepods has shown that chromosome segments that are viable in one population are semilethal in other populations having different genetic backgrounds. His work also shows that the production of the free amino acids, alanine and proline, is regulated in part by structural genes coding for glutamate-pyruvate transaminase and NADP-malic enzyme. However, Burton reported substantial variation in the production of these free amino acids among lines with identical allelic compositions for these two genes. The variation appeared to depend upon the genetic background in which the alleles were found Burton's work shows quite clearly that 'while population divergence may be stochastic, the overall genetic composition of a population is unlikely to be a random set of the allelic variants extant in the species' (Burton, 1990a). Rather, 'the integration of alleles into a population ultimately requires their harmonious interaction with those extant in that population' (Burton, 1990b). Interestingly, both Burton's studies and those of McKenzie and his colleagues show that when coadaptation is disrupted development is also altered. This may be manifested as increased developmental time (Burton, 1990b) or increased fluctuating asymmetry (see McKenzie & O'Farrell, this volume, for references).

The issue is one of integration. To what extent is development dependent upon the interactions among genes (or gene products)? The degree to which evolution involves selection for harmony, as opposed to adaptation, is only beginning to be explored. This exploration will require examining the murky (and, we suspect, nonlinear) realm between genotype and phenotype- i.e. the flux through physiological pathways. Both the experimental work described above, and our own theoretical explorations (Emlen, Freeman & Graham and Graham, Freeman & Emlen this volume) suggest that the generation of coadaptation is important in regulating the production and stability of the phenotype. Measures of developmental stability should prove useful both in this exploration and in conservation biology, because such measures allow one to assess how well the genotype is adapted to its environment, be that environment external or the internal genetic background in which the alleles must work.

Acknowledgements

We would like to thank Walter Rothschild and William Turner for their stimulating discussions, helpful insights, and a good many references. William Turner and Kathy Miglia also supplied two photographs. Kathy Miglia, Umesh Malhotra, Sulada Kanchana, and Veronica Riha provided invaluable help in producing this manuscript. Steven Haak produced the computer images. We would like to thank David Byrd, Mary Tracy, and William Turner for the use of some of their unpublished data which was gathered in conjunction with DCF. Catherine Chamberlin-Graham assisted in the studies of plants in Ukraine.

References

Aono, M. & T. L. Kunii, 1984. Botanical tree image generation. Institution of Electrical and Electronics Engineers Computer Graphics and Applications 4: 10-34.

Burton, R. S., 1987. Differentiation and integration of the genome in populations of *Tigriopus californicus*. Evolution 41: 504-513.

Burton, R. S., 1990a. Hybrid breakdown in physiological response: a mechanistic approach. Evolution 44: 1806-1813.

Burton, R. S., 1990b. Hybrid breakdown in developmental time in the copepod *Tigriopus californicus*. Evolution 44: 1814-1822.

Bush, R. M., P. E. Smouse & F. T. Ledig, 1987. The fitness consequences of multiple-locus heterozygosity: the relationship between heterozygosity and growth rate in pitch pine (*Pinus rigida* Mill.) Evolution 41: 787-798.

Byrd, D. W., 1992. Biochemical assessment of the hybrid zone formed by *Artemisia tridentata tridentata* and *Artemisia tridentata vaseyana* using an automated gas chromatography-data system and statistical methods. Master's thesis. Wayne State University, Detroit, Michigan.

Clarke, G. M. & J. A. McKenzie, 1987. Developmental stability of insecticide resistant phenotypes in blowfly; a result of canalizing natural selection. Nature 325: 345-346.

Clarke, G. M., 1992. Fluctuating asymmetry: a technique for measuring developmental stress of genetic and environmental origin. Acta Zool. Fennica 191: 31-35.

Clarke, G. M., B. P. Oldroyd & P. Hunt, 1992. The genetic basis of developmental stability in *Apis mellifera:* heterozygosity versus genic balance. Evolution 46: 753-762.

Cocho, G. & J. L. Rius, 1989. Discreet aspects of morphogenesis and gene dynamics, pp. 177-190 in Theoretical Biology; Epigenetic and Evolutionary Order from Complex Systems, edited by B. Goodwin and P. Saunders. Edinburgh University Press, Edinburgh.

Cullis, C. A., 1977. Molecular aspects of the environmental induction of heritable changes in flax. Heredity 38: 129-154.

Cullis, C. A., 1983. Environmentally induced DNA changes in plants. CRC Critical Reviews in Plant Sciences 1: 117-129.

Cullis, C. A., 1986. Phenotypic consequences of environmentally induced changes in plant DNA. Trends in Genetics 2: 307-309.

Cullis, C. A., 1987. The generation of somatic and heritable variation in response to stress. American Naturalist 130: S62-S73.

Davies, P. C. W., 1989. The physics of complex organization, pp. 101-111 in Theoretical Biology; Epigenetic and Evolutionary Order from Complex Systems, edited by B. Goodwin and P. Saunders. Edinburgh University Press, Edinburgh.

Davis, T. A., 1962. The non-inheritance of asymmetry in *Cocos nucifera*. J. Genet. 58: 42-50.

Davis, T. A., 1962. Asymmetry and yield in coconut. Experientia 18: 321-322.

Davis, T. A., 1963. The dependence of yield on asymmetry in coconut palms. J. Genet. 58: 186-215.

Desbiez, M. O., M. Tort & M. Thellier, 1991. Control of a symmetry-breaking process in the course of the morphogenesis of plantlets of *Bidens pilosa L.* Planta 184: 397-402.

Duke, S. O., R. N. Paul, Jr. & S. M. Lee, 1988. Terpenoids from *Artemisia* as potential pesticides. ACS Symposium Series 380: 318-334.

Eanes, W. F., 1984. Viability interactions, *in vivo* activity and the G6PD polymorphism in *Drosophila melanogaster*. Genetics 106: 95-107.

Eanes, W. F., 1987. Allozymes and fitness: Evolution of a problem. TREE 2: 44-48.

Eilbeck, J. C., 1989. Pattern formation and pattern selection in reaction-diffusion systems, pp. 31-41 in Theoretical Biology; Epigenetic and Evolutionary Order from Complex Systems, edited by B. Goodwin and P. Saunders. Edinburgh University Press, Edinburgh.

Emlen, J. M., D. C. Freeman & J. H. Graham. Nonlinear growth dynamics and the origin of fluctuating asymmetry (Genetica, this volume).

Fahn, A., 1974. Plant Anatomy. Pergamon Press, New York.

Gilbert, S. F., 1991. Developmental Biology. Third Edition. Sinauer Associates, Inc., Sunderland, Mass.

Goodwin, B. C., 1989. Evolution and the generative order, pp. 89-100 in Theoretical Biology, Epigenetic and Evolutionary Order from Complex Systems, edited by B. Goodwin and P. Saunders. Edinburgh University Press, Edinburgh.

Goss, R. J., 1966. Hypertrophy versus hyperplasia. Science 153: 1615-1620.

Graham, J. H., 1992. Genomic coadaptation and developmental stability in hybrid zones. Acta Zool. Fennica 191: 121-131.

Grahm, J. H., D. C. Freeman & J. M. Emlen, 1993a. Antisymmetry, directional asymmetry and chaotic morphogenesis. (Genetica, this volume).

Graham, J. H., D. C. Freeman & J. M. Emlen, 1993b. Developmental stability: a sensitive indicator of populations under stress. In Environmental risk assessment, ASTM STP 1179, Wayne G. Landis, Jane S. Hughes, and Michael A. Lewis, Eds. American Society for Testing and Materials, Philadelphia pp. 136-158.

Graham, J. H., D. C. Freeman & E. D. McArthur. Selection and dispersal in narrow hybrid zones between two subspecies of big sagebrush (*Artemisia tridentata* ssp. *tridentata* and *A. t.* ssp. *vaseyana*). Submitted to the American Journal of Botany.

Hecks, B., Z. Hejnowicz & A. Sievers, 1992. Spontaneous oscillations of extracellular electrical potentials measured on *Lepidium sativum L.* roots. Plant, Cell and Environment 15: 115-121.

Huberman, B. A., 1989. The adaptation of complex systems, pp. 124-133 in: Theoretical Biology; Epigenetic and Evolutionary Order from Complex Systems. Editors: B. Goodwin and P. Saunders. Edinburgh University Press, Edinburgh.

Hughes, M. B. & J. C. Lucchesi, 1977. Genetic rescue of a lethal 'null' activity allele of 6 phosphogluconate dehydrogenase from *Drosophila melanogaster.* Science 196: 1114-1115.

Jaffe, L. F., 1980. Control of plant development by steady ionic currents. In: Plant Membrane Transport: Current Conceptual Issues. Editors: W. Spanswick, J. Lucas and J. Dainty. pp. 381-388. Biomedical Press/Elsevier, North-Holland.

Jaffe, L. F. & R. Nuccitelli, 1977. Electrical controls of development. Annual Review of Biophysics and Bioengineering 6: 445-476.

Kaplan, D. R. & W. Hagemann, 1991. The relationship of cell and organism in vascular plants. Are cells the building blocks of plant form? Bioscience 41: 693-703.

Kauffman, S. A., 1989. Origins of order in evolution: self-organization and selection, pp. 67-88 in: Theoretical Biology; Epigenetic and Evolutionary Order from Complex Systems. Editors: B. Goodwin and P. Saunders. Edinburgh University Press, Edinburgh.

Kelsey, R. G., 1984. Foliage biomass and crude terpenoid productivity of big sagebrush. Symposium on the biology of *Artemisia* and *Chrysothamus*; Department of Agriculture, Forest Service, Intermountain Research Station. General Technical Report INT-200: 375-388.

Kelsey, R. G., J. R. Stephens & F. Sharizadeh, 1982. The chemical constituents of sagebrush foliage and their isolation. Journal of Range Management 35: 617-622.

Kieser, J. A., 1993. Developmental instability, evolution and the theory of acquisition. Genetica: Special Issue.

Kigel, J., I. Konsens & M. Ofir, 1991. Branching, flowering and pod-set patterns in snap-bean (*Phaseolus vulgaris L.*) as affected by temperature. Can. J. Plant Sci. 71: 1233-1242.

Kojima, K., T. Mukai, H. Suemoto, N. Sueoka & H. Ono, 1955. Studies on the right- and left-handedness of spikelets in Einkorn wheats. VI. Polygenic analysis in a species cross. Proc. Japan Acad. 31: 228-233.

Konsens, I., M. Ofir & J. Kigel, 1991. The effect of temperature on the production and abscission of flowers and pods in snap-bean (*Phaseolus vulgaris L.*). Ann. Bot. 67: 391-399.

Lara-Ochoa, F., 1989. Membrane Phase Transitions, pp. 147-161 in: Theoretical Biology; Epigenetic and Evolutionary Order from Complex Systems. Editors: B. Goodwin and P. Saunders. Edinburgh University Press, Edinburgh.

Leary, R. F., F. W. Allendorf & K. L. Knudsen, 1983. Developmental stability and enzyme heterozygosity in rainbow trout. Nature 301: 71-72.

Leary, R. F., F. W. Allendorf & K. L. Knudsen, 1984. Superior developmental stability of heterozygotes at enzyme loci in salmonid fishes. Amer. Nat. 124: 540-551.

Leary, R. F., F. W. Allendorf & K. L. Knudsen, 1985. Developmental instability as an indicator of reduced genetic variation in hatchery trout. Trans. Amer. Fish. Soc. 114: 230-235.

Leary, R. F., F. W. Allendorf & K. L. Knudsen, 1992. Genetic, environmental and developmental causes of meristic variation in rainbow trout. Acta Zool. Fennica 191: 79-95.

Lerner, I. M., 1954. Genetic Homeostasis. John Wiley & Sons, Inc., New York.

Lindenmayer, A., 1968. Mathematical models for cellular interaction in development, Parts I and II. J. Theoret. Biol. 18: 280-315.

Ludwig, J. A. & J. F. Reynolds, 1988. Statistical ecology. John Wiley and Sons. New York. 337 p.

Markow, T. A. & J. P. Ricker, 1991. Developmental stability in hybrids between the sibling species pair *Drosophila melanogaster* and *Drosophila simulans*. Genetica 84: 115-121.

Mather, K., 1953. Genetical control of stability in development. Heredity 7: 297-336.

McKenzie, J. & K. O'Farrell, 1993. Modification of developmental instability and fitness: Malathion-resistance in the Australian blowfly *Lucilia cuprina*. Genetica: Special Issue.

Meinhardt, H., 1982. Models of Biological Patterns Formation. Academic, New York.

Minzoni, A. A. & F. Alonso-deFlorida, 1989. Time and Space Scales in Neurophysiology, pp. 191-197 in: Theoretical Biology; Epigenetic and Evolutionary Order from Complex Systems. Editors: B. Goodwin and P. Saunders. Edinburgh University Press, Edinburgh.

Mitchison, G. J., 1977. Phyllotaxis and the fibonacci series. Science 196: 270-275.

Mitchison, G. J., 1980. A model for vein formation in higher plants. Proc. R. Soc. Lond. B 207: 79-109.

Palmer, A. R. & C. Strobeck, 1992. Fluctuating asymmetry as a measure of developmental stability: implications of non-normal distributions and power of statistical tests. Acta Zool. Fennica 191: 57-72.

Palmer, A. R. & C. Strobeck, 1986. Fluctuating asymmetry: measurement, analysis, patterns. Ann. Rev. Ecol. System. 17: 391-421.

Paxman, G. J., 1956. Differentiation and stability in the development of *Nicotiana rustica*. Ann. Bot. 20: 331-347.

Pearson, J. E. & W. Horsthemke, 1989. Turing instabilities with nearly equal diffusion coefficients. J. Chem. Phys. 90: 1588-1599.

Prusinkiewicz, P. & J. Hanan, 1985. Lindenmayer systems, fractals, and plants. In: Lecture Notes in Biomathematics. Managing Editor: S. Levin. Springer-Verlag, New York.

Richards, F. J., 1951. The geometry of phyllotaxis and its origin. Symp. Soc. Exp. Biol. 2: 217-245.

Roy, S. K., 1963. The variation of organs of individual plants. J. Genet. 58: 147-176.

Sachs, T., 1975. The control of the differentiation of vascular networks. Ann. Bot. 39: 197-204.

Sakai, K. & Y. Shimamoto, 1965. Developmental instability in leaves and flowers of *Nicotiana tabacum*. Genetics 51: 801-813.

Saunders, P. T. & C. Kubal, 1989. Bifurcations and the epigenetic landscape, pp. 16-30 in: Theoretical Biology; Epigenetic and Evolutionary Order From Complex Systems. Editors: B. Goodwin and P. Saunders. Edinburgh University Press, Edinburgh.

Savageau, M. A., 1989. Are there rules governing patterns of gene regulation?, pp. 42-66 in: Theoretical Biology; Epigenetic and Evolutionary Order from Complex Systems. Editors: B. Goodwin and P. Saunders. Edinburgh University Press, Edinburgh.

Schroeder, M., 1991. Fractals, Chaos, Power Laws; Minutes from an Infinite Paradise. W. H. Freeman and Co., New York.

Souda, M., K. Toko, K. Hayashi, T. Fujiyoshi, S. Ezaki & K. Yamafuji, 1990. Relationship between growth and electric oscillations in bean roots. Plant Physiology 93: 432-436.

Strauss, S. H., 1987. Heterozygosity and developmental stability under inbreeding and crossbreeding in *Pinus attentuata*. Evolution 41: 331-339.

Sueoka, N. & T. Mukai, 1956. Studies on the right- and left-handedness of spikelets in Einkorn wheats. VII. The c-value of the species of *Triticum* and *Aegilops*. Proc. Japan Acad. 32: 191-196.

Tatsum, J., A. Yamauchi & Y. Kono, 1989. Fractal analysis of plant root systems. Annals of Botany 64: 499-503.

Thom, R. An inventory of Waddingtonian concepts. 1989, pp. 1-7 in: Theoretical Biology; Epigenetic and Evolutionary Order From Complex Systems. Editors: B. Goodwin and P. Saunders. Edinburgh University Press, Edinburgh.

Timofeeff-Ressovsky, N. V., 1934. Uber der einfluss des genotypischen milieus und der aussenhebedigunen auf die realisation des genotypes. Genmutation vti bei *Drosophila funebris.* Nachr. Acad. Wiss. Gottingen II. Math-Phys. K1. Fg. 6: 53-106.

Toko, K., S. Iiyama, C. Tanaka, K. Hayashi, Keiko Yamafuji & Kaoru Yamafuji, 1987. Relation of growth process to spatial patterns or electrical and enzyme activity in bean roots. Biophysical Chemistry 27: 39-58.

Toko, K., M. Souda, T. Matsumo & K. Yamafuji, 1990. Oscillations of electrical potential along a root of a higher plant. Biophysical Journal 57: 269-279.

Tornheim, K., 1986. Fructose-2, 6-biphosphate and glycolytic oscillation in skeletal-muscle extract. Federation Proceedings 45: 1840.

Trewavas, A., 1991. How do plant growth substances work? II.

Plant, Cell and Environment 14: 1-12.

Turing, A. M., 1952. The chemical basis of morphogenesis. Philos. Trans. Roy. Soc. London B 327: 37-72.

Van Valen, L., 1962. A study of fluctuating asymmetry. Evolution 16: 125-142.

Vastano, J. A., J. E. Pearson, W. Horsthemke & H. L. Swinney, 1987. Chemical pattern formation with equal diffusion coefficients. Physics Letters A. 1234: 320-324.

Waddington, C. H., 1957. The strategy of genes. Allen & Unwin London.

Ward, J. M. & K. B. Armitage, 1981. Circannual rhythms of food consumption, body mass and metabolism in yellow-bellied marmots (*Marmota flaviventris*). Comp. Biochem. Physiol. 69[A]: 621-626.

Watkinson, A. R. & J. White, 1986. Some life-history consequences of modular construction in plants. Phil. Trans. R. Soc. London B 313: 31-51.

Wegner, T. & M. Peterson, 1991. The Waite Group's Fractal Creations. Waite Group Press. Mill Valley, CA. 315 p.

Weisenseel, M. H., A. Dorn & L. F. Jaffe, 1979. Natural H^+ currents traverse growing roots and root hairs of barley (*Hordeum vulgare L.*). Plant Physiology 64: 512-518.

Weisenseel, M. H. & R. M. Kicherer, 1981. Ionic currents as control mechanism in cytomorphogenesis. In: Cell Biology Monographs, Vol. 8. Editor: O. Kiermayer. pp. 379-399. Springer-Verlag, Berlin.

West, B. J. & A. L. Goldberger, 1987. Physiology in fractal dimensions. Am. Scientist 75: 354-365.

West, B. J. & M. F. Shlesinger, 1989. On the ubiquity of $1/f$ noise. Inter. J. Mod. Phys. B 3(6): 795-819.

West, B. J., 1990. Physiology in fractal dimensions: error tolerance. Ann. Biomed. Eng. 18: 135-149.

West, B. J. & M. Shlesinger, 1990. The noise in natural phenomena. Am. Sci. 78: 40-45.

Zakharov, V. M., 1989. Future prospects for populations phenogenetics. Sov. Sci. Rev. F. Physiol. Gen. Biol. 4: 1-79.

Zakharov, V. M., E. Pankakoske, B. I. Sheftel, A. Peltonen & I. Hanski, 1991. Developmental stability and population dynamics in the common shrew *Sorex araneus*. Am. Nat. 138: 797-810.

Zakharov, V. M., 1990. Analysis of fluctuating asymmetry as a method of biomonitoring at the population level. pp 187-198 In. Krivolutsky (ed). Bioindications of chemical and radioactive pollution. Mir Publishers Moscow/CRC Press Boca Raton.

Zakharov, V. M., 1992. Population phenogenetics: Analysis of developmental stability in natural populations. In V. M. Zakharov & J. H. Graham. Developmental stability in natural populations. Acta Zool. Fennica 19: 7-30.

Zeeman, C., 1989. A new concept of stability, pp. 8-15 in: Theoretical Biology. Epigenetic and Evolutionary Order from Complex Systems. Editors: B. Goodwin, and P. Saunders. Edinburgh University Press, Edinburgh.

T. A. Markow (ed.), Developmental Instability: Its Origins and Evolutionary Implications, 123–139, 1994.

Antisymmetry, directional asymmetry, and dynamic morphogenesis

John H. Graham[1], D. Carl Freeman[2] & John M. Emlen[3]
[1] *Department of Biology, Berry College, Mt. Berry, GA 30149, USA*
[2] *Department of Biological Sciences, Wayne State University, Detroit, MI 48202, USA*
[3] *US Fish and Wildlife, Building 204, Naval Station, Seattle, WA 98115, USA*

Received and accepted 29 April 1993

Key words: fluctuating asymmetry, developmental stability, chaos, nonlinear dynamics, Turing equations

Abstract

Fluctuating asymmetry is the most commonly used measure of developmental instability. Some authors have claimed that antisymmetry and directional asymmetry may have a significant genetic basis, thereby rendering these forms of asymmetry useless for studies of developmental instability. Using a modified Rashevsky-Turing reaction-diffusion model of morphogenesis, we show that both antisymmetry and directional asymmetry can arise from symmetry-breaking phase transitions. Concentrations of morphogen on right and left sides can be induced to undergo transitions from phase-locked periodicity, to phase-lagged periodicity, to chaos, by simply changing the levels of feedback and inhibition in the model. The chaotic attractor has two basins of attraction-right sidedominance and left side dominance. With minor disturbance, a developmental trajectory settles into one basin or the other. With increasing disturbance, the trajectory can jump from basin to basin. The changes that lead to phase transitions and chaos are those expected to occur with either genetic change or stress. If we assume that the morphogen influences the behavior of cell populations, then a transition from phase-locked periodicity to chaos in the morphogen produces a corresponding transition from fluctuating asymmetry to antisymmetry in both morphogen concentrations and cell populations. Directional asymmetry is easily modeled by introducing a bias in the conditions of the simulation. We discuss the implications of this model for researchers using fluctuating asymmetry as an indicator of stress.

Introduction

Fluctuating asymmetry is widely used for evaluating the condition of individuals in natural populations (see Zakharov & Graham, 1992 for reviews). Most practitioners consider it a measure of developmental instability, an organism's inability to buffer random accidents of development. In contrast, antisymmetry and directional asymmetry, two other kinds of asymmetry, are often regarded as a nuisance, because they confound fluctuating asymmetry (Palmer & Strobeck, 1986). Nevertheless, McKenzie and Clarke (1988), Leary and Allendorf (1989), McKenzie, Batterham, and Baker (1990), and McKenzie and O'Farrell (this volume) have recently argued that antisymmetry may sometimes reflect extreme developmental instability. Palmer and Strobeck (1992) have strongly rejected this view. They argue that antisymmetry has an unknown genetic component and thus can not be interpreted as a measure of developmental instability.

The controversy over antisymmetry is actually quite old. Mather (1953), in selecting for increased asymmetry in the sternopleural bristles of *Drosophila melanogaster*, succeeded in selecting for antisymmetry, which he interpreted as an indication of reduced developmental buffering. Van Valen (1962), in comparing the three forms of asymmetry, concluded that both directional asymmetry and antisymmetry were 'developmentally controlled ... and probably normally adaptive,' while fluctuating asymmetry alone represented developmental noise.

He suggested that there was no way of distinguishing the contribution of fluctuating asymmetry to the variance in an antisymmetric distribution.

In a related controversy over antisymmetry, Morgan and Corballis (1978) argued that the direction of handedness in humans, a classic antisymmetric trait, is controlled by the developmental environment. They presented data on familial variation in the direction of handedness and on the tendency for identical twins to be mirror images of one another. Van Valen (1978) rejected Corballis and Morgan's arguments, insisting that the direction of handedness must have a genetic basis, and that the low heritability of the direction of handedness was better explained by a low threshold for reversal under stress.

Palmer and Strobeck (1986), in an influential review, recognized two kinds of antisymmetry: one having a genetic basis and one representing a kind of 'nongenetic environmental noise.' Experimental work in the late 1980s continued to support the hypothesis that antisymmetry can, on occasion, reflect developmental instability. McKenzie and Clarke (1988), for example, observed antisymmetry in sheep blowflies, *Lucilia cuprina*, during the evolution of the flies's resistance to Diazanon. Leary and Allendorf (1989) generated antisymmetry in the mandibular pores of rainbow trout *Oncorhynchus mykiss* by stressing a population of experimentally produced gynogenetic diploids that were derived from a randomly mating hatchery population.

Then, Palmer and Strobeck (1992), in an apparent change of opinion, argued that for both kinds of antisymmetry, 'individuals are genetically or developmentally directed to become asymmetrical.' Thus, antisymmetry can not be used as a measure of developmental instability because it has an unknown genetic component. Furthermore, they prove that a bimodal distribution of $r_i - l_i$ centered on zero can not arise from developmental noise 1) if right and left sides are independent and 2) if both sides have the same frequency distribution. Finally, to explain the antisymmetry observed by McKenzie and Clarke (1988), Palmer and Strobeck (1992) claim that these researchers had inadvertently selected for a gene for antisymmetry.

The controversy is still unresolved. Palmer and Strobeck's criticism of McKenzie and Clarke's (1988) paper cannot be leveled at Leary and Allendorf's (1989) paper. Because the technique of experimental gynogenesis employed by Leary and Allendorf involves a doubling of an activated egg's haploid complement of chromosomes, there is no possibility of inadvertently selecting a gene for antisymmetry. Moreover, the population of rainbow trout from which the eggs were selected normally exhibits fluctuating asymmetry for this trait.

In considering both sides of the argument, we have taken a long hard look at the theoretical basis of developmental stability. It has been more than fifty years since Waddington (1940) established the theoretical foundation for the concepts of developmental homeostasis, homeorhesis, and canalization. He summarized his work on this subject in his book *Strategy of the Genes* (Waddington, 1957). As important as Waddington's work has been (see Thom, 1989), it is also true that much has changed in developmental biology since 1957. Using Waddington's ideas, and the recent research of many of his proteges as a point of departure, we ask whether developmental instability should be reflected only in a normal distribution of asymmetry values. We also reconsider the concepts of developmental stability and fluctuating asymmetry. In this paper, we will argue that transitions from fluctuating asymmetry to antisymmetry, and perhaps even to directional asymmetry, may often reflect severe developmental instability. In making this statement, we implicitly reject the notion that fluctuating asymmetry represents only developmental 'noise.' By exploring the dynamics of Turing's (1952) reaction-diffusion models of morphogensis, we show that antisymmetry, directional asymmetry, and fluctuating asymmetry can arise from nonlinear processes of development.

Kinds of asymmetry

Fluctuating asymmetry

The concept of fluctuating asymmetry can be traced back to research done in the 1920s and 1930s by F. Sumner in the United States, and by W. Ludwig and Soviet expatriates B. Astauroff and N. Timofeéf-Ressovsky in Germany (for the most complete review of the background material see Zakharov, 1987, 1989). As defined by Palmer and Strobeck (1986), fluctuating asymmetry consists of 'minor, nondirectional deviations from bilateral symmetry.' The deviations are measured by the difference between the size (or count) of a structure on the right side and that of the corresponding structure on the left side ($r - l$). For a population exhibiting fluctuating asymmetry, these deviations ($r_i - l_i$) are normally distributed and have a mean of zero. One measure of developmental instability is thus the variance of the deviations, Var ($r_i - l_i$).

Several researchers have suggested explanations for the developmental origins of fluctuating asymmetry. Mather (1953) suggested that fluctuating asymmetry reflects local disturbances during development, which can arise from either 'differences of environment' or 'upsets of cell development.' Later researchers, inspired by the new science of cybernetics, suggested that fluctuating asymmetry was 'analogous to "noise" as used in information theory' (Van Valen, 1962). Reeve (1960), Soulé and Cuzin-Roudy (1982), and Lewontin (1986) believed that fluctuating asymmetry was ultimately reducible to the level of randomness of thermal movements of particles. Palmer and Strobeck (1992) have proposed that fluctuating asymmetry represents variation of 'exclusively environmental origin.'

Directional asymmetry

Directional asymmetry occurs when 'there is normally a greater development of a character on one side of the plane or planes of symmetry than on the other' (Van Valen, 1962). An example of directional asymmetry is the mammalian heart, which shows greater development of its left side. According to Palmer and Strobeck (1986, 1992), an unknown proportion of the asymmetry variance has a genetic basis. Thus, they argue that directional asymmetry can not be used as a measure of developmental instability.

Antisymmetry

Antisymmetry (Timofeéf-Ressovsky, 1934) occurs when most individuals in a population are asymmetric, but it is unpredictable which side of an organism shows greater development. In other words, although the mean deviation may still be zero, the distribution of deviations is no longer normal. In its most extreme manifestation, antisymmetry is characterized by a bimodal distribution of deviations from bilateral symmetry. Mather (1953) has pointed out that antisymmetry requires a negative interaction between sides of a developing organism.

Palmer and Strobeck (1986) distinguished two kinds of antisymmetry: antisymmetry I and antisymmetry II. Antisymmetry I occurs by mixing two genotypes, with each having directional asymmetry in opposite directions. This might also be called polymorphic directional asymmetry. According to Palmer and Strobeck (1992), it has all the disadvantages of directional asymmetry, because it has both a genetic and a nongenetic component. Antisymmetry II, according to Palmer and Strobeck (1986), is a 'form of nongenetic developmental noise in which a character on one side of the body is consistently larger than its partner.'

In contrast to Palmer and Strobeck, we refer to antisymmetry II as fluctuating antisymmetry. Like fluctuating asymmetry, the direction of deviations from perfect symmetry is not heritable. An example of such antisymmetry is the lobster (or fiddler crab) claw. Lobsters (or fiddler crabs) develop a large crusher claw and a smaller cutter claw. Govind and Pearce (1986, 1992) have shown that asymmetry of lobster claws is a result of feedback through the central nervous system. Most of the difference in size (i.e. the dispersion of $r_i - l_i$) here is not due to developmental noise. Even though the original stimulus for one side or the other may have been random, the later growth and tissue differentiation is decidedly nonrandom. The claws have different internal structures and different functions. As we hope to make clear, this may also represent a form of developmental instability, but one that is not due to exogenous or endogenous stress. One might think of such antisymmetry as instability used for adaptation.

Fluctuating antisymmetry may also be produced by endogenous or exogenous stress. The antisymmetry reported by Mather (1953), Leary and Allendorf (1989), and McKenzie and Clarke (1988) may be a manifestation of such antisymmetry, caused by phase transitions leading to a breaking of symmetry in a nonlinear reaction-diffusion system. Before we explain how these processes function, we must discuss several concepts in cybernetics and nonlinear dynamics.

Cybernetics of development revisited

A developing organism might be considered a determinate machine with input (see Ashby, 1956). A determinate machine is one that behaves in the same way as does a closed single-valued transformation. This means that development follows a regular and reproducible course. A machine with input is one in which the transformation varies with the input. In humans, for example, development into a male or female depends on genetic input, the presence or absence of a Y-chromosome. To understand developmental stability, we should consider a population of such machines.

Determinate machines are characterized by variables and parameters. The variables are of two kinds: control and object variables. Control variables are those that are actually regulated. To be regulated, the system must be capable of measuring the variable. Examples of possible control variables of developmental stability are morphogens, chalones, hormones, nerve impulses, etc. Object variables are those variables that are the ultimate object of control. These variables are not easily measured by the system. Examples of object variables include the shape or size of a structure, bilateral symmetry, etc.

Ideally, we would like to measure control variables. They are closer to the regulatory machinery we would like to understand. In most cases, however, we have no idea what the control variable is for a particular structure. Fortunately, object variables are usually easy to measure.

Bilateral asymmetry is the most convenient object variable that we can measure to evaluate developmental stability. Surprisingly, this is not because right and left sides have the same genetic program. Indeed, disregarding somatic mutation, all parts of the organism have the same genetic program. Rather, it is the history of cell lineages on right and left sides that is critical. In general, homologous lineages of cells on right and left sides share similar cytoplasmic determinants and a similar history of cell-cell interactions. They thereby undergo similar patterns of development (see Gilbert, 1991 for examples). In *Caenorhabditis*, however, some nonhomologous cells with dissimilar ancestries also take part in producing a bilaterally symmetrical adult (De Pomerai, 1990). In any case, bilateral symmetry is what mathematicians call an invariant. The symmetry does not change, although the organism itself is passing through a series of changes. This is an especially powerful concept, because it implies that any developmental invariant can serve as a basis for a measure of developmental instability (Graham, Freeman & Emlen, 1993).

The ability of a system, such as a developmental one, to stay within a specific bound is what is meant by stability. Usually, instability is prompted by some external (or internal) disturbance. Under the traditional definition of stability (Cunningham, 1963; Ashby, 1956), a system is stable under displacement if it returns monotonically to its state before the displacement. If it is unstable, it will continue moving away from the original state. Between these two extremes, the system may exhibit neutral equilibrium, damped oscillation, stable oscillation, unstable oscillation, or deterministic chaos.

The traditional definition of developmental stability implicitly assumes a point attractor. In the fluctuating asymmetry literature, that point (r,l) is characterized by $r - l = 0$. Palmer and Strobeck (1992) recognize such an ideal state; they argue that 'in the absence of any ... perturbations, all individuals in a sample should be perfectly symmetrical,' and after perturbation have the 'capacity to ... lower the variance of the difference between right and left sides.' In contrast to such a view, we prefer Zeeman's (1989) less restrictive, and more realistic, definition of stability: the attractor is not a point, but a distribution. In one of Zeeman's models, the trajectory homes in on a point attractor, but never arrives there because of diffusion. He refers to this form of stability as stochastic stability. Although diffusion (random molecular movements) is a likely contributor to developmental 'noise,' we also recognize that diffusion of morphogens, inhibitors, hormones, and other regulatory molecules are normal cellular processes absolutely necessary for normal cell metabolism, cellular differentiation, and pattern formation. Global stability, therefore, may actually require some local instability. West (1987) has taken this argument further, suggesting that chaotic attractors (i.e. strange attractors) may confer both variation and global stability to morphogenesis. We thus accept the notion that zero asymmetry may be impossible for continuous morphometric traits.

Regulation of development

Stability of development implies that some regulation must be occurring. The most common form of regulation in biological systems is feedback, which occurs when information passes in both directions between parts of a system (in this case a developing organism). Right and left sides may be independent of one another (as Palmer & Strobeck, 1992 assumed in their proof on the impossibility of a bimodal distribution resulting from developmental noise). One side may dominate the other; that is, information may flow in only one direction. Or there may be feedback. Feedback is especially important here because it can synchronize morphogen oscillations on right and left sides. It can also produce unstable or chaotic oscillations. As we will show later, a population of organisms exhibiting asynchronous, or chaotic, morphogen oscillations may not have normally distributed asymmetry values.

Which of these models reflects reality in the development of bilateral symmetry? The developmental biology literature generally supports the feedback model, at least for some organisms. For example, Govind and Pearce (1986, 1992), as previously mentioned, discovered that asymmetry of lobster claws involves regulatory feedback through the central nervous system. The claw that receives the most use in its early development becomes the crusher claw; the other claw becomes the cutter. In this situation, however, feedback produces asymmetry rather than symmetry. In addition to this research, Yuge and Yamana (1989) discovered a remarkable interaction between right and left sides of embryos of the frog *Xenopus laevis*. In one experiment, they removed all cells that would eventually develop into the right dorsal side of the embryo, including right dorsal somites and half of the notochord. Surprisingly, the resulting embryos had normal external and internal morphology after this rather severe disturbance. When right dorsal cells were removed, cells migrated from the left side to the right dorsal side to make up for the loss of cells on the right side. The embryos were bilaterally symmetrical. In addition, the notochord was of normal length, but had fewer cells in it. Thus, fewer large cells helped compensate for the initial 50% reduction in prospective notochord cells. Such a response, especially the migration of cells from left to right side, seems impossible without some form of feedback between the two sides. Finally, mice that are mosaics of three zygotes (six parents of three coat colors) still develop into normal bilaterally symmetrical individuals, despite showing a mosaic coat color (Markert & Petters, 1978). Despite these examples, this subject still requires considerable research.

Organisms exhibiting mosaic development, such as the protostomes, exhibit a remarkable degree of cell autonomy, at least up to the 64 cell stage (Gilbert, 1991). One such organism is the nematode *Caenorhabditis*, in which cell interactions begin before the 28-cell stage (Sternberg, 1991). *Caenorhabditis* achieves bilateral symmetry in the adult in a surprising way. While some of its bilateral symmetry derives from homologous cells producing symmetrical lineages on right and left sides, some analogous cells also generate other symmetries (Sulston *et al.*, 1983; Sternberg, 1991). The creation and breaking of symmetry in *Caenorhabditis* entails significant cell interaction (Sternberg, 1991).

As previously mentioned, synchronous oscillations on right and left sides are predicted under the feedback model. Do right and left sides oscillate in phase during development? Or more properly, do morphogens, chalones, and other substances that regulate development oscillate in phase during development? To our knowledge, no one has ever intentionally looked for developmental oscillations involving bilateral parts of an organism. Nevertheless, morphogen oscillations are the rule, rather than the exception. For example, interactions between *Dictyostelium* swarm cells involve oscillations in cyclic AMP concentration (Devroetes, 1989). Plant growth hormone concentrations also oscillate (Wodzicki & Zajaczkowski, 1989). In these two examples, it is the oscillation that carries the information, rather than the concentration of the morphogen.

We should also point out that there is substantial theory predicting the existence of morphogen oscillations. First Turing (1952), and then John Maynard-Smith (1960), proposed that morphogen oscillations might be involved in determining number and location of body segments. Goodwin (1971) later developed the theory further to include two oscillating signals.

The model

Rashevsky-Turing Model

We have used the Rashevsky-Turing Model (Turing, 1952; Smale, 1974) as a simple model of morphogenesis. It was originally conceived as one that models the interaction between two neighboring cells. The model, however, is quite general, and can also be interpreted as either an interaction between two neighboring groups of cells (Turing, 1952) or as an interaction between two distant groups of cells linked via circulatory or nervous system. The model has rather complicated dynamics, giving rise to a chaotic attractor under certain conditions (Parisi *et al.*, 1987). Here, we interpret the model in terms of morphogen concentrations on right and left sides of a developing organism.

The model includes two variables: a, the concentration of an activator, and b, the concentration of its inhibitor. The instantaneous rate of increase of a on a particular side is a function of the concentration of a and b on that side. The instantaneous rate of increase of b on a particular side is a function of the concentration of a and b on that side, plus the concentration of b on the other side as well. The inhibitor can diffuse from side to side, while the activator does not. The inhibitor must diffuse more rapidly than the activator for long-range inhibition to occur (Meinhardt, 1982). This is, however, not a strict requirement for the production of Turing patterns. Recent work has shown that equal diffusion coefficients for activator and inhibitor can still produce Turing patterns (Vastano *et al.*, 1987; Pearson & Horsthemke, 1989). The instantaneous rate of change of a and b on right and left sides is given by the Rashevsky-Turing equations

$$da_r/dt = (k_1 - k_3)a_r - k_2 b_r(a_r/(a_r + K)) + k_5$$

$$db_r/dt = k_3 a_r - k_4 b_r + d(b_l - b_r)$$

$$da_l/dt = (k_1 - k_3)a_l - k_2 b_l(a_l/(a_l + K)) + k_5$$

$$db_l/dt = k_3 a_l - k_4 b_l + d(b_r - b_l),$$

where k_1 is a self-activation constant, k_2 is an inhibition constant, k_3 is a constant expressing the activation of b by a, k_4 is a constant expressing the decomposition of b, k_5 is a control parameter, K is the phenomenological Michaelis-Menten constant, and d is the diffusion (or feedback) constant.

To simulate the changes in the concentrations of a and b over time, we used a standard Euler numerical integration, which converts these ordinary differential equations into finite difference equations. We used a step size of $t = 0.01$ and 2000 iterations for all simulations. This gives results qualitatively similar to those of Parisi *et al.* (1987).

We simulated the three models of right-left interaction by simply varying the parameters in the equations. For the Independence Model without regulation, $k_1 = k_2 = k_3 = k_5 = d = 0$. For the Independence Model with regulation, only $d = 0$; the other constants are greater than zero. For the Feedback Model, $d > 0$. These models were simulated with and without external noise (ranging from $i = \pm 0.0001$ to $i = \pm 0.1$ from a uniform distribution). Noise was added to the concentration of a and b during each iteration. We also set negative values of a and b to zero, if they arose during the simulation. Negative values of a or b rarely arose when noise was low. When noise was very high, they occurred frequently.

Incorporating stress into the Model

Stress is energy dissipative (Alekseeva *et al.*, 1992; Ozernyuk *et al.*, 1992). Thus, modeling stress in a biological system is analogous to increasing friction in a physical system. Winberg proposed that development 'follows the most economical route, with the least expenditure of energy' (as cited by Alekseeva *et al.*, 1992). Thus, under optimal conditions, total specific dissipation of energy is at a minimum. Under stress, dissipation of energy increases; organisms expend more energy to complete a given stage of development. The increase in energy dissipation results from either a loss of gene regulation (see Savageau, 1989) or from changes in the kinetics of enzymatic reactions. Physiological processes require fine tuning, so as to regulate flux through the system. Under stress, we expect that the fine tuning of the system will be disturbed (see Freeman, Graham & Emlen, this volume). Feedback should be disturbed away from optimal levels, because of changes in membrane potentials (d can either increase or decrease); the Michaelis-Menten Constant K should increase (resulting in a decline in the efficiency of the inhibitor). Although the other rate constants (k_1, k_2, k_3, and k_5) may also be reduced by stress, we use constant values suggested by Parisi *et al.* (1987): $k_1 = 10.8$, $k_2 = k_3 = 6$, $k_4 = 3$.

$k_5 = 0.4$ to 2. All simulations consisted of 2000 iterations and a step size of 0.01.

Noise-free morphogenesis

In the absence of noise, and when $a_r = a_l$ and $b_r = b_l$, the concentration of a oscillates, but is always symmetric on right and left sides, as long as the parameters in the model are greater than zero (Fig. 1). If we start with $a_r \neq a_l$ and $b_r \neq b_l$, then a variety of complex dynamics are possible, from phase-locked oscillations to chaos (Fig. 2). The behavior of the

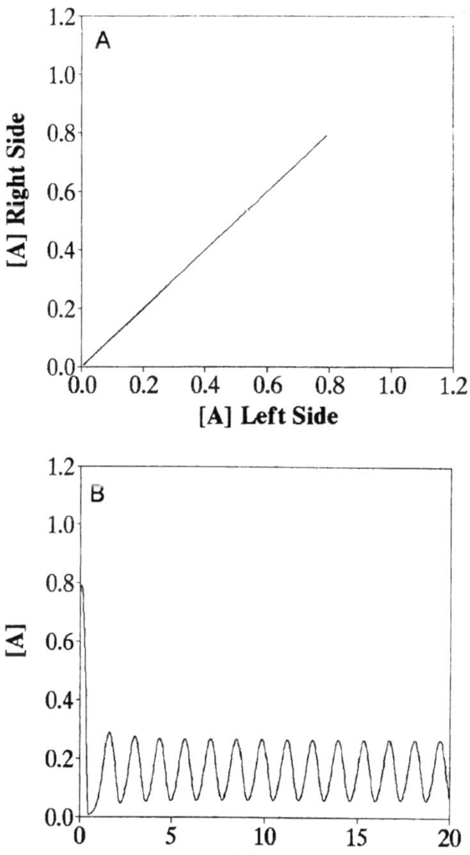

Fig. 1. Behavior of the Rashevsky-Turing equations in the absence of noise, and with $a_r = a_l$ and $b_r = b_l$. This represents a single simulation. A. Phase trajectory of a_r and a_l, with initial conditions $k_1 = 10.8$, $k_2 = k_3 = 6$, $k_4 = 3$, $k_5 = 0.8$, $K = 0.03$, $d = 12$, $a_r = a_l = 0.75$, $b_r = b_l = 0.6$. B. Time series of a_r and a_l. The two curves are superimposed on one another.

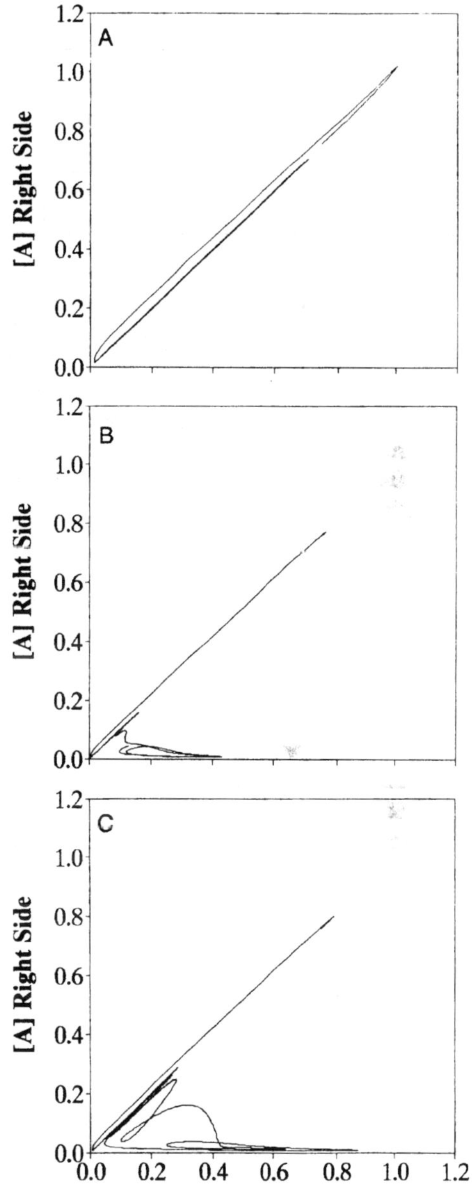

Fig. 2. Behavior of the Rashevsky-Turing equations in the absence of noise, and with $a_r \neq a_l$ and $b_r \neq b_l$. Initial conditions are the same as in Figure 1 except that $a_r = 0.75001$, $a_l = 0.75000$, $b_r = 0.60001$, $b_l = 0.60000$, and A. $d = 1$ (phase-locked periodicity), B. $d = 6$ (phase-lagged periodicity), and C. $d = 12$ (chaos).

Rashevsky-Turing equations under these conditions has already been explored by Parisi *et al.* (1987). We now proceed to models that incorporate noise.

Noisy morphogenesis

Noise is ubiquitous in biological systems, and so the following simulations include noise added to the concentrations of *a* and *b* during each iteration of the model.

Independence Model. Under the Independence Model, and in the absence of any regulation of *a* and *b*, concentrations of *a* on right and left sides undergo a random walk (Fig. 3). Unsurprisingly, $a_r - a_l$ exhibits a mean of zero and a broad (SD = 0.3709) normal distribution (Kolmogorov-Smirnov goodness of fit: D = 0.0434, n = 100, $P > 0.200$; Fig. 4a). With this model, the asymmetry variance increases with time (see Whitney, 1990 for an analytical proof of this conclusion) and with the intensity of the perturbation. Although this is a highly unrealistic model, Figure 4a depicts classic fluctuating asymmetry. Moreover, the variance in $a_r - a_l$ represents pure developmental noise.

A more realistic model involves right-left independence, but with symmetrical regulation of *a* and *b*. This model might represent mosaic development in some protostomes. With only low noise ($i = \pm 0.001$ or less), concentrations of *a* on the two sides stay in phase (if they start in phase) (Figs. 5a and 5b). With increasing noise, however, concentrations of *a* on right and left sides slip out of phase, and asymmetry of *a* increases (Figs. 5c and 5d). Unlike the previous model, this one does not exhibit a random walk of $a_r - a_l$. The reaction system keeps *a* within prescribed bounds. Nevertheless, the asymmetry values have a mean of zero and a

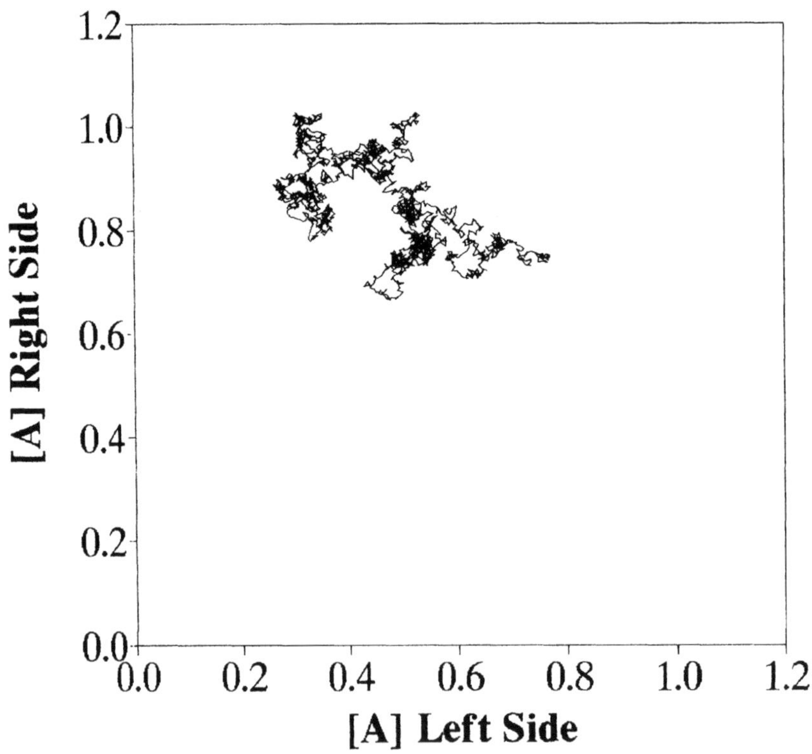

Fig. 3. Phase trajectory of a_r and a_l under conditions of no diffusion ($d = 0$) and no regulation ($k_1 = k_2 = k_3 = k_5 = 0$). All other parameters are the same as in Figure 1, except that noise ($i = \pm 0.001$) is added to *a* and *b* during each step of the simulation.

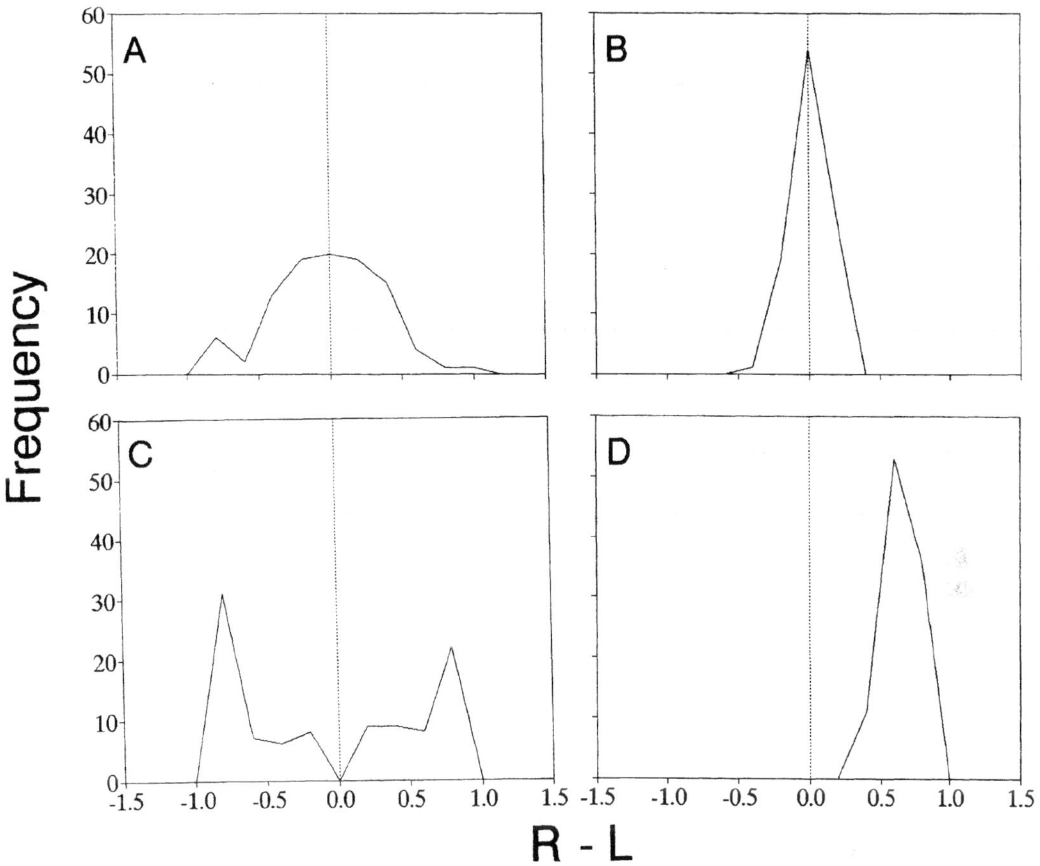

Fig. 4. Asymmetry distributions of $a_r - a_l$. Each graph is based upon 100 simulations. A. No diffusion (d = 0) and no regulation ($k_1 = k_2 = k_3 = k_5 = 0$; $k_4 = 3$; $K = 0.03$; $i = \pm 0.01$), B. No diffusion (d = 0) with regulation ($k_1 = 10.8$; $k_2 = k_3 = 6$; $k_4 = 3$; $k_5 = 0.8$; $K = 0.03$; $i = \pm 0.01$), C. Chaos ($d = 12$; $k_1 = 10.8$; $k_2 = k_3 = 6$; $k_4 = 3$; $k_5 = 0.8$; $K = 0.06$; $i = \pm 0.001$), D. Chaos, with biased control parameter ($k_1 = 10.8$; $k_2 = k_3 = 6$; $k_4 = 3$; $k_{5\,right} = 0.8$; $k_{5\,left} = 0.7$; $K = 0.06$; $i = \pm 0.001$).

narrower (SD = 0.1308) normal distribution (D = 0.040, n = 100, $P > 0.200$; Fig. 4b). This also represents pure fluctuating asymmetry.

Feedback Model. Adding moderate feedback (d = 1) to the model stabilizes the oscillations of a, even with considerable noise (Fig. 6). Asymmetry of a decreases (SD = 0.0804, n = 25 with $i = \pm 0.01$). Increasing feedback beyond d = 1 and increasing the Michaelis-Menten Constant sparks a transition from phase locking to phase lagging and then to chaos in the concentration of a. The situations exhibiting chaos revealed two basins of attraction: right and left dominance (Fig. 7). The locations of these two basins varied, from contiguous to distant depending upon the value of K.

For d = 12 and K = 0.03, the trajectories returned frequently to the vicinity of $a_r = a_l$. With low noise ($i = \pm 0.001$ or less), the trajectory usually stayed in the first basin that it had entered. With increasing noise, the probability of jumping basins of attraction increased (Figs. 8a and 8b).

For d = 12 and K = 0.06, the trajectories did not return to the vicinity of $a_r = a_l$. Consequently, more noise was required to cause the trajectory to jump from one basin to another (Figs. 8c and 8d).

As expected, asymmetry of a also increased with the transition to chaos. For phase-locked and some phase-lagged situations, $a_r - a_l$ maintained a mean of zero and a normal distribution. When a was chaotic, the distribution of a was either normal or bimodal, depending upon the locations of the two basins of attraction. If the basins were contiguous, or closer than the greatest perturbation, and noise was high ($i = \pm 0.01$), then the distribution tended to be normal (SD = 0.4687; D = 0.0576, n = 100, $P > 0.20$). If the basins were separated by a distance greater than the greatest perturbation (say $i = \pm 0.001$), then the distribution tended to be bimodal (D = 0.1833, n = 100, $P < 0.0001$; Fig. 4c). This suggests a mechanism whereby an organism can canalize antisymmetry.

Thus, the simulation revealed a phase transition leading to symmetry breaking in a, which influenced the shape of the asymmetry distributions of a. An important conclusion is that fluctuating asymmetry (i.e. mean of zero and a normal distribution) may represent a combination of random developmental noise and deterministic chaos. Likewise, a bimodal distribution can also represent a combination of developmental noise and deterministic chaos.

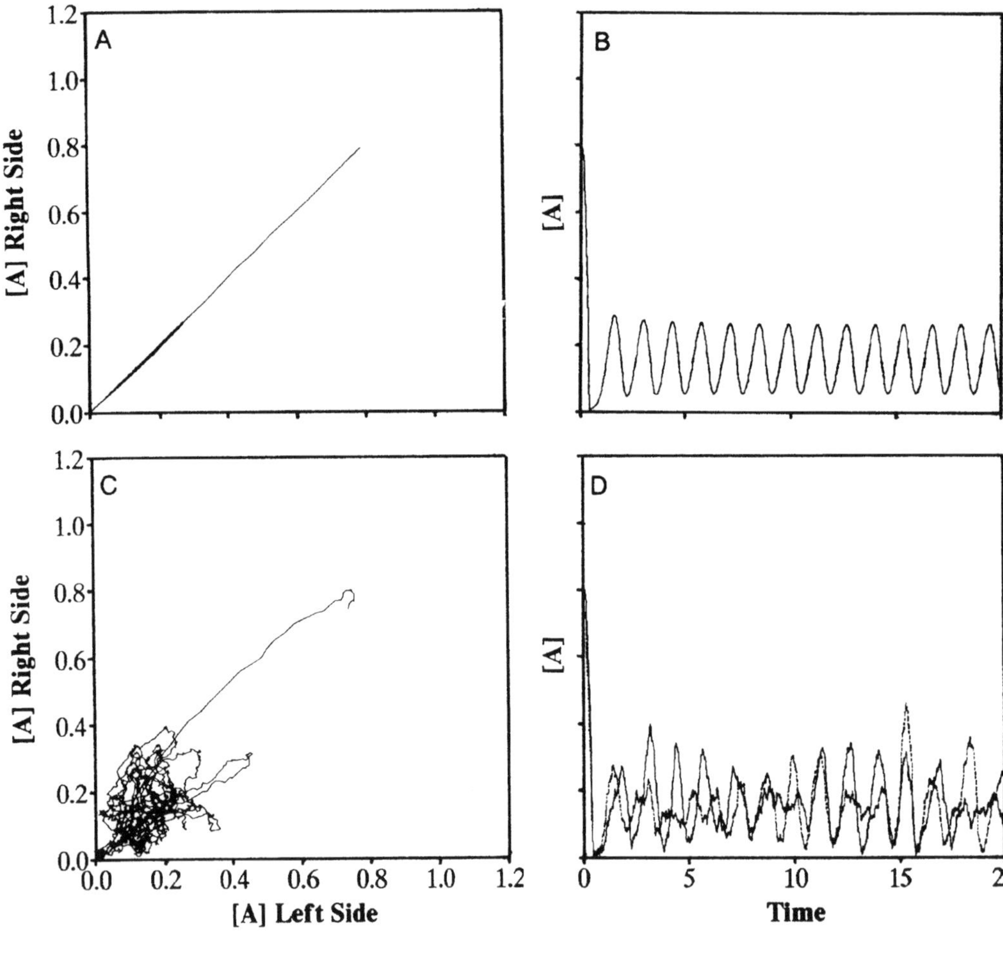

Fig. 5. Phase trajectories under conditions of no diffusion ($d = 0$). All other parameters are the same as in Figure 1, except that noise (i) is added to a and b during each step of the simulation. A. Phase trajectory of a_r and a_l with very low noise ($i = \pm 0.0001$). B. Time series of a_r and a_l with very low noise ($i = \pm 0.0001$). The two curves are superimposed. C. Phase trajectory of a_r and a_l with moderate noise ($i = \pm 0.01$). D. Time series of a_r (solid line) and a_l (dashed line) with moderate noise ($i = \pm 0.01$).

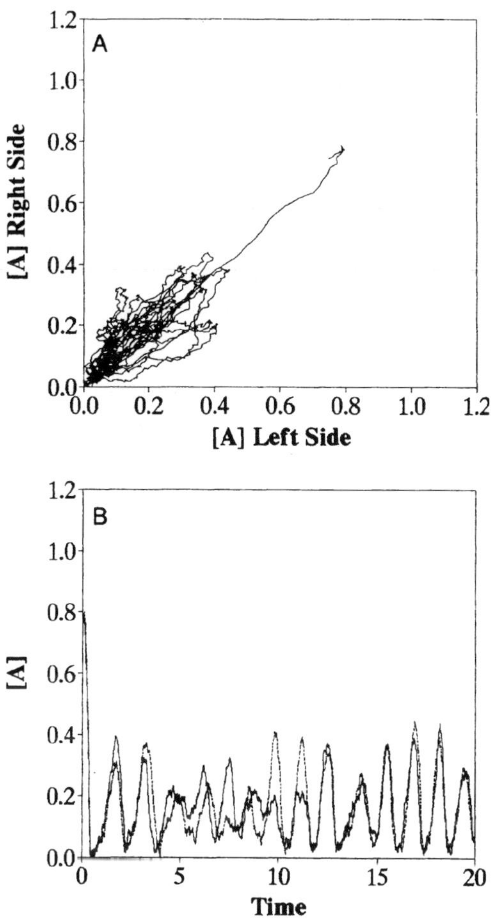

Fig. 6. Behavior of the Rashevsky-Turing equations in the presence of optimal feedback ($d = 1$) and moderate noise ($i = \pm 0.01$). Initial conditions are the same as in Figure 1 except that $d = 1$. A. Phase trajectory of a, B. Time series of a_r (solid line) and a_l (dashed line).

Directional asymmetry. Thus far, the Rashevsky-Turing Model easily accounts for both fluctuating asymmetry and antisymmetry, but what about directional asymmetry? In fact, one can also produce directional asymmetry if there is even a slight bias in the control parameter k_5 on one side of the developing organism. For example, if $k_5 = 0.8$ on the right side and 0.79 on the left side, then most individuals will develop right dominance; a minority will show left dominance. If there is a greater difference in the control parameter, $k_5 = 0.8$ on the right side and 0.7 on the left side, then all individuals will develop right dominance; none will show left dominance (Fig. 4d). Either external or internal asymmetries could conceivably generate this kind of bias.

Dynamic morphogenesis

Our interpretation of the above models is based upon the assumption that chaos is a component of dynamic morphogenesis. Parisi *et al.* (1987) have questioned whether chaos is a 'pathological or even a necessary element' in the development of organisms. Nevertheless, the ubiquity of chaos in physiological processes, from heartbeat to nerve impulses, has been well established (see West, 1990; Glass, 1991). Chaos has also been demonstrated in enzymatic reactions (Olsen & Degn, 1977). The existence of chaos in development has been postulated by numerous theoreticians under the guise of Catastrophe Theory, a branch of mathematics inspired by Waddington (see Thom, 1972; Saunders, 1980). Moreover, a nonlinear dynamic model of development has been recently devised to model cell proliferation and differentiation between cell lineages in *Caenorhabditis* (Bailly *et al.*, 1991).

Several observations, including the existence of antisymmetry and directional asymmetry, support the notion that chaos is indeed a component of development. First, one would predict that handedness in humans and male lobsters should exhibit sensitivity to initial conditions, the hallmark of chaos. Govind and Pearce (1986) found that they could influence the direction of claw development in male lobsters by stimulating one claw with a small brush. Furthermore, Van Valen's (1978) suggestion that handedness in humans has a low threshold for reversal suggests the two basins of attraction we have described.

The ubiquity of developmental thresholds is best explained by the chaotic dynamics of nonlinear systems. In its traditional sense, a threshold occurs when a small increase in the concentration of one substance causes a large increase in the concentration of another substance (Lewis, Slack & Wolpert, 1977). Such thresholds (or transitions, or bifurcations) commonly occur in nonlinear systems that have multiple steady states (Saunders & Kubal,

1989).

Moreover, the steady increase in the information content of an organism during the course of its development from zygote to adult (first postulated by Aristotle, as cited by Ransom, 1981) virtually requires chaotic dynamics. A chaotic attractor provides an efficient means of generating new information (Shaw, 1985).

When developmental stability breaks down, do right and left sides oscillate out of phase or do they exhibit chaos? Wolf, Silk, and Plant (1986) studied developmental growth trajectories in normal and malformed leaves of grape (*Vitis*). The plants having malformed leaves were infected with a virus. The normal leaves showed symmetric growth isopleths on right and left sides of the leaf. The malformed leaves, however, showed random oscillations in their growth isopleths, which changed directions from one day to the next and from one part of the leaf to the next. The malformed leaves became more asymmetric over time.

Finally, *Dictyostelium* occasionally shows chaotic behavior during aggregation of its amoeboid swarm cells. Although wild-type *D. discoideum* exhibits periodic oscillations in cAMP, mutant Fr17 shows chaotic waves of cyclic AMP (Goldbeter &

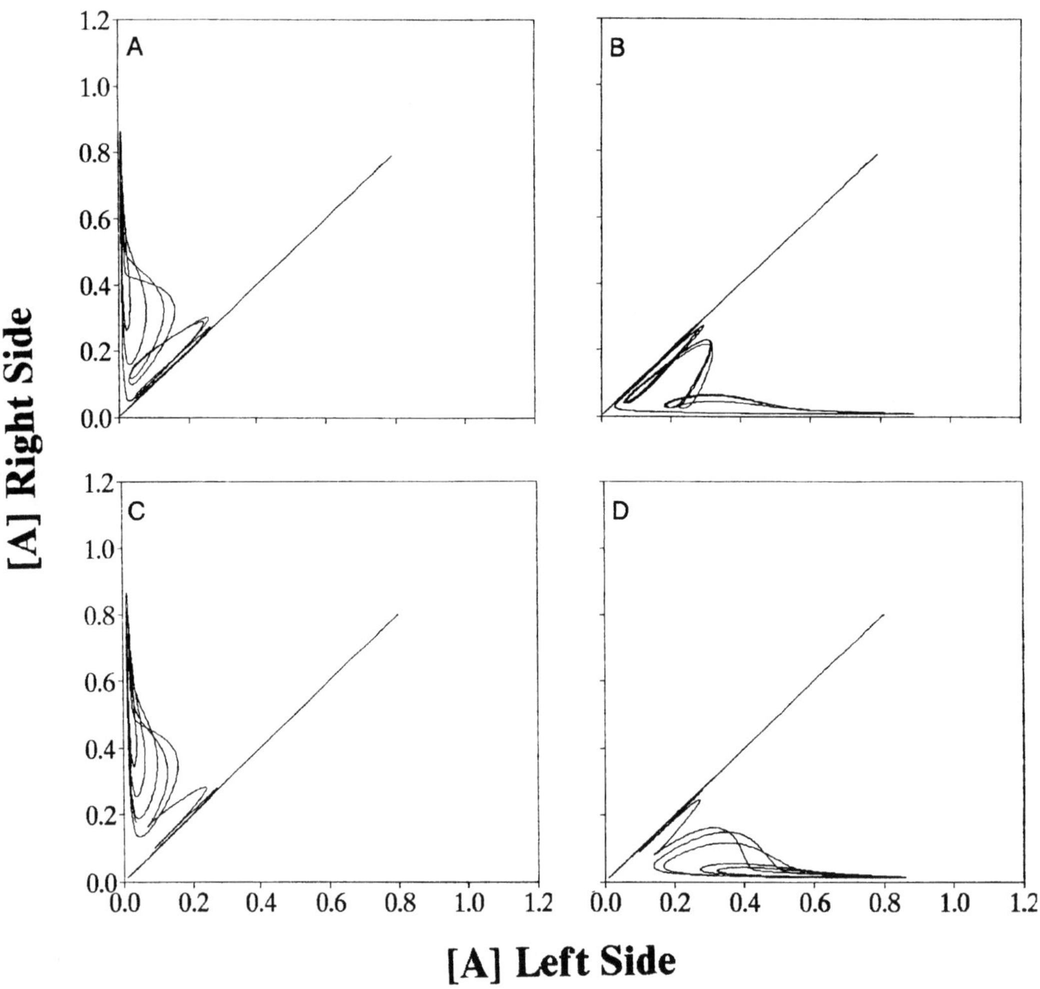

Fig. 7. Behavior of the Rashevsky-Turing equations under conditions of varying feedback and inhibition, and very low noise ($i = \pm 0.0001$). Unless otherwise stated, conditions are as in Figure 1. A. Chaos, right basin, contiguous ($d = 12$, $K = 0.03$); B. Chaos, left basin, contiguous ($d = 12$, $K = 0.03$), C. Chaos, right basin, distant ($d = 12$, $K = 0.06$); D. Chaos, left basin, distant ($d = 12$, $K = 0.06$).

Martiel, 1987). Chaos here is generated by increased feedback in the positive feedback loop that generates cAMP.

But if morphogenesis is indeed actually or potentially chaotic, then how can we account for its ability to generate the same structures over and over again in different individuals? Chaos is defined by sensitivity to initial conditions. If development were globally chaotic, repeatability, as in identical twins, would be impossible.

We postulate that development represents a sort of constrained chaos. The developmental system, which is inherently nonlinear (i.e. potentially chaotic), interacts with the genetic system, which is inherently linear (i.e. predictable in the sense that gene A always codes for polypeptide A). The interaction between the two systems is also nonlinear. Except for transposition and mutation, no new information is generated in the genetic system. In the developmental system, however, gene regulation, flux through enzymatic pathways, cell-cell interactions, and cell movements (all nonlinear processes) generate abundant new information. Linkage of the two systems means that information content can increase, but in the same way each time. An analogy to this view of constrained chaos is a computer

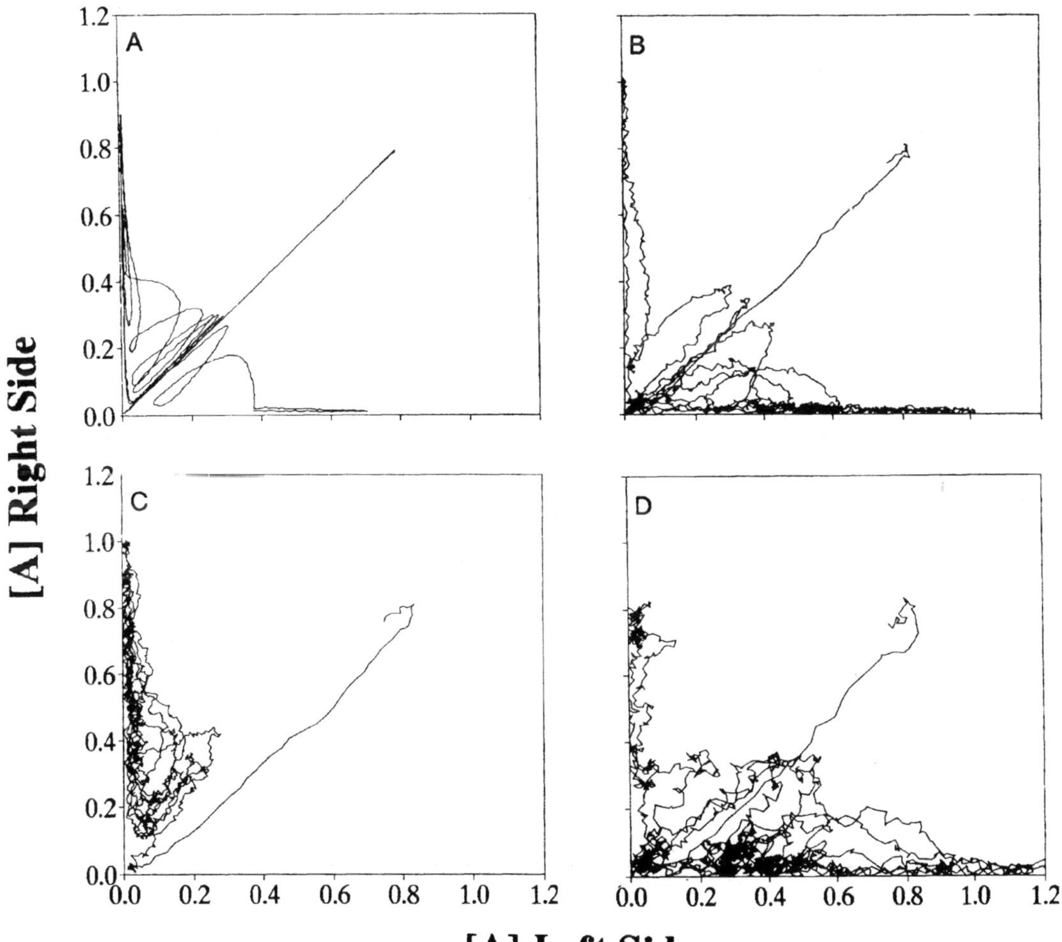

Fig. 8. Effect of increasing noise on basin jumping. A. Basin jumping from right to left with chaos, low noise ($i = \pm 0.001$), and contiguous basins ($d = 12$, $K = 0.03$); B. Basin jumping with chaos, moderate noise ($i = \pm 0.01$), and contiguous basins ($d = 12$, $K = 0.03$); C. Lack of basin jumping with chaos, moderate noise ($i = \pm 0.01$), and distant basins ($d = 12$, $K = 0.06$); D. Basin jumping with chaos, high noise ($i = \pm 0.1$), distant basins ($d = 12$, $K = 0.06$).

generated Mandelbrot Set, possibly the most complex object known. Given the same algorithm, the computer will generate the same figure each and every time.

Directional asymmetry

In our models, we have suggested a possible relationship between directional asymmetry and antisymmetry. Directional asymmetry may be a form of antisymmetry in which there is a bias in the conditions of development. There is much evidence that directional asymmetry is caused by such a bias. Govind and Pearce (1986), for example, showed that it is possible to generate directional asymmetry in a population of male lobsters by stimulating only the right claw. Moreover, there is evidence of directional asymmetry that is produced by environmental influences. *Convolvulus*, for example, always has a right-handed spiral, which coils around supporting sticks and stems. This forces an environmental asymmetry, which then produces directional asymmetry in the leaves. In animals, the asymmetrical environment is the organism itself. In the water snake, *Nerodia*, half scales are more common on the left side than on the right, because of the direction of the coiling of the embryo (Osgood, 1978).

If directional asymmetry is related to antisymmetry, then it should also be possible to generate directional asymmetry by stressing an organism. Graham, Roe, and West (in press) have shown that *Drosophila melanogaster* exposed to 10,000 ppm benzene exhibit significant directional asymmetry for sternopleural bristles. Control flies, in contrast, showed typical fluctuating asymmetry for this trait, and there was no evidence of selective mortality in the flies exposed to benzene.

One would also predict that transition from directional asymmetry to antisymmetry should be possible if the bias is reduced. Normal mice, for example, exhibit directional asymmetry of the visceral organs (i.e. normal *situs*). Hummel and Chapman (1959) and Layton (1976) have described a homozygous strain of mice that displays *situs inversus* of the viscera in 50% of the mice. Morgan and Corballis (1978) have proposed that the absence of the normal gene controlling the direction of *situs* allows 'asymmetry to be determined in a random fashion.'

Conclusion

We are now in the midst of a revolution in the natural sciences. The old viewpoint of Norbert Wiener, Claude Shannon, and Walter Cannon, that of cybernetics, information theory, and linear homeostasis, is being replaced by that of Edward Lorentz, David Ruelle, Ary Goldberger, and Bruce West, that of nonlinear dynamics, strange attractors, and chaotic homeostasis. The new paradigm suggests that random fluctuations may not be transient responses to disturbance. In fact, much minor variation may be a consequence of normal development, which is buffered by parallel nonlinear pathways. Thus, small normally distributed deviations from bilateral symmetry may not reflect disturbance at all! If we ignore non-normal distributions, we may possibly be missing an important response to disturbance.

Part of the recent interest in developmental stability stems from a desire to find a simple correlation with fitness. We hope that we have demonstrated that developmental stability is important, regardless of what it tells us about selection and fitness. The results of our model are not at all intuitive: both antisymmetry and directional asymmetry may be possible responses to either genetic change or stress.

The predictions of our dynamic morphogenetic model are in closer agreement with the data on handedness than are those of the genetic model. The genetic model proposes that handedness is under simple genetic control. With this model, there are genes for directional asymmetry, for antisymmetry, and for bilateral symmetry. Palmer and Strobeck (1992) interpreted McKenzie and Clarke's (1988) Australian blowfly experiment in such terms. Rarely, however, do genes control the direction of handedness (Morgan & Corballis, 1978). When they do influence handedness, they usually influence only the propensity toward asymmetry. According to the dynamic morphogenesis model, random determination of the direction of handedness is another example of sensitive dependence on initial conditions. This model accommodates either genetic or environmental changes leading to a breaking of right-left symmetry. The bias toward right handedness in humans is easily explained by functional asymmetries in the nervous system. This hypothesis does not require ad hoc hypotheses about the existence of specific genes for antisymmetry. It accounts for all of the observed patterns of asymmetry in the literature, and it subsumes Van Valen's reversal hypothesis.

We reject the notion that fluctuating asymmetry reflects only developmental noise. Obviously, some of the variation is due to developmental noise, but a major component of variation is almost certainly due to nonlinear dynamics of developmental processes. In fact, the dynamic processes contributing to the normal variation between right and left sides of a developing organism are most likely the very processes contributing to their stability.

If fluctuating asymmetry represents more than developmental noise, how can one distinguish noise from deterministic chaos? At this moment, we see no simple way of partitioning these components of the total variance, especially for measures of asymmetry. It is currently possible, however, to distinguish noise from chaos in time-series data (Grassberger & Procaccia, 1983; Sugihara & May, 1990). Graham, Freeman and Emlen (1993) discuss the application of time-series analysis to studies of developmental instability.

Palmer and Strobeck's (1986, 1992) prior rejection of antisymmetry and directional asymmetry hinges on their overly restrictive definition of fluctuating asymmetry. They argue that fluctuating asymmetry represents variance that is only of environmental origin. This view has its origin in Waddington's concept of developmental noise. In electronics, noise usually means the Brownian movement of molecules and atoms (Ashby, 1956). But as Ashby (1956) has pointed out, noise in living organisms originates from 'some other macroscopic system from which the system under study cannot be completely isolated.' We argue that only a part of the asymmetry variance is attributable to such noise. We propose that much variation has its origin in the nonlinear dynamics of developing organisms, even in a normal distribution of $r_i - l_i$.

Palmer and Strobeck's (1986, 1992) view also requires unsubstantiated assumptions about the genetic basis for antisymmetry and directional asymmetry. While genes for both forms of asymmetry undoubtedly exist, we have shown that they are not necessary for directional asymmetry or antisymmetry. Palmer and Strobeck's (1992) conclusions follow chiefly from these assumptions. They also assume linearity and no interaction between sides of a developing organism. Both assumptions are highly unrealistic.

We also suspect that there are few evolved feedback loops that ensure bilateral symmetry (see also Freeman, Graham & Emlen, this volume). Phase locking requires only a very weak linkage. Simply connecting parts of an organism together via circulatory and nervous systems is enough for the systems to phase lock if they are sufficiently similar (Bergé, Pomeau & Vidal, 1984; Emlen, Freeman & Graham, this volume). Alternatively, one could postulate the existence of a fluctuating asymmetry demon, analogous to Maxwell's Demon. This demon measures right and left sides to see if they are equal. Of course, there is no evidence for such a demon.

In conclusion, we respond to Palmer and Strobeck's (1992) query: 'whither developmental stability?' We believe that the various symmetries found in living organisms are of vital importance to understanding morphogenesis. And the ways in which these symmetries are broken reveals the evolution of form and structure. As Palmer and Strobeck (1992) suggest, there may be no useful measure of developmental stability, but we believe that fluctuating asymmetry, antisymmetry, and directional asymmetry can all illuminate that most elusive of subjects, the relationship between genotype, environment, and phenotype.

Acknowledgements

We thank Richard Palmer, Robb Leary, and Geoff Clarke for informal discussions that led to our reconsideration of antisymmetry. Richard Palmer also provided many useful comments on the manuscript. Catherine Chamberlin-Graham did much of the literature search, and proofread the final draft. An early version of this paper was presented at the first Biotest Meeting in Moscow, 1990, sponsored by Vladimir Zakharov.

References

Alekseeva, T. A., V. V. Zinichev & A. I. Zotin, 1992. Energy criteria of reliability and stability of development, pp. 159-165 in Developmental Stability in Natural Populations, edited by V. M. Zakharov & J. H. Graham. Volume 191, Acta Zoologica Fennica, Finnish Zoological Publishing Board, Helsinki.

Ashby, W. R., 1956. An Introduction to Cybernetics. John Wiley and Sons, New York.

Bailly, F., F. Gaill & R. Mosseri. 1991. A dynamical system for biological development: the case of *Caenorhabditis elegans*. Acta Biotheoretica 39: 167-184.

Bergé, P., Y. Pomeau & C. Vidal, 1984. Order Within Chaos: Towards a Deterministic Approach to Turbulence. John Wiley and Sons, New York.

Cunningham, W. J., 1963. The concept of stability. American Scientist 51: 425-436.

De Pomerai, D., 1990. From Gene to Animal: An Introduction to the Molecular Biology of Animal Development. 2nd Edition. Cambridge University Press, Cambridge.

Devroetes, P., 1989. *Dictyostelium discoideum*: a model system for cell-cell interactions in development. Science 245: 1054-1058.

Emlen, J. M., D. C. Freeman & J. H. Graham, 1993. Nonlinear growth dynamics and the origin of fluctuating asymmetry, pp. 77-96 in Developmental Instability: Origins and Evolutionary Significance, edited by T. Markow. Kluwer, Dordrecht, The Netherlands.

Freeman, D. C., J. H. Graham & J. M. Emlen, 1993. Developmental stability in plants: symmetries, stress, and epigenetic effects, pp. 97-119 in Developmental Instability: Origins and Evolutionary Significance, edited by T. Markow. Kluwer, Dordrecht, The Netherlands.

Gilbert, S. F., 1991. Developmental Biology, 3rd Edition. Sinauer Associates, Sunderland, Massachusetts.

Glass, L., editor, 1991. Nonlinear Dynamics of Physiological Function and Control. Chaos 1: 1-132.

Goldbeter, A. & J. L. Martiel, 1987. Periodic behaviour and chaos in the mechanism of intercellular communication governing aggregation of *Dictyostelium* amoebae, pp. 79-89 in Chaos in Biological Systems, edited by H. Degn, A. V. Holden & L. F. Olsen. Plenum Press, New York.

Goodwin, B., 1971. A model of early amphibian development, pp. 417-428 in British Society of Experimental Biology Symposium 25, edited by D. D. Davies & M. Balls. Cambridge University Press, Cambridge.

Govind, C. K. & J. Pearce, 1986. Differential reflex activity determines claw and closer muscle asymmetry in developing lobsters. Science 233: 354-356.

Govind, C. K. & J. Pearce, 1992. Mechanoreceptors and minimal reflex activity determining claw laterality in developing lobsters. Journal of Experimental Biology 171: 149-162.

Graham, J. H., D. C. Freeman & J. M. Emlen, 1993. Developmental instability: a sensitive indicator of populations under stress, pp. 136-158 in Environmental Toxicology and Risk Assessment, ASTM STP 1179, edited by G. Landis, J. Hughes & M. A. Lewis. American Society for Testing and Materials, Philadelphia.

Graham, J. H., K. Roe & T. West. Effects of lead and benzene on the developmental stability of *Drosophila melanogaster*. Ecotoxicology (in press).

Grassberger, P. & I. Procaccia, 1983. Measuring the strangeness of strange attractors. Physical Review Letters 50: 346-369.

Hummel, K. P. & D. B. Chapman, 1959. Visceral asymmetry and associated anomalies in the mouse. Journal of Heredity 50: 9-13.

Layton, W. M., Jr., 1976. Random determination of a developmental process. Journal of Heredity 67: 336-338.

Leary, R. F. & F. W. Allendorf, 1989. Fluctuating asymmetry as an indicator of stress: implications for conservation biology. Trends in Ecology and Evolution 4: 214-217.

Lewis, J., J. M. Slack & L. Wolpert, 1977. Thresholds in development. Journal of Theoretical Biology 65: 579-590.

Lewontin, R. C., 1986. Gene, organism, and environment, pp. 273-285 in Evolution from Molecules to Men, edited by D. S. Bendall. Cambridge University Press, New York.

Markert, C. L. & R. M. Petters, 1978. Manufactured hexaparental mice show that adults are derived from three embryonic cells. Science 202: 56-58.

Mather, K., 1953. Genetical control of stability in development. Heredity 7: 297-336.

Maynard-Smith, J., 1960. Continuous, quantized and modal variation. Proceedings of the Royal Society 152: 397-409.

McKenzie, J. A. & G. M. Clarke, 1988. Diazanon resistance, fluctuating asymmetry and fitness in the Australian sheep blowfly, *Lucilia cuprina*. Genetics 120: 213-220.

McKenzie, J. A., P. Batterham & L. Baker, 1990. Fitness and asymmetry modification as an evolutionary process: A study in the Australian sheep blowfly, *Lucilia cuprina* and *Drosophila melanogaster*, pp. 57-73 in Ecological and Evolutionary Genetics of Drosophila, edited by J. S. F. Barker. Plenum Press, New York.

McKenzie, J. A. & K. O'Farrell, 1993. Modification of developmental instability and fitness: malathion-resistance in the Australian sheep blowfly, *Lucilia cuprina*, pp. 67-76 in Developmental Instability: Origins and Evolutionary Significance, edited by T. Markow. Kluwer, Dordrecht, The Netherlands.

Meinhardt, H., 1982. Models of Biological Pattern Formation. Academic Press, New York.

Morgan, M. J. & M. C. Corballis, 1978. On the biological basis of human laterality. II The mechanism of inheritance. Behavioral and Brain Sciences 2: 270-277.

Olsen, L. F. & H. Degn, 1977. Chaos in an enzymatic reaction. Nature 267: 177-178.

Osgood, D. W., 1978. Effects of temperature on the development of meristic characters in *Natrix fasciata*. Copeia 1978: 33-47.

Ozernyuk, N. D., V. I. Dyomin, E. A. Prokofyev & I. M. Androsova, 1992. Energy homeostasis and developmental stability, pp. 167-175 in Developmental Stability in Natural Populations, edited by V. M. Zakharov and J. H. Graham. Volume 191, Acta Zoologica Fennica, Finnish Zoological Publishing Board, Helsinki.

Palmer, A. R. & C. Strobeck, 1986. Fluctuating asymmetry measurement, analysis, patterns. Annual Review of Ecology and Systematics 17: 391-421.

Palmer, A. R. & C. Strobeck, 1992. Fluctuating asymmetry as a measure of developmental stability: implications of non normal distributions and power of statistical tests, pp. 57-72 in Developmental Stability in Natural Populations, edited by V. M. Zakharov and J. H. Graham. Volume 191, Acta Zoologica Fennica, Finnish Zoological Publishing Board, Helsinki.

Parisi, J., B. Röhricht, J. Peinke & O. E. Rössler, 1987. Turbulent morphogenesis of a prototype model reaction-diffusion system, pp. 91-95 in Chaos in Biological Systems, edited by H. Degn, A. V. Holden & L. F. Olsen. Plenum Press, New York.

Pearson, J. E. & W. Horsthemke, 1989. Turing instabilities with nearly equal diffusion coefficients. Journal of Chemistry and Physics 90: 1588-1599.

Ransom, R., 1981. Computers and Embryos: Models in Developmental Biology. John Wiley and Sons, New York.

Reeve, E. C. R., 1960. Some genetic tests on asymmetry of sternopleural chaeta number in *Drosophila*. Genetical Research 1: 151-172.

Saunders, P. T., 1980. An Introduction to Catastrophe Theory. Cambridge University Press, Cambridge.

Saunders, P. T. & C. Kubal, 1989. Bifurcations and the epigenetic landscape, pp. 16-30 in Theoretical Biology: Epigenetic and Evolutionary Order from Complex Systems, edited by B. Goodwin & P. Saunders. Edinburgh University Press, Edinburgh.

Savageau, M. A., 1989. Are there rules governing patterns of gene regulation?, pp. 42-66 in Theoretical Biology: Epigenetic and Evolutionary Order from Complex Systems, edited by B. Goodwin & P. Saunders. Edinburgh University Press, Edinburgh.

Shaw, R., 1985. The Dripping Faucet as a Model Chaotic System. Aerial Press, Santa Cruz.

Smale, S., 1974. A mathematical model of two cells via Turing's equation. Lectures on Mathematics in the Life Sciences 6: 15-26.

Soulé, M. & J. Cuzin-Roudy, 1982. Allomeric variation. 2. Developmental instability of extreme phenotypes. American Naturalist 120: 765-768.

Sternberg, P. W., 1991. Control of cell lineage and cell fate during nematode development, pp. 177-225 in Current Topics in Developmental Biology, Volume 25, edited by H. R. Bode. Academic Press, New York.

Sugihara, G. & R. M. May, 1990. Nonlinear forecasting as a way of distinguishing chaos from measurement error in time series. Nature 344: 734-741.

Sulston, J. E., E. Schierenberg, J. G. White & J. N. Thomson, 1983. The embryonic cell lineage of the nematode *Caenorhabditis elegans*. Developmental Biology 100: 64-119.

Thom, R., 1972. Structural Stability and Morphogenesis. Benjamin, New York.

Thom, R., 1989. An inventory of Waddingtonian concepts, pp. 1-7 in Theoretical Biology: Epigenetic and Evolutionary Order from Complex Systems, edited by B. Goodwin & P. Saunders. Edinburgh University Press, Edinburgh.

Timofeéf-Ressovsky, N. W., 1934. Über den Einfluss des gentypischen Milieus und der Aussenbedingungen auf die Realisation des Genotyps. Nachrichten Göttingen Gesell. Math. Physik. Kl (6) 1: 53-104.

Turing, A., 1952. The chemical basis of morphogenesis. Philosophical Transactions of the Royal Society B 237: 37-72.

Van Valen, L., 1962. A study of fluctuating asymmetry. Evolution 16: 125-142.

Van Valen, L., 1978. The control of handedness. Behavioral and Brain Sciences 2: 320.

Vastano, J. A., J. E. Pearson, W. Horsthemke & H. L. Swinney, 1987. Chemical pattern formation with equal diffusion coefficients. Physics Letters A. 124(6,7): 320-324.

Waddington, C., 1940. Organizers and Genes. Cambridge University Press, Cambridge.

Waddington, C., 1957. The Strategy of the Genes. Allen and Unwin, London.

West, B. J., 1987. Fractals, intermittancy and morphogenesis, pp. 305-314 in Chaos in Biological Systems, edited by H. Degn, A. V. Holden & L. F. Olsen. Plenum Press, New York.

West, B. J., 1990. Fractal Physiology and Chaos in Medicine. World Scientific, Singapore.

Whitney, C. A., 1990. Random Processes in Physical Systems: An Introduction to Probability-based Computer Simulations. John Wiley and Sons, New York.

Wodzicki, T. J. & S. Zajaczkowski, 1989. Auxin waves in cambium and morphogenetic information in plants, pp. 45-64 in Signals in Plant Development, edited by J. Krekule and F. Seidová. Academic Publishing, The Hague.

Wolf, S. D., W. K. Silk & R. E. Plant, 1986. Quantitative patterns of leaf expansion: comparison of normal and malformed growth in *Vitis vinifera* cv Ruby Red. American Journal of Botany 73: 832-846.

Yuge, M. & K. Yamana, 1989. Regulation of the dorsal axial structures in cell-deficient embryos of *Xenopus laevis*. Development, Growth, and Differentiation 31: 315-324.

Zakharov, V. M., 1987. Animal Asymmetry: A Population-Phenogenetic Approach. Nauka, Moscow. (in Russian)

Zakharov, V. M., 1989. Future prospects for population phenogenetics. Soviet Science Reviews F Physiology and General Biology 4: 1-79.

Zakharov, V. M. & J. H. Graham, editors, 1992. Developmental Stability in Natural Populations. Volume 191, Acta Zoologica Fennica, Finnish Zoological Publishing Board, Helsinki.

Zeeman, C., 1989. A new concept of stability, pp. 8-15 in Theoretical Biology: Epigenetic and Evolutionary Order from Complex Systems, edited by B. Goodwin & P. Saunders. Edinburgh University Press, Edinburgh.

T. A. Markow (ed.), *Developmental Instability: Its Origins and Evolutionary Implications*, 141–155, 1994.

Morphological integration of fluctuating asymmetry in the mouse mandible

Larry Leamy
Department of Biology, University of North Carolina at Charlotte, Charlotte, NC 28223, USA

Received and accepted 28 April 1993

Key words: fluctuating asymmetry, morphological integration, mouse mandibles

Abstract

Morphological integration of fluctuating asymmetry (FA) was assessed for nine mandibular characters in random-bred house mice to test the hypothesis that there should be a significant integration or concordance of FAs at the population level, but not at the individual level. FA estimates for each of the nine characters were made for each of 16 subpopulations (two replicates each for mice varying in sex and age) and their correlations indicated a moderate level of integration (index of integration = 0.42). A matrix permutation test of the differences of the correlations within two ('incisor' and 'muscle') developmentally different character groups versus correlations between groups was significant, indicating the presence of morphological integration. Kendall's coefficient of concordance also indicated a significant population asymmetry parameter for FAs within the two character groups. Correlations among individual FA estimates were generally lower than those calculated from subpopulations, the index of integration being 0.21. The matrix permutation test failed to show significant morphological integration among individual FAs, but Kendall's coefficient of concordance was significant, indicating the presence of an 'individual asymmetry parameter' among the nine characters. Principal components analysis and canonical correlation analysis confirmed the overall higher level of integration of FAs among the subpopulations and within the two character groupings.

Introduction

When we measure left minus right side differences (fluctuating asymmetry) in bilateral characters such as various bone dimensions to assess developmental stability, what associations, if any, should we expect among these differences? If fluctuating asymmetry (FA) is a useful measure of developmental stability, then perturbations in the developmental environment affecting specific bilateral characters in a similar way should theoretically be reflected in a significant association, or integration, of FAs for these characters. On the other hand, expression of FA may be character-specific such that characters do not react in the same way to genetic or environmental changes in development (Leary, Allendorf & Knudsen, 1992).

The evidence thus far suggests that there often are significant associations of FA among different characters, but these have been detected primarily at the population, rather than individual, level (Soule & Baker, 1968). Concordance of FA levels among characters across populations implies the presence of a population asymmetry parameter or PAP, and a significant PAP has been demonstrated in many studies (Soule & Baker, 1968; Leamy, 1984, 1992). Attempts to discover a concordance of FAs at the individual level, however, generally have failed (Mason, Ehrlich & Emmel, 1967; review by Palmer & Strobeck, 1986). Yet significant correlations among individual FAs, calculated especially from closely-related characters, occasionally have been found (see Leamy, 1984; Atchley *et al.*, 1984), so perhaps we have not looked hard enough at the right characters.

One way to optimize the search for an 'individual asymmetry parameter', if indeed such a parameter exists, is to estimate FA among characters morphogenetically or developmentally related to each other. Past studies using various dimensions in the rodent mandible (Cheverud *et al.*, 1991) and scapula (Leamy & Atchley, 1984), for example, have shown that characters that are developmentally or functionally related tend to be more highly correlated than characters without these relationships, this being the basic hypothesis of morphological integration (Olson & Miller, 1958; Leamy, 1977; Cheverud, 1982). The murine mandible itself is a complex structure but its genetic and developmental basis is quite well known (see Atchley, 1983, 1992; Atchley & Hall, 1991); thus, it would seem to be an ideal choice for a study on morphological integration of FA.

This paper describes a study primarily aimed at estimating and comparing morphological integration in FA at both individual and population levels among several mandible dimensions in a randombred population of mice. Morphological integration is specifically tested for by comparing the correlations of FA within and between two groups of mandibular characters and by the calculation of measures of concordance. Patterns of integration at each level also are analyzed and contrasted with the use of various multivariate statistical techniques (principal components analysis and canonical correlation). The working hypothesis to be tested is that, based on the literature described above, there should be significant integration or concordance of FA at the population level, but not at the individual level.

Materials and methods

The population

The house mice used in this study were of randombred strain CV1 obtained from Cumberland View Farms in Clinton, Tennessee. This strain originally was derived from inbred strain 101. When the mice were received, they had undergone about 52 generations of random mating. In spite of its isogenic origin, however, previous genetic analyses suggested that this strain contained a significant amount of genetic variation. Thus, heritability estimates for a variety of morphometric characters in CV1 mice averaged 0.40 (Leamy, 1974), a level comparable to that found in other outbred strains.

Single pair matings were made so that eventually 200 families of mice were produced. Each family contained two parents and a maximum of six offspring per litter (those litters containing more than six offspring were reduced to six within two days after birth). Each litter was divided into three sublitters (containing one male and one female where possible) which were sacrificed at three different ages: one month (35 days), three months (90 days), and five months (150 days). Thus eight different subpopulations were generated: five-month-old male and female parents, and male and female offspring one, three, and five months of age. To obtain replicate estimates of FA in each sex and age combination, however, each of the eight subpopulations was divided evenly (mice in families 1 to 100 versus those in families 101 through 200) to produce a total of 16 subpopulations used throughout the analysis.

For several reasons, the numbers of mice in each of these 16 subpopulations varied considerably, and all were less than the maximum possible (100). Some mice died before they could be sacrificed, for example, reducing the total number. Further, not all litters contained six individuals, reducing the numbers of mice especially in the five-month-old sublitters. Some sublitters did not contain one male and one female. Additionally, some bones were broken or lost in the skeletal preparation process and only those individuals with intact right and left mandibles could be used. Finally, statistical outliers were eliminated (see below), further reducing the sample sizes. Altogether, 1028 mice were available for the analysis, the numbers in each of the 16 subpopulations varying from 41 to 80.

The mandible characters

After sacrifice, the skeletons of all mice were prepared by exposure to dermestid beetles. Left and right halves of the mandible in each mouse were separated at the mandibular symphysis and each half was placed in turn on a glass slide on a photographic enlarger. In this way, it was possible to project the mandible image onto a digitizer which was connected to a microcomputer. Twelve points around the periphery of each mandible (Fig. 1) were recorded in x,y space. These points were especially chosen for repeatability (see below), and represent major landmarks which describe the basic shape of the mandible. Euclidean distances between each pair of points were calculated to yield a

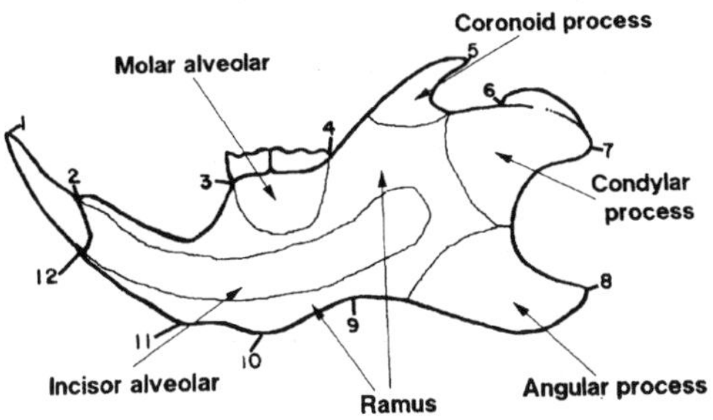

Fig. 1. Outline of a mouse mandible showing the twelve landmark points that were digitized. The nine characters used were: I1, 2-3; I2, 2-11; I3, 2-12; I4, 11-12; M1, 4-5; M2, 4-9; M3, 5-9; M4, 6-7; M5, 7-8.

total of 66 interlandmark distances on both the left and right mandibles in all mice.

A sampling of repeated measures was taken early in the process of recording all points on the mandibles and the repeat measures generally agreed quite well with the original ones. Overall repeatibility was estimated to be around 90% or greater, so that measurement error should have accounted for no more than 10% of the total variation in even the least repeatible mandible character. In addition, it was assumed that whatever measurement error was present was proportional in its contribution to each of the subpopulations and to all covariances of FAs among the 66 characters.

Since use of all 66 mandibular characters would be far too redundant, it was necessary to choose a smaller number of these characters for the analysis. In doing so, it was decided to pick characters from at least two morphologically or functionally different areas of the mandible to facilitate the assessment of morphological integration. Although the murine mandible is quite complex, it does consist of six well-defined morphogenetic units (see Fig. 1): the body of the mandible (ramus), three (coronoid, condylar, and augular) processes, and two (molar and incisor) alveolar components (Atchley & Hall, 1991). The anterior section of the mandible is dominated by the incisor itself and the incisor alveolar component, whereas the posterior section is dominated by the attachment (especially onto the three processes) of a number of muscles (Atchley & Hall, 1991).

It therefore seemed reasonable to choose several characters from interlandmark distances in the anterior (incisor) area and several in the posterior (muscle) area. Thus, characters from the incisor area were chosen from the ten possible interlandmark distances connecting points 2, 3, 10, 11, and 12 (see Fig. 1; point 1 was omitted because it represents the end of the incisor itself rather than the mandible and growth and wear in the incisor is quite variable). Muscle characters were chosen from the 15 possible interlandmark distances connecting points 4-9. An attempt was made to avoid distances all emanating from a single point, since these sorts of colinear measures would generate spurious correlations that would confuse the search for genuine morphological integration (see Cheverud *et al.*, 1991).

Within the incisor and muscle areas themselves, mandibular characters were chosen only if they showed they could be acceptably used to estimate FA. One criterion for acceptability was that the characters chosen should not have significant scaling effects (correlations) with overall size. A second criterion for acceptability was that, especially in the genetically variable CV1 population, distributions of left minus right differences for any characters used should reflect environmental variation only (Palmer & Strobeck, 1992). Thus any contamination of FA by directional asymmetry (DA) or antisymmetry raises the possibility that such characters are unreliable estimators of developmental stability. This is true since, unlike FA, both of these kinds of asymmetry presumably have a genetic basis. DA occurs when one side of a bilateral character is significantly larger than the other (as in the mammalian heart, for example), whereas antisymmetry occurs when the distribution of left minus right differences tends to be bimodal (Palmer & Strobeck, 1992).

Left minus right differences for all characters in each of the 16 subpopulations first were plotted for inspection, and all statistical outliers (Sokal & Rohlf, 1981) were eliminated within each subpopulation. Even after elimination of outliers, however, these differences for some characters showed significant levels of association with overall size in several of the subpopulations. Since associations in these characters were not easily corrected by conventional transformations, they were not chosen for use.

It also was discovered that nearly all of the interlandmark distances exhibited significant DA in at least some of the 16 subpopulations; on the other hand, there was no evidence of antisymmetry. DA may be statistically corrected for, but this does not eliminate any genetic variation for DA which may be present (Palmer & Strobeck, 1992). Fortunately, in this population of CV1 mice, it was possible to test for the presence of genetic variation in these differences by estimating their heritabilities. All heritabilities for the (signed and absolute) differences between sides were estimated by regression of offspring on the mean of the two parents (Falconer, 1989). Differences for some of the characters exhibited statistically significant heritabilities, and where so, those characters also were not chosen for use.

As a result of all of these considerations, nine interlandmark distances were chosen for use in the analysis: four (henceforth denoted I1-I4) in the incisor area, and five (M1-M5) in the muscle area (Fig. 1). Significant skewness or kurtosis in the distributions of left minus right differences for these nine characters was quite rare among the 16 subpopulations, the distributions of these differences being normal in nearly all cases. Although most exhibited some degree of DA (see Table 3 below), none had a significant heritability for left minus right differences. In addition, as assessed by the sequential Bonferroni test (Rice, 1989), correlations of the differences for these nine characters with the mean of the two sides (overall measure of size) also were overwhelmingly non-significant. Thus, it was assumed that these nine characters could be effectively used to assess FA (and thus developmental stability) in the mandible.

Statistical analysis

Preliminary to the FA analysis, the mean of the two sides for each of the nine mandibular characters was calculated and sex and age effects were eliminated in a linear model. Correlations among residuals for all nine characters from this analysis then were calculated for all 1028 individuals. Similarly, correlations were calculated among the means ($n = 16$) of the subpopulations for each of the characters after adjustment of these means for sex and age effects. All correlations were tested for significance by the sequential Bonferroni test (Rice, 1989).

For both (individual and subpopulation) sets of correlation matrices, correlations within the incisor (I1-I4) and muscle (M1-M5) groups of characters were compared with those between groups (I1-I4 with M1-M5) as a test for morphological integration. This was done by comparing the correlation matrix of FA values with a theoretical connectivity matrix consisting of 1s for all within-group correlations and 0s for between-group correlations. A non-parametric matrix permutation procedure (Dietz, 1983) was used for this purpose.

In this procedure, the correlations in rows and columns of either one of the two matrices were repeatedly and randomly permuted, and for each permutation, two test statistics were generated. One test (Mantel's) statistic was simply the sum of the products of the elements of the two matrices; a second was Kendall's tau coefficient (Dietz, 1983). Using either statistic, the null hypothesis of *no similarity* between the two matrices was rejected if the number of such statistics with values equal to or greater than that calculated from the original matrices was less than 5% of all values obtained. Rejection of the null hypothesis of no similarity between these two matrices was taken to be evidence that the within- and between-group correlations differed as would be expected if there is significant morphological integration (see Cheverud *et al.*, 1991).

FA has been measured in many different ways (see Palmer & Strobeck, 1986). For purposes of this analysis, FA was assessed both at the individual and the subpopulation level. At the individual level, differences of left and right sides in each character were calculated for all 1028 individuals. Effects of sex, age, and the interaction of sex and age then were taken out in a general linear model. Residuals from this analysis served as the basic estimates of individual FAs for all characters. Correlations of these values were obtained and all 36 correlation coefficients were tested for significance as before.

Kendall's coefficient of concordance, W (Sokal & Rohlf, 1981), was used to test for associations of the asymmetry levels among the nine characters in all 1028 individuals. This coefficient was calculated by correcting for DA in each character by subtracting the mean of the left minus right differences from each individual difference. Then the absolute values of these FAs were adjusted (in a general linear model) for any sex and age effects. To calculate W, the residuals from this analysis were ranked among all individuals (see Mason, Ehrlich & Emmel, 1967). To test the strength of any such associations within the two groups of characters, coefficients of concordance were also calculated using the absolute, adjusted left minus right side differences from the incisor (I1-I4) characters only, and from the muscle (M1-M5) characters only. Each of these three coefficients was tested for significance (Sokal & Rohlf, 1981). A significant W was taken to imply a concordance of asymmetries among characters in each individual, or the presence of an individual asymmetry parameter (IAP).

FA also was estimated for all characters in each of the 16 subpopulations from two-way unreplicated analyses of variance (see Leamy, 1992). In such analyses, sides (*df* = 1) is a fixed factor accounting for directional asymmetry, individuals (*df* = a-1 where a = the number of individuals) is a random factor that assesses variation among individual mice, and the error variance (essentially the interaction of sides and individuals) provides an estimate of FA. The error variances were adjusted (as above) in a linear model for any differences due to sex, age, or sex by age effects.

It should be mentioned that by using the population method for calculating FA, its relative importance also can be assessed as the percentage contribution of the error variance to the total variance. If the means of the left and right sides for any character are equal, then this relative FA contribution is equal to one minus the correlation coefficient between the sides (Bader, 1965; Leamy, 1984). Similarly, if DA is simply treated as a random variable, its relative importance may be estimated from the percentage contribution of the variance due to sides.

Correlations among (logarithmically-transformed) FAs for each of the nine characters obtained from the 16 subpopulation estimates of FA were estimated and their statistical significance assessed as before. In addition, the matrix permutation test described above was used to test for significant morphological integration. Kendall's coefficient of concordance was used to test for associations of the asymmetry levels among the nine characters in each subpopulation. A significant W was taken to imply the presence of a population asymmetry parameter (Soule & Baker, 1968) whereby overall levels of asymmetry truly differ among the subpopulations. To compare FA associations within and between character groups, coefficients of concordance also were calculated using the subpopulation FAs from the incisor (I1-I4) characters only, and from the muscle (M1-M5) characters only.

Principal components analysis (Harman, 1967) also was used to analyze the covariation in FA among all nine characters. In this procedure, orthogonal factors are constructed, the first several of which are usually sufficient to account for a large percentage of the total variation among all characters. The correlations (loadings) of the variables on these factors are useful in describing the factors and in suggesting patterns of covariation. Principal components analysis was conducted for the correlations generated from individual and subpopulation estimates of FA in the nine characters. The initial solution was subjected to an orthogonal (varimax) rotation so that the final factors could be more easily identified.

Using the principal components analysis procedure, it was possible to estimate an index of morphological integration (I) for the correlations of FAs estimated from both individuals and subpopulations. This index was estimated as follows:

$$I = [\Sigma(\Gamma_i - 1)^2 / (n^2 - n)]^{1/2},$$

where the Γ_i are the eigenvalues from the principal components analysis and n is the total number of characters in the correlation matrices (Cheverud *et al.*, 1983). Values of this index normally range from zero (no integration) to one (perfect integration). Although this index cannot be tested for significance since its sampling variation is unknown (Cheverud, Rutledge & Atchley, 1983; Cheverud & Leamy, 1985), it was useful to compare the I values obtained here from the FAs with those estimated from other studies of integration using various characters in mice and rats. Indices of integration were therefore calculated for all nine characters, for the incisor characters only, and for the muscle characters only, from each correlation matrix.

The relationships between FA values in the incisor and muscle groups of characters (from both individual and population estimates) were quantified using canonical correlation (Green, 1978). In this procedure, a linear combination (canonical variable) is constructed for one group of characters (say the incisor group) such that it has a maximum canonical correlation with a canonical variable constructed for the second set (muscle group). Other pairs of orthogonal canonical variables are similarly constructed, with each successive pair being less correlated. Loadings of the canonical variables are useful in indicating the relative importance of each of the characters to a given variable. Canonical correlations between pairs of variables may be tested for significance to discover how many pairs are useful in describing covariation. Also, squared multiple correlations may be calculated to estimate the proportion of the variation in any character in one group explained by the canonical variable constructed from the characters in the other group (Green, 1978).

Results

The mandibular characters

Before proceeding with the results of the analysis of FA, it is useful to review the means for the nine mandibular characters themselves (mean of the two sides) for each sex and age combination (Table 1). The nine characters differ widely in size, I3 being the smallest (less than 2 mm) and M3 the largest (over 5 mm). Within each sex, mean values for most characters increase throughout the three age groups (especially from one month to three months). Means for the five-month-old parents generally are greater for females, but less for males, compared to their five-month-old offspring. Females have larger means than males in eight of the nine cases for the parents, but in 10 of the 27 cases among the offspring. Analyses of variance showed statistically significant age effects for all nine characters, and significant sex (and sex by age interaction) effects for four of the nine characters.

Although variability for the mandibular characters is not shown in Table 1, its overall level within the subpopulations was low. Coefficients of variation calculated from the error variances in the analyses of variance above varied from 0.51 to 2.00. In general, the four incisor characters were more variable (mean CV = 1.22) than the five muscle characters (mean CV = 0.71).

Correlations among these mandibular characters, calculated from individuals and subpopulation means, are given in Table 2. At the individual level, the correlations vary from −0.65 to +0.76, but most (32 of 36) are positive in sign. They average +0.17, however, which is fairly low. Nonetheless, because of the large population size (1028), 27 of these correlations are significant using the sequential Bonferroni test. Within incisor and muscle groups, the correlations average +0.23 whereas between groups they average +0.11, indicative of the pattern expected for morphological integration. The matrix permutation test of the difference between the within- versus between-groups correlations was significant ($P < 0.05$), suggesting that in fact there is morphological integration of these nine characters.

Correlations among the nine mandibular characters calculated from subpopulation means (Table 2) present somewhat of a different picture. They vary

Table 1. Mean of the two sides for the 9 mandibular characters (I1-I4, M1-M5) in mice of each age (P5 = five-month-old parents; 01, 03, 05 = offspring 1, 3 and 5 monts of age) and sex. These values are rounded to the nearest whole number and represent absolute units calculated from digitizing. For reference purposes, they may be converted to millimeters by dividing by 280.

	Males				Females			
	P5	01	03	05	P5	01	03	05
I1	1069	966	1049	1074	1079	958	1037	1061
I2	1097	987	1083	1100	1106	977	1076	1095
I3	462	398	464	451	482	393	446	444
I4	727	676	706	750	715	670	723	750
M1	1055	918	1054	1064	1102	919	1074	1095
M2	940	865	941	952	955	860	942	950
M3	1608	1403	1590	1627	1659	1406	1606	1651
M4	571	539	560	558	596	530	576	566
M5	1055	921	1063	1109	1034	897	984	1016

Table 2. Correlations of the mandibular characters (mean of the two sides) calculated from individuals (above diagonal) and from subpopulations (below diagonal). * = P <0.05.

	I1	I2	I3	I4	M1	M2	M3	M4	M5
I1		.47*	.34*	-.05	.14*	.15*	.26*	.10*	.17*
I2	.95*		.39*	.40*	.15*	.24*	.19*	.04	.15*
I3	-.43	.50		-.65*	-.02	.10*	.03	-.01	.03
I4	.81*	.89*	-.82*		.15*	.13*	.14*	.05	.11*
M1	.87*	.87*	-.52	-.80*		.21*	.76*	.31*	.14*
M2	.17	.02	-.42	.29	.01		.30*	.09	.32
M3	.78*	.70	-.56	.74*	.76*	.53		.30*	.29*
M4	-.03	.07	-.48	.27	.00	-.06	-.11		.03
M5	-.14	.01	-.49	.32	.08	.32	.22	-.12	

from −0.82 to +0.95, and 13 are negative in sign. Ten of the correlations are significantly different from zero, nine of these ten being positive in sign. Their overall average is +0.16, nearly the same as that calculated from the individual mice. The average of the within- and between-group correlations, however, is the same (+0.16), suggesting that these correlations do not exhibit morphological integration. This was verified by the matrix permutation test, which showed no significant difference between levels of correlations within and between groups.

Directional and fluctuating asymmetry

Significant DA was discovered in most of the mandibular characters, as previously described, and its extent in the nine characters chosen for use is shown in Table 3. As may be seen, all characters except I2 show significant DA in at least some of the 16 subpopulations, the number of significant cases varying from two to ten. In all characters except M1, DA is consistent in its direction throughout the subpopulations; in the M1, however, the left side is significantly larger in some subpopulations and the right side is significantly larger in other subpopulations. In general, the left side tends to be significantly larger than the right side for I4 and the muscle characters, whereas the right side tends to be larger primarily for I1 and I3. In spite of the extent of this DA, however, it should be recalled that there is not significant genetic variation for left minus right side differences in any of the nine characters.

Table 3. Summary of tests for significant directional asymmetry in each of the nine paired mandibular characters for all 16 subpopulations.

Number with left and right sides significantly larger		
	Left side	Right side
I1	0	8
I2	0	0
I3	0	9
I4	10	0
M1	4	5
M2	7	0
M3	9	0
M4	2	0
M5	6	0

All FA estimates for each of the nine mandibular characters made for each of the 16 subpopulations are given in Table 4. These estimates vary widely, and there is considerable variability even between the two replicates within each of the eight sex by age combinations. Coefficients of variation for the FA values in each character are much greater than those for the means of the characters themselves; they range from 18 to nearly 30, indicating considerable variability.

Means of error variances (FAs), sides (DAs) and individual components of variance generated from two-way analyses of variance for each character are shown in Table 5. The relative contribution of each component also is given in the Table in terms of its percentage of the total variation. Mean DA values for the various characters vary from nearly zero to over 600, their relative contribution averaging 5%. FA mean values, on the other hand, are much larger, varying from 578 to over 4300 and contrib-

Table 4. FA values for each of the nine characters estimated in each of the 16 subpopulations (MP5 = five-month-old male parents, F01 - one-month-old female offspring, etc.). Coefficients of variation (CV) also are given for each character.

	I1	I2	I3	I4	M1	M2	M3	M4	M5
MP5	723	1075	2533	4275	1166	634	927	601	748
	703	1575	1929	3478	1128	618	618	955	1157
FP5	607	1461	1449	3829	1530	532	676	856	768
	508	1691	1995	3826	1893	810	1102	834	1202
M01	683	1248	2148	4122	1413	520	940	1462	763
	529	992	2265	3679	1372	750	762	1127	1087
F01	309	937	1599	3057	2534	1077	1858	980	1025
	507	992	2649	4856	1472	694	970	1307	732
M03	599	1570	2230	4274	1457	840	1070	1215	783
	556	1195	2161	3075	1735	1091	1078	814	954
F03	570	1189	2179	4985	1503	1084	1058	527	1102
	576	1168	1965	4952	1536	1061	920	812	1154
M05	477	1844	1213	3885	1459	509	761	751	790
	769	1499	2712	5057	2761	1016	1333	932	789
F05	547	1241	2464	6021	1109	959	970	776	980
	584	1821	2685	5432	2241	1129	970	547	991
CV	19.1	22.0	20.6	19.7	29.5	26.8	28.9	29.3	18.1

Table 5. Mean of the estimates of variance components due to sides (directional asymmetry), individuals, and error (fluctuating asymmetry) and their total for each of the nine mandibular characters. Relative percentage contribution of each of these three factors to the total is given in parentheses.

	Sides		Individuals		Error		Total
I1	47.9	(2.8)	1104.5	(63.8)	577.8	(33.4)	1730.2
I2	3.8	(0.2)	1185.0	(46.8)	1340.4	(53.0)	2529.2
I3	311.3	(9.5)	814.7	(25.0)	2136.0	(65.5)	3262.0
I4	605.6	(9.7)	1338.0	(21.4)	4300.1	(68.9)	6243.7
M1	217.9	(5.4)	2177.8	(53.9)	1644.3	(40.7)	4040.0
M2	156.2	(10.6)	484.1	(32.7)	837.8	(56.7)	1478.1
M3	157.0	(5.7)	1585.1	(57.8)	1000.8	(36.5)	2742.9
M4	14.9	(0.7)	1065.8	(53.7)	906.0	(45.6)	1986.7
M5	55.1	(2.0)	1729.3	(63.5)	939.1	(34.5)	2723.5
Means		(5.2)		(46.5)		(48.3)	

uting from 33% (I1) to nearly 57% (M2) of the total variation in these characters. The average contribution of FA, 48%, is quite large and actually is slightly greater than that due to differences among individuals (46%).

FA correlations

Correlations among the FA values for the nine characters estimated in each individual (above diagonal in Table 6) vary from -0.78 to +0.59, but average just +0.02. Thirteen of the 36 values reach statistical significance ($P < 0.05$), however, so this is good evidence of a fair level of overall integration. Within the incisor and mandibular groups of characters, correlations average +0.07 compared to −0.01 for those between groups, although the matrix permutation procedure showed that the within- and between-group correlations were not significantly different ($P = 0.18$ for Mantel's statistic although 0.064 for Kendall's statistic). Kendall's coefficient of concordance for the FAs is 0.14 for all nine characters, 0.33 for the incisor characters only, and 0.23 for the muscle characters only. All three values are statistically significant ($P < 0.01$), suggesting that in fact there is an individual asymmetry parameter (IAP) for these characters.

Table 6. Correlations of the fluctuating asymmetry values for the nine mandibular characters calculated from individuals (above diagonal) and from subpopulations (below diagonal). * = P <0.05.

	I1	I2	I3	I4	M1	M2	M3	M4	M5
I1		.27*	.31*	-.21*	-.12*	-.11*	-.05	.00	-.04
I2	-.04		.22*	.38*	.04	-.02	.03	.03	-.03
I3	.72	-.26		-.78*	-.13	-.13*	-.05	-.04	-.06
I4	.79*	-.08	.70		.14*	.13*	.06	.04	.03
M1	.05	.08	.31	-.27		.19*	.59*	-.02	.08
M2	-.09	-.25	.43	-.26	.78*		-.01	-.01	.07
M3	-.09	-.27	.34	-.06	.67	.67		.04	.05
M4	.44	.25	.11	.45	-.32	-.22	-.27		-.07
M5	-.58	.14	-.16	-.63	.32	.66	.20	-.08	

Correlations among the FA values for the nine characters, obtained from the 16 estimates in each subpopulation, also are shown in Table 6 (below diagonal). The 36 correlations range from −0.63 to +0.79, averaging +0.12. The seven highest positive correlations, two of which (I1-I4, M1-M2) reach statistical significance, all occur within (incisor or muscle) groups rather than between the two groups. Also, the mean of the within-group correlations is +0.27 whereas the the mean of the between-group correlations is 0.00. The matrix permutation test of within- versus between-group correlations yielded a significant (P = 0.04) Mantel's statistic, although this must be interpreted with some caution since Kendall's statistic calculated in this test was non-significant (P = 0.08). Nonetheless, this result suggests that there may be a real difference in the level of correlations within and between groups, and thus that there is morphological integration among the FAs. This is further suggested by Kendall's coefficient of concordance which was significant (P < 0.05) for the incisor (W = 0.42) and muscle characters (W = 0.34), but narrowly missed significance for all nine characters (W = 0.16).

The indices of morphological integration calculated from both correlation matrices above generally reflect the trends already described. Thus, for the FA correlations calculated from subpopulations, I = 0.42 for all characters, but I is higher in the incisor (I = 0.53) and muscle groups (I = 0.48). For the correlations of individual FAs, I = 0.21, a value one-half that calculated from subpopulation values. The I value for the muscle group is about the same magnitude (I = 0.20), although that for the incisor group (I = 0.41) is higher. In general, therefore, overall morphological integration is higher in the FAs calculated from subpopulations, and is highest within character (especially incisor) groups.

Multivariate analysis of FA

The (first three) rotated factors obtained from principal components analysis of correlations among the subpopulation FA values for the nine characters are shown in Table 7. These three factors account for 81% of the total variation. The first factor has high loadings primarily on I1, I3, and I4, with some contrast from M5. Thus, it primarily is an 'incisor' factor with little contribution from I2. The second factor loads most heavily on M1, M2, M3, and M5 and thus except for M4, is generally a 'muscle' factor. The third factor links I2 and M4, suggesting that their covariation is greater than that between either one and the other characters in their appropriate groups. With the exception of these two characters, however, the principal components analysis does indicate some independence of FA in the incisor versus muscle character groups.

The PCA solution for correlations of individual FAs (also shown in Table 7) is somewhat different. The first three factors account for only 55% of the total variation. The first factor loads heavily on three of the four incisor characters, but is a contrast of I1 and I3 with I4. The second factor primarily links M1 and M3 and the third factor links I2 and I4. Loadings for M2, M4, and M5 are low in all three factors. Nonetheless, this analysis also demonstrates some independence of FA in the incisor versus muscle groups of characters.

The four canonical variables extracted from the canonical correlation analysis of the FAs calculated from subpopulations are shown in Table 8. Only

Table 7. First three rotated factors (and percentage of variation each explains) obtained from an initial principal components analysis of correlations of FAs among the nine characters calculated from individuals and subpopulations.

	Subpopulations			Individuals		
	I	II	III	I	II	III
I1	0.93	-0.02	0.12	0.59	-0.08	0.43
I2	-0.16	-0.07	0.83	0.17	0.06	0.90
I3	0.82	0.46	-0.12	0.92	-0.01	-0.07
I4	0.93	-0.20	0.03	-0.79	0.04	0.58
M1	-0.03	0.88	-0.01	-0.13	0.89	0.02
M2	-0.05	0.96	-0.10	-0.29	0.19	-0.04
M3	0.05	0.77	-0.34	0.02	0.87	0.03
M4	0.44	-0.20	0.67	-0.04	0.02	0.15
M5	-0.60	0.57	0.30	-0.11	0.18	-0.13
%	36.7	30.3	14.0	23.7	16.6	15.2

the first two canonical correlations are statistically significant, although the third nearly reached significance (P = 0.06). Considering the incisor characters, the first canonical variable primarily contrasts I1 and I4 with I3. But from the squared multiple correlation (SMC) values (not shown), this variable is a good predictor only for M5 (SCM = 0.71), in the muscle group, accounting for an average of 30% of the variation of FA in the muscle characters. The second canonical variable for the incisor characters contrasts primarily I2 and I3. It increases the predictability over all five muscle characters from 30% to 49%, and is an especially good predictor for M2. The first canonical variable for the muscle group primarily contrasts M2 and M5 with M2, and explains a considerable amount of the variation in I1 and I4. The second canonical variable for the muscle group contrasts M1 and M5 with M2, and increases the variance explained especially for I2 and I3. The first two canonical variables for the muscle group explain 61% of the variation in the incisor characters. In general, therefore, the first two canonical variables generated from the muscle characters are better predictors of variation in the incisor characters than are the incisor canonical variables predictors of variation in the muscle characters.

The canonical correlation results for the individual FAs also are shown in Table 8. Only the first canonical correlation is significant, which suggests

Table 8. Canonical variables with their associated canonical correlations (CC) for FA in each of the two character groups calculated from individual and subpopulation values. ** = P <0.01.

	Subpopulations				Individuals			
	I	II	III	IV	I	II	III	IV
I1	-0.62	-0.40	1.23	-1.09	-0.60	0.42	0.90	-0.01
I2	0.39	-0.60	0.35	0.70	0.57	-2.46	0.04	-1.08
I3	0.80	0.94	0.52	0.89	0.95	2.42	-0.32	2.83
I4	-0.87	-0.25	-1.20	0.83	-0.27	3.00	0.43	2.88
M1	0.54	-1.10	1.15	0.07	0.76	-0.66	-0.32	-0.48
M2	-0.81	1.95	0.48	-0.58	0.54	0.64	0.03	0.60
M3	0.14	0.06	-0.84	1.04	-0.10	0.01	0.15	0.44
M4	-0.35	-0.33	0.37	0.79	0.19	-0.16	0.96	-0.06
M5	1.14	-0.80	-0.46	0.43	0.14	0.53	0.13	-0.79
CC	1.00**	0.96**	0.79	0.21	0.22**	0.08	0.03	0.03

that FA in the characters in the incisor and muscle groups is somewhat independent. The first canonical variable in the incisor group contrasts I2 and I3 with I1, but explains only 1% of the variation of the FAs in the muscle group. The first canonical variable in the muscle group associates M1 and M2, but accounts for only 2% of the variation of the FAs in the incisor group. Clearly, therefore, the FAs for the characters in the two groups are not very related.

Discussion

Choice of characters

Before the general results as just presented are discussed in any detail, it seems worthwhile to emphasize what perhaps may be obvious: the choice of which characters to use in this study was crucial. Palmer & Strobeck (1986, 1992) have made a particular effort to stress the importance of using characters in these sorts of studies that are reliable estimators of FA. They have outlined several statistical and interpretive problems which may occur if this is not done. As already detailed, a number of potential mandible characters were omitted here since they were considered unacceptable (because of scaling problems, significant heritabilities, etc). In clusion of some of these characters may very well have changed the results of the analysis, but if they did not adequately assess FA (and therefore developmental stability), then these results would have less meaning.

With regard to the mandibular dimensions themselves, skewness or kurtosis was not a problem since distributions of their left minus right side differences were generally quite normal throughout the 16 subpopulations. Scaling was a definite problem with some of the distances, but these distances were simply omitted so that none of the selected characters had to be corrected for scaling effects. Of course this was possible because there were so many original potential distances (66) to choose from, but in other studies the choice of characters may be much more limited.

With all of the various restrictions as already detailed, one consequence of selecting the four incisor characters was that three of them (I1, I2, and I3) emanated from the same point and thus tended toward colinearity (Cheverud *et al.*, 1991). However, inspection of the results throughout the analysis tended to confirm that these three measures (especially I2) were no more highly correlated among themselves than they were with I4, the remaining, non-colinear incisor dimension. Actually, I4 was highly correlated especially with I3, but in a negative direction (see Table 2). In any event, preliminary analyses using other combinations of incisor distances gave comparable results, so it is doubtful if the choice of these three dimensions unduly biased the results.

Given that the various mandibular characters are anatomically quite similar, their correlations were surprisingly low in magnitude (mean of +0.17). Thus the mean of the phenotypic correlations for 19 closely related dimensions in the scapula and humerus in a population of rats, for example, was +0.53 (Leamy & Atchley, 1984). However, the size (means varied from about four to over 30 mm) and variability (CVs varied from about three to ten) of these scapula/humerus dimensions was considerably greater than the mandibular dimensions used here and so perhaps are not comparable. Further, the mean value of +0.17 for the phenotypic correlations of the mandibular dimensions is perfectly in line with the mean genetic and environmental correlations of +0.23 and +0.13, respectively, estimated by Atchley, Plummer and Riska (1985) for 14 mandibular characters in random-bred mice. It may well be that the mandibular dimensions are assessing rather fine shape differences, which, because of their complex developmental origins (Atchley & Hall, 1991; Atchley *et al.*, 1992), account for this rather low overall level of integration.

There is no doubt that morphological integration exists among the mandibular dimensions, at least when assessed among individuals (rather than subpopulations). Evidence for this was seen both from the large number of statistically significant correlations among the characters and from the significant matrix permutation test. It was originally assumed that the correlations within groups would be larger than those between groups, since morphological integration of mouse mandibular characters previously has been demonstrated (see Atchley, Plummer & Riska, 1985; Cheverud *et al.*, 1991). Morphological integration was not as apparent from the correlations calculated from subpopulation means, but it is difficult to interpret what this means. Correlations among the various mandibular characters ordinarily would not be calculated in this manner, but this was done here primarily for completeness since integration of FA was assessed at the subpopulation level.

Directional asymmetry

Perhaps the most important complication in the choice of the mandibular characters turned out to be the presence of significant directional asymmetry in most of the characters in at least some of the subpopulations. By estimating FA by error variances in the analyses of variance, however, any DA actually was eliminated in the 'sides' component in these analyses (Leamy, 1984). DA was present in the individual left minus right differences, and so was statistically corrected for in the calculation of the coefficients of concordance. It was not necessary to correct for DA in the correlations of these individual differences, however, since DA effectively just shifted the mean of the distribution of these differences, and correlations are independent of additive effects.

Beyond these statistical considerations, the more important consideration regarding DA is that it presumably has a genetic basis and thus may contaminate the interpretation of FA, which has a purely environmental origin (Palmer & Strobeck, 1992). As will be recalled, adjustment for any genetic effects of DA was done simply by omitting any characters showing significant heritabilities for right minus left differences. It was fortuitous in this case that parents and offspring were available for estimation of these heritabilities (see Leamy, 1974), although even those heritabilities that reached statistical significance typically were low in magnitude (around 0.10). This suggests that the genetic contribution of DA may not be great in these mandibular characters, but this needs further exploration.

Significant DA in the mandibular characters actually was anticipated since past asymmetry studies in inbred and hybrid mice have shown that it occurs in the lengths or widths of a number of bone characters, including mandibles (Leamy, 1984). The importance of DA in each of these bone characters, however, was slight, amounting to less than 1% of the mean values of the characters themselves (Leamy, 1984). The comparable average for the mandibular characters here (not previously given) was greater than this - 1.9% - with values ranging from 0.6% to 5.0%. Further, DA accounted for an average of 5% of the total variation in the mandibular characters (see Table 5) so its contribution clearly was appreciable.

Perhaps the somewhat small size of the nine mandibular characters used in this study (all less than 6 mm), compared to that of various other bone characters previously used in the inbred and hybrid mice (Leamy, 1984), at least partly accounts for their relatively greater DA. To test this, DA was assessed in the random-bred mice here for the total distance from points 3 to 8 (see Fig. 1). This distance varies from about 7 to 9 mm throughout the subpopulations, and is equivalent to 'mandible length' in the previous study (Leamy, 1984). The absolute contribution of DA to the mean of this distance itself averaged 0.27% over all subpopulations. Since this value is lower than the comparable estimates for all of the nine mandibular characters used here, this suggests that size of a character may play some part in the overall level of DA attained.

Another potential explanation for the DA level seen in the mandibular characters here is that, unlike limb dimensions in which too much DA could not be tolerated (see Leamy, 1984), some significant DA in various skull and jaw dimensions should not unduly affect fitness of these mice. Although it could be argued that it is relatively important to the fitness of these mice that their jaws not show pronounced DA, it is doubtful whether the level of DA seen here in these specific dimensions on the mandible would cause any serious problems. Most of the mandibular dimensions probably define shape patterns (Atchley, Plummer & Riska, 1985; Cheverud *et al.*, 1991) which can accept this level of DA. In any event, the level of DA was far less than that for FA in all of the characters.

Fluctuating asymmetry levels

The level of FA in the mandibular characters (average contribution to the total variance = 48%) was surprisingly high. FA typically has accounted for less than 20% of the total variance in a range of osteometric characters in mice and rats (Bader, 1965; Leamy, 1984, 1986, 1992; Leamy & Atchley, 1985; Parker & Leamy, 1991). On the other hand, FA seldom has been assessed for interlandmark distances on a structure such as the mandible. We therefore have no real benchmark for expectations of FA levels in these sorts of characters. Of course, relative FA values were expressed as a proportion of the total variability in the mandibular characters and this variability was extremely low (mean CV = 1.22). As a comparison, the mean CV for a variety of morphometric characters was about 3.0 in these same random-bred mice (Leamy, 1974) and about 1.5 in inbred and hybrid strains (Leamy, 1982). So perhaps an appreciable amount of relative FA variability can be attributed to the particularly low variability of the mandibular characters themselves.

The considerable FA in the mandibular characters here suggests that it may well be worthwhile to measure both left and right sides so that the mean of the sides can be used as a more accurate measure of these dimensions. If in fact only one side of these characters is used, as has conventionally been done (Atchley, Newman & Cowley, 1988; Cheverud *et al.*, 1991), then considerable unknown variability is introduced. Further, estimates of heritabilities of one side of these characters (for example, Atchley, Plummer & Riska, 1985) should yield lower values than heritability estimates based on the mean of the two sides.

Integration of FA

Given the relatively low level of correlations of the mandibular characters themselves among the 1028 mice, it was somewhat surprising that there was evidence for an 'individual asymmetry parameter', or IAP, in these characters. However, it will be recalled that a large number (13 of 36) of the correlations among FAs for the nine characters calculated for each individual reached statistical significance even using the sequential Bonferroni test. More importantly, Kendall's coefficient of concordance, W, calculated across all individuals also was significant not only for all nine characters, but also for incisor characters only and muscle characters only. Apparently, use of over 1000 individuals was sufficient to detect this trend, whereas it may be missed in samples of much smaller size which are typical in these sorts of studies.

To test the effect of sample size on concordance of asymmetry levels, W values for FA in all nine characters among individuals also were calculated for each of the 16 subpopulations (not shown previously). Treating each subpopulation as independent, six of the 16 W values reached conventional statistical significance. If viewed simultaneously, however, only one reached significance in the sequential Bonferroni test. What this suggests is that with the right sorts of characters, an IAP occasionally may be found within a given population or sample even of size less than 100. But if a number of samples of this size are simultaneously analyzed (as they often are in asymmetry studies), the chance for detecting an IAP is considerably reduced.

The integration level of the FAs among individuals also was demonstrated by the indices of morphological integration. The I value of 0.21 calculated for FA in all nine characters in fact is only slightly less than the comparable value of 0.27 estimated by Atchley, Plummer and Riska (1985) from the covariation among 14 mandibular characters in mice. As will be recalled, this index was even higher (0.41) when calculated from the FAs for the incisor characters alone. Other I values from different static and ontogenetic studies using phenotypic correlations of morphometric characters in mice and rats have varied from 0.36 to 0.65 (Atchley, Plummer & Riska, 1985).

Although an IAP was effectively shown with the individual FA estimates, the matrix permutation test comparing correlations within and between characters groups was not significant. Probably the reason for this is that some of the statistically significant correlations within groups (especially that of −0.78 between I3 and I4; see Table 6) were negative in sign. To test this, the negative correlation of FAs between I3 and I4 was made positive, and the matrix permutation test was rerun and in fact turned out to be significant. The permutation test is directional in that it ranks positive values higher than negative values and this seems appropriate given that we expect morphological integration to associate characters in the same direction (Leamy, 1977; Cheverud, 1982; Cheverud *et al.*, 1991). However, some of the associations of the mandibular characters themselves, especially I3 and I4, were negative (see Table 2), so it seems reasonable that some negative correlations of FA would be found as well.

When variances are used to estimate FAs, as in the subpopulation estimates, directionality is no longer a problem. This is probably at least partly why the tests for morphological integration among the subpopulation correlations of FAs were even more persuasive. Thus, as has typically been found (Palmer & Strobeck, 1986; Leamy, 1992), a PAP was demonstrated by the significant results from Kendall's test and the matrix permutation test. In addition, the index of integration among the subpopulation FAs was moderately high (+0.42) and comparable to that previously found among characters themselves rather than their asymmetries (see Atchley, Plummer & Riska, 1985).

Why is it typically so much easier to discover a PAP rather than an IAP? The results of this study suggest that the signal for FA is weak in the first place and that its detectability at the individual level is considerably complicated by a number of factors, including scaling effects, DA, and measurement error. Thus it would seem that an IAP might be expected only among morphometric characters that developmentally are closely related and for which good estimates of FA have been made. Over a sample or population of even modest size, however, *one* estimate of this signal in each character might be reasonably made by statistics such as variances. The greater the differences (genetic, geographic, etc.) among a number of populations, the greater is the probability that levels of the signals themselves among these populations will be sufficiently different such that significant concordance of FAs might be found.

It is interesting in this regard that a significant PAP was found among the 16 subpopulations here when each simply represented a subsample of the entire random-bred population. Thus FA estimates did not differ because of significant sex or age effects which were taken out and which previously have not been shown to significantly affect FA anyway (Leamy, 1992). Therefore, concordance of FAs must have been generated simply from genetic differences among these subpopulations. A PAP for various morphometric variables previously has been detected for inbred versus hybrid samples (Leamy, 1992), but these genetic differences should have been much larger than those among subpopulations here. In this regard, it would be interesting to see whether a significant PAP could be detected simply among different inbred strains of mice, or perhaps even among samples of one inbred strain.

Multivariate patterns of FA

The multivariate procedures used in this study were primarily useful in depicting patterns of FA assessed both at the individual and population levels. The principal components analysis clearly showed that (a) the overall level of integration (variation explained) was greater at the subpopulation compared to the individual level, and (b) correlations of FAs did tend to cluster within the incisor character group and within the muscle character group. The relative independence of the correlations of FAs between the two groups was especially impressive at the subpopulation level where it will be recalled that one of the two test statistics from the matrix permutation procedure was significant. The principal components analysis results also pointed to a much more localized pattern of covariance of the individual FAs, as has already been discussed.

The canonical correlation analysis revealed that there are some associations between FAs in the two character groups, but they were generally complex, involving various contrasts. The analysis was somewhat more clear cut for individual FAs in that it demonstrated considerable independence of FAs between character groups. Again, this suggests that the effects of individual FAs must be much more localized, even within a bone such as the mandible.

Acknowledgements

I should like to thank Judy Purtell for taking all of the measurements on the mandibles, William Atchley for permission to photograph his drawing of the mandible, Sandy Zane for assistance in preparing Figure 1, Richard Palmer, James Cheverud, William Atchley, and Zhiyi Zhang for helpful discussions on various aspects of this paper, and an anonymous reviewer for excellent revision suggestions.

References

Atchley, W. R., 1983. A genetic analysis of the mandible and maxilla in the rat. Journal of Craniofacial Genetics and Developmental Biology 3: 409-422.

Atchley, W. R., 1992. Genetic and developmental aspects of variability in the mammalian mandible, in The Vertebrate Skull, edited by J. Hanken and B. K. Hall. Univ. Chicago Press, Chicago (in press).

Atchley, W. R. & B. K. Hall, 1991. A model for development and evolution of complex morphological structures. Biol Rev. 66: 669-688.

Atchley, W. R., D. E. Cowley, C. Vogl & T. McLellan, 1992 Evolutionary divergence, shape change, and genetic correlation structure in the rodent mandible. Syst. Biol. 41: 196-221.

Atchley, W. R., S. Newman & D. E. Cowley, 1988. Genetic divergence in mandible form in relation to molecular divergence in inbred mouse strains. Genetics 120: 239-253.

Atchley, W. R., A. A. Plummer & B. Riska, 1985. Genetics of mandible form in the mouse. Genetics 111: 555-577.

Atchley, W. R., S. W. Herring, B. Riska & A. A. Plummer, 1984. Effects of the muscular dysgenesis gene on developmental stability in the mouse mandible. Journal of Craniofacial Genetics and Developmental Biology 4: 179-189.

Bader, R. S., 1965. Fluctuating asymmetry in the dentition of the house mouse. Growth 29: 21-30.

Cheverud, J. M., 1982. Phenotypic, genetic and environmental morphological integration in the cranium. Evolution 36: 499-516.

Cheverud, J. M. & L. J. Leamy, 1985. Quantitative genetics and the evolution of ontogeny. III. Ontogenetic changes in correlation structure among live-body traits in randombred mice. Genetical Research 46: 325-335.

Cheverud, J. M., J. Rutledge & W. Atchley, 1983. Quantitative genetics of development: genetic correlations among age-specific trait values and the evolution of ontogeny. Evolution 37: 895-905.

Cheverud, J. M., S. E. Hartman, J. T. Richtsmeier & W. R. Atchley, 1991. A quantitative genetic analysis of localized morphology in mandibles of inbred mice using finite element scaling analysis. Journal of Craniofacial Genetics and Developmental Biology 11: 122-137.

Dietz, E. J., 1983. Permutation tests for association between two distance matrices. Syst. Zool. 32: 21-26.

Falconer, D. S., 1989. Introduction to Quantitative Genetics, third edition. Longman, New York.

Green, P. E., 1978. Analyzing Multivariate Data. The Dryden Press, Hinsdale, Illinois.

Harman, H. H., 1967. Modern Factor Analysis. Univ. of Chicago Press, Chicago.

Leamy, L., 1974. Heritability of osteometric traits in a randombred population of mice. Journal of Heredity 65: 109-120.

Leamy, L., 1977. Genetic and environmental correlations of morphometric traits in randombred house mice. Evolution 31: 357-369.

Leamy, L., 1984. Morphometric studies in inbred and hybrid house mice. V. Directional and fluctuating asymmetry. American Naturalist 123: 579-593.

Leamy, L., 1986. Directional selection and developmental stability: evidence from fluctuating asymmetry of dental characters in mice. Heredity 57: 381-388.

Leamy, L., 1992. Morphometric studies in inbred and hybrid house mice. VII. Heterosis in fluctuating asymmetry at different ages. Acta Zool. Fennica 191: 111-119.

Leamy, L. & W. R. Atchley, 1984. Morphometric integration in the rat (*Rattus* sp.) scapula. Journal of Zoology (London) 202: 43-56.

Leamy, L. & W. R. Atchley, 1985. Directional selection and developmental stability: evidence from fluctuating asymmetry of morphometric characters in rats. Growth 49: 8-18.

Leary, R. F., F. Allendorf & K. L. Knudsen, 1992. Genetic, environmental, and developmental causes of meristic variation in rainbow trout. Acta. Zool. Fennica 191: 79-95.

Mason, L. G., P. R. Ehrlich & T. C. Emmel, 1967. The population biology of the butterfly, *Euphydryas editha*. V. Character clusters and asymmetry. Evolution 21: 85-91.

Olson, E. & R. Miller, 1958. Morphological Integration. University of Chicago Press, Chicago.

Palmer, A. R. & C. Strobeck, 1986. Fluctuating asymmetry: measurement, analysis, patterns. Ann. Rev. Ecol. Syst. 17: 391-421.

Palmer, A. R. & C. Strobeck, 1992. Fluctuating asymmetry as a measure of developmental stability: implications of non-normal distributions and power of statistical tests. Acta Zool. Fennica 191: 57-72.

Parker, L. T. & L. Leamy, 1991. Fluctuating asymmetry of morphometric characters in house mice: the effect of age, sex, and phenotypical extremeness in a randombred population. Journal of Heredity 82: 145-150.

Rice, W. R., 1989. Analyzing tables of statistical tests. Evolution 43: 223-225.

Sokal, R. R. & J. F. Rohlf, 1981. Biometry, 2nd ed. Freeman, San Francisco.

Soule, M. E. & B. Baker, 1968. Phenetics of natural populations. IV. The population asymmetry parameter in the butterfly *Coenonympha tullia*. Heredity 23: 611-614.

T. A. Markow (ed.), Developmental Instability: Its Origins and Evolutionary Implications, 157–168, 1994.

Relationship between fluctuating asymmetry, morphological modality and heterozygosity in an elderly Israeli population

Gregory Livshits[1] & Peter E. Smouse[2]
[1] *Division of Anatomy and Anthropology, Sackler Faculty of Medicine, Tel Aviv University, Ramat Aviv, Israel*
[2] *Center for Theoretical and Applied Genetics Cook College, Rutgers University New Brunswick, NJ 08903-0231, USA*

Received 19 April 1993   Accepted 28 April 1993

Abstract

We have been checking the following working hypotheses: 1) There is a negative correlation between genetic heterozygosity and fluctuating asymmetry (FA); 2) FA is a measure of developmental stability/instability of the whole organism, i.e. we expect negative correlation between FA and morphological proximity of a set of mass-size variables of an individual to a population centroid; and 3) FA is a measure of character-specific stability in a population, i.e. we expect correlation between magnitude of FA and deviation of an individual from the population centroid of the bilateral characters themselves. For this purpose each individual in a sample of about 200 elderly individuals was assessed for 11 polymorphic blood systems (14 genetic loci) as well as for a set of 26 anthropometric traits: 1) a set of ten mass size variables; and 2) a set of eight pairs of bilateral measurements. Four multivariate measures of morphological centrality were computed, two measures for size and two measures of shape distances from the ith individual to the population centroid for mass-size variables and also for the bilateral variables. A multivariate measure of FA for 8 bilateral pairs was also computed. No relationship was detected between FA and heterozygosity, or between FA and any of the four multivariate deviations. Thus, we concluded that our data do not support the listed hypotheses.

Introduction

The study of fluctuating asymmetry (FA) has a long history. Thoday (1958) claimed that the level of FA could be used as a measure of a developmental buffering capacity for an individual, since differences between bilaterally paired structures could be viewed as resulting from developmental accidents. It has been demonstrated that artificial selection can alter the level of FA (Mather, 1953; Reeve, 1960; Leary *et al.*, 1985), so FA is at least partially under genetic control. The control of FA levels has been postulated by many to be polygenic (Mather, 1953; De Marinis, 1959; Rasmusson, 1960; Ferguson, 1986; Livshits & Kobyliansky, 1989), despite the fact that heritability estimates for single trait FAs are very low (Reeve, 1960; Bailit *et al.*, 1970; Leary *et al.*, 1985; Livshits & Kobyliansky, 1989). There is a growing perception that FA is a non-additive (i.e., dominance or epistasis) feature of the total genome (Goldschmidt, 1955; Thoday, 1958; Livshits & Kobyliansky, 1985, 1989, 1991; Clarke & McKenzie, 1987; Clarke, Oldroyd & Hunt, 1992), rather than being due to any specific locus or to additive genetic effects of polygenic loci (e.g., Soule, 1982).

Lerner (1954) proposed that phenotypic variation and FA, used as measures of developmental homeostasis, would be inversely correlated with heterozygosity. Numerous reports support this contention (e.g., Kat, 1982; Leary & Allendorf, 1989; Quattro & Vrijenhoek, 1989; Blanco *et al.*, 1990; Dubrova *et al.*, 1991a), but conflicting results have also been published (Zink, Smith & Patton, 1985; Zakharov, 1987; Patterson & Patton, 1990; Hartl *et al.*, 1991; Clarke, Oldroyd & Hunt, 1992). The phenomenon is less than universal and the arena is contentious, so it is timely to explore the situation further. Our long term interest is the relationship between heterozygosity and FA in human medical context because FA has been widely used as a predictor of developmental anomalies and morbidity (e.g., Markow & Wandler, 1986; Livshits & Kobyliansky, 1987, 1991; Mellor, 1992).

We have recently developed a multivariate measure of FA and applied it to an assessment of the correlation between FA and departure of the phenotype from morphometric modality on a large Israeli sample (Livshits & Smouse, 1993). We showed that FA is correlated neither with deviations from average size nor with average morphological shape. Moreover, this study showed that if morphometric FA is properly defined and measured, it becomes a measure of developmental noise and shows no discernible pattern at all. For that earlier study, we did not compare FA with heterozygosity per se, because we did not have genetic data for that sample. The major objective of the present study is to rectify that omission, using the same multivariate methodology to examine the relation between FA and heterozygosity in another Israeli sample for which we do have genetic data.

Materials and methods

The population

The present study was carried out during 1984-1988. Our population sample consisted of 208 subjects of both sexes who were 75 years of age or older. The mean age of female participants was 80.85 years, with a range between 75 and 93 years. The mean age of the male participants was 81.67 years, with a range between 75 and 94 years. The sampled individuals were randomly picked from five homes for the elderly, whose managers agreed to participate in the project. The homes involved were: Mishan in Ramat Aviv, Mishan in Holon, Parents' Home in Ramat Chen, Parents' Home in Kfar Shmaryahu, and the Nava Chen Old People's Home in Ramat-Gan, all located in the vicinity of Tel Aviv. In each home, the aims of the research and the manner in which the data would be gathered were explained to the potential participants. We imposed no selection on country of origin, education, or profession of the volunteers, but the ethnic derivation of both parents of each participant was recorded and is shown in Table 1. It is obvious from the Table that the majority of participants of both sexes were of European Ashkenazi origin, mostly from Eastern Europe.

Table 1. Ethnic distribution (according to geographic origin of both parents) in studied elderly sample.

Population group	Males	Females
East Europe	73	84
Central Europe	13	18
Middle East	2	0
North Africa	13	5
Sample Size	101	107

The reason for this predominance of European Jews in the sample is that many of the elderly survivors of the Holocaust were bereft of living relatives. With onset of old age, they elected to live in an Old People's Home. It should be pointed out that many of these survivors receive monetary reparation, enabling them to maintain a relatively high standard of living and affording them the opportunity to stay in such Homes (Aschel & Givon, 1982). People of Mediterranean, North African, or Asian extraction, so-called 'Oriental Jews', are generally less affluent and are thus underrepresented in these same homes. Oriental Jews also differ in cultural habits, especially with regard to elderly parents, whom they keep within the family circle until death. Inevitably, there was a geographic bias inherent in the choice of sample populations, but none in the choice of volunteers within any single home.

Genetic markers

Blood samples were collected from each individual, and several protein enzyme loci were examined through starch gel electrophoresis: acid phosphatase (ACP1), adenosine deaminase (ADA), adenylate kinase (AK1), esterase D (ESD), glutamic pyruvate transaminase (GPT1), glyoxalase I (GLO1), 6-phosphogluconate dehydrogenase (6PGD), and phosphoglucomutase (PGM1). The following blood group loci were also examined: Duffy (using both anty-Fy^a and anti-Fy^b), MNSs (using anti-M, anti-N, anti-S, and anti-s) and Rhesus (using anti-C, anti-c, anti-D, anti-E, and anti-e). With the exception of the MNSs and

Table 2. Allele and haplotype distributions in elderly and adult Israeli populations.

(A) Two-allelic loci — Allele frequencies

System		N	p	q	χ^2 H–W	χ^2 E/A
ADA	Elderly	203	0.9064	0.0936	12.07**	8.96**
	Adult	200	0.8850	0.1150	1.31NS	
AK1	Elderly	205	0.9268	0.0732	3.82NS	3.19NS
	Adult	207	0.9469	0.0531	0.65NS	
ESD	Elderly	203	0.8670	0.1330	4.31NS	7.75*
	Adult	207	0.8941	0.1159	3.56NS	
GLO1	Elderly	203	0.3818	0.6182	0.51NS	1.06NS
	Adult	205	0.3610	0.6390	3.62NS	
GPT1	Elderly	199	0.6633	0.3367	21.02***	6.44*
	Adult	207	0.6087	0.3913	3.38NS	
GPGD	Elderly	203	0.9581	0.0419	1.24NS	4.51NS
	Adult	207	0.9348	0.0652	1.01NS	
PGM1	Elderly	203	0.7586	0.2414	2.56NS	0.95NS
	Adult	206	0.7354	0.2646	7.37*	

(B) Multiallelic loci

ACP1 – Allele frequencies

	N	a	b	c	χ^2 H–W	χ^2 E/A
Elderly	206	0.2524	0.7379	0.0097	1.29NS	0.65NS
Adult	206	0.2476	0.7403	0.0121	4.19NS	

DUFFY – Allele frequencies

	N	F*Y	F ^a^ Y	F ^b^ Y	χ^2 H–W	χ^2 E/A
Elderly	207	0.0412	0.4354	0.5234	12.73**	6.89**
Adult	206	0.0387	0.3986	0.5627	1.55NS	

MNSs, Haplotype frequencies

	N	MS	Ms	NS	Ns	χ^2 H–W	χ^2 E/A
Elderly	207	0.1921	0.2983	0.1461	0.3635	14.56NS	12.17NS
Adult	206	0.2257	0.3350	0.1626	0.2767	4.06NS	

RHESUS, Haplotype frequencies

	N	CDE	CDe	CdE	Cde	cDE
Elderly	203	0	0.4519	0	0.0211	0.1550
Adult	206	0.0049	0.4625	0	0.0157	0.1528

	cDe	cdE	cde	χ^2 H–W	χ^2 E/A
Elderly	0.0383	0.0138	0.3245	6.2NS	9.76NS
Adult	0.0791	0	0.2850	11.42NS	

χ^2 H–W – test for Hardy-Weinberg equilibrium, χ^2 E/A – test for equality of phenotype frequencies in Elderly and Adult populations. NS – Statistically not significant; * = $P < 0.05$; ** = $P < 0.01$; *** = $P < 0.001$.

Rhesus complexes, these loci evolve independently in humans, showing little gametic disequilibrium in population samples. The MNSs and Rh complexes, each representing a set of loci in virtually total linkage, were each treated as multiple-allelic single loci, using haplotype frequencies, as is customary. The distributions of the genetic frequencies for our sample were compared with those of a control sample, 206 adult individuals (Kertzman, 1992), as well as with additional Israeli samples, indicated in context.

Table 2 presents the allele/haplotype frequencies for the two samples. Phenotype frequencies in the two samples were compared, locus by locus, and each locus was also checked against Hardy-Weinberg equilibrium. AK1, GLO1, 6-PGD, PGM1, ACP1, MNSs and Rh showed neither differences among samples nor deviations from Hardy-Weinberg equilibrium. The phenotypic distributions for ADA, GPT1 and Fy loci deviated from Hardy-Weinberg expectations among elderly individuals, while PGM1 deviated significantly among the adult controls. Significant inter-sample differences were detected for the ADA, ESD, GPT1 and Fy loci. To make sure that the observed differences between samples, as well as observed deviations from the H-W equilibrium, are real, we also compared our data with phenotype frequencies reported by other workers for random samples drawn from the same population (Szeinberg & Tomashevsky-Tamir, 1971; Szeinberg *et al.*, 1971; Lahav & Szeinberg, 1972; Golan, Ben-Ezzer & Szeinberg 1977; Kobyliansky *et al.*, 1982). Chi-square tests showed no differences between our adult sample and the published phenotype distributions (not shown). However, when our elderly sample was compared with reference data, significant differences ($P < 0.05$) appeared at three loci, namely ADA, GPT1 and Fy.

At the two enzyme loci, the differences are due to the lower than expected frequency of the heterozygous individuals (0.128 vs 0.170 at ADA locus, and 0.301 vs 0.447 at GPT1). The Fy locus shows the reverse trend, with (aa) and (bb) homozygote frequencies lower than expected, fr(aa) = 0.188 vs $p^2 = 0.224$, and fr(bb) = 0.280 vs $q^2 = 0.315$, respectively (Table 3). It is possible that some selective survival has taken place among the elderly sample. On the whole, our data show limited differences in allele/haplotype frequencies between elderly and mature adult populations. The elderly 'gene pool' may have moved from its 'adult configuration', but it has certainly not moved much.

Anthropometric traits

Each individual in the sample was also assessed for a full battery of 26 anthropometric traits, a set of ten mass-size metrics [stature (STAT), palm length (PALM), Digit-2 length (D2LN), head length (HDLN), head breadth (HDBR), inter-ocular diameter (IODR), head circumference (HDCR), body weight (BOWT), chest circumference (CHCR), bi-iliac diameter (BIDR)] and a set of eight bilateral measurements [right and left ear lengths EL-R and EL-L), right and left ear breadths (EB-R and EB-L), biepicondylic breadths (BP-R and BP-L), bistyloid breadths (ST-R and ST-L), hand breadths (HN-R and HN-L), bicondylic breadths (BC-R and BC-L), bimalleolar breadths (BM-R and BM-L), and foot breadths (FB-R and FB-L)]. With the exception of three traits not measured here, ilio-spinal height (ILHT), tibial height (TIBH) and upper arm circumference (UACR), we used these same measures for the earlier study (Livshits & Smouse, 1993).

Since the 26 anthropometric variables represent correlated features of morphology, an analysis of each variable separately would be highly redundant. We showed that these morphometric traits were related to each other in classic allometric fashion; the traits were non-linearly related and variance increased with the mean (Livshits & Smouse, 1993). To linearize relationships among variables and remove the association between variance and size, purely a scale effect, we first transformed each trait logarithmically. In order to enhance sample size, it would be useful to combine data on males and females. Table 3 lists the means and standard deviations for each of the 26 measurements and their log-transformed analogues, stratified by sex. Even after logarithmic transformation, the males showed significantly larger means and variances for almost all traits. We normalized by subtracting an individual's value from the within-sex mean and dividing by the within-sex standard deviation. Eleven males and nine females had standardized values of some metric more than 3.5 SD from the mean; we removed the outliers, reducing the sample to 188 individuals.

Table 3. Sample averages and standard deviations for 26 morphological measurements among elderly Israeli females and males.

VARIABLES	Females (N = 107)		Males (N = 101)		LOG-transformed Females		LOG-transformed Males	
	Mean	SD	Mean	SD	Mean	SD	Mean	SD
STAT	1506.95	56.33	1622.65	63.48	7.312	0.035	7.391	0.040
PALM	94.45	6.53	103.61	6.70	4.544	0.073	4.643	0.060
D2LN	66.15	5.19	70.52	4.65	4.182	0.075	4.255	0.063
HDLN	179.92	6.16	185.57	6.69	5.192	0.034	5.222	0.035
HDBR	151.43	5.48	155.03	6.86	5.020	0.036	5.041	0.044
IODR	31.47	4.11	31.67	4.08	3.448	0.134	3.440	0.131
HDCR	549.56	23.56	560.11	20.12	6.309	0.044	6.326	0.035
BOWT	615.73	108.44	701.53	96.74	6.410	0.172	6.545	0.139
CHCR	964.00	89.73	981.63	58.72	6.867	0.098	6.888	0.059
BIDR	245.44	18.91	260.74	26.35	5.494	0.076	5.562	0.100
ELR	69.26	5.12	73.42	5.74	4.239	0.075	4.292	0.080
ELL	69.16	4.95	73.48	5.54	4.238	0.070	4.295	0.077
EBR	36.03	3.23	36.23	3.75	3.579	0.086	3.590	0.104
EBL	35.70	3.78	35.92	3.55	3.570	0.102	3.579	0.103
BPR	73.00	3.72	71.64	4.72	4.144	0.058	4.271	0.062
BPL	63.04	4.25	70.63	4.31	4.142	0.064	4.257	0.060
STR	52.73	3.47	58.55	3.61	3.964	0.062	4.069	0.059
STL	52.50	3.74	57.86	3.20	3.957	0.068	4.057	0.055
HNR	74.63	4.19	82.18	4.42	4.308	0.055	4.405	0.052
HNL	73.07	4.58	81.03	4.57	4.285	0.061	4.390	0.054
COR	95.45	7.55	98.88	6.05	4.551	0.078	4.592	0.058
COL	94.79	8.16	98.14	5.83	4.545	0.085	4.584	0.057
MAR	66.34	4.97	73.26	5.18	4.189	0.073	4.290	0.071
MAL	66.21	5.25	72.48	5.39	4.189	0.079	4.279	0.071
FOR	86.08	6.49	90.80	6.35	4.446	0.076	4.507	0.071
FOL	85.18	6.79	89.97	6.04	4.435	0.081	4.497	0.067
AGE	80.85	4.40	81.65	4.39				
HETEROZYGOSITY					0.38	0.17	0.39	0.17

By normalizing, we homogenized the within-sex variances and corrected for average differences in size between males and females, but we also have to be concerned with the question of whether the covariances/correlation structures are the same for the two sexes. In order to test for homogeneity of within-sex covariance matrices, we employed a multivariate Bartlett's test of covariance matrix homogeneity (Anderson, 1987). We compared 26 × 26 covariance matrices for males and females, COV_M and COV_F, having $(N_M - 1) = 89$ and $(N_F - 1) = 97$ degrees of freedom, respectively, with the pooled covariance matrix, COV_T, which has $(N_T - 2) = 186$ degrees of freedom. The formal test criterion is:

$$L = -2r * [(N_T - 2) * \log|COV_T| - (N_M - 1) * \log|COV_M| - (N_F - 1) * \log|COV_F|], \quad (1)$$

where

$$r = \{1 - [\frac{1}{N_M - 1} + \frac{1}{N_F - 1} + \frac{1}{N_T - 2}] * [\frac{2p^2 + 3p - 1}{6 * (p + 1) * (q - 1)}]\} = 0.871, \quad (2)$$

where p = 26 is the total number of traits under examination, and where q = 2 is the number of strata (sexes) being compared. Lambda is asymptotically χ^2-distributed with $p*(p+1)/2 = 351$ df when the null hypothesis of homogeneity is valid. Measured against this standard, Lambda (L) = 350.3 is not significant ($P > 0.10$) and we thus feel free to pool the data for both sexes, after within-sex normalization.

Results

FA and morphological modality

In keeping with the strategy of the earlier paper (Livshits & Smouse, 1993) we conducted a principal components analysis (PCA) of the pooled within-sex covariance matrix for log-transformed metrics. The results are presented for the ten mass-size variables in Table 4. The usual result of such PCA treatment of morphometric variables is a major axis that has positive loadings for all traits and that accounts for a large portion of the total variation, commonly interpreted as a 'size axis', and a series of axes of mixed sign and direction, usually referred to as 'shape axes' (Jolicoeur, 1963; Dubrova *et al.*, 1991b; Livshits & Smouse, 1993). In keeping with that general experience, PC-1 has positive coefficients for all variables and accounts for 37.3% of a total variation. PC-2 explains about 15.2% of the variation and contrasts circumference measurements (CHCR, HDCR, IODR) with longitudinal body measures, (STAT, PALM, D2LN, HDLN). The final eight components each have smaller variance but collectively account for almost half (47.5%) of the variation.

We can partition the deviation of an individual from the population mean morphology into two components: (1) a deviation of an individual's size from mean size, a 'size deviation', and (2) a departure from the average pattern among a set of variables, a 'shape deviation'. The shape deviation represents an individual's morphometric disproportionality, relative to population norm. We (Livshits & Smouse, 1993) have defined size and shape deviations of the ith individual from the morphometric centroid of the populations as:

$$D^2_i = [Z_{i1} - \bar{Z}_1]^2 / V_1 + \sum_{j=2}^{10} [Z_{ij} - \bar{Z}_j]^2 / V_j =$$

$$D_i^2 \text{ (size)} + D_i^2 \text{ (shape)}, \qquad (3)$$

where Z_{ij} and $\bar{Z}_j$ are the jth factor scores for the ith individual and mean of the population, respectively, and the V_js are the respective eigen-values.

We analyzed the 16 (eight pairs) bilateral traits in a similar fashion. An individual's 16 bilateral factor scores were computed, and we present the results of Principal Components Analysis in Table 5. PC-1 has positive loadings for all variables, and is a credible (bilateral) size axis, explaining some 46% of total variation. The next seven components represent bilateral shape differences among individuals, and collectively account for about 47% of the total variation. In general, the R and L elements of each bilateral pair of measures have coefficients that are of the same general magnitude and of the same sign. The 'size and shape' components represent 93% of the total variation, and it is clear that none of the first eight components is related to asymmetry. For the final eight components, an increase in the R element of any bilateral pair is accompanied by a decrease in the L element of the

Table 4. Principal component analysis for size-mass variables. First row represents size loadings, other - shape loadings.

		Size-mass trait loadings									
#	Var	STAT	PALM	D2LN	HDLN	HDBR	IODR	HDCR	BOWT	CHCR	BIDR
1	3.733	.747	763	.603	.706	.508	.100	.666	.729	.529	.456
2	1.524	-.402	-.316	-.378	.261	.485	.406	.447	.057	.391	-.544
3	1.180	.085	.100	.305	.336	.329	-.064	.214	-.553	-.660	-.280
4	.951	-.152	.030	-.256	.169	.054	-.882	.186	-.006	.133	-.008
5	.699	.068	-.013	.364	-.284	.289	-.197	-.298	.214	.119	-.458
6	.567	.082	.142	.061	.222	-.537	.013	.062	.024	.088	-.432
7	.453	-.214	-.396	.400	.065	-.141	-.055	.184	.118	-.022	.121
8	.375	-.360	.335	.181	-.149	-.017	.034	.078	-.201	.167	.047
9	.322	-.192	.027	.031	.367	.050	.023	-.375	.036	.028	.058
10	.195	.162	-.128	.064	.038	.009	-.014	-.052	-.281	.251	.035

Table 5. Principal components loadings and variances for 16 bilateral traits. First row represents size loadings, other rows – the shape loadings.

		Bilateral trait loadings															
#	Var	EL-R	EL-L	EB-R	EB-L	EP-R	EP-L	ST-R	ST-L	HN-R	HN-L	CO-R	CO-L	MA-R	MA-L	FO-R	FO-L
1	7.650	.484	.486	.176	.151	.804	.835	.791	.812	.834	.834	.644	.666	.831	.800	.692	.660
2	2.098	-.576	-.595	-.791	-.788	.153	.090	.132	.141	.072	.054	.160	.182	.121	.082	-.056	-.072
3	1.269	-.466	-.465	.276	.231	.023	.004	.014	.100	.175	.168	-.427	-.429	.077	-.031	.350	.376
4	1.147	-.395	-.364	.383	.435	.101	.075	.103	.044	-.102	-.067	.431	.340	.047	-.038	-.293	-.296
5	1.060	.075	.062	-.024	.031	.243	.151	.297	.245	.184	.148	-.355	-.400	.041	.025	-.487	-.488
6	0.645	.031	-.002	-.079	.055	-.261	-.225	.248	.209	.258	.261	.122	.085	-.302	-.400	-.015	.056
7	0.518	-.014	.055	.005	-.038	.316	.336	.096	.077	-.195	-.185	-.024	.001	-.292	-.316	.051	.157
8	0.465	.032	-.011	.031	-.024	-.148	-.186	.326	.329	-.253	-.307	-.040	-.026	.082	.133	.049	.059
9	0.238	.039	-.017	-.311	.292	.025	.005	.034	-.055	-.046	-.042	.045	-.076	.161	-.102	.033	.019
10	0.210	.050	-.066	-.103	.126	.112	-.093	-.196	.198	-.021	.032	-.081	.101	-.178	.151	-.057	.047
11	0.167	.025	-.049	-.021	.034	.115	-.104	.176	-.201	.114	-.091	.020	-.018	-.158	.138	.014	.027
12	0.140	.062	-.044	.037	-.031	.131	-.114	-.058	.056	.076	-.081	.005	.013	.045	-.101	.194	-.182
13	0.120	-.035	.009	-.051	.059	-.152	.168	.011	.015	.049	-.050	-.046	.036	-.104	.081	.132	-.135
14	0.104	-.170	.173	-.020	.023	.027	-.051	-.010	.008	.102	-.093	-.088	.096	.042	-.027	-.034	.026
15	0.092	.118	-.121	.016	-.004	-.039	.049	.016	-.032	.069	-.069	-.130	.136	.064	-.054	-.063	.044
16	0.077	.021	-.018	.007	-.011	-.033	.041	-.075	.061	.132	-.131	.112	-.104	-.002	.005	-.058	.053

same pair, and vice versa (Table 5). For example, EB-R has a correlation of -0.318 with PC-9, while the correlation of PC-9 with EB-L is +0.299. Correlations of PC-11 with ST-L and ST-R are +0.219 and -0.198, respectively, etc. There is nothing intrinsic to PCA analysis that forces this pattern; the last eight components represent FA, the developmental noise in the system. It is worth a comment in passing that the FA components represent only 7% of the total variation and that smaller PCA components are almost always poorly estimated. Whether we can expect to discern a pattern of FA variation, relative to other interesting variables (among them, morphometric modality and heterozygosity) remains to be demonstrated. To pursue such patterns, we first partition the deviation of an individual's phenotype from the population centroid into three components:

$$D^2_{Bi} = [Z_{i1} - \bar{Z}_1]^2 / V_1 + \sum_{j=2}^{8} [Z_{ij} - \bar{Z}_j]^2 / V_j + \sum_{j=9}^{16} [Z_{ij} - \bar{Z}_j]^2 / V_j = D^2_{Bi}\,(\text{size}) + D^2_{Bi}\,(\text{shape}) + D^2_{Bi}\,(\text{asymmetry}). \quad (4)$$

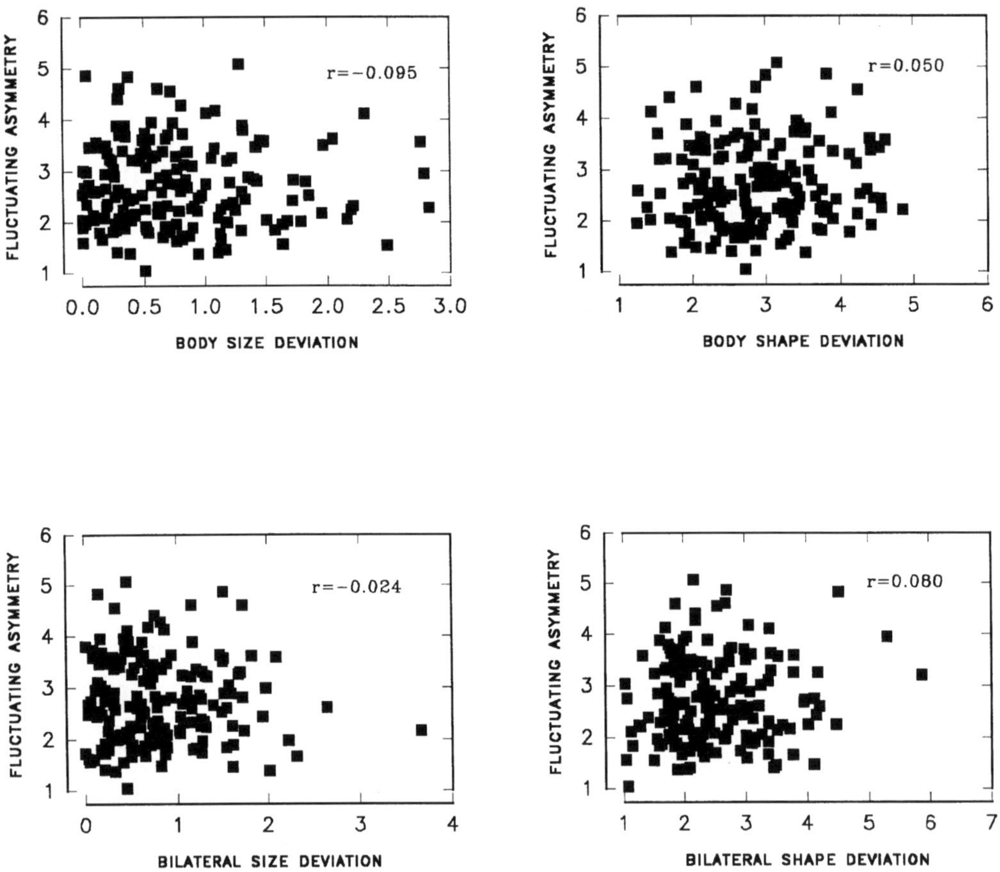

Fig. 1. (a) Relationship between fluctuating asymmetry distance and deviation from central phenotype in size at 10 body mass-size variables; (b) Relationship between fluctuating asymmetry distance and deviation from central phenotype in body shape at 10 mass-size variables; (c) Relationship between fluctuating asymmetry distance and deviation from central phenotype in size at 16 bilateral variables; (d) Relationship between fluctuating asymmetry distance and deviation from central phenotype in shape at 16 bilateral variables.

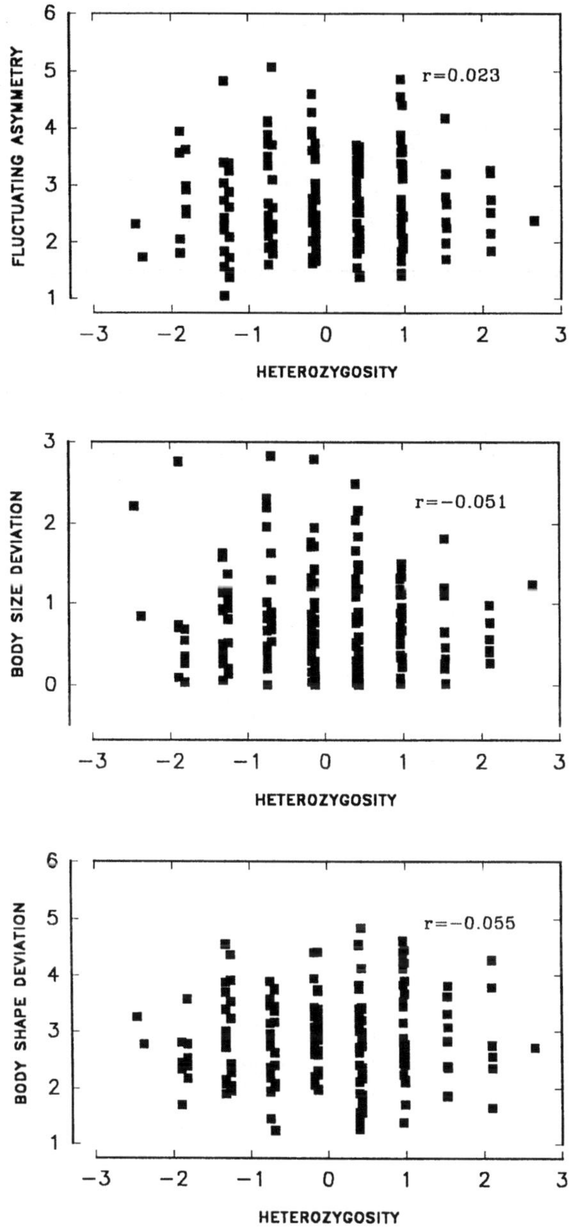

Fig. 2. (a) Relationship between fluctuating asymmetry distance and blood markers heterozygosity. The individual heterozygosity estimates are standardized; (b) Relationship between deviation from central phenotype in body size at 10 mass-size variables and blood markers heterozygosity; (c) Relationship between deviation from central phenotype in body shape at 10 mass-size variables and blood markers heterozygosity.

The third term is a measure of multivariate FA, and is defined naturally as the deviation from zero asymmetry.

The final step of the morphological analysis is to regress an individual's asymmetry distance DBi(FA) against the corresponding mass-size and bilateral measures of 'size and shape' deviations from centrality. The comparisons of FA with the mass-size variables are shown in Figures 1a-1b; FA is correlated with neither. Soule (1982) and Soule and Cousin-Roudy (1982) point out that we should expect an association between the magnitude of FA and the deviation of an individual from the average values of the bilateral characters themselves. The plots of FA against the bilateral variables themselves are shown in Figures 1c-1d; there are no significant correlations (range: -0.024 -0.080).

Heterozygosity, FA and morphology

Numerous studies have reported associations between heterozygosity (H) and univariate FA, as well as between H and univariate centrality; similar tests are in order for our multivariate FA analogue. Figures 2a-2c show plots of H against FA, D^2 (size) and D^2 (shape), respectively. H was computed as the fraction of heterozygous loci in the entire set. More elaborate measures of heterozygosity, such as Smouse's (1986) adaptive distance measure, are possible, but the result is clear enough that H will suffice; there are no significant correlations. The multivariate FA measure was also compared for heterozygous and homozygous individuals at each single locus (not shown); again, no reliable correlations were observed.

The present results are entirely compatible with our comparisons of FA and morphometric modality in adult individuals (Livshits & Smouse, 1993); there is no compelling pattern! In that study, we used the same eight pairs of bilateral traits, along with 13 mass-size variables. We conducted the same sorts of analyses, and reduced the results to plots of multivariate FA against mass-size 'size and shape deviations' and against bilateral 'size and shape deviations'. Collectively, the two sets of results clearly show that FA estimates obtained from morphological traits bear no relation to either morphological modality or to measures of heterozygosity, assessed for a limited number of loci.

Discussion

The predicted correlation of FA and heterozygosity is usually considered from two perspectives: (1) We expect a negative correlation between FA and H individuals sampled from a single panmictic population or from populations connected by regular gene flow, due either to overdominance effects or to the dominance that hides deleterious recessives in the heterozygote (e.g. Soule, 1979; Kat, 1982; Livshits & Kobyliansky, 1984; Leary *et al.*, 1985; Blanco *et al.*, 1990). (2) Theory suggests that we should expect a positive correlation in individuals originating from crosses between populations that have been isolated for many generations, usually attributed to the breakup of differently coadapted gene pools (Zakharov, 1981; Vrijenhoek & Lerman, 1982; Graham & Felley, 1985; Livshits & Kobyliansky, 1985; Markow & Ricker, 1991), a phenomenon sometimes referred to as 'outbreeding depression'.

Theoretical arguments aside, we find no credible support for either claim here. One could argue, particularly in view of departures from panmixia that we described earlier, that one might best study the phenomenon in a panmictic population. Developmental homeostasis may depend less on heterozygosity per se than it does on epistatic and highly coadapted genetic complexes that produce a tightly organized pattern of development (e.g., Futuyma, 1986). The largely fortuitous interactions that emerge from admixture between isolated populations could adversely affect developmental stability, confusing the pattern between FA and heterozygosity. Our sample is not panmictic, probably having suffered some selective attrition, but among the 188 individuals remaining after removal of the 18 outliers, 177 are of Ashkenazi descent. Genetic differences between Ashkenazi populations are small; genetic diversity among them (Nei's Gst) is less than 1% (Livshits, Sokal & Kobyliansky, 1991). In addition, the sampled individuals were first generation immigrants to Israel; they are not themselves admixed. We have no reason to anticipate outbreeding depression, and we should see condition (1) above, not condition (2). The convincing absence of any correlation between FA, on the one hand, and either morphometric modality or heterozygosity on the other hand, has to be viewed as real.

We note that in many other studies where morphological variables are used, correlations between FA and heterozygosity are also not found (e.g. Zink, Smith & Patton, 1985; Patterson & Patton, 1990; Clarke, Oldroyd & Hunt, 1992). In other studies, results are contradictory, i.e., some samples show a relationship and others do not (e.g., Kartavtsev, 1990; Dubrova *et al.*, 1991b). Morphometric variables are subject to considerable ontogenetic changes, due to the effects of growth, obesity variation, aging, labour and physical exercise. It is possible that quantitative morphometrics may not provide the best substrate for studying the relationship between FA and other interesting variables. Be that as it may, we see no relationships here.

On the other hand, such meristic measures as human dermatoglyphics and the number of fin rays or branchial arches in fishes remain virtually unchanged from early embryogenesis until the death of the individual. Published results to date provide almost unvarying support for a correlation between FA and a deviation from modality and for another between FA and H (Mitton, 1978; Jantz and Webb, 1980; Leary *et al.*, 1985; Livshits & Kobyliansky, 1987; Vazquez *et al.*, 1988; Blanco *et al.*, 1990). Dermatoglyphics are also consistent in showing elevated FA in individuals with various genetic anomalies (see review in Livshits & Kobyliansky, 1991). A lack of correlation between FA and morphometric variables or between FA and heterozygosity will not necessarily carry over to the study of dermatoglyphics or other meristic traits. Our only reservation on this point, more a cautionary note than a criticism, is that we need to reexamine multivariate meristic FA, using techniques such as described here, in relation to meristic modality, heterozygosity, and human morbidity.

Acknowledgements

We would like to thank Mr. G. Goldzand for the assistance in the preparation of the manuscript and a pair of anonymous reviewers for helpful comments on a earlier draft of the paper. PES was supported by USDA/NJAES-32102.

References

Anderson, T. W., 1984. An introduction to multivariate-statistical analysis. Wiley and Sons, New York.

Aschel, M. & M. Givon, 1982. Aging among survivors of the Holocaust. Gerontology 21-55: 55-65 (in Hebrew).

Bailit, H. L., P. L. Workman, J. D. Niswander & C. J. Maclean, 1970. Dental asymmetry as an indicator of genetic and environmental conditions in human populations. Human Biology 42: 626-638.

Blanco, G., J. A. Sanchez, E. Vazquez, E. Garcia & J. Rubio, 1990. Superior developmental stability of heterozygotes at enzyme loci in salmo salar L. Aquacul. 84: 199-209.

Clarke, G. M. & J. A. McKenzie, 1987. Developmental stability of insecticide-resistant phenotypes in blowfly: A result of canalizing natural selection. Nature 325: 345-346.

Clarke, G. M., B. P. Oldroyd & P. Hunt, 1992. The genetic basis of developmental stability in Apis mellifera: Heterozygosity versus genetic balance. Evol. 46: 753-762.

De Marinis, F., 1959. The nature of asymmetry and variability in the double bar-eyeless drosophila. Genet. 44: 1101-1111.

Dubrova, Yu. E., I. K. Dambueva, O. N. Kholod, V. D. Prokhorovskaya, E. I. Pushkina & M. L. Blank, 1991a. Influence of mother's heterozygosity on the variation of anthropometric traits in newborns. Genetika (USSR) 27: 2168-2176.

Dubrova, Yu. E., I. K. Dambueva, O. N. Kholod, V. D. Prokhorovskaya, E. I. Pushkina & M. L. Blank, 1991b. Influence of heterozygosity on the variation of anthropometric traits in newborns. Genetika (USSR) 27: 2157-2167.

Ferguson, M., 1986. Developmental stability of rainbow trout hybrids: Genomic coadaptation or heterozygosity. Evol. 40: 323-330.

Futuyma, D. J., 1986. Evolutionary biology. Sinauer associates, inc. Sunderland, Massachusetts.

Golan, R., J. Ben-Ezzer & A. Szeinberg, 1977. Esterase D polymorphism in several population groups in Israel. Hum. Hered. 27: 298-304.

Goldschmidt, R. B., 1955. Theoretical Genetics. Cambridge University Press, Berkeley (L.A.).

Graham, J. H. & J. D. Felley, 1985. Genomic coadaptation and developmental stability within introgressed populations of Enneacanthus gloriousus and E. obesus (Pisces, Centrarchidae). Evol. 39: 104-114.

Hartl, G. H., G. Lang, F. Klein & R. Willing, 1991. Relationship between allozymes, heterozygosity and morphological characters in red deer (Cervus Elaphus) and the influence of selective hunting on allele frequency distrubutions. Hered. 66: 343-350.

Jantz, R. L. & R. S. Webb, 1980. Dermatoglyphic asymmetry as a measure of canalization. Ann. Hum. Biol. 7: 489-493.

Jolicoeur, P., 1963. Bilateral symmetry and asymmetry in limb bones of Martes americana and man. Rev. Can. Biol., 22: 409-432.

Kartavtsev, Yu. F., 1990. Allozyme heterozygosity and morphological homeostatis in pink salmon Oncorhynchus gorbuscha (Pisces:Salmonidae). Genetika (USSR) 8: 1399-1407.

Kat, P. W., 1982. The relationship between heterozygosity for enzyme loci and developmental homeostasis in peripheral populations of aquatic bivalves (Unionidae). The Amer.

Natur. 119: 824-832.

Kertzman, H., 1992. Anthropological and biochemical variables among industrial workers, suffering from hypertension and coronary heart disease. MS Thesis, Tel Aviv University.

Kobyliansky, E., S. Micle, M. Goldschmidt-Natan, B. Arensburg & H. Natan, 1982. Jewish population of the world: genetic likeness and differences. Ann. Hum. Biol. 9: 1-34.

Lahav, M. & A. Szeinberg, 1972. Rec-cell glutamic-pyruvic transminase polymorphism in several population groups in Israel. Hum. Hered. 22: 533-538.

Leary, R. F., F. W. Allendorf, K. L. Knudsen & G. H. Thorgaard, 1985. Heterozygosity and developmental stability in gynogenetic diploid and triploid rainbow trout. Hered. 54: 219-225.

Leary, R. F. & F. W. Allendorf, 1989. Fluctuating asymmetry as an indicator of stress: implication for conservation biology. Trends. Ecol. Evol. 4: 213-217.

Lerner, I. M., 1954. Genetic homeostasis. Oliver & Boyd, London.

Livshits, G. & E. Kobylianskyk, 1984. Biochemical heterozygosity as a predictor of developmental homeostasis in man. Ann. Hum. Genet. 48: 173-184.

Livshits, G. & E. Kobyliansky, 1985. Lerner's concept of developmental homeostasis and the problem of heterozygosity level in natural populations. Heredity 55: 341-353.

Livshits, G. & E. Kobyliansky, 1987. Dermatoglyphic traits as possible markers of developmental processes in humans. Am. J. Med. Genet. 26: 111-122.

Livshits, G. & E. Kobyliansky, 1989. Study of genetic variance in the fluctuating asymmetry of anthropometrical traits. Ann. Hum. Biol. 16: 121-129.

Livshits, G. & E. Kobyliansky, 1991. Fluctuating asymmetry as a possible measure of developmental homeostasis in humans. Hum. Biol. 63: 441-466.

Livshits, G., R. R. Sokal & E. Kobyliansky, 1991. Genetic affinities of Jewish populations. Am. J. Hum. Genet. 49: 131-146.

Livshits, G. & P. Smouse, 1993. Multivariate bilateral asymmetry in human adults. Hum. Biol. 65: 547-578.

Markow, T. A. & K. Wandler, 1986. Fluctuating dermatoglyphic asymmetry and the genetics of liability in schizophrenia. Psych. Res. 19: 323-328.

Markow, T. A. & J. P. Ricker, 1991. Developmental stability in hybrids between the sibling species pair, drosophila melanogaster and drosophila simulans. Genetica 84: 115-121.

Mather, K., 1953. Genetical control of stability in development. Hered. 7: 297-336.

Mellor, C. S., 1992. Dermatoglyphic evidence of fluctuating asymmetry in schizophrenia. Br. J. Psychiatry 160: 467-472.

Mitton, J. B., 1978. Relationship between heterozygosity for enzyme loci and variation of morphological characters in natural populations. Nature 273: 661-662.

Patterson, B. S. & J. L. Patton, 1990. Fluctuating asymmetry and allozymic heterozygosity among natural populations of pocket gophers (Thomomys Bottae). Biol. J. Linn. Soc. 40: 21-36.

Quattro, J. M. & R. C. Vrijenhoek, 1989. Fitness differences among remnant populations of the endangered sonoran topminnow. Science 245: 976-978.

Rasmusson, M., 1960. Frequency of morphological deviations as a criterion of developmental stability. Hereditas 46: 511-536.

Reeve, E. C. R., 1960. Some genetic tests on asymmetry of sternopleural chaeta number in drosophila. Genet. Res. 1: 151-172.

Smouse, P. E., 1986. The fitness consequences of multiple-locus heterozygosity under the multiplication overdominance and inbreeding depression models. Evolution 40: 946-957.

Soule, M. E., 1979. Heterozygosity and developmental stability: Another look. Evol. 33: 396-401.

Soule, M. E., 1982. Allomeric variation. 1: The theory and some consequences. Am. Nat. 120: 751-764.

Soule, M. E. & J. Cuzin-Roudy, 1982. Allomeric variation. 2: Developmental instability of extreme phenotypes. Am. Nat. 120: 765-786.

Szeinberg, A., S. Pipano, Z. Rozansky & N. Ravia, 1971. Frequency of red cell adenosine deaminase phenotypes in several population groups in Israel. Hum. Hered. 21: 357-361.

Szeinberg, A. & S. Tomashevsky-Tamir, 1971. Red cell adenylate kinase and phosphoglucomutase polymorphisms in several population groups in Israel. Hum. Hered. 21: 289-296.

Thoday, J. M., 1958. Homeostasis in a selection experiment. Hered. 12: 401-415.

Vazquez, E., G. Blanco, J. A. Sanchez, E. Garcia & J. Rubio, 1988. Relationship between biochemical heterozygosity and morphological variability in a stock of Salmos salar L. Rev. Biol. Univ. Oviedo 6: 3-15.

Vrijenhoek, R. C. & S. Lerman, 1982. Heterozygosity and developmental stability under sexual and asexual breeding systems. Evol. 36: 768-776.

Zakharov, V. M., 1981. Fluctuating asymmetry as an index of developmental homeostasis. Genetika (USSR) 13: 241-256.

Zakharov, V. M., 1987. Asymmetry of Animals. Nauka, Moscow.

Zink, R., M. Smith & J. L. Patton, 1985. Association between heterozygosity and morphological variance. J. Hered. 76: 415-420.

T. A. Markow (ed.), Developmental Instability: Its Origins and Evolutionary Implications, 169–185, 1994.

Shaping intraspecific variation: development, ecology and the evolution of morphology and life history variation in tiger salamanders

James P. Collins[1], Kim E. Zerba[2] & Michael J. Sredl[3]
[1] *Department of Zoology, Arizona State University, Tempe, AZ 85287-1501, USA*
[2] *Department of Human Genetics, University of Michigan, Ann Arbor, MI 48109, USA*
[3] *Nongame Branch, Arizona Game and Fish Department, 221 West Greenway Road, Phoenix, AZ 85023-4312, USA*

Received and accepted 24 May 1993

Key words: fluctuating asymmetry, heterochrony, intraspecific, morphology, polymorphism

Abstract

The tiger salamander, *Ambystoma tigrinum*, is a geographically widespread, morphologically variable, polytipic species. It is among the most variable species of salamanders in morphology and life history with two larval morphs (typical and cannibal) and three adult morphs (metamorphosed, typical branchiate, cannibal branchiate) that vary in frequency between subspecies and between populations within subspecies. We report morphometric evidence suggesting that branchiate cannibals arose through intraspecific change in the onset or timing of development resulting in the wider head and hypertrophied tooth-bearing skull bones characteristic of this phenotype. We also quantified bilateral symmetry of gill raker counts and abnormalities, then evaluated fluctuating asymmetry as a measure of the developmental stability of each morph. There was a significant interaction between fluctuating asymmetry of developmental abnormalities in cannibals and typicals and the locality where they were collected, suggesting that relative stability of each phenotype could vary among populations. While altered timing of developmental events appears to have a role in the evolution and maintenance of morphs, novel phenotypes persist only under favorable ecological conditions. Predictability of the aquatic habitat, genetic variation, kinship, body size, intraspecific competition and predation all affect expression and survival of the morphs in *A. tigrinum*. This taxon provides an excellent model for understanding the diversity and complexity of developmental and ecological variables controlling the evolution and maintenance of novel phenotypes.

Introduction

Simultaneous occurrence of two or more distinct morphs of a species is a polymorphism if the rarest type is not maintained by recurrent mutation (Ford, 1965). Within the context of the Evolutionary Synthesis, 'conspicuous polymorphisms' (Wright, 1968), such as whole body color, were used to study ecological and genetic factors associated with the origin and maintenance of discontinuous variation in general (e.g., Dobzhansky, 1937), and some kinds of geographic races in particular (e.g., Mayr, 1942). Wright (1978) noted that such polymorphisms are interesting as adaptations and easy to study, but he relegated them to a peripheral role in evolution arguing that they are not typical of the variation on which evolution depends. There may be conditions, however, under which a conspicuous polymorphism could play a significant role in rapid phyletic divergence (Mather, 1955; Alexander & Bigelow, 1960; West-Eberhard, 1986). Stanley (1979: 171-172) elaborated this argument: 'A basic question is how much of the ultimate variability that emerges in adaptive radiation arises as polymorphism within the original species, perhaps later to be fixed by the interposition of breeding barriers... Rapid divergence in this scheme, though preceding reproductive isolation, would still occur almost simultaneously with it on a geologic scale of time, making the described mechanism a form of quantum speciation. Thus, I am in agreement with other authors (Eldredge & Gould, 1972; White, 1978, p. 9) that the quantum morphologic step may commonly predate a population's attainment of formal status as a discrete species.'

West-Eberhard (1986) proposed an 'alternative-adaptation hypothesis' linking conspicuous intraspecific polymorphisms and speciation. Once alternative phenotypes become established in a population, she argued, ecological conditions can then arise that favor only one form. Release of the genome from constraints of accommodating multiple alternatives then facilitates speciation by accentuating divergence from the original population. Ultimately, a new lineage emerges that is characterized by the newly fixed and modified form. West-Eberhard provided several reasons to consider this model. First, speciation via this scheme could be frequent since intraspecific alternative phenotypes are more common than is usually realized as exemplified by protozoa, rotifers, mites, insects, birds, amphibians, mammals, fungi, and higher plants. This hypothesis also provides a clear framework in which to explore the relationship between ecology, development, and evolution. Finally, her model suggests mechanisms for understanding how natural selection can favor the evolution of 'macroevolutionary' changes in design, including phylogenetically important changes caused by heterochrony (Stanley, 1979; Wright, 1982). Heterochrony is phyletic change in the onset or timing of development, so that appearance or rate of development of a feature in a descendant ontogeny is either accelerated or retarded relative to appearance or rate of development of the same feature in an ancestor's ontogeny (Gould, 1977).

The tiger salamander, *Ambystoma tigrinum* (Green) has a distinctive, complex, polymorphic life history which, it is believed, has been strongly influenced by heterochrony (Gould, 1977). This taxon is one of about 113 species of salamanders with a distinct larval stage in the life cycle. In about 42 of these species (37%), some or all adults in a population are morphologically identical to larvae, but larger, and termed adult branchiate morphs. In most of these 42 species, this morph and a mature

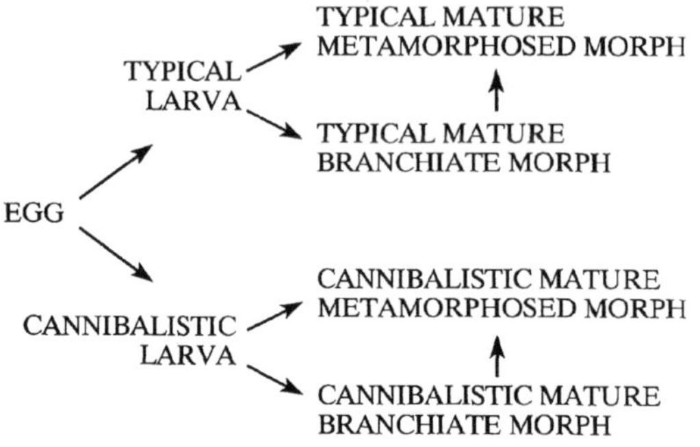

Fig. 1. Summary of variation in life history and morphology in *A. t. nebulosum* or *A. t. mavortium*, subspecies in which cannibalistic larvae and branchiate adults occur regularly.

metamorphosed morph occur in the same population. *A. tigrinum* is unique among these 42 species in that four subspecies each have two juvenile morphs as well (Fig. 1). Some populations of *A. t. mavortium* Baird, *A. t. nebulosum* Hallowell, *A. t. tigrinum* Dunn, and *A. t. melanostictum* Baird have typical larvae and cannibalistic larvae (Powers, 1903, 1907; Gehlbach, 1967; Lannoo & Bachmann, 1984; Rilley, Lauder & Collins, 1992).

Our objective in this study was twofold. First, we sought to analyze the role of changes in developmental timing in the origin of this complex polymorphism. Several arguments suggest mature typical branchiate morphs have evolved via 'intraspecific heterochrony', where juvenile characters are retained by adult ontogenetic stages. We tested the hypothesis that branchiate cannibals also have evolved through delayed differentiation relative to typical larvae. Second, we evaluated the developmental bases of the polymorphism. If the evolutionarily derived cannibalistic phenotype evolved via a disruption of the developmental program of animals with a typical morphology, then cannibals could be viewed as abnormal morphological deviants, or 'phenodeviants' (Lerner, 1954). We used departure from bilateral symmetry as a measure of developmental stability, or homeostasis, of typical and cannibal phenotyes. The cannibals, as phenodeviants, would be expected to exhibit a disruption of bilaterally symmetrical expression of morphometric traits.

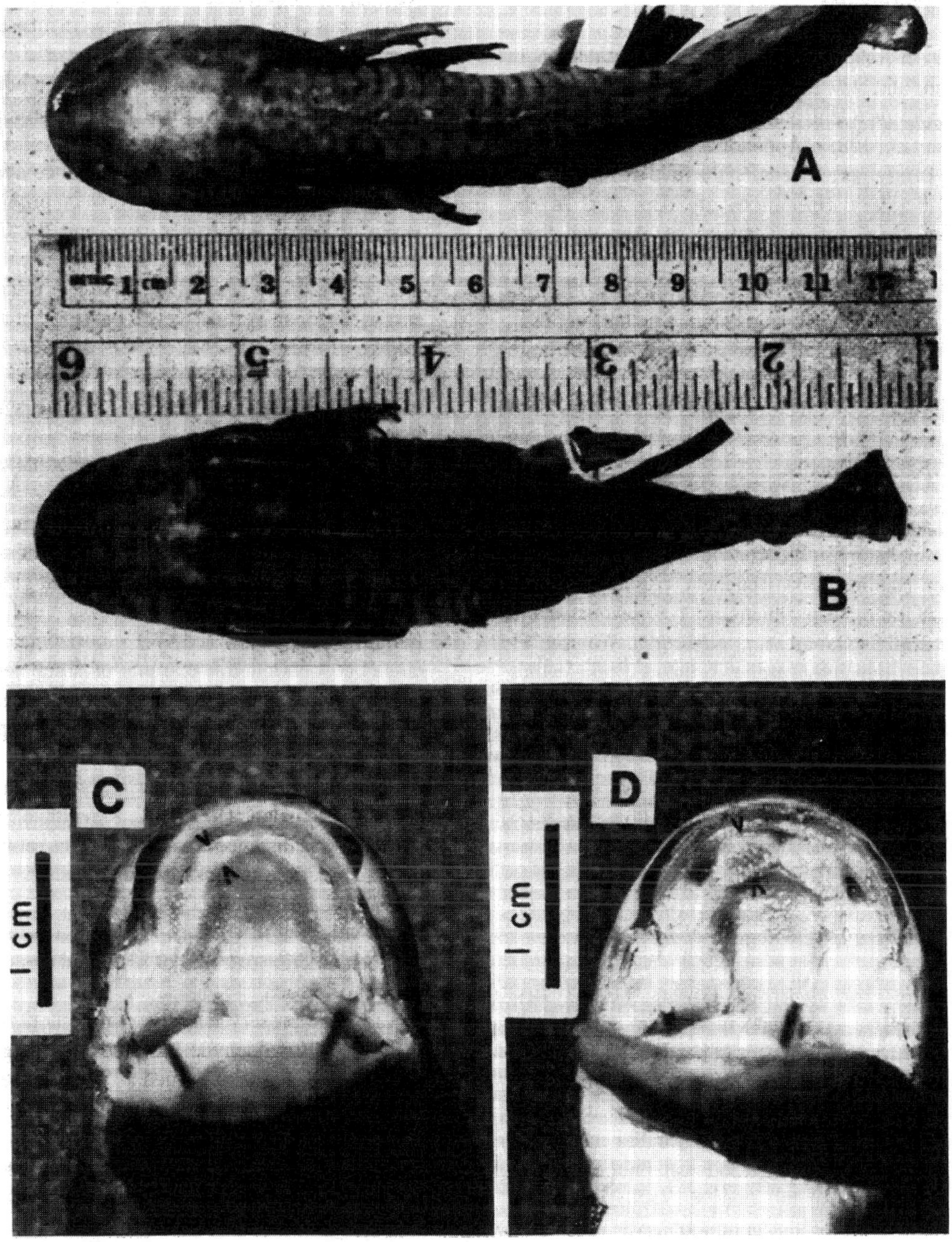

Fig. 2. Views of typical and cannibalistic morphs in *A. t. nebulosum*. Each specimen is the same length ± 2 mm. A, B. Dorsal view of cannibal and typical, respectively. C, D. Enlarged view of vomerine ridge (arrows) in typical and cannibal, respectively.

Materials and methods

Study populations

Ontogeny samples. We sampled ten populations in Arizona and New Mexico, all within the range of *A. t. nebulosum* (Gehlbach, 1965, 1967; Collins, 1981; Jones *et al.*, 1988) [Appendix]. The smallest larvae came from eggs collected in the field and hatched in the laboratory. Specimens are deposited in the Lower Vertebrate collections at Arizona State University (ASU).

We collected 2232 branchiate morphs. We excluded specimens with any sign of metamorphosis, beginning with a reduction in dorsal fin. For large collections we measured from 10 to 20 of the largest specimens and then randomly chose from 25 to 50 other animals. Our resulting sample for analysis was 1104 individuals.

Asymmetry samples. We collected animals from two populations: Old Apache Railroad Tank and Dexter National Fish Hatchery [Appendix]. Old Apache Railroad specimens are *A. t. nebulosum* and those from Dexter are *A. t. mavortium*. We used 19 animals from Old Apache Railroad Tank (cannibals, n = 5; typicals, n = 14) and 22 animals from Dexter National Fish Hatchery (cannibals, n = 11; typicals, n = 11) for analysis. Snout-vent length of larvae varied from 42 to 86 mm. We classified animals by morph using criteria in Powers (1907) and Pedersen (1991).

Description of morphological characters

Ontogeny morphometrics. Powers (1903) first described two branchiate morphs in *A. tigrinum* primarily using differences in head shape and number of vomerine teeth. The larger morph '... is separated from the ordinary type by its much broader and longer head, with parallel instead of converging sides; its enormous gap [sic] of mouth; its flattened or usually concave profile, instead of the even curve of the common larva and adult' (Powers, 1903: 407). He named this the broad-head or cannibalistic morph (Powers, 1907). The vomerine ridge in typical larvae had 1-3 rows of small teeth which were, '... almost indistinguishable points forming little more than a white line;' however, broad-head morphs had 4-6 rows of vomerine teeth (Powers 1907: 263) (Fig. 2).

We quantified differences between morphs using nine measures (Fig. 3); all distances were measured in mm.

1. Width at jaws (JAW) - distance between widest extremes of head at vertical level at proximal angle of upper and lower jaw.
2. Width at mid-parietal (PAR) - distance across head at vertical level midway between proximal angle of upper and lower jaws and posterior margin of eyes.
3. Ocular width of head (OCULAR) - distance across head at vertical level passing through middle of eyes.
4. Nasal width of head (NASAL) - distance between mandibles at vertical level passing through nares.
5. Length of upper jaw (UPJAW) - horizontal distance between anterior extreme of tissue above premaxillary suture and posterior extreme of muscle masses between proximal angle of skull and jaw, and anterior extreme of lowest gill ramus.
6. Length of lower jaw (LJAW) - horizontal distance between anterior extreme of tissue above dentary suture and posterior extreme of muscle masses between proximal angle of skull and jaw, and anterior extreme of lowest gill ramus.
7. Anterior tooth-bearing vomerine projection on left side of palate (VOMER) - width of this structure was distance at a vertical level between median anterior and posterior margins.
8. Mass to nearest mg (WT).
9. Snout-vent length (SVL) - horizontal distance between anterior extreme of skull and posterior margin of cloaca.

Larvae were preserved for varying amounts of time in 10% formaldehyde buffered with magnesium carbonate. We transformed all measurements to natural logarithms before analysis.

Asymmetry morphometrics. We compared total number of gill rakers (hereafter, 'count') and total number of gill raker abnormalities (hereafter, 'abnormalities') for the posterior portion of gill arches 2 and 5 and the anterior and posterior portions of gill arches 3 and 4 for both morphs of salamanders. We quantified counts and abnormalities of gill rakers on preserved animals (10% buffered formalin)

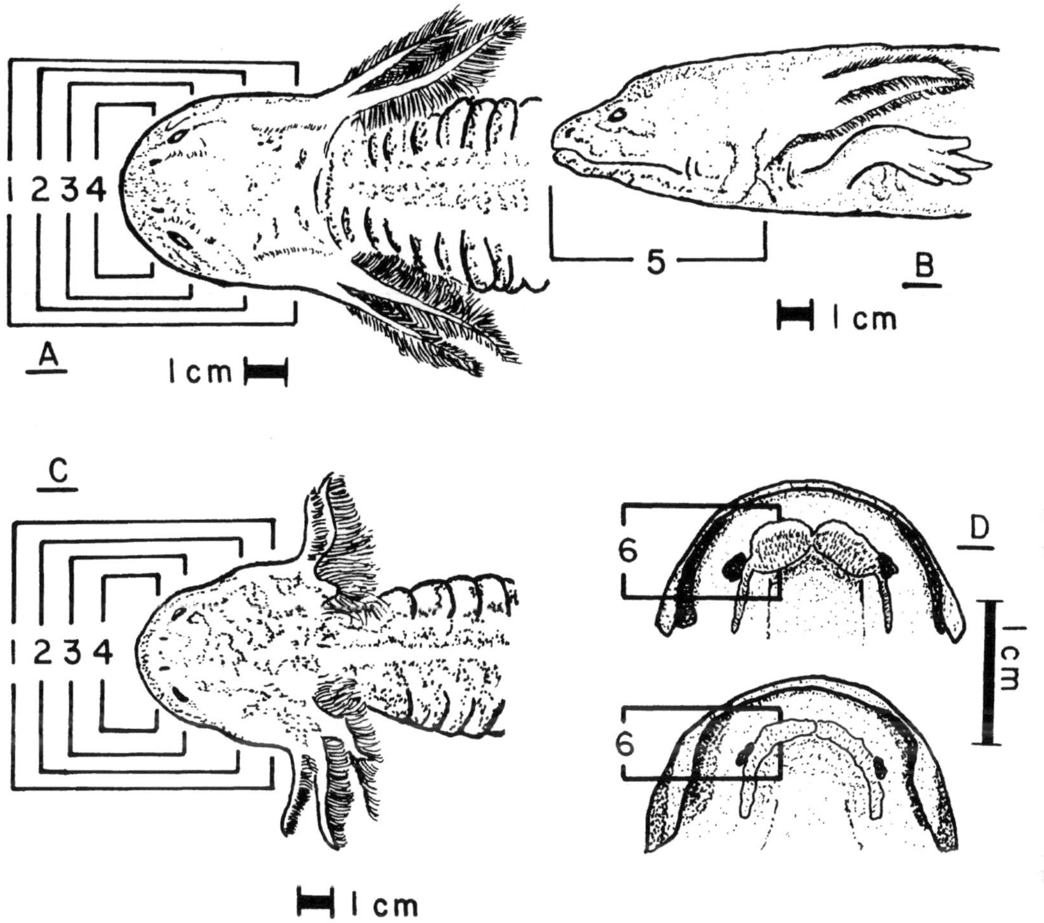

Fig. 3. Illustrations of six characters measured on each specimen. A. Dorsal view of cannibal: 1 = width at jaws, 2 = width at mid-parietal, 3 = ocular width of head, 4 = nasal width of head. B. Lateral view of cannibal, 5 = length of upper jaw. C. Dorsal view of typical larva. D. View of vomerine ridge in cannibalistic (upper) and typical (lower) morph, 6 = width of vomerine ridge.

using a dissecting scope, recording counts and abnormalities for each arch. Development of gill arches is not independent, so we summed the counts and abnormalities of all arches for the left and rigth side for each individual. The signed right (R_i) minus left (L_i) difference was our measure of asymmetry (A_i). We also measured each animal's SVL in mm.

A normal gill raker has a shark-tooth-like shape. When viewed from the side, the shape is pyramidal and the sides may be recurved; it is widest at its base. The edge may be smooth or slightly serrated. Normal rakers do not bend into or out of the buccal cavity. Single gill rakers were identified by relying on the space between respective bases as the means for discriminating between gill rakers.

We classified gill raker abnormalities into eight categories:

1. Fused - two or more gill rakers with bases and tips in such proximity that they are no longer shaped like pyramids.
2. Split - a single gill raker base that has two or more tips.
3. Round - a gill raker base with a very broad or no tip.
4. Branched - a gill raker with a single tip project-

ing vertically and two tips projecting horizontally or nearly so.

5. Bent - one or more gill rakers that bend into or out of the buccal cavity.
6. Spoon-shaped - a gill raker with a midsection wider than the base.
7. Gill raker gaps - spaces in the gill arch where rakers are normally present.
8. Misdirection of gill rakers - a misalignment of gill rakers and gill arch.

Statistical analyses

Ontogeny analysis. We quantified variation in 'size' and 'shape' as follows.

1. Principal Component Analysis (PCA). Using PCA it is common to interpret the first component with coefficients all of the same sign as indicating size variation, while components with positive and negative signs indicate shape variation (Pimentel, 1979; Reyment, Blackith & Campbell, 1984).
2. Somers' Method (SM). This approach constrains conventional PCA by adjusting the correlation matrix for the effects of a theoretical isometric size vector before calculating the remaining eigenvectors (Jolicoeur, 1963). This is done because potential size variation present in apparent shape vectors may mask 'true' shape relationships (Humphries *et al.*, 1981). After this correction for theoretical variation in size, subsequent components summarize information on shape and random variation. This size-constrained method can separate variation in isometric size from variation in shape more effectively than conventional PCA, but there are potential problems with the residual correlation matrix (Somers, 1986). For example, eigenvectors of the residual correlation matrix will be orthogonal to the theoretical isometric size vector only if the average correlation in each row and column of the original correlation matrix is the same (Rohlf & Bookstein, 1987). As suggested by Rohlf and Bookstein (1987) we used Burnaby's Method (Burnaby, 1966) to remove the effects of isometric size from the original correlation matrix so the eigenvectors of the reduced correlation matrix would be orthogonal to isometric size as Somers (1986) intended.
3. Burnaby's Method (BM). This approach allows computation of a size-adjusted PCA by correcting for the size vector of the within group variance-covariance matrix (Burnaby, 1966). Axes produced by this analysis are based on variation orthogonal to the vectors corresponding to variables held constant.
4. Sheared Principal Component Analysis (SPCA). In this method, size is removed using coefficients derived from regression of general size onto principal mean-components centered by group. General size is estimated by the principal axis of the within-group covariance matrix of log-transformed data. Residuals from regressing general size on to the transformed axes approximate a shape-discriminating factor uncorrelated with size within a group, and displaying interpopulation shape differences borne by principal components one and two (Humphries *et al.*, 1981; Bookstein *et al.*, 1985).

Rohlf and Bookstein (1987) discuss limitations of the original formulation of the shear method and propose a superior alternative formulation. We used this Alternative SPCA (ASPCA) for analyzing our data. Rohlf and Bookstein (1987) also note that Burnaby's Method (1966) is the best technique for describing shape since it uses variables clearly orthogonal to a variable representing size, but for most data sets methods 2-4 yield similar results. This was true for our data. For some components we changed signs of all coefficients to facilitate comparisons among methods (Pimentel, 1979).

Asymmetry analysis. In general, we followed the recommendations of Palmer and Strobeck (1986) testing assumptions for all models. Basic statistical analyses follow Sokal and Rohlf (1981); level of significance for all tests was $P < 0.05$.

1. Testing for Size Dependence. We regressed right minus left difference of gill raker count on SVL to examine the relationship between size and asymmetry of counts. Similarly, we examined the relationship between SVL and right minus left difference of abnormalities regressing counts of gill raker abnormalities on SVL.
2. Testing for Directional Asymmetry (DA). We used a two-way mixed model ANOVA [side (fixed) X individuals (random)] to detect differences in DA for counts and abnormalities.
3. Testing for Antisymmetry (AS). We examined AS by testing for departures from normality c

counts and abnormalities using Shapiro-Wilk statistic (W) and examining values of kurtosis (g_2, Sokal & Rohlf, 1981).

4. Testing for Fluctuating Asymmetry (FA). We tested for FA via a Levene's test (ANOVA on the absolute deviations from population means) of the square-root-transformed data.

Results

Testing for ontogenetic change: isolating size and shape components

Eigenvector 1 of the PCA accounted for 97% of the total phenotypic variation; all coefficients were positive, suggesting this is a general summary of variation in size (Table 1). We interpret the positive and negative coefficients for components 2-9 as summaries of shape (allometric variation) and random variation. A Scree Test (Cattell, 1966) suggested that only factors 1 and 2 summarized substantive phenotypic variation. Pearson product-moment correlations between values of the original variables and predicted component scores also supported the interpretation that components 1 and 2 provided a good summary of sample variation (see Pimentel, 1979 and Johnson & Wichern, 1982 for discussion).

Lacking ages for animals, we used SVL as a general indicator of age to study ontogenetic changes in component scores. Body size, as summarized by PCA component 1, varied directly with SVL and identified no distinct groups. We controlled for size in our analysis using two approaches. SM controls for size without assuming there are distinct groups in the sample; BM and ASPCA also control for size, but assume that there are at least two groups, designated before the analysis.

Using all methods, component 2 emerged primarily as change in VOMER relative to the rest of the phenotype (Table 1). SM did not effectively identify groups in our sample as animals increased in SVL. Relying primarily on the characters of Powers (1903, 1907; see also Pederson, 1991 and our Fig. 2), we classified animals as typical or cannibal morphs. Results were similar using BM or ASPCA and are summarized in Figure 4 for ASPCA. Variation in component 2 increased with

Table 1. Component loadings for multivariate analyses. See text for definition of characters/acronyms.

	Unmodified principal components				
Variables	1	2	3	4	5
WT	0.33447	-0.21366	-0.38461	0.41184	0.61291
SVL	0.33489	-0.22592	-0.26325	0.29373	-0.14800
JAW	0.33655	-0.09629	-0.26543	-0.32672	-0.07057
PAR	0.33697	-0.05097	-0.19335	-0.35271	-0.23804
OCULAR	0.33687	-0.05285	-0.16466	-0.35046	-0.24842
NASAL	0.33498	-0.04639	0.53916	-0.42063	0.58110
UJAW	0.33612	-0.09354	0.40749	0.31853	-0.26272
LJAW	0.33609	-0.08865	0.42085	0.30677	0.26480
VOMER	0.31231	0.93268	0.09252	0.13282	0.04757
	Somers' size-constrained PC				
Variables	1	2	3	4	5
WT	0.33333	-0.21647	0.38704	-0.41614	-0.59253
SVL	0.33333	-0.22904	0.26549	-0.29845	-0.12723
JAW	0.33333	-0.10468	0.27339	0.33078	-0.07083
PAR	0.33333	-0.06099	0.19039	0.36066	-0.23739
OCULAR	0.33333	-0.06274	0.16181	0.35803	-0.24753
NASAL	0.33333	-0.05457	-0.53980	0.40655	0.59752
UJAW	0.33333	-0.10166	-0.40740	-0.31290	-0.27382
LJAW	0.33333	-0.09686	-0.42083	-0.30109	-0.27611
VOMER	0.33333	0.92702	0.08991	-0.12744	0.04286

Table 1. Continued.

Variables	Burnaby's adjusted PC 1	2	3	4	5
WT	-0.00573	-0.39842	-0.40563	0.20533	-0.11559
SVL	-0.00103	-0.05167	0.15524	-0.29384	0.68887
JAW	0.00209	0.07820	0.07847	-0.47224	-0.13920
PAR	0.00296	0.13335	0.11831	-0.45466	-0.06949
OCULAR	0.00290	0.13437	0.13162	-0.43154	-0.08214
NASAL	0.00270	0.19163	0.40815	0.08986	-0.62475
UJAW	0.00113	0.15390	0.42749	0.33889	0.21166
LJAW	0.01123	0.16178	0.43580	0.34552	0.19461
VOMER	0.01192	0.84253	-0.48380	0.15504	0.09087

Variables	Alternative PC 1	2	3	4	5
WT	±0.78161	-0.39846	-0.40509	0.20532	-0.11517
SVL	±0.24010	-0.05168	0.15583	-0.29381	0.68880
JAW	±0.22181	0.07822	0.07896	-0.47220	-0.13918
PAR	±0.21556	0.13337	0.11800	-0.45461	-0.06949
OCULAR	±0.21611	0.13440	0.13141	-0.43150	-0.08214
NASAL	±0.20311	0.19164	0.40881	0.08984	-0.62468
UJAW	±0.24141	0.15390	0.42765	0.33885	0.21164
LJAW	±0.24260	0.16178	0.43586	0.34548	0.19459
VOMER	±0.17517	0.84261	-0.48375	0.11500	0.09085

SVL, with cannibals differentiating from typical forms as animals exceeded about 30 mm SVL (Fig. 4, 5B).

Most morphological variation in branchiate salamanders is due to change in size. Variation in component 2 accounts for a small proportion of total phenotypic change (about 2%), but suggests that differences between morphs in pattern of vomerine growth relative to the remainder of the phenotype account for changes in shape. Although the percentage of variation attributable to a component may be small, the component may nonetheless be meaningful biologically (Gauch, 1982). For this reason we also evaluated component 4, since Powers (1907) distinguished typical and cannibal morphs using head width as well as vomerine characters (Fig. 2). Component 4 is a contrast primarily between three measures of head with (JAW, PAR, OCULAR) and two measures of head length (UJAW, LJAW) (Table 1). Of components 3-9, this provided the best summary of changes in cranial characters. No multivariate method completely separated typical and cannibal morphs along this component as SVL increased. Using BM and ASPCA, the best separation occurred for animals from about 30-90 mm SVL when the mean score for component 4 was lower for cannibals compared to typical forms and values for cannibals were most similar to mean values for the smallest size classes of larvae (Figs. 4, 5D). Typical and cannibal salamanders >90 mm SVL were indistinguishable using only component 4.

Testing for FA

Size dependence of FA within samples. Palmer and Strobeck (1986) recommended testing for size-dependence as an essential first step in analyzing for FA. Regression of A_1 on SVL revealed no significant relationship between counts (n = 41, β = 0.32, SE = 0.36, P = 0.36) or abnormalities (n = 41, β = -0.10, SE = 0.10, P = 0.34) and SVL. Size adjustment, therefore, was unnecessary.

Directional asymmetry. A two-way, mixed-model ANOVA [sides (fixed) X individuals (random)] for counts and abnormalities revealed no difference between sides. Therefore, neither counts (ANOVA, $F_{1,40}$ = 0.19, P = 0.67) nor abnormalities (ANOVA, $F_{1,40}$ = 0.11, P = 0.74) exhibited significant DA.

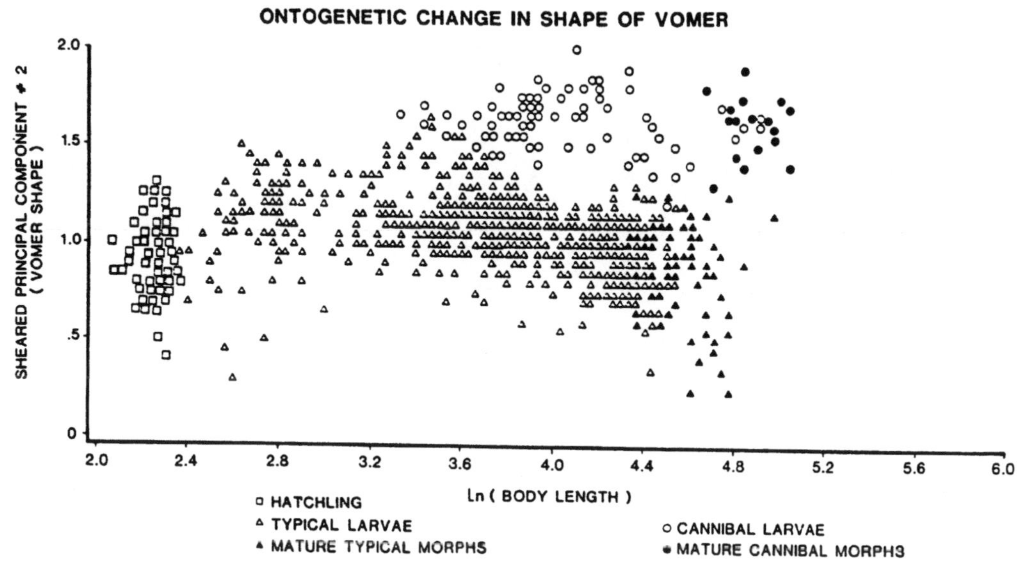

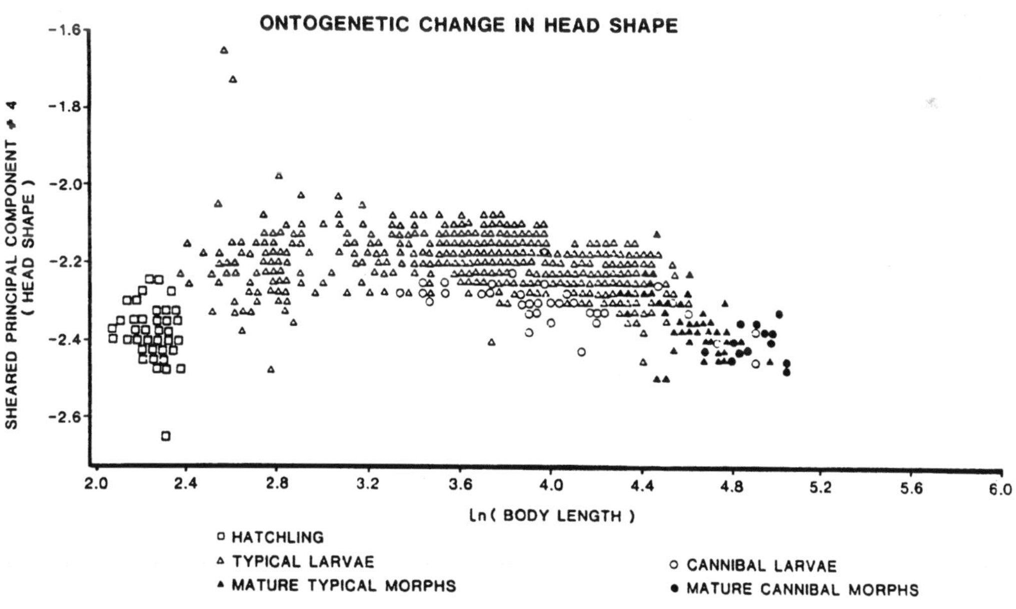

Fig. 4. Relationship to SVL of component scores for components 2 and 4 using ASPCA.

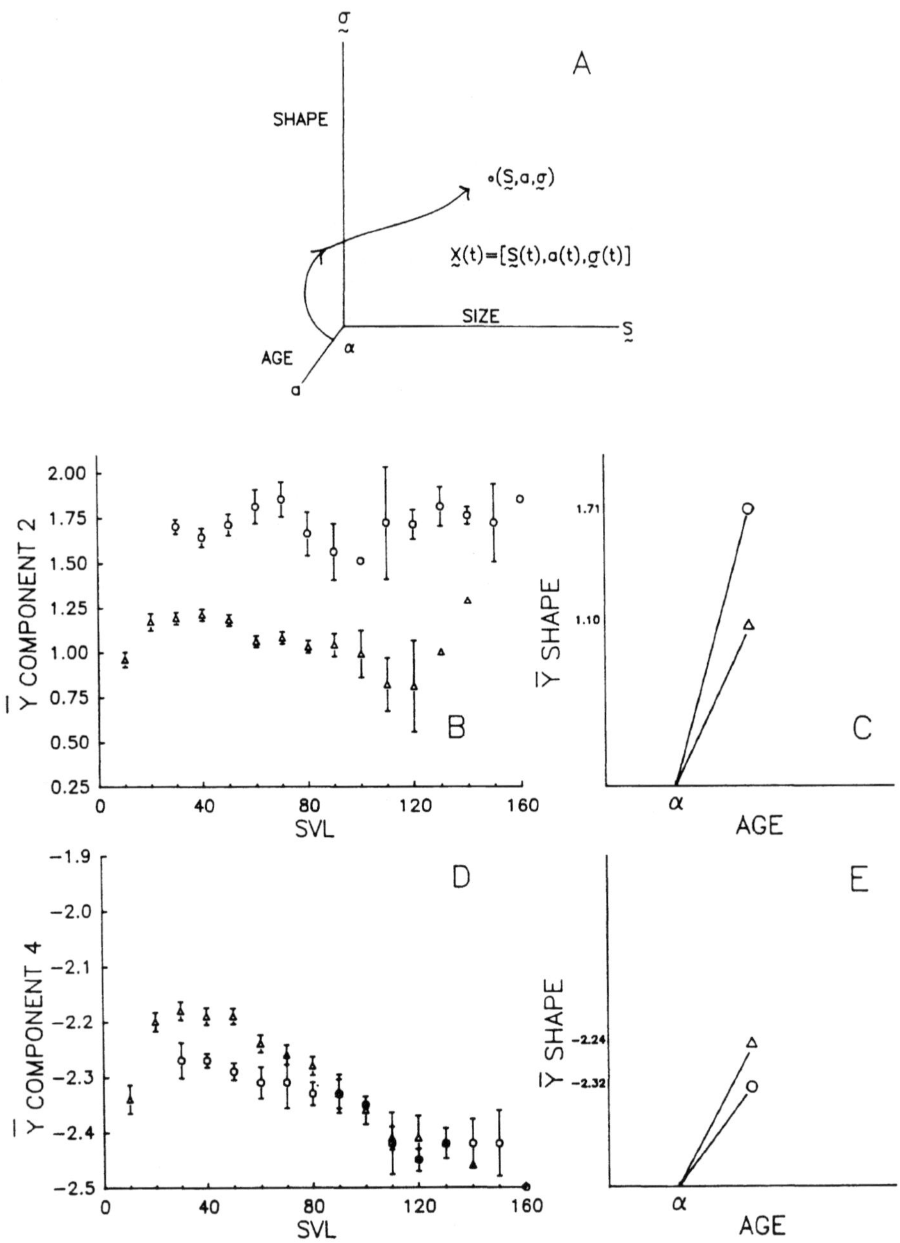

Fig. 5. Ontogenetic trajectory describing growth of an ideal individual using size (S), a measure of shape (σ), and age (a) (after Alberch *et al.*, 1979). B. Mean component 2 scores [± 2 SE] for typical forms (Δ) and cannibals (O) using BM. C. An 'intraspecific descendant' (O) produced by a positive perturbation in mean shape growth rate relative to its 'intraspecific ancestor' (Δ). D. Mean component 4 scores [± 2 SE] using BM. Symbols as in B. E. An 'intraspecific descendant' (O) produced by a negative perturbation in mean shape growth rate relative to its 'intraspecific ancestor' (Δ).

Table 2. Sample sizes (n), means, and standard errors (SE) for A_i by locality and morph: $A_i = | R_i - L_i |$, where R_i = character state (counts or abnormalities) on right side and L_i = character state on left side; Apache = Old Apache Railroad Tank; Dexter = Dexter National Fish Hatchery; C = Cannibal morph; T = Typical morph.

Locality	Morph	n	Mean $A_i \pm$ SE
A. Counts of gill rakers			
Apache	C	5	2.20 ± 1.20
	T	14	1.64 ± 1.02
Dexter	C	11	0.64 ± 0.89
	T	11	0.91 ± 0.67
B. Number of gill raker abnormalities			
Apache	C	5	0.40 ± 0.68
	T	14	1.57 ± 1.32
Dexter	C	11	2.09 ± 4.94
	T	11	1.55 ± 1.88

Antisymmetry. AS can be detected by examining the distribution of A_i for platykurtosis. All groups were normally distributed [Shapiro-Wilk statistic (W), all $P > 0.10$], except abnormalities for cannibals from Old Apache Railroad Tank. Examination of g_2 for this distribution indicated a leptokurtic, not platykurtic distribution.

Fluctuating asymmetry. A 2 X 2 factorial ANOVA of the absolute deviations from population means

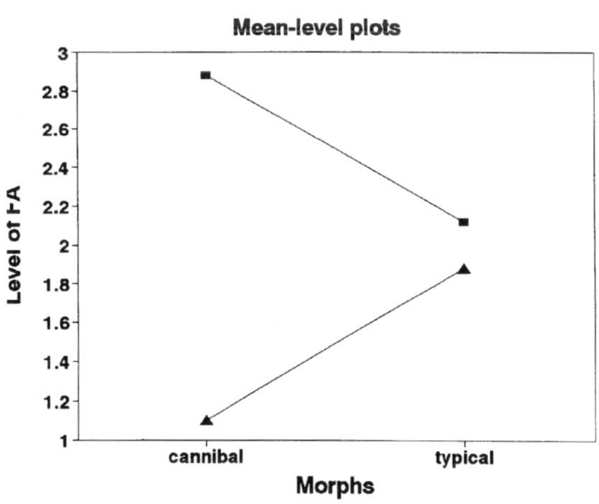

Fig. 6. Mean levels of fluctuating asymmetry relative to morph: ■ = Dexter National Fish Hatchery, ▲ = Old Apache Railroad Tank.

(Table 2) [Levene's Test] revealed no significant differences in the main effects of locality (conditional on morph) and morph (conditional on locality), or interactions of these factors with counts of gill rakers. There were differences between localities in gill raker abnormalities that depended on morph [locality X morph, SS = 5.15, $F_{1,37} = 4.60$, $P = 0.04$; Fig. 6]. Dexter cannibals had higher levels of FA than those from Old Apache Railroad Tank, while typical morphs from the latter had higher levels of FA than Dexter typicals.

Discussion

Development and the evolution of multiple morphs

Heterochrony is phyletic change in onset or timing of development, so that appearance or rate of development of a feature in a descendant ontogeny is either accelerated or retarded relative to the appearance or rate of development of the same feature in an ancestor's ontogeny (Gould, 1977). Any hypothesis concerning heterochrony is therefore a comparative statement requiring a phylogenetic context (Fink, 1982). Complementary arguments from paleontology and ontogeny provide such a context for understanding morphological change in adult *A. tigrinum*.

Paleontology. From Lower Permian to Upper Jurassic, no intermediate types are known between Paleozoic amphibians and primitive, true urodeles (Estes, 1981). The earliest authenticated salamanders are Upper Jurassic, with the main living suborders of salamanders established by end of the Cretaceous (Hecht, 1957). Nonetheless, the simple life cycle of many modern salamanders - eggs laid in water, development of an aquatic larva, and metamorphosis to a terrestrial juvenile/adult - appears to be the primitive life cycle.

Basic features of the amphibian life cycle are first seen in ancient Labyrinthodontia (Szarski, 1962; Baird, 1965), with later forms, especially dissorophids, showing ecological and morphological differences between life cycle phases (DeMar, 1968; Bolt, 1979). Some taxa in all extant salamander families have a larval stage. Recent larvae are phenotypically similar to labyrinthodont amphibians (Szarski, 1957; Baird, 1965), yet share various derived characters (Regal, 1966; Edwards, 1976; Duellman & Treub, 1986). This suggests the larval stage in contemporary salamanders is plesiomorphic (Wiley, 1981), while some larval characters in Recent salamanders are apomorphic relative to labyrinthodont ancestors. Recent species of salamanders in which all members mature and breed in a larval-like form are, therefore, presumably paedomorphic (*sensu* Gould, 1977). In a general sense, the phenotype of ancestral larvae is retained by later ontogenetic stages in some species of modern urodeles; e.g., *Ambystoma mexicanum*. Among salamander taxa, however, there is variance in degree of paedomorphosis exhibited in the terminal stages of development (e.g., Alberch *et al.*, 1979; Duellman & Treub, 1986).

Ontogeny. This standard for recognizing phylogenetic apomorphy complements the out-group criterion (Wiley, 1981). The ontogenetic criterion assumes ontogenetic transformation toward a character reflects phylogenetic development of that ontogeny. Apomorphic characters of a clade will go through ontogenetic stages of development recognizable as plesiomorphic or embryonic characters of that clade's more primitive relatives (Wiley, 1981). Use of this criterion is the object of considerable discussion and debate (Nelson, 1978; Wiley, 1980; Stevens, 1980; Kluge, 1984, 1985; Kluge & Strauss, 1985; DeQueiroz, 1985), but on balance we agree with Wiley's (1981) conclusion that it can be a powerful phylogenetic tool, especially in studies of salamanders.

Among primitive salamanders, dissorophids first showed signs of the separation between larval and metamorphosed life cycle phases that characterize many Recent forms. In Recent species of salamanders with a larval stage, failure to metamorphose and maturation as a branchiate morph is commonly interpreted as reflecting heterochronic change within the lineage. The reasoning supporting this conclusion can be illustrated using *Ambystoma*. Phylogenetic analyses using anatomical (Tihen, 1958; Regal, 1966; Brandon, 1972, 1977; Edwards, 1976) and electrophoretic (Shaffer, 1984a) characters indicate a close relationship among taxa of *Ambystoma*, even though the adult morph in natural populations is always a branchiate morph (e.g., *A. dumerilii*), a metamorphosed morph (e.g., *A. granulosum*), or either (e.g., *A. tigrinum*). Several characters distinguish mature branchiate *Ambystoma* (Ambystomatidae) from phenetically similar taxa like *Necturus* (Proteidae) (Edwards, 1976). Therefore, the most parsimonious hypothesis explaining mature branchiate morphs in species like *A. dumerilii* is evolution via heterochrony.

Evidence of heterochronic change can also come from an ontogenetic transformation in which the presumed ancestral morphology is recovered from the hypothetical derived organism. Evidence for heterochrony in salamanders comes from mature branchiate morphs transforming spontaneously or by thyroxine treatment. Thus 'missing' adult, metamorphosed, ancestral characters are 'discovered' via an ontogenetic transformation. In other words, assuming development is orderly, the 'recovered' character states are interpretable as terminal stages of the hypothetical ancestor (Nelson, 1978; see Shaffer, 1984b concerning the *A. tigrinum* group).

The mature branchiate morph in *A. tigrinum* appears to be one of two instances of evolution of morphological variation via 'intraspecific heterochrony'; the cannibal morph appears to be a second instance. Cannibalism occurs in other salamander taxa (Crump, 1992), but the combination of wide head and hypertrophied, recurved teeth (Pedersen, 1991) associated with this behavior in *A. tigrinum* is unknown in other species. This suggests the typical morph, which largely eats invertebrates and shares many morphological characters with larvae in other taxa, is plesiomorphic and the cannibal morph is apomorphic. Our results extend Pedersen's (1991) conclusion that differential rates of development between the skull (dentigerous bones like the vomer) and the dentition have played a central role in the evolution of this morph from a typical larval-like form.

Alberch *et al.* (1979) used alteration of growth parameters (Fig. 5A) to classify heterochronic changes into two major classes, paedomorphosis and peramorphosis. In the first class, negative perturbations in growth rate and time at which growth ceases during ontogeny produce descendant species shaped like juvenile stages in the presumed ancesta species. Peramorphosis complements neoteny as a second class of heterochronic change. Positive changes in growth rate or time at which growth ceases produce descendants called peramorphs, literally 'shapes beyond', with a shape transcending ancestral shape. Evolution of the cannibal morph apparently involved analogous intraspecific changes.

The wide head of cannibals relative to the presumed ancestral head shape seen in typical morphs places cannibals at one extreme of variation in component 4 and suggests this aspect of their phenotype evolved by delayed development (Fig. 4). Animals can be classified as cannibal or typical morphs after about 30 mm SVL. From 30-90 mm SVL, cannibals develop among those animals in a population with a head shape most similar to hatchlings, which is relatively U-shaped. This suggests that as animals develop, cannibals express a head shape most characteristic of the earliest stages in ontogeny.

Pierce *et al.* (1983) used univariate statistics to analyze variation in 12 cranial characters as a function of size in *A. t. mavortium*. Large head size in cannibals was not simply a consequence of larger body size. For sizes they examined (SVL = 90-120 mm), cannibal head size increased less with SVL than head size of typical forms, which is consistent with our finding that among the largest specimens cannibals and typicals are generally indistinguishable based only on head shape (Fig. 4). They hypothesized that either 1) head size is initially larger in cannibals than in typicals; or, 2) head growth in cannibals initially shows strong positive allometry with SVL, and then isometry or negative allometry in later developmental stages. Our conclusion is similar; namely, cannibals retain the relatively wider head shape characteristic of early ontogenetic stages (Fig. 5D). Very large typical morphs eventually exhibit heads as wide as cannibals because of their increase in size.

Our results also address two points concerning the typical/cannibal polymorphism raised in Reilly, Lauder and Collins (1992). First, gape width is a continuously distributed character at a given SVL (Fig. 4); however, since larval cannibals are the largest, or among the largest, animals in a population at any time (Collins & Holomuzki, 1984), head width is often bimodal among larvae and unimodal for animals >90 mm SVL [animals mature at 76-108 mm SVL (Collins, unpubl. obs.)]. Thus, differences in gape size between typicals and cannibals are greatest early in ontogeny when this difference confers the greatest advantage on animals trying to feed on similarly-sized conspecifics.

The second major conclusion of Reilly, Lauder and Collins (1992) is that hypertrophied vomerine tooth patches distinguish cannibal phenotypes throughout their ontogeny (Fig. 4). This is the primary morphological trait affording cannibals a performance advantage when feeding on conspecifics, and as predicted by Reilly, Lauder and Collins (1992), would be expected to increase as cannibal morphs grow in size and manipulation of relatively large prey becomes important (Fig. 5B). This trait shows an exaggerated morphology relative to vomerine development in typical morphs. In this case, a second form of developmental change is inferred. Differences between typical and cannibal larvae are at the extremes of variation in vomer shape (Fig. 4), a pattern suggesting accelerated development (Fig. 5B,C). Three processes could cause this pattern. A developmental rate change relative to typicals will alter the final morphology of the character, as will changes in the timing of developmental onset or cessation, even if the developmental rate of cannibals is similar to that of typicals. We cannot distinguish among these processes since we cannot recognize larvae <30 mm SVL that will develop as cannibals, hence we cannot describe differences in growth trajectories for presumptive cannibals and typicals (see Emerson, 1986).

Our argument concerning the evolution of the cannibal morph underscores the fact that morphogenesis is complex. Cannibal forms in tiger salamanders illustrate how parts of this morph's phenotype evolved through contrasting developmental processes, acceleration and retardation, relative to the hypothetical ancestral form seen in typical morphs.

Developmental stability is the ability of an organism to develop according to its developmental program despite adverse environmental conditions (Waddington, 1957). We reasoned that animals expressing the evolutionarily-derived, cannibal phenotype might be less capable of buffering themselves against a wide range of environmental conditions and thus would be more likely to exhibit levels of FA. Following Soulé and Cuzin-Roudy (1982), phenotypically extreme individuals should be more asymmetrical for bilateral traits than phenotypically 'normal' individuals. Cannibals, due to their derived features, are among the morphologically extreme individuals within a population (Fig. 4).

We found no significant evidence that morphs differed in levels of SA, AS, or FA in counts of gill rakers, abnormalities conditional on locality, or among localities conditional on morph. There was, however, a significant interaction between morph X locality in FA of developmental abnormalities, suggesting that relative stability of each phenotype could vary among populations. This could be due to differences in selection pressures between localities; alternatively, since the two localities were different subspecies of *A. tigrinum*, this result might reflect differences in the developmental stability of morphs in the respective subspecies. Our results do not provide straightforward support for the prediction that cannibals are less stable developmentally than typicals, but the suggestion that the developmental stability of each phenotype varies with locality is interesting relative to a model like West-Eberhard's (1986).

West-Eberhard (1986) argued that intraspecific alternative adaptations could precede speciation. A prediction of this model is that polymorphic taxa like *A. tigrinum* might exhibit microgeographic and macrogeographic variation in traits like maturation as a branchiate typical or cannibal morph, traits that might ultimately be fixed as differences between taxa. This pattern appears to be present in populations of *A. tigrinum* in Arizona on the Kaibab Plateau. Populations there do not produce mature, branchiate morphs and rarely produce cannibals, whereas populations in the White Mountains commonly produce cannibals and mature, branchiate morphs (Berna, 1990; Pfennig, Loeb & Collins, 1991; Ohlandt, 1992). The evolution of morphological differences through heterochronic change would begin with variation among individuals in development. Thoday (1953) showed that directional selection can cause deterioration of the systems responsible for developmental homeostasis. The differences in developmental stability between salamanders from Arizona (Old Apache Railroad Tank) and New Mexico (Dexter) may be an indication of the kind of developmental variation that would precede the evolution of the macrogeographic differences consistent with West-Eberhard's model. Further evidence that the kind of gene-regulatory flexibility posited by West-Eberhard's (1986) model may occur in *A. t. nebulosum* comes from a study of relative enzyme activities of the products of duplicate loci in each of three enzyme systems in mature branchiate morphs and in mature branchiate morphs forced to metamorphose (Begun & Collins, 1992). The production of such morphs may involve independent regulation of duplicate genes. The results of Begun and Collins (1992) suggest that such regulatory flexibility may be common in the tiger salamander genome.

Ecology and the evolution of multiple morphs

If morphological variation is attributable to 'intraspecific heterochrony', there is a special potential for rapid population and even phylogenetic divergence (Gould, 1977; West-Eberhard, 1986). Developmental differences among individuals and populations would be selected, however, within the context of the population or community in which the animals are found. What selective conditions would favor the persistence of novel phenotypes? Under what conditions might Lerner's (1954) phenodeviants be favored? Tiger salamanders provide some insight into the complexity and diversity of these conditions. Expression of cannibalism is related to density of conspecifics (Collins & Cheek, 1983) and their degree of consanguinity (Pfennig & Collins, 1993; Pfennig, Sherman & Collins, 1993). Eating conspecifics has the advantage of increasing growth rate, but has the disadvantage of increasing exposure to potentially deadly diseases (Pfennig, Loeb & Collins, 1991). Sympatric tadpoles may enhance expression of the cannibal morph, but tadpoles alone cannot induce the morph (Loeb, Collins & Maret, 1993). Multiple variables are clearly affecting expression of cannibalism, suggesting that once a novel phenotype appears in a population, a number of factors must be satisfied if it is to persist.

Phenotypes that are quite different ecologically are being maintained within populations of tiger salamanders, especially in the western U.S.A. Branchiate and metamorphosed forms live in different habitats, while typical and cannibal forms, living in the same habitat, feed largely on different trophic levels (Collins & Holomuzki, 1984; Holomuzki & Collins, 1987). Salamanders and dytiscid beetles are the top predators in most montane, lentic communities in Arizona. In theory, weak interspecific competition could favor substantial intraspecific niche diversification (Roughgarden, 1972; Vrijenhoek, 1984), perhaps even speciation (Seger, 1985). Likewise, intraspecific competition for resources, for example mates, food, or space, can favor the evolution of alternative phenotypes within populations, and perhaps eventually speciation (Knerer & Atwood, 1973; Sage & Selander, 1975; Rosenzweig, 1978; Kirkpatrick, 1982; Seger, 1985; Milligan, 1986; West-Eberhard, 1986; Meyer, 1987). Our results show how relatively small quantitative changes in a phenotype can still lead to significant qualitative differences in behavior and ecology.

Geographically variable conditions also add complexity to understanding the evolution of phenotypic variants. Species richness of fishes, principal competitors/predators of salamanders, is low throughout the ranges of the most highly polymorphic subspecies of tiger salamanders (*mavortium*, *melanostictum*, and *nebulosum*) in the arid western U.S.A. and Mexico, especially compared to the eastern U.S.A. (Minckley, Hendrickson & Bond, 1986). The only reported instance of cannibal larvae in *A. t. tigrinum*, primarily an eastern subspecies, is a population in Iowa at the western edge of its range (Lannoo & Bachmann, 1984). Reduced interspecific competition and intense intraspecific competition could have favored the evolution of the complex intrapopulation variation in morphology and life history found in tiger salamanders in the western United States. Several models now suggest that polymorphisms like the one in *A. tigrinum* may have characterized the ancestors of Mexican taxa like *A. mexicanum*, which are comprised of only adult, branchiate morphs. Within such a scenario extreme phenotypes, Lerner's (1954) phenodeviants, may have a special role in cladogenesis.

Acknowledgements

We thank L. Allison, M. Douglas, S. Reilly and W. L. Minckley for commenting on earlier drafts. The Arizona Game and Fish Department contributed use of equipment to MJS for analyzing asymmetry data. J. Alcock and J. Roth collected many of the morphometric measurements. A grant from the National Science Foundation (BSR-8919901) to JPC provided financial support.

References

Alberch, P., S. J. Gould, G. F. Oster & D. B. Wake, 1979. Size and shape in ontogeny and phylogeny. Paleobiology 5: 296-318.

Alexander, R. D. & R. S. Bigelow, 1960. Allochronic speciation in field crickets and a new species, *Acheta veletis*. Evolution 14: 334-346.

Baird, D., 1965. Paleozoic lepospondyl amphibians. Amer. Zool. 5: 287-294.

Begun, D. J. & J. P. Collins, 1992. Biochemical plasticity in the Arizona tiger salamander (*Ambystoma tigrinum nebulosum*). J. Heredity 83: 224-227.

Berna, H., 1990. Ecology and life history of the tiger salamander, *Ambystoma tigrinum nebulosum* Hallowell, on the Kaibab Plateau. M.S. thesis, Arizona State University.

Bolt, J. R., 1979. *Amphibamus grandiceps* as a juvenile dissorophid: evidence and implications, pp. 529-563 in Mazon Creek Fossils, edited by M. H. Nitecki. Academic Pr., NY.

Bookstein, F. L., B. Chernoff, R. L. Elder, J. M. Humphries, G. R. Smith & R. E. Strauss, 1985. Morphometrics in Evolutionary Biology. Spec. Publ. no. 15. Acad. Nat. Sci. Phil.

Brandon, R. A., 1972. Hybridization between the Mexican salamanders *Ambystoma dumerilii* and *Ambystoma mexicanum* under laboratory conditions. Herpetologica 28: 199-207.

Brandon, R. A., 1977. Interspecific hybridization among Mexican and United States salamanders of the genus *Ambystoma* under laboratory conditions. Herpetologica 33: 133-152.

Burnaby, T. P., 1966. Growth-invariant discriminant functions and generalized distances. Biometrics 22: 96-110.

Cattell, R. B., 1966. The Scree Test for the number of factors. Multivar. Behav. Res. 1: 245-276.

Collins, J. P., 1981. Distribution, habitats and life history variation in the tiger salamander, *Ambystoma tigrinum*, in east-central and southeast Arizona. Copeia 1981: 666-675.

Collins, J. P. & J. R. Holomuzki, 1984. Intraspecific variation in diet within and between trophic morphs in larval tiger salamanders (*Ambystoma tigrinum nebulosum*). Canadian Journal of Zoology 62: 168-174.

Collins, J. P. & J. E. Cheek, 1983. Effect of food and density on development of typical and cannibalistic salamander larvae in *Ambystoma tigrinum nebulosum*. Amer. Zool. 23: 77-84.

Crump, M. L., 1992. Cannibalism in amphibians, pp. 256-276 in Cannibalism: Ecology and Evolution among Diverse Taxa, edited by M. A. Elgar and B. J. Crespi. Oxford University Press, Oxford, UK.

DeMar, R., 1968. The Permian labyrinthodont amphibian *Dissorophus multicinctus*, and adaptations and phylogeny of the family Dissorophidae. J. Paleontology 42: 1210-1242.

DeQueiroz, K., 1985. The ontogenetic method for determining character polarity and its relevance to phylogenetic systematics. Syst. Zool. 34: 280-299.

Dobzhansky, Th., 1937. Genetics and the Origin of Species. Columbia Univ. Pr., NY.

Duellman, W. E. & L. Treub, 1986. Biology of Amphibians. McGraw-Hill, NY.

Edwards, J. L., 1976. Spinal nerves and their bearing on salamander phylogeny. J. Morph. 148: 305-328.

Eldredge, N. & S. J. Gould, 1972. Punctuated equilibria: an alternative to phyletic gradualism, pp. 82-115 in Models in

Paleobiology, edited by T. J. M. Schopf. Freeman, SF.

Emerson, S. B., 1986. Heterochrony and frogs: the relationship of a life history trait to morphological form. Amer. Natur. 127: 167-183.

Estes, R., 1981. Teil 2. Gymnophiona, Caudata. Handbuch der Paläoherpetologie. Gustav Fischer Verlag, Stuttgart.

Fink, W. L., 1982. The conceptual relationship between ontogeny and phylogeny. Paleobiology 8: 254-264.

Ford, E. B., 1965. Genetic Polymorphism. M.I.T. Pr., Cambridge.

Gauch, H. G., Jr., 1982. Noise reduction by eigenvector ordination. Ecology 63: 1643-1649.

Gehlbach, F. R., 1965. The herpetology of the Zuni Mountains region, northwestern New Mexico. Proc. U. S. Nat. Mus. 116: 243-332.

Gehlbach, F. R., 1967. *Ambystoma tigrinum* (Green). Tiger salamander. Catalogue of American Amphibians and Reptiles: 52.1-52.4.

Gould, S. J., 1977. Ontogeny and Phylogeny. Belknap Pr., Cambridge.

Hecht, M. K., 1957. A case of parallel evolution in salamanders. Proc. Zool. Soc., Calcutta, Mookerjee Memor. 1957: 283-292.

Holomuzki, J. R. & J. P. Collins, 1987. Trophic dynamics of a top predator, *Ambystoma tigrinum nebulosum* (Caudata: Ambystomatidae), in a lentic community. Copeia 1987: 949-957.

Humphries, J. M., F. L. Bookstein, B. Chernoff, G. R. Smith, R. L. Elder & S. G. Poss, 1981. Multivariate discrimination by shape in relation to size. Syst. Zool. 30: 291-308.

Johnson, R. A. & D. W. Wichern, 1982. Applied Multivariate Statistical Analysis. Prentice-Hall, Englewood Cliffs, NJ.

Jolicoeur, P., 1963. The multivariate generalization of the allometry equation. Biometrics 19: 497-499.

Jones, T. R., J. P. Collins, T. D. Kocher & J. B. Mitton, 1988. Systematic status and distribution of *Ambystoma tigrinum stebbinsi* Lowe. Copeia: 621-635.

Kirkpatrick, M., 1982. Quantum evolution and punctuated equilibria in continuous genetic characters. Amer. Natur. 119: 833-848.

Kluge, A., 1984. The relevance of parsimony to phylogenetic inference, pp. 24-38 in Cladistics: Perspectives on the Reconstruction of Evolutionary History, edited by T. Duncan and T. Stuessy. Columbia Univ. Pr., NY.

Kluge, A., 1985. Ontogeny and phylogenetic systematics. Cladistics 1: 13-27.

Kluge, A. & R. E. Strauss, 1985. Ontogeny and systematics. Ann. Rev. Ecol. Syst. 16: 247-268.

Knerer, G. & C. E. Atwood, 1973. Diprionid sawflies: polymorphism and speciation. Science 179: 1090-1099.

Lannoo, M. & M. Bachmann, 1984. Aspects of cannibalistic morphs in a population of *A. t. tigrinum* larvae. Amer. Midl. Natur. 122: 103-109.

Lerner, I. M., 1954. Genetic Homeostasis. Dover, N.Y.

Loeb, M. L. G., J. P. Collins & T. J. Maret, 1993, in press. The role of prey in controlling expression of a trophic polymorphism in *Ambystoma tigrinum nebulosum*. Functional Ecology: 29 pp. ms.

Mather, I., 1955. Polymorphism as an outcome of disruptive selection. Evolution 9: 52-61.

Mayr, E., 1942. Systematics and the Origin of Species. Columbia Univ. Pr., NY.

Meyer, A., 1987. Phenotypic plasticity and heterochrony in *Cichlasoma managuense* (Pisces, Cichlidae) and their implications for speciation in cichlid fishes. Evolution 41: 1357-1369.

Milligan, B. G., 1986. Punctuated evolution induced by ecological change. Amer. Natur. 127: 522-532.

Minckley, W. L., D. A. Hendrickson & C. E. Bond, 1986. Geography of western North American freshwater fishes: description and relationships to intracontinental tectonism, pp. 519-613 in The Zoogeography of North American Freshwater Fishes, edited by C. H. Hocutt & E. O. Wiley. Wiley, NY.

Nelson, G., 1978. Ontogeny, phylogeny, paleontology, and the biogenetic law. Syst. Zool. 27: 324-345.

Ohlandt, G., 1992. Microgeographic variation in life history of *Ambystoma tigrinum nebulosum* in Arizona. M.S. thesis, Arizona State University.

Palmer, A. R. & C. Strobeck, 1986. Fluctuating asymmetry: measurement, analysis, patterns. Ann. Rev. Ecol. Syst. 17: 391-421.

Pedersen, S. C., 1991. Dental morphology of the cannibal morph in the tiger salamander, *Ambystoma tigrinum*. Amphibia-Reptilia 12: 1-14.

Pfennig, D. W., M. L. G. Loeb & J. P. Collins, 1991. Pathogens as a factor limiting the spread of cannibalism in tiger salamanders. Oecologia 88: 161-166.

Pfennig, D. W. & J. P. Collins, 1993. Kinship affects morphogenesis in cannibalistic salamanders. Nature 362: 836-838.

Pfennig, D. W., P. W. Sherman & J. P. Collins, 1993, in press. Kinship and cannibalism in polyphenic salamanders. Behavioral Ecology: 43 pp. ms.

Pierce, B. A., J. B. Mitton, L. Jacobson & F. L. Rose, 1983. Head shape and size in cannibal and noncannibal morphs of the tiger salamander from west Texas. Copeia 1983: 1006-1012.

Pimentel, R. A., 1979. Morphometrics. Kendall/Hunt, Dubuque.

Powers, J. H., 1903. The causes of acceleration and retardation in the metamorphosis of *Ambystoma tigrinum*: A preliminary report. Amer. Natur. 37: 385-410.

Powers, J. H., 1907. Morphological variation and its causes in *Ambystoma tigrinum*. Stud. Univ. Nebraska 7: 197-274.

Regal, P. J., 1966. Feeding specializations and the classification of terrestrial salamanders. Evolution 20: 392-407.

Reilly, S. M., G. V. Lauder & J. P. Collins, 1992. Performance consequences of a trophic polymorphism: feeding behavior in typical and cannibal phenotypes of *Ambystoma tigrinum*. Copeia 1992: 672-679.

Reyment, R. A., R. E. Blackith & N. A. Campbell, 1984. Multivariate Morphometrics. 2*nd* ed. Acad. Pr., NY.

Rohlf, F. J. & F. L. Bookstein, 1987. A comment on shearing as a method for 'size correction'. Syst. Zool. 36: 356-367.

Rosenzweig, M. L., 1978. Competitive speciation. Biol. J. Linnean Soc. 10: 275-289.

Roughgarden, J., 1972. Evolution of niche width. Amer. Natur. 106: 683-713.

Sage, R. D. & R. K. Selander, 1975. Trophic radiation through polymorphism in cichlid fishes. Proc. Nat. Acad. Sci. USA 72: 4660-4673.

Seger, J., 1985. Intraspecific resource competition as a cause of

sympatric speciation, pp. 43-53 in Essays in Honor of John Maynard Smith, edited by P. J. Greenwood, P. H. Harvey & M. Slatkin. Cambridge Univ. Pr., Cambridge.
Shaffer, H. B., 1984a. Evolution in a paedomorphic lineage. I. An electrophoretic analysis of the Mexican ambystomatid salamanders. Evolution 38: 1194-1206.
Shaffer, H. B., 1984b. Evolution in a paedomorphic lineage. II. Allometry and form in the Mexican ambystomatid salamanders. Evolution 38: 1207-1218.
Sokal, R. R. & F. J. Rohlf, 1981. Biometry, 2nd ed. Freeman, SF.
Somers, K. M., 1986. Multivariate allometry and removal of size with principal components analysis. Syst. Zool. 35: 359-368.
Soulé, M. E. & J. Cuzin-Roudy, 1982. Allometric variation. 2. Developmental instability of extreme phenotypes. Amer. Natur. 120: 765-786.
Stanely, S. M., 1979. Macroevolution. Freeman, SF.
Stevens, P. F., 1980. Evolutionary polarity of character states. Ann. Rev. Ecol. Syst. 11: 333-358.
Szarski, H., 1957. The origin of the larva and metamorphosis in Amphibia. Amer. Natur. 91: 281-301.
Szarski, H., 1962. The origin of the Amphibia. Q. Rev. Biol. 37: 189-241.
Thoday, J. M., 1953. Homeostasis in a selection experiment. Heredity 12: 401-415.
Tihen, J., 1958. Comments on the osteology and phylogeny of ambystomatid salamanders. Bull. Florida State Mus. 3: 1-51.
Vrijenhoek, R. C., 1984. Ecological differentiation among clones: the frozen niche variation model, pp. 217-231 in Population Biology and Evolution, edited by K. Wöhrmann & V. Loeschcke. Springer-Verlag, Berlin.
Waddington, C. H., 1957. The Strategy of the Genes. MacMillan, NY.
West-Eberhard, M. J., 1986. Alternative adaptations, speciation, and phylogeny (a review). Proc. Nat. Acad. Sci. USA 83: 1388-1392.
White, M. J. D., 1978. Models of Speciation. Freeman, SF.
Wiley, E. O., 1980. Phylogenetic systematics and vicariance biogeography. Syst. Bot. 5: 194-220.
Wiley, E. O., 1981. Phylogenetics: The Theory and Practice of Phylogenetic Systematics. Wiley, NY.
Wright, S. W., 1968. Evolution and the Genetics of Populations. Vol 1. Genetic and Biometric Foundations. Univ. Chicago Pr., Chicago.
Wright, S. W., 1978. Evolution and the Genetics of Populations. Vol 4. Variability within and among Natural Populations. Univ. Chicago Pr., Chicago.
Wright, S. W., 1982. Character change, speciation, and the higher taxa. Evolution 36: 427-443.

Appendix: Localities of specimens used

Arizona

A. t. nebulosum - Apache Co.: Big Meadows Tank #1, ASU 17709-17725[a] (n = 247); Big Meadows Tank #2, ASU 17619 (n = 11); Old Apache Railroad Tank, ASU 19400, 26136 (n = 19). Coconino Co.: Baker Lake, ASU 17612 (n = 172); Limestone Pasture Tank, ASU 17620, 19369-71, 19373 (n = 697); Myrtle Lake, ASU 19374-75 (n = 119). Greenlee Co.: Ackre Lake, ASU 17622-23 (n = 16). Navajo Co.: Lower Cottonwood Tank, ASU 17821, 17878-80, 18543, 19365-66 (n = 209); Snow Tank, ASU 17615-18, 18538, 18542, 18570-71, 18602-03, 18630, 19363-64, 19376 (n = 724).

New Mexico

A. t. nebulosum - Catron Co.: Five Springs Tank, ASU 19368 (n = 8); Lake Erin, ASU 19367 (n = 29).

A. t. mavortium - Chaves Co.: Dexter National Fish Hatchery, ASU 27913 (n = 22).

[a] Arizona State University collection of Lower Vertebrates catalog number.

T. A. Markow (ed.), Developmental Instability: Its Origins and Evolutionary Implications, 187–201, 1994.

Persistence of subtle departures from symmetry over multiple molts in individual brachyuran crabs: relevance to developmental stability

Adam K. Chippindale[1,3] & A. Richard Palmer[1,2]
[1] *Department of Zoology, University of Alberta, Edmonton, Alberta, T6G 2E9, Canada;*
[2] *Bamfield Marine Station, Bamfield, British Columbia, V0R 1B0, Canada*
[3] *Present address: Dept. of Ecology and Evolutionary Biology, University of California, Irvine, CA 92717, USA*

Received and accepted 24 May 1993

Key words: fluctuating asymmetry, growth, morphology, Crustacea, ontogeny, scaling

Abstract

Fluctuating asymmetry (FA) is generally viewed as a population-level character. It is described by some measure of the variance of the difference between the right and left sides for a collection of individuals. Very little is known of the developmental origins of FA, despite the fact that FA is widely used to estimate *developmental* stability. We present a novel technique for examining the growth trajectory of the asymmetries that give rise to FA, and we explore two sample data sets for the brachyuran crab *Hemigrapsus nudus*. We have traced the fate of these small, random deviations from perfect symmetry through three successive molts of individual crabs. Invertebrates that molt, and hence grow in discrete steps, provide an easily preserved record of their growth. Model II regressions of measurements from one molt versus the previous molt can help describe the stability of subtle departures from symmetry over time. Although any number of different patterns may occur, we identify four general cases: a) asymmetries vary at random throughout growth (random determination), b) asymmetries remain unchanged in sign and magnitude (constant asymmetries), c) asymmetries increase in proportion to character size and hence increase with growth (size-dependent asymmetries), and d) asymmetries persist, but are reduced in magnitude (damped asymmetries). Data from ten *Hemigrapsus nudus*, measured for between 21 and 28 metrical, limb-segment characters over three successive molts, yielded associations most similar to our pattern 'b', although some subtle departures in the direction of pattern 'c' were also observed. Persistent asymmetries accounted for 26% and 20% of the variance among asymmetries between molts 1 and 2, and molts 2 and 3 respectively. Thus, in spite of large and rapid increments in the external size of the crab, these subtle asymmetries tended to persist in both direction and magnitude, from molt to molt. This result suggests either i) that individual crabs have a genetic predisposition towards asymmetry in a particular direction but contribute to a continuous and normal distribution of *R–L* differences at the population level, or ii) that these subtle asymmetries arose at some earlier ontogenetic stage and were preserved through growth. Either interpretation has important ramifications for the study of FA. The first suggests that under some circumstances FA may not provide a valid measure of developmental instability, because subtle departures from symmetry in an individual may have a genetic basis. The second implies that subtle departures from bilateral symmetry are not 'corrected' as an individual grows.

Introduction

Fluctuating asymmetry (FA) has been used extensively as measure of developmental stability (Palmer & Strobeck, 1986; Leary & Allendorf, 1989; Parsons, 1990; Zakharov, 1992). The basis of this inference seems straightforward: if the expression of morphological characters on opposite sides of an individual organism is controlled by the same genes, then small, random deviations from perfect bilateral symmetry that give rise to FA should provide a measure of the fidelity of development (developmental stability). Studies of FA over the past 40 years generally fall into two categories: 1) those using FA as a measure of extrinsic environmental stress such as temperature, food deprivation, pesticides, pollution, etc. (Zakharov, 1992) and 2) those using FA as a measure of intrinsic, or 'epigenetic', stress resulting from intense selection, inbreeding depression or hybrid disjunction (Palmer & Strobeck, 1986).

FA refers to one of several patterns of bilateral variation (VanValen, 1962; Palmer & Strobeck, 1992). FA is generally described statistically as a variance of signed, or a mean of unsigned, differences between sides of a sample of individuals (Palmer & Strobeck, 1986). Thus FA is defined as a statistical attribute of a sample of individuals and is not an attribute of an individual organism (VanValen, 1962). To try to avoid confusion, we will use the phrase 'subtle asymmetries' to refer to those departures from symmetry in a given individual that, when examined for a sample of individuals, would yield a pattern of ideal FA. Subtle asymmetries may potentially arise as a result of genetic, environmental, or genotype by environment effects on development (Palmer, Strobeck & Chippindale, 1993), although most appear to arise from developmental noise (Zakharov, 1992).

Although some studies have explored genetic aspects of FA (Mather, 1953; Leamy, 1984; Coyne, 1987; Livshits *et al.*, 1988; McKenzie & Clarke, 1988; Tuinstra, Dejong & Scharloo, 1990), studies exploring the ontogeny or persistence throughout growth of subtle asymmetries are conspicuously absent from a literature on *developmental* stability (see, however, Price *et al.*, 1991). When are subtle asymmetries determined during ontogeny? How is their magnitude related to body size or stage of development? Do they actually fluctuate as an individual grows? Are subtle departures from bilateral symmetry corrected during growth?

We present a novel approach for examining subtle asymmetries ontogenetically or throughout growth. Quite simply, if measurements are nondestructive, the asymmetry of a single character on a single individual may be tracked through time. We applied this method to a growth series of a common shore crab of the northeastern Pacific, *Hemigrapsus nudus* (Dana, 1851). Crabs represent ideal animals for such studies because a) they are readily maintained in the laboratory, b) they have a large number of easily measured, bilaterally symmetrical characters, c) they grow by molting in discrete steps with large size changes from one molt to the next (Hartnoll, 1982), and d) they produce molts that are sufficiently durable to yield a permanent record of external morphology throughout growth.

If subtle asymmetries are determined genetically, or are determined early in development, the sign and approximate magnitude of a character's right-minus-left difference should persist over time. If subtle asymmetries arise due to random processes acting continuously and independently throughout ontogeny or growth, either the sign or the magnitude (or both) should vary as the individual grows, particularly when individuals grow in discrete steps of large magnitude, as they do in most crustaceans.

Materials and methods

Collection and maintenance of crabs

Individual *Hemigrapsus nudus* were collected on May 24, 1988 from a cobble beach on the east side of Dixon Island (125°07′N, 48°51′W), near Bamfield, British Columbia, Canada, where they occur intertidally in great numbers. To minimize potential problems associated with sexual dimorphism and size-scaling of metrical characters, only female crabs between 9 and 13 mm carapace width were collected. Individual crabs were isolated in plastic freezer containers with plastic mesh sides that were held partially immersed in running seawater at the Bamfield Marine Station for approximately 14 months (from late May 1988 to mid-August 1989). During this time, animals were held under ambient laboratory conditions (natural photoperiod, temperatures ranging from 7.3-11.3 °C) and were fed *ad libitum* on a diet of various foods obtained fresh from the original collection site (e.g. *Ulva* and other seaweeds, mussels, and barnacle encrusted rocks, from which they scrape encrusting diatoms) and pieces of fish. This laboratory environment was designed to minimize stresses crabs normally experience in the intertidal zone, including salinity and temperature fluctuations, aggressive interactions, and foraging on rugged substrata.

Twice weekly, cages were cleaned and any molts were removed for measurement.

Measurement protocol
Thirty females were collected originally, but only ten were measured completely because: 1) several crabs died, 2) exoskeletons were incomplete on some individuals, and 3) the large numbers of measurements required per molt (28 × 2 characters each measured five times) made the measurement of more individuals impractical.

A total of 28 limb segment lengths, and one correlate of general body size (carapace width) were measured on each molt. The proximo-distal length of four limb segments, the merus (three different measurements on all but the clawed limb), carpus, propus, and dactylus, was measured on each of the five periopods. The three merus measurements were treated as independent traits in our analyses, since their asymmetry values were not significantly correlated: no indication of any association was detected when individuals were pooled and the characters were plotted against one another in pairs (e.g. merus a vs. merus b, merus b vs. merus c, and merus a vs. merus c) and least-squares regression was performed.

Subtle asymmetries in metrical characters that give rise to FA are small and random, and therefore indistinguishable from measurement error without replicate measurements (Palmer & Strobeck, 1986). Great care was taken to reduce the contribution of measurement error to our estimates of the between-sides variance. Limb segments were measured at approximately 80× magnification using a dissecting microscope fitted with a camera lucida aligned over a 500 DPI resolution digitizing tablet linked to a Macintosh computer. Each intact limb was oriented so as to be viewed perpendicular to the plane of maximum traverse (defined by the proximo-distal and dorso-ventral axes). Individual segments were centered in the field of view and measured once prior to moving to the next segment on the same limb. Care was taken to orient characters so that both endpoints of linear measurements were in the plane of focus before measuring. This avoided error due to refocussing between points, which would alter the calibration of the graphics tablet, and also avoided parallax errors. To further reduce the impact of measurement error, this sequence of measurement was repeated independently five times (i.e. one set of measurements was completed on all limbs for a single molt before beginning a second set of measurements). Thus we obtained a very accurate estimate of measurement error that incorporated errors arising from a) aligning limbs on the measurement stage, b) relocating character endpoints, and c) random fluctuations of the electronic equipment. Carapace widths were also measured five times with the digitizing arrangement described above only at 40× magnification.

Test for preservation artifacts
Estimates of error due to preservation artifacts, such as distortion due to fixation or drying, were made by measuring a set of representative characters on two fresh individuals (seven characters on one and 13 on the other), which were baked for 24 h at 60-70 °C and remeasured. These two crabs were not part of the group of ten selected for complete measurement, although the characters measured were a subset of those used in the main analysis.

Measurement error and tests for departures from ideal FA
Tests for the significance of the between-sides variance relative to measurement error were conducted following the analysis of variance (ANOVA) procedure outlined in Palmer and Strobeck (1986). Before computing means and differences for each segment on a limb in all subsequent analyses, replicate measurements were averaged. Thus, each right-minus-left difference for a given limb segment was a difference between two averages, not an average of five differences.

FA is defined as a normal distribution of *R–L* differences about a mean of zero (VanValen, 1962). Because other patterns of statistical variation of *R–L* may render characters inappropriate for studies of developmental stability (Palmer & Strobeck, 1992) it was important to test for departures from ideal FA. Unfortunately, the small number of individual crabs measured in the present study (ten) greatly reduced our ability to detect departures from ideal FA (e.g. antisymmetry and directional asymmetry) in our samples. Further, since one of our goals was to determine whether subtle asymmetries varied at random during ontogeny, we could not assume *a priori* that they were random to enlarge our sample sizes to 30 by including the three molts for each individual crab. Fortunately, we had access to a larger data set collected by two students as part of a graduate course co-taught by one of us (A.R.P.) at the Friday Harbor Laboratories, Friday Harbor, Washington, in 1986. In this study of FA in *Hemigrapsus nudus*, relatively few (five) characters were measured three times on a large number (80) of individuals of both sexes collected from the field. The measurement procedure was similar to that outlined above, but data were collected using a lower resolution graphics tablet (200 DPI) attached to an Apple II+ computer. We analyzed the data for females only.

Tests for persistence of subtle asymmetries over successive molts

To test for the persistence of subtle asymmetries over successive molts, we treated each individual crab as a 'sample' of from 21 to 28 characters. The question posed was: do characters that exhibit a larger than average asymmetry in one molt of an individual crab exhibit a larger than average asymmetry in the subsequent molt? For each crab we regressed the departure from symmetry (*R–L*) in one molt against the departure from symmetry in the previous molt for all characters measured for that crab. Thus, for each crab we obtained two regressions: molt 2 vs molt 1 and molt 3 vs molt 2.

Data of this type do not conform to the assumptions of a least-squares linear regression analysis, where the X-axis variate is measured without error and all error variation is assumed to parallel the Y-axis (Sokal & Rohlf, 1981). In other words, in this analysis, there is no legitimate reason to define 'dependent' and 'independent' variables, nor is there any expectation for different error variances along the two axes. Both axes were *R–L* differences, measured in identical fashion, and were of similar magnitude. To complicate matters further, the appropriate form of Model II regression for such data remains in dispute (LaBarbera, 1989). We thus estimated two alternate descriptors of the slope of our bivariate scattergrams: reduced-major-axis (RMA, Sokal & Rohlf, 1981, pp. 550) and the principal or major axis of a bivariate scattergram (MA). RMA slopes were obtained by computing a least-squares linear regression slope and dividing it by its correlation coefficient (LaBarbera, 1989). MA slopes were obtained according to the procedure outlined in Sokal and Rohlf (1981, pp. 596). Thus for each molt interval, we obtained ten independent RMA slopes and ten independent MA slopes, one per crab.

To guard against the possibility that one individual crab exhibiting unusually large and persistent asymmetries would bias our conclusions about the persistence of asymmetries, we obtained an average overall slope by computing the arithmetic mean of the ten slopes obtained independently for each crab. In other words, we restricted our degrees of freedom in the final analysis by treating each regression for an individual crab as a single datum. For comparison, we also computed a single RMA and MA slope for the data from all crabs pooled for each molt interval. Finally, because correlations between molts were very low for some individuals, yielding slopes of low certainty, we re-computed the mean RMA and MA slope using only crabs exhibiting significant or nearly significant correlations between characters from molt to molt.

Paired-t tests, tests for skew and kurtosis of distributions of *R–L*, Mann-Whitney U Tests, analyses of variance and regression analyses were conducted with Statview II (Ver. 1.03, Abacus Concepts).

Results

Tests for preservation artifacts

The 20 characters measured to test for preservation artifacts shrank an average of 0.026 (± 0.0033, SE) mm when dried, corresponding to a shrinkage of 0.53% (± 0.069). Differences between sides did not appear to be altered significantly by the harsh drying process. The average difference between sides (mean $|R\text{–}L|$) was 0.060 ± 0.0106 mm before drying and 0.064 ± 0.0101 mm after drying, and this change was not significant ($P = 0.070$, paired-T test). This test of preservation artifacts was liberally abusive: no crab used in the main analysis was exposed to such strong desiccation before measurement. No significant asymmetrical distortion due to baking and drying was observed, hence crab exoskeletons appeared durable enough to make preservation artifacts negligible relative to other sources of error.

Table 1. Proximo-distal length, and statistics describing frequency distribution shape of the differences between sides of limb segments of the posterior-most pair of walking legs in female shore crabs, *Hemigrapsus nudus*, from San Juan Is., Washington, USA.

	Statistic	Limb segment			
		Merus	Carpus	Propus	Dactyl
	N	40	40	40	40
a)	(*R*+*L*)/2:				
	Mean (SE)	6.97 (0.124)	3.91 (0.074)	2.76 (0.044)	3.45 (0.063)
b)	(*R*+*L*):				
	Mean	0.028	-0.013	0.012	0.017
	(SE, *P*◇)	(0.0311, 0.37)	(0.0253, 0.61)	(0.0307, 0.70)	(0.0309, 0.59)
	Kurtosis (*P*)§	0.804 (0.27)	-0.14 (0.85)	•2.082 (0.012)	1.384 (0.06)
	Skew (*P*)§	0.45 (0.23)	-0.584 (0.12)	0.605 (0.11)	-0.271 (0.47)
c)	|*R*-L|:				
	Mean (SE)	0.14 (0.022)	0.125 (0.016)	0.148 (0.019)	0.143 (0.021)
d)	ANOVA*:				
	Error MS	0.0185	0.0127	0.0167	0.0248
	Interaction MS (*P*)	•0.058 (<0.001)	•0.038 (<0.001)	•0.057 (<0.001)	•0.059 (<0.001)
	Sides MS (*P*)	0.0468 (0.37)	0.0106 (0.60)	0.008 (0.71)	0.0235 (0.59)
	σ^2_i (approx. df)	0.013 (17.7)	0.009 (17.0)	0.013 (19.0)	0.011 (12.6)
e)	FA5**	0.039	0.025	0.037	0.038
f)	Size dependence of FA†				
	Slope	-0.01	-0.04	0.03	0.07
	(SE, *P*)	(0.029, 0.70)	(0.034, 0.22)	(0.072, 0.72)	(0.053, 0.19)

◇ *P* = probability that mean (*R*–*L*) differs from zero.

§ SE values for skew and kurtosis are based on sample sizes (Sokal & Rohlf, 1981), and correspond to 0.374 and 0.733 for skew and kurtosis respectively for an *N* of 40.

* Analysis of variance procedure from Palmer and Strobeck (1986) for testing the significance of the between-side variance, directional asymmetry, and for partitioning the between-sides variance (σ^2_i) from the interaction MS. df- degrees of freedom.

** Index 5 of Palmer and Stobeck (1986): $\Sigma\ (R\text{–}L)^2/N$.

† Regression analysis of the dependence of character asymmetry, |*R*–*L*|, on character size, (*R*+*L*)/2.

• Indicates a statistically significant probability.

Measurement error, departures from ideal FA and effects of character size

For the preliminary sample of 40 female *Hemigrapsus nudus* from San Juan Island (Table 1d) the between-sides variance (Interaction MS) was significantly larger than measurement error for all four limb segments. Thus we were justified in analyzing patterns of variation in the between-sides variance in greater detail.

For this larger sample of crabs, all characters appeared to exhibit ideal FA (Palmer & Strobeck, 1992). None of the mean right-minus-left differences were significantly different from zero (Table 1b), thus there was no evidence for directional asymmetry. In addition, frequency distributions of right-minus-left differences exhibited no significant skew (Table 1b). One trait (propus length) exhibited significant kurtosis and a second (dactyl length) nearly significant kurtosis (Table 1b). In both cases kurtosis values were positive, signalling the presence of leptokurtosis (Sokal & Rohlf, 1981). There was no evidence of antisymmetry (platykurtosis) or other peculiar departures from ideal FA (Palmer & Strobeck, 1992).

The between-sides differences were also significantly larger than the measurement error for 28 characters measured on individual crabs in the pre-

Table 2. ANOVA test for significance of non-directional asymmetry (= interaction) relative to measurement error for 28 limb segments measurement five times on a single individual *Hemisgrapsus nudus*, from Barkley Sound, to illustrate the relative magnitudes of measurement error and between-sides variation. Analyses for other inidividual crabs yielded similar results.

Source of variation	df	MS	F	P
Character	27	21.090	2343.3§	<0.0001
Sides	1	0.048	5.3§	0.030
Interaction	27	0.009†	237.5	<0.0001
Error	224	3.96×10^{-5}		

§ MS tested over Interaction MS, see Palmer and Strobeck (1986).

† the between-sides variance, $\sigma^2_i = (0.00941 - 3.96 \times 10^{-5})/5 = 0.00187$ (Palmer & Strobeck, 1986).

sent study (Table 2). Following the procedure outlined in Palmer and Strobeck (1986), the variance between sides (due to all non-directional asymmetries; σ^2_i of Palmer & Strobeck) was 0.00187. Thus the between-sides variance was 47.3 times greater than the variance among repeated measurements of the same limb segment (3.96×10^{-5}).

The smaller sample of crabs from which we obtained multiple molts also yielded no evidence of directional asymmetry. None of the 28 characters measured on the first or third molts (N = 10 per character) exhibited statistically significant directional asymmetry when subject to a sequential Bonferroni criterion which assures a table-wide probability of P = 0.05 (Rice, 1989; P = 0.05/56, data not shown). Furthermore, none of the six first-molt limb segments and only one of six third-molt limb segments (merus a) exhibited a barely significant

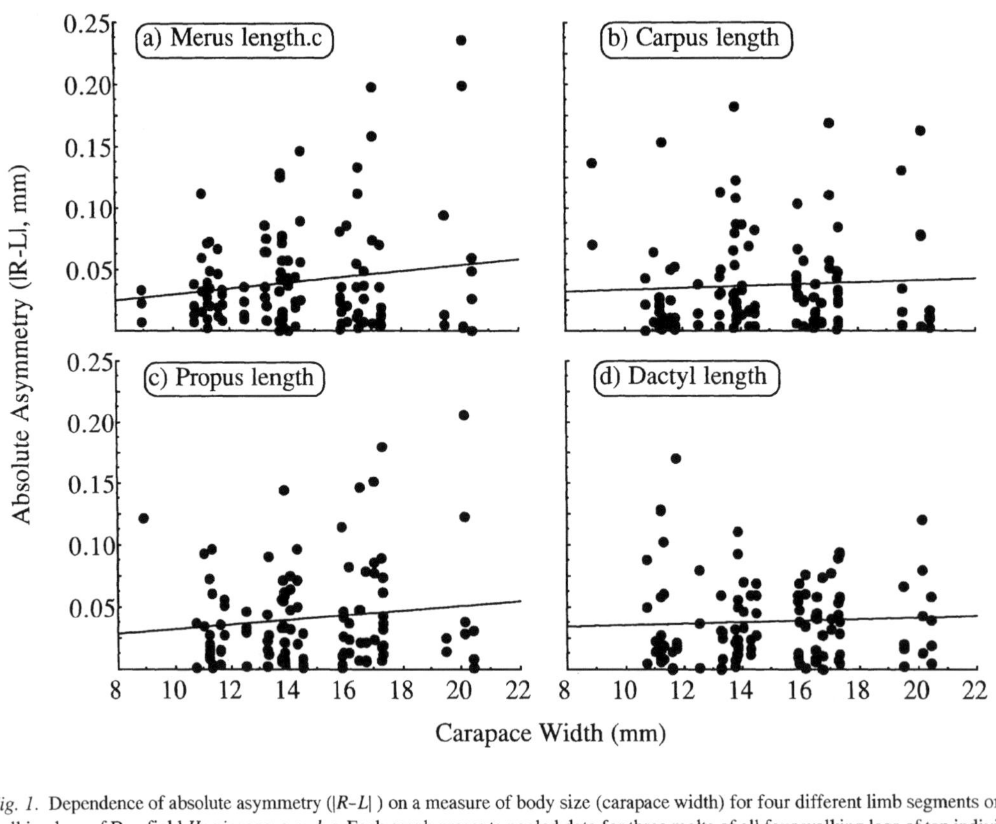

Fig. 1. Dependence of absolute asymmetry (|R–L|) on a measure of body size (carapace width) for four different limb segments on th walking legs of Bamfield *Hemigrapsus nudus.* Each graph presents pooled data for three molts of all four walking legs of ten individu crabs (N = 120). Each point corresponds to one limb segment on a single limb of a single molt. Least-square linear regression equatio are: a) $Y = 0.0024\ X + 0.0058$, $r^2 = 0.026$, $P = 0.086$; b) $Y = 0.0007\ X + 0.0269$, $r^2 = 0.003$, $P = 0.59$; c) $Y = 0.0019\ X + 0.013$ $r^2 = 0.016$, $P = 0.19$; d) $Y = 0.0006\ X + 0.0300$, $r^2 = 0.0022$, $P = 0.62$.

Table 3. Body size, average values for signed (*R* – *L*) and unsigned |*R–L*| asymmetries, and percent asymmetry values {100 × |*R–L*|/ [(*R* + *L*)/2]} for limb segments of the shore crab *Hemisgrapsus nudus* from Bamfield, BC. Entries in section (a) were computed on between 21 and 28 different limb segments on individual molts. All values in mm. SE = standard error.

Crab #	Molt #	Carapace width	(R-L) mean	(R-L) SE	|R-L| mean	|R-L| SE	% asym.	n
a) *Individual molts*								
1	1	11.00	0.017	0.0102	0.041	0.0092	1.49%	21
	2	13.85	0.029	0.0110	0.048	0.0055	1.09%	28
	3	15.94	0.011	0.0084	0.034	0.0086	0.67%	24
2	1	11.20	-0.011	0.0064	0.025	0.0045	0.75%	28
	2	13.74	0.002	0.0060	0.022	0.0046	0.56%	27
	3	16.49	0.002	0.0061	0.022	0.0043	0.42%	24
3	1	11.20	-0.011	0.0072	0.026	0.0036	0.81%	28
	2	14.03	0.000	0.0091	0.041	0.0055	0.93%	28
	3	16.52	0.016	0.0117	0.047	0.0044	0.85%	27
4	1	11.75	-0.013	0.0092	0.034	0.0076	0.88%	26
	2	14.31	0.014	0.0070	0.029	0.0068	0.65%	28
	3	17.28	0.012	0.0073	0.027	0.0051	0.48%	26
5	1	11.33	-0.021	0.0094	0.036	0.0052	1.09%	27
	2	13.87	-0.013	0.0129	0.055	0.0070	1.22%	28
	3	16.76	0.007	0.0066	0.025	0.0083	0.46%	28
6	1	13.79	0.036	0.0166	0.059	0.0048	1.37%	27
	2	17.01	0.066*	0.0169	0.076	0.0135	1.62%	26
	3	20.13	0.055	0.0200	0.079	0.0129	1.34%	28
7	1	12.52	0.015	0.0057	0.025	0.0163	0.63%	26
	2	17.30	0.016	0.0094	0.040	0.0040	0.71%	28
	3	20.46	0.027	0.0113	0.039	0.0066	0.58%	28
8	1	11.63	0.004	0.0053	0.021	0.0100	0.59%	28
	2	14.49	0.000	0.0093	0.033	0.0032	0.76%	28
	3	17.33	0.002	0.0120	0.045	0.0068	0.78%	25
9	1	13.25	0.007	0.0093	0.038	0.0070	0.74%	28
	2	16.13	-0.008	0.0094	0.040	0.0071	0.75%	26
	3	19.50	-0.003	0.0131	0.045	0.0055	0.65%	26
10	1	10.73	0.018	0.0068	0.027	0.0104	0.78%	25
	2	13.34	0.010	0.0108	0.041	0.0051	1.03%	26
	3	15.93	0.014	0.0076	0.031	0.0066	0.59%	28
b) *Average for each molt*								
Molt 1 mean		11.6	0.002		0.033		0.88%	256
	SE	0.08	0.0031		0.0023		0.065	
Molt 2 mean		14.5	0.010		0.042		0.92%	266
	SE	0.11	0.0035		0.0024		0.052	
Molt 3 mean		17.4	0.016*		0.042		0.73%	262
	SE	0.12	0.0037		0.0029		0.049	
c) *All molts pooled*								
	Mean	14.5	0.010*		0.039		0.84%	790
	SE	0.10	0.0020		0.0015		0.032	

* $0.01 < P < 0.05$ according to sequential Bonferroni criteria (Rice, 1989).

directional asymmetry (P = 0.033, standard Bonferroni criterion) when data were pooled across all walking legs (*N* = 40; data not shown). We are thus unable to eliminate completely the possibility that some directional asymmetry was present. However, if it was present, it was very subtle, since we were only able to detect it in one of the 96 comparisons (56 + 40) when subject to sequential Bonferroni criteria.

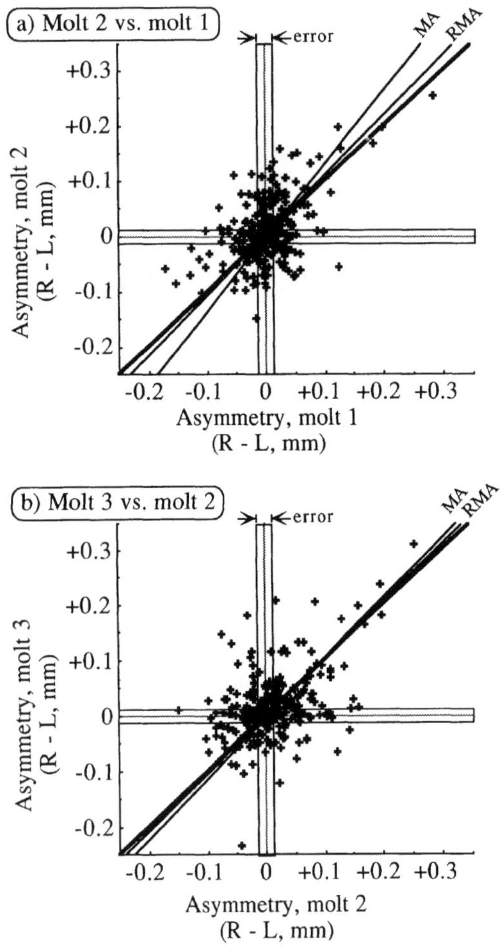

Fig. 2. Persistence of asymmetries in limb segments between successive molts in individual shore crabs *Hemigrapsus nudus*: a) molt 2 vs. molt 1, b) molt 3 vs. molt 2. Data were pooled across all limb segments for all crabs to compute MA and RMA slopes. The heavy solid line indicates perfect persistence of asymmetries. The finer lines indicate observed slopes. MA-major axis slope, RMA-reduced major axis slope. See Tables 4a and 4b for actual slopes. N = 259 and N = 258 in a and b respectively. Error-95% confidence intervals to illustrate the magnitude of measurement error ($\pm$ 2 $\times$ $\sqrt{MS}$ error from ANOVA, Table 2).

Finally, we detected no association between the magnitude of asymmetry and character size, either among individuals for a particular character (Table 1f), or among leg segments (compare values in Table 1c or 1e to those in 1a) in the sample of 40 females from San Juan Island. Although limb segments differed in size by more than a factor of two, estimates of FA varied by only a factor of 0.2 (Table 1c) to 0.5 (Table 1e), depending on the index used. Hence we did not need to correct for size effects on FA (Palmer & Strobeck, 1986).

A similar analysis of each limb segment measured on three successive molts in Barkley Sound *H. nudus* also revealed no significant association between |*R*–*L*| and carapace width when examined across all molts of all crabs (Fig. 1).

Change in overall asymmetry of successive molts

When averaged across all characters for a given molt, differences were observed among molts of individual crabs as well as among crabs (Table 3). One individual exhibited a significant right bias in molt 2 when averaged across all characters (crab #6). This crab also exhibited the highest variation of *R*–*L* for all three molts; however, it also exhibited a significant right skew compared to other crabs.

The absolute magnitude of deviations from symmetry averaged 0.039 mm across all molts of all crabs (Table 3c). Expressed as a relative measure, the absolute asymmetry averaged over all molts was 0.84% of character size (Table 3c). As crab size increased from molt 1 to molt 2, average absolute asymmetry increased significantly from 0.033 to 0.042 mm ($P < 0.001$, Mann-Whitney U Test), but from molt 2 to molt 3 average absolute asymmetry remained unchanged (Table 3b). As a consequence, the level of relative asymmetry (scaled by character size) declined by nearly one quarter between molts 2 and 3, a decline that was also significant statistically ($P < 0.001$, Mann-Whitney U Test). Note that non-parametric tests are more appropriate for these comparisons because frequency distributions of |*R*–*L*| are truncated normal, and hence seriously skewed to the right.

Table 4. Growth increments and three statistical descriptors of the associations between subtle asymmetries in successive molts of the Bamfield sample of laboratory-grown *Hemigrapsus nudus*. Growth increments were computed as the percentage increase in width of the carapace (difference between successive molts over smaller molt X 100). *a)* and *b)*: association between (R–L, mm) in successive molts for 21 to 28 characters measured on individual crabs. *c)* Mean of associations between (R–L, mm) in successive molts computed using only crabs from a and b that exhibited a nearly significant LLR regression (P < 0.1; mean of values for crab # 1, 3, 4, 6, 7 and 8 from section *a* and crab # 3, 5, 6 and 10 from section *b*). *d)* As for c except means computed using only crabs that exhibited a significant LLR regression (P < 0.05; mean of values for crab # 3, 4, 6, 7 and 8 from section *a* and crab # 5, 6 and 10 from section *b*) LLR regression. N = number of characters per crab (or # characters X # crabs). LLR-least squares linear regression statistics (P indicates significance level of regression). RMA = statistics for reduced major axis (Sokal & Rohlf, 1981, pp. 550; P = indicates significance of departure from an expected slope of 1.0). MA = statistics for principal or major axis of a bivariate scattergram (Sokal & Rohlf, 1981, pp. 596; P = indicates significance of departure from an expected slope of 1.0). SE = standard error. r = correlation coefficient.

Crab	Growth Incr.		LLR						RMA		MA	
#	(%)	N	slope	SE	r	P	Ave.X	Ave.Y	Slope	P	Slope	P
a) Molt 2 vs Molt 1												
1	25.0	21	0.49	0.253	0.40	0.070	0.017	0.032	1.21		1.567	
2	18.5	27	0.13	0.181	0.14	0.478	-0.012	0.002	0.92		0.555	
3	20.2	28	0.78	0.197	0.61	<0.001	-0.011	0.000	1.27		1.473	
4	21.8	26	0.35	0.131	0.48	0.014	-0.013	0.014	0.73		0.538	
5	18.3	27	0.42	0.253	0.32	0.108	-0.020	-0.017	1.33		2.280	
6	23.4	25	0.75	0.131	0.77	<0.001	0.038	0.068	0.98		0.971	
7	38.2	26	0.72	0.324	0.41	0.037	0.015	0.014	1.74		3.155	
8	24.6	28	1.27	0.236	0.73	<0.001	0.004	0.000	1.75		2.096	
9	21.7	26	0.26	0.196	0.26	0.193	0.005	-0.008	1.00		1.461	
10	24.3	25	-0.05	0.321	0.03	0.882	0.018	0.007	-1.54†		-28.42†	
Mean	23.6		0.51				0.004	0.012	1.21	0.110	1.57	0.081
SE	1.60		0.108				0.0051	0.0065	0.118		0.283	
Pooled		259	0.60	0.063	0.51	<0.001	0.003	0.010	1.17	0.008	1.35	0.014
b) Molt 3 vs Molt 2												
1	15.1	24	-0.04	0.159	0.05	0.807	0.023	0.011	-0.75†		-0.089†	
2	20.0	24	0.18	0.191	0.20	0.356	0.003	0.002	0.92		0.648	
3	17.8	27	0.42	0.241	0.33	0.095	0.001	0.016	1.28		1.996	
4	20.8	26	0.13	0.203	0.13	0.526	0.012	0.012	1.00		1.013	
5	21.0	28	0.20	0.094	0.39	0.042	-0.013	0.007	0.52		0.257	
6	18.4	26	0.97	0.156	0.78	<0.001	0.066	0.062	1.23		1.302	
7	18.3	28	0.32	0.229	0.26	0.175	0.016	0.027	1.21		1.969	
8	19.6	25	0.20	0.266	0.16	0.459	0.003	0.002	1.29		3.621	
9	20.9	24	0.48	0.285	0.34	0.107	-0.009	-0.001	1.42		2.518	
10	19.4	26	0.31	0.141	0.41	0.036	0.010	0.013	0.76		0.324	
Mean	19.1		0.32				0.011	0.015	1.07	0.500	1.52	0.202
SE	0.6		0.086				0.0070	0.0059	0.098		0.371	
Pooled		258	0.47	0.058	0.45	<0.001	0.011	0.015	1.04	0.536	1.07	0.606
c) Based on crabs where LLR Regression P < 0.1 only (N = 10 crabs)												
Mean	23.0		0.62				0.011	0.023	1.15	0.220	1.37	0.230
SE	1.9		0.105				0.0079	0.0076	0.129		0.286	
d) Based on crabs where LLR Regression P < 0.05 only (N = 8 crabs)												
Mean	23.4		0.67				0.012	0.022	1.12	0.476	1.26	0.472
SE	2.2		0.128				0.0099	0.0095	0.162		0.349	

† excluded from computation of means across crabs in a) and b) because of extremely low r.

Persistence of subtle asymmetries over successive molts

Subtle asymmetries detected in individual limb segments appeared to persist over successive molts for all ten crabs over both molt intervals (Fig. 2, Table 4). For each of the ten crabs examined, segments that exhibited a departure from symmetry in one molt tended to exhibit a departure in the same direction in the subsequent molt. In all, 18 out of 20 slopes for the molt to molt comparisons were positive (Table 4). The two slopes that were negative were also those that had the lowest correlation coefficients by nearly a factor of three. The probability of obtaining nine out of ten slopes positive due to chance, as observed both for molt 1-2 (Table 4a) and for molt 2-3 (Table 4b), would simply be the binomial probability $P = 0.002$.

In addition, the magnitude of the subtle asymmetries appeared to be approximately the same from molt to molt. The mean RMA and MA slopes for both molt intervals were significantly greater than zero, but neither was significantly different from 1.0 (Table 4a,b; two-tailed t-test). When data were pooled across all limb segments and all crabs, to estimate a single RMA and MA slope for each molt interval, both the RMA and MA slopes were significantly greater than 1.0 in the first molt interval (Table 4a) but were not significantly greater than 1.0 in the second molt interval (Table 4b). Finally, when mean slopes were computed using only data for crabs that exhibited nearly significant ($P < 0.1$) or significant ($P < 0.05$) statistical associations from molt to molt based on LLR regressions, neither the mean RMA nor MA slopes differed significantly from 1.0 (Table 4c,d).

Discussion

Confirmation of fluctuating asymmetry

Before measures of the between-sides variance [e.g. Var $(R-L)$ or mean $|R + L|$] can be interpreted as measures of developmental stability, the between-sides variation must be shown a) to be larger than the measurement error (Palmer & Strobeck, 1986) and b) to conform to ideal FA (Palmer & Strobeck, 1992). For a distribution of $R-L$, ideal FA corresponds to a mean of zero with normally distributed variation (VanValen, 1962). For both sets of data we examined, the larger preliminary sample of *Hemigrapsus nudus* from the San Juan Islands (Table 1) and the smaller sample of crabs from Barkley Sound (Table 2), we confirmed that the subtle departures from symmetry that give rise to FA were significantly larger than our measurement error. In addition, these tests also revealed no statistically significant directional asymmetry (Tables 1 & 2), or antisymmetry (platykurtosis) or skew (Table 1). Three of the four limb segments for which we had adequate sample sizes exhibited leptokurtosis and one exhibited platykurtosis (Table 1) however, only one of these kurtoses was statistically significant (leptokurtosis for the propus, Table 1). As discussed elsewhere (Palmer & Strobeck, 1992), the biological significance of departures from normality in the direction of leptokurtosis is unclear, and we attribute no biological significance to it here.

When examining our pooled data, we did detect hints of a right-bias to departures from symmetry, but these became apparent only when pooling measurements together, either at the level of individual crabs (Table 3a), at the level of molts (Table 3b), or pooling the data for all molts for all crabs (Table 3c). These P values, however, need to be interpreted with great care for two reasons. First, when pooled across multiple characters for an individual crab, as in Table 3a, a significant departure of mean $(R-L)$ indicates that more *characters* exhibited a side-bias than would be expected due to chance. In other words, it does not necessarily mean any individual characters exhibited directional asymetry, it only means that an individual crab exhibiting a side-bias for one character was more likely to exhibit a side-bias in the same direction for another character. Second, when pooling across all crabs (Table 3b), or across all molts of all crabs (Table 3c), a significant bias could arise because of a lack of independence. If, by chance, six of the ten crabs we studied exhibited a slight right bias correlated across several characters, then pooling a large number of measurements based upon this small number of crabs would give the appearance of a statistically significant right bias to the entire sample. Finally, even when all data were pooled, the observed directional asymmetry was only 0.010 mm which was roughly one-quarter the magnitude of the average absolute difference between sides (0.039 mm, Table 3c). An average right-bias of this small a magnitude is very unlikely to have given rise to any of the patterns we observed between molts. Thus, although we can not reject entirely the possibility that some slight directional asymmetries were present, the data least subject to analytical bias (Table 1) did not reveal any, and we suspect the apparent directional asymmetries revealed in the pooled data (Table 3) were statistical artifacts of pooling.

In summary, we conclude that the departures from ideal FA we did observe were not only small in magnitude relative to the patterns of between-molt variation we were studying, but appeared to have arisen as a statistical byproduct of pooling data or of conducting a large number of tests. Thus for the purpose of studying between-molt variation, we felt confident that all characters exhibited ideal FA. The presence of a persistent, and significant right-bias (in one of three molts) in a single individual (Crab #6, Table 3) when all characters were pooled suggests the tantalizing possibility that the symmetry of some individual crabs may be disrupted very early in development, leading to a subtle, body-wide asymmetry. Additional study will be required to determine if body-wide asymmetries are more widespread.

Potential biases in the analyses of persistent subtle asymmetries

Another potential concern that arises in the sort of sampling scheme employed here is the independence of characters. If certain characters display directional asymmetry or antisymmetry of sufficient magnitude, they may exert a significant leverage on the slope of regressions of $R–L$ between molts. For example, if claw measurements were included as independent data points in an estimate of overall FA for an individual crab, but claws displayed antisymmetry, they might 'lever' the regression towards a positive slope, even though all other limbs exhibited FA.

To test for possible biases, we repeated our analyses on subsets of the data with certain groups of characters excluded. In one set of analyses, we excluded characters grouped by limb (e.g. all cheliped measurements, or all measurements of walking limb 1, etc.). In a second set, we excluded groups of serially homologous characters (e.g. all measurements of the dactyl on all limbs, or all measurements of the propus on all limbs, etc.). For all 20 of these alternate analyses, the average slope of the least squares regression was not altered qualitatively (i.e. the sign was the same and the slope was similar). Thus neither particular limbs, nor particular limb segments appeared to exert a significant leverage in our analyses.

In addition, we conducted our analyses of persistent subtle asymmetries in such a way that a bias due to one crab would not exert an undue influence on the overall pattern. Hence, although one crab exhibited an overall right bias across all characters (Crab #6, Table 3), we computed an average of the nine individual RMA and MA slopes obtained from single crabs. This procedure yielded similar slopes to those obtained when all data were pooled to compute a single slope.

Our conclusion that subtle asymmetries persist over successive molts does not appear to result from biases in our data or analyses.

Measuring the persistence of subtle asymmetries during growth

FA is normally measured for a sample of individuals at a particular time or place, or under a particular set of conditions in the laboratory (Zakharov, 1992). It is thus a statistical descriptor of a sample, and not normally considered as an attribute of an individual organism. Some individuals may be more developmentally stable than others, and thus exhibit smaller departures from symmetry for a suite of characters. To obtain a multivariate measure of developmental stability, one would have to follow a procedure similar to that proposed by Zhivotovsky (1992). Studies of the origin of FA during development and the dynamics of its change in development and growth are conspicuously lacking. These considerations are of particular importance to the application of size scaling as well as to the use of FA for monitoring the condition of natural populations.

These questions have probably been overlooked for obvious methodological reasons. Because 'individuals' do not exhibit FA that can be quantified, except in cases where clone-mates or sub-components of modular organisms can be used to assess the asymmetry variance for a genotype (Palmer & Strobeck, 1986) or where individuals replace struc-

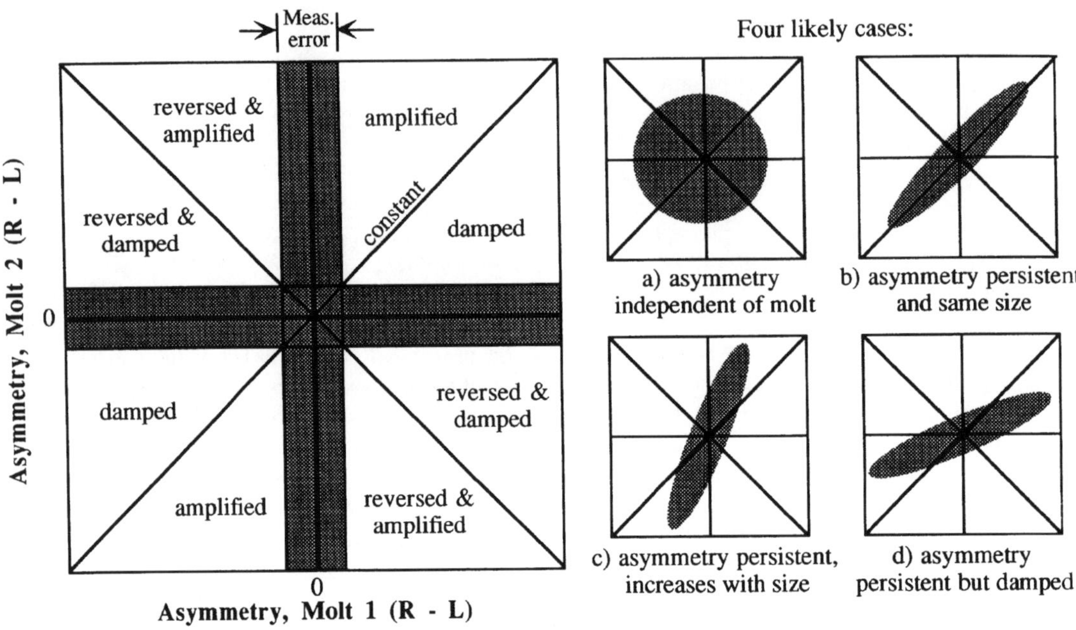

Fig. 3. Key for interpreting associations of subtle asymmetries between molts. Plots of *R–L* differences from two successive growth stages will reveal whether departures from symmetry persist, and whether they are amplified or damped. Many patterns of association are possible, but four seem more likely than others: a) subtle asymmetries are independent from molt to molt (random determination; this pattern would also obtain if the between-sides variation was due exclusively to measurement error), b) subtle asymmetries persist at the same magnitude from one molt to the next (constant asymmetries), c) subtle asymmetries persist but are amplified in subsequent molts because characters are larger (size-dependent asymmetries), and d) subtle asymmetries persist but are damped in subsequent molts (damped asymmetries).

tures repeatedly throughout their life (e.g. bird feathers that are molted; Møller & Höglund, 1992), it is difficult to extract information about the developmental origins of subtle asymmetries for individuals from typical sampling schemes employed to study FA. The method for tracking the progress of asymmetries we have proposed here is limited primarily to laboratory growth studies because data must be collected from the same individual at different times during growth. Given this limitation, however, the method is quite easily applied to a wide variety of organisms. We chose crabs because they grow in discrete steps with easily preserved molts. Skeletal variation in vertebrates could also be measured throughout growth rather easily by using measurements from X-ray plates.

One encouraging implication of our study is that subtle asymmetries in metrical characters in crabs appeared to arise early in development and persist at a more or less constant magnitude throughout growth. In this fashion, they are comparable to meristic characters that often appear to be fixed early in development (Taning, 1952; Wooten & Smith, 1986; but see Strawn, 1961, and Angus & Schultz, 1983, for evidence to the contrary). In addition, because the subtle asymmetries we observed did not increase significantly with character size, FA estimates from pooled samples containing a size range of crustaceans may not be seriously confounded by size-dependent variation (Palmer & Strobeck, 1986).

Interpreting persistent subtle asymmetries

Where subtle asymmetries may be followed over time, as in our study, we may ask how persistent such asymmetries are from one time interval to the next. Although many possibilities exist (Fig. 3), four seem more likely than others. First, departures from symmetry at one time may be completely independent of those at a later time (Fig. 3a). In other words, subtle departures from symmetry could vary continuously and independently throughout growth. This pattern might arise in crustaceans, for example, where body size increments are large and occur in discrete increments at molting. The exoskeleton is shed completely, and the animals inflate dramatically with fluids before the new cuticle hardens (Hartnoll, 1982). This pattern would also arise, of course, where apparent subtle asymmetries were simply the result of measurement error, which further emphasizes the importance of separating measurement error from estimates of FA for metrical characters. Second, subtle departures from symmetry may become fixed early in development and then persist in the same direction and magnitude throughout subsequent growth (e.g. asymmetry is independent of character size; Fig. 3b). Third, subtle departures from symmetry may be fixed early in development, but increase in magnitude throughout subsequent growth as characters become larger (e.g. asymmetry is a constant proportion of character size; Fig. 3c). Fourth, subtle departures from symmetry that arise at any time during development may be damped or corrected during growth (Fig. 3d).

We should also point out here that care must be taken when computing slopes of model II regression where correlations of asymmetries between growth stages are very low. As the correlation coefficient goes to zero, the expected value of the RMA slope approaches 1.0, because the RMA slope is merely the ratio of the standard deviations of Y and X (Sokal and Rohlf 1981, pp. 550). The MA slope, on the other hand, will become very sensitive to the location of individual data lying just outside the bivariate distribution (e.g. see individual 10, Table 4a). We avoided these problems by repeating analyses using only cases where the correlations were significant or nearly so (Table 4c,d).

Persistence of subtle asymmetries in Hemigrapsus nudus*: Relevance to developmental stability*

Our conclusions are based on a small sample of crabs and therefore somewhat limited. However, the small departures from bilateral symmetry exhibited by *Hemigrapsus nudus* were, on average, preserved in both direction and magnitude in spite of large increases in overall body size accomplished in rapid leaps of growth. These patterns of molt to molt variation were thus most similar to case (b) of Fig. 3. Even though our measurement error was quite small (Fig. 2), these patterns were also unlike anything that could have been generated by measurement error, where a random scatter of points would be expected in the molt vs. molt plots (as in Fig. 3a).

We offer two hypotheses to account for the persistence of these subtle (<1% of character size) asymmetries through growth. First, individual crabs may have a genetic predisposition towards asymmetry that manifests itself as a continuous distribution of $R-L$ differences when viewed at the level of a *sample* of individuals, but appears as persistent asymmetries over successive molts when viewed at the level of an individual. Second, developmental errors may have been fixed at some earlier stage of growth, and simply persisted throughout subsequent growth. At present we do not prefer one hypothesis over the other, and our data do not allow us to distinguish between them.

The first hypothesis is very difficult to assess since the published patterns yield contradictory conclusions. Attempts to select for a bias towards one side in characters that initially exhibited FA have been repeatedly unsuccessful in *Drosophila* (Maynard Smith & Sondhi, 1960; Coyne, 1987; Tuinstra, Dejong & Scharloo, 1990). These studies suggest that deviations from symmetry do not have a heritable basis. Nonetheless, strongly asymmetrical structures have evolved repeatedly in otherwise symmetrical organisms (Ludwig, 1932; Neville, 1976), suggesting that heritable variation for subtle deviations from symmetry must have existed at some point. In addition, (Palmer, Strobeck & Chippindale, 1993) have also suggested that certain patterns of normally-distributed variation in $R-L$ may signal the presence of heritable variation.

The second hypothesis implies that a kind of morphological 'templating' may be at work: asymmetries tend to carry over from one molt to the next because the hydrostatic expansion of parts between molts preserves small irregularities under the stable conditions of the laboratory environment. We took individuals from the wild, where they experienced harsh and unstable conditions, and placed them in a controlled environment that interfered little with subsequent growth. It is puzzling, however, that if such 'templating' is in operation, the asymmetries do not scale significantly with overall body size. All observed RMA and MA slopes were greater than 1.0, but in only one instance were they significantly so (Table 4a). The average percent increase in carapace width between molts was approximately 20%, thus if subtle asymmetries increased isometrically with body size they should have increased by roughly 20% (i.e. exhibited a slope of 1.2 between successive molts). In most cases, however, the residual variation in our data was sufficiently large that we could not distinguish between a slope of 1.0 and one of 1.2. Thus we cannot say with certainty that these subtle asymmetries did not scale with overall body size.

Perhaps most puzzling is that *Hemigrapsus nudus*, a species known to autotomize and regenerate entire limbs, seemed unable to correct for asymmetries presumably reflecting developmental noise (see expected pattern in Fig. 3d). A crab can regenerate an entire limb after two or three molts (McVean, 1982), yet subtle asymmetries persisted through two successive molts. Perhaps there is a threshold for asymmetry, above which correction can occur, and below which asymmetry persists through growth. Some exciting possibilities for future research arise from this supposition: Do such subtle asymmetries reproduce themselves in regenerated limbs? How do the other limbs compensate for the loss of a limb?

Acknowledgements

This work was supported by NSERC Operating Grant 55-47180 to ARP. We thank Jeanne Ferris for superb technical support in the maintenance of the experimental animals, and L. David Smith, C. Gratto and R. Helling, who collected the preliminary data on *H. nudus* as part of a student project for the Functional Morphology and Evolution course at the Friday Harbor Laboratories in 1986. R. E. Lenski and C. Klingenberg provided helpful comments on statistical methods. We also thank the staff of Bamfield Marine Station for assistance and use of the research facilities.

References

Angus, R. A. & R. H. Schultz, 1983. Meristic variation in homozygous and heterozygous fish. Copeia 1983: 287-299.

Coyne, J. A., 1987. Lack of response to selection for directional asymmetry in *Drosophila melanogaster*. J. Hered. 78: 119.

Hartnoll, R. G., 1982. Growth, pp. 111-196 in The Biology of Crustacea, edited by L. G. Abele. Academic Pr., London.

LaBarbera, M. L., 1989. Analyzing body size as a factor in ecology and evolution. Ann. Rev. Ecol. Syst. 20: 97-117.

Leamy, L., 1984. Morphometric studies in inbred and hybrid house mice. V. Directional and fluctuating asymmetry. Amer. Nat. 123: 579-593.

Leary, R. F. & F. W. Allendorf, 1989. Fluctuating asymmetry as an indicator of stress in conservation biology. Trends Ecol. Evol. 4: 214-217.

Livshits, G., L. Davidi, E. Kobyliansky, D. Ben-Amital, Y. Levi & P. Merlob, 1988. Decreased developmental stability as assessed by fluctuating asymmetry of morphometric traits in preterm infants. Amer. J. Med. Gen. 29: 793-805.

Ludwig, W., 1932. Das Rechts-Links Problem im Teirreich und beim Menschen. Springer, Berlin. 496.

Mather, K., 1953. Genetical control of stability in development. Heredity 7: 297-336.

Maynard Smith, J. & K. C. Sondhi, 1960. The genetics of a pattern. Genetics 45: 1039-1050.

McKenzie, J. A. & G. M. Clarke, 1988. Diazinon resistance, fluctuating asymmetry and fitness in the Australian sheep blowfly, *Lucilia cuprina*. Genetics 120: 213-220.

McVean, A., 1982. Autotomy, pp. 107-132 in The Biology of Crustacea, edited by D. E. Bliss. Academic Pr., New York.

Møller, A. P. & J. Höglund, 1992. Patterns of fluctuating asymmetry in avian feather ornaments: Implications for models of sexual selection. Proc. Roy. Soc. Lond. B245: 1-5.

Neville, A. C., 1976. Animal Asymmetry. Arnold (Inst. Biol. Stud. Biol.), London. 60.

Palmer, A. R. & C. Strobeck, 1986. Fluctuating asymmetry: measurement, analysis, patterns. Ann. Rev. Ecol. Syst. 17: 391-421.

Palmer, A. R. & C. Strobeck, 1992. Fluctuating asymmetry as a measure of developmental stability: Implications of non-normal distributions and power of statistical tests. Acta Zool. Fenn. 191: 57-72.

Palmer, A. R., C. Strobeck & A. K. Chippindale, 1993. Bilateral variation and the evolutionary origin of macroscopic asymmetries. Genetica (this volume).

Parsons, P. A., 1990. Fluctuating asymmetry: An epigenetic measure of stress. Biol. Rev. 65: 131-145.

Price, T., E. Chi, M. Pavelka & M. Hack, 1991. Population and developmental variation in the feather tip. Evolution 45 518-533.

Rice, W. R., 1989. Analyzing tables of statistical tests. Evolution 43: 223-225.

Sokal, R. R. & J. F. Rohlf, 1981. Biometry. Freeman, San Francisco, CA. 859.

Strawn, K., 1961. A comparison of meristic means and variance of wild and laboratory-raised samples of the fishes, *Etheostoma grahami* and *E. lepidum* (Percidae). Texas J. Sci. 13 127-159.

Taning, A., 1952. Experimental study of meristic characters in fishes. Biol. Rev. 27: 169-193.

Tuinstra, E. J., G. Dejong & W. Scharloo, 1990. Lack of response to family selection for directional asymmetry in *Drosophila melanogaster*: Left and right are not distinguished in development. Proc. Roy. Soc. Lond. B 241: 146-152.

VanValen, L., 1962. A study of fluctuating asymmetry. Evolution 16: 125-142.

Wooten, M. C. & M. H. Smith, 1986. Fluctuating asymmetry and genetic variability in a natural population of *Mus musculus*. J. Mammal. 67: 725-732.

Zakharov, V. M., 1992. Population phenogenetics: Analysis of developmental stability in natural populations. Acta Zool. Fenn. 191: 7-30.

Zhivotovsky, L. A., 1992. A measure of fluctuating asymmetry for a set of characters. Acta Zool. Fenn. 191: 37-77.

T. A. Markow (ed.), Developmental Instability: Its Origins and Evolutionary Implications, 203–220, 1994.

Bilateral variation and the evolutionary origin of macroscopic asymmetries

A. Richard Palmer[1,2], Curtis Strobeck[1] & Adam K. Chippindale[1,3]
[1] *Department of Zoology, University of Alberta, Edmonton, Alberta T6G 2E9, Canada*
[2] *Bamfield Marine Station, Bamfield, British Columbia V0R 1B0, Canada*
[3] *Present address: Dept. of Ecology and Evolutionary Biology, University of California, Irvine, CA 92717, USA*

Received 5 April 1993 Accepted 27 April 1993

Abstract

Given that characters exhibiting macroscopic asymmetry have evolved in a wide variety of taxa, heritable variation for bilateral asymmetry must have arisen at some point in their history. The recognition that heritable variation may underlie some statistical asymmetries not only raises concerns about the incautious use of statistical estimates of FA in studies of developmental stability, but it suggests some intriguing questions about the possible evolutionary origins of macroscopic asymmetries. First, we developed an additive model of bilateral variation based on some simple assumptions about the developmental control of bilateral variation. Second, using a new approach for studying statistical asymmetries, we conducted an analysis of bilateral variation in eight metrical traits of a harpacticoid copepod (*Tigriopus californicus*) to search for novel forms of statistical asymmetries. The model we developed revealed three independent statistical asymmetries of potential evolutionary significance: *a*) a previously unrecognized form of asymmetry (referred to here as *normal covariant asymmetry*), *b*) antisymmetry, and *c*) directional asymmetry. Because each pattern of variation would seem to require different amounts and kinds of developmental-genetic information [*a*- only negative feedback between sides (*bilateral inhibition*), *b*- both *bilateral inhibition* and average departure from symmetry (*bilateral offset*), *c*- *bilateral inhibition*, *bilateral offset*, and a consistent overdevelopment of one side or the other (*side-bias control*)], those requiring less information would seem more likely to represent earlier stages in the evolution of macroscopic asymmetries. Our analysis of bilateral variation in *Tigriopus* revealed no evidence for any form of statistical asymmetry other than fluctuating asymmetry. However, a significant positive covariation between sides, even after correction for body size variation, suggested that factors influencing relative limb length (whether genetic or environmental) affected both sides equally rather than one side at the expense of the other. Finally, we note that certain statistical asymmetries (directional asymmetry, any form of covariant asymmetry) may render characters unreliable for estimating developmental stability because, unlike pure fluctuating asymmetry, they may signal a genetic component to asymmetry variation.

Introduction

Symmetry and departures from symmetry

Although the external body form of most animals is bilaterally symmetrical, departures from symmetry may take a variety of forms (Ludwig, 1932). Sometimes these departures are quite conspicuous, as in the case of gastropods, male fiddler crabs, American lobsters, and flatfishes (Neville, 1976). More commonly, departures from symmetry are quite subtle and require careful measurement to detect (Palmer & Strobeck, 1986). No character, strictly speaking, will exhibit perfect bilateral symmetry, except perhaps by chance, since the mechanisms guiding development simply do not have that kind of precision. Even for characters that are bilaterally symmetrical on avarage, the right and left sides of any given individual will depart subtly from symmetry at some level of measurement. To what extent might such subtle variation provide the raw material from which macroscopic asymmetries evolve?

The biological significance of subtle, statistical departures from bilateral symmetry has stimulated much interest (Palmer & Strobeck, 1986; Zakharov, 1992). Underlying this interest is a belief that one form - fluctuating asymmetry (FA), or small, random departures from bilateral symmetry - can be used to describe 'developmental stability'. In its simplest expression, perfect symmetry provides a convenient reference against which to measure developmental precision. The less able an organism is to buffer itself against disturbances during development, the more likely one or more of its characters will depart from symmetry. In some organisms, FA may increase in response to extrinsic environmental stresses (Leary & Allendorf, 1989; Zakharov, 1992) or to intrinsic genetic stresses (Parsons, 1990). In others, FA appears to decrease with increasing heterozygosity, suggesting a connection between heterosis and developmental stability (Mitton & Grant, 1984; Allendorf & Leary, 1986; Palmer & Strobeck, 1986).

In the extensive literature on subtle, statistical departures from symmetry, however, an important observation is often overlooked: macroscopically asymmetrical structures have, in fact, evolved in a wide variety of taxa (Ludwig, 1932; Neville, 1976). Presumably, such macroscopic asymmetries arose initially as subtle departures from symmetry that had a heritable basis and were subsequently amplified by natural selection. This view of asymmetry raises three questions to which we address this paper. 1) What forms might evolutionarily significant variation in asymmetry take? 2) Of alternative forms, which is the most likely starting point for the evolution of macroscopic asymmetries? 3) What are the implications of such an evolutionary perspective for studies of developmental stability based upon variation in FA?

Components of bilateral variation

An additive model of bilateral variation

The following model attempts to incorporate some simple developmental processes that influence bilateral variation and describe how they contribute to the patterns of variation observed between sides. Building upon the notation of Palmer and Strobeck (1986), let the right and left members of a bilateral trait-pair for the ith individual of a particular sample be

$$R_i = \mu + s_i + (D + \delta_i)A_i/2 + r_i = \mu + s_i + DA_i/2 + \delta_i/2 + r_i, \text{ and} \qquad (1)$$

$$L_i = \mu + s_i - (D + \delta_i)A_i/2 + l_i = \mu + s_i - DA_i/2 - \delta_i/2 + l_i. \qquad (2)$$

where the terms are defined as in Table 1. One half of $D + \delta_i$ is added to the right side and subtracted from the left so that variation in these terms has no affect on average trait size. Note also that because δ_i has a mean of zero and is normally distributed, its contribution to the variation in R or L is not influenced by *side bias*, A_i, which may only take values of $\pm$ 1. Hence both equations simplify to those on the right. Given these components of bilateral variation, the expected means and variances for each side will be:

$$\mathrm{E}(R_i) = \mu + D\,\mathrm{E}(A_i)/2 = \mu + D\,[p - q]/2, \qquad (3)$$

$$\mathrm{E}(L_i) = \mu - D\,\mathrm{E}(A_i)/2 = \mu - D\,[p - q]/2, \text{ and} \qquad (4)$$

$$\mathrm{Var}(R_i) = \mathrm{Var}(L_i) = \sigma_s^2 + (D/2)^2\,\mathrm{Var}(A_i) + \sigma_d^2/4 + \sigma_e^2 = \sigma_s^2 + D^2pq + \sigma_d^2/4 + \sigma_e^2. \qquad (5)$$

The covariance between sides will be

$$\mathrm{Cov}(R_i,L_i) = \mathrm{E}\{[R_i - (\mu + D[p - q]/2)][L_i - (\mu - D[p - q]/2)]\}.$$

Substituting for R_i and L_i from equations (1) and (2), and rearranging yields

$$\mathrm{Cov}(R_i,L_i) = \mathrm{E}\{(s_i + DA_i/2 - D[p - q]/2 + \delta_i/2 + r_i)\,(s_i - DA_i/2 + D[p - q]/2 - \delta_i/2 + l_i)\}$$

$$= \mathrm{E}\{\sigma_i^2 - (DA_i/2 - D[p - q]/2)^2 - (\delta_i/2)^2\}$$

$$= \sigma_s^2 - D^2\,pq - \sigma_d^2/4. \qquad (6)$$

Table 1. Description of variables for an additive model of bilateral variation where there is no variation in body size.

Symbol*	Description	Effect on distribution of bilateral variation	Interpretation of developmental origin
μ	A positive, arbitrary constant defining the mean size of a trait *for a given body size*	n.r.	n.r.
s_i	A random normal deviate (mean 0, variance σ_s^2) added to both sides of individual i	Results in positive covariation between sides	Reflects environmental or genetic factors that increase or decrease development of both sides equally (*positive covariation*)
D	A positive, arbitrary constant representing the fixed difference between sides	Results in directional asymmetry or antisymmetry depending on A_i	Reflects presumably genetic factors that determine average magnitude of departure from symmetry (*bilateral offset*)
δ_i	A random normal deviate (mean 0, variance σ_d^2) added to the fixed difference between sides (D) in individual i	Results in continuous negative covariation between sides	Reflects environmental or genetic factors that influence the extent of bilateral offset in a given individual (*offset variation*)
A_i	A binomial random variable that takes the values +1 or −1 with probability p and $q = 1 - p$ respectively, and indicates whether the fixed difference between sides (D) is added to the right (+1) or subtracted from the right (−1) of individual i	Influences the direction of directional asymmetry if p = 1 or p = 0, or the shape of the frequency distribution of ($R - L$) if $0 < p < 1$	Reflects environmental or genetic factors that influence the direction of departure from symmetry (*side bias*)
r_i and l_i	Random normal deviates (mean 0, variance σ_e^2) added to the right and left side respectively of individual i	Results in residual variation between sides	Reflects non-genetic factors that cause random, independent variation in final trait size (*developmental noise*)

* as in equations 1 & 2; n.r. – not relevant.

The difference between sides, obtained from equations 1 and 2 is

$$R_i - L_i = DA_i + \delta_i + r_i - l_i, \tag{7}$$

and the expected mean and variance of this difference will be

$$E(R_i - L_i) = DE(A_i) = D\,[p - q] \text{ and} \tag{8}$$

$$\begin{aligned} \mathrm{Var}(R_i - L_i) &= D^2\,\mathrm{Var}(A_i) + \sigma_d^2 + 2\sigma_e^2 \\ &= 4D^2 pq + \sigma_d^2 + 2\sigma_e^2. \end{aligned} \tag{9}$$

Equation 9 could also have been derived by recognizing that

$$\mathrm{Var}(R_i - L_i) = \mathrm{Var}(R_i) + \mathrm{Var}(L_i) - 2\,\mathrm{Cov}(R_i, L_i). \tag{10}$$

Note that according to this formulation, variation in the difference between sides (equation 9) includes a contribution due to *bilateral offset* (D), variation in *side bias* (pq), and *offset variation* (σ_d^2). Hence, only if these terms are zero will the variation in ($R_i - L_i$) reflect pure *developmental noise* (σ_e^2).

A graphical view of bilateral variation

Conventional forms of statistical asymmetries. Frequency distributions of $R - L$ are traditionally used to view different forms of statistical asym-

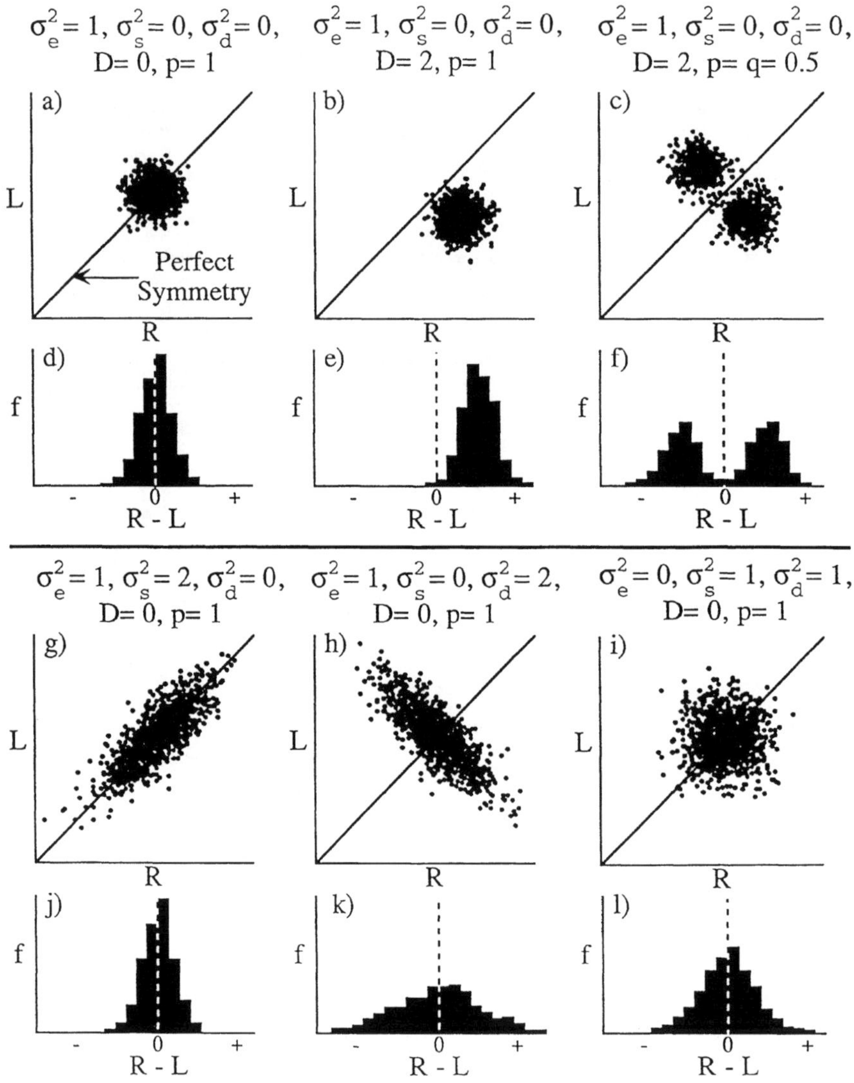

Fig. 1. Computer simulations of bilateral variation using the developmental model described in equations 1 and 2. Variables above each figure are defined in Table 1. Each form of asymmetry is presented as a pair of figures: *scatterplots* of *R* versus *L* (upper figure of each vertical pair), and *frequency distributions* of *R* - *L* (lower figure of each vertical pair, dashed vertical line corresponds to perfect symmetry), (a,d) fluctuating asymmetry. (b,e) directional asymmetry, (c,f) antisymmetry. (*g,j*) positive covariation between sides at a given body size. (h,k) negative covariation between sides at a given body size (normal covariant asymmetry). (i,l) apparent FA in the absence of *developmental noise* (σ_e^2), but where *positive covariation* sides (σ_s^2) and *negative covariation* between sides (σ_d^2) are nonzero and equal.

metries. However, once general size effects have been factored out (see *Scaling variation to a common body size* below), plots of *R* versus *L* reveal more information about bilateral covariation. No such covariation is present in either pure fluctuating asymmetry or pure directional asymmetry (Fig. 1a,b). Of the three currently named forms of statistical asymmetries (Fig. 1a-c), only antisymmetry signals covariation unambiguously (Fig. 1c). All three of these asymmetries may be distinguished by knowing the mean and shape of the frequency distribution of $R - L$ (Fig. 1d-f; Van Valen, 1962).

A novel form of statistical asymmetry. An important point to which we wish to draw attention is that negative covariation between sides need not be expressed only as a bimodal or platykurtic distribution of $R - L$ (Fig. 1c,f). Negative covariation between sides may also be continuous (Fig. 1h). Such covariation would correspond to *offset variation* in our formulation (σ_d^2 in equation 9).

Unlike *positive covariation* between sides (σ_s^2), negative covariation *does* contribute to the variance of $R - L$ (equation 9), even though the frequency distribution of $R - L$ may still exhibit what appears to be pure FA (mean zero, normal; Fig. 1k,l). This may be visualized most easily by considering the shape of the frequency distribution that would result from bilateral variation about the line of perfect symmetry when viewed from the origin. Increasing the *positive covariation* between sides (e.g. σ_s^2 in Fig. 1g) will have no effect of the frequency distribution of $R - L$ (Fig. 1j), whereas increasing the negative covariation between sides (e.g. σ_d^2 in Fig. 1h) will increase the variance of the frequency distribution of $R - L$ (Fig. 1k). Hypothetically, a normal frequency distribution with mean zero could arise even in the absence of any *developmental noise* ($\sigma_e^2 = 0$), so long as σ_d^2 was not zero (Fig. 1i,l).

Terminology for different forms of negative covariation between sides. In keeping with the tradition of naming asymmetries based on their pattern of bilateral variation, as opposed to the presumed mechanism giving rise to asymmetry, we suggest that all forms of negative covariation between sides be referred to as *covariant asymmetry*. However, because the term *antisymmetry* is so deeply entrenched in the literature, we suggest retaining it to describe the case illustrated in Figure 1c,f, even though this is just a special case of covariant asymmetry. We suggest *normal covariant asymmetry* to describe the case illustrated in Figure 1h,i. Skewed frequency distributions of $R - L$ would correspond to another form of antisymmetry (*skewed covariant asymmetry*) where $p \neq q$, and where neither p nor q nor $D = 0$ (see equation 9 and Palmer & Strobeck, 1992).

Distinguishing among different forms of covariant asymmetry

Although directional asymmetry may be detected easily by a variety of analytical procedures (Palmer & Strobeck, 1986), forms of covariant asymmetry are more difficult to detect. As a first step, any form of covariant asymmetry will be revealed as a statistically significant sides X individuals interaction in the ANOVA procedure outlined by Palmer and Strobeck (1986), so long as an estimate of measurement error is available. However, this ANOVA procedure cannot distinguish among forms of covariant asymmetry, hence some additional analysis is required.

Skewed covariant asymmetry and antisymmetry. Both skewed covariant asymmetry and antisymmetry will appear as departures from zero of the third or fourth statistical moments of frequency distributions of $R - L$ (skew and kurtosis). Both are thus readily detected by conventional tests for departures from normality [see Shapiro, Wilk & Chen (1968) and Palmer & Strobeck (1992) for discussion of such tests].

Normal covariant asymmetry: Positive covariation *between sides absent.* Normal covariant asymmetry is more difficult to detect than skewed covariant asymmetry or antisymmetry because the frequency distribution of $R - L$ in this case is indistinguishable from that of pure FA (compare Fig. 1k or 1l with 1d). Where *positive covariation* between sides is absent ($\sigma_s^2 = 0$) and where *developmental noise* is small ($\sigma_e^2 < \sigma_d^2$), *offset variation* (σ_d^2 can be detected as a statistically significant negative correlation between sides (Fig. 1h). Note, though, that if the *offset variation* is small relative to *developmental noise* ($\sigma_d^2 < \sigma_e^2$), then the negative correlation may be difficult to detect statistically for routine sample sizes.

Normal covariant asymmetry: Confounding effects of positive covariation *between sides.* Unfortunately, average character size $[(R_i + L_i)/2]$ may vary even where overall body size does not (e.g. relative arm length may still vary among humans of the same body size). When character size varies in this manner (i.e. where *positive covariation* between sides (σ_s^2) is greater than 0), the detection of *offset variation* becomes problematical. An extreme example should make this clear. Consider a single sample where factors contributing to *positive covariation* between sides (σ_s^2) equal those contributing to *offset variation* (σ_d^2), as in Figure 1g and 1h respectively, but where *developmental noise* is absent ($\sigma_e^2 = 0$). If each of these patterns of covariation is bivariate normal, then the joint distribution will appear to be pure FA (Fig. 1i and 1l)!

Regrettably, the confounding effect of *positive covariation* between sides is not simply an empirical problem. If all three variances (σ_s^2, σ_d^2, and σ_e^2) are non-zero, it is not possible to partition them out precisely. For example, the distribution illustrated in Figure 1h could result either from a) a combination of positive and negative covariation ($\sigma_d^2 > \sigma_s^2 > 0$, $\sigma_e^2 = 0$), b) a combination of negative covariation and *developmental noise* in the absence of *positive covariation* ($\sigma_d^2 > \sigma_e^2 > 0$, $\sigma_s^2 = 0$), or c) some combination of all three variance components (e.g. $\sigma_d^2 >> \sigma_e^2 = \sigma_s^2 > 0$).

Normal covariant asymmetry: A pragmatic approach. The major practical concern for studies of the evolution of bilateral asymmetry is whether the magnitude of normal covariant asymmetry, and its presumed genetic underpinnings, is large enough to respond to selection. For studies of FA, the concern is whether normal covariant asymmetry is large enough to confound inferences about developmental stability. Thus, although it may never be detectable when slight, a pragmatic approach to this problem would be to ask, as illustrated below with *Tigriopus*, whether the *offset variation* (σ_d^2) is large relative to *positive covariation* between sides (σ_s^2) or to *developmental noise* (σ_e^2), among individuals of comparable body size. Continuous negative covariation between R and L will only be apparent statistically where the *offset variation* (σ_d^2) *is greater than positive covariation* between sides (σ_s^2).

A case study of statistical asymmetries in a harpaticoid copepod

To test for the presence of various alternative forms of statistical asymmetries, we analyzed data collected as part of a larger study of geographic differentiation in the common, high-tidepool harpacticoid copepod *Tigriopus californicus* (Chippindale & Palmer unpublished). Burton (Burton & Feldman, 1981; Burton, 1987) has reported extensive divergence at allozyme loci among local populations and we wished to determine if developmental stability was reduced in hybrids between geographically distant populations. Except for its rather small size, *T. californicus* is an ideal animal for developmental-genetic studies because of the ease with which it may be bred and raised in the lab.

Methods

Collection, culture, and preparation. Individuals of *T. californicus* were collected from three sites over a two year period. Laboratory populations were initiated with 125-150 mature females in the early stages of ovigery. Both parents and offspring were reared on a mixed diet of the cultured unicellular green alga *Isochrysis* sp. (Provasoli, Shiraishi & Lance, 1959) and powdered commercial fish food (TetraMin) (Burton, pers. comm.). The cultures were maintained in two liters of sterilized, normal-strength seawater within four liter glass containers left open to the air and held at room temperature. Before initiating the cultures, females were rinsed four times with sterilized, aerated seawater.

When a sufficient number of individuals in a culture had reached their adult moult [copepodid VI (C-VI); approximately 2-3 weeks after hatching] 40 were removed and killed in a dilute formalin solution. Only females were selected because the presence of ovisacs indicated both the sex and the state of maturity of individuals unambiguously. From these individuals, the right and left first antennae, the right and left maxillipeds, and the right and left third thoracic limbs were dissected from the body with modified 00 insect pins in a full strength solution of PVA lactophenol (Clarke, Brand & Whitten, 1986). Limbs were generally removed as pairs to aid with positioning. After removal, the limbs were oriented on a glass slide perpendicular to the plane of articulation as viewed from the posterior, except for the first antennae which were oriented as viewed dorsally. They were subsequently covered with a cover slip, and the mounting medium was allowed to harden. These limbs were chosen for their ease of mounting and because they included clearly defined and resolvable characters for measurement.

The culture we analyzed (initiated from San Juan Is., Washington, USA) was selected because it contained the largest number of individuals having a complete set of replicated measurements. Because of the small size of these copepods (< 2 mm total body length), limbs were occasionally damaged during dissection and hence could not be measured.

Measurement and analysis. A total of eight metrical traits were measured: three on the first antennae (Ant 1, Ant 2, and Ant 3), two on the maxilliped (Mxp 1, Mxp 2), and three on the third thoracic limb (Thc 1, Thc 2, Thc 3). The dimensions measured were: Ant 1- anterior margin of first article, Ant 2- anterior margin of second article, Ant 3- posterior margin of second article, Mxp 1- lateral margin of first article of exopodite, Mxp 2- medial margin of first article of endopodite, Thc 1- lateral margin of first article of exopodite, Thc 2- medial margin of first article of endopodite, Thc 3- from the central lobe of the proximal end to the distal-most tip of the third article of the endopodite. For Ant 1, Ant 2, Ant 3, and Mxp 1, measurements were made from the base of the article at the proximal joint to the distal-most edge of the article adjacent to the next joint. For Mxp 2, Thc 1, and Thc 2, measurements were made from the base of the article to the insertion point of the pronounced, solitary, articulated seta adjacent to the distal-most edge of the article.

All measurements were obtained by viewing the limbs under a compound microscope at 400X magnification and, via a camera lucida, digitizing the dimensions with a 500 × 500 dpi resolution graphics tablet (MacTablet, Summagraphics Corp.) attached to a microcomputer.

Measurement error. Measurement error was estimated by measuring each trait three times independently. In other words, one set of 16 measurements was completed (eight limbs, two sides) on all individuals before beginning a replicate set, hence measurement error included both digitizing error as well as the error in relocating landmarks.

Because differences between sides in studies of statistical asymmetries are often very small (1-2% of character size), these differences must be shown to be significantly larger than would be expected due to the error in measurement. Measurement error will, of course, give the same pattern of between-sides variation as FA. Following the procedure outlined in Palmer and Strobeck (1986), the magnitude of nondirectional asymmetry (which includes FA and all forms of covariant asymmetry) was tested relative to measurement error with a mixed-model, two-way analysis of variance (ANOVA; sides = fixed effect, individuals = random effect, repeat measures = replicates). This same analysis also yielded significance levels (relative to nondirectional asymmetry) for directional asymmetry and body size/character size variation among individuals.

Body size effects and departures from normality. To determine whether the between-sides variance varied with character size for a given trait, we examined scatterplots of R versus L. Variation appeared to be independent of size for each character examined (see Fig. 3 below), so no transformation was necessary to homogenize the variance before computing the conventional descriptors of statistical asymmetries.

To detect departures from normality, frequency distributions of $R - L$ were tested for skew and kurtosis. These tests will detect skewed covariant asymmetry and antisymmetry. To detect normal covariant asymmetry, we tested for negative correlations between R and L after factoring out body size variation (see next section). Replicate measurements were averaged for each side before conducting these tests.

Scaling variation to a common 'body size'. We realize that 'body size' does not correspond to any given entity that we may estimate, because it will always be a function of the particular, usually arbitry, subset of characters measured. We use the term here for convenience, even though we mean only the positive covariation among all the traits we examined. Over the rather narrow size range we examined, the between-sides variance in a particular character did not increase with the mean (see

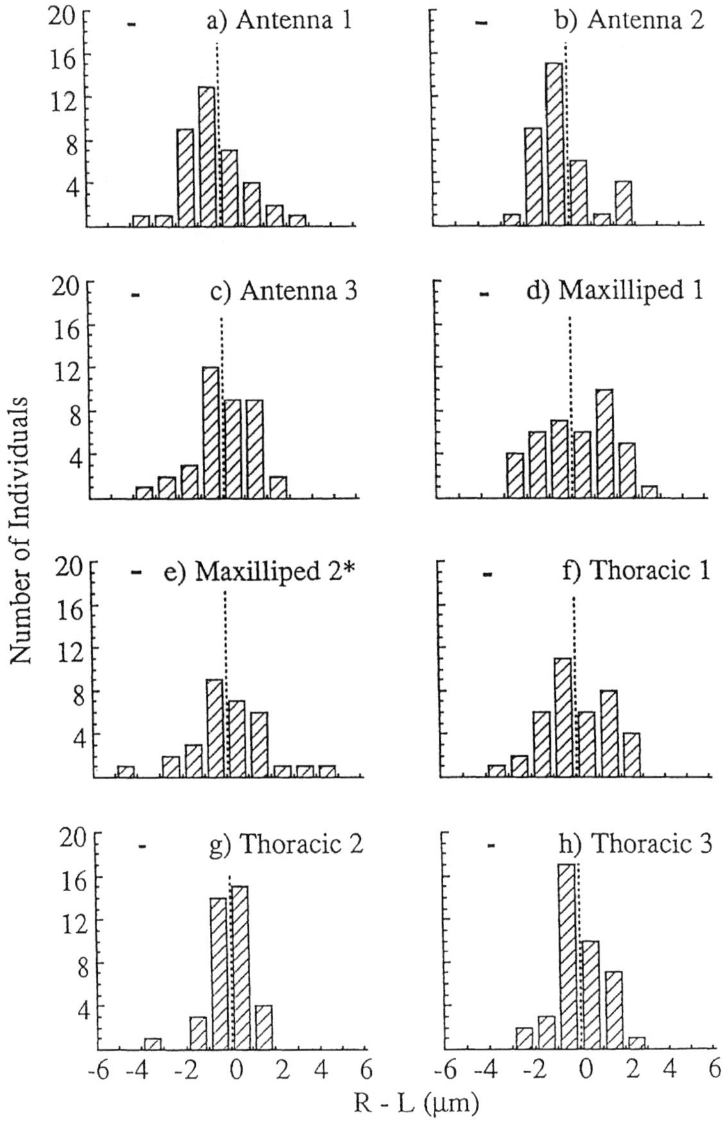

Fig. 2. Frequency distributions of $R - L$ for eight traits in C-VI stage females of the marine harpacticoid copepod, *Tigriopus californicus* Differences were computed after the three replicate measurements were averaged for R and L. Mean measurement error ranged from 0.2(to 0.31 µm among the eight characters, as indicated by small horizontal bars in each figure. Extreme values indicated by † in Fig. 3 were excluded from these distributions to improve the clarity of presentation. Dashed line indicates perfect symmetry. * Note that for Mxp 2 the units of deviation ranged from −12 to +12, but have been plotted to the same scale as the other characters for simplicity. The following five extreme values, some of which lie outside the bounds of these figures, were eliminated as statistical outliers using Grubb's tes (Sokal & Rohlf, 1981): Ant 2: −7 and −8, Mxp 2: −15, Thc 1: +12, and Thc 2: −4.

Table 2. Significance tests for nondirectional asymmetry, body size/character size variation, and directional asymmetry for eight traits in the harpacticoid copepod *Tigriopus californicus*.

	Measurement error			Sides × Individuals (= nondirectional asymmetry)			Individuals (= both body & character size)		Sides (= directional asymmetry)	
Trait	df	MS($=\sigma_m^2$)	σ_m^2/σ_i^2	df	MS	P*	MS	P†	MS	P†
Ant 1	152	0.194	0.225	37	2.78	< 0.001	52.31	< 0.001	2.42	0.36
Ant 2	152	0.280	0.146	37	6.04	< 0.001	33.32	< 0.001	18.14	0.09
Ant 3	152	0.145	0.180	37	2.57	< 0.001	24.11	< 0.001	2.69	0.31
Mxp 1	156	0.113	0.090	38	3.88	< 0.001	240.68	< 0.001	1.90	0.18
Mxp 2	128	0.240	0.024	31	29.78	< 0.001	379.63	< 0.001	4.68	0.69
Thc 1	156	0.194	0.072	38	8.30	< 0.001	51.77	< 0.001	6.44	0.38
Thc 2	148	0.119	0.310	36	1.27	< 0.001	32.81	< 0.001	0.58	0.50
Thc 3	160	0.162	0.285	39	1.87	< 0.001	62.16	< 0.001	0.0002	0.99

Note. Results are from a mixed model, two-way ANOVA with replication, following the procedure of Palmer and Strobeck (1986) ['Sides' = fixed effect, 'Individuals' = random effect, 'Sides × Individuals' = remainder whose expected MS is $\sigma_m^2 + M\sigma_i^2$ (where M = number of replicate measurements per side, three in this study)]. Computations were carried out prior to any correction for body size variation and prior to removal of outliers.

σ_m^2 = measurement error variance, σ_i^2= nondirectional asymmetry variance [includes contributions due to FA and all forms of covariant asymmetry; from Table 3*B* of Palmer and Strobeck (1986)]. The ratio σ_m^2/σ_i^2 indicates the magnitude of measurement error relative to the between-sides variation.

df - degrees of freedom. The degrees of freedom for 'Individuals' is the same as that for 'Sides × Individuals', that for 'Sides' is one. MS - mean squares from ANOVA; P = exact probability; * tested over 'Measurement error' MS; † tested over 'Sides × Individuals' MS.

Fig. 3 below) so a linear transformation to remove 'size effects' seemed justified.

To test for normal covariant asymmetry, body size variation had to be factored out in order to examine the covariation of R and L about the parametric mean character size (μ in equations 1 and 2) for a given body size. Because some *positive covariation* between sides will be body-size independent (i.e. for some individuals, both limbs could be proportionally larger or smaller for the same body size), it was inappropriate to adjust each side simply by subtracting, or dividing by, the mean for each individual $[(R_i + L_i)/2]$. Furthermore, such a correction would merely create a perfect negative correlation between R and L (any departure of R_i in one direction from $[(R_i + L_i)/2]$ would precisely equal the departure of L_i in the opposite direction) and thus obscure any *offset variation*.

Body size was factored out by converting measurements for each side into deviations from a model II regression against a general size metric. To obtain this size metric, the data for each of the eight traits were first transformed to standardized normal deviates (SND) by subtracting the overall mean for the trait ($\Sigma[R_i + L_i]/2N$) and dividing by its standard deviation $[VAR(R_i, L_i)^{-2}]$. This transformation was necessary because traits with a larger mean size had larger variances (Table 3). By converting each variate to an SND, the contribution to body size of each trait was thus weighted equally, independent of overall trait size. In addition, in contrast to a principal components analysis, this procedure permitted us to use all individuals in the analysis because the absence of one or more characters would not bias the size metric for that individual.

The size metric for a particular trait was computed as the arithmetic mean of the SNDs of all the *other* traits. In other words, for Antenna 1 of individual 1, the body size metric was computed as the mean SND for the remaining seven traits for individual 1. This resulted in a series of body size metrics for each individual, one metric for each trait. One advantage to this approach was that the body size metrics for a particular trait were thus independent of the variation in that trait.

Finally, body size was factored out of each trait via reduced major exis (RMA) regression of each side separately against the appropriate size metric

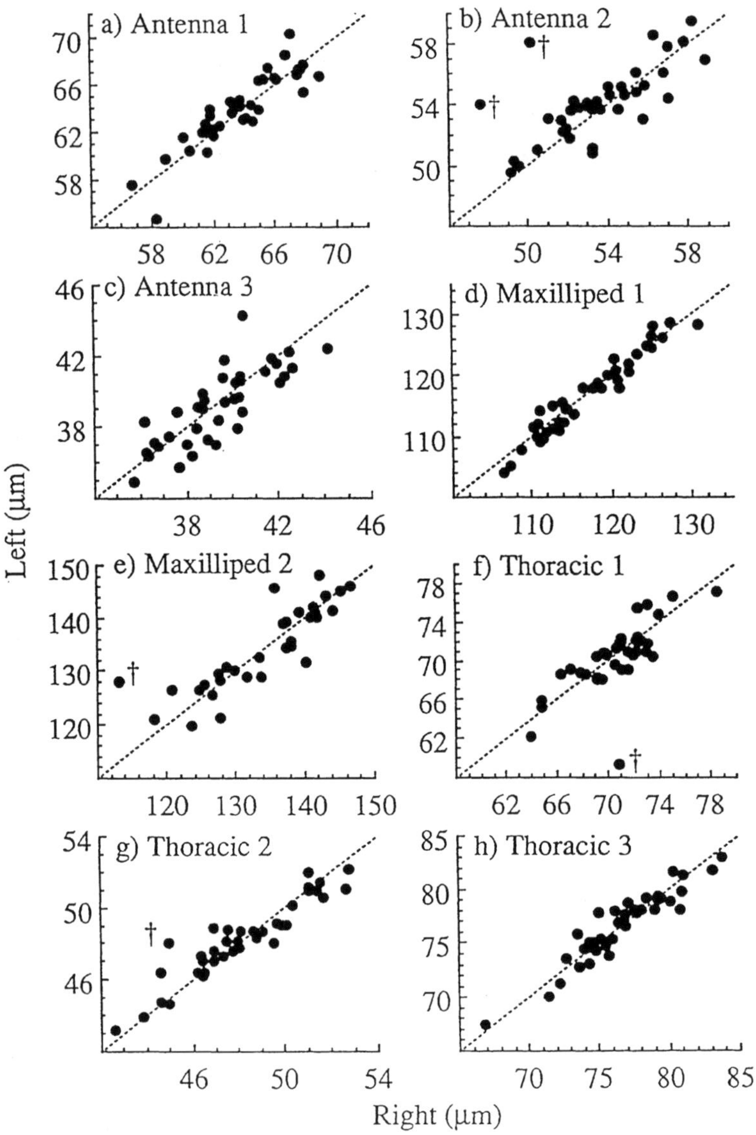

Fig. 3. Scatterplots of *R* versus *L*, uncorrected for body size variation, for eight traits in C-VI stage females of *Tigriopus californicus*. The mean measurement error ranged from 0.20 to 0.31 μm among the eight characters. Points indicated by † were excluded as extreme values from statistical analyses of asymmetry variation. Dashed line indicates perfect symmetry.

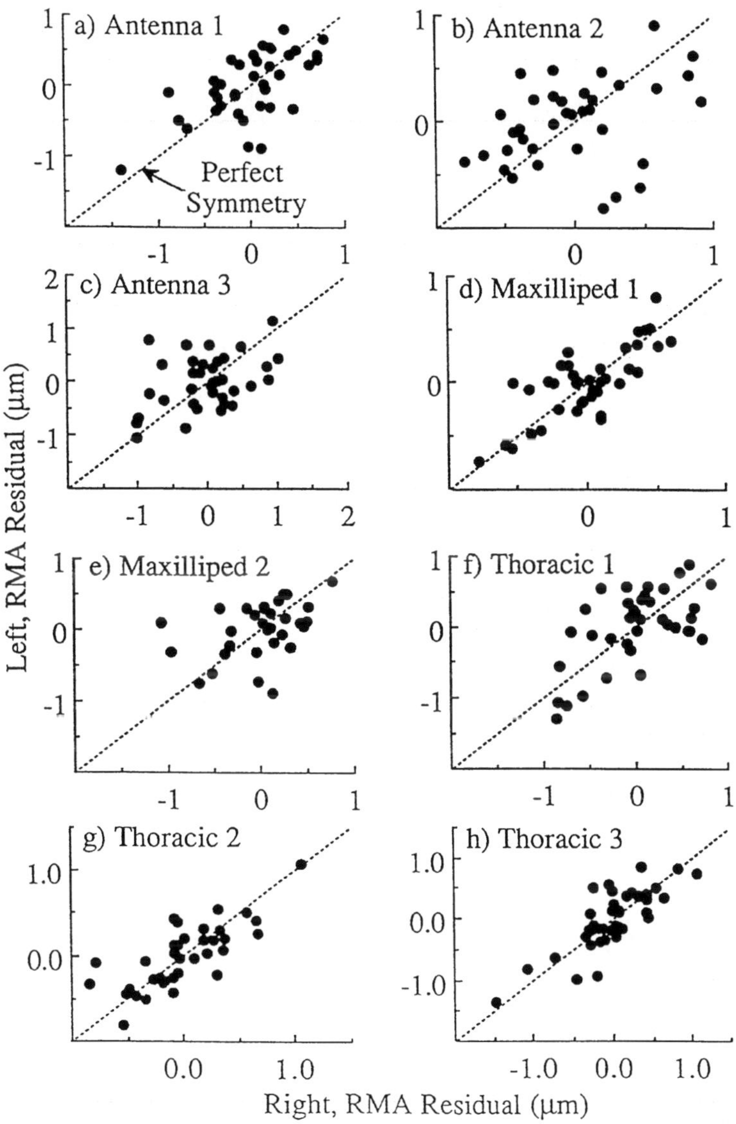

Fig. 4. Scatterplots of *R* versus *L*, after correction for body size variation, for eight traits in C-VI stage females of *Tigriopus californicus*. Data for each side are expressed as deviations from a model II regression on body size (see *Scaling vairation to a common body size* in text). Dashed line indicates perfect symmetry.

Table 3. Average character size, and statistics describing frequency distribution shape of $R - L$ in the harpacticoid copepod *Tigriopus californicus*.

		$(R + L)/2$	(R - L)						Other Indices**	
Character	N	Mean (SE)	Mean (SE)*	P	Skew (SE)	P	Kurtosis (SE)	P	σ_i^2	FA5
Antenna 1	38	63.8 (0.48)	-0.21 (0.221)	0.36	0.58 (0.383)	0.13	0.53 (0.750)	0.48	0.87	1.85
Antenna 2	38	53.9 (0.38)	-0.56 (0.325)	0.09	-1.54 (0.383)	<0.001	4.48 (0.750)	<0.001	1.92	4.23
Antenna 2 †	36	53.9 (0.39)	-0.20 (0.213)	0.35	0.89 (0.393)	0.026	0.14 (0.768)	0.85	0.73	1.63
Antenna 3	38	39.3 (0.33)	0.22 (0.212)	0.31	-0.66 (0.383)	0.088	0.69 (0.750)	0.35	0.81	1.71
Maxilliped 1	39	116.9 (1.01)	0.36 (0.257)	0.18	-0.30 (0.378)	0.43	-0.87 (0.741)	0.24	1.26	2.64
Maxilliped 2	32	134.1 (1.41)	-0.31 (0.788)	0.69	-1.05 (0.414)	0.012	2.45 (0.809)	0.003	10.01	19.33
Maxilliped 2§	31	134.5 (1.38)	0.16 (0.654)	0.81	-0.24 (0.421)	0.58	0.93 (0.821)	0.26	6.54	12.85
Thoracic leg 1	39	70.6 (0.47)	0.33 (0.377)	0.38	2.75 (0.378)	<0.001	11.46 (0.741)	<0.001	2.69	5.50
Thoracic leg 1§	38	70.7 (0.46)	0.04 (0.239)	0.88	-0.04 (0.383)	0.92	-0.76 (0.750)	0.31	1.02	2.12
Thoracic leg 2	37	48.2 (0.38)	-0.10 (0.151)	0.50	-1.00 (0.388)	0.011	1.89 (0.759)	0.014	0.38	0.84
Thoracic leg 2§	36	48.3 (0.39)	-0.02 (0.130)	0.89	-0.35 (0.393)	0.38	0.33 (0.768)	0.67	0.26	0.59
Thoracic leg 3	40	76.7 (0.51)	0.002 (0.176)	0.99	-0.14 (0.374)	0.72	0.44 (0.733)	0.55	0.57	1.21

Note. Computations were carried out prior to any correction for body size variation. The total number of individuals examined was 40, but some structures were missing or damaged on some individuals.
N - number of individuals measured; P - exact probability from a T-test comparing the observed values for the mean, skew, and kurtosis of the distribution of R - L versus the expected value of zero; * For comparison with other FA indices, Var($R - L$) will be N $(SE)^2$; ** σ_i^2 = non-directional asymmetry variance computed from ANOVA, and FA5 = index 5 =$[\Sigma(R - L)^2]/N$ [from Tables 3b and 1 respectively of Palmer and Strobeck (1986)]; for a truly normal distribution these indices are related as follows: Var($R - L$) = FA5 = $2(\sigma_i^2)$; † re-computed excluding two extrema identified by Grubb's test for outliers (Sokal and Rohlf 1981); see Fig. 3). § re-computed excluding one extremum identified by Grubb's test for outliers (see Fig. 3).

for that trait (see LaBarbera (1989) for a nice overview and evaluation of various model II regression techniques). The size-standardized values for each side were computed as deviations perpendicular to the major axis of the RMA regression. Before performing the size-standardizations, five measurements were eliminated as anomalous (see Results).

All statistical analyses were conducted with Statview II (Ver. 1.04, Abacus Concepts).

Results

Measurement error and tests for normality. Average measurement error ranged from 0.20 to 0.31 μm (0.17 to 0.64% of mean character size) among the eight characters, and was independent of trait size. In addition, measurement error was only 17% as large as the nondirectional asymmetry variation on average (mean $\sigma_m^2/\sigma_i^2 = 0.17$), hence nondirectional asymmetry was highly significant for all characters ('Sides × Individuals', Table 2). We were thus justified in pursuing a more detailed analysis of this nondirectional asymmetry, which includes FA and all forms of covariant asymmetry (Palmer & Strobeck, 1986).

Directional asymmetry and departures from normality. Only one of the eight traits (Ant 2) exhibited a suggestion of directional asymmetry (P = 0.09, Table 2), but this was due to a single extreme specimen (Table 3, Fig. 2 legend). In the original data, four traits exhibited distributions of $R - L$ that departed significantly from normality (Ant 2, Mxp 2, Thc 1, and Thc 2; Table 3). A closer inspection of frequency distributions revealed that one extreme value (or in one case two; see Fig. 2 legend) was responsible for the apparent departure from normality. All five extrema were found to be significant outliers by Grubb's test (Sokal & Rohlf, 1981). When removed, none of the characters exhibited significant departures from normality except Ant 2 which exhibited a marginally significant skew (P = 0.026, Table 3). In summary, after eliminating a small number of extreme values (5 of a total of 301), the variation in $R - L$ for all traits was normally distributed about a mean of zero (Fig. 2). Thus no traits exhibited any statistical indication of

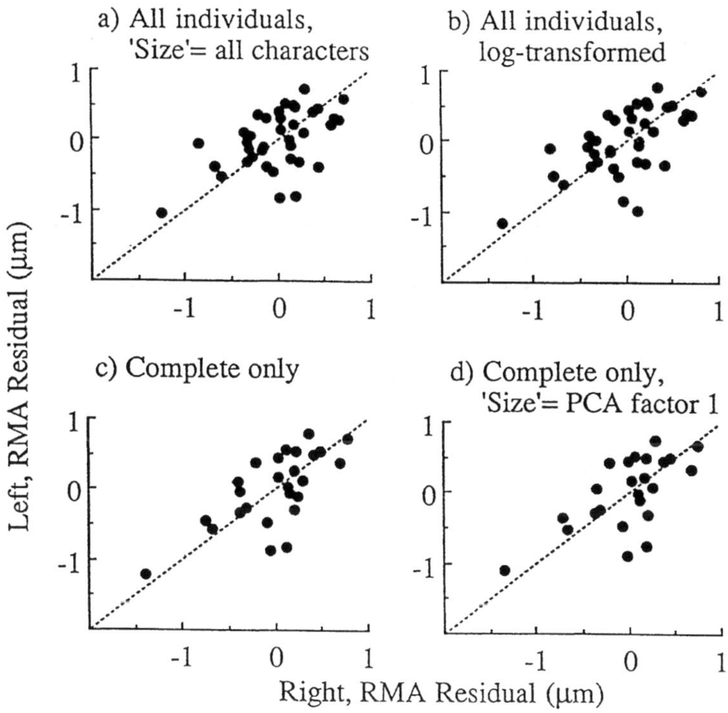

Fig. 5. Scatterplots of *R* versus *L* for Ant 1 of *Tigriopus californicus* only, illustrating the effects of using four differnt procedures to scale out body size variation (compare with Fig. 4a): (a) same procedure as for Fig. 4 except that 'size' was computed as the mean of *all* eight traits instead of all traits excluding Ant1, (b) same procedure as for Fig. 4 except that the raw data were log transformed before beginning the analysis, (c) same procedure as for Fig. 4 except that only individual with a complete set of measurements were used in the analysis (N = 26 vs. N = 31 – 40), (d) same procedure as for 5c except that scores for factor one from a principal components analysis were used as the body size metric.

directional asymmetry, skewed covariant asymmetry, or antisymmetry.

Tests for normal covariant asymmetry. In the original data, all eight traits exhibited highly significant positive correlations between *R* and *L* (Fig. 3). Some of this variation, however, was due to variation in overall body size, whereas some was due to *positive covariation* between sides at a given body size. When body-size was factored out, the predominant covariation remaining was still due to *positive covariation* between sides (Fig. 4). Hence, although we cannot exclude factors contributing to *offset variation* (σ_d^2 in equation 5) as the source of the variation perpendicular to the line of symmetry, the *Tigriopus* data revealed that this asymmetry was at least two to five times less than the *positive covariation* between sides at a given body size (σ_s^2 in equation 5).

To assess the sensitivity of this conclusion to our choice of procedure for scaling out overall size variation, we repeated the analysis four different ways on Ant 1. Although subtle differences were apparent (Fig. 5), the relative magnitudes of *positive covariation* between sides and *offset variation* did not appear to be sensitive to scaling procedure.

Discussion

Although conspicuously asymmetrical morphological traits occur in many otherwise bilaterally symmetrical organisms (Ludwig, 1932; Neville, 1976), the sequence by which they evolve from symmetrical precursors is unknown. Rather than from sudden large mutations, macroscopic asymmetries seem most likely to have originated as small, statistical departures from symmetry that had a heritable basis. If true, studies of developmental stability that assume such statistical departures reflect only developmental noise must be aware of this potential concern.

Implications for studies of developmental stability based on fluctuating asymmetry

Studies of FA presume that differences between sides arise due to non-genetic *developmental noise* (σ_e^2 in equation 9). If individuals exhibit heritable variation for departures from symmetry, then differences in the extent of asymmetry variation among populations could arise via selection or drift. Furthermore, as outlined above, descriptions of FA based on frequency distributions of $R - L$ will not reveal certain forms of negative covariation between sides that may have a genetic basis (e.g. normal covariant asymmetry). These observations suggest caution when using statistical estimates of FA as a measure of developmental stability.

A note on terminology. The terms *developmental noise* and developmental stability are sometimes used as converses of each other in the FA literature, even though they refer to fundamentally different phenomena (Wright, 1920; Mather, 1953; Van Valen, 1962; Palmer & Strobeck, 1992). *Developmental noise* [= random developmental variability, or developmental disturbance of Zakharov (1989), p. 12] refers to the totality of non-genetic perturbations that deflect a developing system away from some ideal growth trajectory under a particular set of environmental conditions. In studies of FA, *developmental noise* refers to those non-genetic processes causing departures from the ideal of perfect symmetry (see *FA: Only developmental noise?* below). *Developmental noise* would seem unlikely to respond to selection.

Developmental stability, on the other hand, refers to the capacity of developing systems to resist or correct for the effects of *developmental noise*. Intrinsic phenomena influencing developmental stability (heterosis, genomic balance) do seem capable of responding to selection (Mather, 1953; Kindred, 1965; Rendel, 1967; Levin, 1970). When considering bilaterally symmetrical characters, *developmental noise* thus refers to the origin of deviations from symmetry whereas developmental stability refers to the capacity to correct such deviations.

Fluctuating asymmetry: The conventional metric for developmental stability. Fluctuating asymmetry has assumed prominence as a measure of developmental stability because of the intuitively appealing premises upon which this inference is based. Consider the paired first antennae of a copepod. In a sample of copepods, both the right and left antennae will exhibit variation in relative length. Some of this variation in an individual antenna will have a genetic basis. The remainder will have arisen as a product of perhaps several levels of environmental effects (see next section). By subtracting the length of the left antenna from that of the right, both the genetic and the environmental effects *common to both* will cancel out (Fig. 6*a*). For characters that exhibit a normal distribution of right-minus-left whose mean is zero, the variance in this distribution is routinely interpreted as a measure of developmental stability (Zakharov, 1992).

Fluctuating asymmetry: only developmental noise? Underlying all studies that relate levels of fluctuating asymmetry (FA) to levels of developmental stability is a common and crucial assumption: the subtle departure from symmetry in a given individual must *not* have been promoted by genetic factors (Palmer & Strobeck, 1992). For traits exhibiting FA, the difference between the right and left members of a bilateral trait-pair in an individual is assumed to have been caused exclusively by *developmental noise* or 'accidents during development' (Wright, 1920; Timoféeff-Ressovsky 1934; Mather, 1953; Waddington, 1957; Zakharov, 1989). Such accidents during development could arise, or be amplified, by non-genetic effects at several levels including: i) those at the level of the whole organism (e.g. extreme physical stress), ii) those having a greater impact on one side (e.g. injury, differential use), and iii) random cellular and sub-cellular noise [e.g. differences in cell division rates, thermal noise (Lewontin, 1983)].

The evidence cited most commonly to support the assumption that FA reflects only *developmental noise* is the apparent inability, under artificial selec-

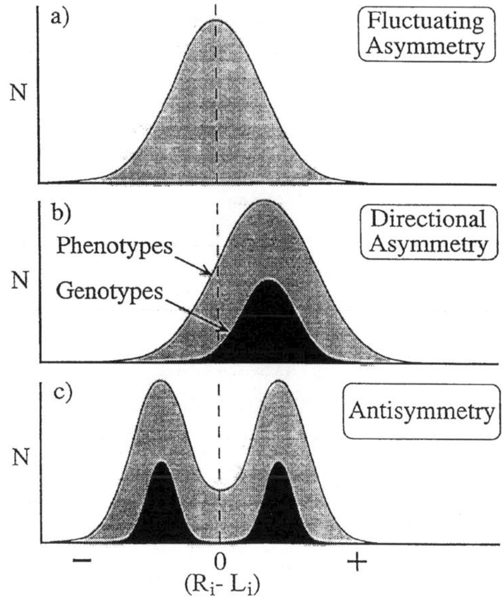

Fig. 6. Contributions of genetic and non-genetic factors to three forms of bilateral variation: (a) fluctuating asymmetry, (b) directional asymmetry, and (c) antisymmetry (from Palmer and Strobeck, 1992). Solid frequency distributions symbolize genetically induced variation, and stippled frequency distributions symbolize total phenotypic variation (including *developmental noise*). The exclusively stippled region under the curve is intended to symbolize environmentally induced variation only. Note that two aspects of these curves are not technically correct. First, the area under each should be the same (the area under any frequency distribution is one). Second, if environmental variation is added to underlying genetic variation, the combined distribution should be broader with a lower peak. We use this heuristic convention a) because we wish to emphasize that some subset of the total phenotypic variation has a genetic basis, and b) because it avoids introducing potentially distracting elements to the figures.

tion, to increase the extent of directional asymmetry (Palmer & Strobeck, 1986). But how strong is this evidence? Are there other ways in which gene action may cause subtle departures from symmetry and consequently compromise FA as a measure of developmental stability?

Directional asymmetry and all forms of covariant asymmetry: subtle asymmetries that may have a genetic basis. To draw conclusions about the extent to which stabilizing mechanisms can counter the effects of *developmental noise*, one must assume that no genes promote the departure of bilateral characters from symmetry. Should such genetic variation exist, differences among natural populations, or among laboratory lines exposed to selection, could arise due to changes in gene frequency at these loci influencing symmetry rather than due to differences in overall developmental stability.

If, for example, the antenna on the right side is larger on average than that on the left, the difference between them will, of course, not be zero (directional asymmetry, Fig. 6b). Within a sample, the variation in directional asymmetry seems likely to have at least some genetic basis (solid curve) although no studies to our knowledge have confirmed this by attempting to select for increased asymmetry of a trait already exhibiting some directional asymmetry.

If, alternatively, the antenna on one side is generally larger than that on the other, but the side which is larger varies randomly among individuals in a sample, then antisymmetry will result (Fig. 6c). As in the case of directional asymmetry, antisymmetry also seems likely to have at least some genetic basis (solid curve, Fig. 6c). See McKenzie and Clarke (1988) for very nice evidence for an antisymmetry polymorphism in the sheep blowfly, *Lucilia cuprina.* Similarly, Mather (1953, p. 332) suggests that some of the increased between-sides variance he obtained in response to selection for decreased developmental stability may in fact have been due to an increase in antisymmetry in his selected lines.

The difference between directional asymmetry and antisymmetry is not profound. It lies only in the *consistency* of the direction of departure for symmetry, not in its magnitude (note that the only difference between Fig. 6b and 6c lies in the value for the parameter p; see also equation 9). The only difference between them lies in how the side to be overdeveloped is specified: does overdevelopment always occur on the same side (directional asymmetry) or can it occur on either side (antisymmetry)?

Suggestions for studies of developmental stability. In many studies of FA a key goal is to compare differences in the levels of developmental stability in different populations. This is true in field studies, where differences in some measure of the variance of $R - L$ are used to infer different levels of environmental or genetic stress (Leary & Allendorf, 1989; Parsons, 1990; Zakharov, 1992). It is also true in laboratory studies examining the effect of selection on developmental stability (Mather, 1953; MacKay, 1980; Leamy & Atchley, 1985). Such studies are particularly vulnerable to the confounding effects of genetically based asymmetries. However, experimental studies with genetically randomized treatment groups, such as those examining the effects of particular environmental stresses on developmental stability within the lifespan of an individual, should be less vulnerable to the confounding effects of genetically based asymmetries.

Because all forms of negative covariation between sides contribute to the variance in $R - L$ (equation 9; see also *A novel form of statistical asymmetry* above), and because the negative feedback implied by such negative covariation (see *Hypothetical pathways* below) seems likely to have a genetic basis, traits exhibiting various forms of covariant asymmetry should be given less weight or eliminated from consideration when drawing inferences about levels of developmental stability. Elsewhere we discuss in some detail the implications of other departures from normality for studies of developmental stability and suggest caution when drawing conclusions based on characters exhibiting such patterns of variation in $R - L$ (Palmer & Strobeck, 1992).

The same cautions should be exercised when traits are found to exhibit statistically significant directional asymmetry, since some of the between-sides variation may have a genetic basis (previous section). Figure 6b also makes clear why a 'correction' for directional asymmetry, by shifting the mean of the frequency distribution of $R - L$ to zero, is inappropriate in studies of developmental stability: such a correction is purely statistical and will not eliminate the fraction of the between-sides variation that may be genetically based.

Table 4. Presumed developmental-genetic factors contributing to different forms of statistical asymmetries.

Form of statistical asymmetry	Side-bias control (G/E)	Bilateral offset (G/E)	Bilateral inhibition (G/E)	Developmental noise (E)
Fluctuating asymmetry (see Figs. 1a,d)				X
Normal covariant asymmetry (see Figs. 1h,k)			X	X
Antisymmetry (see Figs. 1c,f)		X	X	X
Directional asymmetry (see Figs. 1b,e)	X	X	X	X

G - influenced by genetic factors; E - influenced by environmental or non-genetic factors (see *FA: Only developmental noise?* in text).

The evolutionary origins of bilateral asymmetry

Hypothetical pathways. Purely statistical analyses of patterns of between-sides variation cannot, of course, indicate whether such variation has a heritable basis or not (e.g. see Lewontin, 1974), or what developmental mechanisms are involved. Whether a particular pattern of between-sides variation has a genetic basis, for example, requires an explicit attempt to measure heritability as in Mather (1953) and McKenzie and Clarke (1988). More studies of this type would be immensely informative about the potential evolutionary origins of bilaterally asymmetrical characters.

Although our model is only a simple one, we think three forms of bilateral variation may signal the presence of distinctly different heritable variation upon which natural selection could act: normal covariant asymmetry, antisymmetry, and directional asymmetry (Fig. 1h, 1c, and 1b respectively). However, some seem more likely than others as the starting point for the origin of macroscopic asymmetries because they would seem to require less developmental-genetic information.

Off all the major patterns of statistical asymmetries (Fig. 1, Table 4), fluctuating asymmetry (Figs. 1a, 6a) would seem to be the only one not likely to have a heritable basis (but see comments under *Normal covariant asymmetry: Confounding effects of positive covariation* above). Herein lies its appeal for studies of developmental stability (Palmer & Strobeck, 1986; Leary & Allendorf, 1989; Parsons, 1990; Zakharov, 1992). The remaining three would all seem to require some kind of genetic influence (Table 4).

On developmental grounds, we feel that normal covariant asymmetry (Fig. 1h) is the more likely first step in the evolution of macroscopic asymmetries. It would seem to require the least developmental-genetic information, because it could arise simply via some form of *bilateral inhibition* between sides (Table 4). Such inhibition is well documented in the development of conspicuous claw asymmetries in snapping shrimp (Mellon & Stephens, 1978) and lobsters (Govind, 1981), where it seems to involve action by the central nervous system. Whether asymmetry is regulated this way in other bilaterally asymmetrical organisms, or by some more complex developmental mechanism (Brown & Wolpert, 1990) remains to be seen.

Antisymmetry (Fig. 1c, 6c) would seem to be the next most likely precursor to macroscopic asymmetries. In addition to *bilateral inhibition*, however, some developmental mechanism would be required to specify the *bilateral offset*, or average departure from symmetry (Table 4). Finally, directional asymmetry (Fig. 1b, 6b) may be the least likely precursor of macroscopic asymmetries because it would seem to require an additional mechanism for consistently specifying which side was to become overdeveloped (*side-bias control*, Table 4).

This view of statistical asymmetries suggests several predictions that could be tested via a formal comparative study.

1. Among statistical asymmetries that depart from pure FA, the rank order of prevalence should be: normal covariant asymmetry > antisymmetry > directional asymmetry.
2. The number of independent evolutionary origins of macroscopic directional asymmetry should be fewer than those of either normal covariant asymmetry or antisymmetry.
3. Directional asymmetry should have evolved more frequently via ancestors exhibiting antisymmetry than directly from symmetrical ancestors (see also Mather, 1953, p. 334).
4. Among taxa with a clear plane of symmetry, macroscopic asymmetries in novel characters should arise more frequently in taxa with preexisting asymmetries in other characters, because a mechanism for the control of asymmetry would already exist.

Existing data. Our own study of statistical asymmetries in *Tigriopus* revealed no evidence for directional asymmetry or any form of covariant asymmetry. However, the observed residual *positive covariation* between sides in all traits, even after correcting for body size variation (Fig. 4), suggests that the genetic or environmental factors affecting limb development tend to affect both sides equally rather than each side independently or each side at the expense of the other.

Regrettably, although most studies of statistical asymmetries report whether directional asymmetry was detected in the characters examined, few report frequency distributions of $R - L$ or test for statistical departures of such frequency-distributions from normality (but see Ross & Robertson, 1990). Furthermore, as outlined above, such frequency distributions would not reveal normal covariant asymmetry even if it did exist. As a consequence, very little can be said about the prevalence of different forms of statistical asymmetries, in spite of the extensive literature on FA. We can only hope that future studies of FA will include a more detailed presentation and analysis of data. Without such information, generalizations about the prevalence of different patterns of statistical asymmetry will be slow in coming.

Acknowledgements

This research was supported by NSERC Operating Grants A7245 (ARP) and A0502 (CS). We thank the director and staff of the Bamfield Marine Station for their logistical support, and Mikhail Mina and an anonymous reviewer of an earlier paper for stimulating our thoughts on covariance. Legitimate grumblings by two anonymous reviewers prodded us to reorganize this paper substantially, and comments by Günter Wagner, Leigh Van Valen and Louise Roth all helped us clarify the presentation.

References

Allendorf, F. W. & R. F. Leary, 1986. Heterozygosity and fitness in natural populations of animals, pp. 57-76 in Conservation Biology. The Science of Scarcity and Diversity, edited by M. E. Soulé. Sinauer Assoc., Inc., Sunderland, MA (USA).

Brown, N. A. & L. Wolpert, 1990. The development of handedness in left-right asymmetry. Development 109: 1-9.

Burton, R. S., 1987. Differentiation and integration of the genome in populations of the marine copepod *Tigriopus californicus*. Evolution 41: 504-513.

Burton, R. S. & M. W. Feldman, 1981. Population genetics of *Trigriopus californicus*. II. Differentiation among neighboring populations. Evolution 35: 1192-1205.

Clarke, G. M., G. W. Brand & M. J. Whitten, 1986. Fluctuating asymmetry: A technique for measuring developmental stress caused by inbreeding. Austr. J. Biol. Sci. 39: 145-153.

Govind, C. K., 1981. Does exercise influence the differentiation of lobster muscle?, pp. 215-253 in Locomotion and Energetics in Arthropods, edited by C. F. Herreid & C. F. Fourtner. Plenum Pr., New York.

Kindred, B., 1965. Selection for canalization in mice. Genetics 55: 635-644.

LaBarbera, M. L., 1989. Analyzing body size as a factor in ecology and evolution. Ann. Rev. Ecol. Syst. 20: 97-117.

Leamy, L. & W. Atchley, 1985. Directional selection and developmental stability: evidence from fluctuating asymmetry of morphometric characters in rats. Growth 49: 8-18.

Leary, R. F. & F. W. Allendorf, 1989. Fluctuating asymmetry as an indicator of stress in conservation biology. Trends Ecol. Evol. 4: 214-217.

Levin, D. A., 1970. Developmental instability and evolution in peripheral isolates. Amer. Nat. 104: 343-353.

Lewontin, R. C., 1974. The analysis of variance and the analysis of causes. Am. J. Human Genet. 26: 400-411.

Lewontin, R. C., 1983. Gene, organism and environment, pp. 273-285 in Evolution from Molecules to Men, edited by D. S. Bendall. Cambridge Univ. Press, New York, NY.

Ludwig, W., 1932. Das Rechts-Links Problem im Tierreich und beim Menschen, Springer, Berlin. 496.

Mackay, T. F. C., 1980. Genetic variance, fitness, and homeostasis in varying environments: an experimental check of the theory. Evolution 34: 1219-1222.

Mather, K., 1953. Genetical control of stability in development. Heredity 7: 297-336.

McKenzie, J. A. & G. M. Clarke, 1988. Diazinon resistance, fluctuating asymmetry and fitness in the Australian sheep blowfly, *Lucilia cuprina*. Genetics 120: 213-220.

Mellon, D. F. & P. T. Stephens, 1978. Limb morphology and function are transformed by contralateral nerve section in snapping shrimp. Nature 272: 246-248.

Mitton, J. B. & M. C. Grant, 1984. Associations among protein heterozygosity, growth rate, and developmental homeostasis. Ann. Rev. Ecol. Syst. 15: 479-499.

Neville, A. C., 1976. Animal Asymmetry. Arnold (Inst. Biol. Stud. Biol.), London. 60.

Palmer, A. R. & C. Strobeck, 1986. Fluctuating asymmetry: measurement, analysis, patterns, Ann. Rev. Ecol. Syst. 17: 391-421.

Palmer, A. R. & C. Strobeck, 1992. Fluctuating asymmetry as a measure of developmental stability: Implications of non-normal distributions and power of statistical tests. Acta Zool. Fenn. 191: 57-72.

Parsons, P. A., 1990. Fluctuating asymmetry: An epigenetic measure of stress. Biol. Rev. 65: 131-145.

Provasoli, L., K. Shiraishi & J. R. Lance, 1959. Nutritional idiosyncrasis of *Artemia* and *Tigriopus* in monoxenic culture. Ann. N.Y. Acad. Sci. 77: 250-261.

Rendel, J. M., 1967. Canalization and Gene Control. Academic Pr., New York, NY (USA).

Ross, K. G. & J. L. Robertson, 1990. Developmental stability, heterozygosity, and fitness in two introduced fire ants (*Solenopsis invicta* and *S. richteri*) and their hybrid. Heredity 64: 93-103.

Shapiro, S. S., M. B. Wilk & H. J. Chen, 1968. A comparative study of various tests for normality. J. Amer. Stat. Assoc. 63: 1342-1372.

Sokal, R. R. & J. F. Rohlf, 1981. Biometry. Freeman, San Francisco, CA. 859.

Timoféeff-Ressovsky, N. W., 1934. Über der Einfluss des genotypischen Milieus und der Aussenbedingungen auf die Realisation des Genotypes. Mach. Ges. Wiss. Göttingen, Math.-Physik. Klasse, Fachgruppe 6 1: 53-106.

Van Valen, L., 1962. A study of fluctuating asymmetry. Evolution 16: 125-142.

Waddington, C. H., 1957. The Strategy of the Genes. George Allen Unwin, London.

Wright, S., 1920. The relative importance of heredity and environment in determining the piebald pattern of guinea pigs. Proc. Natl. Acad. Sci. USA 6: 320-332.

Zakharov, V. M., 1989. Future Prospects for Population Phenogenetics. Harwood Academic Publishers GmbH, New York. 79.

Zakharov, V. M., 1992. Population phenogenetics: Analysis of developmental stability in natural populations. Acta Zool. Fenn. 191: 7-30.

T. A. Markow (ed.), Developmental Instability: Its Origins and Evolutionary Implications, 221–227, 1994.

Evolution, developmental instability and the theory of acquisition

J. A. Kieser
Department of Oral Pathology, School of Dentistry, Witwatersrand University, Johannesburg

Received and accepted 23 April 1993

Key words: evolution, fluctuating asymmetry, phylogeny

Abstract

Despite major advances in the study of molecular and morphological evolution a substantial rift still exists between these two fields of endeavour. Phenotypic alteration through evolution results from a reallocation of resources which has as its origin the interplay between the production capability of the genes on the one side and the acquisitional need of the phenotype on the other. This process of allocation is coordinated through the environmental arena and is subject to mechanical, biological and economical constraints. Differences in the rates of morphological change at any level (molecular, cellular, organismal or population) depend on the level of environmental challenge, on the availability of variability and on the economics of supply and demand. Short run changes in response to severe environmental stress will be sudden and energetically expensive and will rely on stress-induced unmasking of genetic variability and loss of canalization. Long run changes will be gradual, energetically less costly and less dependent on genetic correlations.

Introduction

Molecular evolution, that discipline traditionally allied to heredity and population genetics, has through its own development remained relatively independent of those fields usually associated with morphological evolution (paleontology, systematics and morphometrics). There are three principal reasons for this dis-union. Firstly, molecular genetic parameters are not easily applied to the interpretation of morpho-evolutionary patterns (Service & Rose, 1985; Mitchell-Olds & Rutledge, 1986; Riska, 1989). Secondly, the spectacular rise of molecular biology coincided with, and was fertilized by the neutralist view of evolution (King & Jukes, 1969; Kimura & Otha, 1971; Nei, 1975), resulting in a preoccupation which prevented a closer association with systematic evolutionary schools of thought (e.g. Hennig, 1966; Gould & Lewontin, 1979; Eldredge & Gould, 1972; Sneath & Sokal, 1981). Thirdly, attributes of segregation and recombination of allozymes channeled molecular biological thought towards concerns of allelic frequencies and heterozygosities which in turn focused thinking onto population geneties and away from morphological phylogenics (Avise *et al.*, 1987, but see De Salle *et al.*, 1992). Moreover, this preoccupation with nucleotide sequences led to estimates of mutation rates during evolution, an interest which failed to address directly the problem of morphometrics. This is because hitherto undisclosed differences in temporal and spatial patterns of genetic expression during embryogenesis, rather than the relatively few nucleotide differences between species, are responsible for morphological diversity (Alberch *et al.*, 1989).

Consequently, a comprehensive theory that explains how developmental processes generate and integrate morphological change during ontogeny and phylogeny into fully functional, adapted and reproducing structures has not been achieved (Maynard-Smith, 1983; Arthur, 1987; Atchley & Hall, 1991; Cowley & Atchley, 1992; Hall, 1992). Realization of these difficulties has resulted in a call for a new discipline devoted to evolutionary developmental biology (Arthur, 1987; Hall, 1992), which would explain the evolution of complex traits in terms of underlying processes of growth and development in addition to the molecular perspective. In what follows, a paradigm is presented which unites genes (as initiators of production) and phenotypes (as acquisitors of production) in a system governed by economic laws. An attempt is then made to outline the possible effects that environmental stress might have on such a system.

The theory of acquisition

Natural selection acts on the phenotype. This action is not limited to the adult organismal phenotype, but also acts on the embryonic phenotype, tissue phenotypes, germ cell phenotypes, developmental cascades and behavioural patterns (Parsons, 1984). To become a phenotype something has to be acquired: macromolecules, cells, developmental paths, behavioural traits or adult forms have to be attained. From whence come these acquisitions? Quite simply, the phenotype, in whatever form it may be visualized, is acquired from the actions of those genes which initiate its production. Genes are units which initiate production or supply, while phenotypes are units of acquisition or demand. While genes will tend to maximize their chances of perpetuation (*vide* Dawkins, 1976, 1982, 1986), phenotypes are driven towards survival and hence, an optimization of their utility (Burian, 1983).

As stated earlier, evolutionary models tend to ignore those developmental, physiological or behavioral processes that govern the middle ground between DNA and the phenotype (Riska, 1989). Phenotypes show observable differences both over time and over different environments. Whether these changes are due to drift in the absence of selection (Wright, 1931; Lande, 1977; Lynch & Hill, 1986; Turelli, Gillespie & Lande, 1988), to correlated changes in unselected traits (Reeve, 1950; Cheverud, 1984; Clark, 1987) or to selection driven shifts in the occupation of adaptive peaks (Wright, 1982; Via & Lande, 1985; Lande, 1986; Felsenstein, 1988; Zeng, 1988), the inescapable fact is that the phenotype is altered. The alteration comes from acquisition or loss, that is, from a reallocation of resources within the organism. This allocation is geared to the fulfilment of the needs of the demand side (the phenotype) and it has as its origin the production capability of the supply side (the genes). Hence we conclude that evolutionary change, in whatever form or from whatever cause, reduces to an interplay between supply and demand.

We now see that, while evolution is governed by chemical, physical and biological laws and constraints (which have been well documented), it is also governed by economic constraints and laws (which have not) - see, for instance, Eldredge (1989) and Liem (1990).

Allocation, scarcity and the arena

Allocation is the assignment of physical, physiological or behavioural features to specific uses. The central problem of evolution is the allocation of resources to meet the needs of a given living organism in a given environment. The problem exists because of three crucial qualities of all living things: Firstly, their needs are unlimited. Secondly, the resources available to produce those features needed are limited. Finally, many features have alternate uses. These three characteristics give rise to the basic need for choice. Evolution determines how the available but finite resources of living organisms are to be allocated among competing needs. As we have seen, biosystems may be reduced to acquisitors of production and initiators of production. The space within which initiators and acquisitors interact is referred to as the environmental market place, or simply, the environmental arena.

The arena has no fixed location. It could be intra or extracellular, it could be within an organ system or within an environmental niche. It is simply defined as a place where acquisitors and initiators exchange information.

Within the arena there are three important conditions: (a) Initiators and acquisitors are very numerous, and the acquisition or initiation of production of a single unit is incredibly small in relation to the aggregate volume of exchange. (b) Although resources are often limited, there is a flow of resources directed to an outcome from which will be derived the greatest mutual benefits for acquisitor (phenotype) and initiator (gene). Inefficient acquisitions will be eliminated by the arena and will be replaced by efficient ones. (c) Resource allocation decisions are made without central direction.

Trade-offs and the arena

In living systems where parts of cells or collections of cells specialize in their activities there has to be a close interaction of costing between the constituent parts of the organism. Costing is necessary because resources are limited; hence any shift in allocation to one component would be at the expense of other components (Parsons, 1993). This paradigm had provided the physiological substrate for a number of trade-off models based upon the subdivision of various processes into competing components of energy (Koehne & Bayne, 1989; Silby & Calow, 1989; De Kruijf, 1992). In the present volume, Emlen, Freeman and Graham argue forcefully for the existence of physiological mechanisms which would, during periods of stress, draw energy off less important developmental pathways, to assure continued supplies of energy to those which are more critical.

Some processes or behaviours are energetically so expensive that they require nearly all of the available energetic resources of the animal (Morgan, 1987; Tanaka, 1989; Parsons, 1991). There is little doubt that the availability of energetic resources beyond maintenance needs interacts with the intensity of environmental adversity to determine the rate of evolution of the organism (Sheldon, 1990; Parsons, 1991). Biological development is thus controlled by the system of trade-offs which results from millions and millions of arbitrations made by cells and cellular collections. The key to the entire process is found in the role of the environmental arena which performs the crucial function of providing the signals that determine the allocation of individual resources.

Every second of every hour, decisions are made for millions of cells within each living organism concerning the production and acquisition of resources. Few of these decisions are motivated by a desire to contribute to a common good, or to make the organism work better, but rather by considerations of self-interest. The arena, and its attendant selective pressures, will coordinate these decentralized decisions so that the whole system is sensitive to the needs of the individual units that compose it.

The arena fulfills its function of coordinating organismal change without any organism having to understand how it works. For instance, the hare does not need to know how many eagles live within its home range, nor how many kilojoules the eagle requires per predatory swoop, or how many chicks it has to feed. What it needs to know is information about the second-to-second state of its particular environment. By responding to such signals as sound or sight or hunger or sex, the organism helps to make the ecosystem fit together, and its phenotype will tend to adapt to what the environment demands of it. Some gambits will work and others will not. Like baseball, evolution is a game of inches.

Time

In any given environment, organisms will acquire over successive generations those phenotypes that best fit them to their surroundings (Parsons, 1992). Change in the environment will result in an expansion of phenotype variation as individuals evolve to occupy new morphological space. The pattern of diversity depends on selective pressures operating on the phenotype (the unit of acquisition or demand) and also on the underlying genotype variation or production potential.

An understanding of evolution hinges on the realization that the entire process is governed by biological as well as economical rules and that the latter probably outweigh the former. Phylogenetic change represents the match of need and ability, of production and acquisition, of supply and demand. However, one of the principal rules of economics is that the responsiveness of the quantity of phenotypic change supplied to a change in demand increases steadily with time.

To illustrate this point, one may subdivide time into three hypothetical periods: instantaneous time, short run and long run. Here, an instantaneous period may refer to a situation where there is simply not enough time to alter production or to reallocate resources. A short run period refers to the span over which production may be varied within the constraints of the initiator's production facilities. An organism has a more or less fixed number of organs and a given bauplan. It is possible for some structure to alter, but such changes remain constrained by that organism's production capabilities. The long run time period refers to the span which is long enough to alter everything that needs to be changed in response to a shift in demand. New cells may be

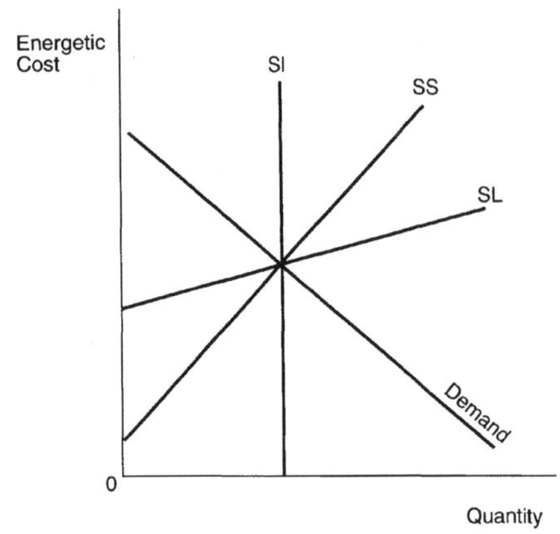

Fig. 1. The economic supply curves relevant to three hypothetical time periods: instantaneous (SI), short run (SS) and long run (SL).

manufactured, new musculo-skeletal systems created, or different biochemical pathways established.

Figure 1 shows a traditional triad of supply curves relevant to each of these time periods. The instantaneous supply curve (SI) occurs when time is so short that production cannot be increased or decreased. In the short run (SS) time is only allowed for some response to changes in the environment, while in the long run (SL) there is sufficient time to allow for full and gradual adaptation to the acquisitor's needs.

Economics teaches that with an increase in demand, the longer the time allowed, the greater the quantity supplied and the less the change in cost. Hence, the slower the phenotypic change, the less it costs. One may thus hypothesize that during periods of gradualistic change or stasis energy expenditure on new acquisitions will be relatively low. As soon as the tempo of phenotypic change increases, the cost also increases. We also know that short run responses of characters to selection differ qualitatively from those in the long run. In the short run, the responses are dependent on genetic correlations between characters, while in the long run they are only determined by the fitness functions of directional and stabilizing selection, independent of genetic correlations (Zeng, 1988).

In summary, change or response in the short run is energetically costly and heavily dependent on genetic correlations. Slower, more gradual change costs less and is less dependent on genetic correlations.

Short run response and developmental instability

In the previous section, an approach for the study of evolution through the concept of acquisition was introduced. In this section, we will examine the problem of short run change and how it relates to the processes of canalization and developmental stability.

All organisms have an innate ability to buffer themselves against those genetic and environmental disturbances that they might encounter during their development. This ability to produce consistent phenotypic end-products is referred to as canalization (Waddington, 1942; Lerner, 1954; Zakharov, 1981; Kieser, Groeneveld & Preston, 1986). Canalization, or autoregulation, works in two ways - firstly, it masks the underlying genetic variability of the organism (Fraser, 1959; Alberch, 1982) and secondly, it colimates the morphogenic programme into a stable creod. That is, it buffers the development against environmental noise (Waddington, 1957).

During periods of environmental uniformity, phenotypic variation will be limited by the process of stabilization which may act either externally on mutational deviants or phenotypic peripherals, or internally in favour of deepened canalization. Stabilizing selection, for instance, will maintain the integrity of harmoniously collaborating genes by eliminating any change in the genetic constitution of the organism either as a result of gene flow or mutation. Such a regimen will eventually result in uniform phenotypes across a range of genotypes within a population (Clarke & McKenzie, 1992). The net result of the long run response is the production of a stable phenotype which is well canalized and is maintained at relatively little energetic cost. In contrast, periods of rapid change in the environment will subject the phenotype to severe stress which in turn will result in decreased canalization and unmasking of underlying genetic variability (Fraser, 1959, 1963; Levin, 1970; Hoffman 1983; Kieser, 1987).

The severely challenged phenotype has to respond or perish. Stress-induced loss of canalization has been demonstrated in otolith production of diet-stressed laboratory rodents (Erway, Hurley & Fraser, 1970), in pentamerous corolla generation in Linanthus (Huether, 1968, 1969) and in increased fluctuating asymmetry of paired organ systems (Siegel & Doyle, 1975; Siegel & Smookler, 1973; Valentine, Soule & Smollow, 1973; Pankakoski, 1985; Kieser, Groeneveld and Preston, 1986; Kieser & Groeneveld, 1991; Livshits & Kobyliansky, 1989; Clarke, Oldroyd & Hunt, 1992).

Any major escalation of environmental stress will require an excess expenditure of energy which ultimately may become incompatible with survival (Parsons, 1990, 1991). Hence a population under stress may, in spite of newly uncovered variability, find that it is incapable of redirecting its energetic resources into the acquisition of a more favourable phenotype. Internal energetic constraints imposed by the exposure to severe environmental challenges may thus preclude rapid evolutionary change. In these cases, the primary targets of stress selection in natural populations will be physiological (see Parsons, this volume). It follows logically that major morphological adaptations would be most likely from generalist animals which have high levels of residual energetic resources. In summary, a short run response to a severe environmental challenge will be energetically costly, heavily dependent on energetic resources and will be facilitated by developmental instability resultant from a lessening of canalization.

Punctuation and gradualism

The foregoing exposition may now be applied to one of the most fiercely fought evolutionary arguments of the past two decades, that of the concept of punctuated equilibria as opposed to the Darwinian theory of gradualism. First conceived by Eldredge and Gould (1972) as a direct empirical extension of the synthetic theory, then seen as a nonadaptive response, only to be subsequently recouched in strong adaptationist terms (*vide* Gould, 1979, 1980, 1982; Ruse, 1988), the concept of punctuated equilibria viewed the course of evolution as one of abrupt changes interspersed by long periods of stasis.

Against this supposition of episodic change is set the more traditional notion of phyletic gradualism which stresses the slow transformation of one species into another (Gingerich, 1976; Rhodes, 1983, 1984). A new unifying perspective was offered by Kieser and Groeneveld (1985) who proposed a threshold catastrophe model for the coexistence of these two opposite views. We now propose that the events surrounding speciation, to which the foregoing relates, might be expanded to include any event of acquisition at any level of phenotypic change (cellular, molecular or organismal).

The rate of evolution depends upon the interaction of the intensity of the environmental challenge on the one side, and the available energy beyond maintenance levels on the other. Ecologically marginal populations could be expected to contain genotypes adapted to counteracting environmental stress (Parsons, present volume). Since genotypes selected for extremes of one environmental stress may be resistant to other stresses (Hoffmann & Parsons, 1992), it follows that major phenotypic shifts could be expected in those marginal populations exposed to fluctuating, mildly stressful environments. In these populations, persistent stress resistance selection will 'pre-adopt' the organism by increasing its available energy resources. One would thus expect a series of minor gradual changes, which increase the available energy, followed (if needed) by a dramatic response as the organism employs its energy towards a major, punctuated shift. The above might explain the observation by Sheldon (1990), who showed that Ordovician trilobites showed the most rapid evolution during periods of fluctuating, slowly changing physical environments, with stasis prevailing under highly fluctuating, rapidly changing environments.

Differences in the rates of phenotypic change (at whatever level of organization) will depend on the severity of the environmental challenge, and the availability of energy. Hence, short run, punctuated, responses in answer to environmental stress will be sudden and expensive and will rely on stress-induced unmasking of genetic variability and loss of canalization. Long run changes, in contrast, will be gradual and energetically less costly.

Acknowledgements

This paper is dedicated to Alex Fraser who gave me the idea back in 1985. I thank Therese Markow for the invitation to submit this manuscript and John Emlen, Carl Freeman, John McKenzie and Peter Parsons for their useful comments on its contents. An anonymous referee savaged and thus vastly improved, the text. Glynny Kieser edited and typed the text.

References

Alberch, P., 1982. Developmental constraints in evolutionary processes, pp. 313-332 in Evolution and Development, edited by J. T. Bonner. Springer, Berlin.

Alberch, B., D. Bray, J. Lewis, M. Raff & J. F. Watson, 1989. Molecular Biology of the Cell. Garland, N.Y.

Arthur, W., 1984. Mechanisms of Morphological Evolution. Wiley & Sons, Chichester, U.K.

Arthur, W., 1987. Theories of Life: Darwin, Mendel and Beyond. Penguin, Marmondsworth.

Avise, J. C., J. A. Arnold, R. M. Ball, B. Eldredge, T. Lamb, J. E. Neigel, C. A. Reeb & N. C. Saunders, 1987. Intraspecific phylogeography: the mitochondrial DNA bridge between population genetics and systematics. Ann. Rev. Ecol. Systematics 18: 489-522.

Atchley, W. R. & B. K. Hall, 1991. A model for development and evolution of complex morphological structures. Biol. Rev. 66: 101-157.

Burian, R. M., 1983. Adaptation, pp. 287-314 in Dimensions of Darwinism, edited by M. Grene. Cambridge University Press, Cambridge.

Cheverud, J. M., 1984. Quantitative genetics and developmental constraints of evolution by selection. J. Theor. Biol. 110: 155-172.

Clark, A. G., 1987. Genetic correlations in the quantitative genetics of evolutionary constraints, pp. 25-45 in Genetic Constraints on Adaptive Evolution, edited by V. Loeschcke. Springer, Berlin.

Clarke, G. M. & G. A. McKenzie, 1992. Coadaptation, developmental stability and fitness of insecticide resistance genotypes in the Australian sheep blowfly, *Lucilia cuprina*: a review. Acta Zool. Fennica 191: 107-110.

Clarke, G. M., B. P. Oldroyd & P. Hunt, 1992. The genetic basis of developmental stability in *Apis mellifera*: heterozygosity versus genetic balance. Evolution 46: 753-762.

Cowley, D. E. & W. R. Atchley, 1992. Quantitative genetic models for development, epigenetic selection and phenotypic evolution. Evolution 46: 495-518.

Dawkins, R., 1976. The Selfish Gene. Oxford University Press, Oxford.

Dawkins, R., 1992. The Extended Phenotype. Oxford University Press, Oxford.

Dawkins, R., 1986. The Blind Watchmaker. Longman, London.

De Kruijf, H. A. M., 1991. Extrapolation through hierarchical levels. Comp. Biochem. Physiol. 100c: 291-299.

DeSalle, R., J. Gatsey, W. Wheeler & D. Grimaldi, 1992. DNA sequences from a fossil termite in Oligo-Miocene amber and their phylogenetic implications. Science 257: 1933-1936.

Eldredge, N., 1989. Macroevolutionary Dynamics. McGraw-Hill, New York.

Eldredge, N. & S. J. Gould, 1972. Punctuated equillibria: An alternative to phyletic gradualism, pp. 82-115 in Models in Paleobiology, edited by T. M. J. Schopf. Freeman, Cooper, San Francisco.

Erway, L., L. S. Hurley & A. S. Fraser, 1970. Congenital ataxia and otolith defects due to manganese deficiency in mice. J. Nutrition 100: 643-654.

Felsenstein, J., 1988. Phylogenies and quantitative characters. Ann. Rev. Ecol. Syst. 19: 445-471.

Fraser, A. S., 1959. Survival of the mediocre, pp. 74-80 in Evolution of Living Organisms, edited by G. W. Leeper. Melbourne University Press, Melbourne.

Fraser, A. S., 1963. Variation of scutellar bristles in *Drosophila*. I Genetic leakage. Genetics 48: 497-514.

Gingerich, P. D., 1976. Paleontology and phylogeny: patterns of evolution at the species level in early Tertiary mammals. Am. J. Science 276: 1-16.

Gould, S. J., 1979. Episodic versus gradualist dogma. Science and Nature 2: 5-12.

Gould, S. J., 1980. Is a new and general theory of evolution emerging? Paleobiol. 6: 119-130.

Gould, S. J., 1982. Darwinism and the expansion of evolutionary theory. Science 216: 380-387.

Gould, S. J. & R. Lewontin, 1979. The spandrels of San Marco and the Panglossan paradigm: a critique of the adaptationist programme. Proc. Roy. Soc. Lond. (B) 205: 581-598.

Hall, B. K., 1992. Evolutionary Developmental Biology. Chapman & Hall, N.Y.

Hennig, W., 1966. Phylogenetic Systematics. University of Illinois Press, Urbana.

Hoffmann, A., 1983. Paleobiology at the crossroads: a critique of some paleobiological research programs, pp. 241-272 in Dimensions of Darwinism, edited by M. Grene. Cambridge University Press, Cambridge.

Hoffmann, A. & P. A. Parsons, 1991. Evolutionary Genetics and Environmental Stress. Oxford University Press, Oxford.

Huether, C. A., 1968. Exposure of natural genetic variability underlying the pentamerous corolla constancy in *Linanthus androsaceus*. Genetics 60: 123-146.

Huether, C. A., 1969. Constancy of pentamerous corolla phenotype in natural populations of Linanthus. Evolution 23: 572-588.

Kieser, J. A., 1987. Epigenetic canalization and phenotypic change: a minimax model. Med. Hypoth. 22: 105-110.

Kieser, J. A. & H. T. Groeneveld, 1985. A threshold model of punctuated gradualism. Med. Hypoth. 17: 219-225.

Kieser, J. A. & H. T. Groeneveld, 1991. Fluctuating odontometric asymmetry, morphological variability and genetic monomorphism in the cheetah *Acinonyx jubatus*. Evolution 45: 1175-1183.

Kieser, J. A., H. T. Groeneveld & C. B. Preston, 1986. Fluctuating dental asymmetry as a measure of odontogenic canalization in man. Am. J. Phys. Anthropol. 71: 437-444.

Kimura, M. & T. Ohta, 1973. Mutation and evolution at the

molecular level. Genetics 73: 19-35.

King, J. L. & T. H. Jukes, 1969. Non-Darwinian evolution: random fixation of selectively neutral mutations. Science 164: 788-798.

Koehn, R. & B. L. Bayne, 1989. Towards a physiological and genetical understanding of the energetics of the stress response. Biol. J. Linn. Soc. 37: 157-171.

Lande, R., 1977. Statistical tests for natural selection on quantitative characters. Evolution 31: 442-444.

Lande, R., 1986. The dynamics of peak shifts and the pattern of morphological evolution. Paleobiol. 12: 343-356.

Lerner, I. M., 1954. Genetic Homeostasis. Oliver & Boyd, London.

Liem, K. F., 1990. Key evolutionary innovations, differential diversity and symecomorphosis. pp. 147-170 in Evolutionary Innovations, edited by M. H. Nitecki. University of Chicago Press, Chicago.

Levin, D. A., 1970. Developmental instability and evolution in peripheral isolates. Am. Naturalist 104: 343-353.

Livshits, G. & E. Kobyliansky, 1989. Study of genetic variance in the fluctuating asymmetry of anthropometric traits. Ann. Hum. Biol. 16: 121-129.

Lynch, M. & W. G. Hill, 1986. Phenotypic evolution by neutral mutation. Evolution 40: 915-935.

Maynard Smith, J., 1983. Evolution and development, pp. 33-46 in Development and Evolution, edited by B. Goodwin, C. N. Holder & C. C. Wylie. Cambridge University Press, Cambridge.

Mitchell-Olds, T. & J. J. Rutledge, 1986. Quantitative genetics and natural plant populations: a review of the theory. Am. Natur. 127: 379-402.

Morgan, K. R., 1987. Temperature regulation, energy metabolism and mate-searching in rain beetles (*Pleocoma* spp.), winter-active endothermic scarabs. J. Exp. Biol., 128: 107-122.

Nei, M., 1975. Molecular Population Genetics and Evolution. Elsevier, N.Y.

Pankakoski, E., 1985. Epigenetic asymmetry as an ecological indicator in musk rats. J. Mamm. 66: 52-57.

Parsons, P., 1984. Colonizing species: a probe into evolution via the organism Endeavour 8: 108-112.

Parsons, P., 1990. The metabolic cost of multiple environmental stresses: Implications for climatic change and conservation. Tree 5: 315-317.

Parsons, P., 1991. Evolutionary rates: stress and species boundaries. Ann. Rev. Ecol. Syst. 22: 1-18.

Parsons, P., 1992. Evolutionary adaptation and stress. Evol. Biol. 26: 191-223.

Parsons, P., 1993. Stress, extinctions and evolutionary change: From living organisms to fossils. Biol. Rev. (in press).

Reeve, E. C. R., 1950. Genetical aspects of size allometry. Proc. Roy. Soc. Lond. (B) 137: 515-518.

Riska, B., 1989. Composite traits, selection response and evolution. Evolution 43: 1172-1191.

Rhodes, F. H. T., 1983. Gradualism, punctuated equilibrium and the Origin of the Species. Nature 305: 269-270.

Rhodes, F. H. T., 1984. Darwin's gradualism and empiricism - a reply to Gingerich. Nature 309: 116.

Ruse, M., 1988. Molecules to men; evolutionary biology and thoughts of progress, pp. 97-128 in Evolutionary Progress, edited by M. H. Nitechi. Chicago University Press, Chicago.

Service, P. M. & M. R. Rose, 1985. Genetic covariation among life-history components: the effect of novel environments. Evolution 39: 943-943.

Sheldon, P. R., 1990. Shaking up evolutionary patterns. Nature 345: 772.

Sibly, R. M. & R. Calow, 1989. A life-cycle theory of response to stress. Biol. J. Linn. Soc. 37: 101-116.

Siegel, M. I. & W. J. Doyle, 1975. The effects of cold stress on fluctuating asymmetry in the dentition of the mouse. J. Exp. Zool. 193: 385-389.

Siegel, M. I. & H. H. Smookler, 1973. Fluctuating asymmetry and audiogenic stress. Growth 37: 35-39.

Sneath, P. H. M. & R. R. Sokal, 1981. Numerical Taxonomy. W. H. Freeman, San Francisco.

Tanaka, K., 1989. Energetic cost of web construction and its effect on web relocation in the web-building spider *Agelena limbata*. Oecologia 81: 459-464.

Turelli, M., J. H. Gillespie & R. Lande, 1988. Rate tests for selection on quantitative characters during macroevolution and microevolution. Evolution 42: 1085-1089.

Valentine, D. W., M. E. Soule & P. Smollow, 1973. Asymmetry analysis in fishes: a possible statistical indication of environmental stress. Fish. Bull. U.S.A. 71: 357-370.

Via, S. & R. Lande, 1985. Genotype-environmental interaction and the evolution of phenotypic plasticity. Evolution 39: 505-522.

Waddington, C. H., 1942. Canalization of development and the inheritance of acquired characters. Nature 150: 563-565.

Waddington, C. H., 1957. Strategy of the Genes. Allen & Unwin, London.

Wright, S., 1931. Evolution in mendelian populations. Genetics 16: 97-159.

Wright, S., 1982. The shifting balance theory and macroevolution. Ann. Rev. Genet. 16: 1-19.

Zakharov, V. M., 1981. Fluctuating asymmetry as an index of developmental homeostasis. Genetica 13: 241-256.

Zeng, Z. B., 1988. Long-term correlated response, interpopulation covariation and interspecific allometry. Evolution 42: 363-374.

T. A. Markow (ed.), Developmental Instability: Its Origins and Evolutionary Implications, 229–236, 1994.

Appearance, fixation and stabilisation of environmentally induced phenotypic changes as a microevolutionary event

Vladimir M. Zakharov
Institute of Developmental Biology, Russian Academy of Sciences, 26 Vavilov St., Moscow 117808, Russia

Received 20 May 1993 Accepted 24 May 1993

Abstract

Certain phenotypic changes originally induced by direct environmental effects (which can be modelled experimentally) and later reflected by corresponding changes in the genotype (revealed by differences in the reaction of individuals from different populations under the same environmental conditions) highlight one of the main trends in microevolutionary processes. During that process, a decrease of initially increased levels of stochastic variation marks the stabilisation of development in a new environment as a change in optimal developmental conditions.

Concordance of interpopulation phenotypic differences with experimentally established dependence on developmental conditions and climatic conditions within habitats signifies the role of environment (by replacing the modification response within the limits of the reaction norm with a corresponding change in the reaction norm). Disturbances of this concordance suggest that some traits of microphylogenesis are playing a role.

Introduction

The study of the appearance of phenotypic differences at the intra-, interpopulation and interspecies levels is one of the most important tasks in population biology and microevolution. To estimate the role of chance fluctuations and directional changes, as well as to distinguish modifications within the limits of the genotypic reaction norm and phenotypic changes caused by alteration of the reaction norm, seem to be of particular importance in elucidating the mechanisms of population and evolutionary processes. In spite of the numerous studies of phenotypic differences and mechanisms of their appearance, to date joint consideration of all such processes has not been done.

It has been established that only the analysis of phenotypic variability from the viewpoint of general developmental homeostasis, including the characterisation of both random fluctuations and directional changes, allows us to fully reveal the genotypic reaction norm and changes therein and to describe all such processes in terms of ontogenetic events. For this it is necessary to distinguish and analyse two aspects of general developmental stability: *stability*, as the ability for stable development under given conditions and *canalisation*, as the ability for stable development under different conditions. It has been shown that random variability of development as a characteristic of developmental stability makes an important contribution to phenotypic variation. Experimental evaluation of the canalisation and plasticity of development allows us to visualise directional phenotypic changes both within the reaction norm and related to changes in the reaction norm itself (Zakharov, 1989, 1992).

The task of this study is to consider phenotypic changes from the perspective general developmental stability. It undertakes to study separately each of the main aspects of developmental homeostasis, developmental stability *per se* and canalisation or plasticity, as well as their parallel study for the same processes in an attempt to estimate the evolutionary significance of each of them and to understand more about the mechanisms of evolutionary transformations.

Developmental stability: chance phenotypic fluctuations

To characterise developmental stability, (defined as the ability of an organism to undergo the same development under given conditions), the phenotypic variation caused by developmental disturbances should be estimated. Analysis of developmental stability is then reduced to estimating the value of the variability, which is stochastic in the sense that it can not be directly explained by genotypic or environmental differences (Astauroff, 1930). Random variability of development, or 'developmental noise' (Waddington, 1957), is a form of variability which satisfies the above requirement.

In order to estimate the phenotypic expression of random developmental variability it is necessary to remove genotypic and environmental differences. This is possible only in exceptional cases, for example when analysing highly homozygous strains, clones, or monozygotic twins. Studies of intraindividual variability which compare different homologous structures are adequate for this purpose.

Minor non-directional differences between two sides of the body, called fluctuating asymmetry, are widespread (Van Valen, 1962). This phenomenon can be defined as asymmetry only by phenomenology, but in essence it demonstrates the imperfection of symmetry and asymmetry as realised in living organisms, since it is observed even in the case of directional asymmetry (Zakharov, 1987). Dissimilarity in expression of a given character on the left and on the right side, observed in the case of fluctuating asymmetry, cannot be explained by either genotypic or environmental differences, since its development is ensured by the same genotype under identical environmental conditions (Mather, 1953; Thoday, 1953-1958; De Marinis, 1959; Reeve, 1960; Palmer & Strobeck, 1986).

This kind of phenotypic variability reflects developmental noise, the imperfection of the ontogenetic processes. The latter naturally depends on both environmental conditions and the genotype. The level of such noise is minimal under certain environmental conditions, those defined as optimal for the genotype under study. Increased levels of developmental noise may also be caused by genotypic imbalance occurring as a result of crosses of various essentially different forms, mutations, etc. (Zakharov, 1989; Clarke & McKenzie, 1992a; Graham, 1992), but usually is only detected in some particular cases.

In general, the level of developmental noise corresponds to the level of developmental stability: it is minimal under optimal conditions and increases when conditions change from this optimum. Thus, the increase of the level of developmental noise proves to be a non-specific response to any changes from the optimum.

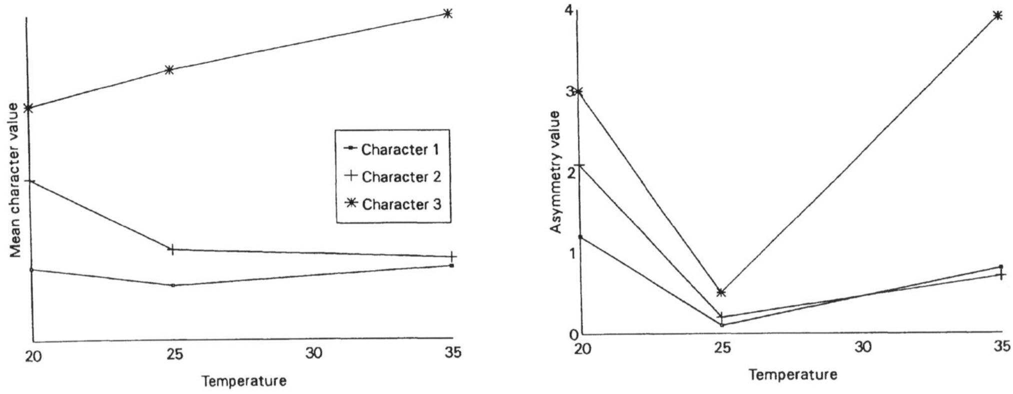

Fig. 1. Asymmetry value and mean character value for pholidosis characters as a function of temperature of egg incubation in the sand lizard *Lacerta agilis*. All characters, in spite of different dependence on temperature (nothing for character 1, negative for character 2, and positive for character 3) have the same pattern of asymmetry varion.

This effect can be easily modelled experimentally in the first generation developing under suboptimal conditions and has been demonstrated in insects, fish and reptiles for both natural (temperature, Fig. 1) and artificial (DDT) environmental changes (Valentine & Soulé, 1973; Zakharov, 1987, 1989; Clarke & McKenzie, 1992b; Leary, Allendorf & Knudsen, 1992).

In nature some changes in the level of developmental stability can be observed in the same population when environmental conditions are changed from the optimal and as a result of various kinds of human activities. Commonly, such phenotypic changes prove to be reversible.

Changes in the value of developmental stability indices with time have been analysed for different species of insects, fish and reptiles (Mason, Erlich & Emmel, 1967; Zakharov, 1981, 1989). These analyses carried out on different populations suggest a high temporal stability of analysed characteristics in most of the populations despite marked interpopulation differences. At the same time, when conditions of development proved to be different, changes in the level of developmental stability were found not only between different generations but also between individuals of the same generation, which differed by time of hatching.

In order to estimate a possible anthropogenic effect on the temporal dynamics of developmental stability in some species of fish and mammals (Valentine, Soulé & Samollow, 1973; Zakharov & Yablokov, 1990), the same population was sampled before and after the onset of intensive pollution related predominantly to the increase in the level of DDT and PCB. The results obtained suggest that developmental stability is markedly affected by anthropogenic pollution of the environment.

Similar differences on a larger scale can be found between different populations which differ essentially in environmental conditions. Serious disturbances of developmental stability occur at the ecological periphery of the species range.

On the whole, analysis of interpopulation differences suggests that many populations of the species existing under relatively optimal conditions despite some interpopulation differences are characterised by high developmental stability. Developmental stability is disturbed at the ecological periphery of the species range in populations existing in suboptimal conditions (Soulé, 1967; Soulé & Baker, 1968; Vrijenhoek & Lerman, 1982; Zakharov, 1981, 1989; Kat, 1982).

Despite the failure of certain attempts to detect the effect of anthropogenic impact on developmental stability (mainly due to methodological problems), many results indicate disturbance of developmental stability in polluted areas (Valentine, Soulé & Samollow, 1973; Valentine & Soulé, 1973; Ames, Felley & Smith, 1979; Zakharov, 1981, 1987, 1990; Jagoe & Haines, 1985).

The most crucial question is whether such differences can withstand the pressure of natural selection and consequently be found in nature. All available population data obtained support the conclusion that some temporal and spatial changes in the degree of developmental stability, to a certain critical level, prove to be not so crucial and can be seen in natural populations.

Canalisation and plasticity: directional phenotypic changes

To characterise canalisation of development (the ability to produce similar developmental outcomes under different environmental conditions) experimental analysis is needed. The main task of such experiments is to demonstrate the canalisation and plasticity of development in the offspring of natural populations and to compare experimentally demonstrated dependence of changes on developmental conditions, with an actual picture of intra- and interpopulation differences. Both highly canalised characters, which have invariant values, and plastic characters, showing certain changes, can be revealed (Zakharov, 1989).

Two main strategies of development can be realised in different species for different characters. The first is the ability to change under environmental influences: *developmental plasticity* (Thoday, 1953). The second is developmental invariability even under the influence of specific environmental conditions: *canalisation* or self-regulation of development (Schmalhausen, 1949). These strategies demonstrate two paths of developmental adaptation (Waddington, 1957). Selection of any strategy appears to be determined by a number of causes, including specific features of the genotype and the possibility of stabilisation, as well as the adaptive value of a certain character. Those

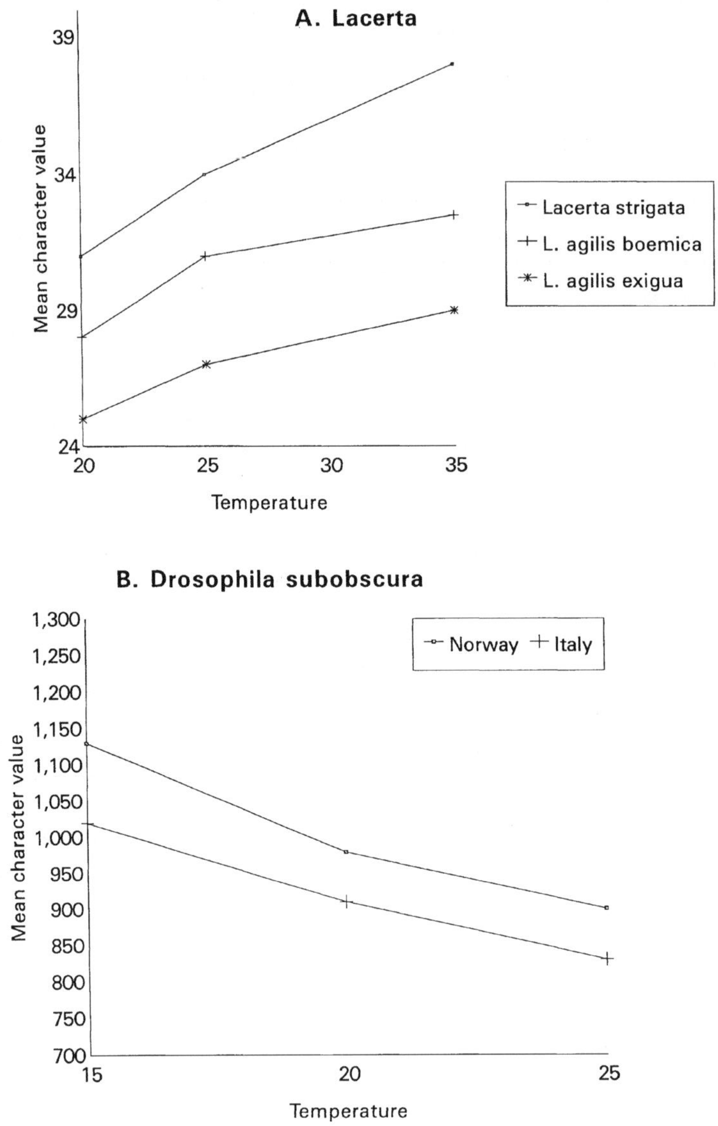

Fig. 2. The mean character value as a function of temperature of egg incubation for number of femoral pores in two species of lizards of the genus *Lacerta* (A) and for the wing length in two populations of *Drosophila subobscura* (B).

characters of the organism which are vitally important under different environmental conditions will invariably be preserved within the range possible for a given species. For other characters, they will certainly change under some environmental conditions and, in this case, the creode will be directionally altered. Finally, certain characters which do not influence viability will markedly change under some conditions due to particular features of their morphogenesis.

In studying developmental plasticity, we should concentrate on the mean value of the character in the group of individuals under study and its possible directional change. This can be caused by alterations in environmental conditions or by genotypic changes under the same conditions. Many such directional phenotypic changes are specific, i.e. they can be seen only under certain environmental conditions (Fig. 1). They do not allow the characterisation of such conditions as optimal (which is possible when studying developmental stability). This opens up possibilities for revealing specific changes which might be indicators of strictly defined developmental conditions (see, e.g., Hubbs, 1922; Fox, 1948; Taning, 1952; Lindsey, 1954, 1962; Fox, Gordon & Fox, 1961; Osgood, 1978). At the same time they prove to be similar for many populations of a species and even for different closely related species (Zakharov, 1992). Such changes can be easily experimentally modelled in the first generation exposed to the changed environmental conditions (Fig. 1).

In nature, directional phenotypic changes can be found as a result of differences in environmental conditions both in time and space. They can appear in populations from different parts of the species range under certain changes in environmental conditions. In general, such phenotypic changes often correspond to the experimentally established dependence on changes in environmental conditions, e.g. changes modelled in experiments under increased temperature are observed in the southern part of the species range.

Analysis of some populations from a major part of the species range reveals that general interpopulation differences agree fairly well with the climatic patterns of corresponding regions and experimentally revealed dependence of the characters on temperature (Zakharov, 1987, 1989). This does not mean that all phenotypic changes may be explained by climatic patterns, but these results illustrate the possibility of estimating the nature of certain phenotypic differences in natural populations and highlight the role of the environment in determining some patterns of phenotypic variability.

However, this does not mean that all variability is exhausted by phenotypic changes within the limits of the reaction norm of the genotype. The differences revealed appear to be related to genotypic changes and it is now possible to determine their causes. The discovered concordance suggests that these differences could initially arise as a modification response to specific environmental conditions.

Experimental analysis may also provide the answer to whether the observed phenotypic differences are accompanied by genotypic changes or not. As an example, Figure 2 provides data on the influence of temperature on one phenotypic character of lizards from different populations of two closely related species. The temperature dependence of this character is the same in both of two subspecies of *Lacerta agilis* and in *L. strigata*. Although these dependence curves run parallel, the values for every temperature regime are markedly different for each species. The value attained in the widespread subspecies *L. a. exigua* only at an increased incubation temperature is found for the southern subspecies *L. a. boemica* at low temperature. A similar picture was obtained when comparing the southern subspecies of the sand lizard with the closely related southern species *L. strigata* (Zakharov, 1987, 1989).

A similar situation can be seen for *Drosophila subobscura* (Fig. 2) (V. M. Zakharov, D. Sperlich, unpublished data). The dependence of wing length on temperature is the same for both northern and southern populations of the species, but at each temperature the values are significantly different.

These data suggest the possible mechanisms underlying development of these intra- and interspecific differences. Changes in the creode, initially under the influence of changes in the conditions of development and then at the expence of rearrangement of the genotype, are considered to be a possible pathway of evolutionary transformation known as the 'Baldwin effect' (Schmalhausen, 1949; Waddington, 1957; Shiskin, 1992). When comparing the starting and terminal points of this evolutionary process, phenotypic changes in the creode (within the reaction norm) are replaced by genotypic changes (changes of the reaction norm).

However, in some cases experimentally established dependence of phenotypic characters on developmental conditions may not correspond to actual interpopulation differences (McPhail, 1963; Osgood, 1978). This might be due to the fact that the populations being compared belong to different groups (e.g. subspecies). To reveal relationships between populations and their groups, microphilogenesis, the task is to distinguish phenotypic changes due to genetic differences, rather than modifications within the limits of the reaction norm of the genotype, which can not be directly explained solely by differences in current environmental conditions (Zakharov, 1987, 1989).

Parallel study of developmental stability and canalisation: chance fluctuations and directional changes in the phenotype

In cases when phenotypic characters under study are either well or poorly canalised, developmental stability is maximal only under certain environmental conditions. This is true even if the creode is invariable within the range of all possible environmental conditions appropriate for the species (Fig. 1). It indicates that these two characteristics are independent, i.e., directional changes of a character and its random fluctuations are two different processes uncorrelated with each other.

Comparing different creodes (or channels) commonly shows that they are not equivalent with respect to developmental stability. Only one of them is characterised by a minimal level of developmental noise that corresponds to the optimum. There could be more than one such creode but these very seldomly occur and only under very special conditions (Waddington, 1957). Commonly different creodes can be equivalent in this respect in cases of genetically determined differences in the reaction norm when both are more or less equally adapted for the particular environment, such as male and female, genetically determined morphs, etc. (Zakharov, 1992).

Comparison of genetically determined phenotypic changes induced by changes in environmental conditions seems to be important. Various phenotypic changes within the limits of the reaction norm due to changes in environmental conditions correspond to the disturbance of developmental stability. The same phenotypic differences caused by changes in the reaction norm correspond to high levels of developmental stability. Such creodes are equivalent with respect to developmental stability. This means that there are the same phenotypic changes, but genetically fixed and stabilised.

This was revealed in our experiments with lizards (genus *Lacerta*): the value of the character which is accompanied by disturbance of developmental stability in *L. agilis* is attained in *L. strigata* at a high developmental stability (Zakharov, 1989).

The establishment of such genetically determined differences consists of three major steps. Changes in a reaction norm, or natural reaction of the genotype under various environmental conditions, as an adaptive response, or simply specificity of the morphogenetic and other processes following environmental changes, are of primary importance for such processes.

Originally, when phenotypic change is induced directly by the alteration of the environmental conditions from the optimum, disturbance of developmental stability takes place. This can be easily demonstrated experimentally in the first generation developing under sub-optimal conditions (Fig. 1). At this stage all processes including both direct and random phenotype changes occur within the limits of the reaction norm. Then, in a population living under the same environmental conditions for many generations phenotypic changes may prove to be genetically determined, but they still have a low level of developmental stability. Disturbance of developmental stability is obviously expressed in indices of developmental noise in a sample from the periphery of the species range (Zakharov, 1989), in spite of the clear significant difference in the reaction under the same environmental conditions employed in experiments (Fig. 2). At this stage, alteration in the reaction norm is manifested as genetically fixed but still unstable directional changes in the phenotype as a result of some processes of genetic assimilation of the modification (Waddington, 1957).

The disturbance of developmental stability, in spite of clear genetic changes, can take place in such situations for two reasons. First, such environmental conditions still are sub-optimal for the genotypes under study. Second the disturbance of general genetic coadaptation, as a result of certain changes in genotype, causes disturbance in developmental stability. Which of these causes is important for the maintenance of the observed disturbance of developmental stability at this stage and how universal is their relationship needs further investigation. Despite the fact that genetic coadaptation is needed to maintain high levels of developmental stability we have some indirect evidence that the adaptation to the new environment in such cases seems to be of primary importance.

First, although selection for developmental stability may proceed for insignificant phenotypic deviations, such as fluctuating asymmetry (Mather, 1953; Reeve, 1960), it is more often related to maintaining the general coadaptation of the genome. This ensures high developmental stability in nature even in the case of directional selection. If pronounced directional selection can take place in this situation, when the necessary effect is attained the required level of genomic coadaptation should be restored. Only in some particular cases, for a rather short time such as several generations, can it be possible to see decreased genetic coadaptation as a result of selection, especially if it is necessary for survival under extremely severe conditions. Such a decrease in developmental stability has been shown in the first generations of flies developing resistance to an insecticide (Clarke & McKenzie, 1992a).

Second, phenotypic changes revealing the disturbance of developmental stability at the ecological periphery of the species range can be reversible to a certain degree, as can be seen at optimal environmental conditions in laboratory experiments (Zakharov, 1989). Disturbance of developmental stability in populations existing for a long time at the ecological periphery of the species range is evidenced by the absence of essential genotypic changes providing biochemical and physiological changes which allow adaptations to sub-optimal conditions. It was established experimentally that the optimal range of the environmental conditions for many populations of the species from different parts of the species range is similar (Zakharov 1992).

However, it is possible to become adapted to the new environmental conditions. In such cases the previously sub-optimal conditions become optimal. Visible differences in the level of developmental stability between sympatric populations of closely related species is indirect evidence of such an opportunity. The actual situation depends on many things such as general genotypic capacity (its ability to provide needed changes), strength of natural selection, etc.

Finally, changes in a reaction norm manifest as directional phenotypic changes are accompanied by high levels of developmental stability. This situation corresponds to development of individuals with high levels of genomic coadaptation under optimal conditions. Thus we can see genetically fixed and stabilised changes in phenotype as a result of adaptation for the new environment and coadaptation of the genome. This can be seen by differences between the sympatric populations of the closely related species. What are sub-optimal environmental conditions for one particular species prove optimal for the other and such directional changes in phenotype are characterised by high developmental stability.

Comparing four species of dragonfly of the genus *Sympetrum* in sympatric populations, produced two groups with respect to their level of developmental stability (Zakharov, 1981). Species belonging to groups differing in developmental stability proved to be different in both taxonomy and distribution.

A similar comparison was made for two species of lizards of the genus *Lacerta* in sympatric populations in Dagestan. *L. strigata* was characterised by a higher level of developmental stability, as compared with *L. agilis* (Zakharov, 1987). The zone of sympatry for *L. agilis*, being one of the most widespread Palaearctic species of reptiles, is the geographic and ecological periphery of the range, while these conditions appear to be optimal for *L. strigata*, a narrow specialist for this environment. This new stabilisation is a result of adaptation, which allows these new conditions to be considered as optimal, with the restoration of originally disturbed genetic coadaptation.

Conclusions

The data presented provide evidence that the study of developmental homeostasis allows the estimation of the role of chance (developmental instability) and directional (canalisation and plasticity) phenotypic changes in evolutionary transformation as well as the elucidation of their mechanisms. The data also illustrate the importance of the pathway of the evolutionary transformation through fixation and stabilisation of originally environmentally induced phenotype changes.

Detailed study of each of these two aspects of general developmental homeostasis during the evolutionary processes, including the precise experimental analysis of the appearance of differences in optimal developmental conditions as well as control of genetic coadaptation, seems to be an important task for the future development of this concept.

In this study, evolutionary processes have been described from the morphogenetic point of view. Consideration of the same processes from biochemical and physiological viewpoints can provide an opportunity for a better understanding of the mechanisms of changes in optimal conditions in adaptation to new environments.

References

Ames, L. J., J. D. Felley & M. H. Smith, 1979. Amounts of asymmetry in Centrarchid fish inhabiting heated and non-heated reservoirs. Trans. Am. Fish Soc. 108: 489-4905.

Astauroff, B. L., 1930. Analyse der erblichen Storungsfalle der bilateralen Symmetrie in Zusammenhang mit der selbstandigen Variabilitat ahnlicher Strukturen. Z.Indukt. Abstamm. Vererb. 55: 183-262.

Clarke, G. M. & J. A. McKenzie, 1992a. Coadaptation, developmental stability, and fitness of insecticide resistance genotypes in Australian sheep blowfly, Lucilia cuprina: a review. Acta Zool. Fennica 191: 107-110.

Clarke, G. M. & J. A. McKenzie, 1992b. Fluctuating asymmetry as a quality control indicator for insect mass rearing processes. J. Econ. Entomology 85: 2045-2050.

De Marinis, F., 1959. The nature of asymmetry and variability in the double Bar-eyeless *Drosophila*. Genetics 44: 1101-1111.

Fox, W., 1948. Effect of temperature on development of scutellation in the Garten snake (*Thanophis elegans atratus*). Copeia 4: 252-262.

Fox, W., C. Gordon & M. H. Fox, 1961. Morphological effects of low temperatures during the embryonic development of the Garten snake (*Thanophis elegans*). Zoologica 46: 57-71.

Graham, J. H., 1992. Genomic coadaptation and developmental stability inhybrid zones. Acta Zool. Fennica, 1992, 191: 121-132.

Hubbs, C., 1922. Variations in the number of vertebrae and other meristic character of fishes correlated with the temperature of the water during development. Am. Nat. 56: 360-372.

Jagoe, C. H. & T. A. Haines, 1985. Fluctuating asymmetry in fishes inhabiting acidified and unacifidied lakes. Can. J. Zool. 83: 130-138.

Kat, P. W., 1982. The relationship between heterozygosity for enzyme loci and developmental homeostasis in peripheral populations of aquatic bivalves (Unionidae). Am. Nat. 119: 824-832.

Leary, R. F., F. W. Allendorf & K. L. Knudsen, 1992. Genetic, environmental, and developmental causes of meristic variation in rainbow trout. Acta Zool. Fennica 191: 79-96.

Lindsey, C. C., 1954. Temperature-controlled meristic variation in the paradise fish (*Macropodus opercularis*). Can. J. Zool. 32: 87-98.

Lindsey, C. C., 1962. Observations on meristic variation in nine-spine stickleback (*Pungitius pungitius*), reared at different temperatures. Can. J. Zool. 40: 1237-1247.

Mason, L. G., P. R. Erlich & T. C. Emmel, 1967. The population biology of the butterfly *Euphydryas editha*. V. Character cluster and asymmetry. Evolution 21: 85-91.

Mather, K., 1953. Genetical control of stability in development. Heredity 7: 297-336.

McPhail, J. D., 1963. Geographic variation in North American nine-spine sticklebacks (*Pungitius pungitius*). J. Fish. Res. Board Can. 20: 27-44.

Osgood, D. W., 1978. Effects of temperature on the development of meristic characters in *Natrix fasciata*. Copeia 1: 33-47.

Palmer, A. R. & C. Strobeck, 1986. Fluctuating asymmetry: measurement, analysis, patterns. Ann. Rev. Syst. 17: 391-421.

Reeve, E. C. R., 1960. Some genetic tests on asymmetry of sternopleural chaeta number in *Drosophila*. Genet Res. 1: 151-172.

Schmalhausen, I. I., 1949. Factors of Evolution: The Theory of Stabilizing Selection. (in Russian) Nauka, Moscow.

Shiskin, M. A., 1992. Evolution as a maintenance of ontogenetic stability. Acta Zool. Fennica 191: 37-42.

Soulé, M., 1967. Phenetics of natural populaitons. II. Asymmetry and evolution in a lizard. Am. Nat. 101: 141-160.

Soulé, M. & B. Baker, 1968. Phenetics of natural populations. IV. The population asymmetry parameter in the butterfly *Coenonympha tullia*. Heredity 23: 611-614.

Taning, A., 1952. Experimental study of meristic characters in fishes. Biol. Rev. Camb. 27: 169-193.

Thoday, J. M., 1953. Components of fitness. Symp. Soc. Exp. Biol. 7: 96-113.

Thoday, J. M., 1958. Homeostasis in a selection experiment. Heredity 12: 401-415.

Valentine, D. W. & M. Soulé, 1973. Effect of p.p′-DDT on developmental stability of pectoral fin rays in the grunion, *Leurethes tenuis*. Natl. Mar. Fish. Serv. Fish. Bull. 71: 921-925.

Valentine, D. W., M. Soulé & P. Samollow, 1973. Asymmetry analysis in fishes: a possible statistical indicator of environmental stress. Natl. Mar. Fish. Serv. Fish. Bull. 71: 357-370.

Van Valen, L., 1962. A study of fluctuating asymmetry. Evolution 16: 125-142.

Vrijenhoek, R. C. & S. Lerman, 1982. Heterozygosity and developmental stability under sexual and asexual breeding systems. Evolution 36: 768-776.

Waddington, C. H., 1957. The Strategy of the Genes. Allen & Unwin, London.

Zakharov, V. M., 1981. Fluctuating asymmetry as an index of developmental homeostasis. Genetika (Beograd) 13:241-256.

Zakharov, V. M., 1987. Animal asymmetry: a population-phenogenetic approach. (in Russian) Nauka, Moscow.

Zakharov, V. M., 1989. Future prospects for population phenogenetics. Sov. Sci. Rev. F. Physiol. Gen. Biol. 4: 1-79.

Zakharov, V. M., 1990. Analysis of fluctuating asymmetry as a method of biomonitoring at the population level, pp. 188-198 in Bioindications of Chemical and Radioactive Pollution. Mir, Moscow.

Zakharov, V. M., 1992. Population phenogenetics: Analysis of developmental stability in natural populations. Acta Zool. Fennica, 191: 7-30.

Zakharov, V. M. & A. V. Yablokov, 1990. Skull asymmetry in the Baltic Grey seal: effects of environmental pollution. Ambio 5: 266-269.

T. A. Markow (ed.), Developmental Instability: Its Origins and Evolutionary Implications, 237–246, 1994.

Parental age and developmental stability in *Drosophila melanogaster*

Jennie Wakefield, Kristie Harris & Therese Ann Markow*
Department of Zoology, Arizona State University, Tempe, AZ 85287-1501, USA
** Author for correspondence*

Received 19 May 1993 Accepted 24 May 1993

Abstract

Two strains of *Drosophila melanogaster*, one outbred, recently derived from nature, and the other created by intensive directional selection on phototactic behavior for 19 years, were used to test the hypothesis that developmental stability is influenced by parental age. Three characters were examined: sternopleural bristle number, wing length, and wing area. The results do not support any relationship between parental age, either young or old, and developmental stability in offspring.

Introduction

Developmental stability is defined as the ability of an organism to buffer itself from environmental disturbances and develop according to its ontogenetic plan (Waddington, 1957). Quantification of developmental stability is typically measured via assessments of fluctuating asymmetry (FA), or the random differences between the right and left sides for any of a variety of characters in bilaterally symmetrical organisms.

The last 15 years have seen a resurgence of interest in developmental stability and fluctuating asymmetry across a wide range of approaches to the study of biology. For example, decreases in developmental homeostasis have been linked to homozygosity, to heterozygosity associated with introgression and its breakup of coadapted gene complexes, and to a variety of nongenetic stressors, leading conservation biologists to use measures such as FA as indicators of endangerment (Leary & Allendorf, 1989).

Parental age is one nongenetic stressor reported to influence developmental homeostasis in a number of organisms including fruit flies (Parsons, 1962) and humans (Livshits *et al.*, 1988). Parsons (1962) observed that in *Drosophila*, offspring often exhibit increased fluctuating asymmetry for sternopleural bristles when parents were very young or very old; he attributed this relationship to differences in the quality of eggs produced by age-extreme females, producing a maternal effect on developmental homeostasis of offspring. Advanced maternal age was found to be associated with increased asymmetry in human infants, possibly as a secondary effect conditioned by changes in length of gestation (Livshits *et al.*, 1988), but an increase in developmental instability has not been seen in the offspring of extremely young mothers.

In many natural populations of organisms, reproduction is occurring simultaneously by individuals of a variety of ages. In others, population age structure may be such that the mean reproductive age may vary as a function of the time at which a population is sampled. If developmental stability can vary with parental age, and parental age can vary temporally or spatially, studies seeking to evaluate the presence of various stressors may be easily confounded by parental age effects. With the increased employment of developmental stability as an indicator of population vulnerability by conservation biologists, it is important that all sources of developmental instability be carefully documented.

In order to ensure that such confounding effects have been sufficiently monitored, we repeated the earlier experiments of Parsons (1962) on parental age and fluctuating asymmetry in *Drosophila melanogaster*, but with an expanded design. In addition to measuring developmental stability for sternopleural bristle number, we added two wing variables: a standard measure of wing length (Markow & Ricker, 1991) and a measure of wing area. We also employed comparison of a pair of strains of greatly contrasting genotype. One was a mass-reared wild-type strain, derived three generations earlier from a mass collection (n > 100) in the wild. The other strain had been intensely selected for photopositive behavior in a maze since 1972, had been through several bottlenecks, and contained little genetic variation (Markow, 1975). We also extended the assessment of progeny through a maternal age of 35 (photoselected) and 39 (wild-type) days, respectively, rather than 30 days, in order to increase the likelihood of detecting effects associated with greatly advanced age.

Materials and methods

Two strains of *Drosophila melanogaster* were utilized in this procedure: ASU390, a strain collected from a natural population breeding in citrus in Tempe, Arizona in March, 1990 and PHOTO+, a laboratory strain artificially selected for positive phototaxis for 20 years, hence relatively homozygous (Markow, 1975).

Ten male and ten female virgins from each strain were removed from uncrowded cultures and placed into collection vials containing a banana medium and yeast. Two such vials were prepared for each strain, each containing ten female/male pairs. These parent flies were transferred to fresh vials on a daily basis (between 20-28 hours from time of last transfer) for a total of 35 days (PHOTO+) and 39 days (ASU390), at which time breeding capacity was minimal.

The eggs laid in these vials were closely monitored and during the first instar larvae stage, 40 larvae per vial were removed from the substrate with a small spatula and transferred to a fresh vial labeled by maternal age (in days). In this fashion, even though female productivity declined with age, density during larvae development was kept constant. Larvae were collected daily for days 1-9, and every other day thereafter until eggs were no longer viable.

At three days post-eclosion, 15 male and 15 female offspring (per maternal age day) were collected at random and anesthetized. Counts for left and right sternopleural bristles were obtained using a dissecting microscope. Both wings were then removed, mounted on projection slides and labeled for 1), sex, 2) maternal age of eggs, and 3) wing side, R or L.

Upon completion of wing mounting procedure, two wing measures (Fig. 1) were obtaining using an IBM PC and BIOSCAN image analysis system.

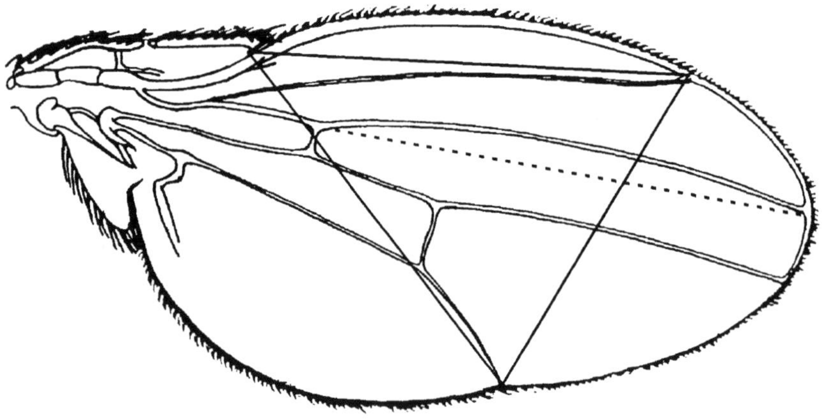

Fig. 1. The length and area measures made on Drosophila wings.

Slides were placed under a light source and camera connected to the computer. Wings were magnified 25× and the images viewed on the computer screen. Images were consistently oriented with the long axis parallel to the horizontal in order to minimize distortion as a function of rotation. Linear measure in mm at 25× was calibrated using a micrometer; all wing data were measured against this calibration. End points were manually placed at the appropriate wing junctures using an electronic mouse.

Both the linear and area measures were calculated and recorded in mm (and mm^2 respectively), as per calculation standard. Twenty random wings were remeasured at the end of the procedure to ensure consistent endpoint placement and calibration invariance and differences were found to be less than 0.5%.

A total of three bilateral characters was thus measured for each animal: sternopleural bristle number, wing length, and wing area. Each character was quantified by the formula R+L / 2, or an average of both sides, and it was these measures that were used to assess differences among strains and sexes for the characters themselves. Developmental stability was measured by calculating fluctuating asymmetry measures, R–L. Scale effects were tested for by seeking correlations between the R–L values and trait magnitude (R+L / 2). As none were found, subsequent analyses were performed on untransformed fluctuating asymmetry values (Palmer & Strobeck, 1986).

Results

Strain means for the three traits, wing length, wing area, and sternopleural bristle number, are shown in Table 1. Before evaluating the impact of parental age in the two strains, we first asked whether the strains and sexes differed in their mean wing and bristle measures. Data were analysed using the General Linear Models (ANOVA) procedure from SAS (SAS, 1985) on the ASU mainframe computer. The F values were significant at $p < 0.0001$ for each trait: wing length ($F = 837.92$, df 3, 1370), wing area ($F = 212.45$, df 3, 1361) and bristle number ($F = 91.43$, df 3, 1362). Both strain and sex were significant factors. Males of both strains have significantly smaller wings, in both the length and the area measures, than females, but males did not always show the smallest values. Bristle number was significantly lower in PHOTO+ males compared to females, while in ASU 390, females had lower mean bristle number. Greater values for all traits were observed in the PHOTO+ strain.

Fluctuating asymmetry values for each measure are reported in Table 2. In no case was any correlation observed between trait magnitude and degree of fluctuating asymmetry. Thus as mentioned above, subsequent statistical analyses were conducted using untransformed data. Wing length asymmetry (strain $F = 7.79$, df 1, 1370) and sternopleural bristle number asymmetry (strain $F = 7.72$, df 1, 1362) were significantly greater in flies from the PHOTO+ strain. PHOTO+ males showed even greater bristle asymmetry than females (sex $F = 7.36$, df 1, 1362) no other significant strain or sex differences emerged.

As expected for a strain having undergone prolonged directional selection, the PHOTO+ strain was characterized by slower development rate and an earlier decline of reproductive capacity (35 days) than the wild type strain (39 days). After day 31, a decrease in the number of eggs laid per day and a decrease in viability of eggs made it difficult to obtain 40 first instar larvae/vial.

The impact of parental age on character magnitude is seen in plots (Figs. 2, 3) of the mean of each character against the age of the mothers at the time of oviposition. Both linear and non-linear equations were used in attempting to find the best fitting

Table 1. Wing length and area (in mm and mm^2) and sternopleural bristle numbers by strain and sex. Standard errors are in parentheses.

Variable	N	Photo + females	Photo + males	390 females	390 males
wing length	330	1.4128 (0.0024)	1.2740 (0.0021)	1.3801 (0.0032)	1.2616 (0.0025)
wing area	326	0.5374 (0.0018)	0.4347 (0.0015)	0.5013 (0.0025)	0.4188 (0.0019)
bristle number	325	11.6385 (0.0699)	11.1167 (0.0689)	10.3062 (0.0605)	10.4592 (0.0578)

Table 2. Wing length and area (in mm and mm²) and sternopleural bristle numbers by strain and sex. Standard errors are in parentheses. WL = wing length, WA = wing area, SP = sternopleural bristle number.

Variable	N	Photo + females	Photo + males	390 females	390 males
WL FA	330	0.0117 (0.0005)	0.0107 (0.0005)	0.0099 (0.0004)	0.0099 (0.0005)
WA FA	326	0.0100 (0.0004)	0.0081 (0.0004)	0.0088 (0.0004)	0.0079 (0.0004)
SP FA	325	1.1661 (0.0559)	1.3545 (0.0597)	1.0112 (0.0445)	1.1268 (0.0490)

Table 3. Linear and non linear regression ANOVAs for average character size on parental age. Plots and equations for best fitting lines are presented in Figures 2 and 3.

				Linear			Quadratic		
Trait	Strain	Sex	DF	Mean square	F	P	Mean square	F	P
WL	Photo +	F	1	0.00000286	0.005	0.9460	0.00000107	0.002	0.9670
		M	1	0.00036174	0.510	0.4833	0.00088102	1.285	0.2705
WA	Photo +	F	1	0.00002311	0.060	0.8096	0.00000320	0.008	0.9286
		M	1	0.00018238	0.521	0.4790	0.00034737	1.015	0.3257
SPB	Photo +	F	1	0.10356899	0.294	0.5935	0.12506723	0.356	0.5572
		M	1	0.65281122	1.502	0.2345	0.22032737	0.483	0.4951
WL	390	F	1	0.00257279	2.384	0.1369	0.00199543	1.805	0.1928
		M	1	0.00058442	0.875	0.3597	0.00018946	0.276	0.6044
WA	390	F	1	0.00217405	3.153	0.0896	0.00170974	2.406	0.1351
		M	1	0.00389986	0.970	0.3354	0.00011091	0.263	0.6132
SPB	390	F	1	0.02521376	0.211	0.6504	0.01336028	0.111	0.7418
		M	1	0.02735592	0.241	0.6283	0.04879880	0.434	0.5170

Table 4. Linear and non linear regression ANOVAs for fluctuating asymmetry on parental age. Plots and equations for best fitting line are presented in Figures 4 and 5.

				Linear			Quadratic		
Trait	Strain	Sex	DF	Mean square	F	P	Mean square	F	P
WL FA	Photo +	F	1	0.00000718	1.974	0.1753	0.00000586	1.583	0.2229
		M	1	0.00000446	0.850	0.3674	0.00000273	0.512	0.4820
WA FA	Photo +	F	1	4.613^{-10}	0.000	0.9916	1.369^{-7}	0.034	0.8562
		M	1	1.618^{-7}	0.033	0.8579	0.00000254	0.530	0.4750
SPB FA	Photo +	F	1	0.00489296	0.086	0.7722	0.01491442	0.265	0.6124
		M	1	0.18299524	2.277	0.1469	0.10527840	1.250	0.2769
WL FA	390	F	1	3.20975^{-7}	0.076	0.7859	0.00000281	0.681	0.4181
		M	1	0.00022931	3.991	0.0583	0.00001568	2.582	0.1223
WA FA	390	F	1	0.00000535	1.371	0.2541	0.00000411	1.038	0.3194
		M	1	0.00000177	0.491	0.4909	1.48772^{-8}	0.004	0.9500
SPB FA	390	F	1	0.00878989	0.139	0.7133	0.01159170	0.183	0.6290
		M	1	0.000536773	0.070	0.7145	0.01926944	0.252	0.6209

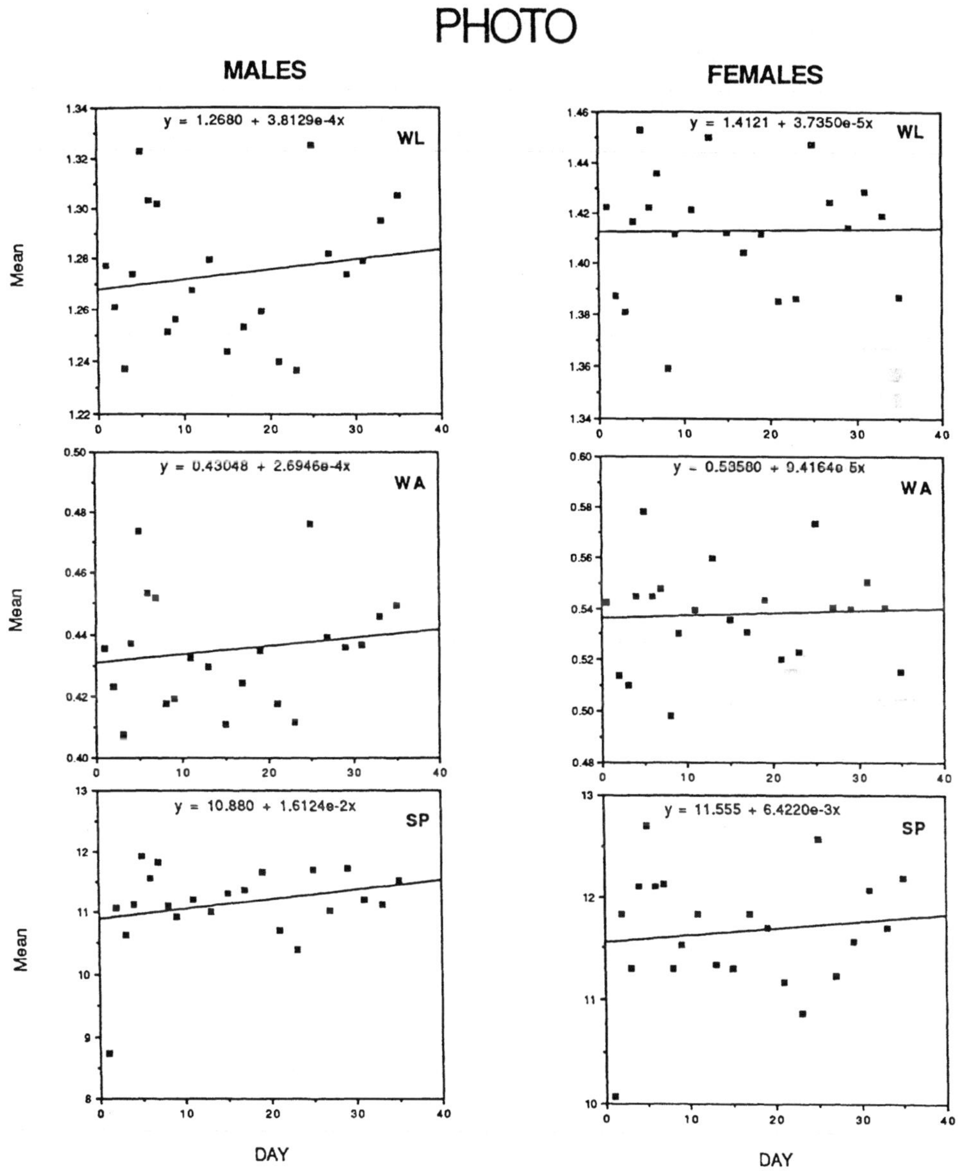

Fig. 2. Regression of trait magnitude on maternal age in the photo positive strain. WL = wing length, WA = wing area, and SP = sternopleural bristle number.

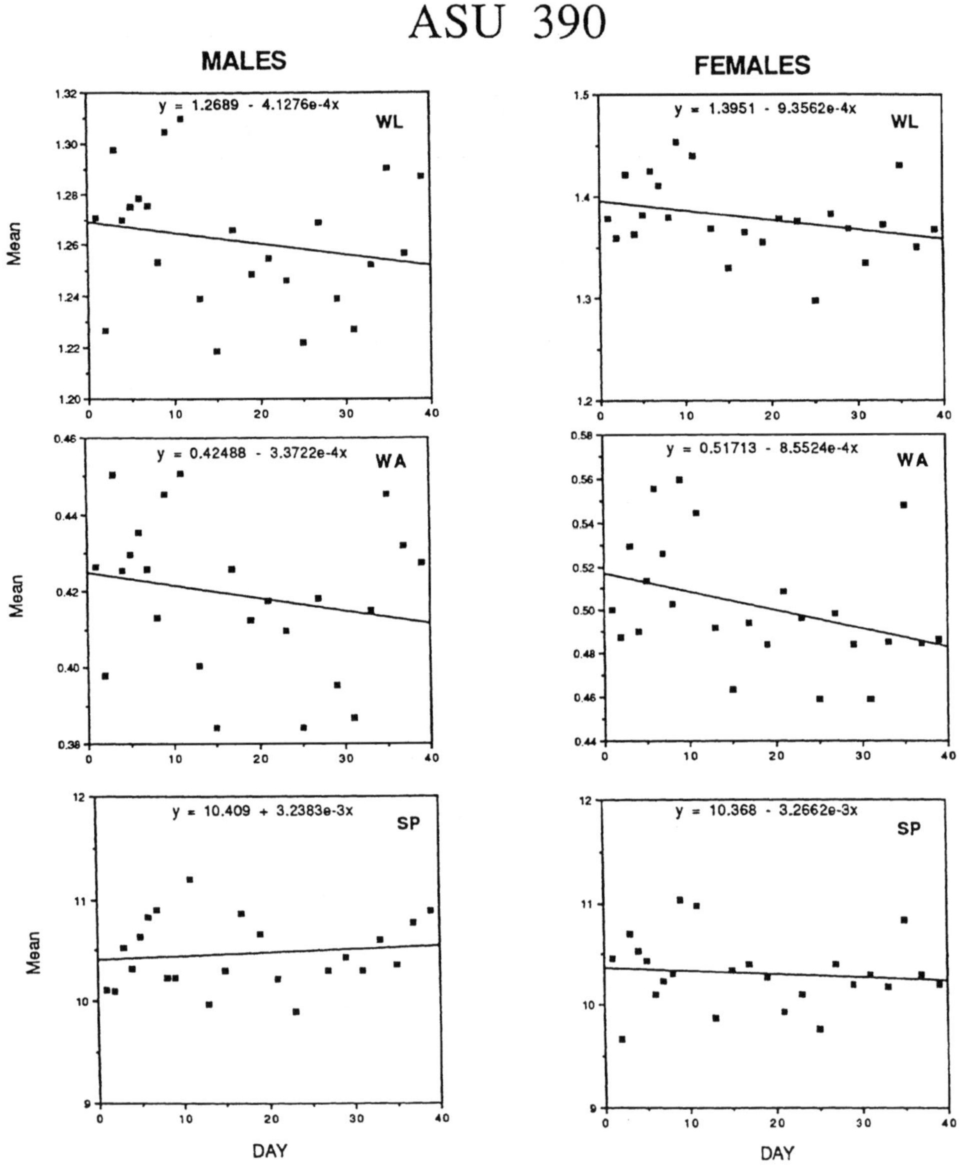

Fig. 3. Regression of trait magnitude on maternal age in the ASU 390 wild type strain. WL = wing length, WA = wing area and SP = sternopleural bristle number.

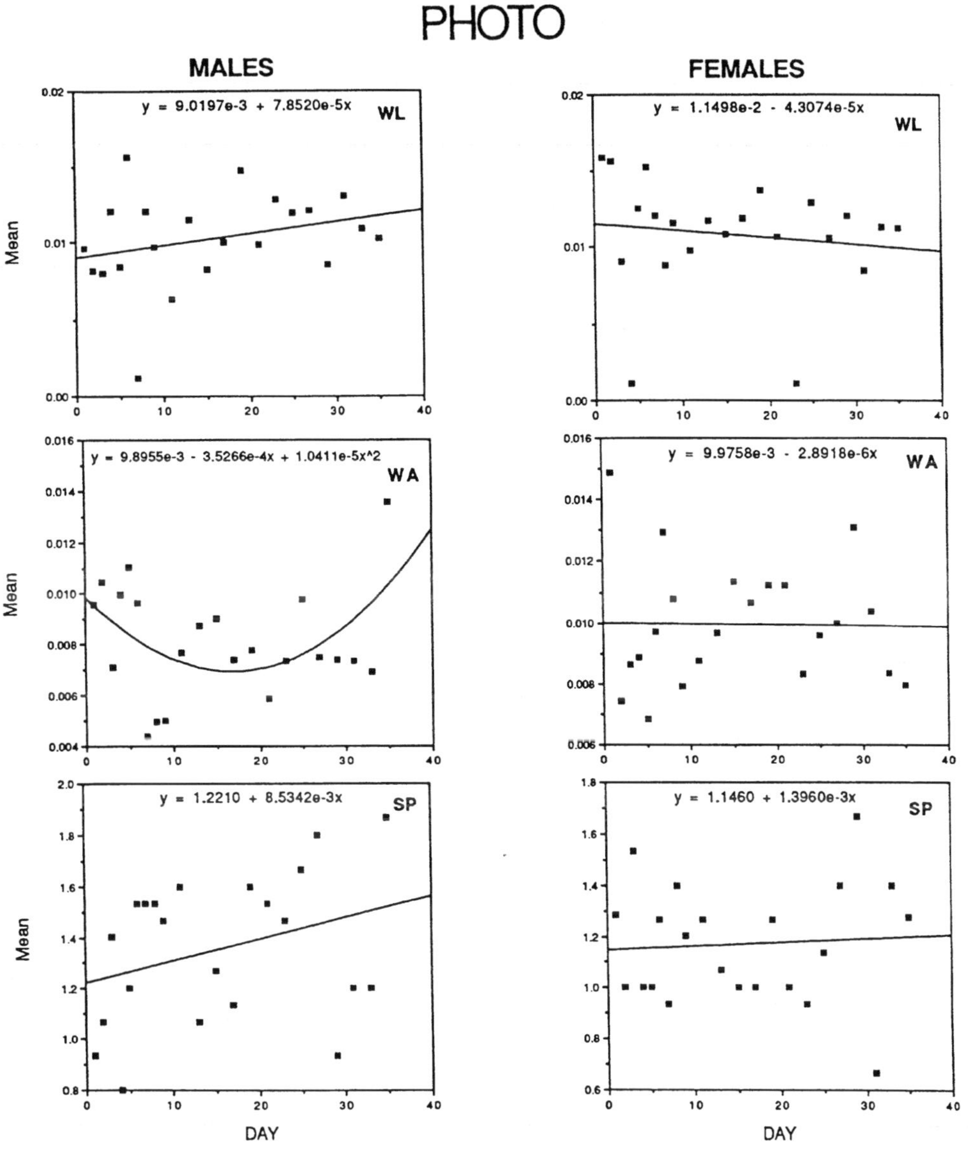

Fig. 4. Regression of trait fluctuating asymmetry on parental age in the photo positive strain. WL = wing length, WA = wing area, and SP = sternopleural bristle number.

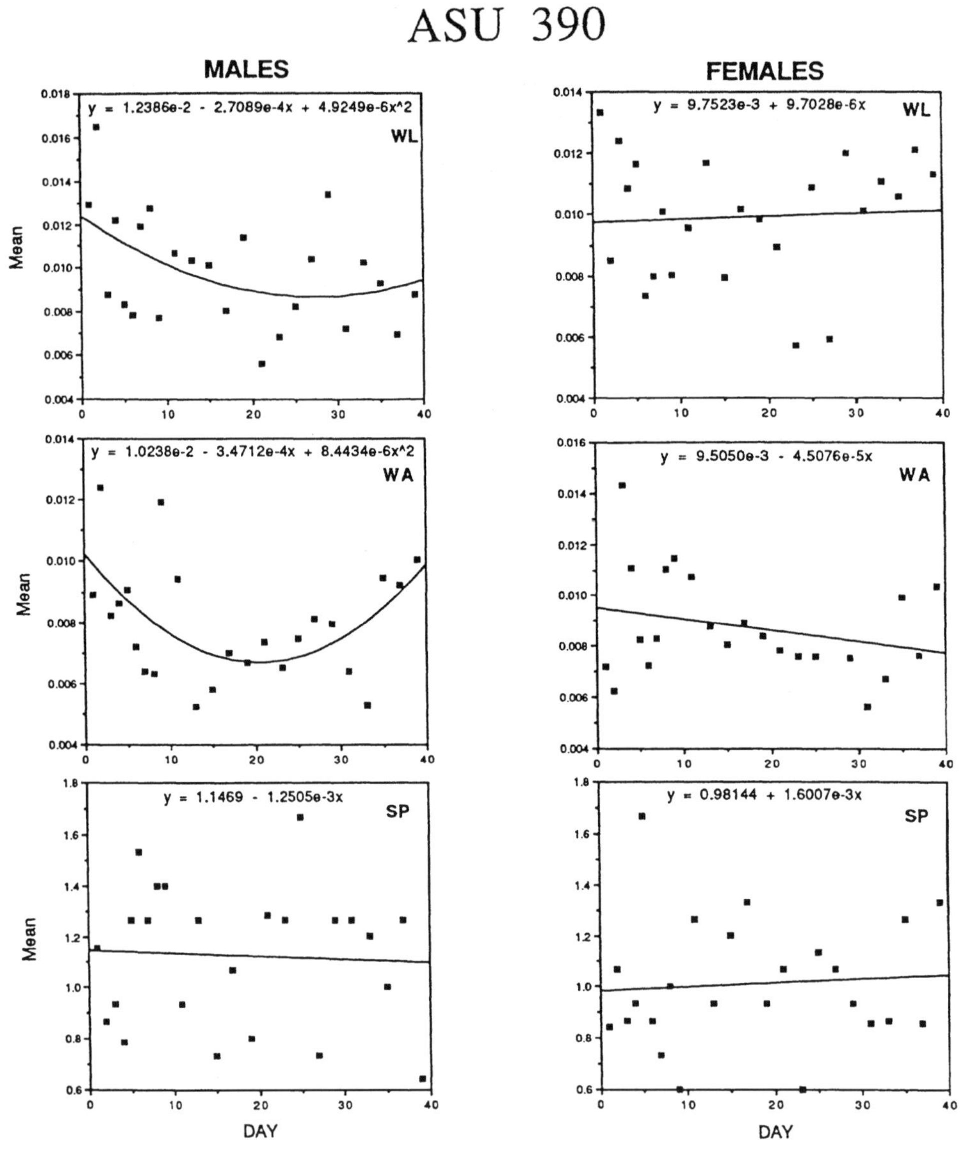

Fig. 5. Regression of trait asymmetry on parental age in the ASU 390 wild type strain. WL = wing length, WA = wing area, and SP = sternopleural bristle number.

relationships. Lines represent the best fit, not necessarily statistically significant ones. Sexes were examined separately to see if they differed in their potential sensitivity to maternal age effects. Figure 2 presents the data for the PHOTO+ and Figure 3 for ASU 390. Analysis of variance revealed that no significant relationships, either linear or quadratic, were present for any of the three traits and parental age (Table 3).

Changes in developmental stability with parental age are presented in Figure 4 for the PHOTO+ strain and Figure 5 for ASU 390. Again, lines are derived from the best fitting equations, regardless of statistical significance. None of the measures of fluctuating asymmetry showed a significant relationship to maternal age in any regression component (Table 4).

Discussion

In our study, parental age was not observed to influence the magnitude of any traits. Parsons (1962) found a significant parental age effect on egg size and argued that this effect was responsible for differences observed in the developmental stability of flies developing from those eggs.

The discussion offered by Parsons (1962) is intuitively satisfying and may well apply in many cases. By observing a different result with different *Drosophila* strains in a different laboratory, we feel that the relationship between parental age and developmental stability is likely to be less general than proposed earlier. For example, Parsons reported, in addition to the findings previously mentioned, that fluctuating asymmetry was even more enhanced in offspring that were incubated at 30 °C relative to those incubated at 25 °C. Temperature is therefore a clear stress factor. However, the temperature of minimal or negligible stress is unknown, and our laboratory temperature of 24 °C may have contributed to the lack of maternal age effect by eliminating a potential confounding interaction.

Unlike Parsons, we did not measure egg size in our study. We did however, investigate the additional FA indicator concerning the degree to which parental age influenced developmental stability of wings and bristles. Several factors in the present study should have enhanced our ability to detect a parental age effect on developmental stability if it existed. First, the upper age limit of parental flies in the present study was greater than in the earlier report. Second, we used a strain of flies, PHOTO+, which was higly inbred and thus should have been highly sensitive to parental age effects. However, no influence of parental age on developmental stability of any trait in any strain was observed. The PHOTO+ strain did yield the expected elevated incidence of developmental instability, but no parental age effect.

Flies from the two strains differed from each other in the magnitude of the three traits. Photostrain flies had significantly higher averages than flies of the wild-caught strain. We had expected that flies from ASU 390 would be larger, due to their recent derivation from nature and concomitant high levels of genetic variability and heterozygosity. The larger trait magnitudes attained by photostrain flies may reflect their longer developmental times. Photostrain adults emerged 1-3 days later than did imagoes from strain 390.

Our observations are not that surprising in view of the report by Livshits and Kobyliansky (1991) on nongenetic factors that influence developmental stability in humans. These investigators found only a minimal influence of maternal age on fluctuating asymmetry for dermatoglyphic traits. Interestingly, the most significant nongenetic predictor of fluctuating asymmetry in offspring was found to be maternal fluctuating asymmetry. Maternal condition, not necessarily maternal age, may be of more importance among factors capable of influencing offspring developmental and fitness. Further, as noted, the work of Livshits *et al.* (1988) suggests that increased FA in humans as a function of maternal age is secondary to complications of gestational age; exclusion of preterm infants would control for this complication. All of these factors should be more thoroughly investigated across taxa before they can be confidently discussed.

Acknowledgements

We thank Dr. Michael Douglas for his advice and use of his image analysis system, Dr. Gregory Livshits for comments on the manuscript, and Mrs. Ann Bradley for running a fine *Drosophila* kitchen. This work was supported in part of NSF grant *GR BSR8919362* to T. A. Markow.

References

Leary, R. F. & F. W. Allendorf, 1989. Fluctuating asymmetry as an indicator of stress: Implications for conservation biology. Trends in Ecology and Evolution 4: 214-217.

Livshits, G., L. Davidi, E. Kobyliansky, D. Ben-Amitai, Y. Levi & P. Melob, 1988. Decreased developmental stability as assessed by fluctuating asymmetry of morphometric traits in preterm infants. American Journal of Medical Genetics 29: 793-805.

Livshits, G. & E. Kobyliansky, 1991. Fluctuating asymmetry as a possible measure of developmental homeostasis in humans. Human Biology 63: 441-466.

Markow, T. A., 1975. A genetic analysis of phototaxis in *Drosophila melanogaster*. I. Selection in the presence of inversions. Genetics 79: 527-534.

Markow, T. A. & J. P. Ricker, 1991. Developmental stability in hybrids between the sibling species pair, *Drosophila melanogaster* and *Drosophila simulans*. Genetica 84: 115-121.

Palmer, A. R. & C. Strobeck, 1986. Fluctuating asymmetry: Measurement analysis, patterns. Annual Review of Ecology and Systematics 17: 291-421.

Parsons, P. A., 1962. Maternal age and developmental variability. Journal of Experimental Biology 39: 251-260.

SAS users guide: Statistics, 1985. SAS Institute Carey, South Carolina.

Waddington, C. H., 1957. The Strategy of the Genes (MacMillan, New York).

T. A. Markow (ed.), Developmental Instability: Its Origins and Evolutionary Implications, 247–255, 1994.

Developmental variability and the limits of adaptation: interactions with stress

P. A. Parsons
Waite Agricultural Research Institute, University of Adelaide, Glen Osmond SA 5064, Australia

Received and accepted 26 April 1993

Key words: adaptation, environmental stress, fluctuating asymmetry, metabolic cost, species borders

Abstract

1. Little evolutionary change may occur at species borders since the cost of accommodating environmental streses is high. Extreme examples of such stasis include cave animals in stable stressed environments and 'living fossils' in widely fluctuating stressed environments.
2. Variability from the molecular to the organismic level tends to be high under extreme stress. At the developmental level, the fitness of such variants may be low. This means that much developmental variability in natural populations may have little evolutionary significance.
3. Rapid evolutionary change of morphological traits is most likely to be based upon genes acting late in a developmental pathway under conditions which are ecologically and energetically permissive.
4. Although some increases in resistance to temperature extremes have been recorded in laboratory selection experiments, major extensions of extremes in natural populations appear difficult to achieve. The energetic costs of surviving extremes at species borders implies that the evolution of major developmental and morphological shifts is more likely to be a feature of populations of more equable habitats.

Introduction

Evolution and stress

Darwin (1859) viewed evolutionary change predominantly as a gradual process and considered competition to be important in the process of natural selection. Even so, various quotations taken from Darwin (1859) by Hoffmann and Parsons (1991) indicate that he appreciated the impact of climatic stress at the extremes of species distributions. A continuum emerges whereby competition is a factor of consequence in benign environments but less so in more marginal environments where environmental stress is important.

In a summary of the diffuse literature on stress and its evolutionary consequences, Hoffmann and Parsons (1991) emphasize the extreme end of the stress gradient at the limits of resistance where non-reversible damage and mortality occur irrespective of the density of organisms. They conclude that stress is important in many evolutionary and ecological processes, implying high selection pressures associated with rapid shifts in the frequencies of underlying genotypes. This is hard selection in the sense of Wallace (1981) as distinct from soft selection, where the fate of an individual depends upon the phenotype and genotype of its immediate neighbours, and density and/or frequency dependence occurs from competitive interactions. This is consistent with data on natural populations indicating that natural selection is often far more intense (Endler, 1986) than envisaged by many of the architects of the modern synthetic theory of evolution.

However, much of the literature in evolutionary biology does not directly consider stress. This is particularly evident in laboratory studies where experimentally easier more optimal environmental conditions are normally chosen rather than those approaching the critical boundary close to extinction where the effects of stress are most obvious. Even though laboratory studies are difficult under stress conditions, it is evident that stress forms an environmental 'probe', permitting generalizations difficult to perceive under more optimal environments. This is partly because expressed variability tends to be high under extreme stress (Parsons, 1987). But ultimately there will be a limit to adaptive change. This article discusses this limit with respect to adult morphology in the context of interactions with environmental stress.

The limits of adaptation

In the absence of abrupt geographic barriers, three approaches indicate an important role for extreme climatic perturbations in determining limits to species distributions (see detailed discussion in Hoffmann & Parsons, 1991):

1. The utility of bioclimatic systems developed from parameters characterizing annual, seasonal and extreme climatic components for predicting the distribution of various animal and plant taxa.
2. Periods of short-term extreme temperature and desiccation stress in the laboratory are effective monitors of climatic races within species and of the habitats of species in the wild in a number of taxa, including *Drosophila*. In many cases, there are clines which are relatable to local conditions expressed in terms of stress.
3. Historical changes in species distributions frequently track climatic change in the natural fauna and flora.

While these approaches indicate that the limits to species distributions correlate with extreme climatic features of the environment, at more local levels other abiotic factors such as pollutants may modify distributions. Furthermore, associations between abiotic factors and species distributions can be affected by biotic factors, especially competition (Dunson & Travis, 1991), but as the time scale increases towards geological time when speciation events necessarily become more evident, climatic factors dominate as the remaining factors become transient (Vermeij, 1987).

Consistent with these observations is the proposition that physiological constraints restrict range expansions into harsh environments (eg. Root, 1993; Bozinovic & Rosenmann, 1989; Parsons, 1991, 1993a). Under this model, energetic limitations restrict normal physiological processes to the level that range expansions are precluded whereby at species borders, the energy gained from assimilation/respiration should equal the energy lost from maintenance/survival. The potential for genetic adaptation and of ecological range expansions then becomes effectively precluded. The metabolic cost of harsh environments has often been demonstrated (Parsons, 1992a). A recent example comes from polar bears following exposure to oil pollution: metabolic rate increased but the consequence was reduced survival, especially when cold and starvation stresses were superimposed (Hurst, Watts & Oritsland, 1991).

Simplistically, species borders therefore appear determined by an interaction between climatic stress and its metabolic cost. In this sense, geographic ranges are 'species-level' traits arising from the biological properties of the individuals making up these species (even though clinal variation can occur within species under less severe conditions, favoring a reduced proportion of phenotypes resistant to extreme stress in more central populations (Hoffmann & Parsons, 1991; Parsons, 1991)). In fossil marine poikelotherm species Jablonski, Flessa and Valentine (1985) came to this conclusion, and considered that temperature is a major determinant of geographical range. This indicates that species borders defined in terms of ecological extremes may represent limits that are difficult to transcend. Turning to the future, this is a point of consequence in the global warming debate (Parsons, 1989).

Many species therefore should track climatic change. In the fossil *Coleoptera* of the late Cenozoic from the Temperate Zone, Coope (1979) found that species tracked climatic change during glacial-interglacial cycles. In addition, extensive latitudinal shifts in the geographical ranges of Californian Pleistocene and recent mollusk species are associated primarily with climatic change (Valentine & Jablonski, 1991). The likelihood of environmental tracking as a generality in the fossil fauna and flora (Upchurch, 1989) concurs with well-documented historical changes in species distributions with climatic change.

'Living fossils' represent an extreme where observable evolutionary change is apparently arrested over vast time periods, including several mass extinction events. In many cases these organisms live in harsh environments. Stromatolites occur in su persaline regions of searing heat and low precipitation. Horseshoe crabs can function in water varying widely in salinity and can endure great swings in temperature and oxygen levels; their extreme hardiness is confirmed by their extreme tolerance to chemical pollutants (see Ward, 1992). The brachiopod, *Lingula*, and oysters and mussels from the Paleozoic occur in similarly stressful environments. Ward (1992) argues that these habitats are so stressful that predators and competitors cannot survive so that there is a lack of interspecific interactions. Furthermore, the energetic costs of mere survival in these habitats would preclude major evolutionary change so that 'living fossils' become the major occupants. The prerequisite for this extreme form of stress is the continuity of defined stress habitats over geological time, even though the geographic location of these habitats may move as climate changes.

One way of increasing resistance to stress is by a genetic reduction in metabolic rate. This occurred in the insect *Drosophila melanogaster*, where reduced metabolic rate was a correlated genetic response to successful selection for resistance to desiccation (Hoffmann & Parsons, 1989a, b). However, the cost of survival was reduced fitness, since fecundity and behavioral activity levels were lower in the desiccation-resistant lines. Turning to the field, many obligate cave species have a low metabolic rate which is an adaptation to high CO_2 levels at concentrations up to 200 times ambient (Howarth & Stone, 1990), and to survival during long periods of starvation stress in an energy-poor environment. In the process of entering such habitats, substantial evolutionary change can occur as shown by cave adaptation, which is a general and predictable process occurring whenever a suitable system of underground voids exists (Howarth, 1991). Once established in such low-energy but stressed environments, the potential for future faunal changes would presumably become restricted. Parallel situations occur in sections of the deep sea and in infaunal habitats.

The above extreme field examples at the limits of adaptation represented by 'living fossils' and caves are situations of little or no apparent evolutionary change due to restrictions on the availability of metabolic energy whereby the energetic cost of survival dominates. In 'living fossils' stasis occurs because of the energetic costs of existing in widely fluctuating stressful environments, and in caves stasis is a feature of stable but stressed environments (see also Parsons, 1993b).

Stressed environments and variability

At the whole organism level, an association between stress intensity and phenotypic and genetic variability has emerged as a plausible relationship (Parsons, 1987; Hoffmann & Parsons, 1991). A difficulty is that relevant data are often collected for other purposes, making reinterpretation in a stress context difficult. Considering experiments on longevity and survivorship under stress, isofemale strains of *D. melanogaster* from three geographically distinct populations gave intraclass correlations >0.5 for adult longevity exposed to stressful concentrations (where longevity was reduced) of ethanol and acetic acid, and <0.5 for more optimal concentrations (where longevity was increased). Since intraclass coefficients reflect the size of the genetic component of variance relative to the environmental component, this result shows that, under stress, the genetic component of the phenotypic variance increased relative to the environmental component in these populations (Parsons, 1982). Based upon a 4×4 diallel cross among inbred strains of *D. melanogaster*, adult longevity was studied following exposure to five different doses of γ-radiation. At the highest dose, 1.2kGy, the additive genetic variance was substantially increased relative to the dominance and environmental components by comparison with lower doses (Westerman & Parsons, 1973). Similarly, at stress levels causing severe longevity reductions additive genetic variance predominates for desiccation, anoxia, ethanol, ether and phenylthiouria. In some cases, genetic activity implying major genes with additive effects have been located to regions of chromosome arms (Parsons, 1974).

Turning to other species, following the introduction of the rice weevil, *Sitophilus oryzae*, into a toxin-stressed environment, Holloway, Sibly and Povey (1990) have shown increased additive genetic variability of two life-history traits, developmental rate and mortality. Finally, in the American cotton stainer bug *Dysdercus bimaculatus*, the heritability of timing of first clutch calculated from parent-offspring regression increased from close to

0 under benign conditions to 0.3-0.4 under the conditions of desiccation stress of the intensity frequently encountered in nature (Derr, 1980).

Additive genetic variability is therefore expected to be high under stress, defined as the point above which is close to where populations become extinct. Approached by conventional quantitative techniques, comparisons of variability pose substantial difficulties in interpretation (Houle, 1992). Notwithstanding these complexities, extreme stress can be regarded as an environmental probe which increases genetic variability revealing associations among life-history and stress traits that are difficult to perceive under more benign conditions.

Recombination and mutation also tend to be high under extreme stress conditions (Parsons, 1988). Taking recombination as a model, environmental stress should affect morphological variability in a parallel way. In the snake *Natrix fasciatus*, vertebral number and the frequency of gross abnormalities increased at both high and low temperatures for experiments carried out over a developmental temperature range of 18-32 °C (Osgood, 1978). In the sand lizard *Lacerta agilis*, in embryos developed at a range of constant temperatures, 20, 25, 30 and 35 °C, developmental stability was estimated as the number of disturbances per individual; stability decreased at both low and high temperatures (Zakharov, 1989).

An epigenetic process that has attracted recent attention at the population level is the fluctuating asymmetry, FA, of bilateral characters (Palmer & Strobeck, 1986; Zakharov, 1989; Parsons, 1990). For quantitative traits, FA increases away from optimum environments, especially at stress levels approaching lethality. An early example comes from an Oregon-R strain of *D. melanogaster* where FA of sternopleural bristle number was higher at the extreme temperature, 30 °C, than at 25 °C (Parsons, 1961, 1962), and a recent example comes from *Lacerta agilis* in parallel with the observations discussed above (Zakharov, 1989). An increase in FA suggests that normal development is upset at one or more of the molecular, chromosomal and epigenetic levels. There is now sufficient evidence to regard FA as a fitness measure whereby fitness is high when FA is low. In other words, the development of high FA in novel environments which are likely to be extreme implies reduced fitness in these environments.

The most stable state of biological systems corresponds to periods of minimum energy dissipation (Zotin, 1990). An immediate effect of a wide range of abiotic stresses is to increase energy metabolism implying a metabolic cost. Optimum development should therefore occur under environmental conditions corresponding to minimum total energy expenditure; under laboratory conditions this occurs at around 23 °C for each of the three ontogenetic stages (embryos, larvae, pupae) of *D. melanogaster* (Zotin, 1990). Similarly, metabolic efficiency, measured from the amount of energy required from a female to produce an average egg at various temperatures, is higher in the 22-25 °C range than at 18 °C and 28 °C (Arking *et al.*, 1988). Furthermore, maximum reproductive potential occurs in the intermediate temperature range which incorporates the region of minimum metabolic cost. Predictably, discretionary energy available for activities such as behavior, growth and reproduction in endotherms is often maximal at intermediate temperatures where metabolic rate tends to be lowest (Ames, 1980; Huey, 1991).

Free radicals are produced by the body as a normal byproduct of oxygen metabolism. Oxidative DNA damage by these highly reactive chemical species has been proposed to bring about damage to cell structure, physiological decrements, and to accelerate the aging process (Adelman, Saul & Ames, 1988). In this way, aging can be viewed as the result of long-term toxic side effects of normal metabolism (Cutler, 1991). The cell has an elaborate antioxidant defense system made up of enzymes and metabolites that inactivate free radicals. Simplistically, an inverse relationship between antioxidant defense activity and the amount of oxidative damage can be postulated, so that strains with a high antioxidant defense system should survive longer (Arking & Dudas, 1989). Under stressful environments where an immediate response is increased metabolic rate, the expectation therefore is elevated free radical production and an increase in abnormalities of all kinds including developmental ones.

Metabolic considerations therefore indicate that an expected response to environmental extremes is increased variability. While high variability is a prerequisite for rapid adaptation and evolutionary change, the fitness of such variants would normally be expected to be lower than optimal (Fisher

1930), thus limiting the potential for adaptive change in natural populations.

Conditions favoring evolutionary change

The most rapid morphological changes are likely for genes acting late in a developmental pathway because only a small subset of the larger developmental whole is affected. The more tightly integrated the developmental subsets, the more exacting the functional interdependence among structures, so that a genetic change should normally be deleterious. This agrees with Futuyma (1989), who suggested that stasis and punctuation in the fossil record should be most commonly observed in 'specially' adaptive characters such as morphological variants of particular structures, by comparison with 'generally' adaptive characters such as body size. In other words, genetic changes conferring general adaptation could be a small minority of those that occur and survive following natural selection.

Many laboratory selection experiments have been carried out on morphological variants of particular structures. In Figure 1, the response to selection for increased scutellar bristle number in *D. melanogaster* occurred in two periods of rapid response separated by a long period of stasis. Genetic experiments showed that the second period of rapid response was due to selection following recombination between two genes of major effect segregating in the founder natural population (MacBean, McKenzie & Parsons, 1971). More generally, relatively few genes of major effect often underlie responses to selection for morphological traits so that discontinuities and rapid responses to selection as in Fig. 1 are not uncommon (Lee & Parsons, 1968).

Apart from providing a reductionist genetic interpretation of discontinuities, selection experiment results are difficult to apply to conditions in nature, including the fossil record, since continuous directional selection is unlikely. Furthermore, optimal laboratory conditions are needed for the survival of extreme phenotypes which transcend the limits of variation normally occurring in a species in nature. The bizarre morphological characteristics of domesticated pigeon breeds noted by Darwin is an analogous situation, since they occur in ecological situations created by man where bizarre forms not normally surviving in nature are favored. Another

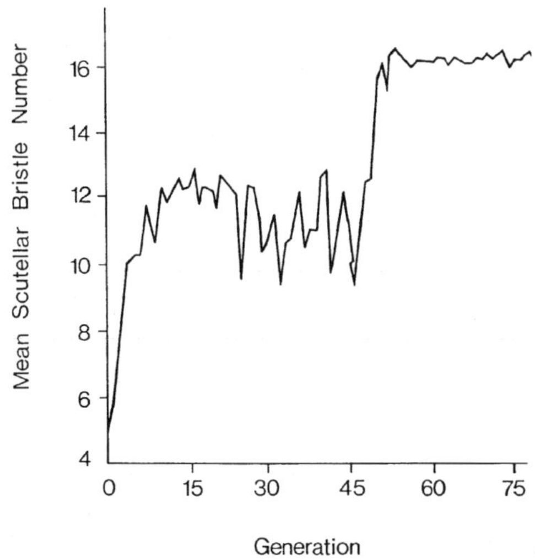

Fig. 1. The response to directional selection for increased scutellar bristle number in *D. melanogaster*. The data are based upon females over 75 generations of selection. Note the two accelerated responses lasting for very few generations in each case and the long plateau or stasis phases. (modified in Parsons, 1983 from MacBean, McKenzie & Parsons, 1971).

example of rapid evolution is the process of 'domestication' in the production of corn from teosinte which is consistent with molecular data indicating that as few as five genes account for 50% to 80% of this morphological change (Doebley *et al.*, 1990). However, fitness under domestication is likely to be reduced on a return to wild conditions, as shown by evidence for lower symmetry of the limb bones of domesticated birds compared with birds taken directly from the wild (Alexander *et al.*, 1984).

An example of rapid morphological change following a resource switch consequent upon drought is the observation of selection leading to large birds in the Galapagos finch (Boag & Grant, 1981). The rainfall returned the next year, indicating that periods of intense selection may occur intermittently in the wild as well as oscillating in direction. Low rates of evolution often found in the fossil record therefore greatly underestimate the rapidity with which populations can change over short periods of time (Parsons, 1992b; Sheldon, 1993).

It is in the colonization of empty islands or other unoccupied habitats and during phases of rapid population increase that organisms commonly diverge from their ancestors, transcending previously established limits of form and behavior. For example, in the butterfly *Melitaea aurinia*, associated with a rapid increase in numbers during 1920-24 'there was a most extraordinary outburst of variation, ... in size, colour, and marking'; this practically disappeared when the numbers became nearly constant in the following four years (letter of E. B. Ford to R. A. Fisher on page 197 of Bennett (1983)). These are relaxed conditions as under domestication, where conventional ecological limitations are reduced or absent, and there would almost certainly be reduced biotic stress from the usual complement of competitors and predators. Indeed in *Melitaea aurinia* above, E. B. Ford comments that parasitism of wild larvae was almost totally absent during the phase of rapid population increase but subsequently reappeared. Vermeij (1987) gives examples in various fish where a wide range of abnormalities survived when introduced into unoccupied habitats. New characters could evolve rapidly so that small populations in geographically isolated regions could become species (Futuyma, 1989). These are situations where adaptive precision is low so that increased diversity of phenotypes would ordinarily be possible (Frazzeta, 1970).

Rapid changes therefore can occur when ecological situations are permissive, especially in previously unoccupied habitats. During the Precambrian-Cambrian transition, substantial changes in ocean chemistry including the release of substantial quantities of phosphorous into shallow waters occurred (Conway Morris, 1987; Brasier, 1990). The influx of phosphorous is important because it acts as a limiting factor in marine productivity as well as being a vital component in cell biochemistry. In this way there would be evolutionary opportunity in an environment of increasing resource abundance, so that life radiated into empty spaces following the evolution of multicellular life. More generally, Liem (1990) emphasizes the likelihood that adaptive radiations depend upon the existence of energetic advantages. Similar arguments appear in Eldredge (1989).

However, any escalation of stress should restrict the evolution of extreme variants, because the energy available above that required for maintenance and survival is reduced. An extreme example comes from obligate cave animals in stressed, energy-poor and anoxic environments. Many such animals show troglomorphy, which in insects consists of a reduction and simplification of body parts such as compound eyes, ocelli and chitinous structures. Various explanations have been proposed (Barr, 1968); Hoch and Howarth (1989) favor energy economy, which is supported by the weak flying ability of these species and more generally their low energy level. In parallel, in marine basins, a diverse calcareous fauna of benthic invertebrates gives way to a low diversity, small, soft-bodied fauna as the oxygen level falls. One of several hypotheses put forward to explain the lack of hard parts in metazoans in low oxygen concentrations is based upon metabolic considerations since the production of the connecting tissue, collagen, is energetically expensive (Rhoads & Morse, 1971).

Major morphological innovation would be likely from small, common and generalist animals when resource availability is not restrictive. From such generalist animals, new forms could occur by crossing functional or metabolic thresholds. Physiological change could precede morphological change. For example, in desiccation-resistant strains of *D. melanogaster* formed by directional selection, no outward morphological changes were correlated with a fall in metabolic rate (Hoffmann & Parsons, 1989a). The primary target of stress selection in natural populations may therefore be physiological. This is consistent with the reductionist approach being taken here and elsewhere (Parsons, 1992a; 1993c). Only when a reaction norm (Schmalhausen, 1949) crosses certain critical functional thresholds would morphological change occur. This may be rapid since a canalization zone (Waddington, 1956) may be disrupted in the process. Following the crossing of such a threshold which could also involve a shift into a new adaptive zone, a period of diversification could follow. The fossil record therefore may lag behind an initial metabolic change, but this would be of little consequence in geological time.

Discussion

Can the limits be extended?

Whatever the environment, ultimately the potential for morphological change will become limited. Selection experiments for increased bristle number have been carried out in *D. melanogaster* to the level that a plateau, or stasis, occurs (Fig. 1). Whe selection ceases, bristle number often declines, indicating that high bristle number is an indicator of a negative correlation with fitness. In some cases, selection lines die out (Mather & Harrison, 1949). Artificial selection therefore may reach a limit because of fitness limitations; this is because genetic combinations not previously exposed to natural selection become exposed.

Assuming FA to be associated with fitness, a selection response is likely to be associated with increased FA, indicating reduced developmental homeostasis. This was found by Thoday (1958) in lines selected for high and low sternopleural bristle number in *D. melanogaster*. In this way FA may assist in monitoring when limits to selection are being approached, since there should be an upper limit to the developmental instability that an organism can tolerate. FA also has a role as a monitor of environmental stress (Section 3 and Parsons, 1992b) as shown in the field by higher FA in ecologically marginal species - border situations (Zakharov, 1989; Parsons, 1990). Since genomic and environmental stresses may be cumulative, or even synergistic, the reduced limits to selection for morphological traits expected under stress could be monitored by FA. Comparative selection experiments for morphological traits under benign and stressful environments therefore would be informative.

There are many examples of plateaus during selection, in some cases followed after some generations by a period of rapid response to a higher plateau as in Figure 1. While it is difficult to conclude that the second plateau in Figure 1 represents an ultimate limit, the variation between generations is much greater in the first than the second plateau, which suggests that major segregating bristle genes were present during the first but became fixed at the time of the second plateau (Parsons, 1983). Such a limit could be extended by migration of genes from another population, or from a major mutant arising *de novo* which may have difficulty in surviving or be initially deleterious (Fisher, 1930; Lande, 1983).

Assuming that climatic extremes are primary in determining species boundaries, can tolerance to extremes be extended? While resistance to environmental stresses such as insecticides has often been increased by selection, stresses directly applicable to climate have been less considered. However, Hosgood and Parsons (1968) found a polymorphism for heat resistance in a natural population of *D. melanogaster*, and Hoffmann and Parsons (1989a, b) rapidly increased desiccation resistance by selection. Selection for increased adult resistance to heat has been successful in insects (White, Debach & Garber, 1970; Stephanou & Alahiotis, 1983; Quintana & Prevosti, 1990; Huey, Partridge & Fowler, 1991) but responses have often been small (Hoffmann & Blows, 1993). Successful selection for increased resistance to heat has been carried out in agricultural animals such as beef cattle (Frisch, 1981). In the bacterium, *Escherichia coli*, replicate lines from a common ancestor previously propagated at 37 °C were propagated for 2000 generations at 32 °C, 37 °C or 42 °C, or a daily alternation between 32 °C and 42 °C. All four experimental groups showed improvement of relative fitness in their own environments in the ranking 42 °C > 32 °C = 32/42 °C > 37 °C, so that among the novel environments adaptation is most rapid at the highest temperature 42 °C (Bennett, Lenski & Mittler, 1992). At 42 °C, mutations favoring fitness improvement are not rare and can spread rapidly through a population (Bennett, Dao & Lenski, 1990). Therefore, some biological systems can evolve in response to changes in environmental factors such as temperature. In this case the most dramatic response was at the extreme temperature, when genetic variability would be expected to be highest. However, adaptation to 42 °C did not much extend the upper limit for population persistence.

Since populations from ecological margins should be selected for resistance to environmental stress, they should contain genotypes useful in countering stressful periods. The possibility of generalized stress resistance associated with reduced metabolic rate means that genotypes selected for extremes of one environmental stress should be resistant to other stresses (Hoffmann & Parsons, 1991; Parsons, 1991). However, even though FA studies indicate the likelihood of much morphological variability at species boundaries, major developmental and morphological shifts are unlikely, especially if metabolically costly. This follows from the point that the extent to which the limits to selection for morphological traits can be modified can be regarded as inversely related to the energetic cost of stress. Bizarre and extreme morphological variants, therefore, would be more likely to survive in less stressful more central habitats where the premium upon survival is reduced compared with the abiotic stresses characteristic of species borders. Much developmental variability in natural populations may therefore have little evolutionary significance. Only when ecological limitations are relaxed would traits persist that would normally be inadmissable. This could lead to a punctuated evolutionary pattern with rapid change at a time of ecological relaxation, but under much lengthier periods of more intense stress, stasis would prevail.

Acknowledgements

I am grateful to J. H. Bennett for drawing my attention to matters on the theme of this paper that are presented in his 1983 book on the correspondence of R. A. Fisher.

References

Adelman, R., R. L. Saul & B. N. Ames, 1988. Oxidative damage to DNA: relation to species metabolic rate and life span. Proc. Natl. Acad. Sci. USA 85: 2706-2708.

Alexander, R. McN., A. Brandwood, J. D. Currey & A. S. Jayes, 1984. Symmetry and precision of control of strength in limb bones of birds. J. Zool. 203: 135-143.

Ames, D., 1980. Thermal environment affects production efficiency of livestock. BioScience 30: 457-460.

Arking, R., S. Buck, R. A. Wells & R. Pretzalff, 1988. Metabolic rates in genetically based long lived strains of *Drosophila*. Exp. Geront. 23: 59-76.

Arking, R. & S. P. Dudas, 1989. Review of genetic investigations into the aging processes of *Drosophila*. J. Amer. Geriatric Soc. 37: 757-773.

Barr, T. C., 1968. Cave ecology and the evolution of troglobites. Evol. Biol. 2: 35-102.

Bennett, A. F., K. M. Dao & R. E. Lenski, 1990. Rapid evolution in response to high temperature. Nature 346: 79-81.

Bennett, A. F., R. E. Lenski & J. E. Mittler, 1992. Evolutionary adaptation to temperature. I. Fitness responses of *Escherichia coli* to changes in its thermal environment. Evolution 46: 16-30.

Bennett, J. H., 1983. Natural Selection, Heredity, and Eugenics including selected correspondence of R. A. Fisher with Leonard Darwin and others. Clarendon Press, Oxford.

Boag, P. T. & P. R. Grant, 1981. Intense natural selection in a population of Darwin's finches (Geospizinae) of the Galapagos. Science 214: 82-85.

Bozinovic, F. & M. Rosemann, 1989. Maximum metabolic rate of rodents: physiological and ecological consequences on distributional limits. Funct. Ecol. 3: 173-181.

Brasier, M. D., 1990. Nutrient flux and the evolutionary explosion across the Precambrian-Cambrian boundary interval. Hist. Biol. 5: 850-93.

Conway Morris, S., 1987. The search for the Precambrian-Cambrian boundary. Amer. Sci. 75: 157-167.

Coope, G. R., 1979. Late Cenozoic fossil Coleoptera: evolution, biogeography, and ecology. Ann. Rev. Ecol. Syst. 10: 247-267.

Cutler, R. G., 1991. Antioxidants and aging. Am. J. Clin. Nutr. 53: 373S-379S.

Darwin, C., 1859. On the Origin of Species by means of Natural Selection. Murray, London.

Derr, J. A., 1980. The nature of variation in life history characters of *Dysdercus bimaculatus* (Heteroptera: Pyrrhocoridae), a colonizing species. Evolution 34: 548-557.

Doebley, J., A. Stec, J. Wendel & M. Edwards, 1990. Genetic and morphological analysis of a maize-teosinte F_2 population: implications for the origin of maize. Proc. Natl. Acad. Sci. USA 87: 9888-9892.

Dunson, W. A. & J. Travis, 1991. The role of abiotic factors in community organization. Amer. Nat. 138: 1067-1091.

Eldredge, N., 1989. Macroevolutionary Dynamics. McGaw-Hill, New York.

Endler, J. A., 1986. Natural Selection in the Wild. Princeton Univ. Press, Princeton.

Fisher, R. A., 1930. The Genetical Theory of Natural Selection. Clarendon Press, Oxford.

Frazzetta, T. H., 1970. From hopeful monsters to bolyerine snakes. Amer. Nat. 104: 55-72.

Frisch, J. E., 1981. Changes occurring in cattle as a consequence of selection for growth rate in a stressful environment. J. Agric. Sci. 96: 23-38.

Futuyma, D. L., 1989. Evolutionary Biology. Sinauer Associates, Sunderland, Massachusetts.

Hoch, H. & F. G. Howath, 1989. Reductive evolutionary trends in two new cavernicolous species of a new Australian cixiid genus (Homoptera: Fulgoroidea). Syst. Entomol. 14: 179-196.

Hoffmann, A. A. & M. W. Blows, 1993. Evolutionary genetics and climatic change: will animals adapt to global warning? pp. 165-178 in Biotic Interactions and Global Change, edited by P. M. Kareiva, J. G. Kingsolver & R. B. Huey. Sinauer Associates, Sunderland, Mass.

Hoffmann, A. A. & P. A. Parsons, 1989a. An integrated approach to environmental stress tolerance and life-history variation. Desiccation tolerance in *Drosophila*. Biol. J. Linn. Soc. 37: 117-136.

Hoffmann, A. A. & P. A. Parsons, 1989b. Selection for increased desiccation resistance in *Drosophila melanogaster*: additive genetic control and correlated responses for other stresses. Genetics 122: 837-845.

Hoffmann, A. A. & P. A. Parsons, 1991. Evolutionary Genetics and Environmental Stress. Oxford University Press, Oxford.

Holloway, G. J., R. M. Sibly & S. R. Povey, 1990. Evolution in toxin-stressed environments. Funct. Ecol. 4: 289-294.

Hosgood, S. M. W. & P. A. Parsons, 1968. Polymorphism in natural populations of *Drosophila* for the ability to withstand temperature shocks. Experientia 24: 727-728.

Houle, D., 1992. Comparing evolvability and variability of quantitative traits. Genetics 130: 185-204.

Howarth, F. G., 1991. Hawaiian cave faunas: macroevolution on young islands. vol I, pp. 285-295 in The Unity of Evolutionary Biology, edited by E. C. Dudley. Dioscorides Press Portland.

Howarth, F. G. & F. D. Stone, 1990. Elevated carbon dioxide

levels in Bayliss Cave, Australia: implications for the evolution of obligate cave species. Pacific Science 44: 207-218.

Huey, R. B., 1991. Physiological consequences of habitat selection. Amer. Nat. 137: S91-S115.

Huey, R. B., L. Partridge & K. Fowler, 1991. Thermal sensitivity of Drosophila melanogaster responds rapidly to laboratory natural selection. Evolution 45: 751-756.

Hurst, R. J., P. D. Watts & N. A. Oritsland, 1991. Metabolic compensation in oil-exposed polar bears. J. Therm. Biol. 16: 53-57.

Jablonski, D., K. W. Flessa & J. W. Valentine, 1985. Biogeography and paleobiology. Paleobiol. 11: 75-90.

Lande, R., 1983. The response to selection on major and minor mutations affecting a metrical trait. Heredity 50: 47-65.

Lee, B. T. O. & P. A. Parsons, 1968. Selection, prediction and response. Biol. Rev. 43: 139-174.

Liem, K. F., 1990. Key evolutionary innovations, differential diversity, and symecomorphosis, pp. 147-170 in Evolutionary Innovations, edited by M. H. Nitecki. Univ. Chicago Press, Chicago.

MacBean, I. T., J. A. McKenzie & P. A. Parsons, 1971. A pair of closely linked genes controlling high scutellar number in *Drosophila*. Theoret. Appl. Genet. 41: 227-235.

Mather, K. & B. J. Harrison, 1949. The manifold effect of selection. Heredity 3: 1-52, 131-162.

Osgood, D. W., 1978. Effects of temperature on the development of meristic characters in *Natrix fasciata*. Copeia 1: 33-47.

Palmer, A. R. & C. Strobeck, 1986. Fluctuating asymmetry: measurement, analysis, patterns. Annu. Rev. Ecol. Syst. 17: 391-421.

Parsons, P. A., 1961. Fly size, emergence time and sternopleural chaeta number in *Drosophila*. Heredity 16: 455-473.

Parsons, P. A., 1962. Maternal age and developmental variability. J. Exp. Biol. 39: 251-260.

Parsons, P. A., 1974. Genetics of resistance to environmental stresses in *Drosophila* populations. Annu. Rev. Genet. 7: 239-263.

Parsons, P. A., 1982. Acetic acid vapour as a resource and stress in *Drosophila*. Aust. J. Zool. 30: 427-433.

Parsons, P. A., 1983. The genetic basis of quantitative traits: evidence for punctuational evolutionary transitions at the intraspecific level. Evol. Theory 6: 175-184.

Parsons, P. A., 1987. Evolutionary rates under environmental stress. Evol. Biol. 21: 311-347.

Parsons, P. A., 1988. Evolutionary rates: effects of stress upon recombination. Biol. J. Linn. Soc. 35: 49-68.

Parsons, P. A., 1989. Conservation and global warming: a problem in biological adaptation to stress. Ambio 18: 322-325.

Parsons, P. A., 1990. Fluctuating asymmetry: an epigenetic measure of stress. Biol. Rev. 65: 131-145.

Parsons, P. A., 1991. Evolutionary rates: stress and species boundaries. Annu. Rev. Ecol. Syst. 22: 1-16.

Parsons, P. A., 1992a. Evolutionary adaptation and stress: the fitness gradient. Evol. Biol. 26: 191-223.

Parsons, P. A., 1992b. Fluctuating asymmetry: a biological monitor of environmental and genomic stress. Heredity 68: 361-364.

Parsons, P. A., 1993a. The importance and consequences of stress in natural populations: from life-history variation to evolutionary change. Amer. Nat. 142: 55-520.

Parsons, P. A., 1993b. Stress, extinctions and evolutionary change: from living organisms to fossils. Biol. Rev. 68: 313-333.

Parsons, P. A., 1993c. Stress, metabolic cost and evolutionary change: from living organisms to fossils, pp. 140-156 in Evolutionary Patterns and Processes, edited by D. R. Lees & D. Edwards. Academic Press, London.

Quintana, A. & A. Prevosti, 1990. Genetic and environmental factors in the resistance of *Drosophila subobscura* adults to high temperature shock. Theor. Appl. Genet. 80: 847-851.

Rhoads, D. C. & J. W. Morse, 1971. Evolutionary and ecologic significance of oxygen-deficient marine basins. Lethaia 4: 413-428.

Root, T., 1993. Effects of global climate change on North American birds and their communities, pp. 280-292 in Biotic Interactions and Global Change, edited by P. M. Kareiva, J. G. Kingsolver & R. B. Huey. Sinauer Associates, Sunderland, Mass.

Schmalhausen, I. I., 1949. Factors of Evolution. Blakiston, Philadelphia.

Sheldon, P. R., 1993. Making sense of microevolutionary patterns, pp. 20-31 in Evolutionary Patterns and Processes, edited by D. R. Lees & D. Edwards. Academic Press, London.

Stephanou, G. & S. N. Alahiotis, 1983. Non-mendelian inheritance of 'heat-sensitivity' in *Drosophila melanogaster*. Genetics 103: 93-107.

Thoday, J. M., 1958. Homeostasis in a selection experiment. Heredity 12: 401-415.

Upchurch, G. R., 1989. Terrestrial environmental changes and extinction patterns at the Cretaceous-Tertiary boundary, North America, pp. 195-216 in Mass Extinctions: Processes and Evidence, edited by S. K. Donovan. Belhaven Press, London.

Valentine, J. W. & D. Jablonski, 1991. Biotic effects of sea level change: the Pleistocene test. J. Geophys. Res. 96: 6873-6878.

Vermeij, G. J., 1987. Evolution and Escalation: an Ecological History of Life. Princeton University Press, Princeton.

Waddington, C. H., 1956. Principles of Embryology. George Allen & Unwin, London.

Wallace, B., 1981. Basic Population Genetics. Columbia Univ. Press, New York.

Ward, P. D., 1992. On Methuselah's Trail: Living Fossils and Great Extinctions. W. H. Freeman, New York.

Westerman, J. M. & P. A. Parsons, 1973. Variation in genetic architecture at different doses of γ-radiation as measured by longevity in *Drosophila melanogaster*. Canad. J. Genet. Cytol. 15: 289-298.

White, E. B., P. Debach & M. J. Garber, 1970. Artificial selection for genetic adaptation to temperature extremes in *Aphytis lingnanensis* (Hymenoptera: Aphelinidae). Hilgardia 40: 161-192.

Zakharov, V. M., 1989. Future prospects for population phenogenetics. Sov. Sci. Rev. F. Physiol. Gen. Biol. Revs. 4: 1-79.

Zotin, A. I., 1990. Thermodynamic Bases of Biological Processes: Physiological Reactions and Adaptations. Walter de Gruyter, Berlin.

T. A. Markow (ed.), Developmental Instability: Its Origins and Evolutionary Implications, 257–267, 1994.

Parasites increase fluctuating asymmetry of male *Drosophila nigrospiracula*: implications for sexual selection

Michael Polak
Department of Zoology, Arizona State University, Tempe AZ 85287-1501, USA

Received and accepted 26 April 1993

Abstract

Fluctuating asymmetry (minor deviations from perfect bilateral symmetry) is manifested by individuals less able to buffer environmental stress during development. I utilized a system of two naturally-occurring parasites of *Drosophila nigrospiracula* to test whether parasitic infection during host development yields elevated degrees of fluctuating asymmetry in two morphological traits of males. This hypothesis has important implications for sexual selection, as it may explain why asymmetric males are often found to be sexually disadvantaged. In my system, nematodes infect larvae and therefore are more likely to disrupt development than mites which only parasitize adult flies. As predicted, nematode-infected male *D. nigrospiracula* had a higher degree of bristle asymmetry than did mite-infested and control (carrying neither parasite) males. There was also a significant relation between nematode number and degree of asymmetry. There was a significant negative relation between nematode load and size of adult males, implicating a causal link between nutritional stress during host development and fluctuating asymmetry. Patterns of wing length asymmetry were inconsistent with those of bristle asymmetry. Nematode-infected males did not differ in wing length asymmetry relative to mite-infested and control males, nor was there a significant relation between nematode number and wing asymmetry. This inconsistency in expression of asymmetry may reflect different intensities of selection operating on each morphological trait.

Introduction

Bilaterally symmetrical traits of animals who experience ideal conditions during development are expected, in theory, to be expressed identically on the two sides of the body (Thoday, 1956; Soulé & Cuzin-Roudy, 1982). Perfect bilateral symmetry is expected because traits on the two sides are the product of the same genotype. Any random departure from perfect symmetry, therefore, has been interpreted to indicate reduced homeostasis or buffering of developmental processes (Thoday, 1956; Waddington, 1957). Fluctuating asymmetry (FA) is a measure of such random deviations from perfect bilateral symmetry in a morphological trait (Van Valen, 1962). Because FA may result from exposure to harmful environmental factors, such as chemical pollutants (Valentine, Soulé & Samallow, 1973), and/or from increased degrees of homozygosity (Soulé, 1982; Leary, Allendorf & Knudsen, 1984; Mitton & Grant, 1984; Leary & Allendorf, 1989; Clarke, 1992), it is, therefore, not surprising that high FA is often associated with reduced fitness at the individual level, and with greater probability of extinction at the population level (Leary & Allendorf, 1989).

High FA is also associated with reduced male reproductive success (RS) (Markow, 1987; Markow & Ricker, 1992; Thornhill, 1992). For example, Møller (1992a) found that male swallows (*Hirundo rustica*) who exhibited reduced FA in their outer-most tail feathers mated earlier in the breeding season and achieved greater annual reproductive success than less symmetrical birds. Møller (1992a) argued that females who discriminate in favour of low-FA males benefit by transmitting to their offspring genes that enhance viability. According to this hypothesis, males vary in the expression of FA in certain characters due to differential heritable ability to cope with environmental contingencies during development. This hypothesis may be subsumed under 'good genes' models of sexual selection, and is currently most often discussed by evolutionary biologists (Balmford & Thomas, 1992; Borgia & Wilkinson, 1992; Møller & Höglund, 1991; Møller, 1992a, b). However, the nature of the environmental stressor(s) which causes FA, and therefore the precise benefits accrued by FA-discriminating females, remain unidentified.

One possible stressor which could increase FA is inferior nutrition during development of a character, especially a character expressed at its physiological limits due to intense sexual selection. Alternatively, parasitic infection, perhaps exacerbated by malnutrition, may be causally related to FA. In fact, Bailit *et al.* (1970) suggest that poor nutrition and disease infection may increase FA in human dentition, and Livshits and Kobyliansky (1991) have suggested that maternal infectious diseases carried during pregnancy could account for high FA in infants.

In the present study I tested whether parasitic infection during development increases FA. Although the role of parasites in sexual selection is currently under intense study (reviewed by Clayton, 1991), any link between parasitic infection and FA is unidentified. A causal relationship between parasitic infection and high FA could be of considerable importance to the study of sexual selection because several hypotheses consequently arise that may explain why asymmetrical males suffer mating disadvantages, as mentioned above. These hypotheses will be discussed in a later section, but two that involve female choice are mentioned here. First, if the ability to resist parasitic infection has a genetic basis, as is often found to be the case (Steinhaus, 1949; Woodard & Fukuda, 1977; Petersen, 1978; Wakelin, 1985; Watanabe, 1987), females who choose symmetrical males as mates could benefit through transmission of resistance to offspring. This possibility parallels the Hamilton and Zuk (1982) hypothesis which has received considerable support, both from inter- and intra-specific studies (Read, 1987, 1988; Ward 1988; Clayton 1991; Zuk, 1991). According to this hypothesis, females discriminate in favour of unparasitized males on the basis of the degree of elaboration and coloration of secondary sexual characters.

According to a second female-choice hypothesis, females may prefer to mate with low-FA males because they gain direct benefits. For example, females may discriminate in favour of symmetrical males to avoid direct transmission of sexually-transmittable parasites to themselves and/or to offspring, or because unparasitized males are better able to provide quality territories or paternal care.

A fundamental assumption of these aforementioned hypotheses is that parasitic infection during development increases host FA. I tested this assumption by examining *Drosophila nigrospiracula* Patt. and Wheeler and two of its parasites, allantonematid nematodes and macrochelid mites. Both parasites co-occur with their drosophilid host in nature. Because nematodes infect *D. nigrospiracula* during its larval stages and mites infest adult flies only (see below), I expected that only nematodes could elevate host FA. Thus, I generated the following predictions from the hypothesis that parasitic infection during development can perturb development and increase FA. First, nematode-infected male *D. nigrospiracula* will exhibit greater FA than mite-infested and control (unparasitized by either parasite) flies sampled from the same population. In the event of a response to infection, I also predicted a positive relationship between the severity of infection and the magnitude of FA.

That both nematode- and mite-parasitized flies will exhibit greater FA than control flies is a prediction of an alternative hypothesis, namely that susceptibility to parasitization is a manifestation of inbreeding depression, and that individuals are more asymmetric because they are more homozygous (Reeve, 1960; Soulé, 1979; Mitton & Grant, 1984; Palmer & Strobeck, 1986). This alternative hypothesis, therefore, assumes no causal link between parasitic infection and FA.

Biology of nematodes

The infective life stage of an allantonematid nematode begins when a fertilized female nematode enters a fly larva. Entry is achieved by direct penetration of the cuticle with the aid of a spear-like stylet and associated pharyngeal gland (Poinar, 1983). These nematodes are nutritionally specialized parasites, and once inside their host, they feed from larval hemocoel by absorbing food directly through their body wall (Welch, 1959; Poinar, 1983). Nematodes achieve maximum size during host pupation at which stage they begin deposition of eggs into the host hemocoel (Poinar, 1983). Damage to flies due to nematode infection can be considerable. Fat bodies, especially those associated with the gonads, are reduced, and fewer eggs are produced by adult flies (Welch, 1959; Jaenike, 1992). For example, all female *D. putrida* infected by nematodes invariably suffer complete sterility (Jaenike, 1992). These findings suggest that nematodes compete with their host for nutritive material, and, as a consequence, the parasites may affect larval biochemical processes, disrupt development, and augment FA. Nutritional deficiencies, or the lack of certain constituents in the diet, are known to affect growth, development, and energy production in insect hosts (Steinhaus, 1949; Sang, 1956). In fact, reduced bioenergetic efficiency is associated with high FA in the sulfur butterfly, *Colias philodice* (Watt, Cassin & Swan, 1983; Leary & Allendorf, 1989).

Biology of mites

Macrocheles subbadius (Berlese) is a cosmopolitan mite associated with several dipteran species (G. W. Krantz, personal communication). Female mites attach to adult flies of both sexes by embedding the entire length of their chelicerae into the fly's abdominal tissue. This mode of attachment suggests that mites actually feed from the host hemocoel rather than assuming an entirely passive phoretic role, as generally believed. In fact, a laboratory study has shown that these mites significantly reduce longevity of both male and female *D. nigrospiracula*, and field work has demonstrated that attached mites have a marked effect on male mating success (Polak & Markow, manuscript in preparation). Thus, these mites exert intense selection on their adult hosts.

Materials and methods

Field collection of hosts

Drosophila nigrospiracula is endemic to the Sonoran desert of North America and breeds in necrotic saguaro cacti, *Carnegiea gigantea*. Male flies gather on the outside skin of these cacti where they court and mate with females who then oviposit in the rotting cactus tissue (Markow, 1988). Samples of male *D. nigrospiracula*, that included nematode-infected and mite-infested males, as well as males unparasitized by either nematodes or mites, were acquired by randomly collecting flies from necrotic saguaro cacti at five different sites within 40 mi of Phoenix, Arizona, USA, between 23 June and 22 October, 1992. Each site was visited 3 - 8 times. On each visit, approximately 25 males were collected from their mating arenas situated on the outer skin of the necrotic saguaro, gently aspirated into 8-dram vials containing banana-agar fly medium, and returned to the laboratory.

Scoring parasites and measuring FA

On the same day of collection, I anesthetized individual males in the laboratory using ether or CO_2 and counted mites attached to each male. Of all mites counted on males, 95% were found attached to the ventral surface of the abdomen near the junction with the thorax, although I occasionally found mites clinging to the neck or face of flies. After counting and removing mites, I measured thorax length of flies to estimate body size (Robertson & Reeve, 1952), removed the front coxae to expose bristles, and counted all bristles, including the sternopleurals, on the second preepisternum of each side of the body (Demerec, 1965). Bristles were counted using a Wild M3Z dissecting microscope. Sternal bristles were not counted. Bristle FA was calculated as the value of the right side minus the value on the left side of the body (R-L).

Using fine forceps, I carefully excised each wing and placed it on double-sided transparent tape adhered to a 35mm glass projection slide. Wings were firmly flattened against the tape using a clean piece of glass. Following the removal of wings, each fly was immersed into 2-3 drops of physiological saline (Shen's solution) on a depression slide and dissected to determine the number of mature nematodes in the body cavity. Dissections were performed under the Wild M3Z microscope. In saline, mature female nematodes, which are distinguishable by their conspicuous sausage-like shape, remained immobile and sank to the bottom of the depression slide. These females were commonly observed to contain eggs. When all dissections were completed, wing measurements were made on digitized images of magnified wings without knowledge of male infection status. The length measured was from the anterior crossvein to the junction of the second longitudinal vein with the outside margin of the wing. Three repeated measures were made of each wing using microcomputer

Optimas software (BioScan Inc.). As for bristle FA, wing-length FA was calculated as the wing length value on the right minus the value on the left (R-L). Because wing length was strongly related to thorax length (slope = 35.6, $F_{(1, 159)}$ = 259.2, P = 0.0001, R^2 = 0.63), I used wing length as an estimator of male body size throughout the study.

Three of the males dissected contained hundreds of larval nematodes distinguished by their thread-like appearance and sustained whip-like movements. Because I was unable to identify mature nematodes among the writhing mass of juveniles, measurements made on these three males were not included in analyses requiring exact counts of female nematodes. In addition, occasional samples (N = 4) were comprised exclusively of males uninfected by nematodes. Measurements made on males from these samples were excluded from all statistical analyses. Males (N = 2) that were parasitized by both types of parasite were also discarded.

Statistical analyses

Van Valen (1962) distinguishes three modes of asymmetry, directional asymmetry (DA), antisymmetry (AS), and fluctuating asymmetry (FA). DA and AS are products of genetical or developmental processes, while FA reflects environmentally-induced perturbations that occur during an organism's development. Thus, in order to evaluate true FA, one must obtain a measure independent of DA and AS. For a trait exhibiting true FA, variation of signed (R-L) values should be normally distributed with a mean equalling zero (Van Valen, 1962; Palmer & Strobeck, 1992). Departure from mean zero identifies DA, while non-normal distributions may reflect several population phenomena (Palmer & Strobeck, 1992), including the presence of AS (handedness) (e.g. Yamaguchi, 1977). To evaluate whether each trait used in this study exhibited true FA, I calculated means, 95% confidence intervals of means, skewness, and kurtosis for signed (R-L) values of each trait.

Another factor confounding analyses of asymmetry is the presence of an association between the magnitude of a trait and the degree of asymmetry (Palmer & Strobeck, 1986). Thus, in order to detect size dependence in bristle and wing length FA, I regressed trait asymmetry (absolute value of R-L) on the magnitude of each trait. In cases where size dependence was found, I scaled out the effect of size by dividing FA by the average size of the trait (R+L/2).

To assess difference in bristle FA among the three experimental groups of males, I conducted a Kruskal-Wallis test on absolute values of (R-L) calculated separately for each male. Wing length FA for each group was obtained from a two-way analysis of variance in which 'side of body' is one factor (fixed) and males the other (random) (Palmer & Strobeck, 1986). Because three repeated measurements of each wing were taken in this study, this analysis allows the variance due to measurement error to be subtracted from the interaction (sides × males) variance. The resultant variance, divided by three (number of replicate measures made on each wing, was the estimate of FA for each male group. The Bartlett's and F_{MAX} tests for homogeneity of variances were used to assess differences among groups of males. As an alternative method of distinguishing FA among groups, I calculated wing length FA as the absolute value of (mean R - mean L) for each male within groups, and tested for group differences by Kruskal-Wallis analysis of variance.

To determine whether there was an association between parasite load and FA of each trait, I performed regression analyses of scaled FA on parasite number. Assumptions of these analyses were evaluated by inspecting plots of the residuals versus the independent variable (FA) (Neter, Wasserman & Kutner, 1990).

Results

Mean parasite loads of those males that were parasitized were 1.9 nematodes (SE = 0.24, N = 24, range 1-5) and 3.1 mites (SE = 0.32, N = 51, range 1-10). Mean bristle number and mean wing length for each group of males are presented in Table 1. From one-way ANOVA, neither mean bristle number ($F_{(2, 199)}$ = 0.62, P = 0.54) nor mean wing length ($F_{(2, 175)}$ = 0.22, P = 0.81) was significantly different among the three groups of males (Table 1). However, bristle number showed a two-fold greater coefficient of variation than wing length, and values were repeatable among groups (Table 1). These coefficients are independent of the mean and thus serve to compare relative amounts of variation between traits (Sokal & Rohlf, 1981).

Table 1. Means (SE, N) and coefficients of variation (CV) in both bristle number and wing length measures of parasitized and control (uninfected) groups of male *Drosophila nigrospiracula*. FA represents absolute values of R-L.

	Nematodes	Mites	Control	Pooled sample
Bristles				
Per side	18.13 (0.40, 27)	18.51 (0.29, 51)	18.63 (0.19, 124)	18.55 (0.147, 202)
CV (per side)	11.44	11.34	11.35	11.31
FA	2.36 (0.26, 27)	1.29 (0.171, 51)	1.45 (0.108, 124)	1.54 (0.87, 202)
Wings				
Length (mm)	1.56 (0.02, 24)	1.58 (0.015, 44)	1.58 (0.08, 109)	1.58 (0.007, 177)
CV (length)	6.12	6.19	5.21	5.57
FA	0.012 (0.001, 24)	0.011 (0.001, 44)	0.011 (0.001, 109)	0.011 (0.001, 177)

Bristle number and wing length FA met the requirements of reliable indicators of developmental instability (Palmer & Strobeck, 1992). For pooled samples of males, the 95% confidence interval of each (R - L) mean included zero, and no significant skewness was detected in the (R - L) distribution of either trait (Table 2). Although bristle FA did not exhibit kurtosis, wing asymmetry was significantly leptokurtic (Table 2). Leptokurtosis indicates that fewer values fall within intermediate regions relative to a normal distribution (Sokal & Rohlf, 1981).

There was no relation between the magnitude of bristle FA and character size (bristles per side) (slope = 0.08, $F_{(1, 200)}$ = 0.40, P = 0.53), nor was there a relation between bristle FA and male thorax length (slope = - 0.004, $F_{(1, 178)}$ = 0.02, P = 0.88). However, wing length FA exhibited some size dependence; there was a negative and marginally insignificant relation between wing length FA and wing length (slope = - 0.03, $F_{(1, 176)}$ = 3.61, P = 0.058).

There was a significant negative relationship between male thorax length and nematode load (Fig. 1), suggesting that nematodes make considerable nutritional demands during larval development of flies (the relation between thorax length and mite load was insignificant (slope = - 0.22, $F_{(1, 19)}$ = 0.72, P = 0.40). I tested whether fluctuating asymmetry in the two characters was greater in nematode-infested males than in mite-infested and control (unparasitized) males. As predicted, bristle FA was significantly greater among nematode-infected males than among mite-infested and control males

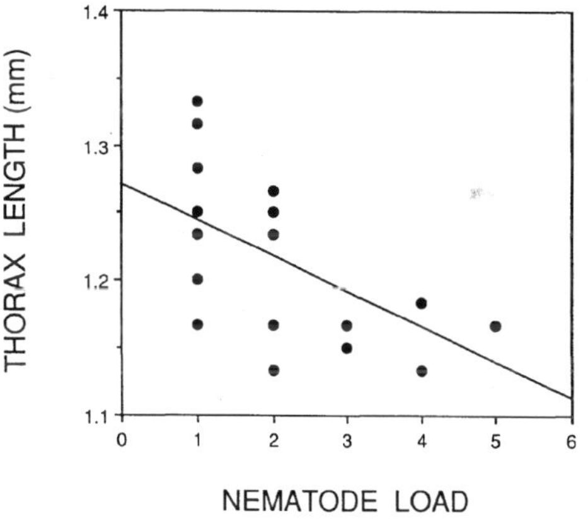

Fig. 1. Negative relationship between thorax length and the number of mature nematodes found within male *D. nigrospiracula*

Table 2. Mean, 95% confidence interval (CI), skewness and kurtosis in bristle number and wing length asymmetry (signed R-L) of male *Drosophila nigrospiracula*.

Trait	Mean FA	N	95% CI	Skewness (P)	Kurtosis (P)
Bristles	0.144	202	-0.130, 0.417	0.037 (> 0.5)	-0.264 (> 0.2)
Wing length	-0.002	177	-0.004, 0.0001	0.008 (> 0.5)	1.29 (< 0.001)

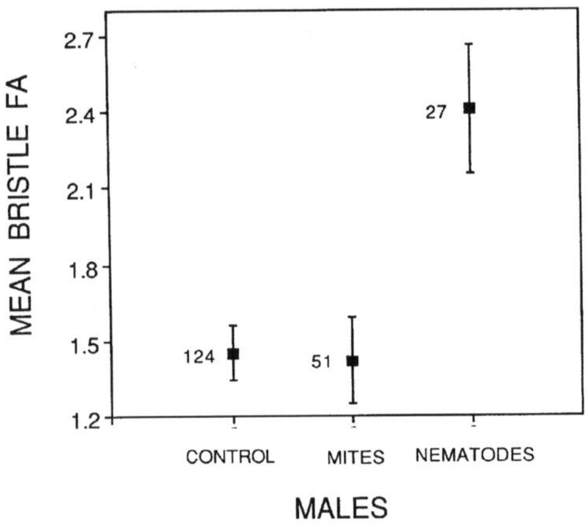

Fig. 2. Mean absolute FA values for thoracic bristles in three groups of male *D. nigrospiracula*. Nematode-infected males were significantly more asymmetric than mite-infested and control (unparasitized) males (Kruskal-Wallis H = 13.29, P = 0.0013). Means are presented in Table 1. Sample sizes are provided in the figure.

(Fig. 2). Moreover, the response of FA to nematode infection was dose-dependent; among males harboring nematodes, scaled bristle FA was significantly related to the number of mature nematodes within flies (Fig. 3), and, as predicted, scaled FA was unrelated to the number of attached mites (Table 3, Fig. 3).

Results of two-way analyses of variance computed on wing length asymmetry are shown in

Table 3. Results of regression analyses of bristle number and wing length FA (scaled for character size) on nematode and mite loads in male *Drosophila nigrospiracula*.

	Slope	F (df)	P	R^2
Bristle FA				
nematodes	0.03	5.46 (1, 20)	0.03	0.21
mites	0.003	0.56 (1, 49)	0.48	0.01
Wing length FA				
nematodes	-0.002	0.09 (1, 20)	0.76	0.01
mites	-0.001	0.35 (1, 42)	0.55	0.01

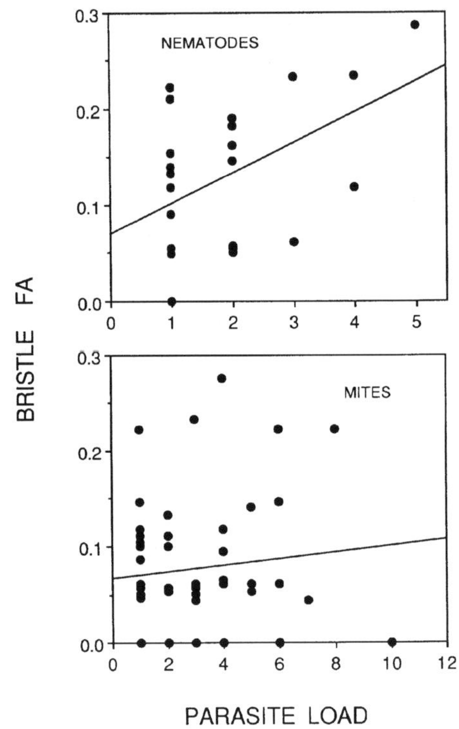

Fig. 3. Relationship between absolute value of bristle FA and number of nematodes and mites parasitizing male *D. nigrospiracula*.

Table 4. Mean squares (ms, × 10^5) from two-way analyses of variance of wing length values from parasitized and control groups of male *Drosophila nigrospiracula*. These means squares were used to calculate the error ms as a percentage of interaction ms (non-directional asymmetry), and the variance component representing FA.

	Error	Interaction	Error/Interaction (× 100)	Fa*	df**
Nematodes	0.29	24.84	1.2%	8.18	40
Mites	0.34	31.81	1.1%	10.49	42
Control	0.31	28.67	1.1%	9.45	106

* Bartletts's and F_{max} tests for homogeneity among FA variances, B = 0.544, 0.90 < P < 0.95, and F = 1.28, P > 0.05, respectively.
** Degrees of freedom associated with FA calculated by the Satterthwaite formula.

Table 4. No difference in FA among the three groups was detected by either the Bartlett's or F_{MAX} test (Table 4). Because the relative contributions of measurement error were small relative to FA for each male group (< 2% for each group, Table 4), I re-calculated FA as the absolute value of (mean R - mean L) for each male, and scaled these values to uncouple FA from character size. Consistent with two-way ANOVA, there was no difference in scaled FA among groups (Kruskal-Wallis H = 2.38, P = 0.30). Among males parasitized by nematodes and mites, there was no relation between scaled wing length FA and severity of parasitization by either parasite (Table 3).

Discussion

Parasites, FA and sexual selection

Results of the present study support the novel hypothesis that parasitic infection during host development augments host developmental instability, and hence FA. Bristle FA was highest among male *D. nigrospiracula* infected with allantonematid nematodes relative to individuals burdened with an adult-infesting mite and unparasitized by either nematodes or mites. Moreover, FA in males was dose-dependent; there was a significant relation between nematode number and the magnitude of FA. Below, I discuss how my findings may be used to explain why asymmetrical males of certain species suffer reduced reproductive success.

One alternative to the hypothesis of a causal mechanism connecting elevated FA with parasitism is that individual flies harbor more parasites due to inbreeding depression and are therefore also more asymmetrical. However, there are two lines of evidence that cast doubt on this alternative. First, there was no relation between FA and the probability of mite infestation, and second, two-generation inbred (sib-mated) *D. nigrospiracula* did not exhibit higher bristle FA than outbred, control lines of flies (Polak, unpublished data).

There are two lines of evidence that make it reasonable to expect that, in *D. nigrospiracula*, low FA indicates heritable ability to resist parasitic infection. First, in some drosophilids, infection by allantonematid nematodes can reduce male mating success and female fertility to the point of complete sterility (Jaenike, 1988, 1992). These results are important, because together with an inverse relationship between nematode number and adult host size found in the present study, they suggest that nematodes represent a strong agent of selection in *Drosophila*. Such strong selective pressures make it reasonable to expect an evolved host response against infection. In fact, response by *Drosophila* to nematode infection is complex and includes melanization, encapsulation, and expulsion (Welch, 1963). Furthermore, that resistance arises in populations of other species under selection against pathogens and even insecticides is well documented (Keiding, 1977; Mulla, 1977; Petersen, 1978; Davidson, 1992). For example, within approximately 104 generations of laboratory selection in mosquito colonies, behavioral changes arose that enabled better active defense by larvae against penetration by mermithid nematodes (Woodard & Fukuda, 1977). In selection experiments with the silkworm, *Bombyx mori*, against a cytoplasmic virus, Aizawa, Foruta and Nakamura (1966) found resistance in only 13 generations. In a case involving chemical insecticides, resistance to dieldrin arose in natural populations of the Australian sheep blowfly, *Lucilia cuprina*, within two years of its introduction (Clarke & McKenzie, 1987).

Second, a field study on the prevalence of infection showed that the distribution of nematodes was more aggregated among drosophilid flies than that expected under a random distribution (Jaenike, 1992). This finding can be interpreted as evidence for genetic heterogeneity in susceptibility (Anderson & Gordon, 1982). Also significant is the presence of *Drosophila* parasitized by allantonematid nematodes found in 26 million year old Dominican amber (Poinar, 1984). This finding certainly suggests a prolonged co-evolutionary history between these animals and presents the possibility that co-adaptational cycles have occurred, which could maintain heritable variation for resistance (Clarke, 1979; Hamilton, 1982; Hamilton & Zuk, 1982). In conclusion, because high FA in male *D. nigrospiracula* is associated with nematode infection, and because it is possible that a genetic basis to resistance exists, low FA may reflect a heritable ability to resist parasitic infection.

Although high FA is associated with reduced male reproductive success in several species (Møller, 1992a; Thornhill, 1992), including *Drosophila* (Markow, 1987; Markow & Ricker, 1992), the mechanism(s) responsible for this mating pat-

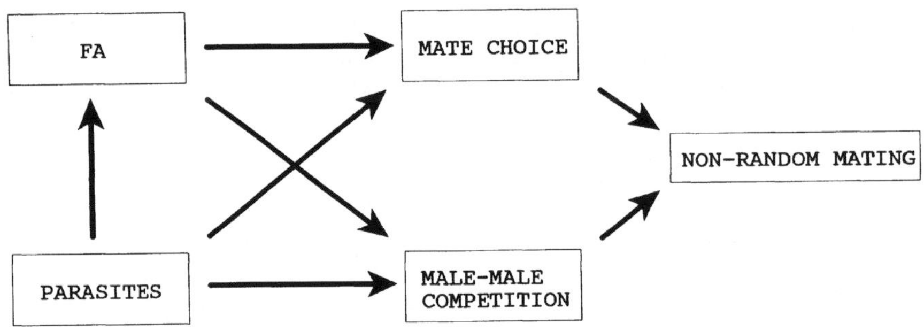

Fig. 4. Flow diagram illustrating how parasite-induced FA may effect non-random mating through mate choice by females and intra-male competition. Under a null model, FA is simply an incidental response to infection and non-random mating is due to parasitism only. Alternatively, FA may be used directly by females to distinguish resistant males or FA may reduce the ability of males to win intra-sexual contests over females. Finally, high FA and parasitism may interact to influcence host sexual selection. See text for further details.

tern remain unidentified and even controversial (Balmford & Thomas, 1992; Borgia & Wilkinson, 1992). One of the most popular hypotheses to explain this mating bias is female choice in favor of symmetrical mates (Møller, 1990, 1992a; Møller & Höglund, 1991; Thornhill 1992). According to this hypothesis, females choose symmetrical mates because symmetry reflects some form of heritable male quality that affects offspring viability. Female choice for low-FA males may thus be subsumed under 'good genes' models of sexual selection (Trivers, 1972; Heisler, 1984; Kodrick-Brown & Brown, 1984; reviewed by Kirkpatrick & Ryan, 1991). Because it is reasonable to apply my findings to other species, I suggest that low-FA males possess greater resistance against parasites, and that females may use observed differences in FA as indicators of male quality, specifically as it relates to male ability to resist parasitic infection. By choosing to mate with low-FA males, females may be more likely to produce parasite-resistant offspring. Hamilton and Zuk (1982) proposed a similar hypothesis, according to which only healthy males can develop the most elaborate and most brightly colored secondary sexual characters. Discriminating females prefer the showiest traits because those males able to produce them are resistant to parasitic infection.

What evidence exists for female mate choice in favor of symmetrical males? Møller's (1992a) experiments, which consisted of manipulating tail length asymmetry of male swallows (*Hirundo rustica*), showed that males with symmetrical tails acquired mates sooner in the breeding season and had greater annual reproductive success than asymmetric males. In the Japanese scorpionfly, *Panorpa japonica*, Thornhill (1992) documented female preference for the pheromone of symmetrical males over less symmetrical individuals. In *P. japonica*, however, it is not known whether females accrue material or genetic benefits from mating with symmetrical males; low-FA males also tend to be more successful at acquiring quality food items on which females feed during mating (Thornhill, 1992).

Although active mate choice for symmetrical mates is currently most often discussed, Fig. 4 illustrates other ways by which parasites and FA may influence host sexual selection. Under a null model, FA is simply an incidental response to parasitism and has no influence on host sexual selection. According to this hypothesis, negative effects of parasites alone account for observed non-random mating of hosts. Such effects are legion. For example, parasites may reduce male attractiveness through their affect on male sexual secondary characters (Hamilton & Zuk, 1982; Read, 1987, 1988; Ward, 1988; Houde & Torio, 1992), courtship display (Kennedy *et al.*, 1987), or because parasites themselves are detected by females (Borgia & Collis, 1989; Spurrier, Boyce & Manly, 1991). Parasites may reduce male competitive ability or disrupt dominance hierarchies (Howard & Minchella 1990), cause their host to voluntarily reduce current reproductive effort (Forbes, 1992), and change hos behaviour to facilitate parasite transmission (Moore, 1984; Edwards & Barnard, 1987).

Alternatively, FA may be used directly by females to discriminate in favour of healthy, unparasitized males, as already discussed (but see below). Non-random mating may also occur if asymmetry weaponry, or characters important for maneuverability, reduce male ability to win intra-sexual contests over females. Finally, the effects of high FA and parasitism may interact, and hence simultaneously influence host non-random mating.

Intensity of selection and FA

Although bristle FA was correlated with nematode load, wing length FA was not. Wing length was, however, associated with a two-fold smaller coefficient or variation than bristle number. This finding has several important implications. First, since stabilizing selection can be expected to decrease the variance of the frequency distribution of a trait (Soulé, 1982; Endler, 1986), these contrasting values of variation suggest that wing length may be under stronger stabilizing selection relative to bristle number. Second, because coefficients of variation are useful also in estimating susceptibility of characters to developmental instability (Mitton, 1978; King, 1985), bristle number is probably less canalized during development relative to wing length. For example, Eanes (1978) showed that heterozygosity in the monarch butterfly is inversely related to morphological variability, due to an heterozygote advantage to canalize morphology during development. Thus, the finding in *D. nigrospiracula* that wing asymmetry was insensitive to parasite-induced developmental stress may be the result of especially strong selection favouring effective wing canalization during development.

The life history of *D. nigrospiracula* requires that individuals disperse long distances to track suitable necrotic cacti, and asymmetric wings may impede effective long-distance dispersal. The strength of selection is probably greater during summer months, when high temperatures, low substrate density and rapid substrate desiccation require continued dispersal attempts of up to 1 km per day (Johnston & Heed, 1976). Thus, selection for wing length symmetry is expected to be strong because asymmetric males may not survive the flight to fresh substrates. This hypothesis is supported by the finding that in *D. mojavensis*, another desert endemic *Drosophila* which feeds on organ-pipe cactus, mated males did not exhibit greater wing length FA than unmated flies (Markow & Ricker, 1992). However, in *D. pseudoobscura*, a widespread generalist species (Begon, 1982; Lakovaara & Saura, 1982), mated males did exhibit greater wing FA than single individuals (Markow & Ricker, 1992).

In the present study, there was a negative relation between absolute wing length asymmetry and wing length. This is a surprising finding because larger individuals should be more asymmetrical for simple allometric reasons and because phenotypically extreme individuals are expected to be more homozygous at the loci controlling body size (Soulé, 1982; Soulé & Cuzin-Roudy, 1982). This pattern of asymmetry is explained, however, if costs of asymmetric wings increase with body size (wing loading), a prediction from the hypothesis that asymmetry impedes effective flight. Research on the aerodynamics of avian fligth shows that asymmetry in tail feathers reduces lift, maneuverability and prey-catching ability of birds (Møller, 1991; and see Balmford & Thomas, 1992).

In summary, my data suggest that traits under very strong stabilizing selection express less observable FA. This idea is supported by two comparative studies showing that traits such as wing length exhibit less FA than which are apparantly under directional selection due to either female choice or male-male competition (Møller & Höglund, 1991; Møller, 1992c). These findings imply that traits under strong stabilizing selection may be less useful to FA-discriminating females than traits under weaker stabilizing, or directional selection. In fact, in modelling the origin of female mating preferences, Heisler (1984) showed that traits under very strong (and very weak) stabilizing selection are least likely to be targets of female choice.

Another consideration is whether FA could ever be the focus of adaptive female choice; the process of either stabilizing natural selection operating on a trait and/or female choice operating on FA should cause rapid trait canalization and reduce asymmetry to dimensions unobservable by females. Thus, the nature and strength of selection, and hence the degree of trait canalization, are important variables that should be embraced by future hypotheses regarding the evolution and maintenance of adaptive mate choice on the basis of fluctuating asymmetry.

Acknowledgements

I am grateful to G. W. Krantz and G. O. Poinar Jr. for identifying the mites and nematodes, respectively. I thank J. Collins for allowing me to use Optimas software in his laboratory, and J. Roth for excellent technical advice. I also thank J. Alcock, E. Davidson, T. A. Markow, J. A. McKenzie, S. Pitnick, R. Rutowski, and R. Snook for valuable comments on the manuscript. This work was partially supported by National Science Foundation Grant BSR 89-19362 to T. A. Markow.

References

Aizawa, K., Y. Furuta & K. Nakamura, 1966. Selection of a resistant strain to virus induction in the silkworm *Bombyx mori*. J. Sericult. Sci. Jap. 30: 403-412.

Anderson, R. M. & D. M. Gordon, 1982. Processes influencing the distribution of parasite numbers within host populations with special emphasis on parasite-induced host mortalities. Parasitology 85: 373-398.

Bailit, H. L., P. L. Workman, J. D. Niswander & C. J. MacLean, 1970. Dental asymmetry as an indicator of genetic and environmental conditions in human populations. Hum. Biol. 42: 626-638.

Balmford, A & A. Thomas, 1992. Swallowing ornamental asymmetry. Nature 359: 487.

Begon, M., 1982. Yeasts and *Drosophila*, pp. 345-384 Vol 3b in The Genetics and Biology of *Drosophila*, edited by M. Ashburner, H. L. Carson & J. N. Thompson. Academic Press, London.

Borgia, G. & K. Collis, 1989. Female choice for parasite-free male satin bowerbirds and the evolution of bright male plumage. Behav. Ecol. Sociobiol. 25: 445-454.

Borgia, G. & G. Wilkinson, 1992. Swallowing ornamental asymmetry. Nature 359: 487-488.

Clarke, B. C., 1979. The evolution of genetic diversity. Proc. R. Soc. Lond. 205: 453-474.

Clarke, G. M., 1992. Fluctuating asymmetry: A technique for measuring developmental stress of genetic and environmental origin. Acta Zool. Fennica 191: 31-35.

Clarke, G. M. & J. A. McKenzie, 1987. Developmental stability of insecticide resistant phenotypes in blowfly; a result of canalizing natural selection. Nature 325: 345-346.

Clayton, D. H., 1991. The influence of parasites on host sexual selection. Parasitology Today 7: 329-334.

Davidson, E. W., 1992. Development of insect resistance to biopesticides. Pesq. agropec. bras. 27: 47-57.

Demerec, M., 1965. Biology of *Drosophila*. Hafner Publishing, New York.

Eanes, W. F., 1978. Morphological variance and enzyme heterozygosity in the monarch butterfly. Nature 276: 263-264.

Edwards, J. C. & C. J. Barnard, 1987. The effects of *Trichinella* infections on intersexual interactions between mice. Anim. Behav. 35: 533-540.

Endler, J. A., 1986. Natural selection in the wild. Princeton University Press, Princeton.

Forbes, M. R. L., 1992. Parasitism and host reproductive effort. Oikos, in press.

Hamilton, W. D., 1982. Pathogens as causes of genetic diversity in their host populations, pp. 269-296 in Population Biology of Infectious Disease Agents, edited by R. M. Anderson & R. M. May. Verlag-Chemie, Weinheim.

Hamilton, W. D. & M. Zuk, 1982. Heritable true fitness and bright birds: A role for parasites? Science 218: 384-386.

Heisler, I. L., 1984. A quantitative genetic model for the origin of mating preferences. Evolution 38: 1283-1295.

Houde, A. E. & A. J. Torio, 1992. Effect of parasitic infection on male color pattern and female choice in guppies. Behav. Ecol. 3: 346-351.

Howard, R. D. & D. J. Minchella, 1990. Parasitism and mate competition. Oikos 58: 120-122.

Jaenike, J., 1988. Parasitism and male mating success in *Drosophila testacea*. Am. Nat. 131: 774-780.

Jaenike, J., 1992. Mycophagous *Drosophila* and their nematode parasites. Am. Nat. 139: 893-906.

Johnston, J. S. & W. B. Heed, 1976. Dispersal of desert-adapted *Drosophila*: The saguaro-breeding *D. nigrospiracula*. Am. Nat. 110: 629-651.

Keiding, J., 1977. Resistance in the housefly in Denmark and elsewhere, pp. 261-302 in Pesticide Management and Insecticide Resistance, edited by D. L. Watson & A. W. A. Brown. Academic Press, New York.

Kennedy, C. E. J., J. A. Endler, S. L. Poynton & H. McMinn, 1987. Parasite load predicts mate choice in guppies. Behav. Ecol. Sociobiol. 21: 291-295.

King, D. P. F., 1985. Enzyme heterozygosity associated with anatomical character variance and growth in the herring (*Clupea harengus* L.). Heredity 54: 289-296.

Kirkpatrick, M. & M. J. Ryan, 1991. The evolution of mating preferences and the paradox of the lek. Nature 350: 33-38.

Kodrick-Brown, A. & J. H. Brown, 1984. Truth in advertising: The kinds of traits favored by sexual selection. Am. Nat. 309-323.

Lakovaara, S. & A. Saura, 1982. Evolution and speciation in the Drosophila obscura group, pp. 2-52 Vol 3b in The Genetics and Biology of *Drosophila*, edited by M. Ashburner, H. L. Carson & J. N. Thompson. Academic Press, London.

Leary, R. F. & F. W. Allendorf, 1989. Fluctuating asymmetry as an indicator of stress: Implications for conservation biology. Trends Ecol. Evol. 4: 214-217.

Leary, R. F., F. W. Allendorf & K. L. Knudsen, 1984. Superior developmental stability of heterozygotes at enzyme loci in salmonid fishes. Am. Nat. 124: 540-551.

Livshits, G. & E. Kobyliansky, 1991. Fluctuating asymmetry as a possible measure of developmental homeostasis in humans: A review. Hum. Biol. 63: 441-466.

Markow, T. A., 1987. Genetic and sensory basis of sexual selection in *Drosophila*, pp. 89-95 in Evolutionary Genetics of Invertebrate Behavior, edited by M. D. Huettel. Plenum Publishing.

Markow, T. A., 1988. Reproductive behavior of *Drosophila melanogaster* and *D. nigrospiracula* in the field and in the laboratory. J. Comp. Psychol. 102: 169-173.

Markow, T. A. & J. P. Ricker, 1992. Male size, developmental stability, and mating success in natural populations of three *Drosophila* species. Heredity 69: 122-127.

Mitton, J. B., 1978. Relationship between heterozygosity for enzyme loci and variation of morphological characters in natural populations. Nature 273: 661-662.

Mitton, J. B. & M. C. Grant, 1984. Associations among protein heterozygosity, growth rate, and developmental homeostasis. Ann. Rev. Ecol. Syst. 15: 470-499.

Møller, A. P., 1990. Fluctuating asymmetry in male sexual ornaments may reliably reveal male quality. Anim. Behav. 40: 1185-1187.

Møller, A. P., 1991. Sexual ornament size and the cost of fluctuating asymmetry. Proc. R. Soc. Lond. B 243: 59-62.

Møller, A. P., 1992a. Female swallow preference for symmetrical male sexual ornaments. Nature 357: 238-240.

Møller, A. P., 1992b. Swallowing ornamental asymmetry. Nature 359: 488.

Møller, A. P., 1992c. Patterns of fluctuating asymmetry in weapons: evidence for reliable signalling of quality in beetle horns and bird spurs. Proc. R. Soc. Lond. B 248: 199-206.

Møller, A. P. & J. Höglund, 1991. Patterns of fluctuating asymmetry in avian feather ornaments: implications for models of sexual selection. Proc. R. Soc. Lond. B 245: 1-5.

Moore, J., 1984. Altered behavioural responses in intermediate hosts - an acanthocephalan parasite strategy. Am. Nat. 123: 572-577.

Mulla, M. S., 1977. Resistance in Culicine mosquitoes in California-Countermeasures, pp. 239-260 in Pesticide Management and Insecticide Resistance, edited by D. L. Watson & A. W. A. Brown. Academic Press, New York.

Neter, J., W. Wasserman & M. H. Kutner, 1990. Applied Linear Statistical Models. Richard D. Irwin, Boston.

Palmer, A. R. & C. Strobeck, 1986. Fluctuating asymmetry: measurement, analysis, patterns. Annu. Rev. Ecol. Syst. 17: 391-421.

Palmer, A. R. & C. Strobeck, 1992. Fluctuating asymmetry as a measure of developmental stability: Implications of non-normal distributions and power of statistical tests. Acta Zool. Fennica 191: 57-72.

Petersen, J. J., 1978. Development of resistance by the southern house mosquito to the parasitic nematode *Romanonermis culicivorax*. Environment. Entomol. 7: 518-520.

Poinar, G. O. Jr., 1983. The natural history of nematodes. Prentice-Hall, Englewood Cliffs, NJ.

Poinar, G. O. Jr., 1984. First fossil record of parasitism by insect parasitic Ty lenchida (Allantonematidae: Nematoda). J. Parasit. 70: 306-308.

Read, A. F., 1987. Comparative evidence supports the Hamilton and Zuk hypothesis on parasites and sexual selection. Nature 327: 68-70.

Read, A. F., 1988. Sexual selection and the role of parasites. Trends Ecol. Evol. 3: 97-102.

Reeve, E. C. R., 1960. Some genetic tests on asymmetry of sternopleural chaeta number in *Drosophila*. Genet. Res., Camb. 1: 151-172.

Robertson, F. S. & E. Reeve, 1952. Studies in quantitative inheritance. I. The effects of selection of wing and thorax length in *Drosophila melanogaster*. J. Genet. 50: 414-448.

Sang, J. H., 1956. The quantitative nutritional requirements of *Drosophila melanogaster*. J. Exptl. Biol. 33: 45-72.

Sokal, R. R. & J. Rohlf, 1981. Biometry. The principles and practice of statistics in biological research. W. H. Freeman and Co., New York.

Soulé, M. E., 1979. Heterozygosity and developmental stability: another look. Evolution 33: 396-401.

Soulé, M. E., 1982. Allometric variation. I. The theory and some consequences. Am. Nat. 120: 751-764.

Soulé, M. E. & J. Cuzin-Roudy, 1982. Allometric variation. II. Developmental instability of extreme phenotypes. Am. Nat. 120: 765-786.

Spurrier, M. F., M. S. Boyce & F. J. M. Manly, 1991. Effects of parasites on mate choice by captive sage grouse, pp. 389-398 in Bird-Parasite Interactions, edited by J. E. Loye & M. Zuk. Oxford University Press, Oxford.

Steinhaus, E. A., 1949. Principles of Insect Pathology. McGraw Hill, New York.

Thoday, J. M., 1956. Balance, heterozygosity and developmental instability. Cold Spring Harbor Symp. Quant. Biol. 21: 318-326.

Thornhill, R., 1992. Female preference for the pheromone of males with low fluctuating asymmetry in the Japanese scorpionfly (*Panorpa japonica*: Mecoptera). Behav. Ecol. 3: 277-283.

Trivers, R. L., 1972. Parental investment and sexual selection, pp. 136-179 in Sexual Selection and the Descent of Man, 1871-1971, edited by B. Campbell. Aldine Press, Chicago.

Valentine, D. W., M. E. Soulé & P. Samallow, 1973. Asymmetry analysis in fishes: a possible indicator of environmental stress. U.S. Natl. Mar. Fish. Serv. Fish. Bull. 71: 357-370.

Van Valen, L., 1962. A study of fluctuating asymmetry. Evolution 16: 125-142.

Waddington, C. H., 1957. The Strategy of the Genes. McMillan, New York.

Wakelin, D., 1985. Genetic control of immunity to helminth infections. Parasitology Today 1: 17-23.

Ward, P. I., 1988. Sexual dichromatism and parasitism in British and Irish freshwater fish. Anim. Behav. 36: 1210-1215.

Watanabe, H., 1987. The host population, pp. 71-112 in Epizootiology of Insect Diseases, edited by J. R. Fuxa & Y. Tanada. John Wiley & Sons, New York.

Watt, W. B., R. C. Cassin & M. S. Swan, 1983. Adaptations to specific loci. III. Field behavior and survivorship differences among *Colias* PGI genotypes are predictable from *in vitro* biochemistry. Genetics 103: 725-739.

Welch, H. E., 1959. Taxonomy, life cycle, development, and habits of two new species of Allantonematidae (Nematoda) parasitic in Drosophilid flies. Parasitology 49: 83-103.

Welch, H. E., 1963. Nematode infections, pp. 363-392 in Insect Pathology, An Advanced Treatise Volume 2, edited by E. A. Steinhaus. Academic Press, New York.

Woodard, D. B. & T. Fukuda, 1977. Laboratory resistance of the mosquito *Anopheles quadrimaculatus* to the mermithid nematode *Diximermis peterseni*. Mosquito News 37: 192-195.

Yamaguchi, T., 1977. Studies on the handedness of the fiddler crab, *Uca lactea*. Biol. Bull. 152: 424-436.

Zuk, M., 1991. Parasites and bright birds: new data and a new prediction, pp. 317-327 in Bird-Parasite Interactions, edited by J. E. Loye & M. Zuk. Oxford University Press, Oxford.

T. A. Markow (ed.), Developmental Instability: Its Origins and Evolutionary Implications, 269–281, 1994.

Fluctuating asymmetry and sexual selection

A. P. Møller* & A. Pomiankowski
Department of Genetics and Biometry, University College London, 4 Stephenson Way, London NW1 2HE, UK
** Present address: Department of Zoology, Uppsala University, Box 561, S-751 22 Uppsala, Sweden*

Received and accepted 23 April 1993

Key words: fluctuating asymmetry, heterozygosity, reliable signalling, sexual selection, stress

Abstract

Fluctuating asymmetry occurs when an individual is unable to undergo identical development on both sides of a bilaterally symmetrical trait. Fluctuating asymmetry measures the sensitivity of development to a wide array of genetic and environmental stresses. We propose that fluctuating asymmetry is used in many signalling contexts for assessment of an individual's ability to cope with its environment. We hypothesize that fluctuating asymmetry is used in sexual selection, both in fighting and mate choice, and in competition for access to resources. Evidence is reviewed showing that the patterns of fluctuating asymmetry in secondary sexual characters differ from those seen in other morphological traits. Secondary sexual characters show much higher levels of fluctuating asymmetry. Also, there is often a negative relationship between fluctuating asymmetry and the absolute size of ornaments, whereas the relationship is typically U-shaped in other morphological traits. The common negative relationship between fluctuating asymmetry and ornament size suggests that many ornaments reliably reflect individual quality.

Introduction

Sexual selection was originally introduced by Darwin (1871) to explain the evolution of extravagant secondary sexual characters. Two mechanisms were suggested to result in sexual selection: competition between members of the same sex and choice of mates of the opposite sex. While the former has been relatively uncontroversial, mate choice continues to attract much attention from evolutionary biologists because of the existence of several models describing the evolution of mate preferences. First, Fisher's model of sexual selection suggests that secondary sexual characters evolve as a result of the coevolution between mating preference and the sexual character (Fisher, 1930; Lande, 1981; Pomiankowski, Iwasa & Nee, 1991). The individuals of the choosy sex, usually the females, get an advantage because their sons are more likely to inherit an attractive ornament and so have higher than average mating success. There is thus no function of the secondary sexual character other than its attractiveness. Second, handicap models suggest that secondary sexual characters are reliable signals of male genetic quality (Zahavi, 1975, 1987; Andersson, 1982a, 1986; Kodric-Brown & Brown, 1984; Nur & Hasson, 1984; Pomiankowski, 1988). If the costs of producing a larger ornament are greater for males in worse condition, then the optimal ornament expression is an increasing function of quality (Grafen, 1990a; Iwasa, Pomiankowski & Nee, 1991). Another version of the handicap idea suggests that ornaments reliably reflect parasite resistance, since only healthy individuals are able to produce the most extravagant sexual traits (Hamilton & Zuk, 1982). Third, secondary sexual characters may signal characteristics of potential mates that directly benefit the chooser. For example, the condition of the sexual trait may reflect infection by ectoparasites that can be transmitted to the female or her offspring (Freeland, 1976) or the parenting ability of a male (Heywood, 1989; Hoelzer 1989).

The direct benefits hypothesis works best when it is not easy for a female to assess those qualities of a potential mate that directly affect her fitness, but when they can be assessed indirectly through the development of extravagant ornamentation.

There is no consensus about the relative importance of these forces in the evolution of mate preferences. The main problem is that no easily measurable criteria have been set out to distinguish these processes. Here it is suggested that fluctuating asymmetry in secondary sexual characters may provide an answer to this problem. In this paper we provide a general discussion about (i) the patterns of fluctuating asymmetry in ordinary morphological traits and in secondary sexual characters, (ii) the factors that cause fluctuating asymmetry, and (iii) the potential use of fluctuating asymmetry in assessment of quality among conspecifics. We primarily consider signalling in sexual selection but also discuss the role of assessment in other signalling contexts.

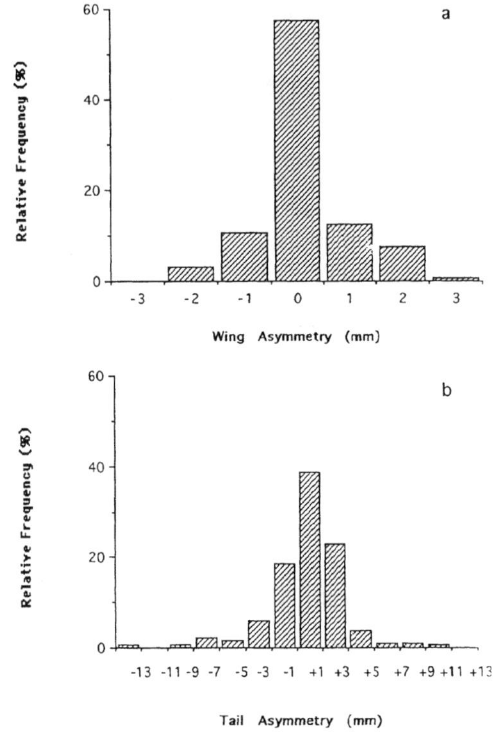

Fig. 1. Frequency distribution of fluctuating asymmetry (right-minus-left value) in wing length (a) and tail length (b) of 181 male barn swallows from Denmark 1991. Four males with more extreme tail asymmetries are not shown.

Fluctuating asymmetry in morphological traits

Fluctuating asymmetry (FA) is one of three kinds of asymmetry demonstrated by bilateral morphological traits, the others being antisymmetry and directional asymmetry. FA occurs when the usual state of a bilateral trait is symmetrical and there is no tendency for one side to have larger character values than the other (Ludwig, 1932; Van Valen, 1962; Palmer & Strobeck, 1986; Leary & Allendorf, 1989; Parsons, 1990). For example, wing length and tail length in the barn swallow (*Hirundo rustica*) demonstrate FA; most individuals are symmetrical, some are slightly asymmetrical and only a few are highly asymmetrical (Fig. 1). The criteria used for determining whether wing length and other morphological characters mentioned in this paper demonstrate FA are that the frequency distribution of left-minus-right character values does not differ from a normal distribution with a mean of zero (Palmer & Strobeck, 1986). This is the case for wing length in barn swallows (test for normality: Kolmogorov-Smirnov one-sample test: $D = 0.06$, $N = 181$, N. S.; one-sample t-test of whether the mean value deviates from zero: $t = 0.23$, d. f. = 180, N. S.). Absolute FA is given by the difference between the numerical values of left and right measures (Palmer & Strobeck, 1986). Asymmetries in the rest of the paper always refer to absolute asymmetries. FA can be measured in paired characters such as wing length, colour patterns (e.g., a breast stripe) or more complex structures (e.g., a peacock's train fan). The inability of an individual to undergo identical development of a trait on both sides of its body is a measure of developmental homeostasis. Individual FA thus measures the ability of individuals to cope with genetic or environmental stress (Jones 1987; Leary & Allendorf, 1989; Parsons, 1990).

Most morphological traits studied demonstrate small degrees of FA, and the absolute magnitude of asymmetry is often negligible (Palmer & Strobeck 1986; Leary & Allendorf, 1989; Parsons, 1990). A recent study of FA in the wing length of 32 different bird species revealed relative values that were most often less than 1% of the absolute size of the trait (Møller & Höglund, 1991). FA in most ordinary morphological traits is thus negligible and c a magnitude that makes detection difficult without direct measurement.

The magnitude of individual FA is often larger for individuals at the tails of the frequency distribution of a character, with relatively smaller values in between (Soulé, 1982; Soulé & Cuzin-Roudy, 1982). The typical pattern of FA when plotted against the absolute size of the trait is U-shaped, as demonstrated by wing length of the barn swallow (Fig. 2). Sometimes a flat relationship of FA and size is reported. This is most likely to be due to small sample sizes that miss individuals with the largest and smallest trait values because of their rarity.

When population FA in a particular morphological character is compared across species the relationship between asymmetry and character size is allometric (Møller & Höglund, 1991; Møller, 1992a). Species with larger traits tend to have higher average levels of FA for both secondary sexual characters and ordinary morphological traits (Møller & Höglund, 1991; Møller, 1992a; Møller & Eriksson, 1993a).

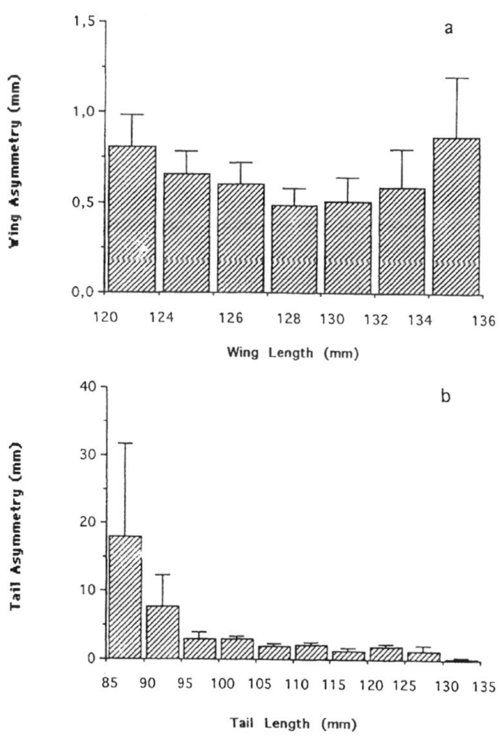

Fig. 2. Numerical value of fluctuating asymmetry in wing length (a) and tail length (b) of 181 male barn swallows from Denmark 1991 in relation to the size of the morphological trait. Values are means + one S.E.

Secondary sexual characters differ from other morphological traits. They are often extravagant traits, bearing highly intricate patterns and striking coloration. Many sexual ornaments are known to be subject to strong directional selection arising from female choice (e.g., Andersson, 1982b; Møller, 1988; review in Møller, 1993a) or male-male competition (e.g., Conner, 1988; Otronen, 1988). In cases where exaggeration has gone to an extreme, sexual traits are likely to be very costly to produce and maintain and may be close to the limits of production. For example, experimental manipulation of the tail feathers of scarlet-tufted malachite sunbirds (*Nectarinia johnstoni*), which are used to attract females, has shown that long tails severely disrupt flying and decrease the efficiency of catching aerial insects (Evans & Thomas, 1992). A similar effect has been reported from the barn swallow (Møller, 1989). Measures of the energetics of courtship, both in frogs (Ryan, 1988) and sage grouse (*Centrocercus urophasianus*) (Vehrencamp, Bradbury & Gibson, 1989), show that display is the most energetically costly act a male performs.

The pattern of FA in ornaments differs from that of ordinary morphological traits. First, the absolute magnitude of FA in secondary sexual characters is considerably larger than in other morphological traits (for example tail length in *Hirundo rustica*: Fig. 1; Møller, 1990a; Møller & Höglund, 1991). Average FA in male feather ornaments of sexually dimorphic birds was five to ten times larger than homologous traits in conspecific females and in both sexes of closely related monomorphic species without ornaments (Møller & Höglund, 1991). Wing asymmetry was of a similar magnitude as tail asymmetry in the species without ornamental feathers (Møller & Höglund, 1991). Absolute FA in ornaments is often so large that it can easily be determined by the naked eye, and it was on average nearly 5 cm in the extravagantly ornamented quetzal (*Pharomachrus mocinno*) (Møller & Höglund, 1991).

Second, individual FA is frequently negatively related to the absolute size of the paired ornament. In a number of bird species, individuals with the largest ornaments tend to have the smallest degree of asymmetry (*Hirundo rustica*: Fig. 2; Møller,

1990a; Møller & Höglund, 1991; Manning & Hartley, 1991). Similarly, individual floral asymmetry is negatively related to the size of floral traits in several plant species (Møller, 1993b; Møller & Eriksson, 1993a). This pattern differs dramatically from that found in ordinary morphological traits. It also fits uncomfortably with the positive allometric relationship found in cross species studies that demonstrates that species with larger traits tend to have higher absolute levels of FA.

Absolute FA in ornaments is not ubiquitously negatively related to the absolute size of the paired ornament. A study of bird taxa with either single or multiple feather ornaments revealed that the average FA was similar in species with single and multiple ornaments. However, the negative relationship between individual FA and ornament size was much more common among single ornament than multiple ornament taxa (Møller & Pomiankowski, 1993). The reason for a flat or U-shaped relationship between individual asymmetry and size of an ornament in many bird species with multiple ornaments is probably that the development of these traits is not costly. These sex traits do not appear to have condition-dependent expression and therefore are interpreted best as representing arbitrary Fisherian traits (Møller & Pomiankowski, 1993).

Another comparative study showed that strong female mate preferences are associated with the pattern of ornament asymmetry (Møller, 1993a). Bird species with a directional female mate preference demonstrated a negative relationship between individual ornament asymmetry and size, while the relationship did not differ significantly from zero among ornamented species without such a preference (Møller, 1993a). This suggests that strong female preferences for the most extravagantly ornamented males are present only if the expression of the secondary sexual character reliably reveals male condition.

Flowers can be considered secondary sexual characters of plants. The patterns of fluctuating asymmetry in flowers are related to the flower preferences of pollinators (Møller, 1993b; Møller & Eriksson, 1993a, 1993b). Flowers visited by pollinators are generally more symmetric than the nearest neighbouring flower not visited by a pollinator (Møller & Eriksson, 1993b). Pollinator preference for symmetric flowers is also apparent in manipulation experiments of individual floral asymmetry (Møller, 1993b). The relationship between floral asymmetry and the size of floral traits is generally negative which indicates that the largest flowers are developmentally most stable (Møller & Eriksson, 1993a). Large, symmetric flowers produce more nectar than small, asymmetric flowers and this relationship may account for the pollinator preference for the most symmetric flowers (Møller, 1993b).

A similar negative relationship between size and degree of FA regularly occurs in weapons used in fighting between males (Møller, 1992a). Weapons are often exaggerated structures and for that reason costly to produce or maintain, and they are therefore likely to show a considerable degree of condition-dependence with the largest characters produced by individuals in prime condition (e.g., Brown, 1984; Andersson, 1986; Bubenik & Bubenik, 1990). Below we consider the causes of FA and how these differences between morphological and sexual traits might be explained.

Causes of fluctuating asymmetry

There exists a vast literature on the causes of FA based on both observation and experimentation. There are two major causes of FA in morphological traits: (i) environmental stress, and (ii) genetic stress. We briefly review each of these causes of FA with particular reference to their importance for sexual selection and assessment.

Environmental stress

FA has been used as a measure of the ability of individuals to cope with different kinds of environmental stress (Jones, 1987; Leary & Allendorf 1989; Parsons, 1990). A large number of differen stressors have been demonstrated to cause FA in morphological traits during development, includ ing extreme temperatures (Parsons, 1962; Siegel & Doyle, 1975; Sciulli *et al.*, 1979), various chemi cals including pesticides (Valentine & Soulé, 1973 Hoffmann & Parsons, 1989a, b), audiogenic stres (Sciulli *et al.*, 1979), and food deficiency either in terms of quality or quantity (Parsons, 1990). Sever restriction of nutrients to mothers during pregnanc causes asymmetry in skeletal traits of offsprin (Sciulli *et al.*, 1979) and in external features such a bristle number in *Drosophila* (Parsons, 1990). It i not known whether nutrient deficiency, or any c the other factors listed above, cause any differential effect on FA in secondary sexual characters compared to ordinary morphological traits.

Pathogens and parasites have also been demonstrated to have debilitating effects on the development of otherwise symmetrical morphological traits. Many human diseases are known to cause distortion of an otherwise symmetrical appearance (e.g. Bailit *et al.*, 1970; Livshits & Kobyliansky, 1991). A recent study of the effects of parasites on development of feathers in birds revealed that not all morphological characters were equally susceptible to parasite loads (Møller, 1992b). The loads of ectoparasites were experimentally manipulated in one year, and the effect of this manipulation on development of individual FA in feather traits was assessed following the subsequent moult. The large tail ornaments in male barn swallows became more asymmetrical than the homologous trait in females as a result of experimentally altered levels of mite infestations prior to the annual moult. Other feather traits, such as wing length and the length of the short, sexually monomorphic, central tail feathers, showed weak, non-significant effects of mites on FA and no difference between the sexes (Møller, 1992b). Different traits are therefore not equally susceptible to the stress caused by ectoparasitism. In barn swallows, mite infections affect asymmetry in secondary sexual characters much more than in comparable non-sexual traits.

Genetic stress

Increased FA has been related to inbreeding and this could either be due to increased levels of homozygosity or to exposure of deleterious alleles among inbred stocks, or both (Lerner, 1954). Subsequent studies demonstrated that homozygous individuals often were developmentally less stable than their heterozygous counterparts (Leary, Allendorf & Knudsen, 1983, 1984, 1985a; Clarke & McKenzie, 1987) and a number of studies have shown similar phenomena at the population level (Lerner, 1954; Soulé, 1979; Kat, 1982; Vrijenhoek & Lerman, 1982; Biémont, 1983; Leary, Allendorf & Knudsen, 1985b; Livshits & Kobyliansky, 1991; review in Mitton & Grant, 1984). This conclusion is based on observations in *Drosophila*, *Mus*, and other animal and plant laboratory and natural populations. Individuals heterozygous for one or more proteins generally develop less FA and have higher fitness as measured by growth, fecundity and various measures of physiological performance (reviews in Mitton & Grant, 1984; Palmer & Strobeck, 1986).

In contrast, crosses between inbred stocks usually yield hybrids with lower FA (Mather, 1953; Thoday, 1956; Reeve, 1960). Hybrids between species or populations that are normally reproductively isolated generally have higher levels of FA than either of the parental species or populations (Tebb & Thoday, 1958; Zakharov, 1981; Graham & Felley, 1985; Leary, Allendorf & Knudsen, 1985). A natural experiment has also been reported in sunfish, where habitat disturbance brought two populations together that were previously isolated. Again, FA increased in the hybrids (Graham & Felley, 1985).

In all the cases mentioned above, it can be inferred that the degree of FA is negatively correlated with fitness. Inbreeding increases the fraction of loci homozygous for deleterious recessives. Outbreeding protects against rare deleterious recessive alleles and may also raise fitness if there is a heterozygote advantage. In contrast, crosses between species disrupt genomic adaptations that have developed since the populations became reproductively isolated and hence reduce fitness. Explicit demonstrations of negative correlations between individual fitness components and individual FA were reported by Thornhill (1992a, 1992b) for mating success and longevity in *Panorpa* scorpionflies, Markow and Ricker (1992) for mating success in *Drosophila*, and Møller (1993d) for mating success, reproductive success and survival in *Hirundo rustica*. Individuals with higher fitness components consistently had less FA.

An apparent exception to this generality is the perturbation to developmental stability caused by the incorporation of a new, advantageous mutant in Australian blowflies (*Lucilia cuprina*) (Clarke & McKenzie, 1987). This mutant confers insecticide resistance. When resistance first evolved it caused greatly elevated FA. But after 20 years of adaptation to the new mutant, FA has returned to the same level found in susceptible, wildtype flies. So even though the fitness increase associated with resistance was initially coupled to increased FA, subsequent selection on resistant flies has ameliorated this developmental instability and probably restored a negative correlation of fitness and FA.

Fluctuating asymmetry and selection

What selective forces can explain the different patterns of FA in ordinary morphological and sexual traits? We suggest that the principal difference lies in the mode of selection. In general most morphological traits are subject to long-term stabilizing selection. The net effect of selection may be directional on the time scale of a few generations. A good example of this is the quick change in body and bill demensions recorded in Darwin's medium ground finch which occurred during a period of drought (Boag & Grant, 1981). But over many generations any net directionality is likely to be small and may even be zero. In sexual characters there may also be a balance of selective forces at equilibrium. In several cases it has been shown that sexual selection for greater ornament size is balanced by viability selection against further increase (e.g., Endler, 1983). But over a longer time-scale we can infer that sexual characters have been subject to strong directional selection and may still be subject to continuing directional selection. The modes of selection affecting ordinary morphological traits and secondary sexual characters differ in four principle ways.

First, while ordinary morphological traits are stabilized by other components of viability selection, sexual characters are stabilized by sexual selection for increased extravagancy opposed by viability selection. This in itself is not an important difference. Second, the two categories of characters differ in terms of the recent history of selection. While secondary sexual characters have been subject to recent dramatic evolutionary changes due to net directional selection, ordinary morphological characters are often of the same size in a genus or a family. In the highly sexually selected birds of paradise, species differ dramatically in male ornamentation but hardly at all in ordinary traits in either sex (Coates, 1990). Similar high variance in sexual characters across species are present in animal genitalia (Eberhard, 1989). Third, the two types of traits probably also differ in the strength of the components of selection. In ornaments, measurements of natural and sexual components indicate that they are very large; in general they are probably much larger than the corresponding natural selection components for ordinary morphological characters. The final difference between secondary sexual characters and other morphological characters concerns the effect of mutation. Mutation pressure on sexual traits that are exaggerated or show great intricacy of design will more often disrupt than improve their attractiveness to females. Mutation bias on ordinary morphological traits is likely to be much weaker (Pomiankowski, Iwasa & Nee, 1991; Iwasa, Nee & Pomiankowski, 1991). This means that sexual ornaments will be subject to continual directional selection because mutation bias creates an excess of poorly ornamented males at equilibrium.

The two forms of selection regimes, directional and stabilizing selection, are predicted to result in marked differences in the nature of genetic and phenotypic variation and asymmetry. The effect of stabilizing selection is to favour individuals with trait values close to the mean. Selection not only acts against individuals that have high or low trait values but also favours modifiers that limit variation in the expression of the trait (Prout, 1962; Milkman, 1970). These modifiers will also reduce FA. A trait that is developmentally stable will suffer fewer and less significant errors in development and so have a lower mean level of FA (Pomiankowski & Møller, 1993).

This simple principle can be extended to explain the U-shaped relationship of size and FA. Three routes may lead to this. First, individuals who produce traits larger or smaller than the selective optimum may suffer greater stress in development because they produce traits that deviate from the optimum size. Hence individuals from the tails of the frequency distribution will show greater variance in development. On average they will have larger differences between left and right sides and higher FA. A second reason is that individuals at the tails of phenotypic distributions tend to be homozygotes while modal individuals tend to be heterozygotes (Soulé, 1982; Soulé & Cuzin-Roudy, 1982). If heterozygosity buffers against a variety of stressors, modal individuals may on average develop smaller degrees of asymmetry than individuals in the tails of frequency distributions. A third possibility is that individuals vary in their ability to control development. Individuals with tight developmental control will have low FA and be distributed closely about the optimum trait size, whereas those with weaker developmental control will have higher FA and greater variation in trait size. The net effect again is low FA in the modal size classes, with higher values at the extremes (Pomiankowski & Møller, 1993).

Under a regime of strong and persistent directional selection for larger character values, as can be generated by sexual selection, the pattern of FA is strikingly different. Strong directional selection for larger size favours existing genes and new mutants that code for higher trait values. Directional selection can also favour relaxation in the developmental control of morphogenesis as this creates greater phenotypic variation and so can aid the response to selection. This is particularly likely when sexual selection disproportionately favours individuals with greater than average values. The effect of these selective forces is predicted to increase FA. First, a trait that is larger than the optimum under natural selection is likely to impose greater stress during development. For example, large secondary sexual characters are more costly to produce and they take longer to produce than small characters (e.g., Bubenik & Bubenik, 1990). Second, selection for weaker control of development will lead to greater variability during morphogenesis and higher FA. This is shown by the greater phenotypic variability of secondary sexual characters which are known to have considerably higher coefficients of variation than ordinary morphological characters (e.g., Alatalo, Höglund & Lundberg, 1988). So as the size and the variability of the trait becomes larger under directional selection, mean FA will increase.

In addition to these effects on the mean, we can predict the relationship between size and FA. Under the most simple assumption, FA in a trait will be directly related to the size of that trait with all other factors merely acting as uncorrelated noise. Hence individuals that grow larger traits will be subject to greater stress and hence on average have higher FA. This predicts that the regression of FA on size will be positive. Such a pattern is likely if the ornament has evolved through Fisher's runaway process and its development is unrelated to condition.

Handicap models of reliable signalling predict a strong correlation between the condition of an individual and ornament size, as only males of high phenotypic or genetic quality can afford to construct large ornaments (Grafen, 1990b; Iwasa, Pomiankowski & Nee, 1991). This prediction is consistent with the finding that the expression of many sexual traits is sensitive to changes in condition (Andersson, 1986; Hill, 1991; Møller, 1991a). The superior condition of individuals with large ornaments may thus mask the increase in stress caused by greater ornamentation. The net effect, if ornament size is mainly determined by male condition, is that the regression of FA on size is negative (Pomiankowski & Møller, 1993). Individuals that produce the largest sexual traits are like to have the lowest FA when the trait reliably reflects condition, because a given level of ornamentation is more costly to low than to high quality individuals.

An old objection to this line or reasoning is that strong directional selection will deplete genetic variation, particularly in traits closely associated with fitness, making female choice for 'good genes' redundant (Maynard Smith, 1978; Taylor & Williams, 1982). However, there are several ways in which genetic variation can be maintained even under regimes of strong selection, for instance frequency-dependent selection, mutational input, and temporal variation in fitness. The heritability of FA has been estimated in a number of selection experiments as well as in laboratory breeding experiments (reviews in Palmer & Strobeck, 1986; Parsons, 1990; Livshits & Kobyliansky, 1991; Thornhill & Sauer 1992). The general conclusion is that there is a small, significant heritability of FA in ordinary morphological traits, while there are no estimates available for secondary sexual characters.

In conclusion, strong directional selection will lead to the evolution of costly secondary sexual characters. These will evolve condition-dependent expression that is reflected in a negative relationship between character size and individual FA. All signals have a morphological basis, and costly signals subject to strong directional selection are likely to demonstrate similar patterns of FA.

On the use of fluctuating asymmetry in assessment

Individual FA in secondary characters and other morphological traits may reflect abilities to cope with stress that may be of importance during assessment to individuals of the same or other species (competitors, predators, parasites, pollinators). We suggest that there are important consequences for the two modes of sexual selection (male-male competition and female choice) and other cases of competition for resources.

Animals are able to discriminate between symmetric and asymmetric features and use asymmetry as a category in classification of objects. Discrimimation experiments on pigeons in operant conditioning chambers reveal that individuals are able to accurately classify novel test stimuli according to degree of asymmetry (Delius & Habers, 1978; Delius & Nowak, 1982; see also Rensch, 1958). This was the case even when the size of the stimuli was held constant. Some animals therefore have the perceptive abilities to discriminate between traits only differing in their degree of individual asymmetry.

In this and the following sections we consider two cases of assessment of FA. First, asymmetries in morphological traits may be of a magnitude that cannot be assessed directly by vision or other sensory modalities. These asymmetries may include those present in ordinary morphological traits such as wing length in the barn swallow (Fig. 1). While such asymmetries cannot be used directly in assessment of conspecifics, they may still represent a useful measure of the ability of individuals to cope with the environment and thus be a useful way to measure quality for the investigator.

Second, asymmetries may be large and quite obvious to human observers and may be used directly in assessment. Many paired secondary sexual characters are positioned next to each other, and this makes direct assessment of asymmetry particularly easy. For example, when a blue peacock (*Pavo cristatus*) spreads its train it is immediately obvious if an eye-spot is missing. It is possible that high quality individuals gain a signalling advantage by placing their paired traits next to each other to make assessment easy. Males of inferior quality will be unable to cheat by making assessment of asymmetries impossible or difficult if females pay attention to asymmetry.

Male-male competition

Fighting ability of males is known to be affected by the ability to acquire sufficient nutrients and to cope with debilitating pathogens and parasites (Freeland, 1976, 1981; Burley, Tidemann & Halupka, 1991; Johnson & Boyce, 1991; Spurrier, Boyce & Manly, 1991; Thornhill & Sauer, 1992). These factors may be reflected by FA in the secondary sexual characters. Conspecifics should be able to assess fighting ability using the size and asymmetry of these traits. In addition, asymmetry in weapons such as claws, spurs, antlers and horns is also informative if asymmetry lessens the effectiveness of these traits in battle. An asymmetrical weapon may be less strong or reveal a weakness that can be exploited (Møller, 1992a).

There might also be direct effects of previous fighting success on the development of FA in secondary sexual characters. Stags are know to develop severe asymmetries in their antlers in response to body wounds acquired during fights in previous years because wounds cause distortion of the ability to symmetrically channel resouces to the right and the left antler during growth (Brown, 1984). Male and female conspecifics will therefore be able to directly assess the previous fighting success of stags from large antler asymmetry.

Female choice

The three major models of female choice make different predictions about the magnitude and the patterns of FA in secondary sexual characters. Handicaps are very costly sex traits, particularly to low quality individuals, and are likely to evolve condition-dependent expression. Only individuals of the highest quality are able to produce the largest ornaments. Handicaps will be subjected to strong directional selection due to mate choice and for that reason demonstrate high levels of population FA. The effect of condition on the expression of secondary sexual characters should be revealed in the relationship between individual FA and the size of secondary sex traits; individuals with the largest secondary sexual characters are predicted to demonstrate the smallest degree of asymmetry.

Fisherian sex traits are sexually attractive characters subject to directional selection due to mate choice and they should also demonstrate high levels of population FA. However, Fisherian traits, contrary to handicaps, are not particularly costly to produce or maintain. Their expression should not be condition-dependent and the relationship between the degree of individual FA and the size of the sex trait is predicted to be flat or U-shaped.

Secondary sexual characters may also signal direct fitness benefits like nuptial gifts, nest sites, territory, absence of contagious parasites, or parenting ability. Direct assessment of these benefits provided by potential partners should be more important than indirect assessment through signals, because direct assessment will be more reliable. Parenting ability is perhaps the only exception to this rule of direct assessment because it cannot be assessed directly until reproduction is in progress. Signals of parenting ability are unlikely to be very expensive, because if they were they might impair the ability to provide parental care. If sex traits signal parenting ability, they are expected to be subject to weak directional selection and demonstrate low to intermediate levels of population FA. The relationship between individual FA and the size of a secondary sexual character will be flat or U-shaped because of the relatively low costs of the sex trait.

Whether mate choice is sensitive to asymmetry in sexual traits is not yet clear. Many studies have demonstrated that females pay attention to the expression of secondary sexual characters and that males with the most extreme ornaments are preferred over less ornamented males (Andersson, 1982b; Møller, 1988, 1990b; Petrie, Halliday & Sanders, 1990; Hill, 1991). These results have usually been interpreted to mean that the size of the ornament is the character assessed by females. However, females might in addition be assessing ornament asymmetry.

There is empirical evidence that females pay attention to FA of ordinary morphological characters. For example, Markow and Ricker (1992) showed that mated male *Drosophila* exhibited less FA than unmated males, although the causal relationship between individual FA and mating status was not verified. Asymmetric male fruitflies may be at a selective disadvantage if morphological asymmetries affected their ability to deliver courtship signals. Similarly, Thornhill (1992a) showed that asymmetric male Japanese scorpionflies *Panorpa japonica* were less successful during fights for nuptial gifts, and asymmetric males had lower mating succes than symmetric males. Asymmetric males also produced pheromones that were more attractive that those of symmetric males (Thornhill 1992b). Again, the causal relationship between asymmetry and mating status could not be verified experimentally.

Individual FA and size of secondary sexual characters are strongly negatively correlated in at least three of the bird species in which an effect of ornament size on mate preference has been inferred: barn swallow (Møller, 1990a), long-tailed widowbird (*Euplectes progne*) (Møller & Höglund, 1991) and blue peacock (Manning & Hartley, 1991). It is thus not possible to determine whether ornament size or ornament asymmetry is the cue used during female choice. Experimental manipulation of the size of the ornament does not resolve this problem because elongation of a tail ornament simultaneously leads to a reduction in the relative FA, while shortening leads to an increase in the relative FA.

A recent experiment on barn swallows involving independent manipulation of both ornament size and FA in ornaments suggested that both characters influenced female choice (Møller, 1992c). Most importantly, there was a significant interaction between ornament size and asymmetry. Males with symmetric tails had higher mating success than males with asymmetric tails. But long asymmetric tails were better than short symmetric tails. This supports a good genes interpretation. It suggests that males are unable to cheat by developing short tails in which the degree of asymmetry is easier to control. The alternative hypothesis that greater success in male-male competition may explain why male barn swallows with symmetrical tails acquired mates sooner during the mating season was ruled out by the fact that male fighting rates or fighting success did not differ between experimental treatments (Møller, 1992d).

Small degrees of asymmetry may also be costly if they impair manoeuvrability in flight either because of effects on the acquisition of resources or on the ability to compete with conspecifics. For example, asymmetry in the outermost elongated tail feathers of barn swallows cause loss of manoeuvrability in a flight maze (Møller, 1991b).

An alternative interpretation of the barn swallow data suggests that females prefer males with the most symmetric ornaments because such males provide better parental care (Balmford & Thomas, 1992; Borgia & Wilkinson, 1992). This was not the case. Males with the most preferred secondary sexual characters contributed relatively less to parental duties than males with less preferred sex traits (Møller, 1993c). Another explanation for the negative relationship between ornament asymmetry and ornament size is that asymmetric sex traits are less costly for males with small ornaments for aerodynamic reasons (Balmford & Thomas, 1992). If this was the case, one should expect tail asymmetry to be positively, not negatively, related to tail length in all aerially insectivorous birds, even those species without tail ornaments. The relationship between tail asymmetry and tail length is neither positive nor negative among sexually monomorphic bird species representing independent evolutionary events of high aerial activity (Møller & Höglund, 1993). This does not support the view by Balmford and Thomas.

Discussion

In the preceding sections we have outlined how fluctuating asymmetry may prove to be an important concept in sexual selection theory. Different models of sexual selection have been suggested to account for the evolution of elaborate secondary sexual characters not directly used in male-male competition, but there is no general consensus concerning the relative importance of these models. The Fisherian and handicap models of female choice predict different correlations between the expression of sexual ornaments and the quality of individuals. We suggest that individual FA may represent a reliable, independent measure of individual quality, because FA is directly caused by the ability of individuals of given genotypes to cope with the environment (including conspecifics, parasites and pathogens). The magnitude and the patterns of individual FA in relation to the size of secondary sexual characters may therefore allow us to determine the relative importance of these two models of sexual selection.

Preliminary comparative analyses of the patterns of FA in secondary sex traits and other morphological traits clearly reveal that population FA in secondary sexual characters is much larger than in other morphological traits. This strongly suggests that ornaments are much more susceptible to disruption of developmental homeostasis than ordinary morphological traits. The strong directional selection that acts on ornaments is the most likely explanation for the high population levels of FA. This interpretation is consistent with earlier experimental data on *Drosophila* and *Rattus* demonstrating that directional selection causes a deterioration of developmental homeostasis as measured in terms of FA (Thoday, 1958; Reeve, 1960; Leamy & Atchley, 1985). Similarly the incorporation of new mutants into the genome is generally associated with increased variability and FA (Fisher & Holt, 1944; McKenzie & Clarke, 1987). In addition, the exaggerated nature of many sexual ornaments and displays means that they are costly traits to produce and maintain. Exaggeration will itself cause stress and lead to an increase in population FA.

Individual FA in ornaments is often negatively related to the mean absolute size of the paired trait, while the common pattern is a U-shaped relationship between individual FA and the size of an ordinary morphological trait (Soulé, 1982; Soulé & Cuzin-Roudy, 1982; Møller & Höglund, 1991; Møller, 1992a). A negative relationship between ornament size and FA may be due to three different factors. First, individuals with the largest ornaments may generally have higher heterozygosity and therefore be less susceptible to the effects of parasites and pathogens and a number of other factors associated with heterozygosity (see section on genetic stress and FA). Second, ornament size may reflect overall quality independently of heterozygosity, and individuals in good condition may be less susceptible to the effects of environmental stress than other individuals. For example, it is well-known that the expression of ornaments is related to condition (Andersson, 1982a, 1986; Nur & Hasson, 1984). Good condition also generally leads to less FA in morphological traits (Parsons, 1990). Third, individuals that frequent more favourable and thus less stressful environments may become more extravagantly ornamented than conspecifics in less favourable environments. It is possible that the negative relationship between FA and size has a non-genetic basis. However individuals with the best genotypes will be selected to occupy the best environments, so it seems likely that environmental and genetic associations are highly correlated.

The absolute size of ornaments may reflect that it is a reliable indicator of individual quality, because the largest ornaments also demonstrate the least FA. The maintenance of reliable signalling reflecting the quality of individuals has been a subject of some controversy. Recent game theory models suggest that costly signals reliably reveal quality because high quality individuals pay a lower cost for a given level of signalling than low quality individuals (Grafen, 1990b). Dawkins and Guilford

(1991) recently stated that current models of reliable signalling assume no cost to the receiver of eliciting and evaluating honest signals, and they use the famous roaring contests in red deer (*Cervus elaphus*) stags as an example of large receiver costs. A receiver will only be able to determine the true quality of an opponent by going through a series of escalating stages of roaring and finally fight, and the cost of evaluating the quality of an opponent may therefore be considerable. Individuals in many signalling systems should therefore, according to Dawkins and Guilford (1991) settle contests by using conventional signals rather than paying the full cost of receiving the costly signal. Such systems would be open to cheating because receivers do not intend to pay the full cost of communication, and Dawkins and Guilford (1991) suggest that many signalling systems will be open to unreliable or dishonest signalling.

We suggest that individual FA may provide a reliable signal of quality that can be perceived without much cost, if directly assessable by vision or other sensory modalities. Cheating will generally not be possible because individual FA directly reflects the ability to cope with genetic and environmental stress including parasites and pathogens, the heterozygosity of the individual and the level of adaptation to local conditions. This will apply to secondary sexual characters that demonstrate large degrees of FA and a negative relationship with character size. Small amounts of FA in non-sexual, morphological traits may not be directly assessed by conspecifics, but FA may still correlate positively with various measures of fighting ability, if asymmetry is a general quality indicator.

Acknowledgements

We thank R. Alatalo, M. Polak and S. Ulfstrand for comments on previous versions of this manuscript. The research was supported by grants from the Swedish NSRC, NERC and The Royal Society.

References

Alatalo, R. V., J. Höglund & A. Lundberg, 1988. Patterns of variation in tail ornament size in birds. Biol. J. Linn. Soc. 34: 363-374.

Andersson, M., 1982a. Sexual selection, natural selection and quality advertisement. Biol. J. Linn. Soc. 17: 375-393.

Andersson, M., 1982b. Female choice selects for extreme tail length in a widowbird. Nature, Lond. 299: 818-820.

Andersson, M., 1986. Evolution of condition-dependent sex ornaments and mating preferences: sexual selection based on viability differences. Evolution 40: 804-820.

Bailit, H. L., P. L. Workman, J. D. Niswander & C. J. MacLean, 1970. Dental asymmetry as an indicator of genetic and environmental conditions in human populations. Human Biol. 42: 626-638.

Balmford, A. & A. Thomas, 1992. Swallowing ornamental asymmetry. Nature, Lond. 359: 487.

Biémont, C., 1983. Homeostasis, enzymatic heterozygosity and inbreeding depression in natural populations of *Drosophila melanogaster*. Genetica 61: 179-189.

Boag, P. T. & P. R. Grant, 1981. Intense natural selection in a population of Darwin's finches (*Geospizinae*) in the Galápagos. Science 214: 82-85.

Borgia, G. & G. Wilkinson, 1992. Swallowing ornamental asymmetry. Nature, Lond. 359: 487-488.

Brown, R. D., 1984. Antler Development in Cervidae. Caesar Kleberg Wildlife Research Institute, Kingsville.

Bubenik, G. A. & A. B. Bubenik. Eds., 1990. Horns, Pronghorns and Antlers. Springer-Verlag, New York.

Burley, N., S. C. Tidemann & K. Halupka, 1991. Bill colour and parasite levels of zebra finches, pp. 359-376 in Bird-Parasite Interactions: Ecology, Evolution and Behaviour, edited by J. E. Loye & M. Zuk. Oxford University Press, Oxford.

Clarke, G. M. & J. A. McKenzie, 1987. Developmental stability of insecticide resistant phenotypes in blowfly; a result of canalizing natural selection. Nature, Lond. 325: 345-346.

Coates, B. J., 1990. The Birds of Papua New Guinea, Volume II Passerines. Dove Publications, Queensland.

Conner, J., 1988. Field measurements of natural and sexual selection in the fungus beetle, *Bolitotherus cornutus*. Evolution 42: 736-749.

Darwin, C., 1871. The Descent of Man, and Selection in Relation to Sex. John Murray, London.

Dawkins, M. S. & T. Guilford, 1991. The corruption of honest signalling. Anim. Behav. 41: 865-873.

Delius, J. D. & G. Habers, 1978. Symmetry: Can pigeons conceptualize it? Behav. Biol. 22: 336-342.

Delius, J. D. & B. Nowak, 1982. Visual symmetry recognition by pigeons. Psychol. Res. 44: 199-212.

Eberhard, W., 1989. Animal Genitalia. Harvard Univ. Press, Cambridge.

Endler, J. A., 1983. Natural and sexual selection on colour patterns in poeciliid fishes. Environ. Biol. Fishes 8: 173-190.

Evans, M. R. & A. L. R. Thomas, 1992. Aerodynamic and mechanical effects of elongated tails in the scarlet-tufted malachite sunbird: Measuring the cost of a handicap. Anim. Behav. 43: 337-347.

Fisher, R. A., 1930. The genetical Theory of Natural Selection. Clarendon, Oxford.

Fisher, R. A. & S. B. Holt, 1944. The experimental modification of dominance in Danforth's short tailed mutant mice. Ann. Eugen. 12: 102-120.

Freeland, W. J., 1976. Pathogens and the evolution of primate sociality. Biotropica 8: 11-24.

Freeland, W. J., 1981. Parasitism and behavioral dominance among male mice. Science, Wash. 213: 461-462.

Grafen, A., 1990a. Sexual selection unhandicapped by the Fisher process. J. theor. Biol. 144: 475-516.

Grafen, A., 1990b. Biological signals as handicaps. J. theor. Biol. 144: 517-546.

Graham, J. H. & J. D. Felley, 1985. Genomic coadaptation and developmental stability within introgressed populations of *Enneacanthus gloriosus* and *E. obesus* (Pisces, Centrarchidae). Evolution 39: 104-114.

Hamilton, W. D. & M. Zuk, 1982. Heritable true fitness and bright birds: a role for parasites? Science, Wash. 218: 384-387.

Heywood, J. S., 1989. Sexual selection by the handicap mechanism. Evolution 43: 1387-1397.

Hill, G. E., 1991. Plumage coloration is a sexually selected indicator of male quality. Nature, Lond. 350: 337-339.

Hoelzer, G. A., 1989. The good parent process of sexual selection. Anim. Behav. 38: 1067-1078.

Hoffmann, A. A. & P. A. Parsons, 1989a. An integrated approach to environmental stress tolerance and life history variation: Desiccation tolerance in *Drosophila*. Biol. J. Linn. Soc. 37: 117-136.

Hoffmann, A. A. & P. A. Parsons, 1989b. Selection for increased desiccation tolerance in *Drosophila melanogaster*: Additive genetic control and correlated responses to other stresses. Genetics 122: 837-845.

Iwasa, Y., A. Pomiankowski & S. Nee, 1991. The evolution of costly mate preferences II. The handicap principle. Evolution 45: 1431-1442.

Johnson, L. L. & M. S. Boyce, 1991. Female choice of males with low parasite loads in sage grouse, pp. 377-388 in Bird-Parasite Interactions: Ecology, Evolution and Behaviour, edited by J. E. Loye & M. Zuk. Oxford University Press, Oxford.

Jones, J. S., 1987. An asymmetrical view of fitness. Nature, Lond. 325: 298-299.

Kat, P. W., 1982. The relationship between heterozygosity for enzyme loci and developmental homeostasis in peripheral populations of aquatic bivalves (Unionidae). Am. Nat. 119: 824-832.

Kodric-Brown, A. & J. H. Brown, 1984. Truth in advertising: the kinds of traits favored by sexual selection. Am. Nat. 124: 303-323.

Lande, R., 1981. Models of speciation by sexual selection on polygenic traits. Proc. Natl. Acad. Sci. USA 78: 3721-3725.

Leamy, L. & W. Atchley, 1985. Directional selection and developmental instability: Evidence from fluctuating asymmetry of morphometric characters in rats. Growth 49: 8-18.

Leary, R. F. & F. W. Allendorf, 1989. Fluctuating asymmetry as an indicator of stress: Implications for conservation biology. Trends Ecol. Evol. 4: 214-217.

Leary, R. F., F. W. Allendorf & R. L. Knudsen, 1983. Developmental stability and enzyme heterozygosity in rainbow trout. Nature, Lond. 301: 71-72.

Leary, R. F., F. W. Allendorf & R. L. Knudsen, 1984. Superior developmental stability of heterozygotes of enzyme loci in salmonid fishes. Am. Nat. 124: 540-541.

Leary, R. F., F. W. Allendorf & R. L. Knudsen, 1985a. Inheritance of meristic variation and the evolution of developmental stability in rainbow trout. Evolution 39: 308-314.

Leary, R. F., F. W. Allendorf & R. L. Knudsen, 1985b. Developmental instability and high meristic counts in interspecific hybrids of salmonid fishes. Evolution 39: 1318-1326.

Lerner, I. M., 1954. Genetic Homeostasis. Oliver and Boyd, Edinburgh.

Livshits, G. & E. Kobyliansky, 1991. Fluctuating asymmetry as a possible measure of developmental homeostasis in humans: A review. Human Biol. 63: 441-466.

Ludwig, W., 1932. Das Rechts-Links Problem im Tierreich und beim Menschen. Springer, Berlin.

Manning, J. T. & M. A. Hartley, 1991. Symmetry and ornamentation are correlated in the peacock's train. Anim. Behav. 42: 1020-1021.

Markow, T. & J. P. Ricker, 1992. Male size, developmental stability, and mating success in natural populations of three *Drosophila* species. Heredity 69: 122-127.

Mather, K., 1953. Genetic control of stability in development. Heredity 7: 297-336.

Maynard Smith, J., 1978. The Evolution of Sex. Cambridge University Press, Cambridge.

Milkman, R., 1970. The genetic basis of natural variation in *Drosophila melanogaster*. Adv. Genet. 15: 55-114.

Mitton, J. B. & M. C. Grant, 1984. Associations among protein heterozygosity, growth rate and developmental homeostasis. Ann. Rev. Ecol. Syst. 15: 479-499.

Møller, A. P., 1988. Female choice selects for male tail ornaments in the monogamous swallow. Nature, Lond. 332: 640-642.

Møller, A. P., 1989. Viability costs of male tail ornaments in a swallow. Nature, Lond. 339: 132-135.

Møller, A. P., 1990a. Fluctuating asymmetry in male sexual ornaments may reliably reveal male quality. Anim. Behav. 40: 1185-1187.

Møller, A. P., 1990b. Male tail length and female mate choice in the monogamous swallow *Hirundo rustica*. Anim. Behav. 39: 458-465.

Møller, A. P., 1991a. Sexual selection in the monogamous barn swallow (*Hirundo rustica*). I. Determinants of tail ornament size. Evolution 45: 1823-1836.

Møller, A. P., 1991b. Sexual ornaments size and the cost of fluctuating asymmetry. Proc. R. Soc. Lond. B 243: 59-62.

Møller, A. P., 1992a. Patterns of fluctuating asymmetry in weapons: Evidence for reliable signalling of quality in beetle horns and bird spurs. Proc. R. Soc. Lond. B 248: 199-206.

Møller, A. P., 1992b. Parasites differentially increase the degree of fluctuating asymmetry in secondary sexual characters. J. evol. Biol. 5: 691-699.

Møller, A. P., 1992c. Females prefer large and symmetrical ornaments. Nature, Lond. 357: 238-240.

Møller, A. P., 1992d. Swallowing ornamental asymmetry. Nature, Lond. 359: 488.

Møller, A. P., 1993a. Patterns of fluctuating asymmetry in sexual ornaments predict female choice. J. evol. Biol. (in press).

Møller, A. P., 1993b. The function of symmetric flowers. Submitted to J. evol. Biol.

Møller, A. P., 1993c. Symmetrical male sexual ornaments, paternal care, and offspring quality. Behav. Ecol. (in press).

Møller, A. P., 1993d. Sexual selection in the barn swallow (*Hirundo rustica*). IV. Patterns of fluctuating asymmetry and selection against asymmetry. Evolution. (in press).

Møller, A. P. & M. Eriksson, 1993a. Patterns of fluctuating

asymmetry in flowers: Implications for sexual selection in plants. J. evol. Biol. (in press).

Møller, A. P. & M. Eriksson, 1993b. Flower asymmetry and sexual selection in plants. Submitted to Ecology.

Møller, A. P. & J. Höglund, 1991. Patterns of fluctuating asymmetry in avian feather ornaments: Implications for models of sexual selection. Proc. R. Soc. Lond. B 245: 1-5.

Møller, A. P. & J. Höglund, 1993. Fluctuating asymmetry and tail length in birds: Do long-tailed individuals always have less asymmetry? ms.

Møller, A. P. & A. Pomiankowski, 1993. Why have animals got multiple sexual ornaments? Behav. Ecol. Sociobiol. 32: 167-176.

Nur, N. & O. Hasson, 1984. Phenotypic plasticity and the handicap principle. J. theor. Biol. 110: 275-297.

Otronen, M., 1988. Intra- and intersexual interactions at breeding burrows in the horned beetle, *Coprophanaeus ensifer*. Anim. Behav. 36: 741-748.

Palmer, A. R. & C. Strobeck, 1986. Fluctuating asymmetry: Measurement, analysis, pattern. Ann. Rev. Ecol. Syst. 17: 391-421.

Parsons, P. A., 1962. Maternal age and developmental variability. J. Exp. Biol. 39: 251-260.

Parsons, P. A., 1990. Fluctuating asymmetry: An epigenetic measure of stress. Biol. Rev. 65: 131-145.

Petrie, M., T. R. Halliday & C. Sanders, 1990. Peahens prefer peacocks with elaborate trains. Anim. Behav. 41: 323-331.

Pomiankowski, A., 1988. The evolution of female mate preferences for male genetic quality. Oxford Surv. Evol. Biol. 5: 136-184.

Pomiankowski, A., Y. Iwasa & S. Nee, 1991. The evolution of costly mate preferences. I. Fisher and biased mutation. Evolution 45: 1422-1430.

Pomiankowski, A. & A. P. Møller, 1993. Fluctuating asymmetry and the strength of sexual selection. ms.

Prout, T., 1962. The effects of stabilizing selection on the time of development in *Drosophila melanogaster*. Genet. Res. 3: 364-382.

Reeve, E. C. R., 1960. Some genetic tests on asymmetry of sternopleural chaeta number in *Drosophila*. Genet. Res., Camb. 1: 151-172.

Rensch, B., 1958. Die Wirksamkeit ästhetischer Faktoren bei Wirbeltieren. Z. Tierpsychol. 15: 447-461.

Ryan, M. J., 1988. Energy, calling and selection. Am. Zool. 28: 885-898.

Sciulli, P. W., W. J. Doyle, C. Kelley, P. Siegel & M. I. Siegel, 1979. The interaction of stressors in the induction of increased levels of fluctuating asymmetry in the laboratory rat. Am. J. Phys. Anthropol. 50: 279-284.

Siegel, M. I. & W. J. Doyle, 1975. The differential effect of prenatal and postnatal audiogenic stress on fluctuating dental asymmetry. J. Exp. Zool. 191: 211-214.

Soulé, M. E., 1979. Heterozygosity and developmental stability: Another look. Evolution 33: 396-401.

Soulé, M. E., 1982. Allomeric variation. 1. The theory and some consequences. Am. Nat. 120: 751-764.

Soulé, M. E. & J. Cuzin-Roudy, 1982. Allomeric variation. 2. Developmental instability of extreme phenotypes. Am. Nat. 120: 765-786.

Spurrier, M. F., M. S. Boyce & B. F. J. Manly, 1991. Effects of parasites on mate choice by captive sage grouse, pp. 389-398 in Bird-Parasite Interactions: Ecology, Evolution and Behaviour, edited by J. E. Loye & M. Zuk. Oxford University Press, Oxford.

Taylor, P. D. & G. C. Williams, 1982. The lek paradox is not resolved. Theor. Pop. Biol. 22: 392-409.

Tebb, G. & J. M. Thoday, 1958. Stability in development and relational balance of X chromosomes in *D. melanogaster*. Nature, Lond. 174: 1109-1110.

Thoday, J. M., 1956. Balance, heterozygosity and developmental stability. Cold Spring Habor Symp. Quant. Biol. 21: 318-326.

Thoday, J. M., 1958. Homeostasis in a selection experiment. Heredity 12: 401-415.

Thornhill, R., 1992a. Fluctuating asymmetry and the mating system of the Japanese scorpionfly, *Panorpa japonica*. Anim. Behav. 44: 867-879.

Thornhill, R., 1992b. Female preference for the pheromone of males with low fluctuating asymmetry in the Japanese scorpionfly (*Panorpa japonica*, Mecoptera). Behav. Ecol. 3: 277-283.

Thornhill, R. & P. Sauer, 1992. Genetic sire effects on the fighting ability of sons and daughters and mating success of sons in a scorpionfly. Anim. Behav. 43: 255-264.

Valentine, D. W. & M. E. Soulé, 1973. Effect of *p,p'*-DDT on developmental stability of pectoral fin rays in the grunion, *Leuresthes tenius*. Fishery Bull. 71: 920-921.

Van Valen, L., 1962. A study of fluctuating asymmetry. Evolution 16: 125-142.

Vehrencamp, S. L., J. W. Bradbury & R. M. Gibson, 1989. The energetic cost of display in male sage grouse. Anim. Behav. 38: 885-896.

Vrijenhoek, R. C. & S. Lerman, 1982. Heterozygosity and developmental stability under sexual and asexual breeding systems. Evolution 36: 768-776.

Zahavi, A., 1975. Mate selection - a selection for a handicap. J. theor. Biol. 53: 205-214.

Zahavi, A., 1987. The theory of signal selection and some of its implications, pp. 305-327 in International Symposium of Biological Evolution, edited by V. P. Delfino. Adriatic Editrice, Bari.

Zakharov, V. M., 1981. Fluctuating asymmetry as an index of developmental homeostasis. Genetiks (Belgrade) 13: 241-256.

T. A. Markow (ed.), Developmental Instability: Its Origins and Evolutionary Implications, 283–298, 1994.

Developmental origins of variation in human hand preference

Ronald A. Yeo & Steven W. Gangestad
Department of Psychology, University of New Mexico, Albuquerque, NM 87131, USA

Received and accepted 28 April 1993

Abstract

Though right-handedness is a prominant characteristic within all human societies, a substantial and stable proportion of individuals are left-handed. Any comprehensive approach to the origin of variation in handedness must account for substantial evidence that left-handedness is associated with reduced fitness, neurodevelopmental disorders, and reduced neuroanatomical asymmetry. In this paper we investigate the hypothesis that developmental instability in early fetal development underlies variation in handedness. In two studies we note an increased incidence of minor physical anomalies (MPAs) and fluctuating asymmetries in both left-handers and extreme right-handers. Moreover, extreme right-handers were more apt to have left-handed parents than moderate right-handers. These data suggest that deviation from moderate right-handedness reflect imprecise expression of a near-universal design due to developmental instability. Preliminary attempts to elucidate the mechanisms underlying developmental instability suggest that both polygenic homozygosity and particular HLA alleles may be important factors. These observations are discussed with respect to current genetic theories of handedness and human evolution.

Introduction

Handedness is a uniquely human phenomenon. While individual members of other species may show characteristic hand or paw preference, there is little or no evidence for a species-wide preference for the left or right sides (Corballis, 1991). Handedness is also intimately linked with the neurobiological substrate for language. Left-handers are more apt to have bilateral or right hemisphere language dominance, while left hemisphere dominance is the rule for right-handers (e.g., Alekoumbides, 1978). These considerations suggest that understanding the origins and development of hand preference may provide some insight into those neural and behavioral features which uniquely characterize our species.

Right-hand preference is quite pervasive among humans. A recent survey of over one million people from the United States revealed that 90% considered themselves right-handed (Gilbert & Wysocki, 1992). Cross-cultural studies show a marked right hand preference in Alaskan Eskimos, Australian Aborigines, the Temne of Sierra Leone, Hong Kong Chinese, Native Americans, Congolese children, and people from Papua New Guinea (see Connolly & Bishop, 1992). Moreover, the incidence of right-handedness appears stable over time. Based on depictions of hand use in works of art, Coren and Porac (1977) estimated a stable right-hand predominance (approximately 90%) for at least 5,000 years.

Hand preference patterns appear to be established quite early in the course of development. Ultrasound scans of fetuses showed that 92% of those who sucked their thumb preferred their right hand (Hepper, Shahidullah & White, 1991), a figure remarkably close to the incidence of right hand preference in adults.

These observations of a pervasive human pattern of right-hand preference and a persistent incidence of left-handedness in a minority of individuals, determined early in neural development, raise the question of how this phenotypic diversity is maintained in the species. Any comprehensive treatment of this issue must consider handedness in the context of its associated traits. It is the rather bewildering array of associations of left-handedness that has generated such widespread interest in handedness within the neurosciences. These associations of left-handedness fall in three major categories. Left-handedness (or when left-handers have been grouped with ambidextrous individuals, 'nonright-handedness') has been associated with diverse indices of reduced fitness, with a host of neurodevelopmental disorders, and multiple neuroanatomical differences.

Reduced fitness

Evidence suggests left-handers have fewer children and are prone to a variety of conditions that could limit fertility. Here are the most salient observations:

1. Left-handers tend to have fewer children (Boklage, 1987; Rife, 1940). In our own data (Gangestad *et al.*, in preparation) we have examined offspring number as a function of parental handedness in 661 individuals older than 20 years of age, partialling out linear and curvilinear effects of age. Regression analyses revealed that whereas right-handers were expected to have 2.03 children by age 45, left-handers were expected to have 1.62, $t(657) = 2.21$, $p < .05$.[1]
2. Left-handers tend to have more spontaneous abortions (Fraser & Rex, 1984, as cited in Boklage, 1987).
3. Left-handers may be older at puberty (Coren, Scarleman & Porac, 1986) and younger at menopause (Leidy, 1990).
4. Halpern and Coren (1991) have reported that left-handers may die at earlier ages, though much controversy surrounds this issue (see letters responding to their 1991 paper in the October 3, 1991, issue of the *New England Journal of Medicine*). They sent letters to the nearest relatives of individuals from two counties in California for whom death certificates had been obtained. Those determined to be right-handers lived an average of nine years longer than the left-handers. Many cross-sectional studies of the relationship between age and hand preference note a decreasing incidence of left-handedness with advancing age (e.g., Fleminger, Dalton & Standage, 1977; Porac, Coren & Duncan, 1980; Smart, Jeffrey & Richards, 1980). It appears unlikely that secular changes in prohibitions against left-handedness can fully account for the aging effect (Coren & Halpern, 1991), raising the possibility that the cross-sectional change reflects earlier death among left-handers.
5. Nonright-handers may suffer more accidents requiring medical attention (Coren, 1989; Daniel & Yeo, 1991) and left-handers may suffer more head injuries producing a loss of consciousness (Daniel, Yeo & Gangestad, 1992).
6. Certain immune disorders, especially those of the thyroid and gut (Bryden & McManus, 1992; Geschwind & Galaburda, 1984) may be more common in left-handers.
7. Among children born with very low birthweight, left-handers may be overrepresented (O'Callaghan *et al.*, 1987).

Neurodevelopmental disorders

Left-handedness has been reported to be more common in a variety of central nervous system disorders that presumably reflect developmental rather than acquired abnormality. These include neural tube defects (Fraser, 1983), autism (Boucher, 1977; Colby & Parkinson, 1977), alcoholism (London, 1990; Nasrallah, Keelor & McCalley-Winters, 1983), psychopathy (Ellis, 1990), stuttering (Bishop, 1990), and schizophrenia (Lishman & McMeekan, 1976). Cleft-palate syndrome is also associated with an elevated incidence of left-handedness (Fraser & Rex, 1984). It should be noted, however, that for all the conditions listed above the effect size (right- vs. left-handers) is rather small and negative results have also been reported (see Bishop, 1990, for a discussion of methodological issues). A related observation is that males, who are more prone to neurodevelopmental disorders in general (Gualtieri & Hicks 1985), are more often left-handed (Annett, 1985).

Neuroanatomy

There are numerous asymmetries in the human brain. In general, the anterior portions of the right hemisphere are larger than the homologous areas of the left, while the posterior regions of the left hemisphere are relatively larger. A variety of asymmetries exist across the left and right perisylvian regions, presumably representing the anatomical substrate for asymmetries in language representation

[1] We also examined interactions of handedness with gender and age. These interactions were not significant and hence the effects of handedness on number of offspring seem not to be age-dependent.

All of these asymmetries are attenuated in left-handers (for reviews of anatomic asymmetries and the role of handedness see Galaburda, 1991; Geschwind & Galaburda, 1987; Witelson, 1980; for reviews of functional asymmetries see Bishop, 1990; Bryden, 1982). That is, left-handers as a group show greater *symmetry* rather than *reversed asymmetry*. Differences in callosal morphology may also be associated with handedness (Habib *et al.*, 1991; Witelson, 1985; 1989), but present data are inconclusive (e.g., Kertesz *et al.*, 1987). When differences are observed, nonright-handers seem to have a relatively larger corpus callosum; this effect may be relatively greater in males (Witelson, 1989).

Genetic theories of variation in handedness

A theory of individual variation in handedness should not only account for the well-known patterns of familial resemblance, which indicate the importance of genetic similarity rather than social learning (Carter-Saltzman, 1980; McManus, 1991; Risch & Pringle, 1985), but also provide some insight into the rich set of associations described above. Among the genetic approaches Annett's 'right shift' theory (Annett, 1985, 1991) is clearly the best known, and along with that of McManus (1985), the most viable. These theories share several features, though for the present discussion we will focus on Annett's. The theory proposes that two alternate alleles influence cerebral dominance, and secondarily, hand preference. In a homozygous state one allele (rs++, for 'homozygous right shift plus') codes for left cerebral dominance and hence right-handedness. In a homozygous state the other allele (rs−−) exerts no bias on cerebral dominance. Chance factors determine cerebral dominance and handedness in these individuals. A majority of heterozygotes become right-handed, but some may become left-handed. Hence, the fact that less than 50% of offspring of two left-handed parents are themselves left-handed (26%, according to a recent meta-analysis [McManus, 1991]) reflects the fact that some proportion of phenotypic left-handers (the heterozygotes) possess genes that code for left cerebral dominance. This single gene model is fairly simple and accounts well for family data, a factor that has surely contributed to its substantial popularity.

A major weakness of both Annett's and McManus' single gene models, however, is that they do not provide a plausible account of the evolutionary basis of variation in handedness. Annett (1985) has been most explicit in her theorizing on this issue. Based on estimated gene frequencies, she has proposed that strong selection pressures favor individuals heterozygous at the single locus influencing cerebral dominance. Different cognitive liabilities are posited for individuals at either end of the continuum of relative hand performance. Deficits in phonological processing are more apt to be seen in those with relatively greater left hand skill (Annett, 1992), while deficits in intelligence and mathematics are more frequent in those with relatively greater right hand skill (Annett & Manning, 1989). Heterozygotes, in the middle portion of the distribution of relative hand skill, are said to be prone to neither of these difficulties, an advantage hypothesized to drive gene frequencies. There is little evidence, however, that variations in phonology, reading, mathematics, or intelligence are linearly related to fitness. If directional selection currently acted on such skills, one would not expect to see significant population variability attributable to additive genetic variance (Fisher, 1959).

Thus, although Annett's single-gene model may adequately account for patterns of familial resemblance, it may be founded on untenable assumptions. In addition to offering no plausible basis for genetic variability, the model also has a limited ability to account for the associations of handedness described above, in particular the associations of neurodevelopmental disorders with left handedness.

Annett (1991) offers two mechanisms to account for these associations: frank brain damage to the left hemisphere and the effects of the rs+ gene on hemispheric development. While perinatal brain damage may account for associations of left-handedness with mental retardation, the rs+ effects are important for individual variation in cognitive skills. The rs+ gene 'prunes certain right hemisphere neurons at a critical stage of cerebral growth. This weakens the capacity of the right hemisphere to discriminate speech sounds in comparison with the left hemisphere and incidentally weakens control of the left side of the body' (Annett, 1991; p. 228). Could the *absence* of this unilateral pruning underlie the elevated incidence of left-handedness with neurodevelopmental disorders? Perhaps so, if the deficient skill requires unilateral control, as may be the case for fluent language production (Helm-Estabrooks *et al.*, 1986). But it is difficult to see how reduced right hemisphere pruning might relate to autism, alcoholism, psychopathy, or schizophrenia.

A very different theoretical approach to accounting for individual differences in handedness has been offered by Bakan (1990). He believes that the primary cause of left-handedness is perinatal anoxic damage to the left cerebral hemisphere, which controls the right hand. Though in principle this approach can account for the association of left-handedness with neurodevelopmental disorders, it has difficulty in accounting for patterns of familial resemblance in handedness and differences in neuroanatomy, which are likely determined long before birth (Witelson & Pallie, 1973). There are other difficulties. A recent meta-analysis (Searleman, Porac & Coren, 1989) determined that less than one percent of the variance in handedness can be attributed to birth stress factors. Along with the observations of Hepper, Shahidullah and White (1991) cited earlier regarding fetal thumb sucking preferences, these findings suggest that frank brain pathology contributes little to normal variation in handedness.

Homozygosity, developmental instability, and handedness

Our approach to variation in handedness emerged from consideration of this question: How is a balanced polymorphism for handedness maintained in the face of considerable evidence for reduced fitness among left-handers? Of the various ways genetic diversity may be maintained - most notably, neutral selection, frequency-dependent selection, diversifying selection, heterozygote selection - the latter seemed the most plausible with regard to handedness. Through heterozygote advantage, variation can be maintained despite differential reproduction across genotypes.

The purported association of homozygosity with developmental instability (Kat, 1982; Leary, Allendorf & Knudsen, 1975; 1983; 1984; Livshits & Kobylianski, 1991; Mitton & Grant, 1984; Soule & Couzin-Roudy, 1982) suggested a specific hypothesis. Perhaps variation in handedness away from the population mean reflects imprecise expression of a universal (or 'near-universal', see below) neural design for left cerebral motor dominance, due to developmental instability at a critical period in fetal development. Both left-handedness and extreme *right-handedness* (which also represents a deviation from the mean) may reflect early developmental instability. We have recently conducted a series of investigations designed to evaluate the hypothesis that developmental instability in early fetal development underlies individual variation in the neural substrate of handedness. Van Valen (1978) first suggested the hypothesis in a brief comment; Markow (1992a, 1992b) has elaborated the idea and provided extensive theoretical support.

In our first two studies of normal, young adults (Yeo, Gangestad & Daniel, 1993) we related handedness variables to two sets of variables reflecting developmental instability: minor physical anomalies (MPAs) and fluctuating asymmetries (FAs). MPAs are physical features, first described by Down (1866), associated with a variety of neurodevelopmental disorders, including schizophrenia (Gualtieri *et al.*, 1982; Guy *et al.*, 1989), autism (Campbell *et al.*, 1978), hyperactivity (Gualtieri *et al.*, 1982; Quinn & Rapoport, 1974; Rapoport & Quinn, 1975; Waldrop & Halverson, 1971), and dyslexia (Cummings, Flynn & Preus, 1982; Steg & Rapoport, 1975; Von Hilsheimer & Kurko, 1979). MPAs reflect deviations in development during the first trimester of pregnancy (Waldrop & Halverson, 1971; Waldrop, Halverson & Shetterly, 1989). Some MPAs probably result from slowed development of given features in time-locked sequences. For example, slowed development in this time-locked developmental sequence may result in the two MPAs of wide-spaced eyes and low ears. During the first trimester of development the eyes gradually move toward each other and the ears migrate upwards from the neck. Other MPAs appear to result directly from disrupted early development (e.g., malformed ears).

Along with the standard assessment of MPAs (Waldrop & Halverson 1971; Waldrop, Halverson & Shetterly, 1989) we have measured several FAs. FAs are deviations in population-typical symmetry in bilateral characteristics, such as foot breadth, biepicondylar breadth, and ear breadth (see Livshits & Kobylianski, 1989, for details). They are consid ered classic markers of developmental instability (e.g., Van Valen, 1962; Parsons, 1990). Schizophrenia (Markow & Wandler, 1986; Markow & Gottesman, 1989; Mellor, 1992), low birth-weight (Livshits *et al.*, 1988), and cleft-palate syndrome (Adams & Niswander, 1970) have each been associated with relatively greater fluctuating asymmetry.

In Study 1 seventeen MPAs were assessed (see Table 1 for a list) and a single FA was measured, asymmetry of the atd angle across the two palms (Penrose, 1954). These variables were standardized and summed to provide a single overall measure of developmental instability. This measure was then related to hand preference (Annett's [1970] Handedness Questionnaire) and relative hand performance, the average difference in seconds (over five trials) between the two hands on Annett's (1985) Peg Moving Task (1985). Linear relationships were observed between developmental instability and both hand preference ($r = .50$, $p < .001$ .001) and hand performance ($r = .47$, $p < .001$), with left-handers showing greater evidence of developmental instability. Importantly, a significant curvilinear (quadratic) relationship was also observed with relative hand performance (beta = .33, $t(84) = 3.44$, $p < .001$). (Similar curvilinear relationships would not be expected to emerge with the hand preference measure, since even a slight consistent skill difference may lead to a complete hand preference). This relationship is shown in Figure 1. Two features of this relationship are noteworthy. First, extreme right-handers, as well as left-handers, showed greater evidence of developmental instability. Second, the minimum of the developmental instability curve falls near the median of the frequency distribution for relative hand skill.

Study 2 sought to replicate these effects and extend them to a broader array of fluctuating asymmetries. Table 1 (part B) shows the set of fluctuating asymmetries included in Study 2. The same MPA evaluation was conducted and a total developmental instability composite score was created from all MPAs and FAs. Results replicated Study 1. Linear relationships were observed between developmental instability and both hand preference and hand performance. A curvilinear relationship was also found, as shown in Figure 2 (beta = .45, $t(107) = 3.28$, $p < .002$). Again, extreme right-handers, as well as left-handers, showed greater developmental instability. The minimum of the developmental instability function again fell near the median of the distribution for relative hand skill.

Table 1 presents the individual correlations of FAs and MPAs with handedness measures when the samples from Studies 1 and 2 are combined. It

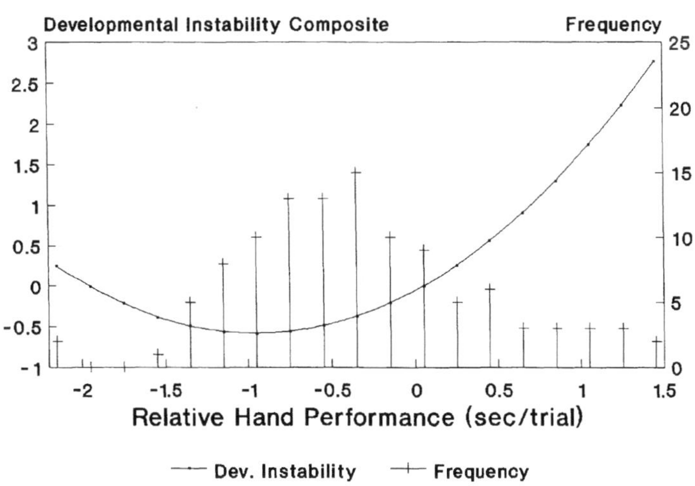

Fig. 1. The relationship between the developmental instability composite and relative hand performance values: Study 1. Curve is based on multiple regression of developmental instability composite on linear and quadratic components of relative hand performance.

Table 1. Correlations of individual MPAs and FAs with hand preference and linear and quadratic components of relative hand performance.

Item	Hand preference	Linear component Hand performance	Quadratic component Hand performance
I. Minor physical anomalies			
Electric hair	.20	.13	.13
Multiple hair whorls	.21	.15	.06
Abnormal head circumference	.05	.00	.10
Epicanthic folds	.08	.15	.04
Wide spaced eyes	.12	.13	-.02
Malformed ears	.09	.03	-.04
Low-seated ears	.10	.06	.02
Adherent ear lobes	.10	.06	.02
High palate	.11	.09	.00
Tongue furrows	.09	.03	-.05
Smooth or rough tongue	.12	.08	-.06
Curved fifth finger	.19	.10	-.02
Single Palmar crease	-.01	.15	.00
Longer 3rd toe	-.05	-.03	-.04
Gap between first 2 toes	.20	.05	.04
II. Fluctuating asymmetries			
atd angle	.16	.15	.29
Foot width	.03	.03	.03
Ankle width	.30	.22	.04
Hand width	.05	.08	.13
Wrist width	-.17	.05	-.11
Elbow width	.08	-.01	.09
Ear width	.08	.07	.11
Ear length	.28	.18	.08

Note. No subjects showed two particular MPAs, soft and pliable ears and partial syndactlia of the two middle toes.

can be seen that the overall relationship of handedness to the developmental instability composite score was not carried by a small subset of particular items. The percentage of items that correlated in the predicted direction with hand preference, hand performance, and the curvilinear component of hand performance were 87% (21 of 24), 96% (23 of 24), and 75% (18 of 24), respectively.

In a third study we evaluated differing predictions of the single-gene theories and the developmental instability hypothesis with respect to familial transmission patterns (Gangestad & Yeo, in preparation). According to both approaches, left-handed parents should produce relatively more left-handed offspring than right-handed parents. The developmental instability hypothesis additionally predicts that, because they possess genotypes predisposing developmental instability, left-handed parents should also produce relatively more extreme right-handed offspring. By contrast, the single-gene models of Annett (1985) or McManus (1985) simply do not yield this prediction. Indeed, Annett's model predicts that left-handed parents will produce relatively few extreme right-handers.

In a sample of 234 individuals, parental handedness, as assessed through subjects' report, was examined as a function of relative hand skill on the Peg Moving Task. Parental handedness was treated as a continous variable with three levels: 0 = no left-handed parents, 1 = one left-handed parent, and 2 = two left-handed parents. As predicted by the developmental instability hypothesis, the quadratic component of relative hand performance significantly predicted parental handedness, once the linear effect was controlled for, beta = .18, $t(233) = 2.46$, $p < .02$.

To examine the separate contributions of maternal and paternal handedness for this curvilinear effect, we performed PROBIT analyses treating the probability of having a left-handed mother or father as dependent measures and the linear and quadratic aspects of relative hand performance as predictors. Results yielded a significant curvilinear relationship between relative hand performance and maternal handedness, $t(222) = 2.05$, $p < .05$. While in the predicted direction, this curvilinear relation was not significant for paternal handedness.

Figure 3 presents the relationship between parental handedness and subjects' relative hand performance. Note the correspondence of the minimum of this function for parental handedness with the minima for developmental instability revealed in Figures 1 and 2. These results provide rather clear support for the developmental instability hypothesis and appear incompatible with single-gene theories.

Results of these studies have three major implications for theories of individual variation in cerebral dominance and handedness. First, variations in early fetal development appear to be quite important for variations in the neural substrate for handedness. MPAs are clearly associated with the first trimester of development. FAs probably result from variability in growth rates over longer periods

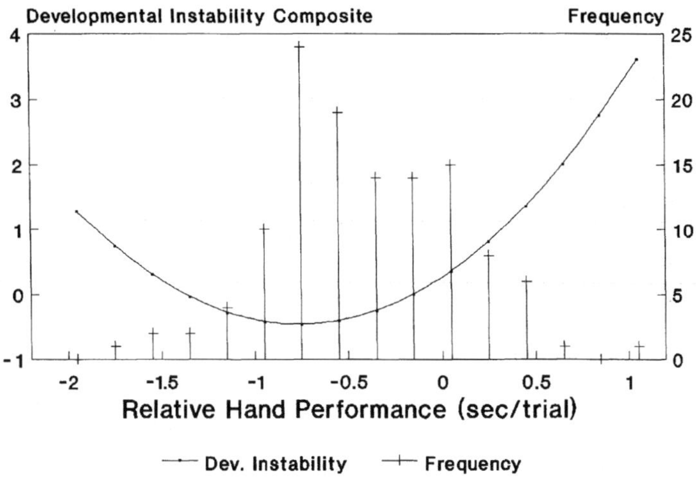

Fig. 2. The relationship between the developmental instability composite and relative hand performance values: Study 2. Curve is based on multiple regression of developmental instability composite on linear and quadratic components of relative hand performance.

of time, but as they have been associated with perinatal complications (Livshits *et al.*, 1988), variations in growth rates during prenatal development appear to be important determinants of FAs.

Second, the possible association of developmental instability with polygenic homozygosity raises concern about any theory of variation in handedness that posits only a single gene. Other aspects of our data are also inconsistent with single-gene theories. Neither Annett (1985) nor McManus (1985) would predict the association of parental left-handedness with extreme right-handedness in off-

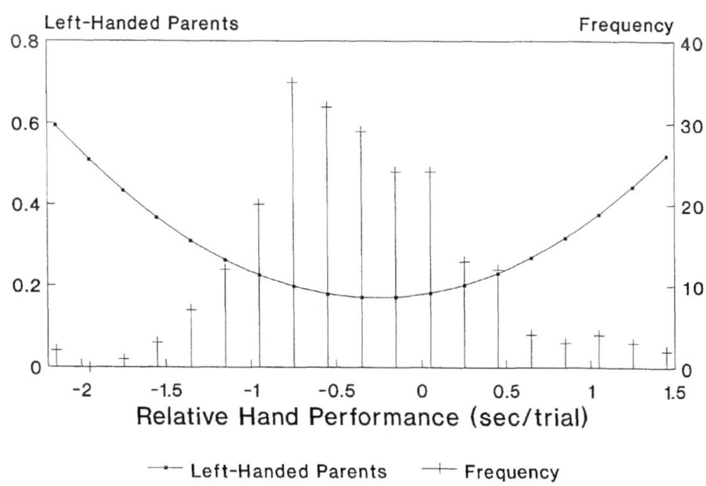

Fig. 3. The relationship between number of left-handed parents and offspring relative hand performance. Curve is based on multiple regression of number of left-handed parents on linear and quadratic components of relative hand performance.

spring. Similarly, neither would predict that extreme right-handers would share with left-handers an elevated incidence of markers of developmental instability.

Third, the theoretical underpinnings of developmental instability may provide a unique account of the association of nonright-handedness with neurodevelopmental disorders. Imprecise expression of the basic neural design, due to developmental instability, may contribute to each of these various disorders. The specific neurobiologic abnormalities underlying these disorders may well be unique, but the factors leading to developmental abnormalities may be quite similar. As noted earlier, many neurodevelopmental disorders show an increased incidence of both MPAs and FAs. Naturally, our findings would suggest that extreme right-handers will similarly show an increased incidence of neurodevelopmental disorders.

We propose that most variation in handedness away from the population mean reflects the influence of early developmental instability. This implies that there is little or no variability in the direct genetic coding for the neurobiologic substrate for handedness. Individuals vary only in the genes that determine the precision with which a design for moderate left hemisphere dominance is expressed. We do not deny the possibility that a small proportion of phenotypic left-handers may have suffered early left hemisphere damage (Satz *et al.*, 1985). In such cases developmental instability need not be a major determinant of handedness. It is perhaps worth noting that the developmental instability approach should not be considered a 'pathological' theory of handedness; no brain damage is hypothesized, only regional variations in fetal growth rates. Such variations may be distinctly non-pathological and help account for occasional reports of 'special talents' among left-handers (Geschwind & Galaburda, 1987).

The phenomenon of situs inversus, in which the organs of the viscera are reversed from their typical asymmetry, presents an interesting contrast to variations in cerebral dominance. Is there an analogous condition in the brain, i.e. a complete reversal rather than imprecise expression to varying degrees? If so, those left-handers with unilateral right hemisphere language dominance (approximately 13% of all left-handers, according to Alekoumbides, 1978) would be likely candidates. If such individuals did not have an elevated incidence of developmental instability markers, they might represent 'reversal' due to some process other than developmental instability. In that case, the basic neural design for moderate left hemisphere dominance would be 'near-universal' rather than 'universal'. Left-handedness, however, is not more common in individuals with situs inversus (Woods, 1986).

The mechanism underlying developmental instability

These three studies lend clear support for the hypothesis that nonmodal handedness is associated with indicators of developmental instability during the prenatal period. But what are the mechanisms that underlie developmental instability? In this section, we consider this question.

Protein homozygosity

As we have noted, a literature that extends back 35 years links developmental instability to widespread genomic homozygosity. In theory, multiple alleles afford multiple molecular forms of a gene's product at a single locus, the result of which is an increased ability to compensate metabolically for environmental perturbations (as reviewed in Markow, 1992a, 1992b). In practice, homozygosity has been found to be associated with both fluctuating asymmetry and abnormal growth rates (see Mitton & Grant, 1984) in a variety of nonhuman species. In light of theory and data, then, it seemed to us reasonable to suspect that the developmental instability underlying variation in handedness might be associated with protein homozygosity. Indeed, it was consideration of the effects of homozygosity on developmental instability that initially led us to examine the effects of developmental instability on handedness variation.

Thus far, we have performed two sets of analyses directly relevant to this speculation. As Soule (1984) notes, individuals with relatively many homozygous alleles with respect to traits that have additive genetic variance tend to be overrepresented at the extremes of the distributions of those traits (see also Falconer, 1981). Individuals who are heterozygous tend to be overrepresented at the middle of the distributions. Livshits and Kobyliansky

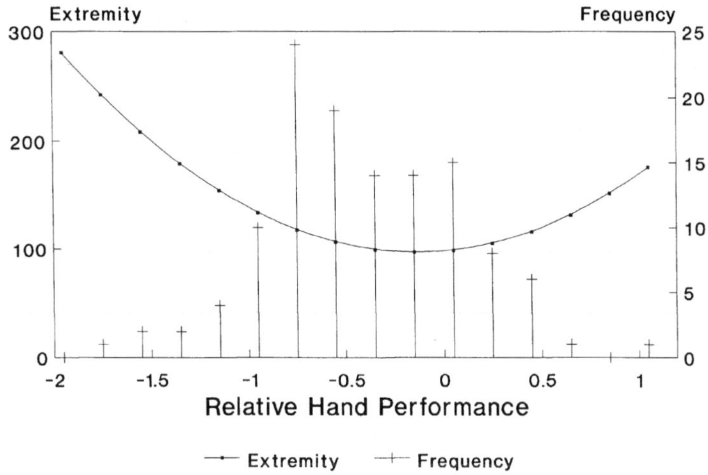

Fig. 4. The curvilinear relationship between phenotypic extremity (summed over seven physical characteristics) and relative hand performance. Curve is based on multiple regression of phenotypic extremity on linear and quadratic components of relative hand performance.

(1987) examined dermatoglyphic asymmetries in samples of individuals with modal and extreme metric characters (e.g., on stature). As expected, those with extreme characters tended to exhibit greater asymmetry.

In our study of skeletal fluctuating asymmetry, we had linear measurements of seven different morphological traits for each of the two sides of the body (again, feet, ankles, hands, wrists, elbows, ear length, and ear breadth). For each measurement, we obtained a measure of extremity by averaging the sizes of the two sides of the body, calculating a mean size for men and women separately, and squaring the deviation of each individual's size from the appropriate mean. Then, to obtain a measure of overall extremity, we averaged these squared deviations across the seven measures for each individual. To the extent that these characters have additive genetic variance (and there is little reason to believe that they do not), these averaged squared deviations should tend to reflect overall degree of homozygosity on genes related to these characters.

Within a sample of 118 individuals, the aggregated measure did not correlate substantially with left hand preference, $r = .02$, *ns*, and correlated negatively with superior left hand performance at a marginal level of significance, $r = -.18, p < .10$. As expected, however, the measure possessed a significant curvilinear relation with relative hand performance, $r = .22, p < .02$, such that individuals with extreme relative hand preference scores tended to possess extreme phenotypes and hence relative homozygosity. The minimum of the regression function predicting extremity from linear and curvilinear components of relative hand performance was a value near ambidexterity. Figure 4 shows this relationship.

More recently, we examined the possible relationship between phenotypic extremity and handedness in a large population of children (Turkheimer *et al.*, in preparation). Table 3 shows means and standard deviations for three physical features (height, weight, and head circumference) as a function of handedness and gender for a sample of almost 6,000 children assessed in the National Collaborative Perinatal Project, a large, multi-center study of perinatal and postnatal development (Broman, Nichols & Kennedy, 1975). The subjects included in Table 3 are those who had measurements taken at each of seven time points (birth, four months, eight months, one year, three years, four years, and seven years). Handedness was determined at age seven by the examining neurologist, who observed hand use across a variety of tasks. Across all three types of measures left-handers tend to show slightly greater variance. Of the 42 specific

Table 2. Height, weight, and head circumference as functions of handedness, gender, and age: Data from the National Collaborative Perinatal Project.

	Right-handers		Left-handers	
	Mean	S.D.	Mean	S.D.
A. *Height* (cm)				
Males (N = 2802 Right-handed, 447 Left-handed)				
Birth	49.86	2.58	49.81	2.58
4 months	62.60	3.02	62.43	3.01
8 months	70.06	2.81	69.77	2.82
1 year	75.22	2.98	74.87	2.96
3 years	94.92	3.81	94.43	4.03
4 years	102.22	4.19	101.74	4.33
7 years	122.09	5.37	121.41	5.59
Females (N = 2957 Right-handed, 333 Left-handed)				
Birth	49.16	2.56	49.12	2.76
4 months	61.14	2.77	61.12	3.15
8 months	68.35	2.73	68.56	2.93
1 year	73.68	2.96	73.60	2.99
3 years	93.95	3.83	93.92	3.83
4 years	101.43	4.23	101.48	4.26
7 years	121.31	5.37	121.19	5.62
B. *Weight (kg)*				
Males (N = 2905 Right-handers, 472 Left-handers)				
Birth	3.22	.49	3.23	.50
4 months	6.59	.94	6.56	.96
8 months	8.81	1.13	8.81	1.21
1 year	10.05	1.18	9.95	1.24
3 years	14.54	1.76	14.47	1.76
4 years	16.54	2.08	16.49	2.08
7 years	23.69	3.77	23.66	4.09
Females (N = 3073 Right-handers, 340 Left-handers)				
Birth	3.11	.48	3.12	.52
4 months	6.10	.85	6.12	.89
8 months	8.19	1.01	8.34	1.07
1 year	9.42	1.14	9.54	1.18
3 years	13.94	1.81	14.15	1.81
4 years	15.96	2.12	16.27	2.17
7 years	23.34	4.15	23.50	4.14
C. *Head Circumference (cm)*				
Males (N = 2802 Right-handers, 447 Left-handers)				
Birth	33.96	1.49	33.97	1.57
4 months	41.32	1.43	41.30	1.50
8 months	44.60	1.42	44.57	1.46
1 year	46.43	1.41	46.37	1.47
3 years	49.69	1.47	49.61	1.54
4 years	50.43	1.47	50.43	1.59
7 years	51.82	1.53	51.81	1.64
Females (N = 2957 Right-handed, 333 Left-handed)				
Birth	33.43	1.47	33.40	1.60
4 months	40.32	1.40	40.32	1.35
8 months	43.45	1.35	43.54	1.29
1 year	45.37	1.41	45.43	1.39
3 years	48.79	1.43	48.82	1.47
4 years	49.61	1.45	49.72	1.49
7 years	50.98	1.50	51.08	1.61

comparisons of variance possible (across measures, time periods, and gender), left-handers show greater variance in 33, right-handers in five, and handedness groups are equal in four. Statistical analyses on variance estimates obtained by partialling out gender revealed significantly ($p < .05$) or marginally significant ($p < .10$) greater variance among left-handers for 10 of 21 comparisons; all but two comparisons were in the predicted direction.

These findings are generally consistent with the speculation that generalized homozygosity underlies the developmental instability associated with handedness. Naturally, our findings are based on just a few quantitative measures, are not fully consistent, and hence require replication. The issue of whether and the extent to which total homozygosity contributes to the developmental instability underlying handedness, then, remains one that future research must more fully address.

Host-parasite coevolution and infection

Recently, we have begun to explore the role of another potential genetic underpinning of developmental instability underlying handedness. Certain sources of environmental perturbation that cause developmental instability are pathogens such as viruses, bacteria, and parasites (Parsons, 1990). Hence, Livshits *et al.* (1988) found that a significant predictor of fluctuating asymmetry across eight different skeletal measures in infants was maternal infection during pregnancy. Schizophrenia, which has been associated with both minor physical anomalies and fluctuating asymmetry, as noted earlier, has been linked to maternal infection during the fifth gestational month in multiple studies (O'Callaghan *et al.*, 1991; Mednick, Machon & Huttenen, 1988; Sham *et al.*, 1992). Experimental work with swallows has shown that manipulated parasite load substantially affects fluctuating asymmetry of their tails (Møller, in press).

Within the last 15 years, substantial theory and research on the evolution of hosts' resistance to pathogens has emerged (e.g., Anderson & May, 1982; Hamilton, 1982; Tooby, 1982). Pathogens, unlike other potential sources of untoward environmental perturbation, are not an unchanging aspect of the environment to which hosts adapt; pathogens themselves evolve. Moreover, because of their rapid reproduction pathogens have an advantage in conflicts with their hosts: They can adapt to them quickly. One defense against pathogens available to hosts is substantial polymorphism. When pathogens adapt to one genetic allele, alternative alleles may be advantaged; when they adapt to the latter, the former may again be advantaged. Host-pathogen coevolutionary cycles result from conflicts between hosts and pathogens. These coevolutionary cycles produce temporal variation in the fitnesses of the host's genetic alleles and thereby lead to the maintenance of genetic polymorphism within the host population. These cycles also maintain additive genetic variation in pathogen resistance within the host population (Anderson & May, 1982; Hamilton, 1982).

These notions suggest a source of the genetic underpinnings of developmental instability: Individuals prone to developmental instability may lack alleles that are associated with pathogen-resistance. Of course, given temporal variation in the fitnesses of alleles subject to host-pathogen coevolutionary cycles, the alleles that are disadvantaged today may be advantaged at some later point in the coevolutinary cycle and, hence, there may be nothing intrinsically disadvantaged about the genetic make-ups of individuals with developmental instability. Indeed, alleles that are advantaged and associated with developmental stability today may be disadvantaged later and therefore associated with developmental instability.

We have begun to explore the possibility that developmental instability underlying handedness is associated with lack of pathogen-resistance by examining the association between HLA alleles and handedness.

Handedness and the HLA system

One genetic system that is probably subject to host-pathogen coevolution is the human leucocyte antigen (HLA) system. HLA antigens code for cell-surface markers that are used by the immune system to discriminate 'self' from 'non-self'. Foreign bodies that do not possess an individual's HLA markers are recognized as 'non-self' and antibodies mobilize to attack them. Because individuals are advantaged by possessing HLA antigens that are uncommon in pathogens, HLA alleles are prime candidates to be subject to host-pathogen coevolutionary cycles. The fact that HLA sites are highly polymorphic, with up to 50 alleles having been

Table 3. HLA allelic frequencies as a function of subject handedness.

Allele or haplotype	Right-handers	Left-handers	t
A1	.26	.33	1.11, NS
B8	.16	.27	1.76, $p < .05$
A1/B8	.13	.24	2.18, $p < .02$
DR3	.12	.33	1.91 $p < .05$

Note. Frequencies reflect proportion of individuals who have particular allele (A1, B8, or DR3) or haplotype (A1/B8); t-tests compared proportions of right and left-handers. For analysis of A1 and B8 alleles and the A1/B8 haplotype, N = 420 right-handers, 45 left-handers. For DR3, N = 212 right-handers, 9 left-handers.

identified at single loci, empirically supports this claim (Tooby, 1982).

Interestingly, certain HLA alleles are associated with a cluster of autoimmune diseases that are relatively prevalent among nonright-handers. The B8 and DR3 alleles are overrepresented within populations of individuals suffering from celiac disease, Graves' disease, Hashimoto's thyroiditis, myasthenia gravis, systemic lupus erythematosus, and juvenile-onset diabetes (Tiwari & Terasaki, 1986), all disorders overrepresented in left-handers (Bryden & McManus, 1992). A third allele, A1, also tends to be associated with these disorders, although generally more weakly. These three alleles are in linkage disequilibrium within Caucasian populations; hence the A1-B8 and B8-DR3 haplotypes are relatively common in whites. Given their association with autoimmune diseases, we suspected that these alleles might also be overrepresented in left-handers.

To assess this hypothesis, we sent by mail an Annett handedness questionnaire to about 1100 persons who had been HLA-typed at a blood bank in Albuquerque. We received 703 questionnaires in return (and thus achieved a return rate of over 60%). Of these, 665 classified themselves as either Anglo or Hispanic. HLA antigen frequencies vary across racial groups and so we included only these two groups for analysis. All of these respondents had been typed at the A and B HLA loci; 221 had been typed at the DR locus.

We divided the sample into subsets of right- and left-handers by splitting it at the midpoint of the handedness questionnaire, and then calculated the percentages of individuals within each category possessing an A1, B8, and DR3 allele. Results are given in Table 3. As can be seen, left-handers were significantly more likely to possess a B8 or DR3 allele than were right-handers, ps $< .05$. The effect on A1 was in the predicted direction but not significant; as we note above, however, this allele is less associated with handedness-related autoimmune diseases than are B8 and DR3. We also examined whether left-handers were more likely to possess an A1-B8 haplotype than were right-handers. As can be seen, this effect was significant, $p < .02$.[2]

Might these alleles currently be associated with less fitness? To assess this possibility, we compared the number of offspring - one component of fitness - of individuals with and without A1, B8, and DR3 alleles, selecting only those individuals 20 years and over and partialling out the linear and curvilinear effects of age as covariates. Numbers of offspring predicted at age 45 on the basis of regression analyses are provided in Table 4. As can be seen, persons who possessed a B8 allele were predicted to have significantly fewer offspring than individuals without a B8 allele, means = 1.66 and 2.01, respectively, $p < .01$. The A1 and DR3 alleles were not significantly associated with reduced family size, although all observed differences were in the expected direction.[3]

An interpretation of these data is that the B8 allele is now associated with lower fertility because it (or, perhaps more likely, a closely linked allele)

Table 4. Multiple regression predictions of offspring number for left- and right-handers by age 45, controlling for linear and quadratic effects of age.

Allele or haplotype	Offspring number with allele	Offspring number without allele	t
A1	1.84	2.00	1.51, NS
B8	1.66	2.01	2.78, $p < .01$
A1/B8	1.63	2.01	2.79, $p < .01$
DR3	2.02	2.04	.09, NS

[2] When we partialled out ethnicity (Hispanic vs. non-Hispanic Caucasians) from these correlations, we found an almost identical effect and hence these relations were not due to a difference between our ethnic populations.

[3] This B8 effect was not moderated by age or gender.

is currently related to greater susceptibility to attack by certain prevalent pathogens, perhaps particularly those that threaten the development of fetuses. Developmental instability associated with handedness may hence be associated with this allele. Naturally, alternative interpretations are possible and, thus, future research should further explore the role of the immune system in developmental instability.

Future directions

Developmental instability appears to be an important determinant of variations in human handedness. We have begun to evaluate specific genetic underpinnings, but much more work needs be done. Our efforts clearly fall short of establishing a comprehensive theory of variations in cerebral dominance and handedness. Important questions remain unanswered at a variety of levels of analysis. In terms of genetic factors, we suspect polygenic homozygosity may be important due to the benefits of protein polymorphism; particular HLA alleles may be important due to disease resistance. Are there other specific genes related to developmental instability? Is homozygosity at any particular locus especially important?

If infection is a determinant of developmental instability, which is more important: individual variation in the risk of becoming infected, or individual variation in the ability to fight off or cope with an infection? Whose immune system is more important, that of the mother or the fetus? In addition to such questions of mechanism, we need to determine the particular time periods during fetal life associated with variations in neural development. The associations of handedness with MPAs and particular dermatoglyphic features (Leche, 1933; Newman, 1934; Rife, 1955) point to the end of the first trimester, but these physical features may represent correlates of even earlier neurodevelopmental events.

What region of the brain is influenced by developmental instability to produce variation in handedness? In part, this depends upon whether one considers asymmetry of hand preference or hand performance to be primary. Though our data reveal that performance measures are necessary to discriminate among right-handers, this may well reflect statistical properties of the variable (i.e., normal distribution as opposed to a J-shaped distribution) rather than any neurobiological primacy. Based on a dissociation of performance and preference in autistics, McManus *et al.* (1992) have suggested that asymmetry or preference emerges first, with skill differences being secondary to differing rates of practice across the two hands. A tendency to *initiate* more behavior with the right hand implicates left hemisphere dopaminergic basal ganglia systems. Dopamine depletion during fetal development may have a selective effect upon the left hemisphere (Craft *et al.*, 1992) and could potentially lead to increased left-handedness.

Should asymmetries in performance or skill emerge as primary, left hemisphere frontal and parietal systems would represent the most probable anatomic locus (Haaland & Yeo, 1989). Galaburda (1991) demonstrated that reduced cortical anatomic asymmetries, such as those seen in left-handers, likely reflect greater cell numbers on the typically smaller side (e.g., right planum temporale). This difference results from different rates of cell division in germinal zones rather than differing rates of neuronal depopulation. Developmental instability may be influencing these very early neural events.

References

Adams, M. S. & J. D. Niswander, 1970. Developmental noise and a congenital malformation. Genetic Research 10G: 313-317.

Alekoumbides, A., 1978. Hemispheric dominance for language: Quantitative aspects. Acta Neurologica Scandinavica 57: 97-140.

Anderson R. M. & R. M. May, 1982. Coevolution of hosts and parasites. Parasitology 85: 411-426.

Annett, M., 1970. A classification of hand preference by association analysis. British Journal of Psychology 61: 303-321.

Annett, M., 1985. Left, right, hand, and brain: The right shift theory. Erlbaum Press, New Jersey.

Annett, M., 1991. Annotation: Laterality and cerebral dominance. Journal of Childhood Psychology and Psychiatry 32: 219-237.

Annett, M., 1992. Phonological processing and right minus left hand skill. The Quarterly Journal of Experimental Psychology 44A: 33-46.

Annett, M. & M. Manning, 1989. The disadvantages of dextrality for intelligence. British Journal of Psychology 80: 213-226.

Bakan, P., 1990. Nonright-handedness and the continuum of reproductive casualty, pp. 33-74 in Left-handedness: behavioral implications and anomalies, edited by S. Coren. Elsevier, New York.

Bishop, D. V. M., 1990. Handedness and developmental disorders. Blackwell Scientific Press, Oxford.

Boklage, C. E., 1987. Twinning, nonrighthandedness, and fusion malformations: Evidence for heritable causal elements held in common. American Journal of Medical Genetics 28: 67-84.

Boucher, J., 1977. Hand preference in autistic children and their parents. Journal of Autism and Childhood Schizophrenia 7: 177-187.

Broman, S. H., P. L. Nichols & W. A. Kennedy, 1975. Preschool IQ: Prenatal and early developmental correlates. Erlbaum, Hillsdale, N. J.

Bryden, M. P., 1982. Laterality: Functional asymmetry in the intact brain. Academic Press, New York.

Bryden, M. P. & I. C. McManus, 1992. Relations between handedness and immune disorders. Journal of Clinical and Experimental Neuropsychology 14: 89.

Campbell, M., B. Geller, A. M. Small, T. A. Petti & S. H. Perris, 1978. Minor physical anomalies in young psychotic children. American Journal of Psychiatry 135: 573-575.

Carter-Saltzman, L., 1980. Biological and socio-cultural effects on handedness: Comparison between biological and adoptive families. Science 209: 1263-1265.

Colby, K. M. & C. Parkinson, 1977. Handedness in autistic children. Journal of Autism and Childhood Schizophrenia 7: 3-9.

Connolly, K. J. & D. V. M. Bishop, 1992. The measurement of handedness: A cross-cultural comparison of samples from England and Papua New Guinea. Neuropsychologia 30: 13-26.

Corballis, M. C., 1991. The lopsided ape: Evolution of the the generative mind. Oxford, New York.

Coren, S., 1989. Left-handedness and accident risk. American Journal of Public Health 79: 1-2.

Coren, S. & D. F. Halpern, 1991. Left-handedness: A marker for decreased survival fitness. Psychological Bulletin 109: 90-106.

Coren, S. & C. Porac, 1977. Fifty centuries of right-handedness: The historical record. Science 198: 631-632.

Coren, S., A. Searleman & C. Porac, 1986. Rate of physical maturation and handedness. Developmental Neuropsychology 2: 17-23.

Craft, S., M. L. Gourovitch, S. B. Dowton, J. M. Swanson & S. Bonforte, 1992. Lateralized deficits in visual attention in males with developmental dopamine depletion. Neuropsychologia 30: 341-351.

Cummings, C., D. Flynn & M. Preus, 1982. Increased morphological variants in children with learning disabilities. Journal of Autism and Childhood Schizophrenia 12: 373-383.

Daniel, W. F. & R. A. Yeo, 1991. Left-handedness and accident rates. American Journal of Public Health 81: 1036-1037.

Daniel, W. F., R. A. Yeo & S. W. Gangestad, 1992. Left-handers suffer more head injuries than do right-handers. Journal of Clinical and Experimental Neuropsychology 14: 77.

Down, J. L. H., 1866. Observations on an ethnic classification of idiots. London Hospital and lecture Reports 3: 259-262.

Ellis, L., 1990. Left- and mixed handedness and criminality: Explanations for a probable relationship, pp. 485-508 in Left-handedness: Behavioral implications and anomalies, edited by S. Coren. Elsevier, New York.

Falconer, D. S., 1981. Quantitative genetics, 2nd ed. Longman, New York.

Fisher, R. A., 1959. The genetical theory of natural selection. Dover, New York.

Fleminger, J. J., R. Dalton & K. F. Standage, 1977. Age as a factor in the handedness of adults. Neuropsychologia 15: 471-473.

Fraser, F. C., 1983. Association of neural tube defects and parental nonrighthandedness. American Journal of Human Genetics 35: 89.

Galaburda, A. M., 1991. Asymmetries of cerebral neuroanatomy, pp. 219-233 in Biological asymmetry and handedness, edited by G. R. Bock and J. Marsh. Wiley, New York.

Gangestad, S. W. & R. A. Yeo, (in preparation). Parental handedness and relative hand skill: A test of the developmental instability hypothesis.

Gangestad, S. W., R. A. Yeo, W. F. Daniel, P. Shaw & R. Thoma, (in preparation). Association of HLA alleles and handedness.

Geschwind, N. W. & A. M. Galaburda, 1987. Cerebral lateralization: Biological mechanisms, associations, and pathology. MIT Press, Cambridge, MA.

Gilbert, A. N. & C. J. Wysocki, 1992. Hand preference and age in the United States. Neuropsychologia 30: 601-608.

Gualtieri, C. T., A. Adams, C. D. Shen & D. Loiselle, 1982. Minor physical anomalies in alcoholic and schizophrenic adults and hyperactive and autistic children. American Journal of Psychiatry 139: 640-643.

Gualtieri, T. & R. E. Hicks, 1985. An immunoreactive theory of selective male affliction. The Behavioral and Brain Sciences 8: 427-441.

Guy, S. D., L. V. Majurski, C. J. Wallace & M. P. Guy, 1989. The incidence of minor physical anomalies in adult males schizophrenics. Schizophrenia Bulletin 39: 571-582.

Haaland, K. & R. A. Yeo, 1989. Neuropsychological aspects of complex motor control, pp. 219-244 in Neuropsychological function and brain imaging, edited by E. D. Bigler, R. A. Yeo & E. Turkheimer. Plenum, New York.

Habib, M., D. Gayraud, A. Oliva, J. Regis, G. Salamon & R. Khalh, 1991. Effects of handedness and sex on the morphology of the corpus callosum: A study with brain resonance imaging. Brain and Cognition 16: 41-61.

Halpern, D. F. & S. Coren, 1991. Handedness and lifespan. New England Journal of Medicine 324: 998.

Hamilton, W. D., 1982. Pathogens as causes of genetic diversity in their hot populations, pp. 269-296 in Population Biology of Infectious Diseases, edited by R. M. Anderson & R. M. May. Springer-Verlag, New York.

Helm-Estabrooks, N., R. A. Yeo, N. Geschwind, M. Freidman & C. Weinstein, 1986. Stuttering: Disappearance and reappearance with acquired brain lesions. Neurology 36: 1109-1112.

Hepper, P. G., S. Shahidullah & R. White, 1991. Handedness in the human fetus. Neuropsychologia 28: 1107-1111.

Kertesz, A., M. Polk, J. Howell & S. E. Black, 1987. Cerebral dominance, sex, and callosal size in MRI. Neurology 37: 1385-1388.

Kat, P. W., 1982. The relationship between heterozygosity for enzyme loci and developmental of aquatic bivalves (Unenidice). American naturalist 119: 824-832.

Leary, R. F., F. W. Allendorf & K. L. Knudsen, 1975. Inheritance of meristic variation and the evolution of developmental instability in rainbow trout. Evolution 39: 308-314.

Leary, R. F., F. W. Allendorf & K. L. Knudsen, 1983. Developmental instability and enzyme heterozygosity in rainbow trout. Nature 301: 71-72.

Leary, R. F., F. W. Allendorf & K. L. Knudsen, 1984. Superior developmental stability of heterozygotes at enzyme loci in salmonid fishes. American Naturalist 124: 540-551.

Leche, S., 1933. Handedness and bimanual dermatoglyphic differences. American Journal of Anatomy 53: 1-53.

Leidy, L. E., 1990. Early age at menopause among left-handed women. Obstetrics and Gynecology 76: 1111-1114.

Lishman, W. A. & E. R. L. McMeekan, 1976. Hand preference patterns in psychiatric patients. British Journal of Psychiatry 129: 158-166.

Livshits, G., L. Davadi, E. Kobyliansky, D. Ben-Amital, Y. Levy & P. Merlob, 1988. Decreased developmental stability as assessed by fluctuating asymmetry of morphometric traits in preterm infants. American Journal of Human Genetics 29: 793-805.

Livshits, G. & E. Kobylianski, 1987. Dermatoglyphic traits as possible markers of developmental processes in humans. American Journal of Medical Genetics 26: 111-122.

Livshits, G. & E. Kobyliansky, 1989. Study of genetic variance in the fluctuating asymmetry of anthropomorphic traits. Annals of Human Biology 16: 121-129.

Livshits, G. & E. Kobyliansky, 1991. Fluctuating asymmetry as a possible measure of developmental homeostasis in humans: A review. Human Biology 63: 441-446.

London, W. P., 1990. Left-handedness and alcoholism, in Left-handedness: Behavioral implications and anomalies edited by S. Coren. Elsevier, New York.

Markow, T. A., 1992a. Genetics and developmental stability: An integrative conjecture on etiology and neurobiology of schizophrenia. Psychological Medicine 22: 295-305.

Markow, T. A. (1992b). Human handedness and the concept of developmental instability. Genetica 87: 87-94.

Markow, T. A. & I. I. Gottesman, 1989. Fluctuating dermatoglyphic asymmetry in psychotic twins. Psychiatry Research 29: 37-43.

Markow, T. & K. Wandler, 1986. Fluctuating asymmetry and the genetics of liability to schizophrenia. Psychiatry Research 19: 323-328.

McManus, I. C., 1985. Handedness, language dominance, and aphasia: A genetic model. Psychological Medicine, Monograph Supplement 8. Cambridge University Press, Cambridge.

McManus, I. C., 1991. The inheritance of handedness, pp. 251-281 in Biological asymmetry and handedness edited by G. R. Bock & J. Marsh. Wiley, New York.

McManus, I. C., B. Murray, K. Doyle & S. Baron-Cohen, 1992. Handedness in childhood autism shows a dissociation of skill and preference. Cortex 28: 373-381.

Mednick, S. A., R. A. Machon & M. O. Huttenen, 1988. Adult schizophrenia following prenatal exposure to an influenza epidemic. Archives of General Psychiatry 45: 189-192.

Mellor, C. S., 1992. Dermatoglyphic evidence for fluctuating asymmetry in schizophrenia. British Journal of Psychiatry 160: 467-472.

Mitton, J. B. & M. C. Grant, 1984. Associations among protein heterozygosity, growth rate, and developmental homeostasis. Annual Review of Ecology and Systematics 15: 479-499.

Møller, A. P., 1991. Parasites differentially increase the degree of fluctuating assymetry in secondary sexual characters. Journal of Evolutionary Biology. (in press).

Nasrallah, N. A., K. Keelor & M. McCalley-Winters, 1983. Laterality shift in alcoholic males. Biological Psychiatry 18: 1056-1067.

Newman, H. H., 1934. Dermatoglyphics and the problem of handedness. American Journal of Anatomy 55: 277-322.

O'Callaghan, E., P. Sham, N. Takei, G. Glover & R. M. Murray, 1991. Schizophrenia after prenatal exposure to 1957 A2 influenza epidemic. Lancet 337: 1248-1250.

O'Callaghan, M. J., D. I. Tudehope, A. E. Dugdale, H. Mohay, Y. Burns & F. Cook, 1987. Handedness in children with birthweights below 1000g. Lancet 7: 1155.

Parsons, P. A., 1990. Fluctuating asymmetry: An epigenetic measure of stress. Biological Review 65: 131-145.

Penrose, L. S., 1954. The distal triradius t on the hands of parents and sibs of mongol imbeciles. Annals of Human Genetics 19: 10-38.

Porac, C., S. Coren & P. Duncan, 1980. Life span age trends in laterality. Journal of Gerontology 35: 715-721.

Quinn, P. O. & J. L. Rapoport, 1974. Minor physical anomalies and neurologic status in hyperactive boys. Pediatrics 53: 742-747.

Rapoport, J. L. & P. O. Quinn, 1975. Minor physical anomalies and early developmental deviation: A major subgroup of hyperactive children. International Journal of Mental Health 4: 29-44.

Rife, D. C., 1940. Handedness, with special reference to twins. Genetics 25: 178-186.

Rife, D. C., 1955. Hand prints and handedness. American Journal of Human Genetics 28: 41-48.

Risch, N. & G. Pringle, 1985. Segregation analysis of human hand preference. Behavior Genetics 15: 385-400.

Satz, P., D. L. Orsini, E. Saslow & R. Henry, 1985. The pathological left-handedness syndrome. Brain and Cognition 4: 27-46.

Searleman, A., C. Porac & S. Coren, 1989. Relationship between birth order, birth stress, and lateral preference: A critical review. Psychological Bulletin 105: 397-408.

Sham, P. C., E. O'Callaghan, N. Takei, G. K. Murray, E. H. Hare & R. M. Murray, 1992. Schizophrenia following prenatal exposure to influenza epidemics between 1939 and 1960. British Journal of Psychiatry 160: 461-466.

Smart, J. L., C. Jeffrey & B. A. Richards, 1980. A retrospective study of the relationship between birth history and handedness at six years. Early Human Development 5: 79-88.

Soule, M. E., 1982. Allomeric variation. 1. The theory and some consequences. American Naturalist 120: 751-764.

Soule, M. E. & J. Couzin-Roudy, 1982. Allomeric variation. 2. Developmental instability of extreme phenotypes. American Naturalist 120: 765-786.

Steg, J. P. & J. L. Rapoport, 1975. Minor physical anomalies in normal, neurotic, learning disabled, and severely disturbed children. Journal of Autism and Childhood Schizophrenia 5: 299-307.

Tiwari, J. L. & P. I. Terasaki, 1986. HLA and disease associa-

tions. Springer-Verlag, New York.
Tooby, J., 1982. Pathogens, polymorphism, and the evolution of sex. Journal of Theoretical Biology 97: 557-576.
Turkheimer, E., P. Hulick, R. A. Yeo & S. W. Gangestad, (in preparation). Left-handedness and growth: Possible differences in variance and stability.
Van Valen, L., 1962. A study of fluctuating asymmetry. Evolution 16: 125-142.
Van Valen, L., 1978. The control of handedness. The Behavioral and Brain Sciences 2: 320.
Von Hilsheimer, E. & J. Kurko, 1979. Minor physical anomalies in exceptional children. Journal of Learning Disabilities 12: 462-469.
Waldrop, M. F. & C. F. Halverson, 1971. Minor physical anomalies and hyperactive behavior in young children, pp. 343-380 in Exceptional Infant: Studies in Abnormality, edited by I. J. Hellmuth. Brunner, Mazel, NY.
Waldrop, M. F., C. F. Halverson & K. Shetterly, 1989. Manual for Assessing Minor Physical Anomalies (1989 revision). Unpublished manuscript, University of Georgia.
Witelson, S. F., 1980. Neuroanatomical asymmetry in left-handers: A review and implications for functional asymmetry, pp. 79-113 in The Neuropsychology of left-handedness, edited by J. Herron. Academic Press, New York.
Witelson, S. F., 1985. The brain connection: The corpus callosum is larger in left-handers. Science 229: 1223-1226.
Witelson, S. F., 1989. Hand and sex differences in the isthmus and genu of the human corpus callosum. Brain 112: 799-835.
Witelson, S. & W. Pallie, 1973. Left hemisphere specialization for language in the newborn: Neuro-anatomical evidence for asymmetry. Brain 96: 641-646.
Woods, R. P., 1986. Brain asymmetries in situs inversus. Archives of Neurology 43: 1083-1084.
Yeo, R. A., S. W. Gangestad & W. F. Daniel, (1993). Hand preference and developmental instability. Psychobiology 21: 161-168.

T. A. Markow (ed.), Developmental Instability: Its Origins and Evolutionary Implications, 299–307, 1994.

Behavioral phenodeviance: a Lerneresque conjecture

Therese Ann Markow[1] & Irving I. Gottesman[2]
[1] *Department of Zoology, Arizona State University, Tempe, AZ 85287, USA*
[2] *Departments of Psychology and Pediatrics, University of Virginia, Charlottesville, VA, USA*

Received 11 May 1993 Accepted 17 May 1993

Abstract

Developmental instability, detectable in morphological and meristic characters, is typically associated with reduced fitness. The actual mechanism by which fitness is reduced in these cases is difficult to identify. We propose that developmental instability also manifests itself at the behavioral level and that when this occurs, behavioral phenodeviance is the result. According to this model, abnormal or phenodeviant behavior compromises fitness. Examples are provided from *Drosophila* and man.

Introduction

Understanding the mechanisms which maintain genetic variation in natural populations is a major goal of evolutionary biology. While mutational and recombinational events generate the original sources of all heritable variation, other evolutionary forces such as drift or directional selection act to reduce genetic variation. Balancing or stabilizing selection, on the other hand, is attributed with being a major variation-maintaining process. While examples of balancing selection are more easily defined for single rather than multiple locus systems, the mechanism of the disadvantage to homozygotes is not usually well understood. Below, we propose a model in which a major phenomenon underlying balancing selection is the 'behavioral phenodeviance' of phenotypically extreme individuals and of homozygotes in general.

In order to develop this concept, we will first review I. Michael Lerner's original concept of morphological phenodeviance and its genetic and developmental basis as presently understood. We will then define the term 'behavioral phenodeviant', going on to exemplify the concept by drawing upon observations in *Drosophila* and man.

Phenodeviants: genetics and developmental homeostasis

Lerner (1954) coined the term 'phenodeviant' to designate the sporadic occurrence of abnormal morphological deviants caused by multigenic Mendelian inheritance in which chance segregation produces a small number of individuals who are homozygous at a certain threshold level of trait-relevant loci. Lerner observed an increase in these morphological abnormalities, such as crooked toes in chickens, as a consequence of processes which increase population homozygosity, i.e. artificial selection and inbreeding.

Lerner linked the appearance of morphological phenodeviants to the reduced ability of homozygotes to display developmental homeostasis or ontogenetic self-regulation. The opposite is true of heterozygotes, whose enhanced ability to self-regulate following developmental disturbances results in a more canalized, evolutionarily pre-programmed phenotype. Homozygosity thus can be seen as eroding the buffering capability of the epigenetic landscape conceptualized by Waddington (1957), leading to a less canalized phenotype. Figure 1 has been adapted from Lerner to illustrate this concept for continuously distributed traits having a polygenic basis. The predicted relationship between the increased number of homozygous loci and reduced developmental homeostasis is clearly seen. The way in which inbreeding, which increases homozygosity, might result in the production of phenodeviants follows from the figure. In addition, by examining this relationship in the broader context of the Gaussian distribution of phenotypes for a quantitative trait, the link between individual developmental homeostasis and populational or genetic homeostasis becomes clear as well: extreme phenotypes found in the tails of the distribution tend to be more homozygous and thus should be less developmentally stable.

An increase in homozygosity, brought about either by inbreeding or by artificial directional selection, is known to reduce fitness. A certain level of homozygosity automatically will be generated by segregation, but is further increased by inbreeding or by artificial directional selection. In natural populations the reduced fitness of homozygous

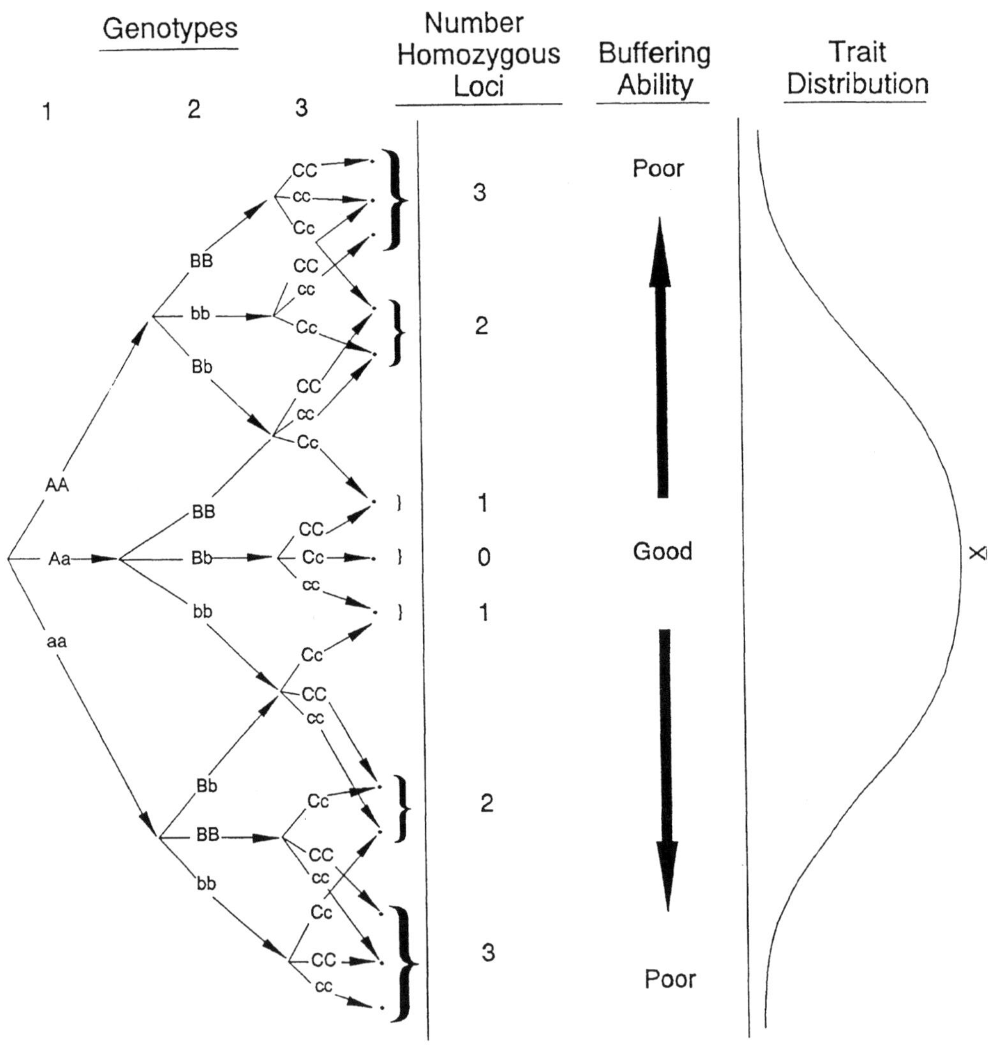

Fig. 1. Hypothetical relationship between three locus genotypes underlying a quantitative trait, phenotype distribution, and buffering ability. Although no dominance or recessiveness exists, upper and lower case letters are used for simplicity. Elements of this figure are based upon Lerner (1954).

phenotypic extremes should result in balancing selection, and thus in the maintenance of heterozygosity and phenotypes which cluster near the population mean.

Several relationships remain unclear in this proposed scenario. One is exactly how, at the physiological level, heterozygosity results in increased developmental stability or buffering. While there is evidence in the literature to support a link between heterozygosity and developmental stability in some systems (Leary, Allendorf & Knudsen, 1984), a number of studies have failed to find this relationship (Patterson & Patton, 1990; Hartl *et al.*, 1991; Zink, Smith & Patton, 1985; Zakharov, 1987), leading Zakharov (1992) to suggest that certain loci, such as those affecting growth rate, are more important than others in their influence on developmental homeostasis. Clarke (this volume) credits reduced coadaptation, rather than homozygosity *per se*, with being the cause of developmental instability in the majority of cases in which there is a genetic component. The issue of mechanisms needs to be addressed empirically across a broad range of taxa.

Another related issue is how impaired developmental stability actually reduces reproductive fitness. The nature of those phenotypic elements underlying reduced fitness are addressed in the model proposed below. A typical effect of inbreeding and artificial selection is the loss of reproductive capacity. However, inspection of the non-reproductive individuals frequently fails to reveal any consistent phenotypic explanation, i.e. reproductive systems are intact and functional, and no gross physical abnormalities are apparent which appear capable of impeding successful reproduction.

It is in these cases that we propose the importance of aberrations in the behavioral phenotype, which usually go undetected because they are rarely the subject of investigation. 'Behavioral phenodeviants' may be defined as sporadically occurring, behaviorally abnormal phenotypes whose inheritance modes are difficult to identity. As with morphological phenodeviants, their incidence may be increased with inbreeding or selection. While their structural and neurophysiological bases may be less developmentally severe, enabling them to reach adulthood, the delicate nature of the relationship between species-appropriate (canalized) behavior and survival and reproduction may make such individuals more vulnerable to exclusion from the pool of eligible mating partners during the mature stages in their lives. While a number of non-genetic stressors (e.g. teratogens such as alcohol or cocaine) may also alter the behavioral phenotype, we confine our discussion to those phenotypes generated by the same genetic mechanisms as the morphological phenodeviants described by Lerner (1954) and discussed by Zakharov (1992), whether these are associated with single or multilocus homozygosity or with loss of coadaptation leading to interlocus interactions such as epistasis.

Mechanisms of behavioral phenodeviance

The most broadly employed measure of morphological developmental instability is fluctuating asymmetry (FA). Simply described, FA is the difference between the right and left sides for paired bilateral traits. Because the deviations from symmetry in FA are random with respect to side, they will be distributed about a mean of zero. The larger the unsigned difference between sides, the greater the developmental instability (but see Palmer, Strobeck & Chippindale, this volume).

What are the potential mechanisms by which developmental instability could reduce fitness? One possibility is that asymmetries measured in bilateral structures examined are the actual cause. We agree with Clarke and McKenzie (1992) that this is unlikely. It is difficult to envision, for example in *Drosophila*, why different numbers of sternopleural bristles on the right and left sides should cause reduced mating success. A similar argument exists for the dermatoglyphic asymmetries in schizophrenic subjects. It is unlikely that finger ridge asymmetry causes reduced fitness. We propose instead, that in these cases, the organ system controlling behavior, the central nervous system, exhibits developmental instability which then goes on to impair the normal behavioral repertoire expected for the species. Thus the FA revealed in ridge counts is simply the echo of endophenotypic CNS developmental instability. This model is schematically presented in Figure 2. The bilateral CNS illustrated is a simplified representation of the hemispheric organization found in man and other vertebrates. However, it can easily be extended to invertebrates, especially those with a segmental somatic

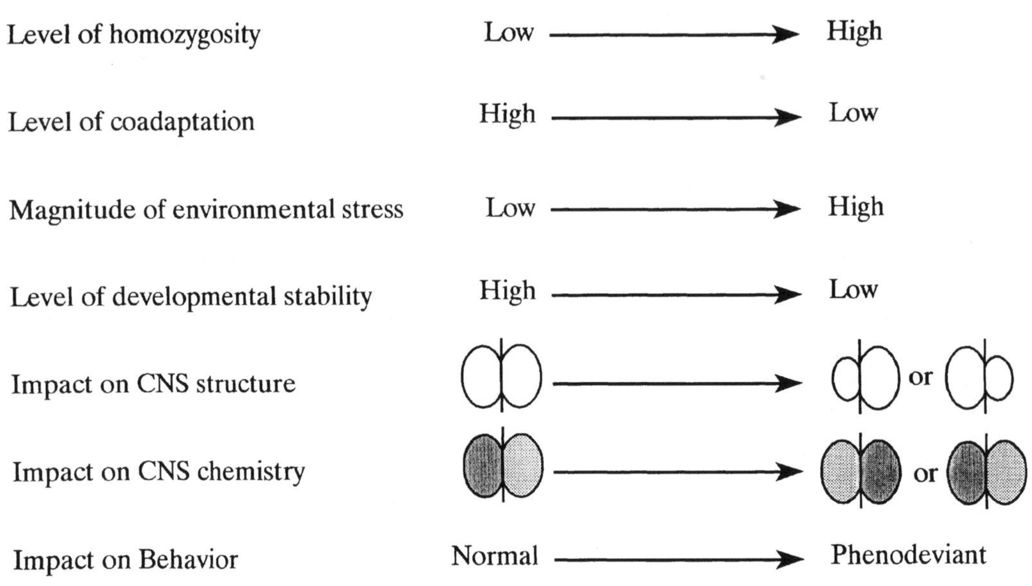

Fig. 2. Hypothetical relationship between genotypic and environmental factors and the neurobiological underpinnings of behavioral phenodeviance.

plan: developmental instability could alter the symmetry of neuronal numbers, chemistry, or connections and thus modify behavior. The quality and quantity of the behavioral abnormalities will depend upon the region and extent of CNS perturbation which in turn will depend upon the genotypic vulnerability, as well as the nature, extent, and timing of environmental results.

Developmental instability and behavioral phenodeviance: predictions

If the existence of behavioral phenodeviants is real and if they actually provide a mechanism of natural or sexual selection, several predictions follow which should be easily testable.

1) Behavioral phenodeviants should be associated with inbreeding in naturally outbreeding populations, and with phenotypic extremes for continuous traits.
2) Behaviorally phenodeviant traits will be heritable but their modes of inheritance will be complex and elusive. In other words, genetic studies will reveal similar patterns as reported for morphological phenodeviants such as crooked toes in chickens (Lerner, 1954), extra crossveins in *Drosophila* (Dubinin, 1948 as discussed in Lerner 1954, 1958), and cleft lip in man (Woolf & Gianas, 1976).
3) Behavioral phenodeviants should have reduced reproductive success compared to other phenotypes in the same natural population.
4) Behavioral phenodeviants as a group should show increased developmental instability in other traits, including the nervous system.

Recognizing the behavioral phenodeviant: testing the predictions

A critical first step in testing these predictions is the identification and measurement of the behavioral phenodeviant. For morphological phenodeviants, a variety of measures are employed for assessment. Some phenodeviant characters, such as crooked toes in chickens, are obvious. Less profound deviation may be quantified along the lines of fluctuating asymmetry (FA; see above) or scales for minor physical anomalies (MPAs; Waldrop & Halvorson 1971). Behavioral phenodeviance as a new concept however, has not yet enjoyed the attention necessary for development of standardized measures. In whatever way they come to be measured, behavioral aberrations will be most obvious in our own species and in species whose normal behavior has been exhaustively or at least intensively studied, such as *Drosophila melanogaster*. As the model and its predictions are new, there have been no *a priori* tests as yet. However, we have been able to pull together a range of *a posteriori* observations which address the predictions.

The first prediction deals with the association between individuals with extreme phenotypes for continuous traits with a multifactorial-polygenic basis and their relative homozygosity. From Figure 1, it is clear that for continuous traits, individuals whose phenotypes lie within the central portion of the distribution should be more heterozygous at the responsible loci, while homozygosity increases in extreme phenotypes. This very basic assumption has yet to be widely tested, but conflicting observations (Leary, Allendorf & Knudsen, 1984; Livshits & Smouse, this volume) underscore the need for critical experiments. In addition, this proposed relationship predicts that individuals in the tail ends of distributions for continuous traits should also exhibit an increase in behavioral phenodeviance. A corollary of this prediction is that behavioral phenodeviance should increase with inbreeding.

Evidence exists in the *Drosophila* literature to support the first prediction. Body size in *Drosophila*, as in other organisms, is a continuous trait with a polygenic basis (Falconer, 1989). In many organisms, large male size is typically associated with increased courtship success, and this tendency is present for *Drosophila* males of a variety of species as well (Markow, 1985, 1987a, 1988; Partridge, Hoffman & Jones, 1987; Partridge, Ewing & Chandler, 1987). However, when male courtship success is tabulated as a function of male size distribution in a population, it is clear that not only the very smallest, but also the very largest males are less successful in stimulating females to mate with them (Markow, 1987a; Markow & Sawka, 1992; Markow & Ricker, 1992). The same is true of inbred strains of *Drosophila melanogaster*. Whether inbreeding is accomplished by sib-mating or by making entire chromosomes isogenic, experiments using both *D. melanogaster* and *D. pseudoobscura* show significant reductions in male courtship success as homozygosity increases (Brittnacher, 1981; Partridge, MacKay & Aitken, 1985; Sharp, 1984; Miller, Glasner & Hedrick, 1992).

Multivariate analysis of courtship behavior reveals striking differences among wild-type *D. melanogaster* males (Markow, 1987b). Unsuccessful phenodeviant males show aberrations in courtship component sequences, levels of performance of various elements within the sequence, and appropriateness of courtship delivery. These aberrations may be thought of as analogous to morphological abnormalities regarded as phenodeviant by Lerner. Furthermore, while we can detect them with sensitive measures, surely these behavioral aberrations must be even more obvious and pronounced to conspecifics.

Planned experiments in humans are not feasible or ethical for the purpose of evaluating our predictions, so we must make do with 'naturalistic observations' that bear on the answers. Both incestuous and consanguineous matings lead to tests for the effects of inbreeding on behavioral phenodeviance (prediction number one). The picture will be confounded by the fact that abnormal personalities will almost always be involved in incestuous matings while consanguineous marriages are often the rule in some populations. It will also be difficult to disentangle the appearance of Mendelizing recessive abnormalities from the simulations of Mendelism (Edwards, 1960) and from the effects predicted by Lerner for polygenic traits under inbreeding. The absence of metabolic diseases in the offspring produced by matings of interest here strengthen the inference that effects are attributable to excessive homozygosity for polygenic traits.

In the largest series of offspring from incestuous matings (Seemanova, 1971), 138 of 161 children born (1933 to 1970) to 141 mothers in Czechoslovakia lived long enough to be evaluated. The incestuous dyads arose from 88 father-daughter, 72 brother-sister, and 1 mother-son unions. A control group was formed from the 95 children born to 46 of the same mothers but with unrelated fathers, that is they are half-sibs with matches for mothering and associated psychosocial variables. Twenty mothers and eight fathers were known to be mentally retarded in the incest group, three mothers were diagnosed as schizophrenics, and 13 fathers as chronic alcoholics (four fathers comitted suicide after the discovery of incest). Only offspring intelligence was evaluated in the behavioral domain: 40 (29%)

of the children of the incestuous matings were severely retarded and 28 (70%) of these children had neither parent retarded, thereby implicating excessive homozygosity as an explanation. No half-sibs were noted to be retarded. Such a sample needs to be examined for a broader range of psychopathology or behavioral phenodeviance.

In a more recent series of incest offspring (from 20 brothers and nine fathers) from British Columbia (Baird & McGillivray, 1982), eight of 21 children were retarded or had conspicuous developmental delays; three of the eight may have had retarded parents.

Costeff *et al.* (1977) examined the families of 904 mentally retarded patients in Israel without significant metabolic disorders. The very high frequencies of first-cousin marriages among Israeli ethnic populations (e.g., 21% Iraqi, 24% Irani, 15% Tunisian) and valid assessment of IQ provided strong evidence for autosomal recessive genes causing severe retardation for only 3.2% of the families. Although not discussed by the investigators, their finding of 19.6% of first cousin marriages among the normal parents of 276 severe 'idiopathic' retardates and of 18.8% first cousin marriages among the normal parents of 165 mild 'idiopathic' retardates is compatible with polygenic homozygosity causing this form of behavioral phenodeviance.

A long-standing tradition of cousin marriages may have the paradoxical effect of reducing the number of abnormalities associated with homozygosity for behavioral deviance if the society uses information about behavior to screen out some potential cousin matings (Saugstad & Odegaard, 1987).

The second prediction of our model is that behaviorally phenodeviant traits will be heritable but the genetic mechanisms will be difficult to identify. Attempts at genetic analysis of behavioral phenodeviants are common in the human genetics literature due to the clinical and social problems presented by affected individuals. For example, schizophrenia has an incidence of about 1% in the general population. Vulnerability to this disorder has a genetic basis which, as yet, is undefined. Gottesman and Shields (1967), in proposing a model for the inheritance of schizophrenia, suggested similarities in occurrence of this disorder and the morphological phenodeviants discussed by Lerner (1954). Genetic epidemiological studies are most consistent with a multifactorial threshold model in which a certain number of genetic factors is required to cross the threshold necessary for the phenotype to be expressed. The threshold may be moved in different directions by either predisposing (stressful) or protective (favorable) environments. All studies to date in which single molecular genetic markers have been sought have been unsuccessful (reviewed by McGuffin, Owen & Gill, 1992), which is exactly what is expected if schizophrenia, like other phenodeviants, is generated by segregation at a number of trait-relevant loci. There is a large and growing body of evidence to show that other major and some of the less severe mental disorders are under an appreciable degree of genetic influence, usually polygenic (McGuffin & Murray, 1991; Gottesman, 1991; Katz & McGuffin, 1993; McGue & Gottesman, 1989). Thus affectation status and severity both imply greater homozygosity at the relevant susceptibility-conferring loci.

Our third prediction deals with the reduced fitness of behavioral phenodeviants. In the *D. melanogaster* examples discussed previously, male behavioral phenodeviants were initially identified because they were at a disadvantage in mating, even though they were phenotypically normal in other respects (Markow, 1987a,b; Markow & Sawka 1992). In these cases, reduced fitness was due to either lack of male mating or reduced fertility in males who did mate.

The marital status and the fertility of psychiatri patients becomes of interest for the evidence it pro vides about human behavioral phenodeviance in a evolutionary context. More than a dozen studies o the marital status and reproductive rates of person suffering from schizophrenia have been conducte since 1935, some better than others in regard t diagnostic validity and control groups (Haverkamp Propping & Hilger, 1982). The studies are in agree ment that the marriage rate of schizophrenics as whole is greatly reduced compared to the genera population, ranging from 30% to 40% for male and 50% to 60% for females. Marital fertility *per s* is not reduced in some studies and is somewh reduced to 70-80% of expected in others. The n effect on fertility of schizophrenics is marked. I the most recent study, using Research Diagnosti Criteria on charts and conducted during the era (effective pharmacological interventions, only 42 of males and 84% of females with a diagnosis of schizophrenia had any offspring (Hilger, Propping & Haverkamp, 1983). For a much larger sample admitted to the Bethlem/Maudsley Hospitals (Slater, Hare & Price, 1971) in London 1952-1966, only 39% (of 1,086) female schizophrenics and 22% (of 1,003) male schizophrenics had any children. An estimate of the relative 'fitness' of those females with schizophrenia who reach age 35, taking into account probability of marriage, age at marriage, and the census data for English women, yields a value of 43%.

Many fewer studies have been conducted on other mental disorders. Research available on the affective disorders usually does not break down the data into bipolar/unipolar/'other' or into psychotic vs. non-psychotic levels of severity; there are many reasons to believe that bipolar disorders can be largely accounted for by a single major locus model and thus are not relevant to the context of phenodeviance. The category of manic-depressive disorder/disease contains a mixture of recurrent severe affective disorders. Manic-depressives' fitness was studied in Norway (Odegaard, 1960) with values of 77% for males and 75% for females (as contrasted with values for schizophrenics in the same period of 36% and 48%, respectively by sex); marriage rates for the group as a whole were close to the non-psychiatric population. Slater, Hare and Price (1971) also computed important data on their large sample of manic-depressives: 79% of males and females were ever married, 58% of males and 61% of females had children, and the relative fitness of females with manic-depression reaching the age of 35 was 73%.

Obsessive-compulsive personality disorder is quite an interesting condition to explore in the context of homozygosity and balancing selective forces. The pervasive pattern of perfectionism, inflexibility, and friendship-avoidance might be expected to lead to lowered fitness; an optimal amount of traits on the same dimensions, however, might actually be adaptive. Slater, Hare and Price (1971) selected obsessional neuroses out of the broader group of 'neuroses' to examine their fertility (229 males and 235 females admitted as inpatients and therefore markedly ill). Only 74% of females and 47% of males were ever married, 52% of females and 34% of males had children, and the estimate of relative fitness of obsessive-compulsive females reaching the age of 35 was 46%, putting them in the same class as schizophrenics who had much more obvious pathology. Both sexes of obsessives had the lowest marital fertility of any groups studied.

If behavioral phenodeviance is actually associated with decreased developmental stability, evidence of developmental instability should be detectable in other characters as well. Waldrop and Halvorson (1971) report an increase in minor physical anomalies in children exhibiting a range of behavioral abnormalities (see also Yeo & Gangestead, this volume). Fluctuating asymmetry or the occurrence of random right-left differences in paired bilateral structures is another, easily quantified measure of developmental instability which has been shown in several studies to be associated with behavioral phenodeviance. This association was first measured by Markow (1987a) for male *Drosophila melanogaster* who were unsuccessful in courting females. Such males were found to exhibit increased sternopleural bristle asymmetry compared to successfully mating males, providing the initial evidence that sexual selection may be mediated by developmental stability. Subsequent studies on other insects (Thornhill, 1992) and a vertebrate (Moller, 1992) have supported this relationship, although Markow and Ricker (1992) report conditions under which the relationship between FA and sexual selection does not exist.

In man, several studies have measured fluctuating dermatoglyphic asymmetry in patients with schizophrenia (Markow & Wandler, 1986; Markow & Gottesman, 1989; Mellor, 1992). In each case, the prediction was supported in that schizophrenic subjects as a group showed significantly greater FA than control subjects. Similarly, Malina and Buschang (1984), upon measuring FA for morphometric traits in mentally retarded males of complex etiology, found significantly greater developmental instability compared to controls. Finally, Durfee (1974) reported that a sample of boys with minimal brain dysfunction (now termed attention deficit disorder/hyperactivity disorder or AD/HD) also showed significantly higher asymmetry than boys without this history.

Our last prediction, that behavioral phenodeviants should show elevated instability in anatomical features, is intimately related to what may actually underlie or exacerbate some abnormal behaviors as discussed earlier and presented in Figure 2. Behavior is rooted in the nervous system. The degree to which the nervous system can tolerate structural or physiological variability has not yet been fully explored, although a number of clinically defined behavioral phenotypes clearly have abnormal neurological correlates. No studies have been reported yet to look *a priori* for developmental instability in the nervous system of any organism. However, a large body of literature exists (reviewed by Markow, 1992) which is consistent with the existence of developmental instability in the CNS of subjects with schizophrenia, including structural, neurochemical, and functional departures from normal symmetry/asymmetry patterns.

The concept of behavioral phenodeviance as a manifestation of developmental instability has the capacity to explain many things. Behavioral phenodeviance may be an important phenotypic link between either balancing selection and the maintenance of genetic variability or selection and the evolution of coadapted gene complexes. The predictions provided here should be subjected to *a priori* testing across a range of taxa and traits. Furthermore, behavioral traits should be included among the metric characters examined by investigators seeking to define the relationship between genotypic and environmental factors and developmental stability.

References

Baird, P. A. & B. McGillivray, 1982. Children of incest. J. of Pediatrics 101: 854-857.

Brittnacher, J. G., 1981. Genetic variation and genetic load due to the male reproductive component of fitness in *Drosophila*. Genetics 97: 719-730.

Costeff, H., B. E. Cohen, L. Weller & D. Rahman, 1977. Consanguinity analysis in Israeli mental retardates. Amer. J. Hum. Genet. 29: 339-349.

Clarke, G. M. & J. A. McKenzie, 1992. Coadaptation, developmental stability and fitness of insecticide resistance genotypes in the Australian sheep blowfly, *Lucilia cuprina*: A review. Acta Zoologica Fennica 191: 107-110.

Durfee, K. E., 1974. Crooked ears and the bad boy syndrome: Asymmetry as an indicator of minimal brain dysfunction. Bulletin of the Menninger Clinic 38: 305-316.

Edwards, J. H., 1960. The simulation of mendelism. Act. Genet. Stat. Med. 10: 63-70.

Falconer, D. S., 1989. Introduction to Quantitative Genetics. Longman, England.

Gottesman, I. I. & J. Shields, 1967. A polygenic theory of schizophrenia. Proceedings of the National Academy of Sciences of the United States of America 58: 199-205.

Gottesman, I. I., 1991. Schizophrenia Genesis-the origins of madness. W. H. Freeman, New York.

Hartl, B., G. Lang, F. Klein & R. Willing, 1991. Relationships between allozymes, heterozygosity and morphological characters in red deer (Cerves elaphus), and the influence of selective hunting on allele frequency distributions. Heredity 66: 343-350.

Haverkamp, F., P. Propping & T. Hilger, 1982. Is there an increase in reproductive rates in schizophrenics? I. Critical review of the literature. Archives of Psychiatry and Neurological Sciences 232: 439-450.

Hilger, T., P. Propping & F. Haverkamp, 1983. Is there an increase in reproductive rates in schizophrenics? III. An investigation in Nordbaden (SW Germany). Archives of Psychiatry and Neurological Sciences 233: 177-186.

Katz, R. & P. McGuffin, 1993. The genetics of affective disorders. Progress in Experimental Personality & Psychopathology Research 16: 200-221.

Leary, R. F., F. W. Allendorf & K. L. Knudsen, 1984. Superior developmental stability of heterozygotes at enzyme loci in salmonid fishes. The American Naturalist 124: 540-551.

Lerner, I. M., 1954. Genetic Homeostasis. John Wiley & Sons, New York.

Lerner, I. M., 1958. The Genetic Basis of Selection. John Wiley and Sons, New York.

Malina, R. M. & P. H. Buschang, 1984. Anthropomorphic asymmetry in normal and mentally retarded males. Annals of Human Biology 11: 515-553.

Markow, T. A., 1985. A comparative investigation of the mating system of *Drosophila hydei*. Animal Behavior 33: 775-781.

Markow, T. A. & K. Wandler, 1986. Fluctuating dermatoglyphic asymmetry and the genetics of liability to schizophrenia. Psychiatry Research 19: 323-328.

Markow, T. A., 1987a. Genetics of Sexual Selection in *Drosophila* in M. Huettl, ed, Genetics of Invertebrate Behavior. Plenum Press, NY.

Markow, T. A., 1987b. Behavioral and sensory basis of courtship success in *Drosophila*. Proc. Natl. Acad. Sci. 84: 6200-6205.

Markow, T. A., 1988. Reproductive behavior of *Drosophila* in the laboratory and in the field. Journal of Comparative Psychology 102: 169-174.

Markow, T. A. & I. I. Gottesman, 1989. Dermatoglyphic fluctuating asymmetry in psychotic twins. Psychiatry Research 29: 37-43.

Markow, T. A. & J. P. Ricker, 1992. Male size, developmental stability, and mating success in natural populations of three *Drosophila* species. Heredity 69: 122-127.

Markow, T. A., 1992. Human handedness and the concept of development stability. Genetica 87: 87-94.

Markow, T. A. & S. Sawka, 1992. Dynamics of male mating success in experimental groups of *Drosophila*. Journal of Insect Behavior 5: 375-383.

McGue, M. & I. I. Gottesman, 1989. Genetic linkage in schizophrenia: perspectives from genetic epidemiology. Schiz. Bull. 15: 281-292.

McGuffin, P. & R. Murray, (Eds.), 1991. The New Genetics of Mental Illness. Butterworth-Heinemann, Oxford.

McGuffin, P., M. Owen & M. Gill, 1992. Molecular genetics of schizophrenia, pp. 25-48 in Genetic Research in Psychiatry, edited by J. Mendlewicz and H. Hippius. Springer Verlag, Berlin.

Mellor, C. S., 1992. Dermatoglyphic evidence of fluctuating asymmetry in schizophrenia. British Journal of Psychiatry 160: 467-472.

Miller, P., J. Glasner & P. Hedrick, 1992. Inbreeding depression and male mating behavior in *Drosophila melanogaster*. Genetica, in press.

Moller, A. P., 1992. Females prefer large and symmetrical ornaments. Nature, Lond. 357: 238-240.

Odegaard, O., 1960. Marriage rate and fertility in psychotic patients before admissions and after discharge. International J. of Social Psychiatry 6: 25-33.

Partridge, L., T. F. C. MacKay & S. Aitken, 1985. Male mating success and fertility in *Drosophila melanogaster*. Genet. Res., Camb. 46: 279-285.

Partridge, L., A. Hoffman & Jones, 1987. Male size and mating success in *Drosophila melanogaster* and *Drosophila pseudoobscura* under field conditions. Animal Behavior 35: 468-476.

Partridge, L., A. Ewing & A. Chandler, 1987. Male size and mating success in *Drosophila melanogaster*: the roles of male and female courtship. Animal Behavior 35: 555-562.

Patterson, B. D. & J. L. Patton, 1990. Fluctuating asymmetry and allozymic heterozygosity among natural populations of pocket gophers (Thomomys bottae). Biological Journal of the Linnean Society 40: 21-36.

Saugstad, L. F. & O. Odegaard, 1987. Inbreeding and the epidemiology of schizophrenia, pp. 466-473 in Human Genetics, edited by F. Vogel and K. Sperling. Springer-Verlag, Berlin.

Seemanova, E., 1971. A study of children of incestuous matings. Human Heredity 21: 108-128.

Sharp, P. M., 1984. The effect of inbreeding on competitive male-mating ability in *Drosophila melanogaster*. Genetics 106: 601-612.

Slater, E., E. H. Hare & J. S. Price, 1971. Marriage and fertility of psychiatric patients compared with national data. Social Biology (Supplement) 18: s60-s73.

Thornhill, R., 1992. Fluctuating asymmetry and the mating system of the Japanese scorpionfly, *Panorpa japonica*. Animal Behavior 44: 867-879.

Waddington, C. H., 1957. The Strategy of the Genes. Macmillan, New York.

Waldrop, M. & C. Halvorson, 1971. Minor physical anomalies: Their incidence and relation to behavior in a normal and deviant sample, in Readings in Development and Relationships, edited by R. Smart, Jr. and M. Smart. Macmillan, New York.

Woolf, C. M. & A. Gianas, 1976. Congenital cleft lip and fluctuating dermatoglyphic asymmetry. American Journal of Human Genetics 28: 400-403.

Zakharov, V. M., 1987. Animal asymmetry: Population-phenogenetic approach. (in Russian) - Nauka, Moscow.

Zakharov, V. M., 1992. Population phenogenetics: Analysis of developmental stability in natural populations. Acta Zoologica Fennica 191: 7-30.

Zink, R. M., M. F. Smith & J. L. Patton, 1985. Associations between heterozygosity and morphological variance. Journal of Heredity 76: 415-420.

T. A. Markow (ed.), *Developmental Instability: Its Origins and Evolutionary Implications*, 309–317, 1994.

Minor anomalies: diagnostic clues to aberrant human morphogenesis

H. Eugene Hoyme
Department of Pediatrics, Steele Memorial Children's Research Center, University of Arizona College of Medicine, Tucson, AZ 85724, USA

Received and accepted 17 August 1993

'It has long been an axiom of mine that the little things are infinitely the most important.'
Sherlock Holmes 'A Case of Identity'

Key words: minor anomalies, dysmorphology, fluctuating asymmetry, birth defects

Abstract

Assessment of the degree of fluctuating asymmetry has been used in a variety of organisms as a measure of genetic and/or environmental stresses encountered during embryonic development. However, fluctuating asymmetry has not been widely used in humans in the diagnosis of congenital anomalies. Rather, assessment of patterns of minor anomalies has been utilized to infer the degree of embryonic developmental instability accompanying either genetic or teratogenic insults.

A minor anomaly is a structural feature seen in less than 4% of the general population, which is of no cosmetic or functional significance to the affected individual. Minor anomalies may or may not have functional or diagnostic significance when taken in the context of the entire child. In dysmorphology, minor anomalies have been useful in three distinct ways. First, some minor anomalies have been external markers of specific occult major anomalies. In addition, the vast majority of malformation syndromes in clinical genetics are recognizable as patterns of minor anomalies. Finally, although 15% of normal newborns have one or more minor anomalies, the finding of three or more minor anomalies is distinctly unusual. The risk of having a major occult abnormality increases proportionately with the number of minor defects present, with three or more minor anomalies signalling a 20% risk of a major occult structural defect.

In summary, just as fluctuating asymmetry may be a marker of abnormal environmental or genetic stress in the developing embryo, the presence of minor anomalies can be utilized to assess developmental instability.

Introduction

Assessment of fluctuating asymmetry has long been utilized to estimate the degree of environmental or genetic stress in the developing embryo. Thoday (1958) asserted that the degree of fluctuating asymmetry present in any organism could be used to determine a developmental buffering capacity for that organism, since structural differences between usually bilaterally symmetric structures could be assumed to result from developmental accidents. Since that time, a number of investigators have confirmed that fluctuating asymmetry may be a useful tool in assessing developmental instability in the embryo (Parsons & Howe, 1967; Parsons, 1990; Siegel, Doyle & Kelley, 1977; Leamy, 1986; Markow & Wandler, 1986).

In the practice of clinical genetics and dysmorphology, fluctuating asymmetry has not been commonly utilized as a marker for assessing abnormal human morphogenesis. Rather, the presence or absence of minor anomalies is utilized to assess the degree of developmental instability in the embryo. This manuscript will explore definitions of minor

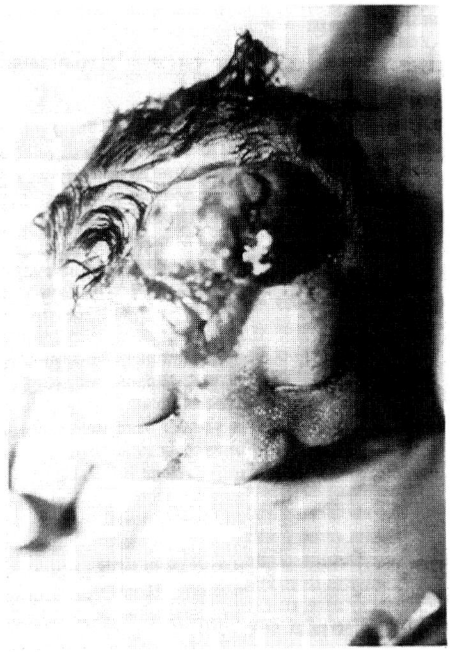

a

Fig. 1. (a) Infant with a major *malformation*, anencephaly, following abnormal embryonic cephalic neural tube closure; (b) Newborn with a major *deformation*, clubbed feet, following abnormal third trimester mechanical constraint accompanying a bicornuate maternal uterus; (c) Child with major structural *disruptions*, amputation defects of multiple digits accompanying amniotic bands.

b

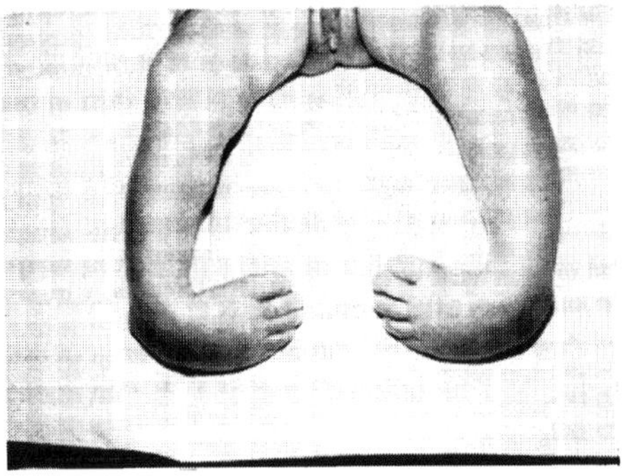

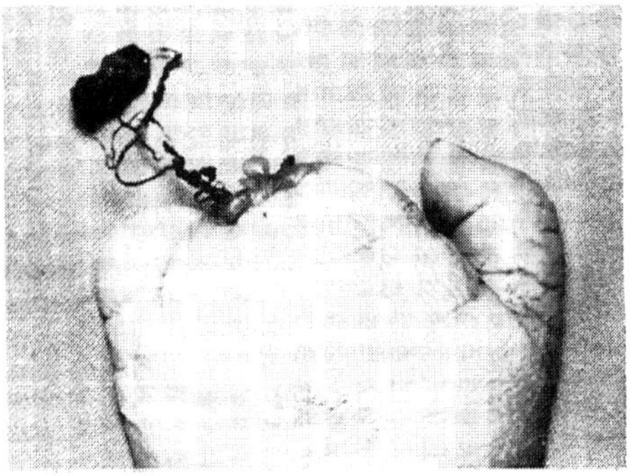

c

anomalies, mechanisms by which they occur, and their utility in diagnostic detection in the practice of clinical genetics and dysmorphology.

Definitions

A minor anomaly is defined as an unusual morphologic feature found in less than 4% of the general population, which is of no serious medical or cosmetic significance to the affected individual (Jones, 1988). This concept has been reviewed thoroughly in other publications. Pinsky (1985) has termed minor anomalies 'informative morphogenetic variants' and Aase (1981) has referred to them as 'microsigns'.

Just as with major structural anomalies (those with serious cosmetic or functional significance), minor anomalies are felt to be caused by a variety of mechanisms (Fig. 1). Minor anomalies may be *malformations* (structural defects arising from intrinsically abnormal developmental processes), *deformations* (abnormal structures resulting from non-disruptive mechanical forces applied to a once normally formed part), or *disruptions* (structural defects arising from destruction of a once normally formed part) (Spranger, Benirschke, Hall *et al.*, 1982). A supernumerary nipple could be considered to be a minor malformation. It is found in less than 4% of the general population, of no cosmetic or functional significance, and follows an intrinsically abnormal developmental process in the embryo. On the other hand, overlapping toes of one foot in a newborn could be considered to be a minor deformation, since they most likely result from intrauterine mechanical forces constraining free movement of the foot, thereby leading to the toes being forced into an overlapping position. Finally, atresia of the appendix could be considered to be an example of a minor disruption. It is well known that intestinal atresia distal to the duodenum follows interruption of arterial blood supply (Louw, 1966). Thus, disruption of blood flow to the appendiceal artery might eventually lead to an atretic appendix. Ascertaining the mechanism by which a minor anomaly has occurred is helpful to the dysmorphologist since it implies the nature of the associated developmental stress during prenatal development. A partial listing of minor anomalies is set forth in Table 1.

The significance of minor anomalies

No individual major anomaly has a high enough incidence in the population for it to be considered a

Table 1. A partial listing of minor anomalies.*

Cranium and scalp	*Sinuses*	*Skin*
Multiple hair whorls	Branchial	Shoulder dimples
Absence of hair whorl	Preauricular	Sacrum dimples
Patent metopic suture	Ear lobe	Dimples over other bones
Metopic fontanel	Helical	Sole crease
Sagittal fontanel	Pilonidal	Horizontal palmar crease (single)
Parietal foramen	*Face and neck*	Bridged palmar crease
Flat occiput	Synophrys	Single crease, finger V
Prominent occiput	Flat bridge of nose	Skin tags
Frontal bossing	Prominent bridge of nose	Hemangioma
Flat brow	Hypotelorism	Nevi
Ears	Hypertelorism	Pigmented spots
Microtia	Nostrils anteverted	Hypopigmented spots
Darwinian point	Long nasal septum	*Trunk*
Darwinian tubercle	Epicanthal fold	Extra nipples
Lack of helical folding	Iris freckles	Single umbilical artery
Bridged concha	Upward palpebral slant	Umbilical hernia
Ear lobe crease	Downward palpebral slant	Diastasis rectus
Ear lobe notched	Short palpebral fissures	Grandular hypospadias
Ear lobe bifid	Cleft uvula	Shawl scrotum
Lop ear	Cleft lip microform	Vaginal tag
Cup-shaped ear	Cleft gum	*Limbs*
Retroverted ear	Long philtrum	Cubitus valgus
Thickened helix	Short philtrum	Tapered fingers
Helix excessively folded	Smooth philtrum	Overlapping fingers/toes
Helix attached to scalp	Microstomia	Broad thumb, great toe
Preauricular tags	Macrostomia	Clinodactyly
Helical pits	Macroglossia	Nails hypoplastic
Preauricular pits	Microglossia	Nails hyperconvex
	Broad alveolar ridge	Increased space, toes
	Micrognathia	Syndactyly, toes 2-3
	Webbed neck	Overlapping digits
	Redundant neck skin	Heel prominent
	Ptosis	

* adapted from Stevenson, Hall & Goodman, 1993

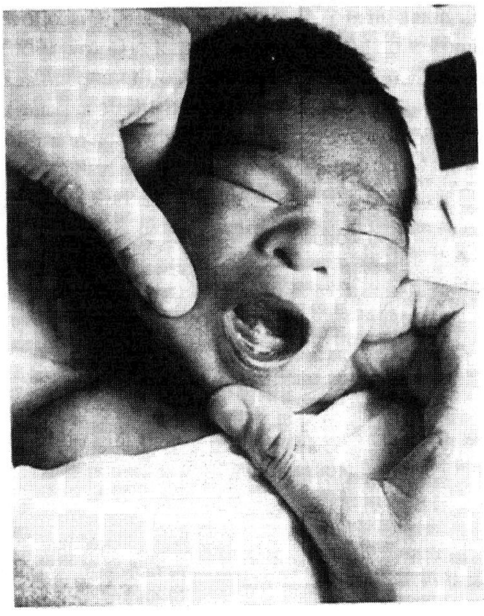

Fig. 2. Newborn native American child with an unerupted mandibular natal tooth, a minor anomaly which is a normal variant in this population subgroup.

variant of normal. However, in some subgroups, certain minor anomalies are so common that they become normal variants. For example, natal teeth, while rare in Caucasian populations, are found in 12% of Native American children from some tribes (Fig. 2) (Ribnik & Hoyme, 1989). The increased incidence of such a minor anomaly in a subgroup of the population could be interpreted as a sign of genetic stress accompanying a high degree of consanguinity.

Minor anomalies have been used clinically in dysmorphology in a variety of ways. First, some minor anomalies may be the external 'red flags' indicating the presence of specific occur major anomalies. In addition, patterns of minor anomalies constitute the majority of the features by which multiple malformation syndromes are defined. Finally, three or more minor anomalies may be a non-specific indicator of an occult major anomaly.

Minor anomalies as markers of related occult major anomalies

A careful dysmorphology examination is essential in the detection of minor anomalies. Seventy-one percent of minor anomalies are present in the craniofacies and the hands (Smith & Bostian, 1964). Particular attention to those parts is, therefore, most likely to be helpful in diagnosing occult major anomalies. A review of some of the minor anomalies associated with specific major anomalies is useful in understanding the concept of minor anomalies as 'red flags' of aberrant morphogenesis.

Preauricular tags are common findings in new-

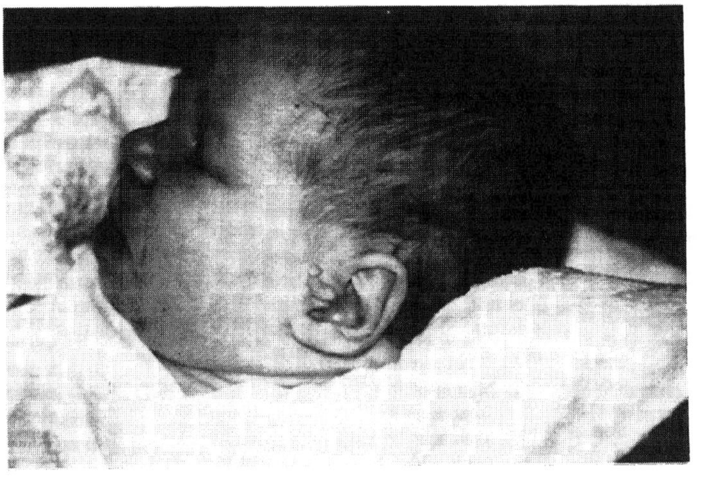

Fig. 3. Infant with congenital hearing loss and multiple preauricular tags.

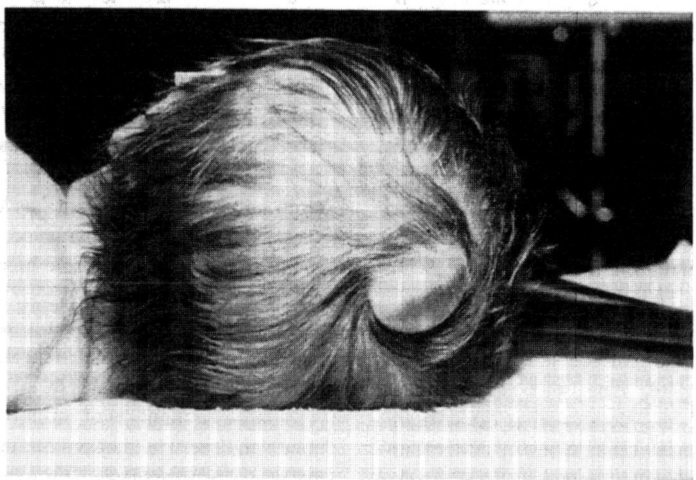

Fig. 4. Child with a small occipital encephalocele. Note the altered hair pattern following aberrant stretch on the fetal scalp at 16 weeks' gestation.

borns (Fig. 3). The first and second branchial arches in the embryo give rise to the hillocks which coalesce to form the external ear, the structures of the middle ear, and parts of the maxilla, mandible and neck. The presence of preauricular tags implies aberrant embryonic first and second branchial arch formation. It is therefore possible that an affected child may have other occult abnormal derivatives of the first and second branchial arches, including malformed middle ear structures. Therefore, the presence of preauricular tags should lead the dysmorphologist to suspect the possibility of hearing loss, either ipsilateral or contralateral to the tags, and should prompt age-appropriate hearing testing (Carey, 1993).

Hair shafts begin to erupt onto the surface of the

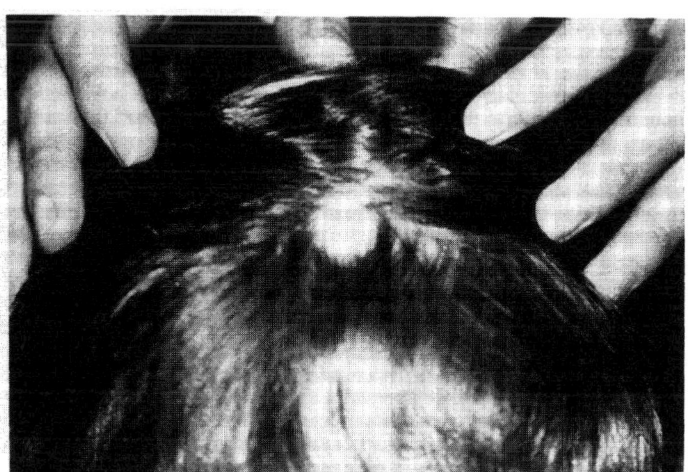

Fig. 5. Scalp aplasia cutis congenita in a child with vitreoretinal degeneration.

fetal scalp at between 16 to 18 weeks of embryonic development. The orientation of the erupting hair shafts is affected by the stretch of the scalp. At 16 weeks of gestation, the vector of brain growth is in a posterior occipital direction, leading to an occipital bulge in the fetus. Hair shafts erupting at this time, therefore, will normally have a concentric pattern around a single posterior parietal hair whorl which represents the point of maximal stretch of the scalp in the occipital region (Smith & Gong, 1973). The infant in Figure 4 had a small occipital encephalocele evident in early embryogenesis. This resulted in abnormal scalp stretch at 16 weeks when the hair shafts erupted. Thus, one sees the hair pattern surrounding and incorporating the encephalocele itself. Similarly, other types of abnormalities in brain growth lead to abnormal stretch on the scalp and either the absence of hair whorls or the presence of multiple unusually placed hair whorls. Aberrant scalp hair patterning, therefore, should lead the clinician to suspect the possibility of early gestational problems in brain growth.

Midline defects overlying the cranium and/or the spine may have particular significance as external markers of occult defects in neural tube closure. The child in Figure 5 was born with an area of aplasia cutis congenita overlying an unusual occipital soft tissue mass. She was later found to have marked vitreoretinal degeneration. Her brother was subsequently diagnosed as being similarly affected. This family manifests features of the Knobloch syndrome (occipital scalp defects and vitreoretinal degeneration), a genetically-determined disorder inherited as an autosomal recessive. In these families, the cutaneous occipital scalp defects are markers of aberrant cephalic neuroectodermal morphogenesis, a more generalized defect associated with the vitreoretinal degeneration in affected patients (Seaver *et al.*, 1993). Similar defects overlying the caudal neural axis may be markers of abnormalities in spinal cord morphogenesis as well (Fig. 6). In the infant depicted, the tail-like appendage overlies an area of tethering of the spinal cord. A variety of cutaneous anomalies over the spine may be associated with tethered cord, including hemangiomas, patches of hair, areas of aplasia cutis congenita, pilonidal sinuses, masses, or unusual scoliosis (Higginbottom, Jones & James, 1980; Hall, Udvarhelyi & Altman, 1981). Congenital nevi in the craniofacial area can also occasionally be markers of abnormal neuroectodermal development. In the epidermal nevus syndrome, craniofacial epidermal nevi are markers of underlying abnormalities in brain development (Fig. 7) (Grebe *et al.*, 1992; Baker, Ross & Baumann, 1987). In the infant pictured, an epidermal nevus extends from the scalp over the forehead and nose and involves the lips, chin, and neck. This infant has seizures and is severely developmentally delayed. A port wine stain in the area of the ophthalmic branch of the trigeminal nerve implies Sturge-Weber syndrome, another neurocutaneous disorder with an external cutaneous marker of abnormal neurologic development (Alexander & Norman, 1960).

FInally, a number of minor anomalies in the hands and feet are markers of abnormalities in embryonic morphogenesis. For example, it is known that the interphalangeal and palmar creases are reflections of the underlying movement of the hand in the embryo and fetus (Popich & Smith, 1970). In children with deficient fetal movement, alterations in creases result. For example, in amyoplasia con-

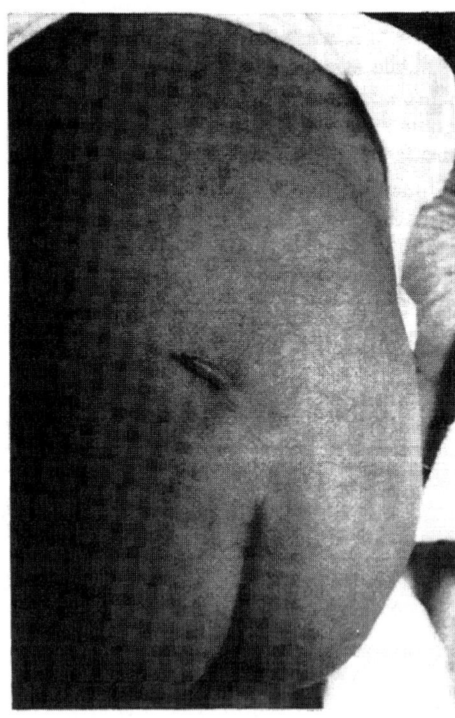

Fig. 6. A tail-like cutaneous lumbar appendage overlying a tethered spinal cord.

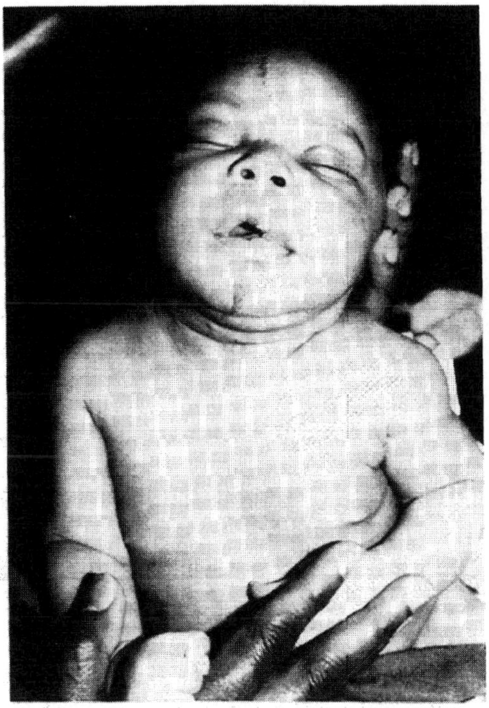

Fig. 7. An extensive midline craniofacial epidermal nevus in an infant with seizures and mental retardation.

genita (a sporadic form of arthrogryposis of unknown etiology) affected individuals frequently have missing interphalangeal creases (Fig. 8) (Hall, 1983). Similarly, lack of movement or decreased movement *in utero* results in skin being more closely applied to the bone or underlying joint than is usually the case. In such instances, unusual dimples result.

Thus, a host of minor anomalies can act as diagnostic clues for the astute clinician in the recognition of occult major anomalies and in determining the nature and timing of embryonic and fetal developmental stresses.

Malformation syndromes as patterns of minor anomalies

The majority of malformation syndromes in dysmorphology and clinical genetics are diagnosed clinically not by the presence of major malformations, but rather by patterns of minor anomalies. Thus, the usual child with Down syndrome does not have obvious external major anomalies. Instead, the diagnosis is suggested by a pattern of minor anomalies (small ears, upslanting palpebral fissures, flat nasal bridge, epicanthal folds, protuberant tongue, short incurved fifth fingers, and a wide space between the first and second toes). In fact, 79% of all malformations detectable in chil-

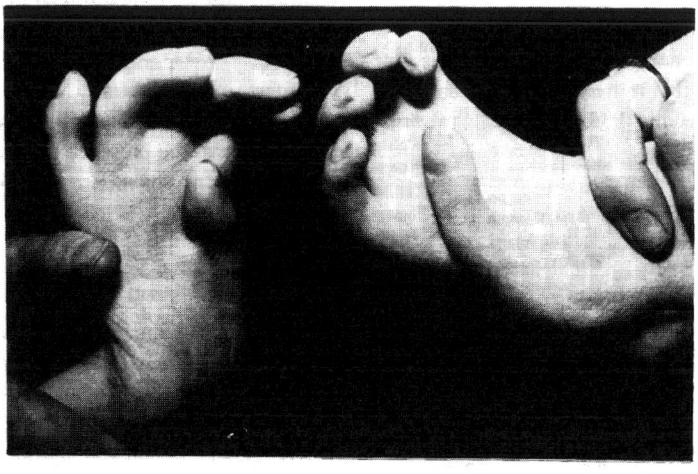

Fig. 8. Absent palmar flexion creases, accompanying lack of embryonic and fetal hand movement in a child with amyoplasia congenita.

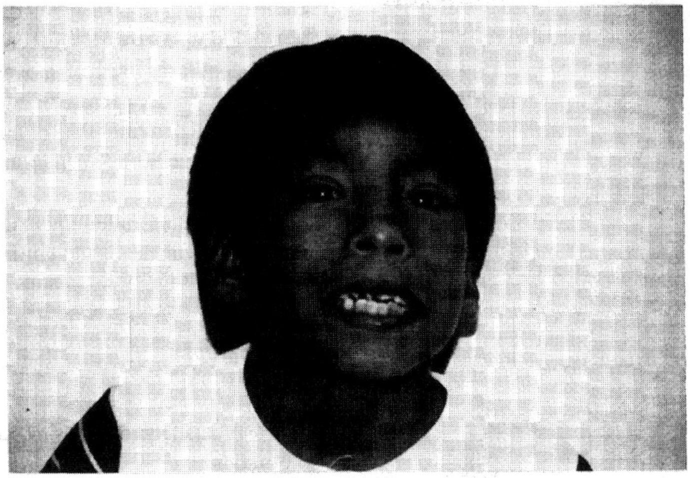

Fig. 9. Child with multiple minor craniofacial anomalies and the fetal alcohol syndrome. Note the short palpebral fissures, epicanthal folds, short nose, and long smooth philtrum with thin upper lip.

dren with Down syndrome are minor anomalies (Smith & Bostian, 1964). Most teratogenically induced abnormalities, such as the fetal alcohol and hydantoin syndromes, are recognizable only through the patterns of minor anomalies present (Fig. 9). In these cases, the minor anomalies reflect teratogenic stresses on the developing organism.

Multiple minor anomalies as indicators of occult major anomalies

Thirteen percent of normal newborns have one minor anomaly. Such infants have a 3% risk of associated major defects. Only 0.8% of normal newborns have two minor anomalies, with an attendant 10% risk of a major malformation; 0.5% of newborns have three or more minor anomalies. A number of investigators have explored the risk of a concomitant major malformation in such children. Although the studies differ, it is likely that there is an approximately 20% chance of a major anomaly in a child with three or more minor anomalies (Table 2) (Marden, Smith & McDonald, 1964; Mehes, 1983; Leppig *et al.*, 1987). Thus, the presence of multiple minor anomalies should lead the clinician to carefully consider occult major anomalies in an affected child. However, additional laboratory investigation of the individual with three or more minor anomalies is dependent on the overall diagnosis of the child and the presence or absence of additional symptoms and signs (Hoyme, 1987).

Table 2. The incidence of major malformations in patients with minor malformations (from various series).

Number of minor malformations	Percent with major malformation		
	Marden *et al.* (1964)	Mehes *et al.* (1983)	Leppig *et al.* (1987)
0	1.4	1.2	2.3
1	2.9	3.8	3.7
2	10.8	12.5	6.7
≥ 3	90.0	26.0	19.6

Summary

Fluctuating asymmetry has been widely used in a number of organisms as a marker of environmental or genetic stresses in the developing embryo. Dysmorphologists and clinical geneticists have traditionally used minor anomalies as clues to the degree of developmental instability associated with genetic or teratogenic stresses in the human embryo. Some minor anomalies may be significant as markers of specific abnormal morphogenetic processes in the embryo. Specific patterns of minor anomalies allow clinicians to diagnose malformation syndromes, thereby leading to correct counseling regarding prognosis and recurrence risks. Finally, the presence of three or more minor anomalies may nonspecifically indicate an increased risk for the presence of occult major anomalies. Clearly, the dysmorphologist must utilize assessment of the degree of fluctuating asymmetry as well as the presence or absence of minor anomalies as clues, not readily apparent to the casual observer, which lead to diagnoses of aberrant human morphogenesis.

References

Aase, J. M., 1981. The dysmorphology detective. Ped Annals 20: 38-43.

Alexander, G. L. & R. M. Norman, 1960. The Sturge-Weber syndrome. Wright and Son, Bristol.

Baker, R. S., P. A. Ross & R. J. Baumann, 1987. Neurologic complications of the epidermal nevus syndrome. Arch. Neurol. 44: 227-232.

Carey, J. C., 1993. Auricular tags, pp. 214-216 in Human malformations and related anomalies, Vol. II. edited by R. E. Stevenson, J. G. Hall and R. M. Goodman. Oxford Univ. Press, New York.

Grebe, T. A., M. E. Rimsza, R. C. Hansen, S. F. Richter & H. E. Hoyme, 1992. Expanding the clinical spectrum of epidermal nevus syndrome: two cases with new findings and literature review. Proc. Greenwood Genet. Center 11: 109.

Hall, D. E., M. D. Udvarhelyi & J. Altman, 1981. Lumbosacral skin lesions as markers of occult spinal dysraphism. J.A.M.A. 246: 2606-2608.

Hall, J. G., 1983. Part I. Amyoplasia: A common, sporadic condition with congenital contractures. Am. J. Med. Genet. 15: 571-590.

Higginbottom, M. C., K. L. Jones & H. E. James, 1980. Aplasia cutis congenita: A cutaneous marker of occult spinal dysraphism. J. Pediat. 96: 687-689.

Hoyme, H. E., 1987. Minor malformations: Significant or insignificant? Ann. J. Dis. Child. 141: 947.

Jones, K. L., 1988. Smith's Recognizable Patterns of Human Malformation, 4th ed. Philadelphia, W. B. Saunders Co., pp. 662-681.

Leamy, L., 1986. Directional selection and developmental stability: Evidence from fluctuating asymmetry of dental characters in mice. Heredity 57: 381-388.

Leppig, K. A., M. M. Werler, C. I. Cann *et al.*, 1987. Predictive value of minor anomalies. I. Association with major malformations. J. Pediatr. 110: 531-537.

Louw, J. H., 1966. Jejunoileal atresia and stenosis. J. Pediatr. Surg. 1: 8-23.

Marden, P. M., D. W. Smith & M. J. McDonald, 1964. Congenital anomalies in the newborn infant, including minor variations. J. Pediatr. 64: 357-371.

Markow, T. A. & J. Wandler, 1986. Fluctuating dermatoglyphic asymmetry and the genetics of liability to schizophrenia. Psych. Res. 19: 323-328.

Mehes, K., 1983. Minor malformations in the neonate. Akademiai Kiado, Budapest.

Parsons, P. A. & W. L. Howe, 1967. Morphogenetic homeostasis in mice. Aust. J. Biol. Sci. 20: 777-784.

Parsons, P. A., 1990. Fluctuating asymmetry: An epigenetic measure of stress. Biol. Rev. 65: 131-145.

Pinsky, L., 1985. Informative morphogenetic variants: Minor cogenital anomalies revisited, pp. 135-170. Issues and Reviews in Teratology, edited by H. Kalter. Plenum Press, New York.

Popich, G. A. & D. W. Smith, 1970. The genesis and significance of digital and palmar hand creases: Preliminary report. J. Pediatr. 77: 1017-1023.

Ribnik, L. R. & H. E. Hoyme, 1989. Natal teeth in Native Americans. Am. J. Human. Genet. 45: A59.

Seaver, L. H., L. Joffe, R. P. Spark, B. L. Smith & H. E. Hoyme, 1993. Congenital scalp defects and vitreoretinal degeneration: Redefining the Knobloch syndrome. Am. J. Med. Genet. 46: 203-208.

Siegel, M. I., W. J. Doyle & C. Kelley, 1977. Heat stress, fluctuating asymmetry and prenatal selection in the laboratory rat. Am. J. Phys. Anthropol. 46: 121-126.

Smith, D. W. & K. E. Bostian, 1964. Congenital anomalies associated with idiopathic mental retardation. J. Pediatr. 65: 189-196.

Smith, D. W. & B. T. Gong, 1973. Scalp hair patterning as a clue to early fetal brain development. J. Pediatr. 83: 374-380.

Spranger, J., K. Benirschke, J. G. Hall *et al.*, 1982. Errors of morphogenesis: Concepts and terms. J. Pediatr. 100: 160-165.

Stevenson, R. E. & J. G. Hall, 1993. Terminology, pp. 21-30 in J. G. Hall & R. M. Goodman (eds.) Human malformations and related anomalies, Vol. I, edited by R. E. Stevenson, J. G. Hall and R. M. Goodman. Oxford Univ. Press, New York.

Thoday, J. M., 1958. Homeostasis in a selection experiment. Heredity 12: 401-415.

T. A. Markow (ed.), Developmental Instability: Its Origins and Evolutionary Implications, 319–334, 1994.

Developmental instability and fluctuating asymmetry in man

F. Clarke Fraser
Departments of Biology, Human Genetics and Pediatrics, McGill University, Montréal, Canada

Abstract

How good is fluctuating asymmetry (FA) as an indicator of developmental instability? Human studies have centered mainly on dental dimensions and dermatoglyphics. Dental studies have provided no convincing evidence that FA shows significant heritability, or is increased by inbreeding or by a stressful environment. It may, however, be increased by teratogenic exposures, and the use of FA as an indicator of teratogenic insult more subtle than overt malformation should be explored.

Dermatoglyphic studies may be confounded by the fact that FA of finger-tip patterns will vary with the frequency of patterns (as will the ridge-counts derived therefrom) irrespective of any changes in developmental stability.

Dermatoglyphic FA is not consistently related to homozygosity, but is increased in certain overt developmental and behavioural disorders, particularly cleft lip and schizophrenia. Whether this results from a genetically determined reduction in developmental stability or from prenatal stress is a question still to be resolved.

In Waddington's metaphorical epigenetic landscape, it is suggested that genetically based canalization determines the topography, and fluctuating asymmetry involves the minor surface features.

Introduction

Much of the literature on developmental stability in humans has centered on the overt results of its failure: major malformations. In experimental animals, developmental stability must have been apparent to experimental embryologists ever since the discovery of embryonic regulation – the fact that when a piece is removed from an amphibian embryo, the rest of the embryo will reorganize itself and still produce a normal organism. From a combination of embryological and genetic approaches, an understanding of regulation, morphogenesis, and how they might fail began to emerge (Waddington, 1957), and gave rise to various concepts that were applied, with varying success, to humans. The following pages will review, briefly, a number of these concepts, ending with the main subject of this paper, fluctuating asymmetry, and its relation to developmental instability.

Embryological concepts relevant to fluctuating asymmetry

Canalization

The idea of developmental stability or homeostasis is implicit in the observation that members of the same species usually develop into individuals that are sufficiently alike that they can be recognized as members of that species. The concept was put on a more formal basis by Waddington (1957), a Scottish developmental biologist who coined the term canalization to refer to the fact that embryogenesis usually proceeded along predetermined developmental pathways, or channels, and resisted attempts by environmental insults or uncoordinated genes to divert the organism or its component organs from their channels. This was represented by the 'epigenetic landscape' where a ball (the embryo or an organ) rolls down a system of valleys, with one main valley and branches, some of which can only be reached by surmounting a threshold. The degree of canalization was represented by the steepness and height of the sides of the channel. A big enough diversion would result in the ball surmounting the threshold and switching into another valley.

A series of elegant experiments showed that the degree of canalization was genetically determined. For example, Alex Fraser (an Australian quantitative geneticist) and others studied a highly invariant character, vibrissa number in the mouse (Fraser & Kindred, 1962). They used the Tabby gene to destabilize vibrissa number, selected lines for high or low vibrissa number by breeding from the appropriate Tabby mice, and showed that there were eventually corresponding changes in the non-Tabby littermates. Thus canalization of a character may conceal a great deal of underlying genetic variation, and once that variation is revealed it can be modified, by selection, to change the epigenetic landscape. The genetic response was specific to the organ system involved.

Genetic assimilation

Waddington (1957) used much the same approach to demonstrate 'genetic assimilation', whereby selection can convert an 'acquired' character into an inherited one. An environmental stimulus, such as heat-shock, is used to produce a phenocopy, such as cross-veinless (*cv*) in *Drosophila*. Selection of flies that produce high frequencies of the *cv* phenocopy eventually results in strains that show the *cv* phenotype without any heat shock, and a normal *cv* gene. Again, the environmental stimulus allows selection to operate on the otherwise hidden genetic variability and thus to modify the epigenetic landscape. It is very clear that developmental homeostasis is under genetic control which, in a highly canalized character, is largely hidden.

Phenodeviants

Another approach to developmental instability was the study of 'phenodeviants'. Michael Lerner, a population geneticist, was struck by the fact that in strains of many species (particularly chickens and mice) with increased levels of homozygosity through selection or inbreeding often had low levels of minor or major malformations (Lerner, 1954). Each strain had its own characteristic type and frequency of these phenodeviants, which did not segregate as Mendelian traits but could be selected for, once present. The genetic basis for some phenodeviants, such as crooked toes in chickens, seemed to be present in many stocks. As one explanation for this phenomenon, Lerner proposed that increasing homozygosity of genes in general would tend to destabilize developmental homeostasis, which would result in the appearance of these minor aberrations. Since the type of phenodeviant differed from strain to strain, there had to be some genetic specificity; it was suggested that a general increase in homozygosity decreased developmental homeostasis, but which type of aberration would appear depended on the genetic background influencing susceptibility of a particular organ. Lerner was not incontrovertibly committed to this interpretation, as opposed to increased homozygosity for genes for susceptibility to the particular trait, but marshalled a lot of evidence in support of it.

The phenodeviant concept was extended to the common human malformations by James Neel, one of the first American medical geneticists (Neel, 1958). He noted that human races differed in the frequencies of individual malformations (cleft lip in Orientals, neural tube defects in Celts, polydactyly in West Africans) but that the total number of malformations was about the same in each race. This suggested (not very strongly) that falling below an obligate level of heterozygosity caused developmental instability and hence the occurrence of malformations. Again, there had to be genetic susceptibility for particular organ systems, and there was not much evidence for a general reduction in developmental stability with increasing homozygosity. More importance was given to the concept that the common malformations had a polygenic basis, the 'polygenic' variation being the distant results of genes having major effects on some other system. The fact that the malformations were so common implied that the said major genes were advantageous, and that the malformations were the price that the population paid for maintaining these genes at optimal frequencies.

Quasi-continuous variation

The concept of phenodeviants overlapped that of 'quasi-continuous' variants, a term coined by Hans Gruneberg, a pathologist mouse-geneticist (Gruneberg, 1952) who built on the idea of Sewal Wright (the great American quantitative geneticist) that non-Mendelian malformations (e.g. polydactyly in the guinea pig) occurred when a polygenically determined trait exceeded a physiologica threshold, not further specified (Wright, 1934). Thus quasi-continuous referred to the separation of a continuous distribution of liability to a malformation into discontinuous parts, affected and unaffected. For example, the size of the third molar in the mouse is normally distributed, but in strains with a small mean size, the distribution is truncated at the lower end and absence of the tooth appears. It appears that there is a critical (threshold) mass of the tooth anlage below which the tooth, instead of just being small, is more likely not to form at all, and the continous distribution of tooth size is separated into quasi-continuous variation, present or absent. Gruneberg amassed numerous examples of this phenomenon among the minor skeletal variants of the mouse and rat (Gruneberg, 1963).

The multifactorial threshold concept

The idea that this model could also apply to major malformations developed concurrently in developmental genetics, mainly in the mouse, and in human population genetics (see Fraser, 1976; 1980b for a review). Susceptibility to teratogen-induced cleft palate in the mouse could be seen in terms of how a continuous distribution of liability relates to a developmental threshold. Liability was manifested by the developmental stage at which the palatal shelves move from vertical to horizontal, so they can meet. The threshold was the latest stage at which the shelves could reach the vertical and still meet. Many factors, both genetic and environmental, were identified that could influence the position of the liability distribution relative to the threshold, and hence the frequency of cleft palate. Strains and crosses in which normal shelf movement was late were more susceptible to teratogen-induced cleft palate than were early-closing strains (Fraser, 1961; 1980a). Being close to the threshold made the embryo's palate unstable. Thus susceptibility or, if you will, developmental instability of the palate was a reflection of the normal, genetically determined, developmental pattern.

It is important to note that the underlying genetic systems were specific to the particular organ system. A strain in which the palate was susceptible to 6-aminonicotinamide, for example, could have a vertebral system that was resistant (Goldstein, Feiner-Pinsky & Fraser, 1963). Differences in susceptibility between pairs of inbred strains could be determined by a small number of genes, but the genes involved were different between pairs of strains, so the total number of loci influencing liability of the palate in an outbred population could be quite large.

Another important implication of the model was that inbreeding would increase the frequency of phenodeviants; increasing homozygosity for the underlying liability genes would increase the variance of the liability distribution, and hence the number of individuals falling beyond the threshold (Newcombe, 1964). But this would relate to the liability genes for that particular system, and not to a non-specific decrease in developmental stability. This would explain why inbreeding would increase the frequency of various malformations, without requiring an increase in non-specific developmental instability.

The problem with the model in humans is that without inbred lines and accessible embryos, it is usually impossible to see the underlying distribution. Nevertheless, it was shown by Douglas Fal coner, a Scottish quantitative mouse geneticist, that assuming polygenically distributed liability to a trait and a threshold separating it into discontinuous parts, heritability could be estimated from the frequency of the trait in the first degree relatives of probands (Falconer, 1965). Surprisingly, perhaps, heritability estimates turned out to be high even when recurrence risks and monozygotic twin concordance rates were relatively low (Smith, 1971).

Concurrently, Cedric Carter, an English medical geneticist, was collecting evidence to show how pyloric stenosis and cleft lip had family patterns consistent with the model (Carter, 1976), now referred to as the multifactorial threshold model, since it should incorporate environmental factors as well as polygenes. The term polygenic, referring to a large number of genes of small and equal effect, was used rather loosely in the human context (Fraser, 1976). The idea of many genes of equal effect is biologically implausible, and there could be a few 'major' genes influencing liability that could produce the same distinguishing characteristics of recurrence risk as the polygenic model, since the characteristics relate more to the presence of a threshold than to the shape of the liability distribution. 'Major' appeared to refer to genes of low penetrance, but nevertheless detectable by complex, computer-assisted analysis of family recurrence rates, and 'poly' to at least a few genes not so detectable. The favoured model now appears to be a 'mixed model' involving some combination of a polygenic component, 'major' genes, phenocopies, and perhaps even familial environmental factors (Fraser, 1989).

An implicit but often ignored aspect of multifactorial threshold models is a stochastic element - individuals with the same genotype, in the same uterus, do not necessarily end up at the same point on the liability distribution. This was elegantly demonstrated by David Kurnit and others (1987) by computer simulation of atrial septum development. They endowed the hypothetical random-walking ancestral cells for the septum with particular properties such as rate of migration, adhesiveness, and mitotic rate, and specified that when two cells adhered, migration and mitosis would stop. With repeated simulations the number of cells reaching the septum area varied because of random differences in the behaviour of the cells. A threshold was represented by assuming that below a certain minimum number of cells the septum would be incomplete. The frequency of septal defect was normally distributed, and the mean could be markedly altered by changing one of the properties, e.g. adhesiveness, which would change the mean of the distribution. An important message is that two genetically identical individuals, or two sides of the same individual in the same environment, can differ in the presence or absence of a threshold trait, simply on the basis of random variation. That is, there will always be a certain amount of fluctuating asymmetry in a system that involves stochastic decisions, and it is hard to imagine a developmental system of any complexity that does not.

Homeostasis and homozygosity

Considerable effort has been put into assessing the role of homozygosity in contributing to variability at the extremes of the distribution of quantitative traits. One must distinguish between a non-specific increase in variation as postulated for phenodeviants, and the increase in homozygosity that occurs at the tails of a polygenically determined distribution of a quantitative trait, for the genes that influence that particular trait. Homozygosity for the quantitative genes that determine one trait does not imply the same for genes that determine another trait, and being at the extreme of the distribution for one trait does not necessarily mean being at the extreme of the distributions for other traits. And if there are more than a few genes involved, the number of individuals who are homozygous for all the relevant genes will be extremely small. So one should be cautious about referring to individuals at the tails of the distribution (e.g. beyond two standard deviations from the mean) as homozygous - most of them will not be.

The literature on the relation of being at the extremes of distributions, homozygosity, variance and fluctuating asymmetry is inconsistent and conflicting. It is well reviewed elsewhere (Van Valen, 1962; Livshits & Kobyliansky, 1985; Livshits & Smouse, this volume) and will not be discussed here.

Fluctuating asymmetry

And so we come to the question of fluctuating asymmetry (FA), how useful it is as an indicator of developmental instability, and its relation to homozygosity.

It would seem reasonable to suppose that if FA is a measure of how successful the two sides of the body were at doing the same thing, it would be a useful indicator of canalization, but the facts do not provide much support for the idea. For example, Waddington (1957) and Reeve (1960) showed that selection for better canalization in Drosophila did not decrease FA.

The use of dermatoglyphics and dental morphometrics to study these questions is reviewed below, and the other literature relating to these questions is well reviewed elsewhere. A few relevant comments are made here.

There is some evidence in the literature that increasing homozygosity contributes to increasing FA, but it is certainly far from a consistent effect. To take one example, there is the study of Parsons and Howe (1967) on minor skeletal variations in inbred mouse-lines. To see if there was developmental homeostasis in minor skeletal anomalies, they counted the frequency of 29 variants on each side, and observed whether the proportions of LL, RL, LR and RR combinations differed from random expectation. A majority of traits showed an excess of LL and RR combinations, indicating that there *was* developmental homeostasis. However, the degree of deviation from random expectation, or strength of homeostasis, differed for different variants, suggesting that each variant had its own

Table 1. Fluctuating asymmetry for dermatoglyphic and dental traits in patients with cleft lip, with or without a family history, and in patients with isolated cleft palate (Adams & Niswander, 1967).

Condition	Trait	N	Subject	FH+		FH-
CLP	atd	24	patients	16.5**		9.6
		67	controls	7.9		7.9
	molar	26	patients	0.039*		0.027
		62	controls	0.022		0.022
CP	atd	17	patients		10.5	
		67	controls		7.9	
	molar	15	patients		0.021	
		62	controls		0.022	

** *significantly different from the controls at the 0.025* or 0.005** level.

genetic system, not that there was a generalized homeostatic effect. Furthermore (though they did not comment on this), the F_1 hybrids did not characteristically show lower frequencies of the variant, or more deviation from random expectation, but rather tended to show frequencies intermediate between the strains or close to the value for one or the other parent strain (dominance) and similar degrees of homeostasis. So in this system, at least, there is no evidence that homozygosity causes a general decrease in developmental stability and FA is not decreased by heterozygosity. Similar conclusions were reached for FA of tooth size in inbred mouse strains and hybrids (Bader, 1965, discussed in the next section).

Table 2. Fluctuating asymmetry for dermatoglyphic and dental traits in patients with cleft lip, with (FH+) or without (FH-) a family history, and their near relatives (Woolf & Gianas, 1976, 1977).

Trait	N	Subject	FH+	FH-
atd	47	propositae	12.9*	6.0
	81	sibs	6.7	5.1
	86	parents	4.2	6.9
	166	controls	5.4	5.4
ab	68	propositae	7.6*	3.5
	90	sibs	5.6*	4.2
	74	parents	6.9*	3.8
	166	controls	3.3	3.3
patterns	63	propositae	1.3*	0.9
	91	sibs	1.3*	1.1
	100	parents	1.2*	1.0
	166	controls	0.9	0.9

* significantly different from the controls at the 0.01 level

FA in people has been approached largely through the study of dental dimensions and of dermatoglyphics; a number of these studies are reviewed in the following sections.

Fluctuating asymmetry in tooth size

What is the significance of fluctuating asymmetry in tooth size? It may be useful, when considering the studies discussed below, to keep in mind the findings of Garn (1966) who addressed this question and concluded that fluctuating asymmetry (1) was markedly related to tooth size *per se*, (2) was greater for the more distal tooth of each morphological class, and (3) showed a correlation within, but not between, morphological classes, so that (4) bilateral asymmetry of teeth is seen to be of purely local origin, and probably post-embryonic in timing. One would predict, therefore, that FA of tooth size would not be strongly related to canalization.

Dental studies relating to FA have addressed these questions:

1. Is dental FA increased by increasing homozygosity?
2. To what extent is dental FA under genetic control?
3. Is dental FA increased by environmental stress?
4. Is dental FA increased in conditions related to overt development instability?

Relation to inbreeding

The one human study that addresses the question of inbreeding directly is that of Niswander and Chung (1965), done as part of a large study of the effects of inbreeding in Japanese children which, in turn, was done as part of a large study of the genetic effects of the atomic bomb explosion. They found no effect of inbreeding on tooth diameter, and a 16% increase in variance in offspring of 1st cousins, which they attributed to environmental factors. To look at asymmetry, they examined 1,076 dental casts of the upper central incisors of children whose parents were unrelated (582), 2nd cousins (117), $1^1/_2$ cousins (78) or 1st cousins (299). There was a suggestion of an increase in intrapair variance with increasing inbreeding, and multiple regression analysis showed a significant ($p < 0.05$) association, but almost all of the increase was in the $1^1/_2$ cousins, with a 'very small increase in the other inbred classes', and '... little more than a suggestion of an over-all inbreeding effect.' The authors do not offer an explanation for the non-conformity of $1^1/_2$ cousins, but point out that they are mostly cross-generation and hence somewhat unusual. Thus, they conclude, the biological importance of this finding is doubtful. Nevertheless, one finds several references to this paper as supporting the relation of inbreeding to increased FA.

Another study (Dibernardo & Bailit, 1978) looked at the same Japanese population to evaluate various environmental effects (see below) but also looked at the effect of inbreeding on asymmetry in the lateral incisors and first molars. They found no correlation, supporting the scepticism of Niswander and Chung.

Others have approached the question of inbreeding on a population basis. A comparison of tooth width in four populations varying in inbreeding and environmental conditions showed that FA was increased in the most inbred, that of Tristan da Cunha (Bailit *et al.*, 1970), but this was also the population with the most severe environment. There was no relation of dental FA within the population to the degree of inbreeding, and heritability was low.

There is some experimental evidence that inbreeding can increase FA. An experiment designed to test response to selection for dental FA in mice did not show any such response, but did show a trend towards an increase in FA in both the high, low and control lines (significant only in the low line), presumed to result from increasing homozygosity brought about by selection in small populations (Leamy, 1986). There was no increase in additive genetic variance for FA.

On the other hand, an extensive study comparing FA of molar width in wild, random-bred, inbred and hybrid mouse strains, gave inconsistent results (Bader, 1965). When FA was measured by correlation of the values of the two sides, the isogenic populations (inbred and hybrid) were more variable than the wild and random-breds. Using intra-individual variance as a measure of FA, the hybrids had a lower variance, on average, than the other three groups (which did not differ), even though they were probably not as heterozygous as the wild and random-bred groups. Furthermore, as with the skeletal variants of Parsons, the F_1s between hybrids were in some cases less variant, in some cases intermediate, and in some cases close to one or the other parental value, and there were differences in ranking between the three molars. Thus there is no evidence that the inbred (homozygous) lines have lower developmental homeostasis than the heterozygous lines.

Thus there is no convincing evidence that increasing homozygosity increases dental FA in people, and only tenuous evidence that it does so in mice.

Heritability of dental FA

It also appears that there is little, if any, evidence for the heritability of dental FA in humans. The only robust contribution to the question is a large and well-documented twin study wich found little or no evidence of a genetic basis for dental asymmetry, though there was strong evidence for genetic determination of tooth size (Potter & Nance, 1976). Heritability estimates of dental FA in two Solomon island populations (Bailet *et al.*, 1970) and Australian aboriginals (Townsend & Brown, 1980) did not differ significantly from zero. The observation by Garn, Lewis and Kerensky (1966) that there is a correlation of FA within but not between morphological classes of teeth, suggests that whatever is influencing bilateral asymmetry is of purely local origin, which argues against a general genetic influence on dental asymmetry. A similar finding is reported in a Colombian Indian tribe (Harris & Nweeia, 1980). Another tantalizing suggestion of a possible genetic influence on dental FA comes from a study of a Paraguayan Indian tribe and a sample of South African caucasoids, both of which showed a *bimodal* distribution of dental FA, as measured by a technique that uses antimeres for all teeth to derive a value for FA for an individual mouth (Kieser, Groeneveld & Preston, 1986). This suggests the existence of a major factor influencing dental FA that could be either genetic or environmental. The authors interpret the finding as suggesting that loss of canalization is individual - specific and does not establish a causal link with environmental stress.

In conclusion, there is no convincing evidence of genetic control of dental FA in general, and fairly convincing evidence to the contrary.

FA and environmental stress

There is, on the other hand, evidence that environmental stress may influence dental FA. This comes from experimental animal studies, observations on human individuals, and comparisons of populations (Leary & Allendorf, 1989). Studies in which pregnant rats were exposed to stressful levels of noise, cold and heat showed an increased FA of teeth, as well as of bone density (Siegel Doyle & Kelley, 1977; Gest, Siegel & Anistransky, 1983). These agents are also known to cause malformations (Edwards & Wanner, 1977; Fraser, 1977).

Few examples are known in humans of dental FA in children that were exposed to environmental stress. Children of alcoholic mothers, who did not have overt features of the fetal alcohol syndrome, nevertheless showed increased ($p < 0.05$) FA in 3 of the 4 maxillary teeth measured, and in 8 of the total of 16 teeth measured (Kieser, 1992). Further studies would be of interest as, if substantiated, this would provide a useful approach to the study of suspected teratogens. One wonders if the dermatoglyphics would have shown a similar effect.

In a study of Japanese children previously referred to (Dibernardo & Bailit, 1978), no relation of dental FA was found with factors thought to be related to prenatal stress, such as maternal age and parity, prematurity, low birth weight and low socioeconomic class, but these are not very sensitive indicators of prenatal stress. Data from the NINDS Collaborative Perinatal Project failed to detect any relation of dental FA to more impressive indicators, such as maternal diabetes, hyperthyroidism and hypertension, as well as short gestation length and birth weight (Smith, Garn & Cole, 1982). There was, however, an increased dental FA in the offspring of diabetic Rhesus macaques (Kohn & Bennett, 1986).

A number of studies have attempted to relate dental FA to environmental stress by comparing populations. For example, dental FA was shown to be higher in a population with severe environmental conditions (Tristan da Cunha), than in Boston caucasians, with a tribe from the Solomon chain intermediate in both dental FA and environmental rigor (Bailit *et al.*, 1970). Skulls from Alaskan Eskimos and Pueblo Indians had dental FA similar to that of Neanderthal man and greater than that of Ohio caucasians, which was considered to support a relationship to environmental stress (Doyle & Johnston, 1977). A lowland forest tribe had a level of dental FA intermediate between Ohio caucasians and prehistoric groups (Pueblo Indians and Eskimos), again supporting the stress hypothesis (Harris and Nweeia, 1980). Similarly the teeth of prehistoric hunters had higher FA than those from a later aboriginal farming group and a modern cadaver population (Perzigian, 1977). In contrast, however, dental FA in a sample of white Michigan children was higher than in those of a presumably more stressed Archaic American Indian population, and the published data from a Boston population (Black, 1980).

Dental FA in children with developmental disorders

Finally there are several studies of dental FA in children with overt evidence of disturbed development. In a U.S. series of cleft lip patients (with or without cleft palate), with a positive family history (undefined), FA of the lower first molar was higher ($p < 0.025$) than that of a comparison group of patients attending a cleft palate clinic for non-cleft condition (Adams & Niswander, 1967). Those with a negative family history (the 'sporadic' group) did not differ from either the 'FH+' or the comparison group, though they were much closer to the comparison group. One would have thought that, since the sporadic group would be diluted with a substantial portion of genetically predisposed cases who did not happen to have an affected relative, that they might have had an FA value closer to that of the 'FH+' group. This question also arises with respect to dermatoglyphic asymmetry (see below).

The isolated cleft palate group, interestingly, did not differ from the comparison group.

A similar Scottish study examined all teeth, except lateral incisors, in patients with cleft lip with or without cleft palate (Sofaer, 1979). All teeth examined had more FA than those of non-cleft orthodontic patients. The difference between FH+ and FH- patients was not so clear. Of the maxillary teeth, FA was higher in the FH+ group for 2 of the 6 teeth examined in the mesio-distal diameter (1 at $p < 0.01$) and 1 in the bucco-lingual diameter, but for the latter the mesio-distal dimension was *less* asymmetrical. For the mandible only two measures were significantly increased, one at $p < 0.01$. For this one (1st molar) the FA of the upper tooth was *lower* in the FH+ group, though not significantly so. Thus it seems clear that dental FA is increased in children with cleft lip, even in regions remote from the cleft. The evidence that this effect is greater in cases with a genetic disposition is not so clear, and needs further investigation, as the answer could be important.

FA was increased in the majority of teeth of children with Down syndrome as compared to a comparison group derived from the literature (Barden, 1980), supporting a previous study of Garn, Cohen and Geciauskas (1970). This finding is not surprising in view of the evidence of generalized developmental instability shown by this and other aneuplodies (Opitz & Gilbert-Barness, 1990). In fact, it is somewhat surprising that differences are significant only at the 5% level.

Dental FA was also somewhat increased in a small sample of children with fragile-X syndrome, though not enough to be of diagnostic significance (Peretz *et al.*, 1988). This is interesting in view of the mosaicism in size of the trinucleotide repeat expansion sometimes seen in these patients (Nelson & Warren, 1993).

Perhaps this section should have begun with reference to the cautionary papers of Smith, Garn and Cole (1982), Greene (1984) and Palmer (this volume). Greene points out that differences between pairs of teeth are small and that, in many studies, sample sizes are small. The measurement error can confound comparisons, even in large populations, and the power of the F-ratio to distinguish small difference diminishes with diminishing sample size. Smith, Garn and Cole use computer simulations to show that sample sizes of several hundred are needed to detect population differences in dental asymmetry. They conclude that 'given sampling limitations and the current inability to link increased human dental asymmetry to defined prenatal stresses, we suggest that fluctuating dental asymmetry is not yet established as a useful and reliable measure of general stress in human populations.'

In conclusion, dental FA is not convincingly related to homozygosity, or to general stress. There is, however, evidence that it can be increased by teratogens, both in experimental animals, and in one case (alcohol) in people.

Dermatoglyphics and fluctuating asymmetry

Studies of FA in dermatoglyphic traits have addressed some of the same questions as those for dental traits but, unfortunately, rarely on the same subjects. Comparisons of two traits in the same samples would have important implications for the generality of the findings with respect to heterozygosity and environmental stress and whether their effects on developmental instability are general or specific to particular organs.

Dermatoglyphics as indicators of FA

How good is dermatoglyphic FA as a reflector of developmental asymmetry? Dermatoglyphics are accessible, and easily recorded, but they have some characteristics that can constitute analytical pitfalls if not guarded against.

There are three basic finger-tip patterns, arches, loops (ulnar or radial) and whorls. For statistical analysis these can be converted into ridge-counts. Arches have the lowest ridge count (0), and whorls the highest. Conversion of patterns to ridge counts makes them into quantitative characters, more amenable to statistical analysis, but this tends to blur pattern differences. To translate patterns into ridge counts, and average the counts over the ten fingers, will provide figures for statistical analysis that will surely have a multifactorial distribution, but are not likely to provide meaningful insights into the nature of inheritance, or stability, of the underlying patterns.

Two palmar patterns have been used in studies of FA. Normally there is a triradius at the base of each of the four fingers, labelled a-d. The number of ridges between the a and b triradii is the a-b ridge count. The *atd* angle is the angle subtended by the a and d triradii and the proximal triradius, normally at the base of the hand. It has long been recognized, but not well documented, that the patterns are related to the shape of the fingers. Broad thumbs often have whorls, and thin fingertips are likely to have arches, as in trisomy E or children exposed *in utero* to diphenylhydantoin (Andermann *et al.*, 1982). It is assumed that the finger patterns are related to the shape of the volar pads, which are related to that of the distal phalanges. The volar pads appear on the fingers at 8 weeks post-fertilization. They begin to regress at 10-11 weeks, at which time the epidermal ridges appear. Thus the patterns are probably determined by 8 weeks and certainly by 10 weeks post-fertilization, earlier than previously thought. Timing of appearance is related to pattern; early ridge formation is associated with whorls and late formation with arches (Babler, 1991).

Each finger has its own characteristic array of patterns (Holt, 1968). Correlation between fingers on the right and left hand is fairly high, and there is some correlation of pattern between adjacent fingers (Micle & Kobyliansky, 1988). Developmental disturbances, particularly aneuplodies, cause in some cases changes in frequencies of patterns that are so characteristic that they can be used diagnostically (Preus & Fraser, 1972).

It seems fairly obvious that FA will vary with the frequency of patterns, quite apart from developmental instability. For example, individuals with nine or ten arches (as often seen in trisomy 18) will obviously have very little FA, either for finger pattern or ridge count. Similarly, 31% of Down syndrome patients have ten ulnar loops, vs. 7% of controls (Preus & Fraser, 1972), which must contribute to a decrease in FA in Down syndrome dermatoglyphics, no matter what the differences in developmental homeostasis. Dermatoglyphic FA varies from finger and pattern to pattern. Variance of right-left differences in ridge count in normal subjects was greatest for digit 2 and least for digit 5 (Jantz, 1979). Similarly, in a sample of 300 normal individuals, the number of discordant patterns was greatest (45%) on digit 2 and least (14%) on digit 5 (Fraser & Garabedian, unpublished). It seems reasonable to suggest that this difference may relate to the fact that the most frequent pattern, ulnar loop, has the highest frequency (83%) on digit 5 and the lowest (36%) on digit 2. There is, indeed, a tendency for FA to increase with decreasing frequency of the pattern (Fraser & Garabedian, unpublished). Thus it is important to recognize that a change in FA of finger-tip patterns, or (more obscurely) ridge-counts, can result from a change in pattern frequency *per se*, irrespective of differences in developmental stability. It follows that an increase in dermatoglyphic FA should not be assumed to represent increased developmental instability without considering whether there are changes in pattern frequencies.

Inheritance of dermatoglyphics

Dermatoglyphic patterns have a very high heritability, at least when they are quantified as ridge counts. The inheritance of specific patterns is not so clear. From analysis of an Israeli sample Slatis (1976) suggested that the basic pattern is all ulnar loops, and that various genes cause deviations from it. Some of these are postulated to show intermediate dominance, so that a heterozygote for the gene for, say, whorls on the thumbs might have a whorl on one thumb and an ulnar loop on the other. This would result in an *increase* in FA in *heterozygotes*, contrary to the idea that heterozygosity decreases variability.

Pertinent to this is the observation of Holt (1968) that the variance of right-left difference was greatest at intermediate ridge count levels, which might be supposed to contain the greatest proportion of heterozygotes in a multifactorial system, again contrary to the hypothesis. This phenomenon was also noted in a study of racial variation (Jantz, 1976). On the other hand, when asymmetry was examined directly in the *a-b* ridge count, there was more asymmetry at the extremes of the distribution (Jantz & Webb, 1980), supporting the hypothesis. Could this be because the extreme values were associated with nearby palmar patterns such as a loop in the third interdigital space, which tend to be asymmetric?

Segregation analysis of *a-b* ridge count suggested a multifactorial basis, with less firm evidence for a major factor (Darlu & Iagolnitzer, 1984). Data from a Louisiana population supported the Israeli findings on inheritance of patterns to some degree (Juberg, Morgan & Faust, 1980), but the sample size was small, and polygenic inheritance was not convincingly ruled out. It is surprising that Slatis' intriguing, though implausible, proposal has not been explored more thoroughly.

Heritability of dermatoglyphic FA

Although dermatoglyphic ridge counts (and presumably patterns) have a high heritability, their fluctuating asymmetry does not. Holt (1968) found that familial correlations in differences between right and left hand total ridge counts were close to 0 and concluded that nearly all variation in asymmetry was environmentally determined. That is, ridge-count is under strong genetic control (Maynard-Smith *et al.*, 1985). Somewhat higher estimates of heritability were found using a method that incorporates the variance between fingers as well as differences between homologous fingers (Singh, 1970). Parent-child heritability estimates ranged from 20 to 40%. Note that combining the data for the two sexes resulted in a very low estimate of heritability. A twin study, using multiple regression analysis, found that, by three measures of asymmetry, FA appears to be largely under environmental control, but with some genetic variation, more so in males. For example, total *R* vs. *L* asymmetry showed a heritability of 0.52 in males and 0.12 (n.s.) in females (Martin *et al.*, 1982). Could this be related to lyonization? Another twin study showed no significant differences in FA of ridge counts between monozygotic twins, dizygotic twins and singletons, showing that heritability was low and that twinning *per se* did not affect ridge count FA (Markow and Gottesman, 1989a).

If the FA of dermatoglyphic patterns shows low heritability, one would not expect that it would be influenced by consanguinity. Nevertheless there was significantly more asymmetry in fingertip pattern and *a-b* ridge count among inbred members of an American Indian band, the Havasupai. However, the samples were small (Markow, in press). If substantiated, this would support the idea that increasing homozygosity decreases developmental stability.

Dermatoglyphic FA in overt developmental disorders

Studies of FA in individuals with overt signs of developmental disorder have involved Down syndrome, cleft lip and behavioral disorders.

Dermatoglyphic FA in Down Syndrome

The increased asymmetry seen in Down syndrome (DS) is widely recognized but not very extensively documented (Opitz & Gilbert-Barness, 1990), although increased variability is (Shapiro, 1983). Shapiro has pointed out that the least buffered traits in the normal population are those most commonly abnormal in DS, suggesting that the DS phenotype results from a generalized genetic imbalance due to trisomy and a consequent exaggerated response to various genetic and environmental factors to which all individuals are exposed. Arguments to the contrary are presented by Epstein (1988).

With respect to dermatoglyphics, DS patients show an increased variance for palmar *atd* angle (Shapiro, 1975). Variance of ridge count is somewhat *decreased* in DS (Shapiro, 1975), as expected from the increase in ulnar loops, as mentioned above. Shapiro suggests that the *atd* angle is more variable because it is more sensitive to environmental influences (Shapiro, 1983), but one wonders if it is simply because the angle is considerably larger in DS, and probably influenced by the presence of hypothenar patterns, more frequent in Down syndrome. The only attempt to look at dermatoglyphic asymmetry specifically showed that the palmar lines were *more* symmetrical in DS patients (Rodewald, Zana & Diegelmayer, 1976). Possibly, all these differences may result from differences in the frequencies of patterns in DS rather than from differences in developmental stability (see above). In any case they do not attest to increased dermatoglyphic FA in Down syndrome.

Dermatoglyphic FA and cleft lip

Several studies have been done on dermatoglyphic asymmetry in patients with cleft lip, with or without cleft palate. Adams and Niswander (1967) showed that patients with familial (but not those with 'sporadic') clefts had increased asymmetry for lower first molar diameter (see previous section), though there was no difference in the mean or variance. An excess of FA in familial cases vs. 'sporadic' cases is also seen in the palmar *atd* angle, and is more convincing than that for molar size (Table 1). As noted previously, one would expect the 'sporadic' group to contain some genetically predisposed individuals who did not happen to have affected relatives, which should bring the mean of the sporadic group closer to that of the familial group, whereas in fact it is very close to the control. It would have been interesting to see whether the two traits were correlated. Still, taken at face value, the data suggest that cleft lip patients with a family history are developmentally less stable in their dermatoglyphic patterns and molar size than individuals with a negative family history or those without clefts.

If this instability has a genetic basis, the near relatives of patients with familial clefts would also show increased asymmetry. The observations of Woolf and Gianas (1976, 1977) provide only partial support for this prediction (Table 2). For *atd* angle, *a-b* ridge counts and finger print patterns, the familial cases had higher FA values, as shown by Adams and Niswander. But for *atd* angles the sibs and parents of the familial cases had values similar to those of the sporadic cases and their controls. The authors suggest that this is because the biological effect is small, and would be diluted in the relatives (Woolf & Gianas, 1976).

On the other hand, for the *a-b* ridge counts and fingerprint patterns, asymmetry *was* higher in the normal sibs and parents of familial cases, and close to the values for the propositae - that is, *higher* than predicted by the hypothesis. Values for the non-familial cases and their relatives are close to the controls, contrary to expectation if this group contains some genetically predisposed cases.

We can only conclude that the data are inconsistent. It seems clear that familial cases have more asymmetry than controls, but the data for sporadic cases and for both sets of relatives do not conform very well with the hypothesis that genetic factors predisposing to cleft lip also increase developmental stability in general. According to the hypothesis, one would expect family history positive probands to have the highest FA, followed by their near relatives, followed by the family history negative probands, with the FH-relatives and controls having least. The diagram below illustrates the findings.

The data are not consistent with the hypothesis. They are more compatible with the hypothesis that FH+ clefts result from a familial environmental teratogen that also causes increased FA, but this seems rather implausible. The alternative is to conclude that either the data for FH+ relatives or FH- probands, or both, are suffering from sampling error, and that further studies may reconcile them.

Dermatoglyphic FA in behavioural disorders

A third area where FA has been involved as a possible indicator of developmental instability is the behavioral disorders, and schizophrenia in particular.

The literature on dermatoglyphic patterns in schizophrenics has been inconsistent, possibly because frequencies differ in different subcategories of schizophrenia, and different studies have included different proportions of sub-samples (Mellor, 1968). If schizophrenia is a multifactorial threshold trait, then patients (being at the extreme of the distribution of liability) should have increased homozygosity which, by Lerner's phenodeviant theory, would destabilize development (Markow, 1992). FA in dermatoglyphic traits was examined in schizophrenic patients as an indication of developmental instability (Markow & Wandler, 1986). There were significant increases over controls in asymmetry of *a-b* ridge count ($p < 0.01$) and fingertip pattern ($p < 0.05$) in schizophrenic patients and (to a lesser extent) schizoaffective patients, but not in those with affective disorders. Also, as expected for a multifactorial threshold

	FH+ probands	FH+ relatives	FH- probands	FH- relatives	Controls
Expectation	+++	++	+	–	–
Observed					
atd	+++	–	–	–	–
a-b	+++	++	–	–	–
patterns	+++	+++	–	–	–
teeth	+++		–		–

trait, there was more asymmetry in the more severe patients, as measured by early age at admission and rate of deterioration. A study of another sample of schizophrenic patients by the same group also demonstrated increased dermatoglyphic FA (Ricker *et al.*, 1989). Further support came from a study of twins in which ridge counts and fingerprint patterns were more asymmetric in twins concordant for schizophrenia than in those discordant ($p < 0.05$) – the concordant twins presumably being more homozygous for schizophrenia susceptibility genes (Markow & Gottesman, 1989b). Study of a larger sample by Mellor (1992) also showed increased FA in schizophrenics. Ridge counts for each pair of fingers in male and female schizophrenics showed increased FA in nine of ten comparisons with controls ($p < 0.02$); fingerprint patterns were more often discordant than in controls (23% vs. 18%; $p < 0.005$) the palmar *atd* angle was more asymmetric (significantly only in females), and the *a-b* ridge count was more asymmetrical ($p < 0.01$). Controls were drawn from the literature. Mellor had some reservations about his findings: the controls were not ideal, and there was some inconsistency between the sexes. Nonetheless, they support the findings in the three studies by Markow and others, and it is reasonable to accept that there are, indeed, signs of developmental instability in schizophrenic patients, as manifested in their dermatoglyphic patterns.

The idea that the 'polygenes' that appear to increase susceptibility to schizophrenia also decrease developmental stability in general seems somewhat implausible, for reasons discussed elsewhere. In view of the increasingly convincing evidence that susceptibility to schizophrenia is increased by prenatal damage (Waddington, 1993), it may be more reasonable to assume that such insults also increase instability in the development of dermatoglyphic patterns. It would be interesting, in this respect, to see whether the near relatives of schizophrenics also show increased FA.

The interesting finding (Bracha *et al.*, 1992) that monozygotic twins discordant for schizophrenia have less similar dermatoglyphics than concordant pairs might also be explained on the basis of prenatal environmental stress.

Dermatoglyphic FA analysis has also been used to study personality differences. Monozygotic twins were selected according to whether their *a-b* ridge counts were very similar or very different on the two hands. The asymmetric pairs showed greater differences in their Minnesota Multiphasic Personality Inventory scales, suggesting that their personalities were less well buffered from environmental noise (Rose, Reed & Bogle, 1987). The authors suggest that the *a-b* ridge count may be the best dermatoglyphic character for the study of FA.

Dermatoglyphic FA as an indicator of homozygosity

What do dermatoglyphics have to tell us about the question of the relationship between increased homozygosity, appearance at the extremes of the distribution and increased developmental instability? I have already referred to Holt's observation that variances of ridge counts were greatest at intermediate values (contrary to the hypothesis), and Slatis' theory that there are specific genes for dermatoglyphic patterns and that heterozygotes may have one pattern on one hand and the other on the other, thus increasing FA, contrary to the hypothesis. Similarly, Jantz (1976) found increased variability in the intermediate finger ridge count values. Conversely, individuals at the extremes of the *a-b* ridge count distribution (and therefore presumably more homozygous, though there was no direct measure of homozygosity) showed more asymmetry than those close to the mean (Jantz & Webb, 1980). Soulé and Cuzin-Rudy (1982) found that in seven human populations variance was, if anything, lower in individuals at both extremes of the ridge count distribution than those more centrally located. Of the 35 comparisons between 'out-groups' and 'in-groups', eight were significant, and for seven the 'in-group' values were higher. Soulé points out that dermatoglyphics differ from the many (non-human) traits that do show increased variance of outliers in that they are the only pattern differences studied and have exceptionally high heritability. Also, the relation of variance to FA may not be simple, as pointed out previously.

Taking a somewhat different approach to the problem, Livshits and Kobyliansky (1987) separated a sample of young adult Israeli males into modal and extreme groups on the basis of five morphometric variables, and examined five dermatoglyphic traits (selected by discriminant analysis from 67 traits recorded). Variance was greater in the extreme groups for all five dermatoglyphic traits, supporting the hypothesis. On the other hand, digital pattern asymmetry increased with increasing heterozygosity at four blood group loci (Kobyliansky & Livshits, 1986), even though the morphological traits showed the opposite trend. It may be significant that the modal and extreme groups differed in the frequencies, as well as asymmetry, of digital patterns (see previous section). The authors comment that the significance of asymmetry in morphological traits is not established unequivocally.

In conclusion, dermatoglyphics can provide a useful way of looking for developmental instability, but can provide confusing results unless interpreted with caution. Dermatoglyphic FA is not consistently related to homozygosity, but it is increased in some overt developmental and behavioural disorders. It may be more useful as an indicator of prenatal stress than of generalized developmental instability.

Conclusion

The evidence discussed above, and a more wide-ranging review by Parsons (1990), would suggest that the developmental instability reflected by increased FA is related more to environmental stress than to changes in the mechanisms underlying developmental homeostasis. Parsons (1990) concludes, convincingly, that FA is a useful trait for monitoring stress in the laboratory and in natural environments. His conclusion that increased FA is a reflection of poorer developmental homeostasis at the molecular, chromosomal and epigenetic level is more debatable.

There is little evidence to support the idea that FA is related to developmental homeostasis in general. Waddington (1957) and Reeve (1960) showed that selection for increased canalization in Drosophila did not increase FA appreciably. Admittedly, homeostatic mechanisms may be somewhat different in organisms showing mosaic (as in Drosophila) rather than regulative (as in vertebrates) development, so this may not be a valid argument. It would be useful to have some data from vertebrates.

There is also little evidence that FA is related to a general developmental buffering capacity, or decreased buffering by increased homozygosity, and its relation to extreme phenotypes seems tenuous (Livshits & Smouse, this volume). Rather, canalization seems to be organ specific, and there is little correlation of asymmetry in different structures of the same organism (Van Valen, 1962). Nevertheless, malformations do occur together more often than a random expectation would predict, and it is not clear how much of the excess is due to syndromes.

There is, on the other hand, convincing evidence from animals and some from people that environmental insults increase FA (Parsons, 1990). Teratologists rarely examine FA while counting malformations in embryos exposed to teratogens; it might be revealing to do this in experiments designed to analyze the interactions of genetic predisposition to liability and environmental insult (Biddle, 1979; Fraser, 1980a).

Further exploration of FA as an indicator of prenatal insults in humans could be profitable, particularly with respect to its possible use in diagnosis, for example, of fetal alcohol syndrome, and it might be useful to consider morphometric traits separately from threshold traits.

One might speculate that FA correlates only poorly with canalization because selection has developed mechanisms of developmental homeostasis for the major features of the epigenetic landscape; they are the geological infrastructure of the landscape, whereas FA applies to the details – the ripples in the sand, the rivulets, the hummocks. The details bear little relation to the major geological structure, but they would reflect the effects of environmental disturbances which, if severe enough, could also affect the landscape.

These ideas could be profitably explored by experimental teratologists, and there are lots of opportunities for further meaningful studies in humans. But there must be careful attention to the choice of controls, and the methods of statistical analysis (Greene, 1984; Smith, 1982; Livshits & Smouse, this volume). Attention must be paid to how FA varies with size and trait frequency. Finger ridge counts should not be used to compare groups for FA unless it can be shown that the groups do not differ in pattern frequency, or some method has been designed to allow for this. Indeed, it would be a challenging problem, for someone endowed with the necessary mathematical skills, to work out how FA would change with changing pattern frequencies, irrespective of changes in developmental stability.

It has not been possible, in this review of fluctuating asymmetry, to even touch on many other interesting and important aspects of developmental instability. These include handedness (Yeo, Gangestad & Daniel, in press) and its relation to twinning, both forms of developmental instability (Boklage, 1987), and neural tube defects (Fraser, 1983). There is also the intriguing suggestion that the sibs of children with a particular malformation have an increased frequency of other malformations (Fraser, Zeizel & Hanson, 1982). There is still much to be learned.

Acknowledgements

I thank Dr. Kenneth Morgan and Dr. Daphne Trasler for a critical reading of the manuscript. Supported by a Medical Genetics Group Grant from the Medical Research Council of Canada.

References

Adams, M. S. & J. D. Niswander, 1967. Developmental 'noise' and a congenital malformation. Genet. Research 10: 313-317.

Andermann, E., L. V. Dansky, F. Andermann, T. M. Loughnan & J. Gibbons, 1982. Minor congenital malformations and dermatoglyphic alterations in the offspring of epileptic woman. A clinical investigation of teratogenic effects of anticonvulsant medication, pp. 235-249 in Epilepsy, pregnancy and the child, edited by D. Janz et al., Ravin Press, New York.

Babler, W. J., 1991. Embryologic development of epidermal ridges and their configurations. Birth Defects: Original Article Series 27: 95-112.

Bader, R. S., 1965. Fluctuating asymmetry in the dentition of the house mouse. Growth 29: 291-300.

Bailit, H. L., P. L. Workman, J. D. Niswander & C. J. Rachean, 1970. Dental asymmetry as an indicator of genetic and environmental conditions in human populations. Hum. Biol. 42: 626-638.

Barden, H. S., 1980. Fluctuating dental asymmetry: a measure of developmental instability in Down syndrome. Am. J. Phys. Anthropol. 52: 169-173.

Biddle, F. G., 1979. Genetic studies of teratogen-induced cleft palate in the mouse. Advances in the Study of Birth Defects. Vol. I. Teratogenic Mechanisms. T.V.N. Persaud, Ed., chap. 6, pp. 85-111, MTP Press Ltd., Lancaster.

Black, T. K., 1980. An exception to the apparent relationship between stress and fluctuating dental asymmetry. J. Dent. Res. 59: 1168-1169.

Boklage, C. E., 1987. Twinning, nonrighthandedness and fusion malformations: evidence for heritable causal elements held in common. Amer. J. Med. Genet. 28: 67-84.

Bracha, H. S., E. F. Torrey, I. I. Gottesman et al., 1992. Second-trimester markers of fetal size in schizophrenia. Amer. J. Psychiat. 149: 1355-1361.

Carter, C. O., 1976. Genetics of common single malformations. British Medical Bulletin 32: 21-26.

Darlu, P. & E. R. Iagolnitzer, 1984. Segregation analysis of ab-ridge count in human palms. Ann. Hum. Biol. 11: 537-544.

Dibernardo, R. & H. L. Bailit, 1978. Stress and dental asymmetry in population of Japanese children. Am. J. Phys. Anthropol. 48: 89-94.

Doyle, W. J. & O. Johnston, 1977. On the meaning of increased fluctuating asymmetry: a cross populational study. Amer. J. Phys. Anthropol. 46: 127-134.

Edwards, M. J. & R. A. Wanner, 1977. Extremes of temperature. In Handbook of Teratology 1. General Principles and Etiology. J. G. Wilson and F. C. Fraser, Eds. Plenum Press, New York.

Epstein, C. J., 1988. Specificity versus non-specificity in the pathogenesis of aneuploid phenotypes. Am. J. Med. Gen. 29: 161-165.

Falconer, D. S., 1965. The inheritance of liability to certain diseases, estimated from the incidence among relatives. Ann. Hum. Gen. 29: 51-71.

Fraser, A. S. & B. M. Kindred, 1962. Selection for an invariant character, vibrissa number in the house mouse. III. Correlated responses. Austral. J. Biol. Sci. 15: 188-206.

Fraser, F. C., 1961. The use of teratogens in the analysis of abnormal developmental mechanisms. In First International Conference on Congenital Malformations. M. Fishbein, Ed., J. B. Lippincott Co., Montreal.

Fraser, F. C., 1976. The multifactorial/threshold concept - uses and misuses. Teratology 14: 267-280.

Fraser, F. C., 1977. Interactions and multiple causes, pp. 445-463, Handbook of Teratology Vol. I, edited by J. G. Wilson and F. C. Fraser. Plenum, New York, Vol. I.

Fraser, F. C., 1980a. Animal models for craniofacial disorders. In 'The etiology of cleft lip and palate', M. Melnick, D. Bixler and E. D. Shields, Eds., Alan R. Liss, Inc., pp. 1-23.

Fraser, F. C., 1980b. The William Allan Memorial Award Address: Evolution of a palatable multifactorial threshold model. Am. J. Hum. Gen. 32: 796-813.

Fraser, F. C., A. Czeizel & C. Hanson, 1982. Increased frequency of neural tube defects in sibs of children with other malformations. Lancet III: 144-145.

Fraser, F. C., 1983. Association of neural tube defects and parental non-righthandedness. Am. J. Hum. Genet. 35: 89A.

Fraser, F. C., 1984. Mapping the cleft-lip genes: the first fix? Am. J. Hum. Gen. 45: 345-347.

Garn, S. M., A. B. Lewis & R. S. Kerewsky, 1966. The meaning of bilateral asymmetry in the permanent dentition. Angle Ortho. 36: 55-62.

Garn, S. M., M. M. Cohen & M. A. Geciauskas, 1970. Increased crown-size asymmetry in trisomy G. J. Dent. Res. 49: 465.

Gest, T. R., M. I. Siegel & J. Anistransky, 1983. Increased fluctuating asymmetry in the long bones of neonatal rats stressed by cold, heat and noise. Am. J. Phys. Anthropol 60

196-199.

Goldstein, M., M. Feiner-Pinsky & F. C. Fraser, 1963. Genetically determined organ-specific responses to the teratogenic action of 6-aminonicotinamide in the mouse. Genet. Res. 4: 258-265.

Greene, D. L., 1984. Fluctuating dental asymmetry and measurement error. Am J. Phys. Anthropol. 65: 283-289.

Gruneberg, H., 1952. Quasi-continuous variation in the mouse. Symposia Genetica, Pavia, III: 215-227.

Gruneberg, H., 1963. The pathology of development. A study of inherited skeletal disorders in animals. Blackwell Scientific Publications, Oxford.

Harris, E. F. & M. Nweeia, 1980. Dental asymmetry as a measure of environmental stress in the Ticuna Indians of Colombia. Am. J. Phys. Anthropol. 53: 133-142.

Holt, S. B., 1968. The genetics of dermal ridges. Charles C. Thomas, Springfield.

Jantz, R. L., 1976. The relation between total finger ridge-count and variability of counts from finger to finger. Genetic implications of racial variation. Ann. Hum. Genet. 40: 221-224.

Jantz, R. L., 1979. On the levels of dermatoglyphic variation, pp. 53-61. Dermatoglyphics 50 years later, edited by W. Wertelecki and C. Plato. National Foundation - March of Dimes. Alan R. Liss Inc., New York.

Jantz, R. L. & R. S. Webb, 1980. Dermatoglyphic asymmetry as a measure of canalization. Ann. Hum. Biol. 7: 489-493.

Juberg, R. C., L. Y. Morgan & C. C. Faust, 1980. The inheritance of digital dermatoglyphic patterns in 54 American Caucasian families. Am. J. Physic. Anthropol. 52: 7-12.

Kieser, J. A., H. T. Groeneveld & C. B. Preston, 1986. Fluctuating dental asymmetry as a measure of odontogenic canalization in man. Am. J. Phys. Anthropol. 71: 437-444.

Kieser, J. A., 1992. Fluctuating odontometric asymmetry and maternal alcohol consumption. Ann. Hum. Biol. 19: 513-520.

Kobyliansky, K. & G. Livshits, 1982. Anthropometric multivariate structure and dermatoglyphic peculiarities in biochemically and morphologically different heterozygous groups. Am. J. Phys. Anthrop. 70: 251-263.

Kohn, L. A. & K. A. Bennett, 1986. Fluctuating asymmetry in fetuses of diabetic Rhesus monkeys. Am. J. Phys. Anthropol. 71: 477-483.

Kurnit, D. M., W. M. Layton & S. Matthysse, 1987. Genetics, chance, and morphogenesis. Am. J. Hum. Genet. 41: 979-995.

Leamy, L., 1986. Directional selection and developmental stability: evidence from fluctuating asymmetry of dental characters in mice. Heredity 57: 381-388.

Leary, R. F. & F. W. Allendorf, 1989. Fluctuating asymmetry as an indicator of stress: implications for conservation biology. Trends in Ecology and Evolution 4: 214-217.

Lerner, I. M., 1954. Genetic homeostasis. John Wiley, New York: 1-134, figs. 1-14.

Livshits, G. & E. Kobylianskey, 1985. Lerner's concept of developmental homeostasis and problem of heterozygosity level in natural populations. Heredity 55: 341-353.

Livshits, G. & E. Kobyliansky, 1987. Dermatoglyphic traits as possible markers of developmental processes in humans. Am. J. Med. Gen. 26: 111-112.

Livshits, G. & P. E. Smouse, 1993. Relationship between fluctuating asymmetry, morphological modality and heterozygosity in an elderly Israeli population. (this volume).

Markow, T. A. & K. Wandler, 1986. Fluctuating dermatoglyphic asymmetry and the genetics of liability to schizophrenia. Psychiatry Research 19: 323-328.

Markow, T. A. & I. Gottesman, 1989a. Dermatoglyphic fluctuating asymmetry in twins and singletons. Hereditas 110: 211-215.

Markow, T. A. & I. Gottesman, 1989b. Fluctuating dermatoglyphic asymmetry in psychotic twins. Psychiat. Res. 29: 37-43.

Markow, T. A., 1992. Genetics and developmental stability: an integrative conjecture on aetiology and neurobiology of schizophrenia. Phychol. Med. 22: 295-305.

Markow, T. A. & J. F. Martin, 1993. Inbreeding and developmental stability in a small human population. Ann. of Hum. Biol. 20: 389-394.

Martin, N. G., J. L. Jinks, H. S. Berry & D. Loesh, 1982. A genetic analysis of diversity and asymmetry in finger ridge counts. Heredity 48: 393-405.

Maynard-Smith, J., R. Burian, S. Kauffman, P. Alberch, J. Campbell, B. Goodwin, R. Lande, D. Raup, L. Wolpert, 1985. Developmental constraints and evolution. Quart. Rev. Biol. 60: 265-287.

Mellor, C. S., 1968. Dermatoglyphics in schizophrenia. Part I. Qualitative aspects. Part II. Quantitative study. Brit. Jour. Psychiatry 114: 1387-1397.

Mellor, C. S., 1992. Dermatoglyphic evidence of fluctuating asymmetry in schizophrenia. Brit. J. Psychiat. 160: 467-472.

Micle, S. & E. Kobyliansky, 1988. Sex differences in the intra individual diversity of finger dermatoglyphics: pattern types and ridge counts. Hum. Biol. 60: 123-134.

Neel, J. V., 1958. A study of major congenital defects in Japanese infants. Am. J. Hum. Genet. 10: 398-445.

Nelson, D. L. & S. T. Warren, 1993. Trinucleotide repeat instability: when and where? Nature genetics. 4: 107-108.

Newcombe, H. B., 1964. The phenodeviant theory, pp. 345-349 in Congenital Malformations. Second International Conference on Congenital malformations, edited by M. Fishbein. Lippincott, Philadelphia.

Niswander, J. D. & C. S. Chung, 1965. The effects of inbreeding on tooth size in Japanese children. Am. J. Hum. Gen. 17: 390-398.

Opitz, J. M. & E. Gilbert-Barness, 1990. Reflections on the pathogenesis of Down syndrome. Am. J. Med. Genet. Supp. 7: 38-51.

Parsons, P. A. & W. L. Howe, 1967. Morphogenetic homeostasis in mice. Aust. J. Biol. Sci. 20: 777-784.

Parsons, P. A., 1990. Fluctuating asymmetry: an epigenetic measure of stress. Biol. Rev. 65: 131-145.

Peretz, B., P. Ever-Hadani, P. Casamassimo, E. Eidelman, C. S. Shellhart & R. Hagerman, 1988. Crown size asymmetry in males with Fra(X) or Martin-Bell syndrome. Amer. J. Med. Genet. 30: 185-190.

Perzigian, J. A., 1977. Fluctuating dental asymmetry: variation among skeletal populations. Amer. J. Phys. Anthrop. 47: 81-88.

Potter, R. H. & W. E. Nance, 1976. A twin study of dental dimension. I. Discordance, asymmetry and mirror imagery. Am. J. Phys. Anthropol. 44: 391-396.

Preus, M. & F. C. Fraser, 1972. Dermatoglyphics and syndromes. Am. J. Dis. Child. 124: 933-943.

Reeve, E. C. R., 1960. Some genetic tests on asymmetry of sternopleural chaeta number in Drosophila. Genet. Res. 1: 151-172.

Ricker, J. P., R. H. Boring, A. E. Harris & T. Markow, 1989. Fluctuating dermatoglyphic asymmetry and its association with positive and negative symptoms in schizophrenia. Schizophrenia Research 2: 73.

Rodewald, A., K. D. Zang & G. Ziegelmayer, 1976. Bilateral symmetry of qualitative dermatoglyphic patterns in the Down syndrome. Z. Morph. Anthrop. 67: 333-344.

Rose, R. J., T. Reed & A. Bogle, 1987. Asymmetry of a-b ridge count and behavioral discordance of monozygotic twins. Behaviour Genetics 17: 125-140.

Shapiro, B. L., 1975. Amplified developmental instability in Down's syndrome. Ann. Hum. Genet. London 38: 429-437.

Shapiro, B. L. 1983. Down syndrome - A disruption of homeostasis. Am. J. Med. Gen. 14: 241-269.

Siegel, M. I., W. J. Doyle & C. Kelley, 1977. Heat stress, fluctuating asymmetry and prenatal selection in the laboratory rat. Am. J. Phys. Anthropol. 46: 121-126.

Singh, S., 1970. Inheritance of asymmetry of finger ridge counts. Hum. Hered. 20: 403-408.

Slatis, H. M., B. M. Katznelson & B. Bonne-Tamir, 1976. The inheritance of finger-point patterns. Am. J. Hum. Gen. 28: 280-289.

Smith, B. H., S. Garn & P. M. Cole, 1982. Problems of sampling and inference in the study of fluctuating asymmetry. Amer. J. Phys. Anthropol. 58: 281-289.

Smith, C., 1971. Recurrence risks for multifactorial inheritance. Amer. J. Hum. Gen. 23: 578-588.

Sofaer, Y. A., 1979. Human tooth-size asymmetry in cleft lip with or without cleft palate. Arch. Oral Biol. 24: 141-146.

Soulé, M. E. & J. Cuzin-Roudy, 1982. Allometric variation. 2. Developmental instability of extreme phenotypes. Amer. Nat. 120: 765-786.

Townsend, G. C. & T. Brown, 1980. Dental asymmetry in Australian aboriginals. Hum. Biol. 52: 661-673.

Van Valen, L., 1962. A study of fluctuating asymmetry. Evolution 16: 125-142.

Waddington, C. H., 1957. The strategy of the genes. George Allen and Unwin Ltd., London.

Waddington, J. L., 1993. Schizophrenia: developmental neuroscience and pathobiology. Lancet 341: 531-536.

Woolf, C. M. & A. D. Gianas, 1976. Congenital cleft lip and fluctuating dermatoglyphic asymmetry. Am. J. of Hum. Genet. 4: 400-404.

Woolf, C. M. & A. D. Gianas, 1977. A study of fluctuating dermatoglyphic asymmetry in the sibs and parents of cleft lip propositi. A. J. Hum. Genet. 29: 503-507.

Wright, S., 1934. An analysis of variability in number of digits in an inbred strain of guinea pigs. Genetics 19: 506-536.

Yeo, R. A., S. W. Gangestad & W. F. Daniel, 1993. Hand preference and developmental instability. Psychology (in press).

T. A. Markow (ed.), Developmental Instability: Its Origins and Evolutionary Implications, 335–364, 1994.

Fluctuating asymmetry analyses: a primer

A. Richard Palmer
Department of Zoology, University of Alberta, Edmonton, AB T6G 2E9, Canada; and Bamfield Marine Station, Bamfield, BC V0R 1B0, Canada

Outline

1. Fluctuating asymmetry and developmental stability: a cursory overview

The developmental stability of an organism is reflected in its ability to produce an 'ideal' form under a particular set of conditions (Zakharov, 1992). The lower its stability, the greater the likelihood it will depart from this 'ideal' form. Ideal forms are rarely known *a priori*. However, bilateral structures in bilaterally symmetrical organisms offer a precise ideal, perfect symmetry, against which departures may be compared (Palmer & Strobeck, 1986). Thus they provide a very convenient method for assessing deviations from the norm, and studying the factors that might influence such deviations.

Subtle departures from symmetry are most commonly described by frequency distributions of right - left (R - L). Such frequency distributions usually exhibit one of three patterns (Fig. 1; VanValen, 1962). Other patterns, however, are theoretically possible (skewed asymmetry, Palmer & Strobeck, 1992; normal covariant asymmetry, Palmer *et al.*, 1993; see Sect. 2.3), so no assumptions should be

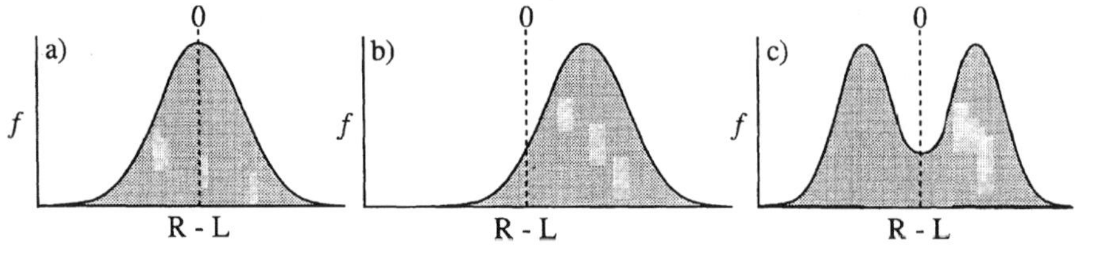

Fig. 1. Three common distributions of R - L in bilateral organisms: a) fluctuating asymmetry (FA), b) directional asymmetry (DA), c) antisymmetry (platykurtic or bimodal).

made about distribution shapes or patterns of covariation between R and L.

Fluctuating asymmetry (= FA, Fig. 1a) is widely used as a measure of developmental stability. It either shows no change or it increases with increasing extrinsic (environmental) or intrinsic (predominantly genetic) 'stress' (see reviews in Palmer & Strobeck, 1986; Leary & Allendorf, 1989; Parsons, 1990; Zakharov, 1992). Deviations from symmetry also appear to correlate with fitness differences, particularly where traits directly affect performance (Thornhill, 1991; Møller, 1994).

A critical assumption underlying studies of FA as a measure of developmental stability is that departures from symmetry in an individual should not have a heritable basis (Palmer & Strobeck, 1992). If departures from symmetry are heritable, then differences in the extent of asymmetry among individuals will have both a genetic and a non-genetic basis. Unless the genetic component can be partitioned out, the variation in R - L can no longer be assumed to describe developmental stability. Debate remains over the reliability of DA and antisymmetry as measures of developmental stability (Palmer & Strobeck, 1992; Graham *et al.*, 1993; McKenzie & O'Farrell, 1993), so caution is advised when encountering these forms of subtle asymmetries in studies of FA.

Most studies of FA attempt to distinguish differences in the between-sides variation among two or more samples. Since differences between sides are often very small (generally <5% and often <1% of trait size), great care must be taken during both measurement and analysis to maximize the likelihood of detecting FA differences among samples.

Regrettably, data from published studies of FA are nearly impossible to compare quantitatively for a variety of reasons outlined below. Where discrepancies exist in the literature, I suspect that these are most commonly due to limited analytical power. I hope that this primer will help illustrate: a) how to conduct FA analyses with the greatest power and precision, b) how to avoid common pitfalls in FA analyses, and c) how to present data in a fashion that allows it to be compared *quantitatively* among studies. The checklist at the end (Sect. 16) should serve as a useful guide.

2. Terminology

2.1 Asymmetry in an individual vs pattern of asymmetry variation in a sample

The terms FA, DA and antisymmetry all refer to patterns of variation in a particular trait exhibited by a *sample* of individuals. One cannot determine to which pattern a departure from symmetry in a single individual belongs without reference to other individuals in the sample. More seriously, one can not conclude that because a particular trait does not depart statistically from FA that DA or antisymmetry are absent for this trait. One can only conclude that, given the sample size used and given the particular precision of measurement, no departure from FA could be detected. For example, small sample sizes or moderate measurement error may obscure statistical departures from FA even though they may exist.

For this reason, some other term would seem advisable to apply to departures from symmetry in an individual. '*Subtle asymmetry*' might be useful as a general term to refer to slight departures from symmetry in an individual since it makes no presumptions of the distribution to which that departure in that individual might belong. This distinction may seem like splitting hairs, but considerable confusion arises when referring to correlations among traits or correlations among samples, and when trying to discuss the heritability of subtle asymmetries (Sect. 14.1).

2.2 *Patterns vs processes*

Care should be taken to use descriptors of patterns of asymmetry variation solely to describe patterns. FA is a *pattern* of between-sides variation in a sample of individuals. It reflects some developmental compromise between two presumably independent but opposing sets of *processes* (Fig. 2) grouped loosely under the headings of:

developmental noise = developmental instability: a suite of processes that tend to disrupt precise development, such as a) small, random differences in rates of cell division, cell growth and cell shape change, b) effects of thermal noise on enzymatic processes, c) small, random differences in rates of physiological processes among cells.

developmental stability: a suite of processes that tend to resist or buffer the disruption of precise development, such as a) negative feedback systems that regulate enzyme activity (both concentration and catalytic rate) within and between cells, b) central nervous regulation of non-contiguous structures, or c) hormonal regulation of non-contiguous structures.

FA may increase either because developmental noise increases, or because developmental stability decreases (Mather, 1953; VanValen, 1962; Palmer & Strobeck, 1992). Thus differences in FA among samples may arise due to differences in one or the other or both of these phenomena; the differences alone can not delineate their relative contribution. The outcome of these processes influences the extent of homeorhesis, or stabilized flow of development, of Waddington (1957) and Zakharov (1992).

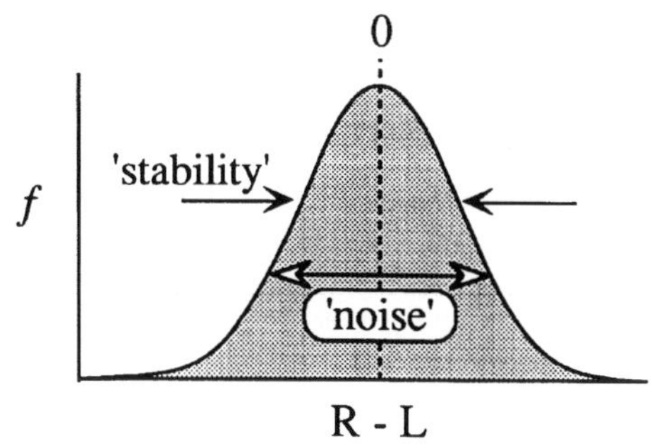

Fig. 2. The magnitude of FA reflects a compromise between two opposing processes: developmental noise and developmental stability.

2.3 *Glossary*

absolute asymmetry: the absolute value of the difference between the R and L sides of a trait in an individual, |R - L|.

antisymmetry: a pattern of bilateral variation in a *sample* of individuals, where a statistically significant difference exists between sides, but where the side that is larger varies at random among individuals (Fig. 1c); detected by statistical tests for departures of frequency distributions of (R - L) from normality in the direction of platykurtosis (broad peakedness, or negative values for kurtosis; Sect. 8); mean R - L normally zero (VanValen, 1962). May render traits unusable for studies of developmental stability (Sect. 8).

bilateral variation: a general term referring to all patterns of between-sides variation in bilaterally symmetrical organisms (FA, DA, antisymmetry, NCA, etc.).

canalization: the ability of a structure to develop along an ideal developmental trajectory under a variety of *different* environmental conditions (Waddington, 1940). A phenomenon distinct from developmental stability, which refers to the ability of a structure to develop along an ideal developmental trajectory under a *particular* set of environmental conditions (Zakharov, 1992). A trait that is highly canalized will exhibit little phenotypic plasticity in response to different growth environments, even though it may still exhibit normal levels of developmental noise.

covariant asymmetry: a general term referring to patterns of bilateral variation that involve negative covariation between sides within a sample (i.e. antisymmetry and normal covariant asymmetry; Palmer & Strobeck, 1992).

developmental instability: see Sect. 2.2.

developmental noise: see Sect. 2.2.

developmental stability: see Sect. 2.2.

directional asymmetry (DA): a pattern of bilateral variation in a *sample* of individuals, where a

Table 1. FA indices (modified from Table 1 of Palmer & Strobeck, 1986; see original paper for references).

Form of size correction	Measure of asymmetry in an individual: Unsigned (= absolute) asymmetry |R - L|	Signed asymmetry (R - L)	Ratio between sides (R/L)
none	**FA1:** mean |R - L|	**FA4:** var(R - L) **FA5:** $\Sigma(R - L)^2/N$	
by individual	**FA2:** mean [|R - L|/((R + L)/2)]	**FA6:** var[(R - L)/((R + L)/2)]	**FA8:** var(log(R/L))
by sample	**FA3:** mean |R - L|/mean((R + L)/2)	**FA7:** var(R - L)/mean((R + L)/2)	

Other indices for single traits:

FA9: 1 - r^2 of correlation between R & L (i.e. % bilateral variation not due to positive covariation).

FA10: $\sigma_i^2 = (MS_{sj} - MS_m)/M$ from a sides x individuals ANOVA (from Table 3 of Palmer & Strobeck, 1986); describes the magnitude of non-directional asymmetry after partitioning out measurement error (see Sect. 6.6 for a full explanation).

Indices based on multiple traits per individual:

FA11: asymmetry in an individual (A_i) = Σ|R - L| for all traits; the index for a sample is $\Sigma A_i/N$ where N = number of individuals in the sample (Leary *et al.*, 1985b).

FA12: asymmetry in an individual (A_i) = total number of asymmetrical traits in an individual, independent of how large the deviation is between sides; the index for a sample is $\Sigma A_i/N$ where N = number of individuals in the sample (Leary *et al.*, 1985b).

FA13: Generalized index of overall fluctuating asymmetry (GFA); a multivariate measure of average deviation from symmetry for multiple metrical traits (Zhivotovsky, 1992).

statistically significant difference exists between sides, but the side that is larger is generally the same (Fig. 1b); detected by statistical tests for departures of mean R - L from zero (Sect. 7; VanValen, 1962). May render traits unusable for studies of developmental stability (Sect. 7).

fluctuating asymmetry (FA): a pattern of bilateral variation in a *sample* of individuals where the mean of R - L is zero and variation is normally distributed about that mean (Fig. 1a); a pattern of bilateral variation that may arise via many processes (Sect. 2.2).

homeorhesis: 'stabilized flow' of a developmental trajectory; a phrase coined by Waddington (1957) and encouraged by Zakharov (1992); refers to the capacity for a structure to develop along an ideal developmental trajectory under a *particular* set of environmental conditions. A phenomenon distinct from canalization.

non-directional asymmetry: a general term useful to refer to all forms of bilateral variation *except* directional asymmetry (i.e. FA, antisymmetry, normal covariant asymmetry).

normal covariant asymmetry (NCA): a hypothetical pattern of bilateral variation in a sample where bilateral traits exhibit continuous negative covariation; on a scatterplot of R vs L for a sample, it would appear as an elliptical distribution of points whose major axis was perpendicular to the line of perfect symmetry (R = L); it may or may not pose a problem for studies of developmental stability depending on how widespread it is (see Fig. 1e of Palmer *et al.*, 1993).

phenodeviant: refers to a trait whose value in an individual lies well outside the 'normal' range for a trait. For each trait described in this manner, one must take care to define precisely, and to defend, what is considered to be 'normal'.

phenotypic plasticity: the capacity for a trait produced by identical genotypes to exhibit different phenotypes under different environmental conditions, the greater the plasticity the lower the canalization.

Population Asymmetry Parameter (PAP): a phrase coined by Soulé (1967) to describe the correlation of asymmetry indices for different traits among populations (i.e. populations with greater asymmetry variation for one trait also exhibit

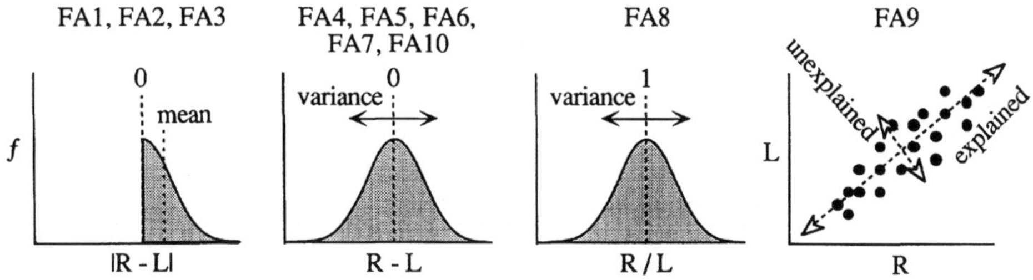

Fig. 3. Graphical illustration of what is being estimated by different FA indices; see Table 1 for actual indices.

greater asymmetry variation for other traits). Soulé believed such correlations arose because differences in developmental stability were reflected in all traits (Sect. 14.3).

signed asymmetry: the difference between the R and L sides of a trait in an individual with the sign included, (R - L); retains information about the direction of the asymmetry.

skewed asymmetry: a pattern of bilateral variation in a *sample* of individuals where the frequency distribution of R - L departs significantly from normality in the direction of skew.

subtle asymmetry: a convenient term to refer to departures from symmetry of a trait in a *single individual*, as opposed to a *sample of individuals* (Sect. 2.1). Avoids much of the confusion that may arise where terms for the pattern of variation exhibited by a sample are applied to deviations from symmetry in an individual (Sect. 14).

3. Indices for describing the level of FA in a sample

3.1 FA indices

Many different indices have been used to describe the level of FA in a sample (Table 1). The two most frequently used ones are **FA1** & **FA4**. A common modification is to (try to!) correct for size-dependence by dividing by the mean trait size either at the level of individuals (**FA2, FA6, FA8**) or at the level of samples (**FA3, FA7**; see Sect. 10 for cautions regarding size correction). Another index (**FA10**) can be obtained from the results of an ANOVA procedure that simultaneously tests for measurement error, size differences among individuals and directional asymmetry (Sect. 6.6).

3.2 Pros & cons of different FA indices

FA1: *Pros:* easily computed; yields a number that is intuitively easy to understand; unbiased estimator of the sample standard deviation; readily used in the recommended ANOVA procedure for testing for differences among 3 or more samples (Palmer & Strobeck 1992); not as sensitive to outliers as FA4; may allow more sophisticated multi-trait tests for differences among samples (Sect. 13.2).

Cons: will be *very* biased if either DA or antisymmetry are present; only 87.6% as efficient as the sample standard deviation (Kendall & Stuart, 1951), hence some statistical power is lost that might be important where sample sizes or differences between samples are small; sensitive to size-dependence of |R - L|.

Recommendation: probably the most generally useful index for moderate to large sample sizes (30+). May be used with smaller sample sizes, but at the sacrifice of some statistical power. Use *only* where DA and antisymmetry are absent (Sects. 7, 8), and where |R - L| does not change with overall size (Sect. 10).

FA2: *Pros:* as for FA1.

Cons: as for FA1 except not biased by size-dependence of |R - L|.

Recommendation: use only where clear evidence exists of a size dependence of |R - L| among individuals *within a sample* (Sect. 10.2).

FA3: *Pros:* as for FA1.
Cons: as for FA1 except not biased by size-dependence of |R - L|.
Recommendation: use only where clear evidence exists of a size dependence of |R - L| *among samples* (Sect. 10.3).

FA4: *Pros:* easily computed; lends itself to the most powerful test for differences between two samples (F test, Lehmann, 1959); more efficient than FA1 for estimating the between-sides variation; not biased by DA.
Cons: more sensitive to outliers than FA1; will be *very* biased if antisymmetry is present; tests for differences among 3 or more samples are more sensitive to departures from normality than FA1 (Palmer & Strobeck, 1992); yields units that are not as intuitively easy to understand as FA1; sensitive to size-dependence of |R - L|.
Recommendation: A useful descriptor of FA particularly when only two samples are to be compared (Sect. 13.1). Use *only* where antisymmetry is absent (Sect. 8), and where |R - L| does not change with overall size (Sect. 10).

FA5: *Pros:* yields an estimate of the between-sides variance with one additional degree of freedom than FA4 because the mean is assumed to be zero (i.e. one degree of freedom is not used to estimate the mean); statistical power is gained that might be important where sample sizes, or differences between samples, are small.
Cons: harder to compute than FA1 or FA4; will be *very* biased if either DA or antisymmetry are present; more sensitive to outliers than FA1; sensitive to size-dependence of |R - L|.
Recommendation: The main value of this index is the one additional degree of freedom it yields over FA4. Where sample sizes are small (on the order of 20 or less), or where differences in FA between samples are small, this index may yield additional needed statistical power. Use *only* where DA and antisymmetry are absent (Sects. 7, 8), and where |R - L| does not change with overall size (Sect. 10).

FA6: *Pros:* as for FA4.
Cons: as for FA4 except not biased by size-dependence of |R - L|.
Recommendation: Use only where clear evidence exists of a size dependence of |R - L| among individuals *within a sample* (Sect. 10.2).

FA7: *Pros:* as for FA4.
Cons: as for FA4 except not biased by size-dependence of |R - L|.
Recommendation: Use only where clear evidence exists of a size dependence of |R - L| *among samples* (Sect. 10.3).

FA8: *Pros:* easily computed; not biased by size-dependence of |R - L|.
Cons: may yield artificial differences among samples if |R - L| does not depend on trait size but trait size varies widely among individuals or samples since the value of this index depends on both the difference between-sides and mean trait size; restricted number of tests for among-sample differences; not widely used in studies of FA; units not as intuitively easy to understand as FA1 or FA4.
Recommendation: Not recommended. Basically equivalent to FA6, but does not yield numbers that are as well defined statistically.

FA9: *Pros:* easily computed; expresses between-sides variance as a proportion of the total variation between sides and among individuals; not biased by DA.
Cons: is *very* dependent on overall trait size variation in the sample since, for a given level of developmental noise, the index decreases as the range of body sizes increases (Angus, 1982); will be *very* biased if antisymmetry is present.
Recommendation: Not recommended. Do not use except in conjunction with other indices.

FA10: *Pros:* this is the *only* index that allows the measurement error variance to be partioned out of the total between-sides variance; not biased by DA; lends itself to the most powerful test for differences between two samples (F test; Lehmann, 1959).
Cons: more difficult to compute than FA1 or FA4; degrees of freedom approximate and depend on the difference between MS_{inter} and MS_{err}; will be *very* biased if antisymmetry is present; yields units that are not as intuitively easy to understand as FA1; sensitive to size-dependence of |R - L|; tests for differences among 3 or more samples are more sensitive to departures from normality than FA1 (Palmer & Strobeck, 1992).
Recommendation: The principle advantage to this index is that the measurement error variance may be partitioned out of the non-directional

asymmetry variance (although taking the average of repeated measurements reduces the impact of measurement error, residual variation due to measurement error still remains, see Sect. 6.6). This yields a more accurate estimate of the true non-directional asymmetry variance, but the degrees of freedom for the estimate depend upon the relative sizes of the non-directional asymmetry and measurement error variances (Palmer & Strobeck, 1986, pg. 408); in other words, the greater the measurement error, the lower the confidence in the estimate of the non-directional asymmetry variance (Sect. 6.1). Probably most useful when accompanied by other indices (e.g. FA1 or FA4).

FA11: *Pros:* easily computed; combines information from multiple traits and may thus yield a more accurate estimate of the overall average asymmetry of an individual where the average deviation from symmetry of all traits is comparable; combines information from all traits into a single index for a sample.
Cons: will be *very* biased if either DA or antisymmetry are present in any trait; obscures differences in variability among characters; yields values that may not be compared quantitatively with other studies; may be biased by one or a few traits where some traits are more variable than others.
Recommendation: Use only a) in conjunction with other indices for individual traits, b) where tests are needed for correlations between the inferred level of developmental stability in an individual (e.g. its 'average' asymmetry over several traits) and some other factor such as fitness (Sect. 15), performance, heterozygosity, etc., or c) where one wishes to pool information from all traits to increase the likelihood of detecting differences among samples. However, see Sect. 14.2 to determine whether differences in FA among individuals are significant statistically or not and Sect. 13.3 for a test that pools information from multiple traits using conventional FA indices.

FA12: *Pros:* easily computed; combines information from multiple traits and may thus yield a more accurate estimate of the overall average asymmetry of an individual; combines information from all traits into a single index for a sample; less likely to be biased than FA11 by one or two extreme values or by one or two extremely variable traits.
Cons: may only validly be applied to meristic traits or traits where one may identify 'phenodeviants' (Sect. 2.3); will be *very* biased if either DA or antisymmetry are present in any trait; obscures differences in variability among characters; yields values that may not be compared quantitatively with other studies; less able to reveal differences in overall average asymmetry among individuals than FA11 because no account is taken of the magnitude of the difference between sides in a given trait.
Recommendation: Use only in parallel with FA11 where concerns exist about differences in the relative variability of different traits.

FA13: *Pros:* combines information from all traits into a single index for a sample even where the average deviation from symmetry of all traits is not comparable.
Cons: difficult to compute; may not be valid for meristic traits where differences between sides are small; will be *very* biased if either DA or antisymmetry are present in any trait; obscures differences in developmental stability among characters; yields values that may not be compared quantitatively with other studies.
Recommendation: Use only in conjunction with other indices for individual traits, and where one wishes to pool information from all traits to increase the likelihood of detecting differences among samples. See also Sect. 13.3 for a test that pools information from multiple traits using conventional FA indices.

3.3 Relationships among FA indices

The indices of Table 1 differ in the degree to which they may be biased by overall trait size variation, by the presence of DA and antisymmetry, and by departures from normality of the frequency distribution of R - L. However, for a truly normal distribution where: i) mean (R - L) = 0, ii) |R - L| does not depend upon overall trait size, and iii) measurement error is absent, the relations among expected values of the indices are:

FA1 = 0.798 * $\sqrt{}$**FA4** = $\sqrt{}$**FA4** / ($\sqrt{(\pi/2)}$) (Kendall & Stuart, 1951)

FA5 = **FA4** (Palmer & Strobeck, 1986)

FA10 = **FA4** / 2 (Palmer & Strobeck, 1986)

FA2, FA3, FA6, FA7, FA8, FA9 - relations to each other and other indices are not defined because values depend either on mean size (all but FA9) or size range (FA9) of traits.

3.4 General recommendations regarding FA indices

i) Every index of Table 1 is sensitive to antisymmetry, hence tests for antisymmetry should be conducted for all traits (Sect. 8). Traits exhibiting antisymmetry should be used only with caution if at all (Sect. 1).
ii) Many indices of Table 1 are sensitive to DA, hence tests for DA should also be conducted for all traits (Sect. 7). Traits exhibiting DA should be used only with caution if at all (Sect. 1).
iii) Never assume |R - L| *depends on* overall trait size. Indices that incorporate size scaling (FA2, FA3, FA6, FA7) should only be used where clear evidence exists that |R - L| depends on overall trait size (Sect. 10). If used indiscriminately, size-scaled indices can generate artificial differences in FA where samples differ in overall size but not in the average absolute difference between sides.
iv) Never assume |R - L| is *independent* of overall trait size. Where |R - L| depends on trait size, artificial differences in FA among samples may arise where samples differ in overall size.
v) Because indices differ in their sensitivity to other factors (outliers, DA, trait size variation, other departures from normality), tabulations of FA indices should include values for at least two FA indices (e.g. FA1, FA4, or FA10) to illustrate the dependence of differences in FA on the choice of index (Sect. 12).
vi Where two or more indices are presented, one should be FA10, because this is the only index that effectively removes the impact of measurement error from the between-sides variance (Sect. 6).

4. Choice of traits

4.1 Pros & cons of meristic vs metrical traits

meristic: *Pros:* may often (but not always!, see Jagoe & Haines, 1985) be measured without error thus avoiding the need for replicate measurements; may be fixed early in development and thus not sensitive body size variation or age (Taning, 1952; Leary *et al.*, 1985a; but see Strawn, 1961 and Angus & Schultz, 1983, for contrary evidence).
Cons: differences between sides are often small (e.g. 1 or 2); quantizing of variation may obscure subtle departures from FA (particularly antisymmetry); where most asymmetrical individuals differ by only 1 between sides, all FA indices depend on how close mean trait size is to an integer value (Sect. 4.2); the degree to which indexes of FA describe the level of developmental noise depends on how individual meristic elements are added to or lost from a group (Sect. 4.2).
Recommendation: Great care must be taken with traits that rarely differ by more than 1 between sides in a given sample (Sect. 4.2). However, when traits are easy to count with little or no error, exhibit moderate variation between sides, and counts are independent (or nearly so) of body size and age, meristic traits can be ideal for studies of FA.

metrical: *Pros:* variation is continuous so the ability to detect differences between sides, or departures from FA, is limited only by measurement precision and accuracy; virtually any trait by used.
Cons: measurement error generates the same pattern of between-sides variation as FA and hence *some* estimate of measurement error is necessary to confirm that between-sides differences are greater than measurement error; between-sides variation often increases with overall trait size so a correction for size-dependence of |R - L| may be needed.
Recommendation: Any use of metrical traits in studies of FA variation should be accompanied by clear estimate of the size of measurement error relative to FA (Sect. 6). Unless measurement error is less than roughly 25% of the between-sides variation, all traits should be measured at least twice. Because measurement error can obscure differences in FA among samples, particular care must be taken to document the contribution of measurement error in studies reporting negative results.

Care must also be taken to test for, and correct

if necessary, a dependence of |R - L| on overall size for each trait (Sect. 10).

4.2 Two idiosyncrasies of meristic traits

Idiosyncrasy I. Although meristic traits may be advantageous in some circumstances for studies of FA, particular care must be taken where most asymmetrical individuals in a sample differ between sides by only 1 (see Swain, 1987, for a general discussion of this problem). Swain notes that in such cases the number of asymmetrical individuals in the sample depends heavily on how close the mean count is to a whole integer value. Consider the following hypothetical situation. If bristles are uniformly spaced along a limb segment, and the spacing between bristles remains constant on average, then even where the uncertainty in bristle position (developmental noise, shaded frequency distributions at segment bases in Fig. 4) remains the same, all measures of FA will depend on the average length of the limb segment and hence the average bristle number. If a limb segment had seven bristles on average, and the length of the segment was such that the average position of the terminal bristle was well in from end of the segment (Pop. A, Fig. 4), then even if the position of this bristle varied, most individuals would still be symmetrical, and only a few would have one fewer or one extra on a given side. If, however, the segment was just a little shorter, and the average position of the terminal bristle was precisely at the end of the segment (Pop. B, Fig. 4), the bristle is much more likely to be missing from one side or the other. In this hypothetical example, the terminal bristle would be missing half of the time and the average bristle count would be 6.5. Hence, even though the underlying developmental instability in terms of placement of the bristle is the same in both cases, the observed asymmetry variation is larger in the for Pop. B, simply because the segment on which it occurs is shorter.

This impact of a non-integer mean on the variance of the between-sides difference is particularly apparent when considering the absolute value of R - L (right-most histograms, Fig. 4). Where the mean count is a whole integer value (Pop. A), a large peak occurs at zero with a small count in the +1 class, but if the mean count is a half-integer value (Pop. B), equal frequencies exist in both the zero and +1 class. Clearly, the computed FA will larger for Pop. B than Pop. A, even though the level of developmental stability is the same for both populations (shaded frequency distributions at segment bases). This arises simply because it is not possible to have 6.5 elements on one side of an individual.

Such concerns are not entirely hypothetical, as illustrated by data for pectoral fin rays and lower gill rakers in rainbow trout (Fig. 5). Should such an association be detected, one solution might be to

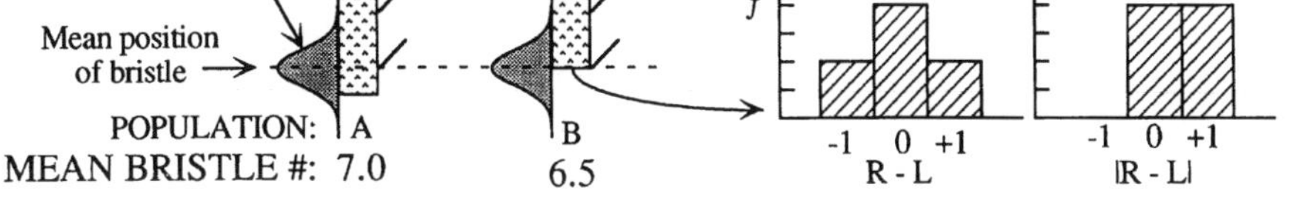

Fig. 4. Effect of non-integer mean on FA variation when sides differ by only 1. In population B, the limb segment is slightly shorter than that for population A. Shaded frequency distributions indicate the hypothetical frequency of occurrence of bristle #7 at various positions near the end of the limb segment for both populations. Cross-hatched frequency distributions indicate the frequency of symmetrical and asymmetrical individuals for both signed (left) and unsigned (right) asymmetry for each population.

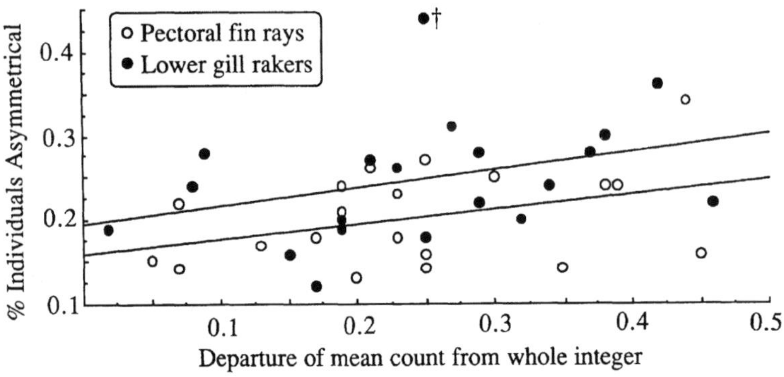

Fig. 5. Effect of deviation of sample mean count from a whole integer value on asymmetry variation among 20 populations of rainbow trout, *Onchorynchus mykiss* (common slope P = 0.0171 ANOVA, † - outlier excluded from analysis; data from Tables 2 & 5 of Leary *et al.*, 1992; three other meristic traits did not exhibit such an association).

convert asymmetry estimates to residuals from the regression of asymmetry against departure of mean count for a trait from a whole integer. Swain (1987) recommends that care be taken when ≤20% of the sample depart from symmetry by no more than 1 and the mean count is a whole integer, or where ≤1% of the sample depart from symmetry by no more than 1 and the mean count is halfway between integer values.

Idiosyncrasy II. Variation in the number of meristic elements in a series can arise via two fundamentally different mechanisms (Scharloo, 1991). In one, elements are added or lost sequentially from terminal positions in a series (e.g. fin rays in fishes). In the second, elements may be present or absent independently at each of several fixed positions in a series (e.g. certain bristles in *Drosophila*). The sensitivity of a particular meristic trait to developmental noise may very well depend on the mechanism by which meristic elements are added or lost from the series, though this has yet to be explored theoretically.

4.3 Single vs multiple traits

As a rule, multiple traits are preferred when testing for differences in developmental stability *among samples* (Leary & Allendorf, 1989; Palmer & Strobeck, 1992). If several traits yield a concordant pattern of variation (e.g. see Zakharov *et al.*, 1991), even if very subtle, one has greater confidence that the pattern is somehow 'real', and not an artifact of something peculiar to a particular trait (Palmer & Strobeck, 1992). Thus, wherever possible, examine patterns of FA variation in several traits (Sect. 12).

Furthermore, when using Levene's test for differences in FA among samples, additional power may be gained by using multiple traits to conduct a two-way factorial ANOVA (traits × samples) on |R - L|, and testing for an overall effect of 'samples' (Sect. 13.3).

4.4 Choose traits that are developmentally independent

Although subtle asymmetries in different traits seem rarely to correlate among individuals *within a sample* (Palmer & Strobeck, 1986), such correlations are occasionally found (Waddington, 1955; Chang *et al.*, 1960; Bader, 1965; Brown & Wolpert, 1990). If the goal of a study is to be able to infer differences in developmental stability *among samples*, the less developmentally correlated the traits the more robust the conclusion. If, however, the goal of a study is to be able to make inferences about the developmental basis of subtle deviations from symmetry *in individuals*, then a comparison of closely connected versus distantly connected traits could be quite informative (Leamy, 1993).

4.5 Choose traits that exhibit 'ideal' FA

In studies of developmental stability, traits that exhibit 'ideal' FA (mean (R - L) = 0, variation = normal) should be selected over those that exhibit

DA or antisymmetry, where possible. For traits that exhibit DA or antisymmetry, some of the between-sides difference in an individual may be due to heritable variation for asymmetry (Palmer & Strobeck, 1992). If so, then the variation in such traits in a sample of individuals will reflect a mix of developmental noise, heritable differences for asymmetry, and effects of genotype-environment interactions for asymmetry. As with any trait whose development is affected by both non-genetic and genetic factors, the relative contribution of these sources can only be assessed via some form of heritability analysis. Debate remains over the reliability of traits that exhibit DA or antisymmetry as indicators of developmental stability (Palmer & Strobeck, 1992; Graham *et al.*, 1993; McKenzie & O'Farrell, 1993), so caution is advised when encountering these forms of subtle asymmetries.

Laboratory studies, for example of the effect of stress on developmental stability, would seem less vulnerable to concerns about DA and antisymmetry since the genetic makeup of experimental groups may be controlled by the investigator. So long as DA and antisymmetry are small, and experimental groups are reasonably homogeneous genetically, differences in developmental stability (however measured) among experimental groups should not be too seriously confounded. However, where DA and antisymmetry are moderate or large, the possibility arises that some of the differences among experimental groups might arise due to genotype-environment interaction (i.e. the effect of genes contributing to DA or antisymmetry may depend on the environment in which the individual develops), in which case differences among groups would not be due to differences in developmental stability. In addition, the between sides variance of traits exhibiting antisymmetry cannot be compared quantitatively with that for traits exhibiting ideal FA.

Concerns about DA and antisymmetry are more serious for studies of developmental stability in natural populations. If deviations from symmetry are heritable in traits that exhibit DA or antisymmetry, then apparent differences in developmental stability among populations could arise due to differences the amount of heritable variation for asymmetry.

Ideally, a preliminary survey should test for departures from FA of R - L in a range of traits. As a precaution, traits exhibiting significant DA (Sect. 7), or antisymmetry or skew (Sect. 8) should be excluded from further study. If this is not practical, see Sect. 9 for some guidelines for avoiding spurious associations.

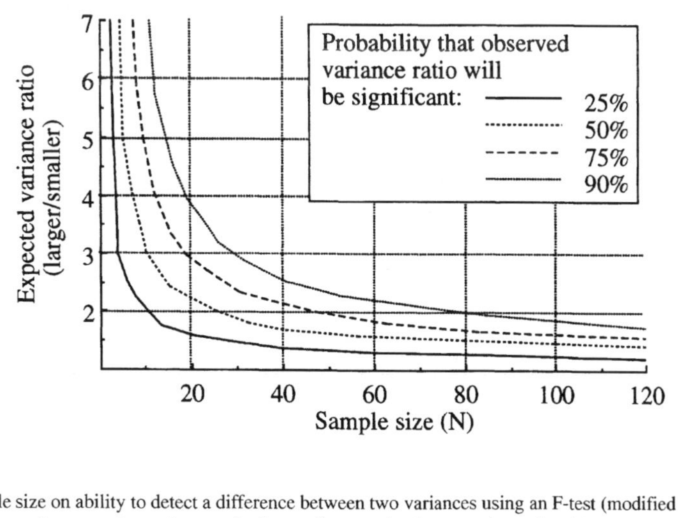

Fig. 6. Effect of sample size on ability to detect a difference between two variances using an F-test (modified from Smith *et al.*, 1982). The smaller the difference between the variances, the larger the sample size must be to be able to detect the difference at a given probability. For example, to detect a true difference between variances of 2 even 50% of the time requires a sample size of at least 20. To detect such a difference 75% of the time, requires a sample size of 40. $\alpha = 0.05$ for all tests.

5. Sample sizes

Tests for differences in FA are effectively tests for differences in variances among samples. Even using an F-test to test for differences in FA between two samples (the most powerful test in this case), the ability to detect, for example, a two-fold differences in the variance depends very much on the sample size: a sample size of 10 will only reveal a significant difference ($\alpha = 0.05$) 25% of the time, a sample size of 25 will only reveal a significant difference 50% of the time, and a sample size of 40 will only reveal a significant difference 75% of the time (Fig. 6). For a detailed discussion of this problem, see Smith *et al.* (1982).

As in any statistical analysis, the recommended sample size depends on the magnitude of the differences in the 'signal' (i.e. FA) among samples. As a rough rule of thumb, N = 30 should be considered a minimum sample size in studies of FA. Where concerns exist about possible departures from normality, N = 40 or 50 would be more desirable since the size of the sample will greatly influence the ability to detect departures from normality (Sect. 8).

6. Measurement error

6.1 Why is measurement error a particular concern in studies of FA?

Measurement error is a particularly troubling concern in studies of FA because the 'data' to be compared among samples are measures of the variation between-sides. Not only can measurement error obscure differences among samples, as in any analysis, but it can also give the illusion that the between-sides variation is large and does not differ among samples. Consider, for example, a bilaterally symmetrical mechanical part machined to within 0.0001 mm on each side. If 50 such parts were given to a student and s(he) was asked to measure the asymmetry of these parts with calipers to a precision of 0.1 mm, the frequency distribution of R - L would, of course, be normally distributed about a mean of zero (e.g. see Fig. 9c). Virtually all of the apparent 'between-sides' variation, however, would simply be due to measurement error, not to variation among individual parts.

More seriously, measurement error not only affects confidence in estimates of FA, as it does for any trait, but it also biases estimates of the between-sides variance (Fig. 7). For a conventional trait

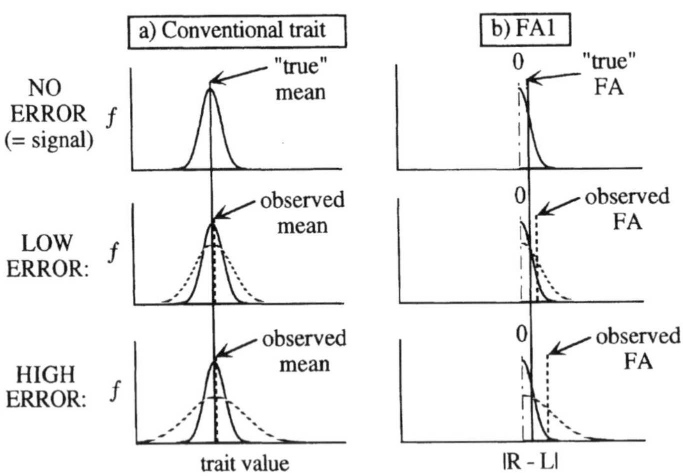

Fig. 7. Impact of measurement error on conventional traits (a) and estimates of FA1 (b). Although only one FA index is illustrated, the impact of measurement error is the same for all indices except FA10. Solid curves and solid vertical lines indicate the 'true' values for a sample. Dashed curves and uniformly-dashed vertical lines indicate the values actually observed when measurement error is added to the 'true' values. Short/long dashed line indicates the expected value for perfect symmetry.

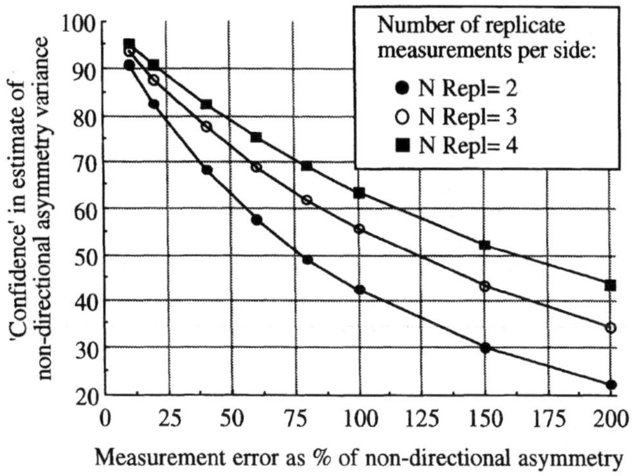

Fig. 8. Impact of measurement error on the 'confidence' with which the magnitude of non-directional asymmetry (FA, antisymmetry, etc.) may be estimated. X axis = 100 * (σ^2_m/σ^2_i) from Table 2b, Y axis = 100 * [(approximate df for σ^2_i)/(df for number of genotypes, J - 1)] from Table 2d.

where one is attempting to estimate the mean, in creasing measurement error simply increases the variance about the mean, but it has no impact on the mean itself (Fig. 7a). However, for FA, where one is attempting to estimate a variance, increasing the measurement error actually increases the variance (Fig. 7b). The greater the measurement error, the greater the impact on the estimate of the between-sides variance (compare middle vs lower panel of Fig. 7b). Fig. 7 thus illustrates why a quantitative estimate of measurement error is essential in studies of FA. Without it, the 'observed FA' contains some unknown mix of 'true FA' and measurement error. Fig. 7 also illustrates why an FA index such as FA10 (Sect. 3.1, 6.6) is particularly valuable. This index formally attempts to separate the fraction of the between-sides variance due to FA from that due to measurement error. All other indices simply lump these two components of the between-sides variance together, and thus yield estimates of FA that are not comparable quantitatively among traits or among studies.

The statistical impact measurement error has on estimates of FA may be more apparent if one examines the effect it has on the effective size of the sample in which one is attempting to estimate FA (Fig. 8). As measurement error increases relative to the underlying non-directional asymmetry, the effective size of the sample with which one estimates non-directional asymmetry decreases. For example, if the measurement error variance is only 25% of the non-directional asymmetry variance (for two replicate measurements per side), the effective sample size is 80% of that actually measured. However, if the measurement error variance is 100% of the non-directional asymmetry variance, the effective sample size for estimating the non-directional asymmetry variance is only 42%.

Because of the bias it introduces into studies of FA, accurate estimates of measurement error are essential.

6.2 *Error in meristic traits*

In principle, meristic traits can be counted without error. In practice, error can arise from several sources. First, unless counting is conducted by an image analysis system, human observers invariably make some mistakes. These may be sufficiently infrequent that they can be ignored. However, when beginning a study, one sample should be counted at least twice to assess how frequent such errors are.

Commonly, establishing the last element in a series of meristic traits that decrease in size may require a subjective assessment (Fig. 9a, b), as may establishing whether a meristic element falls inside or outside an arbitrarily defined area. This subjectivity can lead to an error that is no different from the measurement error of metrical traits. If subjectivity is involved when counting meristic traits, an assessment of the magnitude of 'measurement' error is essential (Sect. 6.7).

6.3 Error in metrical traits

With metrical traits, measurement error is unavoidable and can seriously bias estimates of FA (Sect. 6.1). Measurement error can create the illusion of FA, even where it is negligible (Fig. 9c), or it will artificially inflate estimates of FA (Fig. 9d). Furthermore, Fig. 9d illustrates how meaningless it is to report measurement error as a percent of overall trait size [e.g. as a % of (R + L)/2]. Clearly what matters in studies of FA is measurement error as a percent of between-sides variation (compare solid vs dashed lines with arrows, Fig. 9d).

Ideally, all metrical traits should be measured at least twice. However, where sample sizes are large, or where measurements are difficult to obtain, repeated measurements of all traits may not be feasible. In these cases, tests for the impact of measurement error (Sect. 6.5) should be conducted for each trait on a subsample of at least 30 individuals.

6.4 Quantizing error in image analysis systems

Video image analysis systems (e.g. Morphosys, Jandel) offer a seductively simple, and seemingly more reliable method for quantifying the size and shape of structures. For conventional morphometric studies they can be quite useful. For studies of FA, however, great care must be taken to determine the size of the pixellation error relative to the average difference between sides.

Pixellation error is a problem for studies of FA because most imaging systems have rather low resolution (on the order of 600 × 600 pixels per image). When a structure spans 200 pixels, the pixellation error would seem to be quite small (0.5%). However, since the average difference *between sides* in FA studies is on the order of 1% of trait size, pixellation error can make up a substantial fraction of the average between-sides difference.

6.5 Recommended procedure for conducting repeated measurements

To obtain a totally unbiased estimate of measurement error, repeat measurements should be conducted 'blind' (i.e. without reference to earlier measurements). Furthermore, since subjective 'preferences' of the observer may vary with time, repeat measurements are more reliably conducted one or more days apart. Given the bias that measurement error introduces (Sect. 6.1), the more reliable the estimate of measurement error, the more reliable will be the estimate of FA. Thus, where replicate measurements are to be taken, the most reliable estimate of measurement error will be obtained if one complete set of measurements is finished, and several days allowed to elapse, before the repeat set of measurements is started. Ideally,

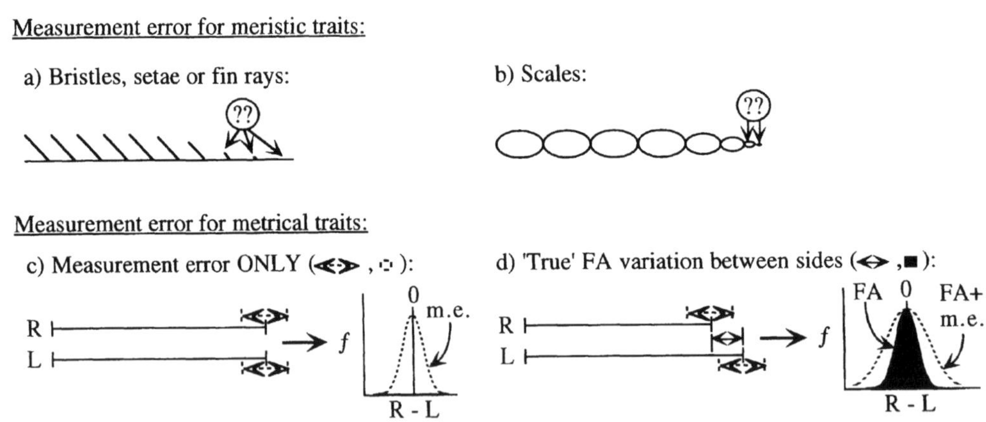

Fig. 9. Examples of measurement error for meristic and metrical traits. (a,b) errors due to subjective decisions about meristic traits, (c) 'artificial' FA created solely by measurement error, (d) a combination of 'true' FA and measurement error. m.e. - measurement error. R - trait on right side, L - trait on left side. Note that measurement error should be expressed relative to FA rather than to trait size [(R + L)/2].

Table 2. ANOVA procedure for testing the significance of FA relative to measurement error. This test also simultaneously tests for the presence of DA and for trait-size differences among individuals.

a) Data layout

Format of data file						Layout of analysis for 1 trait		
Grouping Variables								
Individual	Side	Repl.	Trait 1	Trait 2	etc.	Individual (random)	Side (fixed) Right	Left
A1	R	1	M1	M1				
A1	R	2	M2	M2		A1	M1	M1
A1	L	1	M1	M1			M2	M2
A1	L	2	M2	M2				
A2	R	1	M1	M1		A2	M1	M1
A2	R	2	M2	M2			M2	M2
A2	L	1	M1	M1				
A2	L	2	M2	M2		A3	M1	M1
A3	R	1	M1	M1			M2	M2
A3	R	2	M2	M2				
A3	L	1	M1	M1		etc.		
A3	L	2	M2	M2				
etc.								

*b) Variance components in a mixed model, two-way ANOVA**

Source of variation	Mean squares label	df	Expected mean squares	Mean squares used to test for
Sides	MS_s	$(S-1)$	$\sigma^2_m + M(\sigma^2_i + J/S - 1\ \Sigma\alpha^2)$	Directional asymmetry
Genotypes**	MS_j	$(J-1)$	$\sigma^2_m + M(\sigma^2_i + S\sigma^2_j)$	Size/shape variation §
Side x Genotype	MS_{sj}	$(S-1)(J-1)$	$\sigma^2_m + M\sigma^2_i$	Nondirectional asymmetry †
Meas. error	MS_m	$SJ(M-1)$	σ^2_m	

* From Table 3b of Palmer & Strobeck (1986). S = number of sides, J = number of 'genotypes' (= 'individuals'), M = number of replicate measurements, $\Sigma\alpha^2$ = added fixed variance component due to 'sides' (i.e. directional asymmetry), σ^2_j = added random variance component due to 'genotypes' (= 'individuals'; reflects variation in overall trait size or shape among individuals), σ^2_i = added random variance component due to non-directional asymmetry (reflects variation in the between-sides difference among individuals such as FA and antisymmetry), σ^2_m = random variance component due to measurement error.

** 'Genotypes' refers to genetically unique individuals and may be considered as 'individuals' in most cases. Where clonemates are available, a more detailed analysis is possible that can distinguish between two forms of antisymmetry (see Table 3a of Palmer & Strobeck, 1986).

§ Indicates variation in overall trait size or shape among individuals; this may be useful for testing the effectiveness of size scaling.

† Includes *all* forms of non-directional asymmetry including FA, antisymmetry and NCA.

the time between replicate measurements should be roughly equal to the total elapsed time spanned by one complete set of measurements of all treatment groups.

6.6 *Tests for the significance of FA relative to measurement error in metrical traits*

Palmer & Strobeck (1986) describe a test for determining whether the between-sides variation is significantly larger than the measurement error. The

Table (Table 2 continued)

c) Significance tests of results from ANOVA (abbrev. as in Table 2b)

Significance test	F ratio	Degrees of freedom Numerator	Denominator
Nondirectional asymmetry (i.e. FA if antisymmetry is absent) relative to measurement error	MS_{sj} / MS_m	$(S - 1)(J - 1)$	$SJ(M - 1)$
Directional asymmetry	MS_s / MS_{sj}*	$(S - 1)$	$(S - 1)(J - 1)$
Variation due to overall trait size or shape among genotypes (= individuals)	MS_j / MS_{sj}*	$(J - 1)$	$(S - 1)(J - 1)$

* Note that the 'error' term from this ANOVA yields no information about the variation between sides, it simply reflects measurement error. Only the sides x genotypes (= individuals) interaction contains information about between-sides variation. Hence the test for directional asymmetry asks: is the mean difference between sides greater than expected given the amount of non-directional asymmetry, as reflected in MS_{sj}? The same logic underlies the test for variation due to trait size or shape.

d) Partitioning measurement error variance from between-sides variance (abbrev. as in Table 2b)

The total between-sides variance (MS_{sj}) includes a contribution from both measurement error and from all forms of non-directional asymmetry. A better estimate of the true between-sides variance (σ_i^2) may be obtained by partitioning measurement error out of the sides x genotypes (= individuals) mean squares of the ANOVA results (Palmer & Strobeck, 1986):

non-directional asymmetry = (between-sides MS − meas. error MS) / # replicate meas.

$$\sigma_i^2 = (MS_{sj} - MS_m) / M$$

$$\text{approx. degrees of freedom for } \sigma_i^2 = \frac{(MS_{sj} - MS_m)^2}{\left(\frac{(MS_{sj})^2}{(S-1)(J-1)} + \frac{(MS_m)^2}{SJ(M-1)}\right)}$$

Note that as the magnitudes of measurement error and between-sides variance become more similar, the approximate degrees of freedom for σ_i^2 decrease. In other words, the larger the measurement error, the lower certainty in the estimate of the between-sides variance (see Fig. 8).

test is a simple two-way ANOVA (sides × individuals, Table 2b), and should be conducted routinely as a part of any study of FA. If the interaction variance is not significant (MS_{sj} of Tables 2b, c), then tests for FA differences among samples are not justified.

A crucial point to keep in mind is that this ANOVA procedure tests for the significance of *all* between-sides variation relative to measurement error, including antisymmetry and FA (Table 2c). In other words, this test asks: *Does the difference between sides vary more AMONG GENOTYPES (= individuals) than would be expected, given the size of the measurement error?* Hence the ANOVA procedure can not establish whether the between-sides variation arises due to antisymmetry or to FA. Some other procedure must be conducted to confirm that the observed between-sides variation is statistically indistinguishable from FA (Sect. 8).

If the interaction variance (MS_{sj}) is significantly greater than the measurement error (MS_m), then measurement error may be partioned out to yield a more accurate estimate of the underlying between-sides variance (σ_i^2, Table 2d). Of all the different indices for FA, only FA10 (σ_i^2 of Table 2d) provides an estimate of the between-sides variance after removing the effects of measurement error.

6.7 Tests for the significance of FA relative to counting error in meristic traits

Unfortunately, for meristic traits, it is less clear

Table 3. Tests for significance of directional asymmetry.

	Test	Description of test
a)	one-sample *t*-test	tests for a departure of the mean of (R - L) from an expected mean of zero
b)	paired *t*-test	tests the consistency of the direction and magnitude of (R - L) among individuals
c)	nested ANOVA (individuals within sides)	tests for a significant difference between mean R and mean L in a sample of individuals
d)	factorial ANOVA (Table 2) (sides x individuals)	tests for a significant difference between mean R and mean L in a sample of individuals relative to the between-sides variation *after accounting for measurement error*; also tests significance of non-directional asymmetry and overall trait size variation among individuals.

how to test for the significance of non-directional asymmetry relative to counting error. Where the differences between sides are large, for example an average of four or five, the ANOVA technique (Sect. 6.6) should probably work adequately. However, where the true difference between sides is only one or two, and a low level of scoring error may exist, no formal tests have been suggested. A likelihood ratio or g-test (Sokal & Rohlf, 1981) comparing the counts in each asymmetry class (e.g. -1, 0, +1) vs the counts in each class of deviations between replicates (e.g. -1, 0, +1) might be used to yield an estimate of the statistical significance of the between-sides variation relative to counting error.

7. Directional Asymmetry

7.1 Why test for DA in studies of FA?

Tests for the presence of DA should be conducted in studies of FA for two reasons: 1) the presence of DA artificially inflates the values of certain FA indices (FA1, FA2, FA3, FA5, Table 1; Sect. 3.2), and 2) if a trait exhibits DA, some portion of the between-sides variation may have a genetic basis, hence the between-sides variance may not purely be a product of developmental noise (Palmer & Strobeck, 1992). Although the biasing effect on FA indices may be removed statistically by 'correcting' for DA, the possibility of heritable variation for DA remains and can not be eliminated without a formal heritability analysis. Hence, if other traits are available that do not exhibit DA, these will be preferred because they raise fewer concerns about inferences based on patterns of bilateral variation among samples.

7.2 Tests for DA

Many tests are possible and appropriate for DA (Table 3), since the question is simple: *Is one side significantly larger than the other on average?* One advantage to the factorial ANOVA procedure (Table 3d; Sect. 6.6) is that the significance of DA may be tested at the same time as that of FA relative to measurement error, hence this procedure is the recommended one. Care should be taken to apply the Sequential Bonferroni test (Sect. 11) when conducting multiple tests for DA.

8. Departures from normality (e.g. antisymmetry and skew)

8.1 Why test for departures from normality in studies of FA?

Frequency distributions of R - L may depart from normality in several ways, including skew (a long tail on one side), leptokurtosis (narrow peaked & long-tailed) or platykurtosis (broad peaked & short-tailed, or bimodal). Traits exhibiting significant departures from normality should be viewed with skepticism as valid measures of developmental stability because they do not conform to the pattern of bilateral variation expected due to developmental noise. Because developmental noise is most precisely viewed as the cumulative consequence of subtle, random variation in perhaps several developmental processes that affect each side independently (Sect. 2.2), it should generate perfectly normal distributions of R - L. Although processes considered to be developmental noise may potentially give rise to non-normal distributions of R - L (Palmer & Strobeck, 1992), processes *not* considered to be developmental noise could also give rise to such distributions. Hence, unless alternative explanations can be ruled out, it seems safest to exclude traits exhibiting significant departures from normality from studies of developmental stability.

Tests for the presence of antisymmetry should be conducted in studies of FA for the same two reasons as tests for DA (Sect. 7.1): 1) antisymmetry artificially inflates the values of *all* FA indices (FA1 - FA10, Table 1; Sect. 3.2), and 2) if a trait exhibits antisymmetry, some portion of the between-sides variation may have a genetic basis, hence the between-sides variance may not purely be a product of developmental noise (Palmer & Strobeck, 1992). Unlike DA, however, no statistical 'corrections' for antisymmetry have been suggested. In addition, estimates of the between sides variance based on traits that exhibit antisymmetry can not be compared quantitatively with those that exhibit ideal FA (Sect. 4.5). As for DA, if other traits are available that do not exhibit antisymmetry, these will be preferred because they raise fewer concerns about inferences based on patterns of bilateral variation among samples.

The relevance of skewed or leptokurtic (long-tailed) distributions of R - L to studies of FA is somewhat less clear, though they may also signal the presence of a genetic component to subtle asymmetries (Palmer & Strobeck, 1992). Until more is known about their developmental significance, traits exhibiting significant skew or leptokurtosis should be avoided.

Note, however, that a common source of skew or leptokurtosis in studies of FA is statistical outliers. For example, the R - L difference in one or two individuals in a large and otherwise normally distributed sample may deviate unusually far from zero. Such deviations could arise due to extreme measurement errors, or they might reflect truly unusual deviations from symmetry due to prior injury or trauma. If measurement error is suspected, outlying individuals could simply be remeasured. Alternatively, if injury or trauma are suspected, such individuals can legitimately be excluded from samples used to estimate FA, because injury or trauma are not normally considered 'developmental noise' (Sect. 2.2).

8.2 Tests for departures from normality: general comments

Numerous tests for departures form normality are available but their statistical power varies depending on the form of departure (Shapiro *et al.*, 1968). After a review of prior studies, Palmer & Strobeck (1992) suggested that conventional skew and kurtosis statistics provide the most useful descriptions of and tests for departures from normality because a) they are readily computed by most commercial statistical packages, b) their standard errors are known (Table 4), and c) these two statistics taken together provide additional useful descriptors of the form of between-sides variation. A visual inspection of frequency distributions of R - L may also be particularly helpful when attempting to interpret apparent departures from normality.

As with all such testing, care should be taken to apply the Sequential Bonferroni test (Sect. 11) when conducting multiple tests.

8.3 Tests for departures from normality in either meristic or metrical traits

Since for a truly normal distribution the expected values for both skew (g_1) and kurtosis (g_2) are zero and the degrees of freedom infinite (Sokal & Rohlf, 1981), the significance test is a straightforward one-sample *t*-test (see Table 4 for standard errors and interpretations of departures from normality):

skew $t_s = g_1/s_{g1}$ (compare versus values of t for ∞ df)

kurtosis $t_s = g_2/s_{g2}$ (compare versus values of t for ∞ df)

Unfortunately, the power to detect departures from normality with these statistics drops off radically as sample size decreases, hence other tests may be more useful for small samples (<30; Sect. 8.4).

8.4 Tests for departures from normality in small samples of metrical traits

The *Kolmogorov-Smirnov Test* offers a useful non-parametric test for departures from normality in small samples (Sokal & Rohlf, 1981). Unfortu-

Table 4. Descriptors of departures from normality, their standard errors and interpretation (Sokal & Rohlf, 1981). See Sokal & Rohlf (1981) p. 114 for computation of the skew and kurtosis statistics themselves.

Descriptor	Standard error (s)	Large sample approximation (n > 150)	Interpretation of departure from normality
skew (g_1)	$\sqrt{\dfrac{6n(n-1)}{(n-2)(n+1)(n+3)}} =$	$\sqrt{\dfrac{6}{n}}$	negative = long tail to left (skewed to L) positive = long tail to right (skewed to R)
kurtosis (g_2)	$\sqrt{\dfrac{24n(n-1)^2}{(n-3)(n-2)(n+3)(n+5)}} =$	$\sqrt{\dfrac{24}{n}}$	negative = platykurtic, broad or bimodal peak and short tails positive = leptokurtic, narrow peak and long tails

nately, because of the 'steppiness' of the frequency distribution of R - L for most meristic traits, this test will yield many 'false' positive results if the number of asymmetry categories is small (<10). Thus it is not very useful when applied to meristic traits. For metrical traits, however, the variation in R - L is continuous and hence the test may be applied with more confidence. See Sokal & Rohlf (1981, pp. 716-721) for worked examples of this test.

9. What to do when traits depart from ideal FA

For the reasons outlined above (Sects. 7.1, 8.1), if traits are found to exhibit DA, or departures from normality, some care should be taken to show that inferred differences in developmental stability (however measured) among samples are not confounded by those traits that depart from 'ideal' FA (Sect. 4.5). The principle concern is that both DA and antisymmetry can bias certain FA indices, and thus render them unreliable as measures of developmental stability (Sects. 7.1, 8.1). If necessary, the presence of DA may be removed *statistically* by subtracting (mean (R - L))/2 from the side with the larger mean and adding it to smaller side of all individuals in a sample. However, concerns remain about the validity of inferring levels of developmental stability using such 'corrected' variation in R - L (Palmer & Strobeck, 1992). Regrettably, no methods have yet been suggested for statistically 'correcting' for the presence of antisymmetry or skew in distributions of R - L.

Should traits be found to depart from 'ideal' FA (Sect. 4.5), and should it not be possible to complete a study without incorporating such traits in the final analyses, several questions should be asked to determine if such traits might be biasing any conclusions:

Do traits exhibiting DA, antisymmetry or skew appear to reveal greater differences in developmental stability among samples than those exhibiting ideal FA (Sect. 4.5)?

Do traits exhibiting greater DA, antisymmetry or skew appear to reveal a greater range in developmental stability among samples than those exhibiting lesser DA, antisymmetry or skew?

Does the magnitude of developmental stability correlate with the magnitude of DA, antisymmetry or skew among samples?

If the answer to any of these questions is yes, then it may not be possible to say with certainty that differences in the between-sides variance among samples are due to differences in developmental stability.

10. Size dependence of FA

10.1 Why is size-dependence a concern in studies of FA?

A dependence of asymmetry on trait size can influence inferences made in studies of developmental stability in three ways: 1) if the magnitude of asymmetry depends on trait size within samples, and some samples have a greater size *range* than others, spurious differences in FA may arise among samples (i.e. samples with a greater size *range* will appear to exhibit higher FA), 2) if FA depends on mean trait size among samples, and *mean size* varies among samples, inferred differences in developmental stability among samples may be an artifact of size variation (i.e. samples with a larger mean size will appear to exhibit higher FA), and 3) if FA depends on mean trait size when comparing different traits, inferred differences in developmental stability among traits may be an artifact of size differences. At the same time, arbitrary 'correction' for presumed size-dependence can lead to spurious differences among samples (Sect. 3.4). For example, if samples differ in average body size, and the variance of R – L is independent of trait size, then inappropriate correction for size effects will create artificial differences among samples (i.e. those with larger mean size will appear to exhibit artificially lower FA). Furthermore, differences in overall size among samples may partly reflect differences in general 'condition', which itself is often correlated with FA (Zakharov, 1992). Hence, eliminating all size-dependence of FA among samples could partially or completely obscure associations between FA and condition (Palmer & Strobeck, 1986).

Among related traits the variance often increases with the mean, which has lead to considerable discussion about how best to describe relative variation (Lande, 1977; VanValen, 1978). Thus, individuals should ideally be selected in such a manner that mean trait size does not differ among samples. Where this is unavoidable, some form of size-correction may be necessary. The question thus is: Does asymmetry vary with size? If, for the size range examined, an association is weak or absent using unscaled indices (e.g. FA1, FA4, FA5 or FA10), no size correction is needed. If the unscaled index varies significantly with size, then some correction may be required.

Never assume that size-corrected FA indices remove all size effects. Regardless of what kind of size corrections are considered or employed, there is no substitute for visual inspections of scatterplots of asymmetry versus 'size', both before *and* after conducting any size correction.

10.2 Tests for size dependence of FA within samples

Where size differences exist mainly within samples, analyses of the size-dependence of FA are less clear cut. Note that a regression of (R – L) against some independent measure of body size will yield a zero slope even though the variance of (R – L) clearly depends on size (Fig. 10a). A more appropriate regression is |R – L| against an independent measure of body size (Fig. 10b). If a significant association remains, a scaled asymmetry index may be in order (e.g. FA2, FA6 or FA8 of Table 1).

Scaling |R – L| by (R – L)/2 for the same trait should be done with care. If body size variation within a sample is small, and asymmetry variation large, much of the apparent variation in (R + L)/2 will arise solely from departures from symmetry (Fig. 10c).

10.3 Tests for size dependence of FA among samples

Where size differences exist mainly among samples, a simple linear regression or visual inspection of plots of log(var(R – L)) vs mean((R + L)/2) should reveal any size-dependence. Log (var(R – L)) is preferred here because the variance increases as a square of the average difference between sides.

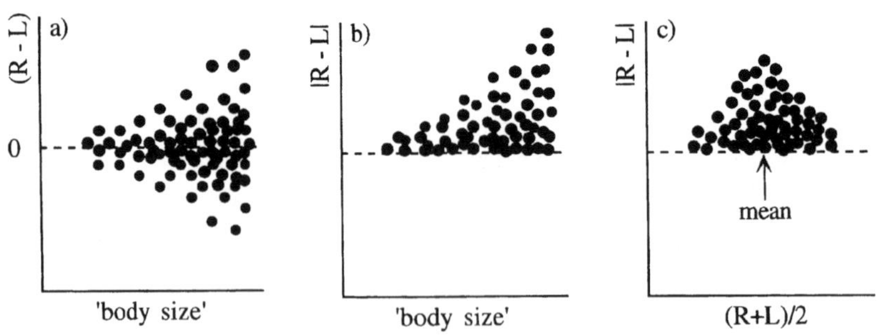

Fig. 10. Examples of size-dependence of asymmetry within a sample. a-b) asymmetry versus some independent measure of 'body size'. c) asymmetry vs. trait size for a single trait exhibiting FA, in the absence of any 'body size' variation (i.e. all apparent variation in trait size is due solely to random variation in each side independently).

Table 5. Illustration of the Sequential Bonferroni procedure to ensure an appropriate table-wide likelihood of 'false' positives (i.e. probability of Type I error at the level of the whole table is at the desired α, normally 0.05); modified slightly from Rice (1989). N = total number of tests in a given table (8 in this example). Note that this somewhat simpler method can yield values or P_r greater than 1.0 where values of P_o are high or N large, hence values of P_r should not be reported.

Test #	Initial Probability P_o	Bonferroni correction	Revised Probability P_r	Significant at P = 0.01	Significant at P = 0.05
1	0.001	$P_0 * N$	0.008	yes	yes
2	0.002	$P_0 * (N-1)$	0.014	no	yes
3	0.007	$P_0 * (N-2)$	0.042	no	yes
4	0.011	$P_0 * (N-3)$	0.055	no	no
5	0.012	$P_0 * (N-4)$	0.048	no	no
6	0.031	$P_0 * (N-5)$	0.093	no	no
7	0.037	$P_0 * (N-6)$	0.074	no	no
8	0.042	$P_0 * (N-7)$	0.042	no	no

Table 6. A recommended 'complete' presentation of descriptive data in studies of FA.

Sample label	Trait	N	(R + L)/2	(R - L)				|R - L| = FA1	FA10§		
			Mean ± SE	Slope ± SE¶	Mean ± SE*	Skew ± SE†	Kurtosis ± SE†	Mean ± SE	MS_m	σ_i^2	df
1	A										
	B		<- - - - - computed AFTER taking average of replicate measurements - - - - ->						(uses replicate measurements to partition out meas. error)		
	C										
	D										
2	A										
	B										
	C										
	D										
etc.											

¶ Slope ± SE from regression of |R - L| or log|R - L| versus some measure of size (either organism size or trait size, Sect. 10.2).

* For mean (R - L), FA4 = N * SE^2.

† SE optional since it may be computed from the sample size (Table 4, Sect. 8.3).

§ MS_m = measurement error mean square, σ_i^2 = non-directional asymmetry, df = approx. degrees of freedom for non-directional asymmetry after partitioning out measurement error (Table 2d, Sect. 6.6).

If a significant association is detected, some correction may be warranted (e.g. FA3 or FA7 of Table 1). However, if body size is also correlated with the factor under investigation (e.g. stress or heterozygosity), such a correction may mask an association with FA (Palmer & Strobeck, 1986).

11. Care when conducting multiple tests

Where a large number of traits or a large number of samples have been analyzed in a single study, care must be taken not to be mislead by 'false' significant results. Recall that a significance level of $P < 0.05$ means that, on average, one out of every 20 tests with completely random data will yield a statistically 'significant' result. The Sequential Bonferroni correction provides a useful guard against such 'false' positive tests. It should be applied to each set of related tests (e.g. all tests for DA or all tests for antisymmetry, etc., in a single study). This procedure is explained and justified in more detail by Rice (1989).

The Sequential Bonferroni test ensures the appropriate table-wide probability of type I error (i.e. 'false' positives) and is very simple to conduct (Table 5):

i) Rank all the original test results from most significant to least significant.
ii) Multiply the lowest original P value (P_o) by the total number of tests in the table.
iii) If the revised P value (P_r) is significant, proceed to the next higher P value in the table.
iv) Multiply the next lowest original P value (P_o) by the number of tests in the table, minus the number of prior Bonferroni tests up to that point.
v) Repeat steps (iii) and (iv) until reaching the first revised P value (P_r) that is *not* significant. All lower ranked tests beyond this test would be considered not significant.
vi) Cease computing revised P values. In other words, even though tests #5 and #8 (Table 5) would appear to be significant at the 0.05 level, these would not be considered significant by the Sequential Bonferroni procedure.

12. Adequate presentation of descriptive data

Anyone who has ever attempted to extract quantitative estimates of FA from the literature for comparison among studies will realize how hopelessly inadequate most presentations of descriptive data are. In some cases, either the number of samples, or the number of traits is too large to permit a full description. However, in most cases, the authors seem mostly concerned with the qualitative patterns of their own study rather than with a quantitative presentation of their data in a way that makes it usable by others. In view of the extraordinary effort required to collect a respectable set of FA data, and in view of how little we understand about the developmental origins and implications of variation in subtle asymmetries, a detailed presentation of descriptive data when results are published should be a high priority.

A detailed presentation presentation of data should ideally include several descriptors for each trait in each sample (Table 6). The more complete the description of bilateral variation, the more useful and informative the comparisons among published studies will be. The logic behind including these statistics includes:

mean ± SE of trait size [(R + L)/2]: to assess dependence of FA indices on trait size among samples and among traits (Sect. 10); also indicates the range of trait size variation within a trait in a given sample.

dependence of |R – L| on size within samples: confirms presence/absence of size dependence of subtle asymmetries among individuals within a sample (Sect. 10).

mean ± SE of the difference between sides (R – L): indicates the presence and significance of any directional asymmetry (Sect. 7), also the SE may be easily converted to FA4.

skew and kurtosis of frequency distribution of R – L: to assess the magnitude of variation of skew and kurtosis among samples and among traits (Sect. 8).

two (ideally three) separate FA indices:

a) FA1 - indicates the average difference between sides, |R - L|, in easy to understand units,
b) FA4 - (the SE of (R - L) may easily be converted to FA4 just by knowing the sample size)
c) FA10 - σ_i^2 provides an estimate of FA after measurement eror has been partitioned out (Table 2d, Sect. 6.6).

estimate of measurement error: critical for knowing what fraction of the bilateral variation is due to measurement error and for comparing values of FA quantitatively among studies (Sect. 6.6).

13. Significance tests for differences in FA

13.1 Between two samples

Where frequency distributions of R - L have been confirmed to be mean = 0, normal (Sects. 7, 8), the most powerful test for the significance of the difference in FA between two samples is an F-test (Lehmann, 1959) using FA4, FA5, FA6, FA7, FA8 or FA10 (Table 1). All these indices are variances, and an F-test is simply a ratio of the larger over the smaller variance. The significance of this ratio need only be looked up in a statistical table for the appropriate degrees of freedom.

13.2 Among three or more samples

Tests for differences in FA are effectively tests of whether variances are significantly heterogeneous among samples. Many tests for heterogeneity of variances are available, but they vary in their statistical power and their sensitivity to departures from normality (VanValen, 1978; Conover *et al.*, 1981). After a detailed comparison of four common tests (Bartlett's, F_{MAX}, Levene's and Scheffé's), Palmer & Strobeck (1992) concluded that Levene's test may be the most practical for studies of FA for the following reasons: 1) it was the least sensitive to departures from normality in the direction of leptokurtosis or platykurtosis and hence more likely to yield valid P values where such departures are small, 2) it was only slightly less powerful than Bartlett's and F_{MAX}, both of which were quite sensitive to departures from normality, and 3) it is extremely easy to conduct on commercial statistics packages, even those that do not formally incorporate the test.

Levene's test is disarmingly easy to conduct: it is simply a one-way ANOVA (i.e. a comparison among samples) conducted with the unsigned asymmetry values, |R - L|. Thus in a data file where |R - L| have already been computed to estimate FA1 or FA2, these same data may be used in an ANOVA, with 'samples' as a grouping variable, to test for differences in FA among samples. Even though these data are not normally distributed (they are truncated normal, and hence skewed to the right; Fig. 3a), this test yields very close to the proper Type I error when no differences exist among samples (Palmer & Strobeck, 1992). However, when all the variation in R - L within samples is normal, Levene's test does lose on the order of 10-15% statistical power compared to Bartlett's and F_{MAX} tests when the differences among variances are slight.

Another potential advantage to using Levene's test is that *a posteriori* tests may be conducted among samples if an overall significant difference is detected. Thus, conventional multiple-comparisons tests can identify subsets of samples among which variation is not significantly different. In addition, where comparisons among subsets of data within a larger analysis are desirable and identifiable *a-priori*, one may conduct a formal analysis of contrasts (Sect. 9.6 of Sokal & Rohlf, 1981). Levene's test thus provides a robust and versatile method for analyzing FA differences among samples.

Table 7. Structure of and results from a test that combines information from several traits to test for differences in FA among samples. It is a modified version of Levene's test for heterogeneity of variances (e.g. FA) among samples: a. Model I, two-way ANOVA of variation in |R - L|. $R_1 - L_1 = R - L$ for individual 1, etc., MS = mean squares, P = probability.

a) Layout of analysis

Sample (fixed effect)	Trait (fixed effect) Trait 1	Trait 2	Trait 3	Trait 4	etc.
A	$\|R_1 - L_1\|$	$\|R_1 - L_1\|$	$\|R_1 - L_1\|$	$\|R_1 - L_1\|$	
	$\|R_2 - L_2\|$	$\|R_2 - L_2\|$	$\|R_2 - L_2\|$	$\|R_2 - L_2\|$	
	$\|R_3 - L_3\|$	$\|R_3 - L_3\|$	$\|R_3 - L_3\|$	$\|R_3 - L_3\|$	
	...	...	...	...	
	$\|R_n - L_n\|$	$\|R_n - L_n\|$	$\|R_n - L_n\|$	$\|R_n - L_n\|$	
B	$\|R_1 - L_1\|$	$\|R_1 - L_1\|$	$\|R_1 - L_1\|$	$\|R_1 - L_1\|$	
	$\|R_2 - L_2\|$	$\|R_2 - L_2\|$	$\|R_2 - L_2\|$	$\|R_2 - L_2\|$	
	$\|R_3 - L_3\|$	$\|R_3 - L_3\|$	$\|R_3 - L_3\|$	$\|R_3 - L_3\|$	
	...	...	...	...	
	$\|R_n - L_n\|$	$\|R_n - L_n\|$	$\|R_n - L_n\|$	$\|R_n - L_n\|$	
etc.					

b) ANOVA results

Source of variation	*MS*	*P*	Interpretation if significant*
Traits (T)	MS_t	P_t	Some traits are repeatably less stable developmentally than others.
Samples (S)	MS_s	P_s	When information from all traits is pooled, the level of developmental stability varies among samples.
Interaction (TS)	MS_{ts}	P_{ts}	Extent of FA variation among samples depends on the trait.
Error (w/in sample)	MS_e		

* All MS tested over MS_e. Assumes that all traits have been shown to exhibit 'ideal' FA prior to ANOVA (Sects. 7, 8).

13.3 Among samples using multiple traits simultaneously

Since FA variation is often quite subtle, and since the FA of individual traits may not differ much among samples, additional statistical power may be gained by combining information about FA variation from several traits. Here too, a variation on Levene's test (Table 7) may offer an easy and effective way to combine information from several traits: a two-way ANOVA (samples × trait) of variation in |R - L| would yield a single test for overall significance of FA differences among samples. Such an analysis would also yield, via the interaction MS, a test of how consistent traits were at revealing among-sample differences in FA (i.e. a significant interaction term would indicate that some traits were better than others at revealing presumed differences in developmental stability among samples, Table 7b).

14. Correlations of subtle asymmetries among traits

When examining correlations among subtle asymmetries, one must keep in mind that the terms FA, DA and antisymmetry refer to attributes of a *sample* of individuals (Sect. 2.1). Since all bilateral traits will depart from perfect symmetry at some level of precision, the only way to recognize the pattern of subtle asymmetry to which an individual belongs is with reference to other individuals in a sample. In addition, the general presumption in studies of FA is that as developmental stability decreases the variance of R - L increases. This does not mean that all individuals, or all traits become more asymmetrical, only that the range of variation in R - L increases. Hence, even a sample with very low developmental stability will still exhibit a frequency distribution of R - L whose mean is zero: i.e. many individuals *by chance* will still not differ significantly from symmetry (Fig. 7a). In other words, for traits exhibiting FA (assuming it arises due to developmental noise), *the expected value of R – L is always zero*. The expected value of |R - L| depends on the variance, but even here many individuals will still not differ significantly from symmetry (Fig. 7b).

Once one recognizes that the expected value of R - L is zero for all traits that exhibit ideal FA (Sect. 4.5), some initially puzzling patterns begin to make more sense.

14.1 Between parents and offspring (heritability)

How does one assess the heritability of subtle asymmetries? Conventional heritability analyses (parent-offspring regressions) of signed R - L will yield no information about the heritability of developmental stability (Fig. 11a) for the simple reason that regardless of the size of the departure from asymmetry of the parents, the *expected value* of signed R - L among the offspring is zero. Thus, although the *range of variation* among offspring should, on average, be greater from parents that were more asymmetrical (curved dashed lines, Fig. 11a), the apparent heritability would be zero (straight dashed line, Fig. 11a). If, on the other hand, subtle deviations from symmetry per se are heritable, we see quite a different pattern: right-biased parents give rise to right-biased offspring, or left-biased parents give rise to left-biased offspring, but the variation among offspring in all the crosses should remain the same (Fig. 11c).

Since an increase in developmental instability is reflected in an increase in the variance of R - L, the proper form of heritability analysis is a regression of some measure of the *variance* of R - L among offspring against some measure of the *variance* of R - L of the parents. Since mean |R - L| is an unbiased estimator of SD(R - L) (Kendall & Stuart, 1951), a regression of offspring |R - L| versus parent |R - L| should yield a reasonable estimate of the heritability of developmental stability (straight dashed line, Fig. 11b). However, a potential problem arises when asymmetries are expressed as absolute values of R - L, since both cases yield a positive slope. In one case, however, this positive slope effectively represents heritable variances (i.e. heritable differences in developmental stability, Fig. 11b), and in the other heritable means (i.e. heritable deviations from symmetry, Fig. 11d).

The different estimates of heritability for (R - L) versus for |R - L| suggest another way in which FA (if due to developmental noise) may be distinguished from DA. For bilateral variation in the form of FA, the estimated heritability of (R - L) should be zero (Fig. 11a) whereas that for |R - L| should be significant and positive (Fig. 11b). On the other hand, for bilateral variation in the form of DA, the estimated heritability should be significant and positive for both (R - L) (Fig. 11c) and |R - L| (Fig. 11d).

More importantly, if a trait exhibiting FA yields a statistically significant heritability of (R - L) (Fig. 11c), this reveals that departures from symmetry in individuals have a heritable basis and hence have *not* arisen exclusively due to developmental

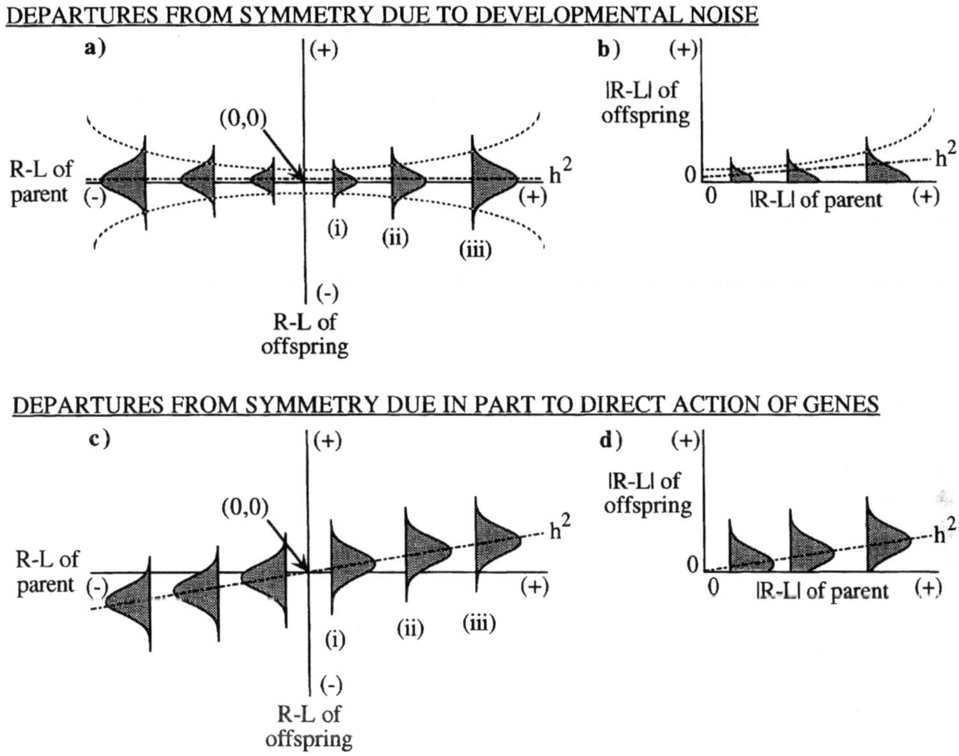

Fig. 11. Asymmetry correlations between parents and offspring for two cases: deviations from symmetry in an individual are due to developmental noise (a,b) or due in part to the action of genes promoting asymmetry (c,d). For each case, expected associations are illustrated using both signed (R - L) and unsigned |R - L| asymmetry. Frequency distributions represent the expected frequency distribution of asymmetry among offspring from a pair of parents. h^2 - estimated heritability (from parent-offspring regression, dashed straight line). Dashed curved lines delineate the range of variation among offspring (a & b only). These figures assume that differences in developmental stability are *not* due to differences in heterozygosity, but are due to some form of heritable variation for developmental stability (e.g. if extremely homozygous parents were homozygous for different alleles, their offspring would be highly heterozygous).

noise. In such a case, FA may *not* be a valid descriptor of developmental stability. Palmer *et al.* (1993) describe a pattern of bilateral variation where continuous negative covariation may exist between sides (normal covariant asymmetry, NCA). Even though NCA will apear to be ideal FA (Sect. 4.5) when viewed as a frequency distribution of R - L, the negative feedback mechanism implied by such negative covariation between sides may have a heritable basis and thus signal the earliest stages of heritable variation for asymmetry. Such variation may set the stage for the evolution of macroscopic asymmetries (Palmer *et al.*, 1993).

Finally, given the truncated-normal shape of the distribution of |R - L|, it is not clear how estimates of the heritability of |R - L| should be compared against those for conventional traits, since heritability analyses assume that the distribution of offspring phenotypes more closely resembles those of Fig. 11c.

14.2 Among individuals within samples: Are some individuals more stable developmentally than others?

Why do we so often not see correlations of asymmetries among traits when examining a sample of individuals, when, at the same time, we often do see correlations among traits when examining different samples? The logic used in Sect. 14.1 should make it clear that one should not expect correlations of (R - L) among traits if developmental noise leads to asymmetries in each trait that are independent of those in every other trait. This would be analogous to the pattern in Fig. 11a: even though (R - L) for one trait may depart substantially from zero, the expected value of (R - L) should still be zero for all other traits (e.g. distribution *iii*).

If signed asymmetries (R - L) do correlate among traits, this may arise via two possible mechanisms. The first would result in some structure-wide or organism-wide DA (analogous to Fig. 11c), where genes or some other factors promoting asymmetry affect several traits in the same individual in the same direction simultaneously. If this mechanism were responsible for the correlation of asymmetries among traits within a sample, such traits would probably exhibit DA when examined independently.

A second mechanism that might account for the correlation of signed asymmetries among traits is developmental noise acting at the level of several traits simultaneously. Unlike the first mechanism, however, all these traits would exhibit ideal FA (Sect. 4.5) when examined independently. Such a pattern would suggest that shared departures from symmetry reflect the action of developmental noise either a) at a very early stage in ontogeny (e.g. early embryo), before developmental independence is achieved among traits, or b) in substances that regulate the development of several traits simultaneously later in ontogeny. In other words, studies of the correlations of signed asymmetries (R - L) among traits may be very informative about the stage at which suites of traits are influenced by developmental noise.

Since unsigned asymmetry |R - L| *does* provide an estimate the variance of (R - L) for traits exhibiting ideal FA (Sect. 4.5), one might expect to see correlations of |R - L| among traits within a sample even if all bilateral variation were due to developmental noise. However, one must keep in mind that for a given trait in an individual organism, the unsigned asymmetry |R - L| estimates the variance of (R - L) with only one degree of freedom (i.e. one knows the observed difference between sides, and one assumes that the true mean of R - L is zero). This should be obvious with a little thought: individuals that are very unstable developmentally will nonetheless be symmetrical for some traits some of the time just due to chance (e.g. distribution *iii* of Fig. 11a). Thus the confidence one has in an estimate of the variance of R - L in an individual (which presumably reflects the level of developmental stability in that individual) is not very high since it is based on a sample with only one degree of freedom. Thus the ability to detect correlations of |R - L| for different traits among individuals within a sample will depend on a) the magnitude of the differences in developmental stability among individuals (e.g. the larger the differences, the greater the likelihood of detecting a correlation) and b) the number of traits measured on each individual (e.g. the larger the number of traits sampled

Table 8. Structure of and results from a test that combines information from several traits to test for differences in FA among individuals. It is a modified version of Levene's test for heterogeneity of variances (e.g. FA) among individuals: a. mixed Model, two-way ANOVA of variation in |R - L|. $R_A - L_A$ = R - L for individual A, etc., MS = mean squares, P = probability.

a) Layout of analysis

Individual (random effect)	Trait (fixed effect) Trait 1	Trait 2	Trait 3	Trait 4	etc.
A	$\lvert R_A - L_A \rvert$	$\lvert R_A - L_A \rvert$	$\lvert R_A - L_A \rvert$	$\lvert R_A - L_A \rvert$	
B	$\lvert R_B - L_B \rvert$	$\lvert R_B - L_B \rvert$	$\lvert R_B - L_B \rvert$	$\lvert R_B - L_B \rvert$	
C	$\lvert R_C - L_C \rvert$	$\lvert R_C - L_C \rvert$	$\lvert R_C - L_C \rvert$	$\lvert R_C - L_C \rvert$	
etc.					

b) ANOVA results

Source of variation	*MS*	*P*	Interpretation if significant*
Traits (T)	MS_t	P_t	Some traits are repeatably less stable developmentally than others.
Individuals (I)	MS_i	P_i	When information from all traits is pooled, the level of developmental stability varies among individuals.
Interaction (TI)	MS_{ti}	P_{ti}	(used as error term for testing MS_t and MS_i)

* MS_t and MS_i both tested over MS_{ti}, since no error MS is available. This analysis assumes no significant interaction between traits and individuals (e.g. the rank order of the underlying instability of traits is the same for all individuals) and assumes that all traits have been shown to exhibit 'ideal' FA prior to ANOVA (Sects. 7, 8).

for an individual, the greater the confidence in the estimate of developmental stability of that individual). The widespread lack of such correlations in most studies of FA (Palmer & Strobeck, 1986) may thus be a result of a) limited differences in developmental stability among individuals within a population or b) examining too few traits.

Kendall's Coefficient of Concordance (Sokal & Rohlf, 1981) is commonly used to test for correlated differences *among individuals* in the levels of asymmetry for several traits. However, the ANOVA test suggested in Sect. 13.3 could also be used here (Table 8). If the effect of 'individuals' was significant (Table 8b), this would indicate that some individuals exhibited consistently higher average asymmetry than others when the information from multiple traits was pooled.

14.3 Among samples

Unlike the situation for correlations among traits within a sample (Sect. 14.2), correlations of FA indices for different traits among two or more samples often are significant (Palmer & Strobeck, 1986). Such corelations among *samples* are the basis of Soulé's (1967) 'Population Asymmetry Parameter'. Correlations of asymmetry indices for different traits among populations are more likely because the average asymmetry in a sample of individuals is a better indicator of average developmental stability of that sample than is the observed asymmetry in a single individual. In other words, with a sample of individuals, a single trait provides a better estimate of the level of average developmental stability than it does for a single individual.

Kendall's Coefficient of Concordance (Sokal & Rohlf, 1981) is also used to test for correlated differences *among samples* in the levels of asymmetry for several traits. Here too, the ANOVA test suggested in Sect. 13.3 would also be appropriate. If the effect of 'Samples' was significant (Table 7b), this would indicate that multiple traits exhibited consistently higher FA in some samples than others.

15. Correlations between asymmetry and fitness: can asymmetry predict mate choice?

Groups of individuals that exhibit differences in fitness sometimes exhibit parallel differences in FA: lower fitness is associated with higher FA (Biémont, 1983; Clarke & McKenzie, 1987; Quattro & Vrijenhoek, 1989). This seductive association has prompted others to test for, and find, an association between the magnitude of subtle asymmetry and mate selection: males more asymmetrical for a particular trait were less preferred by females in both barn swallows (Møller, 1992) and scorpion flies (Thornhill, 1991). For the reasons outlined above (Sect. 14.2), the deviation from symmetry of a single character in an individual should be a very poor predictor of the developmental stability of that individual. So if females are selecting males based on differences in developmental stability (as deduced from subtle differences in asymmetry), and deviation from symmetry in a single trait is a very poor predictor of developmental stability, how can these results be explained?

Significantly, in both cases the traits whose asymmetry correlated with mate choice also directly affected performance. In barn swallows, asymmetry in the outer tail feathers affects maneuverability (Møller, 1991), and in scorpion flies, forewing asymmetry affects foraging efficiency (Thornhill, 1992) and hence (presumably) the volume of pheromone output upon which females based their choice. Hence, the traits examined in these studies were not just arbitrary indicators of developmental stability.

Future attempts to use subtle asymmetries to predict mate choice should either a) measure asymmetry in a large number of traits to ensure an accurate estimate of the developmental stability of individuals, or b) focus on traits whose asymmetry directly affects performance, because one or a few randomly selected traits will be a very poor predictor of developmental stability unless differences in developmental stability among individuals are large (see Sect. 14.2).

16. Checklist for studies of FA

- *Has FA variation been examined in several developmentally independent traits?* Because the 'signal' of FA variation is often very weak, and can sometimes be biased in unexpected ways, correlated patterns of FA variation among several traits can greatly strengthen inferences about levels of developmental stability (Sect. 4).

- *If meristic traits have been used, have concerns about the dependence of FA on the proximity of mean (R – L) to whole integer values been adequately addressed?* (Sect. 4.2)
- *Are sample sizes large enough to be able to detect differences in FA among samples?* (Sect. 5)
- *Has the measurement protocol been described in sufficient detail?* What landmarks were used to measure particular features? Were repeat measurements conducted totally blind (Sect. 6.5)? Did more than one 'observer' take measurements, and if so did the estimate of measurement error include differences among observers? Did any dimensions share a common landmark?
- *Was measurement error estimated accurately?* In other words, did it include *all* possible sources of variation: temporal variation in observer accuracy, among-observer variation if more than one observer took measurements, variation in ease of measurement for live and dead specimens, digitizing or quantizing error (Sect. 6.5)?
- *Is FA significantly larger than measurement error in all traits?* Because measurement error yields the same pattern of between-sides variation as FA, such a test is essential to all studies of FA (Sect. 6). This is perhaps *the* most common (and important!) oversight in studies of FA.
- *Do all traits exhibit ideal FA of R – L* (i.e. mean zero, normal variation)? If not, have those traits departing from ideal FA in due to DA, antisymmetry or skew been eliminated from the analyses (Sects. 7, 8)? If they must be included, have tests been conducted to verify that they have not biased FA indices or biased the conclusions about differences among samples (Sect. 9)?
- *Does asymmetry correlate with trait size?* If so, have the appropriate corrections for such dependence been applied (Sect. 10) and has the size correction been confirmed to eliminate any dependence of asymmetry on size? If not, have the proper FA indices been used (i.e. those that do *not* correct for trait size; Sect. 3)?
- *Where multiple related statistical tests have been conducted, have the P-values been corrected to reflect this fact?* (Sect. 11)
- *Have the data describing bilateral variation among traits and among samples been presented in sufficient detail?* Has more than one index of FA been included with these descriptive data (Sects. 3.4, 12)?
- *Have the proper statistical tests been used to test for differences in FA among samples?* (Sect. 13)
- *If heritability estimates for asymmetries are presented have the analyses distinguished between heritability of developmental stability vs heritability of subtle asymmetries?* (Sect. 14.1)
- *If deviations from symmetry have been used as an index for fitness, have a sufficiently large number of traits been used to obtain valid differences among individuals?* (Sects. 14.2, 15)
- *In the final paper, have terms referring to patterns (which are observed) been kept distinct from those refering to processes (which are inferred)?* (Sect. 2.2)
- *Have the indices used to estimate FA been identified unambiguously in ALL table and figure legends?* (Sect. 3)

Acknowledgements

This 'Primer' was written as a draft document to provide a focus for discussion at the Developmental Instability Symposium organized by Teri Markow in Tempe, Arizona, June 1993. It does not reflect a consensus about either terminology or methods, but reflects my rather personal opinion on a variety of subjects. I thank Teri for all of her efforts to make the symposium a success, and particularly for her persistence about including this document as part of the proceedings. I hope others will find it as useful as I have when trying to answer questions that commonly arise in FA analyses. Curt Strobeck, as always, has patiently helped me avoid making gross statistical errors, those that may remain are my responsibility. L. David Smith and Graeme Taylor both offered thoughtful comments on an earlier draft of the MS. Throughout, this research has been funded by the Natural Sciences and Engineering research Council of Canada (operating grant A7245), and I gratefully acknowledge their sustained support.

References

Angus, R. A., 1982. Quantifying fluctuating asymmetry: not all methods are equivalent. Growth 46: 337-342.

Angus, R. A. & R. H. Schultz, 1983. Meristic variation in homozygous and heterozygous fish. Copeia 1983: 287-299.

Bader, R. S., 1965. Fluctuating asymmetry in the dentition of the house mouse. Growth 29: 291-300.

Biémont, C., 1983. Homeostasis, enzymatic heterozygosity and inbreeding depression in natural populations of *Drosophila melanogaster*. Genetica 61: 179-189.

Brown, N. A. & L. Wolpert, 1990. The development of handedness in left-right asymmetry. Development 109: 1-9.

Chang, K. S. F., F. K. Hsu, S. T. Chan & B. Chan, 1960. Scrotal asymmetry and handedness. J. Anat. 94: 543-548.

Clarke, G. M. & J. A. McKenzie, 1987. Developmental stability of insecticide resistant phenotypes in blowfly; a result of canalizing natural selection. Nature 325: 345-346.

Conover, W. J., M. E. Johnson & M. M. Johnson, 1981. A comparative study of tests for homogeneity of variances with applications to the Outer Continental Shelf Bidding data. Technometrics 23: 351-361.

Graham, J. H., D. C. Freeman & J. M. Emlen, 1993. Antisymmetry, directional asymmetry, and chaotic morphogenesis. Genetica (this volume).

Jagoe, C. H. & T. A. Haines, 1985. Fluctuating asymmetry in fishes inhabiting acidified and unacidified lakes. Can. J. Zool. 63: 130-138.

Kendall, M. G. & A. Stuart, 1951. The Advanced Theory of Statistics. Hafner, London.

Lande, R., 1977. On comparing coefficients of variation. Syst. Zool. 26: 214-217.

Leamy, L., 1993. Morphological integration of fluctuating asymmetry in the mouse mandible. Genetica (this volume).

Leary, R. F. & F. W. Allendorf, 1989. Fluctuating asymmetry as an indicator of stress in conservation biology. Trends Ecol. Evol. 4: 214-217.

Leary, R. F., F. W. Allendorf & R. L. Knudson, 1985a. Developmental instability and high meristic counts in interspecific hybrids of salmonid fishes. Evolution 39: 1318-1326.

Leary, R. F., F. W. Allendorf & R. L. Knudson, 1992. Genetic, environmental, and developmental causes of meristic variation in rainbow trout. Acta Zool. Fenn. 191: 79-95.

Leary, R. F., F. W. Allendorf, R. L. Knudson & G. H. Thorgaard, 1985b. Heterozygosity and developmental stability in gynogenetic diploid and triploid rainbow trout. Heredity 54: 219-225.

Lehmann, E. L., 1959. Testing Statistical Hypotheses. Wiley, New York. 369.

Mather, K., 1953. Genetical control of stability in development. Heredity 7: 297-336.

McKenzie, J. A. & K. O'Farrell, 1993. Modification of developmental stability and fitness: Malathion-resistance in the Australian sheep blowfly, *Lucilia cuprina*. Genetica (this volume).

Møller, A. P., 1991. Sexual ornament size and the cost of fluctuating asymmetry. Proc. Roy. Soc. Lond. B 243: 59-62.

Møller, A. P., 1992. Female swallow preference for symmetrical male sexual ornaments. Nature 357: 238-240.

Møller, A. P., 1994. Sexual selection in the barn swallow (*Hirundo rustica*). IV. Patterns of fluctuating asymmetry and selection against asymmetry. Evolution 48: (in press).

Palmer, A. R. & C. Strobeck, 1986. Fluctuating asymmetry: measurement, analysis, patterns. Ann. Rev. Ecol. Syst. 17: 391-421.

Palmer, A. R. & C. Strobeck, 1992. Fluctuating asymmetry as a measure of developmental stability: Implications of non-normal distributions and power of statistical tests. Acta Zool. Fenn. 191: 57-72.

Palmer, A. R., C. Strobeck & A. K. Chippindale, 1993. Bilateral variation and the evolutionary origin of macroscopic asymmetries. Genetica (this volume).

Parsons, P. A., 1990. Fluctuating asymmetry: An epigenetic measure of stress. Biol. Rev. 65: 131-145.

Quattro, J. M. & R. C. Vrijenhoek, 1989. Fitness differences among remnant populations of the endangered Sonoran topminnow. Science 245: 976-978.

Rice, W. R., 1989. Analyzing tables of statistical tests. Evolution 43: 223-225.

Scharloo, W., 1991. Canalization: genetic and developmental aspects. Ann. Rev. Ecol. Syst. 22: 65-93.

Shapiro, S. S., M. B. Wilk & H. J. Chen, 1968. A comparative study of various tests for normality. J. Amer. Stat. Assoc. 63: 1342-1372.

Smith, B. H., S. M. Garn & P. E. Cole, 1982. Problems of sampling and inference in the study of fluctuating dental asymmetry. Amer. J. Phys. Anth. 58: 281-289.

Sokal, R. R. & J. F. Rohlf, 1981. Biometry. Freeman, San Francisco, CA. 859.

Soulé, M. E., 1967. Phenetics of natural populations. II. Asymmetry and evolution in a lizard. Amer. Nat. 101: 141-160.

Strawn, K., 1961. A comparison of meristic means and variances of wild and laboratory-raised samples of the fishes, *Etheostoma grahami* and *E. lepidum* (Percidae). Texas J. Sci. 13: 127-159.

Swain, D. P., 1987. A problem with the use of meristic characters to estimate developmental stability. Amer. Nat. 129: 761-768.

Taning, A., 1952. Experimental study of meristic characters in fishes. Biol. Rev. 27: 169-193.

Thornhill, R., 1991. Female preference for the pheromone of males with low fluctuating asymmetry in the Japanese scorpionfly (*Panorpa japonica*: Mecoptera). Behav. Ecol. 3: 277-283.

Thornhill, R., 1992. Fluctuating asymmetry, interspecific aggression and male mating tactics in 2 species of Japanese scorpionflies. Behav. Ecol. Sociobiol. 30: 357-363.

VanValen, L., 1962. A study of fluctuating asymmetry. Evolution 16: 125-142.

VanValen, L., 1978. The statistics of variation. Evol. Theory 4: 33-43.

Waddington, C. H., 1940. Organizers and genes. Cambridge Univ. Pr., Cambridge.

Waddington, C. H., 1955. On a case of quantitative variation on either side of the wild type. Zeitschr. Indukt. Abstammun Vererbungslehre 87: 208-228.

Waddington, C. H., 1957. The Strategy of the Genes. George Allen Unwin, London.

Zakharov, V. M., 1992. Population phenogenetics: Analysis of developmental stability in natural populations. Acta Zool. Fenn. 191: 7-30.

Zakharov, V. M., E. Pankakoski, B. I. Sheftel, A. Peltonen & I. Hanski, 1991. Developmental stability and population dynamics in the common shrew, Sorex-Araneus. Amer. Nat. 138: 797-810.

Zhivotovsky, L. A., 1992. A measure of fluctuating asymmetry for a set of characters. Acta Zool. Fenn. 191: 37-77.

T. A. Markow (ed.), Developmental Instability: Its Origins and Evolutionary Implications, 365–387, 1994.

Session 1

Measuring developmental instability

Monday morning June 14

Larry Leamy, Moderator

Leamy:

In general, the topic of the session this morning is measuring developmental instability, and I think this is really an important one, because if we don't know what we're measuring, we don't know what we're talking about. So all the successive workshops topics that come up after this morning, although important, depend upon a thorough understanding of how we might best assess developmental instability.

The literature on developmental instability, especially as assessed by fluctuating asymmetry, is enormous. But the very enormity of this literature, as well as the variety of approaches used in measuring fluctuating asymmetry, is confusing. Rich will try to shed some light on this confusion by reviewing various methods for measuring fluctuating asymmetry and making recommendations regarding these measures.

I think that we need to try to achieve a specific goal by the end of this morning. The goal I have in mind would be basically to develop a protocol of how we believe we should typically proceed in measuring fluctuating asymmetry and/or developmental stability. This should be particularly useful to researchers who perhaps are not as familiar with fluctuating asymmetry or developmental stability and are confused by all the approaches to measurement they discover in the literature.

One important issue regarding developmental stability involves the relative role of genetics in influencing asymmetry estimates. That is, are genetic differences responsible for some of the variation in directional asymmetry and anti-symmetry, and if so, are these differences contaminating our estimates of fluctuating asymmetry? If at least some of the asymmetry we measure has a genetic basis, how important is this (i.e., what might the heritability of asymmetry be?) Typically most investigators measure characters phenotypically and have little knowledge of underlying genetic influences. We tend to view fluctuating asymmetry, at least, as a measure of pure environmental variation, with no genetic control. So beyond the statistical considerations in measuring fluctuating asymmetry, which sometimes can be considerable, we need to know more about its genetic basis. So I hope when we get into the discussion after talks by Rich Palmer first and then by Carl Freeman and John Graham, we can keep our focus on developing a consensus approach to measuring developmental stability. I also hope that as a result of our discussion, we can sharpen our thinking about what we mean by developmental stability as distinguished from other related phenomena.

Palmer:

As someone who lives in an extreme climate in Edmonton, I thought it quite amusing to come to a meeting in Tempe in June. It's rather like having a meeting in Edmonton in January or February, except that here it's about +40 °C, and in Edmonton in the winter it's about −40 °C. Anyhow, it's quite a pleasure to be here, and I'm really grateful to Teri for being given this opportunity to put forward some thoughts that I hope will set the stage for our discussion of methods this morning.

Although we have all come to this symposium because of a common interest in developmental instability, each of us is interested in different aspects. A crucial general point I think we need keep in mind is that, unlike many other phenomena we study, the study of developmental instability is really a study of *variability* as opposed to a study of *means*. Some of the morphological variability we observe arises from the phenomenon we're interested in, developmental instability, but a lot arises from other sources: direct action of genes, direct action of environment, genotype-environment interactions, and of course, measurement error. As I'll try to make a little bit clearer today, certain biases and curious pitfalls in ways of doing analyses really make studies of FA rather different from those of other phenomena.

Two other points will probably be apparent to anyone who has conducted such studies, but I think they are very important to keep in mind. First, the 'signal' for FA is often very small. As a consequence, this signal is easily obscured by observational noise or insufficient statistical power. Second, some of the descriptors of FA are prone to biases that can either create spurious patterns, or obscure existing patterns of interest. Most of what I will say this morning relates to these three issues.

As I gather most of you received this 'Primer' with your registration package, perhaps I should provide a little background. Vladimir commented yesterday when I showed it to him that I clearly didn't have enough to do if I had all this time to sit down and write such a detailed document about issues that should be obvious to anyone who has done these kinds of analyses. Perhaps he is right, but many of these issues weren't obvious to me initially and I suspect others may have been equally puzzled. I was motivated to write this 'Primer' by the hope that we could sit down as a group with a document where at least one, perhaps foolhardy, person laid out their ideas on how to tackle a variety of methodological problems. This document could then provide the focus for a discussion of concrete issues in the hopes that errors or inconsistencies or differences of opinion might be resolved. I would really very much appreciate comments back, should any of you find things that are either wrong or could be clarified. I don't want to give anyone the illusion that this document is written as a bible about how to do studies of FA. I view it merely as a starting point, but it will be for you to judge how close I have come to a thorough and accurate account. I'm not going to try to go through all the sections in detail, but I'd like to draw attention to a few.

One issue that I hope we can address at this workshop is terminology: exactly what do we mean by the particular terms that we use when discussing developmental stability and related issues? One problem I have encountered frequently in the developmental stability/FA literature is that terms for *patterns* of bilateral variation are often used synonymously with those for *mechanisms* or *processes* or vice versa. I think discussion gets very confusing when terms for patterns and processes are used interchangeably, particularly when the mechanisms generating the patterns are uncertain or a matter of dispute. If we wish to convince others outside of this very august group to take studies of developmental stability seriously then we should try to be a little more thoughtful and consistent with our use of terms. I'm the first to admit I've made mistakes when initially trying to wrestle with these phenomena. Terms for mechanisms are all too easy to substitute for those for patterns, but we do so at our peril.

Another terminological problem that I think is a bit more serious and widespread is our lack of a word to describe departures from symmetry *in an individual*. Many people refer to the 'FA' of an individual, or 'directional asymmetry' in an individual, but as I discuss in the 'Primer', these terms describe the pattern of variation observed in a sample of individuals. A deviation from symmetry in an individual, by itself, can tell us nothing about what the pattern of asymmetry variation is without reference to other individuals. A term I have found useful for referring to departure from symmetry in an individual is: *subtle asymmetry*. This term carries no baggage, no biases, no prejudice, about either the patterns to which it might belong, or the mechanisms that might have generated it. I think such a term, if used consistently, might avoid a lot of confusion.

As most of you know, many indices have been used to describe the level of FA. This table (Table 1 of 'Primer') illustrates that among the great variety of descriptors or indices of FA, there really was some order. There is nothing really fundamentally wrong with any of the indices. They are all reasonable ways of describing FA, depending on what the data look like. However, I'd particularly like to draw your attention to this last index (FA10), not just because we developed it, but because I believe that it is quite important in studies of FA to keep track of the contribution of measurement error to bilateral variation. The principal virtue of this index is that, unlike all the other indices, it allows you to partition out the portion of the bilateral variation that is due to measurement error, thereby, hopefully, yielding better quantitative estimate of the 'true' bilateral variation.

Another issue that has troubled me in thinking about studies of FA is the choice of traits: which traits are the most useful descriptors of FA? One easy split in trait types is between meristic traits, which can be counted, and metrical traits, which vary in a continuous fashion. Meristic traits seem appealing because you would think they could be counted without error. But there is a rather sinister weirdness that can arise with meristic traits that I think is widely overlooked. In 1987, D. P. Swain published a short note in The American Naturalist discussing this potential problem. On first reading I, quite frankly, thought the paper was rather trivial. But reading it later, I came to understand the point he was trying to make, and I think it's actually a fairly important one. Swain shows that when most asymmetrical individuals in a sample differ between sides by only one, the estimate of FA depends on how close the mean count is to a whole integer value, for the very simple fact that it is not possible to have, for example, 6.5 elements on one side (Sect. 4.2 of 'Primer').

Regardless of whether meristic or metrical traits are used, opinions vary about whether traits that exhibit directional asymmetry or antisymmetry should be used to infer the level of developmental stability. If statements made in some of the papers submitted for this symposium are any indication of the strength of feeling on this issue, I expect some animated discussion later. In our previous papers, Curt Strobeck and I have suggested that, if at all possible, traits that exhibit directional asymmetry or anti-symmetry should be avoided in studies of developmental stability (Sect. 7 & 8 of 'Primer'). Having said this, however, I continue to be fascinated by the thought-provoking results that John McKenzie has obtained in his genetic studies of sheep blowfly. In both his 1988 paper with Geoff Clark on Diazinon resistance, and his paper with O'Farrell submitted for this symposium on Malathion resistance, John finds that the locus that leads to resistance, when expressed in the absence of a fitness-modifier locus, also results in antisymmetry for three meristic traits. The antisymmetry observed in the Malathion resistance study was quite subtle, and not significant statistically, but all three traits exhibited greater platykurtosis in the resistant-unmodified group compared to normal. In other words, John appears to have uncovered a single locus that induces the expression of antisymmetry. Intriguingly, he also finds that fitness declines in direct proportion to the degree of antisymmetry, as would be expected if antisymmetry were an expression of developmental instability. Here, then, is a case where subtle antisymmetry clearly has a genetic basis, yet, because of its effect on fitness, it also seems to signal a decrease in developmental stability. I think it would be particularly interesting to conduct a selection experiment with this system. Is it possible to increase the average difference between sides more quickly in strains that exhibit antisymmetry than in those that don't? This might tell us a great deal about the developmental basis of departures from symmetry and their potential to respond to selection.

I'd like now to focus attention on the problems that measurement error creates for studies of FA. We have all paid homage to measurement error or uncertainty in the statistics classes we've had, and so we are all aware that measurement error reduces the likelihood of detecting differences. But I think measurement error is a far more serious problem for studies of FA than it is for other studies for several reasons (Sect. 6 of 'Primer'). The most important of these is that, unlike conventional traits where measurement error simply reduces confidence in the mean, measurement error in studies of FA actually *biases* the results. This is because no matter what index is used, we are trying to estimate a variance, the between-sides variance. Unless measurement error has been partitioned out, the between-sides variance always includes a contribution due to measurement error. This is why it is more important to be worried about measurement error in studies of FA than in other studies.

Smouse:
One of the things that has me a bit worried about this whole exercise is that in the normal course of events morphometric fluctuating asymmetry variation will represent on the order of 2.5-5% of the variation, depending on how much the measurement error is. That is disturbing, because the variance of a variance is fourth order. Variances are poorly estimated unless we have very large sample sizes, and quite frequently, we don't have large sample sizes.

Most of our techniques are either intensely tied to normal theory, and quite vulnerable to deviations from it, or to newer non-parametric techniques, probably more robust but less powerful. In the case of multiple characters, we can always do a principal components rotation. That 2.5% of the variation tends to be the leftover part that is poorly estimated and unreliable. The normal practice is to discard (as unreliable) the noise in the system. Here, we are hoping that this noise in the system will send us a signal of some kind. The data are somewhat undeniable at one level, in the sense that they're clearly trying to tell us something, but I find it disturbing that we are trying to extract strong signal from the tiny bit of leftover variance.

Palmer:
My reason for not just throwing up my hands and abandoning the whole problem for the reason that you say, namely that asymmetry variation really is a small fraction of the total variation, is that you can't argue with the data. We have seen patterns of asymmetry variation arise enough times in enough different organisms to suspect that there must be some common explanation for them. But you're absolutely right. We're talking about a phenomenon in which the effects are really quite small. I guess one of the points I was trying to make earlier was that, not only is there a very small signal, but this signal can easily get lost underneath a whole variety of different, rather imprecise, ways of trying to describe it. Nonetheless, in spite of the many different ways that people have described the associations between some measure of fluctuating asymmetry and stress or heterozygosity or whatever, the patterns keep recurring. So, to me, the empirical evidence suggests that this is a phenomenon worthy of farther study.

Kieser:
Surely, if measurement error is a noise, and if one observer only is used, it's a constant noise, an invariable noise, so the mean error variance which is observed here, remains a constant over the experimental program, and could thus be factored out.

Smouse:
If you take 2.5% of variation, and you extract 1% of variation as measurement error, you're now down to 1.5% that you can't account for. I have to wonder whether the phenomenon in question might not represent something else.

Graham:
In a comparison of several populations, you are saying that the fluctuating asymmetry signal is somewhere on the order of 2 percent of the total variance. Where does this number come from?

Smouse:
Greg Livshits and I were looking at human bilateral morphometrics where correlations between the two sides would be on the order of .9 to .95 for a giver character. Moreover, we had several characters that were all highly correlated with each other, never mind the right vs. the left sides. We did a multivariate analysis, and standardized to normal (0, 1) for each character. Now, all characters are on the same scale. Of course they are all still highly intercorrelated. We extracted the principal components and discovered that the first half of the components had both elements of a particular character (right and left) with the same loadings and same signs. The last half of the components, representing 2.5% of the total variation, inevitably had the right side with the opposite sign from the left side but of the same magnitude. We then looked at the eigenvalues (the variances of the components). It always turns out that the first component has the biggest variance, since that is how we define it, and we work our way down; by the time we get down about half-way through the first half of the components, we're down to the point where one would be inclined to say: 'They're so minimally contributory that we might just as well throw the rest away; they are not telling us anything useful; we're probably manipulating the rounding error in the correlation matrix. Even with the sample sizes at our disposal, we don't have all that much precision. On the other hand, we could take just a few more components to be sure we had all the information worth having. By the time we get down to the asymmetry components, the last half of the components, they are just not credible in terms of being believable.' Maybe that's a good definition of the statistical noise in development, what fluctuating asymmetry is supposed to represent. I can certainly live with that, but

to then turn around and wonder what kind of signal that residual noise is going to send on some other criterion makes me more than a little nervous.

Koehn:
I think Peter's missing the point. I think Rich's point was that if we had all done analyses of that type and gotten that result, we wouldn't be here today. But that is in fact not the case. So either the analyses that were done which didn't get that conclusion were flawed by some statistical artifact, or measurement error, or they were not. And that seems to me to be the point of the discussion ... is there or is there not statistical artifacts in these analyses?

Smouse:
Fair enough, so let me be provocative and wonder out loud about whether a lot of what we think we see *is* statistical artifact.

Rich gave us a nice presentation, we had another one from John and Carl, on better ways to do things than are routinely done to avoid some of these problems. I think it's fair to respond that in the literature, most people have not been meeting these current standards of excellent practice. I'll concede that the standard of practice has improved over time, but I cannot but wonder how many of the numerous traditional studies would meet today's criteria of excellence. That does not make the conclusions wrong, of course, but some careful reexamination might be in order.

Leamy:
Peter, let me just say one thing about measurement error. I think it's true, most people have not been taking repeated measures, and we need to do that henceforth. It's just an awful lot of work, which is why we often don't do it. But I have to say that many studies have shown differences in fluctuating asymmetry between inbreds and hybrids, heterozygotes verses homozygotes, and the like. Now, you typically assume there's no reasonable reason to believe that there's going to be more measurement error in inbreds compared with hybrids. That is, in any study with several samples, you assume that any measurement error is going to contaminate these samples to the same extent. And when you get these patterns of fluctuating asymmetry, already described, that seem to hold up pretty well, you have to assume there's really something there, beyond errors of measurement. We do need to factor out measurement error and get more precise estimates of fluctuating asymmetry, but I think the measurement error problem may be not as big as people have thought. That doesn't say that we shouldn't be taking into account measurement error - you're absolutely right, but I think that it may not be as bad as we've been imagining.

Palmer:
That's precisely my point. I don't see how the positive associations reported in many studies could have arisen due to measurement error. I suspect that in many of the studies where the results have been equivocal, or where no differences were reported among samples, this may have been an artifact of measurement error or limited statistical power. But it's really hard for me to imagine how so many people independently could have measured their samples in such a way to create the patterns that have brought us all here.

Soulé:
One could go even further and say that principle components analysis is an elegant way of taking interesting meat and potatoes and spices and turning it into a very uninteresting stew.

Smouse:
There is some fairness in that comment. The point of using the principal components analysis for the purposes of discussion was merely to add context to the 2.5% variation figure. Technical details aside, however, the larger concern is whether the phenomenon is real or merely a consequence of bad statistical practice. Reasonably, I am not averse to a delicious Irish stew. The question that I'm trying to raise, is the following: If we're going to persist in thinking of this asymmetry that we're after as simply the wobble in development, and if we're going to reduce it to the point where it's so small, having taken good measurement technique and everything taken into account that needs to be taken into ac-

count, I find it amazing that we're finding anything, by way of pattern, between that and anything else. I'm not disputing that we do find a pattern, clearly we do, of course we don't always hear about those who don't find a pattern. We all know that problem. I guess I would worry a little bit about whether something is full of meat and interesting; is it as interesting and full of meat as we think it is? I would just like to be confident that when we come out the back end with the result it's a real result and not the fact that we have stone soup and think we've done something.

Vrijenhoek:
I would like to expand on Michael's point. If an unbiased investigator makes the measurements on all the animals, it is reasonable to assume that the errors of measurement or counting are consistent across the groups to be compared. For asymmetry measures it is probably best to use the simplest statistics that make no assumptions about underlying distributions. The sophisticated multivariate approaches of Smouse and Livshits might be elegant, but they remind me of measuring the coastline of Norway with a millimeter ruler. You can get a result, but you probably will lose Norway in the process. That's silly! Perhaps it is better to step back from the data and look for larger patterns of consistency. My experience with *Poeciliopsis* taught me that genetic groups rarely show statistically significant differences in asymmetry for any one character. Instead, it is the consistent pattern of differences across multiple characters that is significant. Often a nonparametric signs test is all that is needed to make the comparisons among groups. I suggest we stop trying to get more sophisticated in our statistical dissections. If the pattern is not apparent, it is probably not there anyway.

Johnson:
I feel a little uncomfortable, especially with the *Fucus* study coming from Seattle, working for an agency that monitors those sites and having some experience with them, I would be very hard-pressed to say that Four Mile Rock is much more polluted that Alki Flats. Four Mile Rock is right near Seattle's water treatment plant. It's probably the cleanest part of the entire Pacific Ocean. It's monitored on an hourly basis. But, local pollution, especially for *Fucus* could occur anywhere. Friday Harbor probably has more small oil spills that even Puget Sound has. So I'm sort of worried about the question, and we see a pattern and there, all of the sudden we have a very polluted place and not so polluted place and a pristine place, where really, they probably are not. But we see a pattern, so we immediately want to label things. And John Emlen is not here, so he may have very good reasons why he chose this.

Freeman:
Can I respond to this? To begin with the levels of pollution were determined independent of the pattern. Levels of pollution were determined by measuring the pollution levels in shellfish, that were collected from exactly the same locations that the *Fucus* came from. The shellfish find the pattern of pollution to be exactly the same as the pattern the *Fucus* found. And that's both PCB's and PAH's.

Johnson:
How did you do the shellfish, though? It is on the rocks.

Freeman:
They did it with mussels. No, John Emlen was in charge of determining pollution levels. We had no knowledge of the pollution levels when we were doing our study. So, in that sense I think we did it right. So, having defended our little study, let me go back to your general point, and say that we expect certain patterns, and when we don't observe those patterns, we tend not to report them. When the desired patterns are observed, the world is well as it should be, and the result is more apt to be reported. John Graham and I had that experience with our hybrid zone. Hybrids are supposed to be unfit, and all of our measures tended to suggest that one of the parents was less fit than the hybrids. We reached this conclusion by counting seeds, and conducting seed germination experiments and reciprocal transplanting experiments. The hybrids were consistently turning out to be more fit than one of parental taxa. Well, our terpene analysis had previously shown that this parental taxa was less developmen-

tally stable than the hybrids. If you're going to be absolutely honest, then you just take nature as it comes. But I also think we ought to report negative studies. And in that regard, one of the things that's always been amazing to me is every time I pollute plants with heavy metals, at least the above ground parts, don't show increased asymmetry. So I'm looking at roots now, which is really difficult work.

Palmer:
I think the study that you prefer to, Carl, looking at the variations in the *Fucus* over that gradient is a useful example of the kinds of things that people have done repeatedly, at least when monitoring natural populations. I guess a question I would have, which is one that can always be raised when populations are sampled over a gradient, is how one separates the effects of pollution, for example, from other factors that also vary over the gradient. As I recall, in Puget Sound many other things change as you move from the San Juan Islands down into the lower Puget Sound, and wave exposure is among them. Have you ever, for example, tried to look at a series of sites that were all at the same pollution levels and ask how *Fucus* responded to wave exposure in terms of your indices? Because, ultimately, as I noted in my talk, and the study of FA is a study of variability, and that variability can be influences by a great variety of different factors. It seems to me that to conclude that the increase in variability is due to pollution requires that other causes of variability be rejected.

Freeman:
Both of us would like to respond. Ok, I have to confess, I'm a terrestrial botanist, so *Fucus* for me is a side project. But I worried about the same thing that you're worried about. In fact, we went back after you and I were up the first time, and we sampled at different depths that would be subject to different amounts of wave action. There is a consistent patterns across depth, but there's also a profound difference among the harbors sampled.

Graham:
The *Fucus* study was a small exercise on one species. Much better examples of an ecotoxicological application are the two Biotest expeditions to Odessa and Novgorod, where we looked at developmental instability of several plants, small mammals, and insects. At Odessa, the first year we went to the Soviet Union for research, the region near the chemical facility looked obviously polluted. Then we went out to our control site, and the farmers were spraying herbicides and insecticides. There were no natural environments for a suitable control! We were upset; we were sure we would not see any pattern. But you can ask Vladimir Zakharov about the Odessa report. In species after species after species, we saw the same patterns, whether they were in goldfish from some of the facility's waste water ponds, or whether they were plants, or small mammals. When we went to Novgorod the next year, it was much easier to find a pristine control some distance from the chemical facility. Almost all of the organisms showed parallel patterns of developmental instability. And I wouldn't have even believed it at Odessa. I said' This isn't going to work; we're not going to see any pattern. It's going to be totally random.' And I was surprised when it turned out not to be so. Different people were studying different organisms, so we weren't in communication with one another.

Palmer:
Well, of course, the larger the number of species that show the same pattern over such a gradient, the greater confidence that the pattern has a common cause. This would be particularly true where organisms with totally different developmental systems were included, such as plants and animals. So yes, some of the people in this room will study the impact of human-caused environmental stress, as you and Vladimir and Carl are trying to do, by using measures of developmental variability. The caution I urge is the following. Envision yourself in court, being cross-examined. How do you, or any of us, for the matter, respond to lawyers who ask us to distinguish between competing explanations? What is the evidence? What is the evidence on which we can agree? Does the evidence really allow us to distinguish between alternative explanations? I think as the number of species in a study increases, particularly where species with totally different traits are included, the stronger a case can be made that the pollution is the culprit. But it would seem

rather difficult to make a convincing case that variability in a single species over a pollution gradient is due to pollution *per se*, as opposed to some other factor that varies in a parallel fashion.

Graham:
We would agree. The *Fucus* study was just an exercise. We were exploring dichotomous fractal branching for the first time. Our attitude was, here is a species that displays dichotomous branching; well, let's look at it.

McKenzie:
I would like to ask an important clarification of Carl. With the fractal index taking inbreeding as being increased stress, the fractal index went down in one study you mentioned. Is there a constant prediction of that index?

Freeman:
No. If you notice with my *Fucus* study, the fractal index went up, and the difference has to do with regulation. With *Fucus*, and with some other plants that we've applied this to, when you stress the plants, you disrupt the feedback mechanism between the growth promoter and the growth inhibitor. This feedback mechanism is responsible for the dichotomous branching. In the case of *Fucus*, stress impacts the level of the inhibitor more than that of the activator. This actually promotes growth. This is true not only for *Fucus*, but it occurs in a number of plants. Only when the stress becomes really, really severe is growth reduced. But initially you disrupt that feedback system, and so it depends on the nature of the feedback system, whether you effect activators more than inhibitors as to which way it goes. But for a lot of things that will be of interest to zoologists - skull sutures or veins in the retinas for example, the dimension generally declines.

Mitton:
I'd like to raise a point, something that I've always found puzzling. We think in terms of individuals varying in their degrees of developmental stability, and so we might imagine individuals that are highly stable over here and highly unstable over there. If you measured a population of individuals that are strung out along this continuum, some individuals would show very little evidence of instability, and others would show very much. But it seems like most of the empirical studies don't detect any correlation in fluctuating asymmetry among characters when they're measured among the individuals within a population. That is, if ten characters on are measured on each individual, examine correlations in fluctuating asymmetry among those ten characters. Most studies report that asymmetry is not correlated among characters. But if you then look for correlations among those ten characters among populations, you find positive correlations, and this led Mike Soulé to define the population variation parameter. If a population tends to be highly variable for character nine, it's highly variable for character seven, highly variable for character three. Well, why doesn't that work within a population? Is it a matter of statistical precision? I mean, if we take these highly unstable individuals over here, and they've got a fluctuating asymmetry for character seven, wouldn't it only make sense that they're more likely to have fluctuating asymmetry for other characters as well? Is it a matter of statistical precision, or is fluctuating asymmetry a rather haphazard measure of developmental stability?

Palmer:
I've been profoundly troubled by this apparent contradiction right from the very beginning. It wasn't until I trying to put my thoughts together for the talk here, that I think I understand why those patterns are not contradictory. I believe the explanation is that, among samples, you're comparing variances, whereas among individuals, you're comparing deviations. There is no necessary correlation between deviations, which arise through completely random processes for each trait in an individual, but there is a correlation between variances when you can estimate the variance for a trait based on a sample of several individuals. Keep in mind that even where a trait is very unstable developmentally and hence some individuals may depart wildly from symmetry, the frequency distribution of signed differences between all sides still has a mean of zero. In other words, there are still a large number of individuals that are symmetrical, or

nearly so, even in, say, a uniformly highly imbred population where all individuals have a low developmental stability. Thus, the expected value of the signed difference between sides for any normally symmetrical trait is zero. It doesn't matter what the variance is. Hence a developmentally unstable individual that exhibits extreme asymmetry for one trait may or may not exhibit extreme asymmetry for other traits, since the expected value of the difference in any other trait is always zero unless the traits are developmentally linked somehow.

Mitton:
If you look in Palmer's handout you can see that there should be a correlation. If you're measuring absolute differences, why aren't the absolute differences in fluctuating asymmetry correlated? If an individual is highly instable, then it should have multiple deviations, and if there are multiple deviations in one group and no deviations in another group, then those deviations will be correlated. Is the slope of the line too shallow to produce correlations that we can measure?

Koehn:
I'm not sure I understand the point. Are individuals unstable, or is the development of characters unstable? In other words, whatever the underlying mechanism of instability, is the mechanism necessarily the same for the development of one character versus another in the same individual?

Mitton:
And that may be the problem, that fluctuating asymmetry is not measuring a generalized instability, but multiple opportunities for error. If different characters develop at different times, then you may expect asymmetries within a suite of characters to be correlated, but correlations among characters from different suites may be much weaker.

Leamy:
A lot of people are still baffled about this problem of why we often find correlations of fluctuating asymmetries among characters between but not within, populations. Rich, I want to hear more of your explanation. This issue, which Jeff has raised, is why we typically find what Michael called a population asymmetry parameter. But we don't find, or it's very hard to find, what I call an individual asymmetry parameter. That is, sometimes you can pick up individual asymmetry correlations within organisms. But these are rare and you have to really look carefully and have a large sample size. In fact, my paper in this workshop addresses that very point, but it is certainly one which would warrant more discussion.

Soulé:
It's been a long time since I thought about these things. I remember thinking I had it figured out about twenty years ago, but since Jeff brought it up, I've been struggling to figure it out again. There's very little evidence for a genetic basis for asymmetry. And if there's not much heritability for asymmetry, then the genetic differences for asymmetry potential among individuals are going to be very little in a population: that is, you and I are not going to have much genetic difference in our proclivity or tendency to be more or less asymmetrical. But, if you compare populations, there may be overall genetic differences in populations, but more important, there are environmental differences between populations, in terms of stressors, for example, such as nutritional differences, or in the amount of toxins in the environment. So, what I'm saying is that unless there were a strong heritability for asymmetry within a population, you wouldn't really expect an individual asymmetry parameter. But if I were in a terrible environment, that is, I was nutritionally stressed, as an embryo or fetus, and you weren't, then I might be a lot more asymmetrical for a lot of characters than you.

Freeman:
I have to agree with everything you said. In addition, I'd like to point out a problem that confronts zoologists more than botanists. Zoologists usually end up with one set of characters on the left side and one set of characters on the right side. We botanists (and there are few of us here) have an advantage in that we have multiple structures on the same individual that we can compare. And when we do that, we do find both significant differences among indi-

viduals within populations and among populations. Sakai and Shimamoto did this in the 50s, and they looked at a suite of traits for leaves, and suite of traits for flowers. They found that the flower traits were all tightly correlated, and they shared, as Richard said, a common developmental mechanism. So you'd expect them to be tightly correlated. The leaf traits were also correlated with each other, but they were not correlated with the flower traits. This type of analysis is easier if you have repeated structures on an individual. Zoologists get a single data point per organism, and it makes it more difficult. In the *Fucus* study, I showed you that fluctuating asymmetry was higher for plants that had a breakdown of self-symmetry, i.e., phenodeviants. These plants also exhibited significantly higher fractal dimensions than non-phenodeviants from the same harbor. I can do this because I have studied repeated parts on the same plant. I can also do it because of the development history. These traits have a common developmental basis, and so they should be correlated. They also have a common developmental time, and so you'd expect correlation. I think you have to look at when things develop and the mechanisms by which they develop before you get worried about whether or not there is a correlation.

Livshits:

Our experience shows that first of all, when the individual is strongly deviated in terms of asymmetry from the population average, it shows that it deviates for several characters. So there is a correlation between the variables when the individual is indeed a strong deviant. In relation to the correlation between different FA variables, our experience shows the following. When we compare the newborn infants suffering from various diseases, in particular cardio-vascular morbidity, against healthy infants, the first group of infants also show very high - relatively very high asymmetry, and they show asymmetry for a number of characters.

Palmer:

I'd like to make another stab at responding to Jeff's earlier question about asymmetry correlations, because it certainly is a puzzling one. He is quite correct that, if you look at the absolute value instead of the signed value of the difference between sides, then you *should* expect a correlation between the asymmetry for one trait and the asymmetry for another on the same individual. This is the pattern illustrated in Fig. 11B of the 'Primer', except that for the present discussion the X-axis would represent asymmetry in one trait, and the Y-axis asymmetry in a second trait in the same individual. So why do we so rarely see such correlations? I still think the answer has to do with the confidence with which we can infer the developmental stability of an organism simply by observing the deviation from symmetry in a single trait. When you have a sample, and Figure 11B illustrates a sample of offspring, you have multiple individuals that yield information about variability. However, consider the situation where you examine only two traits in a single individual. If one trait is highly asymmetrical, what are the possible asymmetries that could be exhibited by the other trait? I believe the frequency distributions in this figure would illustrate the asymmetry potential, to use Michael's term, for that second trait. In other words, as would be the case for all traits exhibiting developmental noise, the second trait still has a reasonable probability of being nearly symmetrical. On average, it should be more asymmetrical if the individual is more developmentally unstable, but one has very little confidence in the information that a single trait yields about developmental stability.

Perhaps I can make this clearer by stating it in the form of an analysis of variance. If we had several individuals, and we wished to know whether the developmental stability differed among these individuals, this could be tested by conducting a two-way ANOVA on absolute asymmetries, with traits as one factor and individuals as the other. In fact, this would be a modified form of Levene's test for differences in variances. The statistical question would be: do individuals differ significantly in their average deviation from symmetry? This is the appropriate question since the average absolute deviation from symmetry should increase as developmental stability decreases. As I see it now, the likelihood of detecting a significant difference among individuals will depend on two things: first, the size of the underlying difference in developmental stability among individuals, and second, the number of traits that are measured. Thus, it seems to me that increasing the number of traits examined would increase the likelihood of detecting a significant difference among individuals in the same way that increasing the number of individuals in a sample would increase the likelihood of detecting a difference in asymmetry for a single trait between two samples. Perhaps this accounts for why such asymmetry correlations are detected so rarely among individuals. If more traits had been included in these studies, perhaps they would have detected average asymmetry differences among individuals.

Carl makes the interesting point that plants, and I should point out, certain clonal invertebrates such as bryozoa, ascidians and corals, offer tremendous potential for studies of developmental stability because they have multiple, developmentally equivalent and (presumably) genetically identical modules whose variation can be measured. I think they allow us to ask this question about correlated asymmetries more precisely. If we find one zooid in a bryozoan colony, or one leaf on a plant, that is highly asymmetrical, what can we infer about the asymmetry of the other zooids in the colony, or the other leaves in the plant, assuming symmetry is the normal state? If the leaf that was measured was representative of the developmental stability of the whole plant, then the other leaves should be more asymmetrical on average (i.e. exhibit a higher average absolute asymmetry). But how likely is it that we can detect differences in developmental stability among separate plants? Clearly, as more leaves are measured, the greater the confidence one has in one's estimate of the developmental stability of one individual plant, and the greater the likelihood that one can detect differences in developmental stability (i.e. correlations among asymmetries) among separate plants. Thus, I think the lack of correlations of asymmetries among individuals arises simply because deviations from symmetry in any one trait in an individual do not yield an estimate of developmental stability for that individual in which one has very much confidence. Hence, either the developmental stability differences among individuals must be quite large, if they are to be detected with a few traits, or one must examine a great many traits to detect statistically significant differences in average absolute asymmetry among individuals.

Zakharov:
I would like to make some comments for the study of correlation for fluctuating asymmetry (FA) of various characters.

Correlation (at least in terms of statistically measured coefficient of correlation) does not really exist in nature. It is just an idea of statisticians. Let me show you what happens when we try to find some correlation: successfully in some cases and unsuccessfully in the others.

As it was originally established by Michael Soulé, correlation between FA measures of various characters does not exist at the individual level, but exists at the population level. Follow-up observations revealed all gradations, from the presence to the absence of such correlation at both levels.

Even comparing a set of populations, we can see correlation for measures of FA in some cases but not in the others. The reason for such differences is very simple. To reveal the correlation it is necessary to have significant heterogeneity in the parameters studied. Even for a hundred populations rather uniform in their developmental stability, we can not predict the presence of any correlation between studied characters. And observations support such an assumption. Otherwise, introduction to the set of studied samples of one seriously stressed population, characterized by developmental instability, will induce the correlation between asymmetry measures of various characters.

The same explanation applies to understanding the situation at the individual level. How can we originate the correlation in measures of FA at the individual level? It is possible to induce such correlation by mixing some individuals from the normal population without any stress and some individuals from the stressed population in one sample. But commonly it is not possible to reveal such a correlation within a normal population. This is because of the specificity of FA as a measure of developmental stability. It mainly depends not on the presence or absence of particular allele, but on disturbance in general genetic coadaptation and deviation of environmental conditions from the optimal range. For such general characteristics as genetic coadaptation it is less probable to see high intrapopulation heterogeneity. The same is true for the environmental conditions. FA heritability studies provide us with another manifestation of the same phenomenon. All attempts to reveal significant heritability of FA for various characters of different invertebrates and vertebrates including humans were unsuccessful.

I know only one exception. D. Hagen (1973) has revealed the high heritability for FA of lateral plátes in crosses in hybrid population of three-spined stickleback, because of serious differences in developmental stability between the two original genetic morphs and their hybrids. It means that if commonly we can not reveal significant heritability for FA in intrapopulation crosses in some particular case, we can originate it in crosses of essentially different genetic forms.

Livshits:
There is one more example. In our study we examined two groups of families. One group of families includes those infants which I just described. They include both term and preterm infants and their parents. The second set of families came from the three common farms called Kibbutzim from the northern part of Israel. We took approximately 80 nuclear families. We used the integration approach, i.e., each individual was assessed for eight characters. No single character showed any remarkable measure of family correlation. However, when we combined, a kind of average measure for these eight characters, both sets of data, both sets of family data, showed similar results. In both cases we got mid-parent/child correlation equal to approximately 0.3. Significant statistically. And the correlation of interaction between parents FA with child combined FA was also approximately the same. Thus when we studied a combined measure of asymmetry by regression of child's FA on parents mean FA, we were able to explain about 60% of the variation, but we got virtually zero for each measurement separately.

One more point which I think is important. Looking for the asymmetry, we may look before we know the biological meaning of the character under the study. For example, linear measurements of the long bone, tibia, show almost no asymmetry, no measurement error with the same caliper that we use for breadth measurements. It is about 1% of the measurement size. On the other hand, ear measurement shows fluctuating asymmetry, including measurement error, of course, comprising about 5% of the ear size. And the question which is raised at this point, do the differences in biological significance of these two characters, or survival significance, contribute in difference in FA values for these two characters. Obviously, that asymmetry in long bones, such as tibia, will make an individual handicapped immediately. At the same time, a difference in ear measurements, as far as we know, within 5%, would hardly imply any significance, even considering sexual selection. So, as an example, when we took the measurements of newborn infants, just born, the long bones, specifically the tibia and femur didn't show the asymmetry almost at all. In absolute terms, it was, I think, less than the asymmetry in ears in the percentage expression. So maybe when we approach the study of the asymmetry, we should also take in to consideration the possible fitness or survival significance of the character which we are going study.

Mitton:
I think I just heard two different answers. One answer was that you do get correlations when you have a severe abnormality. Allendorf's group has shown that individuals with fused vertebrae are also more asymmetric. But then you said that if you look af fluctuating asymmetry of a single character, you're not likely to find anything. However, if you put them all together, you may detect associations. And that suggests that there may be some sort of variation among individuals within a population in the degree of stability that they have – it's just difficult to measure. So let me ask a rhetorical question for people who have data: is there a better way to measure developmental stability? Fluctuating asymmetry is an infrequent strike, and it may not record developmental accidents very well. What if you just measured many morphological characters, and then determine, for each character, whether an individual is towards the center of the distribution for that population, or out in one of the tails. Would standardized distances from the means be correlated among characters? Is that a better way to measure developmental instability?

McKenzie:
It's an interesting point in the sense that in the discussions that we're having, with the fluctuating asymmetry, first, it's interesting that Vladimir made a comment that perhaps he and Richard disagreed about the patterns of asymmetry that gave a measure of 'developmental noise', developmental

stability and instability. I wonder if their discussion could come back to the starting point, the point that we're all measuring variability in a number of different scales, but is there any consensus about exactly what we mean by developmental noise, whether we mean exactly the same thing by developmental noise and instability? If I talk about developmental perturbation, am I talking about any of those things, and do any of the three basic patterns of asymmetry, or if you like, just a gross phenodeviant as an extended instability measure, provide a better measure of whatever it is that we are actually measuring?

Clarke:
Can I just make a comment on one of Rich's earlier comments and Jeff's first comment on the expected correlation between the characters. I really don't expect that for the independent characters there will be a correlation in the level of asymmetry between characters, unless as Rich pointed out, the characters are developmentally related. More importantly, unless the mechanisms responsible for the development of these characters are correlated, then you wouldn't expect that correlation in the level of asymmetry, and of course we don't know anything about such mechanisms, and whether they are the same for every character we're looking at, or whether they're different mechanisms. We don't understand the mechanisms that control stability for any character; there is no reason to suspect that those mechanisms are the same for every character and therefore, we might not expect any correlation in asymmetry between characters. Let's use the data which are there, which would suggest that in a majority of cases such a correlation does not exist, so perhaps there are different mechanisms that control stability in those characters.

Smouse:
Just to pick up on that thought, and as a partial answer for Jeff's question. If you use a multivariate suite and measure from the center, you can determine whether asymmetry for different characters is correlated. Greg and I have done that with the adult sample that we described earlier, and we found very small correlations. We did that because we felt that any one character would yield a single trial (imperfect at best) but that with enough trials (characters), whether they were correlated or not, a larger pattern would emerge. We found no pattern in our study, but more of this sort of thing needs to be done.

Mitton:
I asked a different question. I was asking, if one character has unusual value, do other characters also have unusual values? What you've done is looked at a multivariate distance from a centroid, isn't that correct?

Smouse:
We get the same thing on right vs. left that everybody else gets - the correlations among characters are small, and you can't really make them stick statistically.

Koehn:
But you can't ask this question when you have zero correlation. That's why you can't answer the question, though, that he's asking.

Smouse:
In some sense the question *is* whether the correlation is non-zero. In our case, we have answered it.

Livshits:
Actually, in our case, we didn't compare the incidence of phenodeviants in the extreme individuals vs. modal individuals. In our case we compared multivariate FA among polyextreme individuals vs. morphologically polymodal individuals. We did not observe any reliable pattern.

Koehn:
I want to go back to the point that Jeff made, and I talked about earlier, namely that a correlation among characters will depend very much upon the underlying mechanisms by which the variation is generated. Let's take the example of environmental pollution - say, cadmium or PCB. The question is whether each character that you select in order to measure this instability is the same - is the develop-

mental mechanism of each character the same and are the characters effected the same way by a particular environmental agent or insult, of some kind? And then you have the problem that even assuming that they are, just for simplicity's sake, would each character scale identically on to cadmium concentrations in terms of its asymmetry? Certain characters may have been historically subjected to higher balancing selection; they would therefore have a different genetic interaction term with that environmental insult. So that's why I think the question of 'Why don't they correlate?' is really not very interesting until you understand a little bit more about the developmental mechanism for the characters.

Vrijenhoek:
You can use to your advantage the observation that your characters do not correlate across individuals. Mike Soulé noted this lack of correlation in his lizard studies. The same was true for *Poeciliopsis*. Ideally, you would want eight or nine *independent* measures of asymmetry. Of course, you could have to worry whether the apparent *independence* of characters isn't really a consequence of low statistical power.

Koehn:
There's another important point, actually, that Rich mentioned, that has sort of gotten lost in this. I thought at first it wasn't very important. The measurement error of different characters is different and so are your confidence limits. The point you made about where an individual lies on a deviation is not so important when you're comparing populations, because the error's the same for a given character. But that's not true for different characters. So your conclusion of whether they're independent or not may not be correct, if your error variance is very different. The characters may actually be correlated in some developmental sense, but they wouldn't be in your data.

Graham:
I think this is a good time to bring up a concern of mine. In a typical study, one is measuring several different characters on the same individual. The appropriate statistical approach is a multivariate one, yet no one seems to do that. I was wondering if we could address sequential Bonferroni test, which Richard Palmer mentions in his primer. The sequential Bonferroni test is not a multivariate test; it controls for experiment-wide error rate. I have a feeling that a true multivariate approach would be a bit more powerful, because you would have more degrees of freedom.

Smouse:
Greg and I did MANOVA-type tests, because we were a little bit concerned about whether things were independent, as the data would tend to indicate. We didn't want to make an assumption about it. The MANOVA tests are more powerful than non-parametric alternatives, but they're exquisitely sensitive to non-normality. It doesn't take very much failure of distributional assumption to get you into a lot of trouble with MANOVA, and some of the non-parametric techniques, though not as powerful, are very much more robust to those sorts of deviations. There are trade-offs, as I mentioned earlier, but the jury is not yet on the choice. We need a lot more work on both sets of techniques.

Graham:
We need somebody, maybe Rich, to do another one of those power of statistical effects on MANOVA.

Vrijenhoek:
I'd like to get back to another point that Rich Palmer mentioned earlier. We all accept there is a problem with metric characters. There's an estimation error built in, regardless of the size of the ruler. Meristics have errors too. Rich gave the example of little tiny scales falling off at the end of the tail. From my point of view as a fish biologist, you have to think a little about the characters you measure. You can't measure everything and you don't want to count things willy-nilly. As a student, I remember going to my advisor and asking him to help me key-out some species. The fish were supposed to have 29 or 30 scales on the lateral line, but I keep coming up with 31 or 32. My advisor answered, well, you dummy, you don't count them all. You have to start here, at the pectoral fin, and you stop there at the last scale on the caudal pedun-

cle. All those extra scales are accessory scales – they don't count. In other words, you have to know a little bit about your characters. The old taxonomists knew this when they were keying fish out. There are certain scale characters you can rely on and others that you can't. Often you have to clear and stain the specimen. Many of these characters are buried within tissue and become emergent at various periods in development. It's important to clear and stain so you can see the actual base of the fin rays as they attach to boney processes on the vertebrate. You need to relate your characters to landmarks in the whole organization of the organism. You can't count every twig and branch and scale that you see. You have to know something about development, and you have to think carefully about landmarks and boundaries of the things you count.

Clarke:
I totally agree with Bob; something that's been lost here today is that if you look at the characters first, a lot of the problems that we're looking at, such as directionalized symmetry, anti-symmetry, or whatever, could be eliminated by choosing a large suite of characters, and I think it's something that most of us would do automatically. We're not going to say, I'm going to use this character and this character before you even have a look at firstly, the distributions and look at the patterns they're showing you and saying, well, I'm not going to use this one because it can't tell me anything for whatever reason. You choose your characters first, and then you can believe the signal when it hits you in the face.

Leary:
I chose in my material two sets of gill arches because I can sit down and count that gill arch, and I'll consistently get the same damn number of gill rakers. Ok, I feel comfortable with that character. I count lateral line scales on a rainbow trout – I do it five times, I've got a variance of 30 scales. Same individual! What do I do? I quit counting lateral line scales. I see something that's directional or demonstrates anti-symmetry in my data, I don't use it as a measure of fluctuating asymmetry or as the signal for developmental instability. So I think by being a practical biologist, you can really take care of a lot of this statistical nonsense.

Smouse:
We statisticians earn our living by cleaning up the loose ends. You might be surprised how much nonsense is masquerading as practical biology.

Palmer:
I agree with all these comments. Somewhat to my surprise, in the course of learning more about multivariate morphometric methods, I have discovered that there are no absolute rules that govern what you do, other than, perhaps: 'be careful'. There is still no substitute for thinking carefully about each step: choose traits appropriate to the biological questions you are trying to answer; try to display patterns of variation graphically so you can inspect them for unusual or unexpected features; don't rely on single statistical tests if you feel your data may violate assumptions; etc. etc.

The example Bob gave, however, raises the question of whether particular methodological concerns are 'obvious' or not. Where terminal elements of a meristic trait gradually tail out and disappear, it should be obvious to anyone who has tried to count them that they will be 'noisy'. And as Robb noted, once you discover that a trait is too 'noisy', you simply don't continue to use it. But even so, many people appear to get excited by the phenomenon of FA, and decide to study it themselves without paying much attention to these 'obvious' problems. We may pooh-pooh such individuals, but what is obvious to some may not be obvious to others. Similarly, what may be 'obvious' to us now, may not have been obvious to us before. For example, I now understand the concerns that Swain raised about studying the FA of meristic traits where very few individuals in the sample are asymmetrical: with such traits, the variance will depend to a large extent on how far the mean count is from a whole integer value. To me, this problem is now 'obvious', and I hope for some of you, now, it will also be obvious. But my point here is that, until Swain described his concern in a way that others could understand, this problem certainly was not an 'obvious' one.

I'd be curious to hear your comments on this, Bob. It seems to me to be a nice example of how, once you know something, it becomes common sense. But until it has been pointed out, it may not be obvious at all. We may still choose to use such traits. But at least we will be aware of the potential problem, and can look for associations between average asymmetry and mean count that might be confounding the patterns we observe.

Vrijenhoek:
The problem Swain raised is real for characters that have very few elements. If you look at pelvic fin rays in *Poeciliopsis*, you almost always find seven. Occasionally, an individual will miss a ray on one side and you have six. You almost never get eight. Swain's criticism may be correct in this case. However, sometimes we find the complete absence of one or both of the pelvic fins. Quattro observed this in homozygous populations of endangered topminnows. The absence of one or both pelvic fins seems to be a manifestation of inbreeding depression. Just counting rays and estimating asymmetries might miss this more important phenomenon. You just can't go in there and count things and expect it all to make sense.

Trivers:
I need some guidance here, because, on the advice of Michael Soulé, one of the characters I've measured in lizards was digital lamellae of the largest toe pad because it's a countable character. Now the problem is that towards one end, the lamellae may suddenly get a lot smaller. Do you count only those above a certain threshold or try to count all? Is there any general experience on that question that's useful? When you find a character that starts diminishing in size, do you keep counting that character?

Soulé:
In general you choose a character which has good landmarks, but why does a landmark at each end, which is pretty clear.

Trivers:
But landmarks can move.

Soulé:
For example, with digital lamellae, the a landmark at one end is where the two digits come together. Of course there is a certain amount of arbitrariness in any characteristic, and that contributes to the variance. Hopefully, the bilaterality of the character takes some of that into account, but in any meristic character, you're going to have some problems.

Zakharov:
I would like to mention some important items concerning the main topic of our discussion. I do not know why it would be interesting to study FA of fingerprints or any other characters without understanding the reason it is so important. I think that first of all we should establish that we study FA as a measure of developmental stability. It means that by use of a common morphological approach we study general and fundamental characteristics. This general characteristic can be measured not only from a morphological, but also from a genetic viewpoint. Again by use of a cytogenetic approach, but with emphasis on chromosome abnormalities in somatic cells, we will have a good measure of the condition of developing individuals. Similar measures exist for in physiology: energetic metabolism studies based on the estimation of total energetic cost for individual development provides us with another very useful measure of developmental homeostasis. There are also some other biochemical and immunological tests for developmental homeostasis.

We have already established a general approach based on the estimation of developmental homeostasis from different points of view (including morphological, genetic, physiological, biochemical, and immunological ones) at various structural and functional levels. We have shown that it is possible to combine all these measures of developmental homeostasis for various species in the field and laboratory studies. The results revealed that even independent parameters prove to be correlated in cases of serious differences in such general characteristics as developmental homeostasis. Such experience makes us more optimistic in our morphological studies of FA.

Thus, the main task for our work when we study FA is to measure developmental stability. In general, to estimate the level of developmental stability we need to measure deviations from the norm as a result of disturbance in individual development. But to distinguish them in a whole phenotype diversity is not such an easy task, especially in population studies. When we exclude phenotype variety caused directly by genotypic or environmental differences we shall get a good morphological measure of developmental stability. This is why we are so concerned with the measures of FA - not because it is so important as such, but as one of a few useful measures of developmental stability. It is preferable to use it because it is very simple and clear theoretically and methodologically.

I would like to propose the following statements for our discussion. First, interest of developmental stability study in general. Second, its estimation, for morphological characters, is a measure of deviation from the norm. The ideal model to study this variation is a genetically uniform group of individuals under the same environmental conditions. There is no signal at all in such variation, just noise. But even here we can seriously change its level by stress.

To estimate the level of developmental stability we can use different measures. It is better to use different methods and different characters. I prefer to use not only parametric but also non-parametric methods to evaluate the data. If we have found some real effect it should be revealed by all methods.

I think that an idea to prepare the protocol on how to analyze FA is very useful, but I would like to mention that even in such a case it is necessary always to be very careful. I would like to discuss some examples.

First, symmetry as a measure of developmental stability. Formally, to study FA, we should be concentrating on deviation from symmetry. One zoologist following this approach was amazed when he found increased symmetry under chemical stress for one particular character in roaches. This was the number of pharyngial teeth and the norm for this character for roaches is 6 on the left and 5 on the right side of the body. Under stress we will see the appearance of deviation from this norm and the frequency of symmetrical states of the character (6/6 and 5/5) increases. Formally we should exclude such characters from our analysis, but really it provides us with a useful measure of deterioration of developmental stability under the impact of stress. It illustrates that in general, we should focus on form not symmetry, as the norm and deviation from it are so important for developmental stability studies.

Second, scaling effect in FA study. One example concerns study of two genetic morphs in three-spine stickleback. The White Sea is inhabited by one genetic form with a high number of lateral plates on the body, while fresh water bodies are inhabited by low plates form. Each form is adapted for particular environmental conditions. The level of FA measured as variance of differences between two sides of the body is the same in both cases, meaning that the compared forms are equal with respect to developmental stability. Following the protocol we should use not variance for the absolute values of the difference between the sides, but scaled values. After such transformation the level of FA between compared populations differs by five times. Should we conclude that the level of developmental stability is so dramatically different for these populations?

Third, study of the correlation between two sides of the body. It is not a measure of FA, but of the proportion of FA in total phenotype diversity in a studied group. If we have none of the basic sources of variation in terms of genetic and environmental variability, the correlation between left and right sides will be 0, because all variation will present just developmental noise. When we introduce some genetic or environmental variation the correlation will appear. It can be illustrated by the same example of the stickleback study. Between-sides correlation is rather small for the lateral plates within each form. But the correlation is very high for the pooled sample of individuals of both forms.

Fourth, measurement error. It is an important aspect, but I think that its possible influence commonly might be neglected. We have done such work specially for the wing measurements in *Drosophila.* For 2 of 10 used characters measurement error was too high, i.e. its magnitude was the same as the difference between the sides. By this criterion we have excluded them from the analysis. But all characters, including these two, revealed significant differences in the same direction between the control and stressed populations. It illustrates that if measurement error is the same for all samples under study, all characters proved to be useful. When measurement error is too high, we can loose some small differences, but this is not so dangerous as the creation of some artificial difference that does not really exist.

Finally, I would like to emphasize the most important things that we should establish in our discussion: the importance of study of general developmental homeostasis and estimation of the deviation from the norm as a way to assess developmental stability from the morphological point of view.

Leamy:
How are you going to define the norm, Vladimir?

Zakharov:
From my point of view, it is important for the theoretical and applied study of developmental stability and condition of developing organisms to assess the degree of deviation from the norm - from the optimum. It is clear theoretically, but not so easy to realize in practice. This is why we are so focussed on FA. At least in such a case it is clear what the norm is: it is symmetry. But as I have already demonstrated, even this statement is not universal - in some cases asymmetry could be a norm. Formally we should establish a statistical norm and a functional norm. In some cases it is possible to establish both. But if it is not possible, even a statistical norm is useful. Now we have the results of a parallel study of the optimum by use of various morphological, genetic, and physiological measures. It allows us to be confident that we really can establish the norm and measure the deviation from it. The advantage of FA study is that it allows us to get needed information about developmental homeostasis and the condition of developing organisms the easiest way.

Vrijenhoek:
I would like to return to the question that Bob Trivers asked Mike Soulé a moment ago. How do you pick the characters to study fluctuating asymmetry? What do you measure, and what do you count? I have been pondering these questions in the context of our studies with *Poeciliopsis*. Why did our studies work? It turns out that we used characters that were useful as taxonomic characters. These characters distinguished populations and species of *Poeciliopsis*. My advice to is to read the taxonomic literature on your target animals. Traditional systematists focused their attention on characters that were relatively stable within species and varied between species. They were searching for traits that exhibited developmental stability but were not so rigid that no variation occurred at all. They avoided characters that had a large environmental component.

Leary:
I would like to go back to the Swain argument, and again, I'm not a statistician, but I'm not so certain that a lot of that might not be statistical artifact. When you determine the mean, there's also asymmetry in there, and I'm not so sure that as asymmetry goes up, it may not push you further from a whole integer value. Using an example to clarify my point - say you had a clone of fish, where the actual number of pectoral rays was supposed to be fourteen. Say that was the total number, seven on each side. All of a sudden I get an asymmetrical individual in there. It's got seven on a side and six on a side. Now, all of the sudden, I deviated from a mean integer value of fourteen total, to 13.9 something, so I give that a value of 0.1. Then, I start to do a correlation between departure vs. asymmetry, and I find at times that it's a subtle positive relationship. Maybe all you're really picking up is that the more asymmetrical the trait is in the populations, the more the mean tends to depart from a whole integer value.

Palmer:
Presumably though, as the developmental noise goes up, you should be just as likely to get individuals with a higher count on one side as with a lower count, so the trait mean should stay at 7. Even if the range went from 4 to 10, as opposed to just 6 to 8, I don't see why the mean should change necessarily.

Leary:
But if I'm using the total count, I'm going to start getting odd values in there. I'm not sure. I think - I'm really addressing this to you, Rich - that it may be something you want to go back to. I'm not positive what I'm saying is what's driving these relationships. Sometimes you find it, sometimes you don't. But, in my mind, I think there's a possibility that it may be part of the explanation. In a

sense what you're doing is that your departure from the mean may somewhat be related to how asymmetrical the trait is.

Palmer:
By mean do you refer to mean total count? I find it easier to think about the problem if we talk about the average count on a particular side.

Leary:
I'm saying asymmetry. It has a potential of possibly modifying your mean total count.

Palmer:
True, asymmetries can affect the mean total count in a particular sample, but it still seems to me that if the loss of a single fin ray on one side is as likely as a gain, as would be the case for developmental noise, then these should balance out and have no effect on the overall average.

Leary:
Because I'm going to have in my data set fourteens and twelves and thirteens and fifteens. The odd values can only be created by asymmetry.

Livshits:
You use your average.

Leary:
But really when you get into it, it's not going to be perfectly normally distributed. You're going to have some sampling error in there, and that's going to generate for you a deviation from a whole integer value.

Palmer:
But Swain's point is a simple one, really. When asymmetrical individuals in a sample rarely differ between sides by more than one, the observed between-sides variance will be very sensitive to the mean count on a given side. The further the mean count on a side deviates from a whole integer value, the greater the apparent between-sides variance will be. This seems to be an unavoidable problem since it is simply not possible for an individual to have 6.5 elements on a side; it will have either 6 or 7.

Leary:
But this isn't the problem I had originally when that note came out. I'm not really certain that this may not be a factor involved in why we might see these relationships.

Palmer:
My gut feeling is that it's not.

Leary:
And how do we factor the asymmetry out of the calculation of the mean?

Strobeck:
Suppose you are measuring things that were right at - for your landmark as you're measuring in your scale - were right at a 6.5. And sometimes you'd call that 6 and sometimes you'd call that 7, if there's slight variation in your landmark, which may not have anything to do whatsoever with asymmetry. So my understanding of that paper is that you may end up having asymmetry in development, just because, not of anything really dramatic, but just because the way you're measuring it. Now, does that make sense? That seems to me, once you've measured that asymmetry in that way, you can't tell it from fluctuating asymmetry, because, it's all variable. And so, the main reason why you'd want to standardize the landmarks is because of the changing the numbers there. If you have a large amount of change, then you would be very sure that the asymmetry you're seeing is real, but if you only have one, and the landmark's not right on, then it may be what you're seeing is just variation because you're measuring 6 things that are 6.5 apart. That's why that comes in.

Graham:
It has been a long time since I've read Swain's paper. As I recall, he was working with a threshold

model for meristic traits, assuming an underlying continuity with a normal distribution. This continuity was superimposed on the integer values of the thresholds. He gave the number of fin rays in a fish as an example. Swain suggested that two populations having different means for the continuous variable could have different levels of fluctuating asymmetry, due entirely to the locations of the means with respect to the integer values of the threshold.

Strobeck:
Yes, because essentially maybe because there's a shift in the morph, you're still measuring things that come at 6.5, on average, as opposed to 7. Seven, you're more likely to call absolutely correct, than 6.5. Because of the shift in developmental noise, you'll call sometimes a 7 and sometimes a 6. And that will be much greater than the variance than if you just get 7. And I think you can't tell those apart. And so the variance may be different because of this morphic effect. I think we agree.

Clarke:
In some of those cases when the landmark moves within the character, you would in fact use the shift of the landmark as a measure of asymmetry. This is actually what happens in dragonfly wings where the total number of wing cells in the first rows of the wing is constant, but the position of one of the landmarks actually changes its position within this row of cells in response to stress. So in those cases, to avoid problems of shifting landmarks, use the landmark itself as the character.

Leamy:
I would now like to summarize the talks and discussions we have had on this topic of our workshop – measuring developmental instability.

We started out this morning with a talk by Rich Palmer, much of which was a review of the excellent primer he had earlier distributed. Rich emphasized that his primer was intended to be a draft for his discussion, and it did serve to generate a number of questions and comments later during our discussions. In his presentations, Rich talked about the need for distinguishing patterns from processes; fluctuating asymmetry, directional asymmetry, etc. are patterns, whereas developmental noise, stability, etc. are processes. He also talked about the need to distinguish measures of asymmetry in individuals versus samples and the problems associated with trying to define individual asymmetries. And he talked about the need for frequency distributions of right and left sides more carefully and for future publications to show these distributions in some appropriate form.

Rich quickly reviewed some of the main methods that have been used to measure asymmetries; these included absolute and signed differences between sides, ratios of sides, and an analysis of variance method. He talked about the idiosyncracies of meristic characters and we heard a lot about meristic characters later in our discussions. He talked about perhaps not using characters to show directional asymmetry or anti-symmetry if we really want a reliable measure of developmental stability. We had a lot of discussion about this, and may or may not all agree on how important those contaminants might really be. Rich talked some about measurement error being a problem. But in later discussion, I think we heard that this sort of error may or may not be as much of a problem as we think it is. It depends basically on what sort of experiment you are conducting, and what you intend to measure. It might be more of a problem, for example, if you're looking at correlations among individual characters that exhibit different amounts of measurement error. Rich talked about the potential errors in using meristic characters, which apparently a lot of people think are pretty eror-free. They may not be, given these landmark definitions and problems which Rich pointed out.

Finally, Rich talked briefly about the heritability of asymmetries. I personally wish we could have spent a little more time on this topic, but there are a number of unknowns here. It is not entirely clear, for example, exactly what we're estimating when we estimate heritabilities of asymmetries and what kinds of distributions we really are using. Theoretically, at least, if we want a good measure of developmental noise, it should have no heritability at all; it should be truly environmental. But Vladimir Zakharov has pointed out that phenomena other than bilateral asymmetry used to assess developmental instability, for example, chromosome aberration cases, definitely have a genetic basis. So maybe we're focusing too much just on a narrowly defined and environmentally-based phenomenon, fluctuating asymmetry, when we assess developmental stability.

John Graham next gave a talk which primarily was on the topic of developmental invariance. John pointed out that developmental invariance is a phenomenon exhibited by some aspect in an organism which is invariant with respect to changes in growth and development. Although bilateral symmetry is an example of a developmental invariant, there are other examples such as translational symmetry and rotational or radial symmetry. He then talked about within-individual asymmetry, a form of asymmetry which typically is measured in organisms that have repeated parts. He gave an example of individual asymmetry that could be assessed by counting the numbers of flowers in the inflorescence of various plants. In Russia, John was able to show that within-individual asymmetry, measured as the variance of subumbels per umbel in a species of plant, was greater the closer these plants were to a chemical dump. Another measure of within-individual asymmetry, derived from snails, showed opposite results when assessed near and away from this same chemical dump site. John also talked briefly about the possibility of applying the concept of developmental invariance to allometry.

Carl Freeman then gave a presentation which continued the topic of developmental invariance and which also introduced the use of fractals in the study of developmental stability. Carl explained rotational invariance (radial symmetry), and outlined a study he had conducted which showed that plants at control sites had significantly more leaflets per leaf than plants exposed to herbicides. And the angles between the main veins of Norway Maple leaves, used as another measure of rotational invariance, became more acute in polluted, versus non-polluted, sites in Russia. Carl and John Graham also participated in a study in which they fit a regression to the length of leaves as you move up the stem and discovered that the amount of variation explained by regression was greatest (best fit) in the least polluted site. So clearly we often can expect polluted sites to show greater amounts of within-individual variation. On the other hand, Carl mentioned that within-plant measures of variation, such as numbers of petals per flower, showed no differences in plants exposed, and not exposed to copper treatments.

Carl talked about the advantage of plants for assessing within-individual variation, or what he called self-symmetry. He explained the generation of a model which he depicted by dividing up lines of some structure into three pieces and replacing the middle piece by two pieces. If this same rule is applied many times, a number of structures can be generated, all of which would occupy much more space than the original structure. That is, these new structures possess what are known as fractional or fractal dimensions. Fractal dimensions can be looked at in a number of organisms, and Carl showed how this could be done in a brown alga. Once again, algae taken from relatively polluted areas exhibited different patterns of branching than did those from non-polluted areas. Similar kinds of techniques can be applied to non-morphological characters such as biochemical and/or physiological ones.

After these presentations, we had some spirited discussions of a number of topics generated as a result of these talks. Early in our discussions, Peter Smouse brought up a real concern that I'm sure a lot of people have when dealing with fluctuating asymmetry: right minus left differences in various characters typically account for such a small percentage of the total variation that we wonder if they really are meaningful assessments of developmental stability. Beyond the fact that the relative amount of variation accounted for by such asymmetries varies in different studies (in some it may be 50% or more of the total variation), the consistency of the patterns exhibited by fluctuating asymmetry seemed to be the compelling reason that most of the participants used to justify these studies. Thus statistically we may be dealing with just a small amount of the total variation, but we often find that this variation is significantly different between inbreds versus hybrids, stressed versus non-stressed individuals, and the like.

Some concerns were expressed that asymmetries do not always exhibit these expected patterns and that we need to publish, when possible, 'negative' patterns or results. It was also pointed out that increasing the numbers of characters used often yields more clear-cut results and that where possible, we should look for general trends across multiple characters. Sometimes also, defining what is a 'stressed' individual or a 'polluted' environment is not always easy to do unambiguously, and careful consideration should be given to the experimental design in any study which attempts to assess developmental stability.

We spent a considerable amount of time discussing why we find associations or correlations of fluctuating asymmetries among different characters mostly at the population, rather than individual, level. A number of studies have shown the existence of a population asymmetry parameter, first championed by Michael Soulé, in which there are different levels of overall asymmetries among a suite of characters in certain populations versus others. But we seldom find an individual asymmetry parameter, in which fluctuating asymmetries of different characters are significantly correlated in individuals. Some exceptions occasionally occur in highly asymmetrical individuals, as shown in a study of human infants by Greg Livshits, or among highly correlated characters, as I have shown in mouse mandible characters, but these are relatively rare.

Possible explanations for the prevalence of a population, rather than an individual, asymmetry parameter were offered by several participants. Rich Palmer suggested that we typically measure *deviations* from perfect symmetry when comparing individual asymmetries (but *variances* when measuring population differences) and that there is no necessary association between these subtle, individual deviations in one character versus another. Jeff Mitton suggested that perhaps fluctuating asymmetry is not really assessing some generalized instability factor which can ordinarily be picked up at the individual level. Michael Soulé suggested that since there is little, if any, genetic component to fluctuating asymmetry, we would not expect any significant genetic correlation between character asymmetries in individuals. Carl Freeman suggested that the search for individual fluctuating asymmetry associations is easier in plants which offer the advantage of multiple structures on the same individual, and that in fact such associations have been found.

It seems clear that we do not know enough about the developmental basis for fluctuating asymmetry; which is to say, we still do not understand precisely how fluctuating asymmetry is related to developmental stability. Since we do not know in most cases what the mechanisms are that are controlling the development of the characters that we use in such studies, we have no real expectations for the magnitude of the associations of their asymmetries. We need to know more about the basic units of development which control the characters that we use, and only then might we be better able to predict how these asymmetries may be related.

As Bob Vrijenhoek made clear, we need select our characters carefully in any study of developmental stability. Probably a lot of us start out measuring convenient characters, or characters we are familiar with, and I think we all need to think very carefully in the future about why we are using the characters we have chosen. That is, are these characters correlated with each other and/or is their genetic or environmental basis known? And do we have a sufficient number of characters to analyze them in a multivariate fashion? Some of these same questions have been answered previously by classical taxonomists, and we therefore might do well to select characters chosen by these taxonomists.

In other cases perhaps we should not use fluctuating asymmetry estimates from bilateral characters at all, since other measures of developmental stability are possible. As Vladimir Zakharov has persuasively argued, perhaps the best measures of developmental stability can be made in terms of deviations from the norm. He also pointed out that there are a number of approaches to the assessment of developmental stability other than morphological ones: these include studies which use chromosomal aberration approaches, physiological phenomena, and biochemical data. One advantage of using a wide variety of approaches is that it forces us to think more broadly about what developmental stability is.

Now I want to say a few words about a specific goal I had hoped we might accomplish by the end of this session. You may recall that I started this session suggesting that it might be useful for us to generate a protocol for assessing developmental stability, especially for those investigators not familiar with this area. I had earlier thought that such a protocol would be extremely helpful, but as a result of our presentations and discussions, I am now persuaded that this is probably not realistic. In our discussions about many of the problems we all have encountered in trying to assess developmental stability, it has become clear that many approaches to such assessment are possible and the methods used will necessarily differ among experiments.

Regardless of the approach used, however, there have been many cautions and/or concerns issued today which bear on the planning of any experiments which attempt to assess developmental stability. These include: the selection of appropriate characters, which we spent considerable time discussing today, special problems associated with meristic characters, the standard concerns about eliminating scaling effects, directional asymmetry and anti-symmetry from true fluctuating asymmetry, the potential for some genetical basis for asymmetry, the experimental designs which would best assay pollution and/or stressor effects in the environment, the confusion which sometimes exists between patterns such as fluctuating asymmetry and processes such as developmental stability, and the like. It was suggested that all of us should strive to improve the standards of experiments in this area by carefully taking into account these sorts of cautionary factors.

I believe that the presentations and discussions we have had this morning point quite clearly to some potential prospects for future studies on developmental stability. Perhaps first and foremost, we particularly need studies which would be aimed at assessing the developmental basis of various characters and their asymmetries. Thus, what precisely is the unit (or units) of development which controls a given characters, its apparent asymmetry, its relationship with fitness, and its association, if any, with asymmetries of other characters? Until we understand the developmental processes which produce the characters we like to measure, we probably will not be able to develop more viable hypotheses which would predict associations of asymmetries as well as differences in asymmetry levels between individuals and populations, inbred and hybrids, stressed and non-stressed populations, and the like. For those of us interested in developmental stability from an ecological, genetical or evolutionary standpoint, there is probably much to be gained by collaboration with individuals expert in the modern techniques of developmental biology.

We also need studies which assess the genetic and environmental basis of asymmetry, especially so that we have a better understanding if there is some genetic basis for the asymmetry we observe. It seemed clear today that all of us are not entirely sure what the genetic/environmental profile of asymmetry really is. We need additional, direct estimates of the additive genetic (heritability) and non-additive genetic components of variation for asymmetry. And it would be even more informative if we knew which patterns of asymmetry (fluctuating, directional, anti-symmetry), if any, have a genetic basis. Beyond quantitative genetical approaches, qualitative genetic studies (such as those by McKenzie and Clarke) which assess asymmetry changes associated with specific genetic loci, also should be quite helpful in discovering pathways which directly affect fitness and developmental stability.

Finally, I believe that studies which assess developmental stability need to be broadened to include more organisms, especially various plants which would seem to offer excellent prospects for intra-individual variability estimates. And we need to broaden our studies as well to include suites of characters which can be subjected to multivariate approaches. We need to have some understanding, for example, of how asymmetry sorts out, if it does, in the components generated from principal components analysis. And we can certainly always use carefully conducted studies which directly link asymmetry levels to fitness, however fitness may be defined. Only when we have a sufficient number and variety of well-conceived and executed experiments will we be better able to clarify our understanding of how we can best measure developmental stability.

T. A. Markow (ed.), Developmental Instability: Its Origins and Evolutionary Implications, 389–406, 1994.

Session 2

Developmental genetic origins

Monday afternoon June 14

Thomas Kaufman, Moderator

Trivers:
Are there any data available on the typical heterozygosity at the regulatory loci we're talking about?

Kaufman:
No, that was something that we really need to talk about in this discussion. That again is a very important question because these regulatory regions are bloody huge. And it's really not known how much variability exists. People focus on the protein. And what you find is that these proteins have a small amount of variability, not large. You don't find fast and slow alleles, particularly. But you do see wonderful changes in these proteins between species that are really neat and tell you what's important and not so important in the protein. You see essentially no change in the homeodomain. The actual binding motif, that just doesn't change. That doesn't change over 500 million years of evolution. But other motifs in the protein do change.

Ok, so now that's sort of genes and how genes work in the process, in very broad strokes. Many of you have asked a lot of very critical questions, you know, I don't want to be up here forever, nor do you want me to be up here forever.

So, one last thing I'd like to address is the idea of phenocopies. Because I think that they're important to the people who are interested in environmental effects, how they might be mediated. And it also brings up another topic that I think is going to be talked about next, windows of opportunity. There is a gene in mammals called *phocomelia*. Mutations at their locus cause small flipper like appendages. Back in the sixties there was a rash of births in Europe with this phenotype. They were born to mothers who had taken thalidomide during their first trimester and were not 'mutants' but phenocopies. What is likely the case for those cases was that the drug influenced the functioning of the pathway the *phocomelia* gene happens to be in. Note that the effect is caused during first trimester during the critical times of limb development in utero. Subsequent to that, the fetus would probably be relatively immune to the intake of the drug. So there's a time at which development would be critically affected. In a less personal example, we know that, for instance in *Drosophila*, you can phenocopy the bithorax phenotype with ether vapors. However, an added important item to take home from that case is that the effect is very genotype dependent. There are some genotypes, wild-type strains, that are completely immune to phenocopy. There are other strains that are very highly sensitive, and in these, 20-30% of the animals treated as embryos will show the mutant phenotype. So somehow, ether vapors are capable of interfering with the function of the homeotic gene. How is not known. Another thing that demonstrates the window of opportunity and signalling amongst cells, which is probably very important to all of this, is the way in which bristles become developed. I notice that some people are very interested in chaetotaxy and how that can be used for assessment. This cartoon shows how the bristle develops. There's actually a set of cells which give rise to the external bristle, a socket cell, a nerve cell, and a support cell. All four are involved as a clone in the formation of the structure. Now the cell, the progenitor cell, that gives rise to this group of cells, is surrounded by other cells, and we know that there's an interaction amongst the surrounding cells that do not form bristles and this cell which specifies which progenitor will become a bristle. This process involves a couple of genes, *Notch* and *Shibire*. You can, with mutants in these genes, affect that process. This is not a phenocopy, but it is an excellent example for windows of opportunity. The genotypes of the flies shown in the top and bottom parts of this figure thing are temperature sensitive of either the *Notch* gene, which is an epidermal growth factor gene and is important to the development of the nervous system, and then, at the top, is a temperature-sensitive mutation in a gene called *Shibire*, which is Japanese for paralyzed. If one takes adult flies that at 22 degrees are happy; you put them in 29 degrees, they all fall to the bottom of the vial, paralyzed. Both mutations also have effects on the development of the bristles and the eye in a very similar way. If you take an animal, going along happily at 22 degrees and you pulse it to 29 degrees and bring it back down to 22, for very short periods of time, during specific times in the life cycle, the pupal and larval stages, you get very, very different kinds of effects on the development of external morphology of the animal. So, for instance, in a very specific window in a third larval instar you can affect the development of the eye, so that you can produce a scarring across the eye. And depending on whether you do it early in this period, or late in this period, you get the scar in a different place in the eye. Also, during these other times, you can change that relationship of the external cells and that internal cell that are going to make a bristle. If you pulse an early point that overlaps this eye-scarring period, you get the eye scar, but you also see that all bristles are duplicated. If you pulse later, you remove all bristles. These two genes do exactly the same thing. Early in the sensitive stage, essentially every cell in this field is fated to make a bristle instead of cuticle and bristle, while later, all cells make cuticle.

What this demonstrates is that there are very discrete times in development during which you can affect, very strikingly, the development of specific structures. And, I would argue that the correlation and non-correlation of defects, like cleft-lip and pallet and dentition, that when you don't see correlation, you probably have an effect that occurs late, and therefore it's in a gene that's being regulated and expressed only in a subset of the total number of structures that it affects. If you get an effect early, you're going to see more of a correlation. So time, and time in development and place, are very, very important here. It tells you a lot about what you see and what you get.

McKenzie:
Just to reinforce a point you made about the discrete time of gene action. For a variety of reasons we have been interested in producing diazinon-resistant mutants using mutagenesis and selective screening. The mutants have elevated the levels of, but one has turned out to be temperature sensitive with respect to asymmetry. The period of time that determines whether or not it shows a normal level of resistance asymmetry, or a lower level, is a period of about 2 hours in the development cycle. So the temperature shock period is very precise.

Kaufman:
What we're talking about here is two three hour periods. Where is it, embryonic?

McKenzie:
At the pre-pupae to pupae interface.

Kaufman:
So its when the imaginal disks are also actually going through metamorphosis. You can also get these temperature effects very early when the cells are set aside. The imaginal cells are specified in the embryo, and they inherit that fated state, and then during differentiation it's revealed. So you're affecting things right at the time that fate is actually being revealed.

Vrijenhoek:
For me to come up here and talk about development after Thom Kaufman makes me feel like a motor-scooter at a Harley Davidson convention. One benefit of my knowing so little about development is that I can say perfectly silly things with a straight face. I intend them only to provoke thought and discussion.

When I first started working on fluctuating asymmetry in *Poeciliopsis*, I really didn't see any good reasons why allozyme-based estimates of heterozygosity should be correlated with anything, much less developmental stability. Other than some general ideas like those of Michael Lerner's (1954) about developmental homeostasis, we had little to go on. However, Lerner's evidence was based on inbreeding, not allozyme heterozygosity. Unless allozyme heterozygosity was negatively correlated

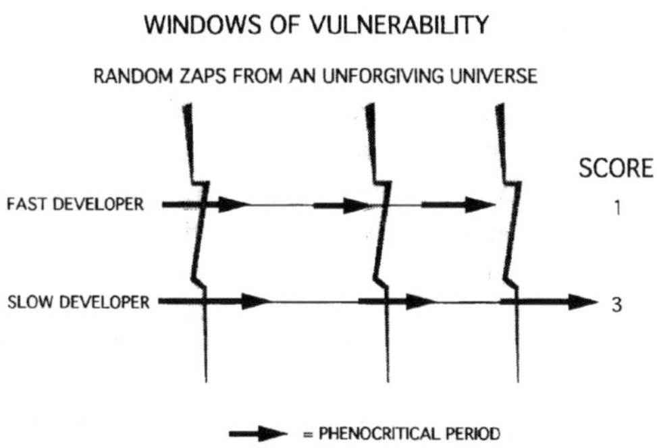

Fig. 1. From: Vrijenhoek, R. C., 1985. Animal population genetics and disturbance: the effects of local extinctions and recolonizations on heterozygosity and fitness, pp. 265-285 in The Ecology of Natural Disturbance and Patch Dynamics, edited by S. T. A. Pickett and P. White. Academic Press, New York, NY.

with inbreeding in nature, the relationships observed by Soulé and others made little sense.

In developing what has been called the 'windows of vulnerability' model, I was influenced greatly by a paper the late Robert W. Harrington, Jr. (1968) wrote entitled 'Delimitation of the thermolabile phenocritical periods of sex determination in the ontogeny of the hermaphroditic fish, *Rivulus marmoratus*.' I was impressed as a beginning graduate student that this title had more big words than I had ever seen in so small a space. Harrington worked with perfectly homozygous fish strains. With heat shocks at the appropriate time of development, he could induce primary males. He also found some morphological abnormalities associated with specifically timed stresses. He described phenocritical periods or windows of vulnerability in the development of certain traits. It occurred to me that if factors that could disturb development occurred randomly at some background rate, then fast developers would expose these open windows for shorter periods of time than slow developers. I have illustrated (Fig. 1) such disturbances as occurring randomly. The slow developer is going to accumulate more hits in its windows than the fast developer. Papers by Lefteri Zouros and Dick Koehn helped me connect this process to allozyme heterozygosity (Zouros *et al.*, 1980) (Koehn & Shumway, 1982). They found that multi-locus heterozygotes of marine bivalve mollusks grew faster. Perhaps the fish I studied also developed faster. Thus, more heterozygous individuals might escape the developmental insults by having smaller windows of vulnerability. I put this idea forward in a review article on 'Animal population genetics and disturbance...' (Vrijenhoek, 1985). At about the same time Allendorf and Leary found corroborative data in trout. Allozyme based heterozygosity was correlated with growth rate and growth rate was correlated with developmental stability.

We have since tried to test this hypothesis with a panel of laboratory synthesized clones of *Poeciliopsis* (Leberg, Wetherington, and Vrijenhoek, unpublished data). As predicted, we found a weak correlation between the average growth rate of a clone and its developmental stability, but the relationship was not statistically significant. The postnatal growths we measured might only be weakly correlated with developmental rate in these fish. Unfortunately, we could not measure developmental rate as the fish are viviparous. More work needs to be done with this question. One thing this study did reveal is a strong and significant genetic component to the variance in developmental stability among the clonal strains. This finding might have some bearing on questions regarding the heritability of developmental stability. In this case it has nothing to do with heterozygosity as all the fish were essentially the same in this regard - they were hybrids.

Koehn:
Did you say what asymmetry was in this paper?

Vrijenhoek:
I don't think windows of vulnerability apply to all kinds of characters. They are probably not important for characters that are size or growth dependent, characters that accrue throughout the whole lifetime of the organism. In this case you can imagine different kinds of processes. Fast growers might be at a disadvantage for characters that continue to change as the organism grows; it might get thrown off track much more severely than a slow grower if it's perturbed early in development. On the other hand, mammals appear to exhibit compensatory growth for some characters. If you slow down growth of one bone, or one part of the body, it can catch up later on. By the time you have an adult organism, it somehow ends up having its bones

roughly the right size. I have no idea how an organism does that. I'm not sure anybody has any good ideas - maybe Larry has some ideas. There's quite a bit of evidence about this phenomenon of compensatory growth in mammals. I don't think we understand it very well. It's certainly worth discussing. How do the right side and the left side communicate? Now, we had an example a while ago, Livshits talked about the tibia in human beings. Perhaps bone development depends on feedback processes. Once an organism is out and walking on its legs, that feedback process may work to equalize growth.

Livshits:
When do we get these characteristics? After you're born. Well, ok, but I spoke about newborn infants. They didn't use their limbs, and yet, if you compare the asymmetry in length of the tibia on the one side, and of the breadth of the same bone, the length has very small asymmetry, while the breadth has relatively high asymmetry. The question is: What kind of mechanism may regulate this symmetric development of the length of two bones?

Vrijenhoek:
The point I'm trying to make is that we have to think about the characters. We have to think about potential feedback processes once the organism is hatched and out interacting with the world. We have to think about what time in the life history of the organism we are going to measure the characters. It's not a simple process. Just going out and measuring a bunch of characters does not seem like a reasonable procedure in this kind of research. We need to talk to developmental biologists who seem to have information about these things.

Kaufman:
One of the issues, genetic mechanisms underlying developmental stability, is going to be discussed next. Robb has some informal comments in this regard.

Leary:
I really hadn't prepared anything formal to say on the relationship between heterozygosity and asymmetry. Everything I'm going to say here is off the cuff. So because of that I thought I'd start with a story, a story of my pre-fossil days.

I started off as a graduate student with Fred Allendorf. And, what I was interested in was obtaining some heritability estimates of meristic characters of rainbow trout, and also investigating how environmental parameters might influence counts of the characters. And about that time, there were a few papers coming out in the literature showing negative associations between heterozygosity at allozyme loci at the population level and population levels of fluctuating asymmetry of morphological characters. And, I was thinking to myself, well, the data are teasing, but there's a real problem. Populations evolve, they do not develop. The hypothesis was a developmental one; more heterozygous individuals will have increased developmental stability. So I said to myself, I've really got some material here to test the hypothesis, as long as I flip the fish over on its right side. Most of you are not ichthyologists, but, conventionally ichthyologists always count characters on the left side of the fish. I had five bilateral characters and I had left counts, I had allozyme data from 50 loci, and said, 'Well, Fred, what I'm going to do is flip these fish over and count the bilateral characters on the right side. And then we're going to look for a correlation between individual asymmetry and individual heterozygosity. And I'll bet my butt - I didn't have any money then, as a graduate student, but I could bet my butt - that we'll find nothing. I thought we'd publish a little note pointing out the flaws in these studies and say, really at the individual level they tell us nothing about what's going on.' So I called up the allozyme data in the computer and the asymmetry data, and let the computer do the correlation - minus .40. We still published the paper, but took a different position; these data support the hypothesis. Ok, now we repeated this type of analysis in this population and in many others, and the story is that there appears in rainbow trout to be a negative, but certainly a weak negative, association between heterozygosity at protein loci and levels of fluctuating asymmetry. We interpret this to mean that yes indeed, more heterozygous individuals tend to have increased developmental stability. Now, since then, there have been a lot of other papers published in many other organisms claiming that they also have shown this association. Unfortunately, a lot of those papers still have the same original flaw that we were trying to point out, that many, many, of them are done at the population level. Again, populations don't develop. Individuals develop. There are also many papers that demonstrate that inbred individuals tend to be developmentally less stable than non-inbred individuals. Now, in one sense you can interpret these data as supporting the hypothesis. But, I really don't think they level strong support for the hypothesis that heterozygous individuals in random mating populations have increased developmental stability. What these data do say is that reduced developmental stability is simply another consequence of inbreeding depression. Whether that's due to reduced heterozygosity, or to increased homozygosity of deleterious recessives which are normally not expressed in random mating populations, we have no way of knowing.

Now there are a few other reports out there. I can recall one on fruit flies by Biémont (I believe, I'm not very good with French) where he pretty much found the same association that we did. There are a few other ones that claim they have or have not found the association. However, I still don't think those data are appropriate to test the hypothesis because often what they call a population is ill-defined. Often what you'll find if you read through the grey lines is they have mixed individuals from, what I'd call in a population genetic sense, different populations. Individuals from these different populations not only have a different evolutionary history, they are also often subjected to different environments. One example might be lumping human beings from Eastern European cultures together and calling them a population. They are a population, I guess in the social sense that they all live in the same area now, but historically, they did not, they had different evolutionary backgrounds, and different environmental backgrounds. So, in sum, where does that leave us about the association between heterozygosity and developmental stability? I think that question is still a very open one. I have nothing more to say, so I'll open it up to Thom to lead what hopefully will be a good discussion.

Trivers:
How many studies are there now that have correlated between heterozygosity and asymmetry within populations?

Leary:
In a correct fashion? As far as I know, three. Two give positive results, one on mice gave negative results, and that's it.

Trivers:
The second positive one was one what?

Leary:
On fruit flies - *Drosophila melanogaster*. So the two model organisms are fruit flies and rainbow trout.

Mitton:
Robb, wait a minute. You're losing the audience here. Some things aren't being said quite explicitly. What Robb means by not having the correct experimental design is that if you're not measuring fluctuating asymmetry, then you're not measuring anyting to do with developmental stability. I disagree with that.

What I and others have done is to collect a group of individuals and divide that group of individuals, collected at the same time at the same place, into groups, heterozygous or homozygous for a single gene. Variances or coefficients of variation are then used to compare the levels of morphological variation in heterozygotes and homozygotes. Rob does not consider this to be correct design. It is not the way he prefers to do it. But that doesn't mean that it's not valid. These studies often report that morphological variation is lower in heterozygotes than in homozygotes. They've been published in *Nature* and the *American Naturalist* and *Evolution*. I believe that morphological variation measured this way reflects developmental stability.

Leary:
There's an alternative explanation of that. If you have additive genetic variation affecting the trait, you expect heterozygotes to be more centrally oriented.

Mitton:
Only, if the genes that you're measuring directly influence that character.

Leary:
Or are marking chromosome sections that are. That gets us down to the mechanisms. If there's additive variance and the genes you are looking at are directly or indirectly through linkage involved in that variance then you expect exactly the association you're looking at.

Koehn:
Do you know the mechanism for the explanation of this observation in salmon?

Leary:
No.

Koehn:
So knowing the mechanism is not, then, a prerequisite for interpreting the data.

Leary:
We have estimated the additive variation for asymmetry in rainbow trout and it's zero. So in this sense, I'm free of the additive variation interpretation. If you had heritability estimates for your morphological characters, and you knew they were zero, I would say the approach is valid. But, in all the studies of such type I'm aware of, heritability estimates for the traits are unknown. And so when we see an association between variance and protein heterozygosity, how do we interpret it? It is simply a reflection that we're marking portions of the genome by luck, of course, that are involved in additive genetic variation for the trait?

Koehn:
People have actually done some work on that question, that I know of, in both theoretical and, Lefteri Zouros can tell us about some empirical data on that question, right?

Biddle:
I hear some argument about the correct and incorrect. And I'm lost. Can I ask our speaker what is the correct experimental design for detecting what?

Leary:
What do I view as developmental stability?

Biddle:
What's the question?

Leary:
Ok, the question is, is heterozygosity positively associated with developmental stability? So, in order to answer that...

Koehn:
Heterozygosity at a series of allozyme loci?

Leary:
Or any. The first problem is obtaining some estimate of individual heterozygosity. The second problem is estimating individual developmental stability. With the technologies available at the time the best way of getting a heterozygosity estimate for individuals in a randomly mating population, of course, was to use allozymes. The next problem is to get a developmental stability estimate, and in order to do that, you want to measure deviations from a developmental target. And, in my mind, an appropriate choice would be characters that demonstrate fluctuating asymmetry. We assume the developmental target is equal left/right. So, when you have a left/right difference, that's a deviation that indicates less stable development. So we correlate the amount of right/left differences on the individual with how heterozygous it was at that suite of allozyme loci analyzed.

Strobeck:
There are several people that I don't think probably have read your paper. So, if you can say precisely right minus left squared, absolute the value of right minus left...

Leary:
Ok, in our material it was just the number of characters at which the individual was asymmetric. I feel this was valid. They were all meristic characters. Asymmetry differences, by and large, were a plus one or a minus one; in other words, binary data, zero or one. Thus we felt the number of asymmetry characters would be a sufficient estimate of the amount of asymmetry on that individual.

Hedrick:
Is the heterozygosity that you measured from allozymes indicative of the heterozygosity of the developmental genes that Thom was talking about? Is that the underlying mechanism that you would think was effecting developmental stability? That's the context I think we're talking about here. Is that the connection you would make? If not, then why not?

Leary:
The data we have right now don't allow us to say what is the cause of the association. That is still open. I do believe, and this is only specific to the rainbow trout, we can rule out that it's due to differences in levels of inbreeding, because we can find the association among full sibs, and they have, by definition, the same inbreeding coefficient. But, what remains is the question of whether the associations due to heterotic effects of the loci we're examining. Or are they due to heterotic effects of chromosomal segments? And personally, while we have no evidence for this, I think that we're seeing heterotic effects of linked chromosomal segments. We have some circumstantial evidence that suggests, at least in rainbow trout, that you probably have big blocks of chromosomes in which you have practically no recombination; in other words, supergenes fortuitously containing some of our markers.

Vrijenhoek:
Robb, I think you're being unnecessarily cryptic in your comments. At the risk of directing the slings and arrows, I'd like to pin you down on a couple of your initial statements. You indicated that you got started in this business because you found some problems with studies that compared population levels of heterozygosity and found population levels differences in asymmetry. You wanted to see if you could find this within a population among individuals. That raises a couple of interesting questions. Would heterozygosity as estimated through half a dozen or so polymorphic enzyme loci be any more informative about genomic heterozygosity of individuals than would heterozygosity estimated from the same number of polymorphic loci be informative about heterozygosity differences between populations? What are the allozymes or the markers in this case telling us about individual genomic heterozygosity and about population levels of heterozygosity? Secondly, as I recall from your initial papers, you were critical of between population studies specifically because the environment in which the population grew was not controlled. Different populations experience different levels of environmental stress, and perhaps that is what we're looking at, not developmental stability per se. That's a perfectly valid criticism. It's worth exploring further, although I think we all understand the problem. Getting back to the first question, what do really think about the predictive value of allozymes for estimates of genome heterozygosity among individuals within populations vs. between populations?

Leary:
My guess is that between populations is better.

Vrijenhoek:
Why?

Leary:
Usually because you'll find a much broader array of differences.

Koehn:
The biggest variance of heterozygosity is among genes. That is an empirical fact. We know that heterozygosity varies among genes by 2-3 orders of magnitude, but heterozygosity variation that's among populations is actually fairly small. And among individuals as well.

Leary:
Is that a valid argument if you're looking at individuals, you have the same suite of allozyme loci?

Koehn:
Well, it certainly bears on this question, though.

Leary:
My only argument on between or the among population comparisons is that the hypothesis is a developmental one. And it's the environment that really bothers me.

Vrijenhoek:
You said that populations don't develop, but individuals within populations do develop. So what do you mean when you say that populations don't develop? I don't understand your statement.

Leary:
I have a population with a very low level of allozymic heterozygosity. Why is that; why does that population have a low level of allozymic heterozygosity? We don't know. Maybe, and this is where I see problems, maybe it has a low level of allozymic heterozygosity because it tends to be maintained by small effective population sizes. Why is it being maintained by small effective population sizes? Maybe for that taxon, it's in a stressful environment. And because it's in a stressful environment, individuals have reduced developmental stability, they have high asymmetry or whatever is the appropriate measure that we choose to use. And then when we do our correlation analysis, we come out with this negative correlation. Certainly the individuals in that population are less heterozygous than the ones in the more heterozygous population, but we're not picking up, what I would call, a real association between heterozygosity and developmental stability. What we're picking up is an association between environmental stress and developmental stability. And in practically all of the among population comparisons that have been done, this can't be ruled out.

Vrijenhoek:
I accept that, but you can rule it out in certain conditions. If you have a control organism that's genotypically uniform across your environments you use that organism to rule it out. We were lucky enough to have a clone in our studies of *Poeciliopsis*. If you don't have a convenient clone, you could find another syntopic species that doesn't show a gradient of heterozygosity and use it as a control for variance in environmental effects.

Leary:
But that has not always been done, Bob.

Vrijenhoek:
I don't think it's so hard to design these controls.

Leary:
No. But, my point is that a lot of the data that are in the literature saying that they support the association don't really allow you to make a conclusion, because there are other interpretations of it.

Trivers:
This discussion suggests that it would be wise to keep the two categories of evidence separate. And that's something you are perhaps intending to stress. In other words, if we have a population that is inbred for quite a while, if it shows greater fluctuating asymmetries, the fact that it's been inbred for a long time tends to eliminate the hypothesis that deleterious recessives being expressed is causing the greater fluctuating asymmetry, because by logic those should have been modified by now. By contrast, as you point out, associations within a population between heterozygosity and fluctuating asymmetry can be explained by reference to either the intrinsic advantage of heterozygosity or the effect of deleterious recessives. So I think it is worth keeping the two kinds of evidence quite separate because they probably do not support a single general principle.

Koehn:
It seems to me one of the questions that we haven't

addressed - I thought we were going to, as it is an interesting one, having now seen how some developmental systems are genetically controlled, at least as Thom Kaufman described them - is how might heterozygosity at those kinds of systems or variations of heterozygosity produce the kinds of phenotypic effects we are concerned with? I mean, it seems to me that's the question. Or, you're asking the question: 'How is it that heterozygosity of a series of allozyme marker loci, are linked to these genes that produce these effects?' That is, in effect, the same question, because now you have to explain the action of these linked genes. So, you ultimately come down to the genetic control of development and how variability at these developmental systems can effect this phenotype. We haven't addressed that point, really.

Leary:
We need to demonstrate that there is an association between heterozygosity and developmental stability, and I guess, I'm not as comfortable with feeling that it's as universal a phenomenon as others.

Koehn:
Well, let's forget about universality. We'll just take your paper and go with that.

Leary:
Once you've convinced yourself that the association is there, then it's appropriate to ask, what is the cause and effect? And I would be happy to go onto that.

Zouros:
We have, again, to think in terms of the mechanism involved. Either we have to assume the ranking of a few hundred individuals in terms of their heterozygosity at a dozen, at the most, allozyme loci tells you something about the ranking of these individuals in terms of total heterozygosity (a question that has been raised perpetually in the past) or that by pure luck, you happened to hit on the loci that are most responsible for bilateral asymmetry. It seems to me your answer is that the former is true and that the reason for this is because the genome of you fish is organized in linkage blocks (rather large, I am afraid) that resist recombination and which harbor, here and there, deleterious recessive genes. Once you make these assumptions everything follows, we all know that. But, what is the evidence for these linkage blocks? Surely, people have looked for them quite carefully in *Drosophila* and mouse and could not find them. Even chromosomal inversions do not appear to obey the model you envisage. Trout could be different, but until we have some good evidence for that we cannot consider this explanation of your results as more likely than other explanations, of which there is no shortage.

Leary:
Can I just interrupt? Maybe this would be helpful for others in the room to point out the possible causal mechanisms, one being the loci you're looking at are heterotic themselves. Another explanation would be the loci you're looking at are actually causally having no effect on the phenotype, but they are linked to a locus or loci that have a heterotic effect on the phenotype. Another possibility, and this is the one I prefer at this point, at least in the trout, is that the loci are in heterotic blocks of genes, and really you're masking deleterious recessives. I'd gladly add other possibilities if people in the room want to propose them.

Smouse:
Let me comment a little bit on this, because there's something that some of us are aware of here that others are probably not. In a population without any inbreeding, that is to say in a large panmictic population, Mitton and Pierce and Chakraborty showed over a decade ago that a small sample of allozymes randomly chosen from the genome would convey virtually no information about the overall heterozygosity of the genome. I later showed that the only reason your example of allozymes shows anything at all is that the sampled loci are included in the total. If you compare the sampled loci with everything else, you do really get nothing. Now, that's true as long as all loci are segregating independently. There is one case where there is a correlation, and Dave Foltz has very forcefully pointed this out, as has Zouros. If there is inbreeding in the popula-

tion, that is to say, a range of inbreeding levels among individuals, it will turn out that the homozygosity or heterozygosity of your marker loci *are* correlated with those of the rest of the genome. The degree to which that is true reflects two things: (a) how closely the loci of interest are linked, and (b) which alleles are homozygous in your marker set. Unlinked loci are minimally correlated, but correlated, tightly linked loci are very correlated. Rare allele homozygotes are more inbred for the rest of the genome than are common allele homozygotes. The associative overdominance hypothesis derives from these two types of correlation, and it is a very credible alternative. It's interesting that the full-sib families showed strong correlations among even unlinked loci. One of the things that's always true in full-sib families, whether the population is in linkage equilibrium or not, is that you will get such correlations, and you will certainly get big blocks of chromosomes that don't have a chance to randomize in the progeny of a single cross. People have looked at this in conifers and have found that if you just take outbred material, you don't see much of an effect. If you do full-sib families and look within a family, the slope gets really steep, and you start picking up the effect. That is to say, we may have credible markers for whole blocks of the genome within a family and it may not be an accident that things work out well in this full-sib situation.

Clarke:
I'm going to throw a spanner in the works here. We've done some work on haplodiploids, for the reason that in haplo-diploid there shouldn't be any deleterious recessives in the genome as they have been eliminated through exposure to selection in the haploid part of the life cycle. So you inbreed the diploid part of the life cycle, increase heterozygosity up to 75%, and find out what happened to the level of asymmetry. You find that there's not difference between the completely outbred material and the material which is highly homozygous.

Koehn:
No difference in what way?

Clarke:
The level of asymmetry is the same in outbred and in highly inbred material. So in this system the level of homozygosity does not appear to have any effect on the level of asymmetry. What is interesting, and as I was just thinking about what Thom said today, is that you do find that haploids are always more unstable than the diploids, and we're just doing some tests now to see how wide-spread that is across haplo-diploid systems. If we're looking at the level of variability at developmental loci, perhaps the fact that having two copies in the diploid is making them more stable, and the fact of having a single copy in the haploid is making it less stable.

Trivers:
Even though you're saying when they're redundant they're no less stable, as in your inbred diploids?

Clarke:
The haploids are always more unstable, even more unstable than highly homozygous diploids.

Leary:
Now that Peter was on the blocks, I think that, as an explanation for why we consistently find this association, even in individuals of these fish taken from random populations, is we know in these fish that there's extremely strong interference to crossing over, and when you have a single cross-over effect on a chromosome during meiosis, that's all you get. We know that the cross-overs occur from the centromere outwards. So when you get a cross-over here, there's no recombination out here. And as far as we know, this always occurs. In rainbow trout, we don't know how big these are, but distal ends of chromosomes never have cross-over events; they're blocks of genes. Three of our allozyme markers are in distal chromosomal segments. In two of the three, we have never seen a crossover event in the regions. In the other one, you can see a crossover event 5% of the time. So that's why I find, in the rainbow trout, the last as the most attractive hypothesis right now. We have a mechanism in which we can build up very big blocks of genes. And we know that we have some markers in these blocks. Furthermore, we have no recombination in

males. Crossovers never occur in males. They only occur in females.

Koehn:
Do you see equilibrium in your marker loci? (No.) Are any of them linked that you know? (Yes.) I figured as much. Are they in linkage equilibrium? Those two are? How close are they?

Leary:
Ten centimorgans. They are and they aren't. I'll have to qualify that. In some recently founded populations they are. I think that's just demonstrating a founding effect. If you look in very old populations or natural populations, and you have suitable frequencies, they tend to be in equilibrium. And also, they both tend to be fairly close to the centromere, in the region where we're getting a lot of crossover.

Koehn:
So you might actually suspect that they were out further toward the end and would scramble faster?

Leary:
No, if they were out toward the end they would stay linked.

Koehn:
I thought you had more recombination further out?

Leary:
No, recombination begins at the centromere and works outward, and then once you get a recombination event, then bang, you get no others. So it's out here where you're getting your blocks of genes. We'd love to, but we do not have any two markers in the same distal end of a chromosome. Our prediction, of course, would be significant linkage disequilibrium and possibly the construction of heterotic blocks. So maybe in rainbow trout, the data are more convincing for an association between heterozygosity at protein loci and developmental stability, not just because of better experimental design, but also because, fortuitously, it's an organism which would tend to show the strongest association. And *Drosophila* is very similar in that respect.

Armstrong:
To bring up the point previously discussed concerning the hierarchy of developmental genes, it seems possible that if some of these linkage groups are independently diverging haplotypes which produce trans-acting proteins, then homozygosity could provide twice as much regulatory interaction between the homozygous haplotypes.

Clarke:
Homozygous or diploid?

Armstrong:
This would only be valid for diploid organisms because haploid organisms may have different developmental trajectories due to processes such as dosage compensation. In other words, if a haplotype represents a specific developmental trajectory, and an individual is homozygous, then it will have the potential for more negative feedback between the two haplotypes. This concept then provides support for John's model which illustrates that with increasing negative feedback there is increasing chaos which would result in increasing fluctuating asymmetry.

Vrijenhoek:
Getting back to the ideas about heterozygosity and developmental rates - perhaps development works like metabolic flux. Kacser and Burns (1981) described a model in which heterozygotes exhibit dominance with respect to flux in an enzyme pathway. Perhaps heterozygotes exhibit developmental dominance. Alternatively, for a complex developmental pathway, heterozygotes might be intermediate in rate, but that intermediacy itself could lead to more rapid and efficient overall development. Imagine an individual that is homozygous for genes that affect developmental rate. Some genes in a pathway are fast and perhaps the next genes are slow. This might result in a herky-jerky movement through the pathway. If heterozygotes are average

they will have a smoother flux. A sequence of starts and stops might have a lower harmonic mean rate of development than a sequence composed of average genotypes. Perhaps increased heterozygotes dampens the oscillations in development.

Armstrong:
Especially if they are linkage groups that have evolved under different selective regimes. Then you may have multiple developmental options to recover in the varying selective pressures which some of the diverse linkage groups may be familiar with.

Smouse:
Just before the break, a thought hit me, and this is a question for John Graham on the one hand and Thom Kaufman on the other. In John's model, the key to the chaotic dynamics is the feedback compensation from one side of the organism to the other. If you get overcompensation you're into chaotic dynamics. Is there developmental precedent for regulation of that feedback between the two sides, Thom? You have said that the two sides of the fly cuticle are basically just two different worlds, and they have nothing to do with each other, but with vertebrates, it might be otherwise. What is known about that?

Kaufman:
The external cuticle of the fly is indeed independent from side to side. The exception to that would be effects brought about through the endocrine system, where you've got a circulating molecule that can now act on both sides. And you can, with, for example ecdysone, transcend the boundary. You can create a situation and get a differential responsiveness in the cuticle deposition pathway, which would respond to ecdysone or juvenile hormone. Several of those kinds of studies have been done, say by Wigglesworth, where he's created ligatures that would block communication, but you have to have a molecule that has the capability of moving from side to side. The thing is that with a sort of standard compartmentalized developmental program that you see in a fly, you don't get that in the hard structural parts. They just don't move. They're there - they can respond differently to a circulating molecule, but that's rare. Another possible exception would be cross wiring by the nervous system. But the pathways of developmental patterning are for the most part autonomous.

Leary:
What about in vertebrates?

Kaufman:
In vertebrates, again, I would think there could be hormonal influence on limb development, but you can get autonomy in mosaics of limb structure development, and you can see left/right differences. It's not the case that if you've got deficiency in the right forelimb that you generally see in the kind of systems that we look at that there's been a compensatory change in the other side. You don't get that kind of structural feedback. You just don't see that.

Mitton:
In preparation for this meeting I generated two new analyses of the relationship between single locus allozyme genotypes and developmental stability. What I wanted to know was, can you reliably detect developmental differences with just a single locus? One study was on honeybees, the other was on killifish. Two observations focused my attention on malate dehydrogenase - MDH - in honeybees. First of all, MDH is variable in all geographic areas, an unusual pattern for a species with relatively little generic variation. Second, some physiological studies we've done in the lab showed that MDH heterozygotes and homozygotes use different amounts of oxygen while hovering. We measured symmetry of wings and found that the heterozygotes are slightly more symmetrical than homozygotes. Furthermore, analyses of variance and discriminant function analysis revealed that the wings of heterozygotes and homozygotes have slightly different shapes. There are some problems with that study; it hasn't been replicated, and all individuals were from the same hive, so it's conceivable that the MDH is in linkage disequilibrium with other genes. But the second study doesn't share that problem. Many years ago, I reported that allozyme heterozygotes in killifish had lower morphological

variation than homozygotes. These fish were taken from a very large population - perhaps 10 to 15 thousand individuals - so linkage disequilibrium should not be a problem. The difference between heterozygotes and homozygotes was most apparent for the gene coding for lactate dehydrogenase, or LDH. A year ago in *Science* there was a paper by Powers and others that really hammered home the fact that LDH has a profound impact upon physiology of killifish. They know these things about LDH. 1) There's kinetic variability, so that the different genotypes have biochemical differences. 2) They know that within a few hours of when sperm meets eggs, the LDH is active, and within 16 hours, there are dramatically different levels of the downstream metabolite. Within 16 hours of fertilization, concentrations of lactate differ by about 60%, causing metabolism to differ by a factor of two among genotypes. The differences among genotypes occur early in development, and therefore have the opportunity to influence development. Once again, heterozygotes tended to be less variable than homozygotes. In addition, I found that the determinant of the correlation matrix tends to be higher in heterozygotes suggesting more regular relationships among the morphological characters in heterozygotes. I found one striking difference in that dorsal fin rays and scales below the lateral lines are positively correlated in homozygotes and negatively correlated in heterozygotes. I found the same result in fish from a separate locality. I don't believe for a second that lactate dehydrogenase directly influences the number of fin rays or directly influences scale counts. But the work of Powers and his colleagues showed that LDH genotypes influence the developmental environment. Within 16 hours, one genotype has 60% more lactate, a major source of carbon for the developing embryo, and respiration rates differ accordingly. When we look for the morphological consequences of the physiological variation, they're easy to find. I see this as clear evidence that enzymes can influence development.

Livshits:
Quick question. Did they find any difference between the different alleles in LDH or did you find any difference in MDH, functional difference?

Mitton:
The functional differences are very, very well known at LDH in killifish. I am not aware of kinetic studies of the MDH genotypes in honeybees.

Livshits:
Anyway, we can consider LDH locus as highly functional in this context, so maybe MDH alleles are also functionally different. I raise this question just because if there is a connection between asymmetry and heterozygosity, it implies that different alleles are not equal in their function, while most studies in population genetics the protein loci, enzymatic loci are mostly neutral.

Mitton:
Depending on who reviewed the literature.

Thornhill:
I am thinking about the connection between heterozygosity and developmental quality, fitness in general. It would seem that genetic diversity in populations is very important as a defence against parasites and a lot of comparative evidence that supports this. It could follow that you could think about heterozygosity as within individual genetic diversity, and in an environment where there are a lot of parasites, the individual, who is highly heterozygous, might have an advantage, because inevitably, the parasites are going to be more adapted to one of those alleles than another and a heterozygous locus. So heterozygosity could give individuals, back-up enzymes and so forth - common enzymes are knocked out by the parasites, and I think with enough time a reasonable argument could be made that heterozygote advantage may relate to the general problem that all organisms face - the big problem - and that is adapting to disease organisms.

Koehn:
I've been sitting here debating whether I should launch into this, but as long as everyone has their favorite idea on the table, I might as well get mine out too. I haven't worked on this problem of asymmetry, but I've thought about it. I've thought about

it a lot, because I wrote a lot, as you all know, on growth rates and heterozygosity. I've often been perplexed as to whether these two phenomena were actually mechanistically connected to one other. What we know about the effect of allozyme heterozygosity on growth rate, and, of course it's true in salmon as well, is that there are average growth rate differences associated with heterozygosity. We know that those average growth rate differences are due to differences in energy metabolism because they're due to differences in the cost of maintenance metabolism. For a particular feeding regime, individuals with a lower basal metabolic demand have more energy for growth, and therefore grow faster, and heterozygotes do that. We've measured that directly and we know that metabolic difference in cost is due to differences in proteins in the turnover costs of the protein pool, and that's been done on several occasions, primarily by the Plymouth group. We know that somehow the differences in the effects of these genotypes is on the efficiency of maintaining these protein pools. And we also know that there's some evidence that inbreeding has a similar effect on these costs; certainly disease states are known to have a similar effect on these costs. That is, they enhance the cost of maintenance, and therefore, any other biological process that depends upon energy cannot occur as effectively. Thus, the energy differences of this sort are correlated with heterozygosity. And I've often wondered, believing naively that there was a lot of evidence that related heterozygosity to developmental stability, if this and my work shared the same underlying mechanisms. And then hearing Thom describing the mechanisms of development, I couldn't help but think whether there was some relationship between the cost of the maintenance of these systems, and the degree to which they are subject to environmental perturbation. We also know that the lower the energy status of an animal - plants too, for that matter - the more susceptible it is to individual environmental stress. Peter Parson's done an enormous amount of work on that. So one could envision this complicated sequence-dependent process of gene products, all of wich involve energy, all of which involve control, being differentially susceptible to environmental insult. I wouldn't say these were 'homeostatic properties' - I don't know what the hell that means - that's what I trying to get at - I'm trying to get a mental conception of what that means in terms of a developmental mechanism. Why should inbred individuals be less 'homeostatic' in this kind of metabolic sense than heterozygotes? And the reason why I have never jumped into this discussion, is that I can not conceptually relate these two phenomena to one another. I'm curious whether anybody else can. I mean, you mentioned this metabolism thing, and I've talked about this balanced pathway business. But I really couldn't describe the model, a sort of a concept that would tie these things together - I can't do it - and yet somehow, it seems to me that they ought to be related.

Kaufman:
This is something that I wonder about - the energy argument, that is how much energy an organism uses or expends to perform certain tasks. I could give you an example where the fly does something in such a bizarre way, that, it's so baroque, that you never would have invented it to do this that way. And it obviously utilizes a hell of a lot of energy to do it that way, and yet it does it that way anyway. There's a far simpler way that it could have solved this developmental problem, and if you were going to sit down with a T-square and a pencil and design it, you would have done it in a more simple and straightforward fashion. But evolution doesn't choose to do things that way. When you start making energy-use, energy-conservation arguments about organisms, I think you, with respect to development, you start running into problems.

Koehn:
Actually there's an interesting corollary of my presentation in the salmon literature. If you look at when different genotypes hatch, at hatching, in other words what's coming out of embryogenesis, you find that heterozygotes are not faster developing in embryogenesis, they're actually intermediate between different homozygotes. Since heterozygotes grow faster because of higher metabolic efficiency, why didn't they come out faster? Well, it turns out there is no metabolic maintenance cost in embryogenesis. So development is not a metabolic efficiency issue, it's a metabolic rate issue. You have variations in metabolic rate among genotypes, and the faster the metabolic rate of a genotype, the

faster they at hatch. Efficiency becomes important at hatching, basically because you've now got a metabolic maintenance cost at that point. This is not an argument about what *should be* happening on the basis of energy, it is an attempt to interpret what is happening in development on the basis of known energy costs during development.

Freeman:
I want to hazard a possible explanation for you, and I'm sort of hesitant to do it, because the number of real geneticists that sit around here. But one of the things that's always bothered me about heterozygosity is that the enzyme uses is the same, and the product that's produced is the same. The only thing that's different is the machine that makes the product. This led me to believe that it had to do with rates and flux. If you read Kacser and Burns' papers on flux, they assume a dynamic equilibrium. That is everything moves towards some point. But if you've ever sat down and tried to model a chain of enzymes with first or higher order kinetics, you'll soon recognize it's very difficult to get things to reach a point. They oscillate. They don't reach this dynamic wonderful point unless the critter dies. It seems to me that one of the things that heterozygosity might do is genetic oscillations that are not identical. If you have two different alleles, perhaps these oscillations may not be in phase. They may be offset. And one of the things that we learned about heartbeat in the last few years is that if your heartbeat becomes regular, you're going to have a serious cardiac problem in the near term, the near term being a few weeks to a month. It's the irregularness of it, it's the complexity of the pattern, the little bumps that ride the big bumps that keep you going, and if you have the waves so they're not perfectly superimposed when you deal with that flux, then it may be a more efficient system for preserving some substrate flux. The complexity may confer stability in the same sense that it does with the heart. And so we might want to look at it in that kind of vein, where you're starting to model flux. But if you're going to do that, the first notion you have to shed is the notion that everything is going to move toward the point.

Graham:
May I add something to Carl's comments that may answer your question? If you look at the oscillations of chemical species in a metabolic pathway, and if you look at them in phase space, you will find that they describe a chaotic attractor. There is much literature on this. The interesting thing about strange attractors is that they are resistant to perturbation. If you perturb the trajectory a little bit, the trajectory is still on the attractor. Bruce West has proposed that morphogenesis is resistant to perturbation because metabolic pathways, enzymatic reactions and developmental pathways involve chaotic dynamics. Now that is not to say that development is globally chaotic, but that chaos occurs within the system.

Koehn:
Just to respond to this point. I've read all of Kacser and Burns, and I've written a lot of papers about it, and so I'm aware of their arguments. And I agree with you that it's somewhat an oversimplification of the way metabolism actually behaves. And I think the suggestion you make is a perfectly reasonable one, but you have the obligation, it seems to me, when you make a hypothesis, to make it consistent with the data that's known. And the data that we know is that we have differences in protein turnover. It is not therefore sufficient to speak simply of 'chaotic dynamics' or 'chaos' generally. You have an obligation, I think, to be less nebulous.

Freeman
That's not what I said. What I am suggesting is that if and only if, the waves are out of phase, then the system may be the more stable to perturbation. Now, do I know whether things are out of phase in metabolic pathways, I don't. But, I don't know anybody who can do those experiments. But the assumption that most people make, that they exist in dynamic equilibrium, is both over-simplistic, and totally unsupported with everything we know about kinetics and complex systems. So the model that I'm proposing, while I don't have all the data for it, is based on the first principles of physics and chemistry, and I think it is more defensible than a model postulating that everything strives for a dynamic equilibrium.

Kaufman:
There seems to be, it seems to be in my naive view of all of this, is that it's reasonable to me to think that a heterozygous individual would be more stable developmentally, but again, I'm very naive with respect to this, and I have a very simplistic reductionist view of how the genome works to create the people that are sitting around this table, and other organisms. So is there any evidence that says that homozygosity is good or that heterozygosity is bad?

Freeman:
Sure. There are thousands and thousands of plants which are obligate self pollinators. Cleistrogamous plants pollinate themselves before flowers open, they're homozygous at virtually every locus. Further, they're remarkably successful, and they count for one of your major agricultural exports from this country, soybeans. They've been very successful. Now, what happens with respect to developmental stability, I don't know. It's very difficult to get enough crosses, and I actually have someone who's sitting down there with jeweler's lenses, trying to emasculate these flowers, at the moment, so I can answer this question for you. And she's not proficient enough yet.

Kaufman:
I know one case where outcross heterozygosity would be very bad, and that's when we start mobilizing the transposable elements of the genome. That wreaks havoc. I think if somebody looked at a hybrid dysgenic cross they would find a set of dramatic problems that are enormous. Therefore, in my mind that is one place where it would be the case.

Vrijenhoek:
We compared fluctuating asymmetry in all-female hybrids of *Poeciliopsis* and one of their sexually reproducing ancestors (Vrijenhoek & Lerman, 1982). The hybrids reproduce naturally in clonal fashion and they were believed to be heterotic rather than disgenic as Thom Kaufman has just suggested. We found no overall differences in developmental stability between the hybrid clones

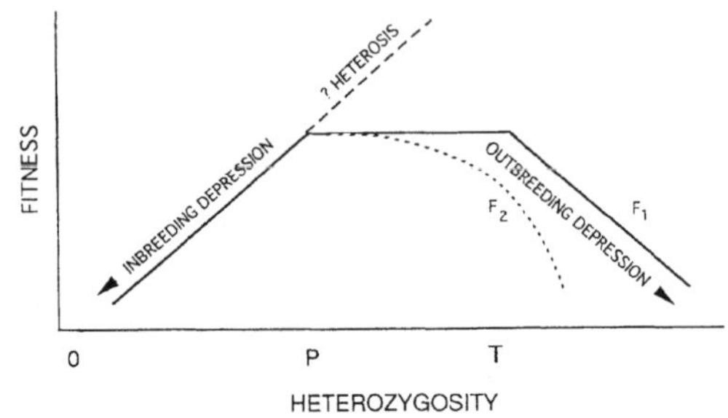

Fig. 2. From: Vrijenhoek, R. C. & S. Lerman, 1982. Heterozygosity and developmental stability under sexual and asexual breeding systems. Evolution 36: 768-776.

and the coexisting sexual species, as long as the sexuals had some genetic diversity. However, the hybrids were significantly more stable than a co-occurring founder population of sexuals that had essentially no heterozygosity.

We developed a simple graphical model to explain these relationships (Fig. 2). Levels of heterozygosity vary along the horizontal axis and developmental stability along the vertical. Imagine a large, numerically stable, panmictic population with a heterozygosity level of *P.* For a typical outcrossing species, a rapid reduction of heterozygosity through inbreeding, bottlenecks, or founder events should also result in a loss of fitness, and that might be manifested in a loss of developmental stability. This decline is mostly due to increased homozygosity for slightly deleterious recessive alleles that accumulate in the large ancestral population.

What happens in interracial and interspecific hybrids? Crosses between 'inbred' populations of strains with heterozygosity below *P* would result in an increased heterozygosity in the resultant population to the plateau at point *P*. If heterosis is due to simple dominance, then the fitness lost during inbreeding is recovered upon outcrossing. If overdominance plays a role, hybrids between different populations at point *P* might temporarily lead to increased fitness beyond the threshold level described for a single population. Unisexual species that could clone the F_1 genotype might permanently fix this heterosis. Nevertheless, we did not observe this for developmental stability in *Poeciliopsis*. The unisexual hybrids, although heterozygous at more than 50% of their loci, were no better than a typical outcrossing sexual population with 4 to 5% heterozygosity. However, the unisexuals were considerably better than the founder population of sexuals that had lost heterozygosity. The existence of the highly heterozygous unisexuals at a plateau of developmental stability suggested that simple dominance was sufficient to explain the relation ship between heterozygosity and developmental stability in these fish.

However, dominance can only extend its benefits so far. If the hybridizing species are different enough (point *T*) coadapted genes from each species might not work well in the hybrid context and developmental stability will decline. This decline might be accelerated in F_2's and backcross progeny if recombination has disrupted epistatic interactions. Graham and Felley (1985) found such a decline in developmental stability associated with hybridization in banded sunfish from New Jersey. They did not invoke transposable elements though.

One thing we don't understand is how broad the plateau might be for maximal developmental stability. How different must two populations or species be to get a decline due to hybridization? I suspect the answer differs immensely across taxa. In some plants that show strong local adaptation to soil conditions the plateau might be very narrow (Waser and Price 1989). However, for many fishes, hybridization between species and even genera does not appear to be that maladaptive. This problem deserves more study.

Soulé:
There's an interesting parallel to the ecology where intermediate levels of disturbance in ecological systems are supposed to generate maximum diversity. Let's not carry this analogy further.

Kaufman:
Well, we're approaching the witching hour, and Teri says that I need to sum up. However, the sort of hubris that I would have to exhibit by attempting to sum up in a roomful of experts in a field that I know nothing about is beyond me. What I'd like to do, I made this overhead, so you're going to have to look at it, and the bar is almost open, so -

Let me begin by saying that I'm convinced that you're seeing something real, and it has an underlying developmental cause, and that there are genes involved in this process, and I don't think that's just an article of faith on my part, that genes are involved in development. It seems to me that one of the points that needs to be looked at very carefully involves the selection of the morphological traits that you're going to use in any analysis. It is essential that you have an understanding of the underlying mechanism that results in that morphology, because if you're just measuring some random noise in the morphology that you're looking at, it's really not going to tell you much. But if you're looking at something where you can define an underlying mechanism and how that can be influenced by genetic or environmental factors you are on the path to cause-effect relationships. It seems to me we've come up with a few highly conserved mechanisms for how development proceeds and is controlled, and that these involve a series of signalling events and a series of responsive events in cells. The signalling can either be from cell one to cell two, or they can be external influences. By external, I mean outside of the organism. So, let's first deal with the cell to cell interaction. Cell 1 produces by some mechanism that we can ignore right now a signal, which is a ligand. That ligand then interacts with a receptor molecule on cell 2. That receptor molecule has an external domain and an internal domain, which is in the cytoplasm of cell 2 - this internal domain quite often has a kinase-like function, so the external domain gets tickled and then causes the activation of this internal protein which then has consequences inside cell 2. Likewise this external signal in case 2, which can be something as complicated as a hormone or something as simple as an ion, can be sensed by the cell in a receptor that can either be external or internal. That event then has consequences on transcription factors which can interact with the regulatory elements in the genome that each factor targets. The result being that cell 2 responds to produce new products which either can be new receptors, new transcription factors, new structural proteins, a new enzyme, or a new signal, and therefore result in a differentiation event that changes morphology. On the diagram the places where you can alter this scenario are shown in red.

What we do in the developmental biology labs is simply wipe out a component. Clearly, however, you're not going to find anything running around out there in the real world with a wiped out control step because it's dead. But the kinds of things that I think you're looking at are rather more subtle changes. For example, you can change the concentration of the ligand that's produced here; not remove it, just change its concentration. That will change the magnitude of the target cell response. There are also multiple elements in these target sites in the genome. If you now change the number of these sites by mutation, you can subtly change the effect of the level of transcription independent of the amount of the transcription factor. This will in turn be reflected in how strong the cellular response would be. Also relevant to the kinase functions is the size of the ATP and GTP pools, and these are relevant to the metabolic state of this cell, how much phosphate there is available to carry out these reactions. You can then make subtle changes in the binding coefficients of these transcription factors, by changing the protein signals, very subtly. You can change the number of binding sites in targets which will change the responsiveness of this gene and how strong the response is. What I would say would be very important to any analysis is that you have to understand at what point in this maze your morphology is being controlled and changed for each test example. One must also understand that environmental influences can impinge on many of the stages outlined. Any molecule that's being dumped out there by some unscrupulous giant of industry potentially can interfere with these steps directly or indirectly. For example, analogs to any of these signals that can combine irreversibly with the receptors and block their action. You can find things that change ion concentrations or that change the phosphate pools inside of cells. All of these things can have very strong effects on the developmental process. The other thing that you have to be very careful about is looking at what point in the ontogenic pathway the blockage takes affect. The earlier, the broader the ramifications. So, I would say that based on this kind of thing, look carefully at the characters you're going to assess, understand the point and time in development when the defect is being incurred by the animal or plant.

The other thing relates to the question of heterozygosity. I think one of the things that came out of this meeting is my naive feelings that heterozygosity should, in general, be good, for many of the reasons that were stated around the table. But, perhaps in some cases, we need to have a better measure of heterozygosity. I think that a lot of the complaining that I heard had to do with, ''are you really sure that just because you're fast/slow, and you're making this heteromorphic dimer, or what have you, if that isn't necessarily telling you that the whole genome is heterozygous.'' That was a problem that I, at least I think I heard. Now that's my naive view, and my very sort of reductionist approach to all of this, and if anybody else wants to say anything, or you want to throw something, that's fine by me. I can duck.

References

Harrington, R. W., Jr., 1968. Delimitation of the thermolabile phenocritical period of sex determination and differentiation in the ontogeney of the normally hermaphroditic fish, *Rivulus marmoratus Poey.* Physiol. Zool. 41:447-460.

Koehn, R. K. & S. E. Shumway, 1982. A genetic/physiological explanation for differential growth rate among individuals of the American oyster, *Crassostrea virginica* (Gmelin). Mar. Biol. Lett. 3:35-42.

Lerner, I. M., 1954. Genetic Homeostasis. Oliver and Boyd, London.

Vrijenhoek, R. C., 1985. Animal population genetics and disturbance: the effects of local extinctions and recolonizations on heterozygosity and fitness, p.p. 265-285 in The Ecology of Natural Disturbance and Patch Dynamics, edited by S. T. A. Pickett and P. White. Academic Press, New York, NY.

Zouros, E., S. M. Singh & H. E. Miles, 1980. Growth rate in oyster: an overdominant phenotype and its possible explanations. Evolution 34:856-867.

T. A. Markow (ed.), *Developmental Instability: Its Origins and Evolutionary Implications,* 407–419, 1994.

Session 3

Human health and phenotypic variation

Tuesday morning June 15

F. Clarke Fraser, Moderator

Koehn:
Would you say your statement on the lack of evidence for a relationship between asymmetry and genotype, as you put it, is a function of negative data or no data?

Fraser:
Unconvincing data.

Koehn:
My impression is that there just haven't been really very many thorough studies.

Fraser:
Yes. Actually they do involve the problems that Rich brought up yesterday in faults in design and whatnot.

Livshits:
The question which I want to ask at the beginning of my talk is what kind of finding or evidence should we get in order to prove that FA has indeed something to do with developmental instability? In my view we should take individuals who suffer from congenital disorders whose development we know is disturbed and compare their FA with that of healthy controls. This was done in several studies. For example, our first slide shows Peretz *et al.* (1988) dental asymmetry data in clinically normal individuals vs. individuals who suffer from fragile (X) syndrome and from Down's syndrome. As seen in the figure, healthy individuals consistently show lower FA in both the mandible and the maxilla than do affected individuals. The second slide compares intraclass correlation coefficients between antimeric teeth in probands with Down's syndrome and in normal subjects, obtained by Barden (1980). Again, individuals with Down's syndrome exhibit increased FA: their correlation coefficients were lower in 10 of the 14 teeth measured.

Along the same lines are the observations made by Malina and Buschang (1984) on morphometric FA in etiologically heterogenous groups of mentally retarded individuals (slide 3) and in Markow and Wandler (1986) on dermatoglyphic asymmetry in schizophrenic subjects.

Our team in Tel-Aviv University for may years has monitored physical growth and motor development of newborn children. We have compared numerous characteristics, including morphometric FA in preterm and normal term infants. Preterm infants have a very high percentage of various morbidity categories. They therefore were considered as a group with decreased developmental homeostasis. The next slide (5) shows that FA in infants and their parents decreases with the increase of the infant's gestational age (Livshits *et al.*, 1988). Slide 6 confirms this observation on a sample gathered by Steve Fields (visiting from Ann Arbor) and us ten years later. The one year follow-up study of these infants demonstrates that all preterm infants somehow compensated their FA, i.e., it decreased with the age while term infant's FA virtually did not change. The following slide compares measurement error (ME) in infants of various gestational age and the respective FA. One can see that ME is independent of gestational age, but on average is of a comparable magnitude with FA estimates. This means that all morphometric FA estimates may not be reliable. To explain the consistency of the results we assumed that measurement error is normally distributed around each measurement. That is, the mean estimate and the variance in a group with higher FA will be higher than in a group with smaller FA value, yet the individual values will be biased.

As a result of large measurement error one can assume that morphometric traits *per se* might not be very reliable and this is the major reason for the observed uncertainty. Therefore other traits such as dermatoglyphic characters that have negligible measurement error in relation to asymmetry might be more appropriate. Several studies involving dermatoglyphic variables were performed at our department. In particular, Kobyliansky, Berjerano and Bat-Miriam-Katznelson compared finger patterns and ridge counts asymmetry in Turner's, Down's and Klinefelter's syndromes as well as in patients with cleft lip/palate, cervical endometrial carcinoma and cystic fibrosis vs. healthy control. The next two slides provide the results according to sex, and show that many FA measures are significantly higher among affected individuals. However, not all the samples and not all the variables show differences.

Koehn:
Greg, do you have any idea of to what degree these characters are statistically independent?

Livshits:
Some of these traits are correlated, however, this was not taken into consideration, nor was the Bonferonni correction applied. It makes our decision about the relationship between FA and developmental stability even more problematic.

Fraser:
You wouldn't really expect it for cystic fibrosis, would you?

Livshits:
Well, I would, because cystic fibrosis, although monogenic is still a severe genetic anomaly in development. My idea here was to check whether the FA depends on definite genetic abnormalities, and to what extent, in the chain: chromosomal aberrations, polygenic (cleft palate) and monogenic mutation (cystic fibrosis).

Fraser:
It is not a developmental anomaly in the usual sense.

Livshits:
What will you call it then, if it is not a developmental anomaly? It is a lethal mutation - this is the genetic catastrophe. Thus, of course, in the broad sense I do consider it as a developmental anomaly.

Summing up the facts provided above, I am not sure that I can accept today the idea that FA does reflect developmental stability. If we do accept this idea because we have many facts that suggest a relationship, I believe we should probably change the concept of FA. My suggestion is that the appearance of FA has a probabilistic, i.e., stochastic character, rather than a deterministic one. The probabilistic approach could explain inconsistencies in general, i.e., why sometimes we get results consistent with theory, but sometimes not. We propose a stochastic model which consists of two stages in FA development. The first stage is actually a latent stage where symmetric development of bilateral structures occurs or asymmetry is comparable with the population mean. As the slide shows, the second stage begins at the conditional time zero. At this moment, with probability equal to q, appeared first deviation from the typical (or ideal) symmetry. I believe that this appearance is genetically determined. From this point, the following development of asymmetry occurs with the probability p, or the probability of no changes is 1 - p. Thus this hypothesis states two stages in FA development and each stage is probabilistic.

Fraser:
I think we have to distinguish clearly between morphometric traits and threshold traits, between growth disorders and developmental disorders - morphogenetic disorders, so let's keep that in mind.

McKenzie:
I'm really following your point about cystic fibrosis; the expectations you would have from that. I mean to all organisms, life's a terminal disease. I'm just wondering, with respect to that measure of cystic fibrosis as a developmental character,

whether it's a reasonable character, or more importantly, with cystic fibrosis sufferers, are they born earlier, or are they born later than non-sufferers? Is there information known at this point?

Livshits:
Well, I don't have much information at this point, the information which I showed, are just previous ones. We'll get much more detailed in the very near future, maybe in a month or two.

Gottesman:
I'm the coauthor of a paper in your handed-out material with Teri Markow that attempts to relate fluctuating asymmetry and fingerprints to the Lernerian ideas of phenodeviance, and to make an indirect suggestion for these pieces of evidence being related to increased homozygosity as underlying the liability to schizophrenia. I've never been invited to a meeting on cold fusion before, and I hope I haven't been invited to one yet!

I know why I am here today; it's because almost forty years ago at the beginning of my career as a psychologist, I was 'adopted' by the *Drosophila* geneticist Sheldon Reed because of our shared interests in the genetics of mental retardation and other human behaviors. It was through Sheldon Reed that I was personally introduced to both Dobzhansky and to Lerner, and I was encouraged by both of them to learn more about genetics, and to, in effect, become one of the founders of behavior genetics in the United States. I know there is such a field, because, among other things, Peter Parsons and Lee Ehrman had written a text book about behavioral genetics as had John Fuller and William Thompson as early as 1960. It's not going to be a simple problem to just go from what we know about *Drosophila* to what we would *like* to know about our own species, although yesterday I certainly ran into a lot of enviable shopping lists for pieces of information that I wish we knew at the human level. 'Nature abhors a homozygote' is a thought of mine from in 1965; that was as a consequence of my having been exposed to Lerner's data on anomalies in chicken morphology. So we're going from *Drosophila* to chickens to schizophrenia all in one fell swoop. I felt that I had the burden, as one of the few behavioral geneticists in this crowd, to try to elevate behavior to the same plane as morphology as a trait worthy of genetic interest. If any of you have seen the June (1993) issue of *Scientific American* you'll realize that this has not been as easy struggle; it's been an uphill battle, and we have many detractors even today. In regard to schizophrenia, I believe that facts are quite clear, that not only is schizophrenia familial, but the basis for the familiality is in large part, genetic. If you want to take the trouble to calculate the heritability of the liability to this complex trait, which we have reason to believe is associated with a multi-factorial polygenic basis, the heritability of that liability is approximately 70%. This is not very different from the heritability of the liability for variation in IQ; that domain also has had an awkward history. There are some plain facts about the risks of schizophrenia in the relatives of schizophrenics, and if you model them, you'll find that you have some interesting evidence for more than one model underlying schizophrenia. I want to select the twin data, because twins are really the closest we can come to any effort at experimentation in human beings where we have the equivalent of a clone. The trouble is we just have a bunch of individual clones, two twins in each clone, and those are the identical twins. It's clear from post-war studies done on schizophrenia using a reasonable diagnosis - I don't want to get into that very much, I'd like to avoid it - but these are studies that I have faith in, and the total sample of post World War II studies; they all reach the same conclusions, namely that the identical twin concordance rate is at least three times higher than the fraternal twin rate (48% vs. 16%). This all fits together with the idea that one or more among these models will do the job of explaining the data points. While I've eliminated those studies that seemed to have a high loading of phenocopies - phenocopies, for us, are the individuals who show their symptoms of schizophrenia only after they were addicted to alcohol, only after they had been shot in the head during WWI or WWII, etc., and I removed those kinds of studies to get a cleaner sample, and then McGue and I did model fitting and found that out of all possibilities, we probably could exclude a single major locus model. The mixed model, which we had a lot of faith in because nobody's ever tested it before, turned out to be not all that good. Multi-factorial oligogenic, and multi-factorial polygenic; I can't currently distinguish between them, but I prefer multi-factorial oligogenic as being one of the major models that underlies the etiology of schizophrenia. Recent efforts have tried to show that a single major locus does indeed account for the transmission of schizophrenia across generations, but it turns out not to be so. Using a marker on chromosome 5 (originating from observing a chromosomal abnormality in a pedigree with an affected nephew-uncle pair), the first study was, of course, a positive one. Since that time we've had just a continuous series of negative findings. This atmosphere of negative findings for linkage studies of schizophrenia has led to the incorrect conclusion that genes have nothing to do with schizophrenia. That's a very wrong way of looking at the world. The data from more than 60 years of family and twin studies show it is certainly highly heritable, but, it isn't going to lend itself to explanation with a single major locus. That pushes us more toward the interests in multi-factorial polygenic models, and those things that may indirectly support it, such as fluctuating asymmetry and increased homozygosity leading to phenodeviance. When I analogized from Lerner's chickens to schizophrenia, saying, 'Isn't this a phenodeviant?', he readily agreed to that idea and hoped that I could support it eventually. My resolution to the issue of how schizophrenia is transmitted, my ecumenical effort to reduce hostility amongst warring factions is to recognize heterogeneity. But these partitions of the 'Bosnian territory' are due to my efforts at computer simulation with different kinds of models, and our paper will give you some of those are references. Now, one of the interesting things that turned up in the research I've been doing all these past decades is reminiscent of one of the things that Clarke Fraser had in his paper. He said 'an implicit but often ignored aspect of multi-factorial threshold models is the stochastic element; individuals with the same genotype in the same uterus, do not necessarily end up on the same point on the liability distribution.'

Well, I've been very fortunate in being able to have access to a wonderful set of data from Denmark, where, they have a national twin register that goes back to all twins born since 1870, that's a 123 years ago. And because of the age of that twin register, and because of my working with Margit Fischer first, and now Aksel Bertelsen, using this national register of twins, and matching it against the national psychiatric register. The Danes have had national socialized health insurance for a long time, were able to find all those twins who are schizophrenic and their co-twins, and because of the age of the sample, were able to find the children of those twins, and then to look at the children of those twins as a function of whether or not their identical twin mother or father did or did not have schizophrenia. The results of looking at the offspring of identical and fraternal twins as a function of whether or not the identical twin parent did or did not have schizophrenia provides a concrete example of what Clark Fraser was talking about, the effect of stochastic processes that lend to these co-twins of schizophrenics actually being normal; there are only six of them, I'm afraid, because it's not a very common phenomenon, and then the data set are further reduced because I had to look only at those who had children. And you can tell from our paper (Markow and Gottesman) that the fertility of schizophrenics is greatly reduced. We can infer from the fact that these normal monozygotic co-twins of schizophrenics produce children who have schizophrenia at the same high rate as these probands who have schizophrenia, that these co-twins indeed have the genetic liability for schizophrenia, but because of stochastic processes, both prenatal and perinatal and postnatal, I'm assuming – did not have schizophrenia themselves, but they transmitted the liability to their offspring who were unlucky and they developed schizophrenia at this rate of 17.4%, which is 17 times the population rate. The lifetime risk of developing of schizophrenia is 1%. For control information we found that the fraternal schizophrenic twins have transmitter schizophrenia at the same high rate to their children, 17.4%. These rates are higher than we usually see (13%), because in Denmark we used a narrower definition leading to probands with severe cases of schizophrenia. The nieces and nephews of the fraternal schizophrenic twins and they rate we observed in other studies for nieces and nephews, 2% or so. And this all fits together for supporting both multi-factorial polygenic transmission models and to show you that there can be latent genotypes that remain unexpressed.

Shifting to another set of data, again, rather exotic as far as human genetics data go, but which speaks to the issue of the generation of homozygosity which should generate more phenodeviance. Those few studies where we can look at the children where both the mother and the father had schizophrenia, and these people at least had children - you should know that most schizophrenics who have children have their children before their first episode of schizophrenia. After their first episode, they tend not to have any more children, and in fact, they have a very high probability of the dissolution of the marriage. But if you wait long enough, if you wait from 1923 to the end of the 1980s, you can pick up a number of these couples from different countries (Soviet Union, Norway, Sweden, Germany, and Denmark). These children who are unlucky enough to have both the mother and the father who suffer from schizophrenia have tremendously high rates of schizophrenia - 46% averaged. Those rates are much higher than you would find from assuming that it's not multi-factorial polygenic; we get an exponential increase when both parents are affected vs. one parent affected. The same is true when both parents are manic depressive. There are reasons to believe (from the literature) that manic depressive disorder is more likely to be due to a single major locus. So it may not be too surprising that we do have high rates when both the mother and the father have manic depressive disorder - 30% to 80%. And one other indicator of the possible effect of homozygosity comes when we pick out those few individuals where the mother has either schizophrenia or manic depressive disorder and the father has the other disorder. It's interesting that these disorders seem to segregate, schizophrenia, schizoaffective, and affective disorder - and they don't blend or add together to give you one very severe psychosis that we don't yet have a name for. The children segregate into these three categories. That also speaks to another point that was raised in Clark Fraser's paper and other remarks in the last few days. Just because you are an outlier on one dimension doesn't give you any predictive power for saying that individual will be an outlier on other dimensions. So the homozygosity is relatively trait specific.

I want to leave room for Teri to comment on how we got together. It was partly accident, and partly the result of my having been exposed throughout my career to *Drosophilists* so I knew that when one talked to me, they had something worth saying. And because I never throw out any data, and because we've used fingerprints in order to diagnose zygosity in the old days, I had fingerprints from twins who were schizophrenic. Teri, do you want so say just a little bit about how we played with Lerner's ideas and just a word or two about the homozygosity and phenodeviant story from your point of view.

Markow:

Our intent was to test the prediction that vulnerability to schizophrenia was a function of a polygenic threshold. The prediction was that if schizophrenia or vulnerability to it is caused by polygenes with a threshold, individuals past the threshold manifest this condition and should also then be more homozygous at the relevant loci. This situation would predict an increase in developmental instability as evidenced by FA in dermatoglyphics. So, we looked at FA in a number of different samples of schizophrenic patients and found that the evidence was consistent with these predictions. The idea for this study came from earlier work by Charles Woolf on cleft lip (with or without cleft palate).

Koehn:

So if I understand you're saying in the sample of schizophrenics that measures of asymmetry in dermatoglyphic data is higher than it would be in non-schizophrenics, but you don't really have any direct evidence that schizophrenics are actually more homozygous.

Markow:

You're absolutely right, Dick. It's only inferential.

Fraser:

So what do you think is happening, are the genes for schizophrenia also upsetting the dermatoglyphic patterns?

Markow:

I don't think specifically. I think it's probably a more general vulnerability in this situation.

Smouse:

Teri, what do you know about asymmetry within or

about differences among individuals who are concordant or discordant co-twins? I should say, by way of preface, that the narrow sense heritability of total ridge count is on the order of 99%. For all practical purposes, it's totally genetic under normal circumstances. I'm not sure that a corresponding figure is available for ridge count asymmetry.

Markow:
Well, we've looked at different sets of twins that were concordant and discordant. We looked at another category where one twin was schizophrenic and another twin had a different mental disorder. And the idea was that twin sets which were concordant would have a higher genetic loading and that loading would be reflected as a higher homozygosity in those individuals as well. So concordant twins were predicted then to have a higher asymmetry than discordant twins. The data supported our predictions.

Fraser:
So twinning itself..

Markow:
Twinning itself has no apparent effect on the dermatoglyphic characters that we were looking at.

Fraser:
Did you look at familial vs. sporadic?

Markow:
Did we look at familial vs. non-familial cases? We haven't looked, in that sense, like the cleft-lip studies that Charles Woolf did. We hope to do that in the future.

Parsons:
Are you going to look at other disorders, like manic depressives for homozygosity?

Markow:
We've looked at manic depressives for elevated FA and we don't see it, as would be expected if affective disorders are monogenic.

Wakefield:
I have two points. The first is, from what I've read in the literature of genetics and brain morphology, in schizophrenia in my own work, it seems to me that the intermediate mechanism between genetics and brain morphology, seems to be, for the most part, a slow development in general, which is interesting considering what John and Robb talked about yesterday, and if it is in fact the case that slow development is what is allowing these individuals to be vulnerable to some insult, then homozygosity would, in fact, fit into this model, because one of the generic by-products of increased homozygosity is developmental delay or developmental retardation. Of course, that's not the only way you can get it, and there are other genetic mechanisms which would lead to developmental delay, and this might account for the fact that there would appear to be more than one genetic causes of schizophrenia, so that in one family set, you might find some abnormality in chromosome number 5, you wouldn't necessarily expect to find it in any other family. But the intermediate mechanism may, in fact, be the same. Let me go to the second point first, with regard to ridge counts. And that is the, if ridge counts are going to be, if they're going to depart from the number of ridges you would get as determined by the genetic propensity for it, then the disruption's going to have to occur during the second trimester which is when the cells migrate to the fingertips, and in fact the cellular level abnormalities that you see in the brains of schizophrenics occur during the second trimester. So you would expect, knowing one bit of that information, you would expect to find another.

Koehn:
To ask a question on the first point, are schizophrenics typically pre-term? Is there any evidence that there's a relationship between term and the occurrence of schizophrenia?

Wakefield:
That I don't know, and that's one thing I didn't look

at, but they have a host of other neuropsychological traits that would all fit together only if you have them developing slowly. For example, they all seem to have language acquisition that occurs quite late compared to normal children.

Koehn:
I'm trying to get a handle on the developmental rate. When in development do the rate differences occur.

Wakefield:
It would have to be quite early in embryogenesis, because, in fact, these things have to occur during the second trimester. So, it's unclear to me whether during the first trimester your development is slow, but certainly during the second, it would have to be. If it's the case that it's a generic slowing of development, it would be across the life-span, or at least up until adolescence and those changes that you see there, and the hormonal effects, might lead to the fact that you get schizophrenia onset during adolescence or sometime following that.

Gottesman:
In general, because we're dealing with a multi-factorial polygenic disorder, what you're talking about now are those things from the environment that make it multi-factorial. No one schizophrenic will have a lot of the other contributors.

Koehn:
But there are two ways you can have a disruption when you try to develop: one environmental, and one is genetic.

Wakefield:
Right, but, for example in psychological data of delayed language acquisition and the fact that the lateralization seems to be unusual, that shows up in the family members as well. And there's probably something you couldn't account for strictly by environment.

Fraser:
I think we need to move along. There's a large body of data on the relationship of twinning, HLA, handedness and congenital malformations, as manifestations of developmental instability, which we have not had time to get into.

Koehn:
Could I just ask, because I'm not familiar with HLA alleles, did you say that B8 alleles have a higher association with left-handedness?

Gangested:
The B8 alleles?

Koehn:
Yes. Was associated with left-handedness? B8 genotype mothers have fewer children?

Gangested:
Right.

Smouse:
People have been looking for HLA associations with just about everything over the years; somewhere along the line, someone had asked that fertility question. Did they get the same result?

Gangested:
I have looked for that association in the literature and I haven't been able to find it. And I there are whole volumes, and I have not found positive or negative reports in that regard. I would think that the data are out there.

Markow:
It gets complicated with HLA because there's a high incidence of recurrent spontaneous abortions among parents who share HLA types for both Class One and Class Two, and because the system is so polymorphic and there are so many alleles it's been really difficult to address those kinds of questions. But what about other HLA alleles? There are other

Class One and Class Two alleles, significantly associated with auto-immune diseases, and are you saying that those are not associated with handedness?

Gangested:
Well it turns out that handedness is not associated with auto-immune diseases. So that for instance B27 which is associated with ankylosing spondylitis. Handedness is not associated with that. Handedness is not associated with MS, which is associated with certain alleles.

Mitton:
Can you give some sort of indication how strong this association is, between B8 and handedness?

Gangested:
Yeah, it's not an extremely strong association. I think it was something like 33 percent or 27 percent of the left-handers have the B8 allele, while 16 percent of the right-handers have it.

Zouros:
What's the frequency in the population of B8 alleles?

Gangested:
Somewhere about 20 percent.

Zouros:
It's a common allele?

Gangested:
It's a fairly common allele.

Clarke:
Just to take the point a bit further, what's the level of association between handedness and the disease and the association between the disease and the actual allele? Is the relationship between the allele and handedness lower than the other associations or higher or about the same?

Gangested:
The association between the disease and the alleles is higher than what it is between the alleles and handedness.

Clarke:
And handedness and the disease?

Gangested:
Handedness and the disease, I think that the handedness and the disease is probably the case that it is lower than what it is between handedness and the allele.

Soulé:
While Dr. Goldberg is arranging her slides I would like to ask Dr. Livshits a question. Your data, in fact, suggest that, extreme individuals in some cases were more asymmetrical and in some cases they weren't. Among Israeli's in general, they weren't; you found no correlation between asymmetry and whether or not the individuals were at the ends, at the tails of distribution for morphological characteristics. But among Poles there was significant effect, that is extreme Poles are more asymmetrical than straight Jews.

Livshits:
But the characters are different.

Soulé:
Well, for one thing, the Jews are coming from all over the world and belong to many different ethnic groups, but even more importantly, you probably have recent immigrants with vastly different nutritional histories and environmental histories mixed in with residents hetero that have been there for generations. So, it's a much more heterogeneous population environmentally than the Poles, and maybe that could be part of the problem.

Livshits:
Well, I find that it could be true in general. However, our sample for Ashkenasi trends is also a homogeneous sample. Actually, the individuals included in this sample came mostly from Poland, and they settled in Israel after the second World War. Thus, I do not believe that they are genetically or environmentally more heterogeneous than the Polish sample which showed the relationship.

Goldberg:
Just in case you don't know, and I know that not everybody knows, this is what idiopathic scoliosis looks like. If you have a school scoliosis screening awareness program in your area, this is the sort of thing they would show you. Most of them look much less deformed, like this one. It is quite subtle, but you can see if you look that this shoulder is higher and this scapula is further out and the waist is asymmetrical. The big problem for a clinician is will she stay like this, or will she end up like the first girl with gross asymmetry and trunk imbalance? Obviously, she has a lot of asymmetry - if you divide her down her median saggital plane, her left and right side are not remotely mirror images. But what we don't know is whether this asymmetry is in the sense that you speak of asymmetry, or whether it is an asymmetry only in the sense that a left femur fracture is. I think it must be the former, because there is not a random right-left distribution, it runs in families in the same sort of way that handedness does (I mean that it obviously has a familial tendency but the exact pattern has proved elusive), and there are other associated by non-vertebral asymmetries. People have found differences in arm and leg length (Burwell *et al.* in Scoliosis and Growth, Ed. Zorab P.1977), breast size (Normelli *et al.*, Spine. 9: 749-752. 1986), teeth irregularities (Pecina *et al.*, Int. Orth. 15: 57-59. 1991), deep peripheral sensation i.e., vibration and joint-position sense (Cook *et al.* Clin. Orthop. 213: 118-124. 1986). I myself found that there is an association between scoliosis convexity and hand and foot preference and that girls with adolescent idiopathic scoliosis were more uniformly lateralised than normal controls. (Goldberg & Dowling, Spine. 15: 61-64. 1990. and 16: 84-87, 1991).

The underlying anomaly is the way the spine is held, not straight when viewing in AP projection, but bent sideways. In the thoracic spine it is perhaps 5 times as often convex right as left. Also, the right curves by my count are more likely than the left to be severe. On the other hand, the left curves, if they are severe enough to come to a surgeon's attention, are thought to be much more likely to have a detectable underlying pathology. In fact, the left curve here is not in an adolescent but developed when she was still a child. This brings me to another odd thing: although scoliosis may complicate many childhood conditions, especially those of the neuromuscular system, the typical patient with a spinal deformity is an adolescent girl with a right rib hump and no demonstrable disease or disability. It is best illustrated for me by one girl who never came back for her check-up because she was so busy winning awards for gymnastics that her family felt there couldn't really be anything wrong with her back, and she had progressed from mild to surgery in the interim.

The cause of this type of scoliosis, the most frequent, is still unknown. It can occur at any time during growth, but has peaks in infancy to a certain extent, less so in the US for unknown reasons, and adolescence. The infantile form has a tendency to be convex left, the childhood a random right-left distribution and in adolescence, a strong tendency to be convex right. I have looked at patterns of idiopathic scoliosis and found that there is an association between handedness and curve convexity, and between consistency of lateralization and the presence of deformity. I have arrived at the point where I am saying either idiopathic scoliosis is a genetically controlled asymmetry, or it is only asymmetrical in the sense of the left femur fracture, and if it is the former, I should find some more major asymmetries in other non-skeletal systems, and if the latter, I won't. I can't give you any results because I am still in the middle, it's too early. I came here for advice and information, really.

Koehn:
Do these individuals ever have children?

Goldberg:
Yes.

Koehn:
They do. So, you said this runs in families. Has no one made a serious attempt to study the variability, or associated with other genetic traits or some other measure of genetic variation?

Goldberg:
Yes, scoliotics have children, though their marriage rate is said to be reduced. They have psycho-social problems from being found with deformity during adolescence. There have been some family studies, the first was Ruth Wynn Davies in Edinburgh (Wynne-Davies, J. Bone & Joint Surg. 50-B: 24-30. 1968), and some more since, but no formal genetic studies are complete, as far as I know, though there may be some in progress. The general incidence is somewhere just below 1%, but is significantly higher among first and second degree relatives of known scoliotics.

Fraser:
We really should move on. Dr. Keiser is a dentist who spends a lot of his time wandering around the world looking at teeth and skulls of various kinds and has some very interesting observations about fluctuating asymmetry.

Kieser:
My interest is in the developmental instability of the unborn infant, as expressed later in life in the teeth. The adverse effects that maternal alcohol abuse has on the developing fetus is well documented. In the craniofacial region, teratogenic effects of alcohol include a constellation of defects, collectively known as the fetal alcohol syndrome (FAS). Lower levels or less frequent drinking has been associated with less noticeable effects, known as fetal alcohol effect (FAE). The thrust of our research has been directed at those children who fall beneath the thresholds of FAS or FAE (beneath A in Fig. 1). What is the evidence for developmental destabilization by ethanol abuse in the pre-threshold region?

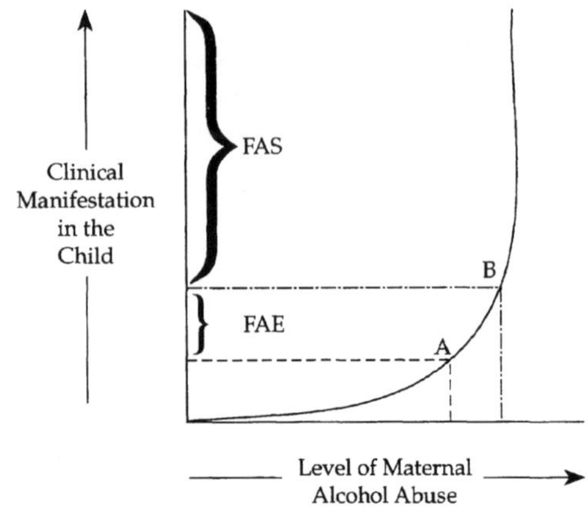

Fig. 1.

As a measure of developmental stress, I compared fluctuating dental asymmetry in 112 non-FAS children of alcoholic mothers with a control group of 120 normal children. Length (mesio-distal) and width (bucco-lingual) measurements were taken on three key teeth in the dental arch - the canine, first incisor and first molar teeth. Using student t-tests to evaluate differences in the levels of rescaled dental asymmetry, we were able to show there were significant differences in the magnitudes of antimeric asymmetry between the children of alcoholic mothers and the control group. It is well known that ethanol has a definite teratogenic effect on dento-facial development. However, ethanol-induced deficiencies are often obscured by subsequent repair and growth with the result that early embryopathy is lost by subsequent remodelling. Given that tooth germ completion of the teeth considered occurs around week 30 intra-utero, we conclude that the stress of prenatal exposure to alcohol affects the soft tissue stage of odontogenesis and that subsequent calcification results in elevated fluctuating dental asymmetry, the magniture of which bears testimony to subthreshold destabilization of development.

The exact teratogenic effect of ethanol is not known, but it appears that direct toxicity via acetaldehyde and fetal anoxia are major factors. Whether binge drinking or chronic alcohol consumption causes the greater stress is the focus of our continued research.

Having thus demonstrated the usefulness of fluctuating dental asymmetry as an early-warning system in the children of alcoholic mothers, we conducted a similar investigation into dental asymmetry in the offspring of parents who smoked. While some studies had linked prenatal exposure to tobacco smoke to such conditions as decreased birth weight, increased perinatal mortality and childhood cancers, others have not. It seemed useful to search for evidence of subthreshold effects of indirect exposure to tobacco smoke on the same basis as our FAS-study.

In this case, we contrasted the results of three experimental groups (a) those whose mothers only smoked, (b) those whose fathers only smoked, and (c) those in whom both parents smoked, with a control group in which neither parent had smoked. Our results showed very clearly that, while exposure from a single parent did not result in developmental stress, exposure to tobacco smoke from both parents increased developmental instability as measured by fluctuating dental asymmetry.

Fraser:
It is curious that you find that it works through the father and not through the mother? How can that be?

Kieser:
There's evidence for the deleterious effect of smoking on the spermatogenesis of the male. Male smokers have been shown to have significantly fewer and less motile sperm than non-smokers. Infants whose fathers smoked have also been shown to have significantly lower birth weights than those whose fathers did not smoke at all.

Koehn:
Would another possible explanation of that be through the maternal effects of the paternal smoking on the fetus? Is the mother exposed to a long-term effect of smoking which has a physiological effect on the fetus, indirectly?

Kieser:
There's a lot of noise there, I think it's connected.

Koehn:
But he didn't say how long the mother smoked.

Kieser:
In the last table, in the mother who smoked for five years, in excess of five years, still no effect.

This is a very clean data set. It was nice to get away from alcoholic mothers. Because alcoholic mothers smoke, neglect their children, beat their children, don't feed their children, abuse substances, God knows what. There was very little noise in this study. These people were non drinkers, non-alcoholics.

Livshits:
The first study showed correlation between fluctuating asymmetry and the alcohol consumption. Is that right? But also in those children who are not FAS children, right? Then the second question is, do you know anything about the fluctuating asymmetry among those children who have this FAS?

Kieser:
Oh, I see what you mean. No, I haven't done that yet. There are not many of them. They're difficult to get a hold of. They're very difficult to manage, so if you're going to take impressions of them, which is how we measure them; it's very, very difficult.

Koehn:
Can I ask one quick question? You presented data on three teeth. Will you take data for more than three teeth?

Kieser:
There are three developmental fields within the dental arch: an incisor field, a canine field and a molar field. Within each field the first member is the most stable or key tooth. We figured that asymmetry detectable as the most stable teeth would give us the most robust indication of developmental instability.

Fraser:
Let's get back to continue the discussion. I hear a lot of reference to homozygosity at the ends of the distribution and instability, and I'd like to hear

some discussion about that. Of course, only the very last section in the tail of the distribution is totally homozygous, and what we're really talking about is total numbers of plus or minus genes for that particular target. Anyway, who has some ideas about what is the mechanism that makes increased homozygosity at the tails of the distribution turn into increased asymmetry, if, indeed, it does? I suggest that one of the main reasons is the fact that inbreeding increases the variance of the total distribution and therefore brings more people close to the threshold, but that's not enough, is it?

Smouse:
I don't want to address that one particularly, but I just wanted to make a passing comment on the two psychiatric disorders. It's quite fascinating that with manic depressive illness, which is clearly monogenic, you don't see the effect in the tails of the distribution, and you wouldn't expect it for that particular disorder, because polygenic homozygosity is not really the issue here. From schizophrenia, however, which clearly is polygenic, and where you would expect a buildup of homozygosity in the two tails, relative to what would happen in the middle of the distribution, we get an increase in asymmetry. Perhaps we could get someone to comment on that from who actually did the work. That is what I heard as it went by. Is that a reasonable recap?

Gottesman:
Very reasonable.

Fraser:
But what's the mechanism?

Smouse:
For polygenic disorders, when you're out in the tail, you can expect a higher degree of homozygosity, or fraction of homozygosity than you will near the middle of the distribution, and you see the buildup of the asymmetry from the middle outward. For a monogenic disorder, and manic depressive was the one that was provided, polygenic tails are absolutely irrelevant as an expression of the disorder, and you don't see increasing asymmetry, in fact. I thought that was particularly striking. It makes perfect sense, and to return to the homozygosity issue, for some things we're going to see an effect and for others we are not.

Kieser:
Linking up with what Peter's just said about it, we heard earlier that many patients who show cleft lip show an increase in the fluctuating asymmetry of the teeth. Now we have a condition in Africa which is very prevalent in the black ethnic group around there, which is called gigantiform cementoma; it's a huge outgrowth and it's inherited as a simple dominant with incomplete penetrance. We've looked at the fluctuating asymmetry in the teeth of these people and found no elevated levels, which seems to point away from a genetic basis of fluctuating asymmetry. The sample size is very small, however.

Zouros:
If alcohol and smoking affects tooth morphology and if asymmetry is to some degree genetically determined, then we should conclude the basis of this determination is most likely polygenic. If every exogenous factor we bother to study is shown to affect asymmetry, then it's got to be polygenic, and if it is polygenic, it would follow that asymmetry will correlated with multilocus homozygosity. Is that a fair statement? Do we all agree that because fluctuating asymmetry appears to be correlated with all kinds of environmental abuses it most likely is polygenic?

Kieser:
No.

Vrijenhoek:
It has a large environmental component, is that what you're suggesting?

Soulé:
That assumes that it's a genetic trait. And I would argue that the heritability studies of fluctuating

asymmetry suggest that what we're measuring is noise and not genes. That is, you've got environmental variation, you have genetic variation, and then you have what's left over, that is, random events during development. And what these insults are doing, these exogenous insults such as alcohol and tobacco, are increasing the randomness in development, increasing the noise, but fluctuating asymmetry itself is not genetic. I would argue as an extreme null hypothesis, but not too extreme.

Fraser:
The presentation and discussion in this session addressed, argued about, and sometimes skirted around several issues. They considered some of the human data that relate to the questions of (1) whether fluctuating asymmetry (FA) simply represents developmental 'noise' or is an expression of diminished canalization; (2) whether FA of a trait is increased by increasing homozygosity and, if so, whether this is a result of a general increase in developmental stability as proposed by Lerner, or an increase in homozygosity of the (poly)genes controlling that particular trait; (3) is the corollary, that individuals at the extremes of the distribution of a trait will have increased FA, supported by evidence.

If FA is, indeed, a measure of developmental noise, as seems likely, an increase in FA could represent either a decrease in canalization (developmental homeostasis) rendering the embryo more sensitive to disruptive environmental factors, or simply an increase in these environmental factors, to which the embryo is exposed. There is some evidence to support the latter interpretation, but not to rule out the former.

There is considerable experimental evidence that increased homozygosity may (thought not always) increase developmental instability, but attempts to relate increased FA to inbreeding in human populations have been largely unsuccessful, perhaps the degree of inbreeding is not great enough to reveal what would probably be (as in experimental animals) a small increase in FA.

It is well established, however, that FA is increased in persons with certain developmental and behavioral disorders. In particular, FA is increased in the dermatoglyphics and teeth of children with cleft lip, and the dermatoglyphics of schizophrenics (but not of patients with affective disorders). This could represent either a general destabilizing effect of increased homozygosity for underlying polygenes or an effect of prenatal insults increasing liability to these multifactorial traits. An increase in FA in the near relatives of these patients would support the former interpretation, but the evidence for this is inconsistent. The question needs further investigation.

It is also well established, in experimental animals, that exposure to certain teratogens increase FA, though not enough attention has been paid to this by teratologists. There is some evidence for this effect in the sibs of children with the fetal alcohol syndrome, and in children of heavy smokers, and of diabetic mothers. This needs further exploration, as FA might be useful as a more subtle indicator of prenatal insult than is an increase of overt developmental anomalies.

Finally, the polygenic basis for quantitative, or multifactorial, threshold traits implies increased homozygosity *for the genes controlling the trait* in individuals at the tails of distribution. If increased homozygosity does indeed decrease developmental homeostasis one would expect that individuals at the tails of the distribution for a quantitative trait would have more FA than those with more central values. Apart from the intriguing finding that FA is increased not only in the strongly left-handed, but the strongly right-handed, attempts to demonstrate such an effect in humans have been equivocal. It was suggested that we may need more sensitive, accurate and rigorous methods of measuring FA. In particular, it would be helpful to measure FA in several traits in the same individual, and to recognize that FA of dermatoglyphic patterns (and ridgecounts) will vary with the frequency of pattern irrespective of any change in developmental instability.

There was no time to consider the intriguing epidemiological evidence for factors causing a general increase in developmental instability, or the molecular basis for developmental homeostasis and its failure. There is a lot of scope for further workshops.

T. A. Markow (ed.), Developmental Instability: Its Origins and Evolutionary Implications, 421-436, 1994.

Session 4

Population biology and conservation

Tuesday afternoon June 15

Philip W. Hedrick, Moderator

Soulé:

What kinds and degrees of fluctuating asymmetry are you likely to see as a function of developmental stage? Here's kind of a flow chart in which I attempt to address this issue. You can break down developmental stages into before birth and after birth, basically - before fledging and hatching and after fledging and hatching - early and late. And then there's right and left feedback. Yesterday we talked a bit about feedback during development and after development. The question is, does the right side know what the left side is doing? And, if the right side knows what the left side is doing, then there's the potential for feedback, either negative feedback or positive feedback during development or after birth. In some cases the answer is yes; in some cases the answer is no. For example, on the fingers of a lizard, I doubt that the scales developing on the right forefinger know how many scales are developing on the left forefinger. There's no way of passing that information across the fetus or the embryo. A possible example of negative feedback comes from owls. One external auditory meatus of an owl is usually much smaller than the external auditory canal of the owl, presumably to help the owl stereophonically determine distance of prey. Crab claws might be another example, but I'm not sure how early in development asymmetry begins, and whether there is any true negative feedback in the determination of size of crab claws. There can also be positive and negative feedbacks late in development. For example, if one side of the heart hypertrophies, the other side of the heart may know that the first side is undergoing enlargement and may also hypertrophy as a compensation. So, that's an example of possible positive feedback during late ontogeny. Negative feedback might be a case where one side of the body is wounded, and the other side compensates in same way.

It's important to consider these sorts of developmental interactions when thinking about which characters to use in a study of fluctuating asymmetry. So, first of all, here's a quiz. What traits best manifest fluctuating asymmetry as a response to disturbance or instability? Now there are basically two categories of disturbance; genetic disturbance or environmental disturbance. We've heard about two kinds of genetic disturbance; disturbance by major genes or major chromosome mutations, such as Down's Syndrome. The other category of genetic disturbance we've talked about is relative homozygosity. So, individuals who are relatively homozygous may be more subject to developmental perturbations than individuals who are relatively heterozygous. Now, what traits best manifest fluctuating asymmetry as a response to greater disturbance? We certainly don't want to look at traits that are effected late in development by the environment. So developmental stage is relevant. Also, canalization is relevant, though there are degrees of canalization which lead to peculiar paradoxes. Certainly we don't want to use arm number in human beings as a trait to look at fluctuating asymmetry, but in fish, as the work of Robb Leary and Fred Allendorf suggests, even though the number of fin rays is highly canalized, the heritability is almost one. The trait is canalized to a narrow range or zone, but within a zone; within that zone it's highly heritable - almost one. It is hard to say if this 'zone' of canalization is the rule in vertebrates. In any case, we want to avoid traits where there's the possibility of feedback during development or later in life.

The second question is: what traits best manifest the association of homozygosity with fluctuating asymmetry? Recall that we talked quite a bit this morning about the possibility that relatively homozygous individuals at the tails of a phenotypic distribution, both the upper tail and the lower tail, are likely to be more homozygous. If they're more

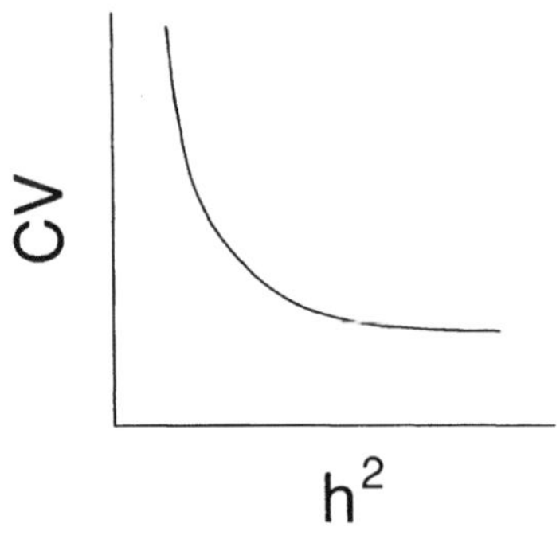

Fig. 1.

homozygous, they might be more asymmetrical. So, what traits best manifest the kind of distribution where the homozygotes are likely to be at the tails of phenotype distribution. It would be those which have a relatively high heritability. In 1982, in The American Naturalist, I published evidence for a relationship, a negative relationship, between the coefficient of variation and heritability. So, it would be desirable to pick traits that have low coefficients of variation, as long as they aren't so canalized that there's no heritability at all. And, we probably want characters that are fixed, fixed phenotypically early in development, so they're not effected by the environment later in life. Meristics often fit this description. Certainly things like scale number in reptiles and dermatoglyphics in humans are fixed very early in development and not subject to environmental perturbation.

Mitton:
I think somebody expressed the opinion and you just told us that one of the better characteristics to look for in the trait to measure fluctuating asymmetry is one with high heritability. I think somebody else mentioned that you would want traits with low heritability. Can you clarify this?

Soulé:
If you want to look for differences between homozygous and heterozygotes, you want high heritability. If you just want a trait that becomes asymmetrical under stress such as pollution, I'm not so sure. It may not make so much difference.

Livshits:
If the trait under consideration is highly heritable, I think the pollution may not have an effect, especially if the inheritance is by one major locus. I think that if a character is variable under environmental pressure, it may be preferable to take it as a marker of the environment or pollution.

Soulé:
Yes, that raises the question, if a character is highly heritable in the laboratory, is it highly heritable in the field? And that's difficult to answer. In general though, I agree.

Moller:
I've been wondering about this relationship you've shown between CV and heritability. A number of people have shown that generally, secondary sexual characters have extremely large coefficients of variation, but also relatively large heritabilities. So I think you have to say that this is for non-sexual characters. I'd like to see some data on the heritability of secondary sexual characteristics.

Soulé:
Well, that contradicts the curve in my figure, and perhaps this curve is only true for meristics of certain kinds. It's based on tooth number and morphometrics of skeletal traits, so maybe it doesn't apply, as you say, to secondary sexual characteristics. We're just beginning to figure out what's going on with regard to the relationship of heritability to the typology of characteristics. That's a good point.

Hedrick:
OK, last comment.

Leary:
Most of our meristic traits have very high heritability, but we can really modify them quite a bit through environmental alterations. I don't really know if that's of real concern.

Johnson:

I wanted to discuss our ongoing population conservation studies on chinook salmon (*Oncorhynchus tshawytscha*) in the Pacific Northwest. This study was funded by Bonneville Power Administration (BPA) and is included in our 1992 Annual Report to BPA 'A Genetic Monitoring and Evaluation Program for Supplemented Populations of Salmon and Steelhead in the Snake River Basin' (Project Number 89-096). The fluctuating asymmetry portion of the project was conducted by Orlay Johnson, Robin Waples, Amy Cook, and Kathleen Neely at the National Marine Fisheries Service's (NMFS) Northwest Fisheries Science Center (NWFSC).

The purpose of the program is to evaluate the long term effects of outplanting hatchery-reared fish on natural and wild populations. A natural population is one that includes hatchery and wild fish. The project is based on studies conducted by Betty Landrum (1966) at the NWFSC and by Robb Leary and his co-workers at Montana (see Leary *et al.* this volume).

Juvenile chinook salmon were collected in 1989 and 1990 from selected wild, natural, and hatchery populations in Idaho and Oregon within the Snake River basin. The experimental design involves monitoring genetic and meristic charactes in yearly samples. Seven bilateral meristic characters (e.g., fin ray and gill raker counts on right and left sides of an individual) were identified that show promise as indicators of asymmetry. Several indices of asymmetry developed from these paired characters are being evaluated as indicators of developmental stability, which has been shown to be correlated with levels of genetic variability in other fish species. No consistent patterns in fluctuating asymmetry values (FA) were detected in hatchery versus natural or wild populations, but population asymmetry analysis (PAP) revealed lower overall levels of FA in the Grande Ronde and Imnaha Rivers compared to the Salmon River. While not statistically significant, it was also found that hatchery stocks had consistently more negative mean asymmetry values (Ai) in both years of the study.

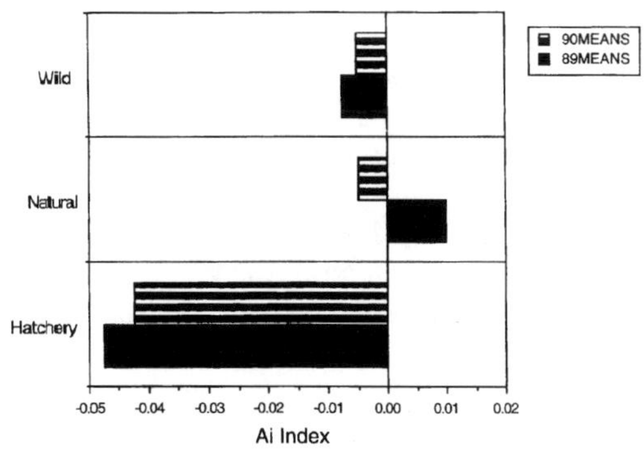

Fig. 2. Mean directional asymmetry (Ai) of chinook salmon by type of management. 'Natural' represents collections sites with mixed wild and hatchery populations.

Electrophorectic analysis of the fish revealed genetic variation at 35 gene loci, but no correlation was found between the level of asymmetry and the level of heterozygosity within populations or individual fish.

The study is continuing with analysis of chinook data from 1991 and steelhead data (*O. mykiss*) from 1989-1991.

Koehn:

In growth-rate studies that we've been involved in, which are sort of analogous to this and maybe you can relate it now, we published a significant point that people tend to miss. We looked at rank order of phenotypic effects by individual genes in a replicated design, and found that there were certain genes that had phenotypic effects and certain genes that didn't and this was a consistent result. There was really no difference in heterozygosity between those genes that had an effect and those that did not. So, the lesson was that the luck of the draw, whatever that means, in selecting allozyme loci could very much affect your answer. You had a lot of loci. (Yes, we did). But the fact is that if only a few genes have effects, say, two or three or four loci out of the total studied. If we dilute those small effects, with a bunch of genes that are not causally related to the phenotype(s) of interest, you will not see an overall effect. Have you decomposed your data, locus by locus, or tried various ways of analyzing it other than just as overall relationships with heterozygosity?

Johnson:

Now, that is something that we have considered, but no, we haven't done that. That's one of the things that I will do when I get back. Any suggestions would be certainly welcome.

McKenzie:

Well, I belong to the group who weren't anticipating talking at all. But, Phil and Teri decided the time had come to lighten the day up by presenting a village idiot, and I was obviously pristinely designed for the part, so here I am.

In terms of the areas of interest that I've found from the conference, I just want to talk briefly about blowflies and their relationship to these. First of all, John Graham's overhead seems to be the continuum between antisymmetry, fluctuating asymmetry and directional asymmetry, with the potential to interchange with those particular components. I think it's something that Vladimir has been going on about for a long period of time. If I understand him correctly, he's arguing that there is not necessarily the one Godhead, so to speak, in terms of there being a single measure that applies to all situations. I think Richard's point in terms of pattern/process is very important in that regard as well. One of the things, if you like, the potential continumm, not perhaps in developmental instability or developmental stability, but more in terms of developmental perturbation. I guess, in a sense, whatever developmental perturbation means, it is a term that removes some of the emotional baggage that runs with other terms that I can use.

The second point that I want to talk about in the context of the blowfly, is the idea that Thom Kaufman had, in really a wonderful talk yesterday. I think that in lots of ways we missed an opportunity in not seizing on the ground that Thom provided us. He provided in the course of an hour an overview of developmental genetics that was just superb. I wouldn't say it was a simplified overview, but an overview that, at least, gave us a feel of the way things can act, and I've represented his final diagram with a series of simplified axis of potential points in terms of cell-cell communication with different genetic, and perhaps environmental, effects that come into play. And finally in a sense that relates to each of those, the individual asymmetry parameter (IAP) vs. the population asymmetry parameter (PAP). Bob pointed out windows of opportunity where genes can act, which I think can come back into this line and that ongoing rubric that flows through this particular conference, the importance of micro and macro effects in evolutionary phenomena. I'm fascinated in this sense by this. There's rock group called 'Ten Years After'. I look at Michael Soulé and sort of think 'twenty years before' or whatever, because I can remember back when Michael was touring Australia in the early 70s when I was a young graduate student, and then soon after with Max Whitten. We were talking about the influence of major effects like insecticide resistance; whether or not you would expect to get changes in asymmetry under those circumstances? Well, Whitten and I were certainly lazy and it took about 7-8 years before we got somebody young, naive and enthusiastic like Geoff Clarke to actually go out and count bristles of particular resistance phenotypes to see whether or not these things happened. And in very brief summary, the sheep blowfly system, as it stands with respect to both frequency of resistance and fluctuating asymmetry is as follows.

Two chemicals, both organophosphorus chemicals, malathion and diazinon; susceptible individuals of both genotypes; the genotypes show basal levels of fluctuating asymmetry in both chemical resistance groups. In the case of diazinon-resistant individuals, when resistance first evolves, asymmetry is first anti-symmetric, and there is an elevation in the asymmetry levels. A modifier has been selected, subsequent to resistance development; that modifier enhances fitness and also decreases asymmetry basal levels, and importantly, the asymmetry we observe at that point is fluctuating asymmetry. So, with respect to John's diagram, we've gone along that particular route, fluctuating asymmetry, antisymmetry to fluctuating asymmetry. For malathion-resistant phenotypes fluctuating asymmetry of baseline levels is observed for susceptibles. The resistant individuals show an increased level of fluctuating asymmetry. In this case the asymmetry levels are partially dominant, in terms of the effect of the resistant allele, where in the case of diazinon resistance it was totally dominant. And, once again, with the same modifier, presumptively selected for reasons I won't go into now, for malathion-resistant phenotypes, the modifier returns the malathion-resistant individuals to basal levels of fluctuating asymmetry. Fitness correlation with asymmetry in both instances gives quite a good correlation between levels of asymmetry and the key component, that's been consistent throughout the studies, of developmental time. So, depending on the asymmetry levels you have, which in turn are genetically related, they influence the developmental target you observe. Well, what about this modifier allele. It, more and more convincingly, is becoming seen as an allele of the *Scalloped wings* locus. Now, interestingly, *Scalloped wings*, is homologous to *Notch* in *Drosophila*, and those of you who have any knowledge of the developmental literature will be aware that *Notch* has a wide variety of a potential pleiotropic effects; there's no shortage of ways in which, potentially *Notch* can actively influence interactions between cells, bristle formations and the like. Importantly, with respect to the resistance gene, we know the biochemistry of the resistance mechanism in each case. In the case of diazinon resistance, it's an allelic substitution at a single genetic locus, *Rop-I*; the product is a carboxyl esterase, E3. That particular esterase provides resistance to a wide range of different organophosphorus chemicals (OPs). In the case of malathion resistance, which in fact turns out to be closely linked to the *Rop-1* locus, it's a specific esterase that acts only upon malathion, it doesn't influence other Ops. So quite different modes of action, in one sense, and yet, in terms of the impact of the modifier, it's having a similar effect on both of those, which is just intriguing, characters being determined at a specific time; I'll refer to again. But, in a sense, I want to come down to the idea of the mechanisms, and then go back to those three original cartoons that were placed on the overheads. I guess I perceive what's happening from an asymmetry viewpoint is, in a sense, the resistance gene acting as a genetic switch. We're all conscious of what happens under disruptive selection with the segregation of either a supergene or perhaps an environmental switch, switching a phenotype into different phenotypic routes.

There's another possible analogy. I think in some ways, what the resistance gene is doing is acting as a gate that switches the developmental pathway one way or the other. And if that's a general working model, how can we interrelate these resistant gene products against a producent of a neurogenic locus such as *Notch*? Well, the basic requirement that we would have, clearly, is that there must be synchronous expression between the products of the *Rop-1* locus, the *Rmal* locus and the modifier locus. Because if we can't get synchronous expression, then looking at mechanisms and interaction potentially just becomes far too complex. I think in thinking about this we've had a fortuitous event through our mutagenesis studies, to try and mimic the type of resistance that evolves in the field in the lab, and ultimately, to predict the type of resistance that might occur to novel pesticides. Diazinon will work and stand as the control, in that regard. We've identified the stage that seems to be critical to bristle pattern, having produced a temperature-sensitive mutant, so that we know the stages in terms of the changes in asymmetry pattern. They seem to be at the pre-pupal/pupal interface. This is the sort of window, if you like, that Bob had referred to. So we have unlimited opportunities where we can look at these particular components. Now under those circumstances, as I said, *Notch*, is in no way short of options. It potentially could be active at each of those stages. But interestingly, there is convincingly more and more evidence that the esterases, and particularly, these associated with resistance, show significant similarities with cell adhesion molecules. If we have cell adhesion molecules, I think the possible target where this interaction is going to be will be in a signaling process from one cell to another. So I'd like to believe, as a result of thinking of asymmetry parameters as a continuum, linking with the sort of Kaufman view of the universe, which I said was an excellent summary of where developmental genetics is at, we, at least in this modifier system, have the possibility of looking at a mechanism of where this particular gene-product interaction is.

It seems to me, that yesterday we had a very intersting debate about whether we'd expect correlations within individuals compared to correlations across populations for particular characters. If we are taking this view of specific times of action of particular genes, potentially for particular characters, then to me, at least, it's not at all surprising that we don't commonly end up with correlations within individuals. Because quite subtle differences in developmental windows, may, in fact give rise to quite marked differences. However, on average, across the population, we have those events that are going to be observed between individuals. Therefore it doesn't strike me as being surprising that a PAP doesn't turn out to be correlated with an IAP or that we end up with that particular character as being a reasonable measure of asymmetry within a particular population. I'll just stop at that point.

Smouse:
One of the things that's interesting here is that when you go from the wild type to the resistant mutant, for all practical purposes instantaneously, by imposition of the insecticide, you discover that the asymmetry distribution is itself heritable. You have a modifier that has been naturally selected, so for a while there was at least some additive variance for asymmetry. I gather that no one has done a before and after heritability study, but when you impose a major perturbation, either environmentally induced or selectively forced, you move the gene pool way off center from whatever it was before. The same genes at the other loci were probably present in their initial (equilibrium) frequencies, but all of a sudden they matter because the situation has changed drastically. Once the gene pool converges on a new stable equilibrium, there is no additive variance left. I think we sometimes have a tendency to think that things are heritable or not, without thinking about the context, and this is a beautiful example of a case where it really does matter.

McKenzie:
Jules Kieser has made a very interesting comment in terms of evolution being like baseball, a game of inches. And as I said to him this morning, we've got to allow for the fact that every now and again, home runs get hit. So there are those quantal jumps, if you like, within the structure, it seems. I think the situation in diazinon resistance in *Lucilia* is very unusual in the insecticide resistance literature. Farmers kept using diazinon on the order of 20 years after resistance evolved, for the very simple reason that there was nothing else available. And it provided reasonable levels of protection. Now, under that circumstance, the evolutionary opportunity is a very rare one, if not unique. And it's very much, I think, akin to the evolution of dominance. You've set up a circumstance, where resistant homozygotes are sufficiently common, so you select for the modified, and 'correctly' adapted phenotype. The modifier is presumably there initially in a low frequency and having some sort of effect. And interestingly, we have been able to find, you know, that being susceptible and carrying the modifier might lead to an absolutely marginal disadvantage, but close enough to neutral, I would have thought. Under those circumstances, it's only when you've altered the goalpost, so to speak, that you can select for that particular gene.

Zouros:
My talk ought come after Robb Leary's talk, because it raises issues that came out of his talk. It is really a matter of debate whether heterozygosity is correalted with developmental stability, but if we were to look at another aspect of development, growth rate, we would probably agree that a correlation does exist, at least in some groups of organisms such as plants and mollusks, as some people present have shown. The correlations have been extensively discussed in the literature, so I shall not say here anything more than to remind you that until very recently, heterozygosity was measured exclusively by scoring enzyme loci. Now we can do something more; we can score heterozygosity at anonymous DNA loci. The question then is: Are we to see the same correlation when two sets of loci, one made of allozymes and another of single RFLP loci, are scored in the same set of animals? Since the DNA polymorphisms are neutral (or so we would like to think), proponents of the view that the allozyme correlations are there because allozyme heterozygosities selected for would predict that RFLP heterozygosity will not correlate with a fitness-related characters. Opponents of this view would predict that RFLP heterozygosity will correlate because allozymes and RFLPs are equally likely to act as neutral markers of associations with deleterious alleles at loci that cause the correlation.

The experiment was done by Dr. Grant Pogson in my laboratory. Pogson examined about 250 juveniles from an age cohort of the deep sea scallop *Placopecten magellanicus* for seven enzyme loci and found a reasonable correlation (significant at the 3% level) between degree of heterozygosity and shell height. He then selected colonies from a cDNA library made from scallop adductor muscle that revealed single locus RFLP polymorphisms when used as probes against digests of total DNA from individual scallops. He scored the same animals for eight such polymorphisms. When the new set of polymorphisms was combined with the old, the new degree of heterozygosity (now ranging from zero to 12), was not correlated with shell height; the correlation vanished. When the slopes of the allozyme and the RFLP regressions against shell height were compared, they were different. Proponents of the view that allozyme heterozygosity is causing the correlation should rejoice; opponents, like myself, should repent.

This result has forced me to re-address the basic distinction of hypotheses offered as explanations for the heterozygosity-fitness correlation. Robb Leary mentioned a few of them; there are many more. But all fall into one or the other of the following two categories. There are hypotheses that treat allozymes as the agents of the correlation and hypotheses that treat allozymes as markers of other genes that are the agents of the correlation. This is a sharp and important dichotomy. Both views are well represented in this workshop; I can go around the room and point to the devotees of the one or the other. I will deal exclusively with one version of the hypothesis that allozymes act as markers, referred to by many as *associative overdominance*. According to associative overdominance, the apparent overdominance at a scored locus is the result of genotypic correlations between the locus unscored selected loci. It is known from some time that these correlations are caused mainly by two factors acting alone or in concert: linkage disequilibrium in the gametic pool and non-random union of gametes. In turn, each one of these will have its own causes, but I will not go into this. Instead, I will put out the question, we all agree that these processes can provide a qualitative explanation of the heterozygosity-fitness correlation, but can they explain these correlations quantitatively? In other words, what sort of inbreeding coefficients and linkage disequilibria do we need to explain the empirical observations?

To answer this question, I modelled associative overdominance as follows. Let *B* be the wild allele at the agent by unscored locus (the locus that affects the character) and *b* be a fully recessive deleterious allele. Let A_1, A_2, ..., A_n be alleles at the marker locus, and p_1, p_2..p_n be their frequencies in the population. Then assume that in the gametic pool A_i co-occurs with *b* with the probability a_i, so that the summation of $p_1 a_i$ gives the frequency of *q* of *b*. These a_is will, of course, change from generation to generation. But I am not interested in the long-term outcome of the process. My interest is to account for the observations we obtain from a single sample representing one generation, using the information contained in that sample alone. Now assume that gametes unite to form zygotes with a probability of inbreeding *f*. Standard, population genetics allows one to write down expressions for the frequency of genotypes A_iA_iB-, A_i, A_ibb, A_iA_jB and A_iA_jbb in terms of p_i, p_j, a_i, a_j and *f*. If the phenotypic values for *B*- and *bb* are 1 and 1 - s, respectively, then one can obtain the mean phenotypic value of A_iA_i and A_iA_j as functions of p_is, a_is, *f* and *s*. Going one step further, on can pool all homozygotes at the marker locus into a 'homozygote' class and all heterozygotes into a 'heterozygote' class and come up with an expression for the expected phenotypic value fpr each of them. Subtracting the former from the latter gives the phenotypic depression, *D*, due to associative overdominance.

At this point we may return to the empirical studies. They give us estimates of *D* (say the difference of mean shell height of heterozygotes from homozygotes in the study I mentioned in the beginning of this presentation; but the phenotypic trait can, of course, be anything from developmental and morphological characters to viability and fertility), or p_i (the allele frequencies at the marker locus) and *f* (as the fixation index). But we don't know the a_is and it might be difficult to obtain reliable estimates for them. So we cannot compare the observed depression with the expected; the associative overdominance hypothesis remains intestable. Yet, there are a few results that come out of this analysis and they all agree with what the qualitative models of associative overdominance had predicted. Take, for example, one marker locus at a time. The analysis that if you compare homozygotes for a rare allele with homozygotes for a common allele, the mean depression will be higher in the former. Then take two loci at a time. The analysis says that *D* should be larger for the locus with the higher expected heterozygosity.

More importantly, one can partition *D* into two components, one due to inbreeding and one due to linkage disequilibrium, and ask what values of *q*, the frequency of the deleterious allele, one should have to explain the results if *s*, the effect of homozygosity for the deleterious gene, takes certain values. I will use as an example a study from my lab published in *Evolution* in 1988 in which shell height and heterozygosity were correlated in the common blue mussel, *Mytilus*. The mean fixation index (f) was 5%, and in six out of seven loci the difference of the mean shell height of heterozygotes exceeded that of homozygotes. When this difference (D), the inbreeding coefficient (f), and the expected heterozygosities (calculated from the observed allozyme frequencies, p_i) were plugged into the equations, and then s was allowed to vary so that the contributions of inbreeding and linkage disequilibrium to associative overdominance be both positive, it was found that the resulting frequency, q, of the deleterious gene was unrealistically high. There was no case where q was allowed to be less than 8% and these lowest permissible values of q paired with very high values of s almost close to the maximum value of one. With lower, and more realistic, values for s, the values of q increased rapidly. The model constrains the load at a locus to remain the same as s and q change, and it turned out that this load was around 1%, which might not be far from reality. This is, however, small comfort. The important thing is that the associative overdominance has its problems, problems to account for empirical observations like the ones with RFLP heterozygosities that I mentioned in the beginning, and problems theoretical, as it requires that the deleterious genes associated with the scored markers have strong effects on the character, or else that their frequencies in the population are very high.

Hedrick:
We have a case of a really honest scientist; he's bent on proving himself wrong.

Koehn:
I think it's fantastic, that the theory you developed must be rejected, because since you rejected the idea on the basis of experiment, it's even more important thant the theory you developed shouldn't fit correctly.

Can I just ask a serious question. You said the inbreeding was about 5%, but as you know, the inbreeding coefficient varies dramatically among these loci.

Zouros:
Yes, electrophoretic estimates for the inbreeding coefficient may vary dramatically from locus to locus. For me this is bothersome, but other people would say it's okay because each locus has its own history in the population and, therefore, it ought to have its own fixation index. But unless Walhund effects occur every generation and the gene frequency differences between mixing populations vary dramatically from locus to locus, these are very large Hardy-Weiberg deviations to expect for neutral markers. My feeling is that like some people accuse others of inventing phantom genes in order to explain the heterozygosity-fitness correlation, so some people may accuse others of inventing phantom inbreeding. It is possible that there is no inbreeding in these mollusk populations with highly varying fixation indices and strong heterozygosity correlations, and that huge mortality effects in the first life stages cause the excesses of homozygosity. It doesn't necessarily have to be selection on the allozymes. But, I repeat, it is a difficult problem, and we don't have at present a satisfactory explanation for it.

Hedrick:
Are there any other comments before the next group of speakers are going to talk about sexual selection; does anyone have any comments in general about mechanisms of fluctuating asymmetry or the underlying basis of what we're seeing. Peter?

Smouse:
Just one last comment on the allozyme vs. DNA marker contrast. If it turns out that we have to adopt a position that some of these allozymes really do matter for fitness or asymmetry, then we also have to adopt the position that we shouldn't be using them as random markers of genomic homozygosity, particularly if we're talking about FA. If Elefterios' results are general, and this is the first study of its kind, hopefully to be repeated for several different organisms, we may have to back off a bit on which of these loci we've been using as random neutral markers.

Soulé:
I agree, and I've been worried about allozyme estimates of heterozygosity for a long time, because they didn't fit my expectations. I have some gopher data from Jim Patton, and I've been analyzing that for a number of years, there were some real anomalies in that his populations that had too much heterozygosity for the amount of morphological variation involved and some that had too litle heterozygosity for the amount of morphological variation involved. So we've gone back to fingerprinting, DNA fingerprinting, some of those populations, and indeed we find that fingerprint results are much more highly correlated with the morphological variation, than the results are.

Hedrick:
Well, why don't we go on; we've got about half an hour until our break, so the next group of people were the ones that wanted to talk about sexual selection and mate success.

Moller:
Ok, actually I'm not going to talk about sexual selection; I'm going to talk about stabilizing selection, and how that might relate to fluctuating asymmetry. I think it was Rich Palmer who started out by saying that we should be very careful about measures of fluctuating asymmetry, because we have measurement errors, and sometimes they're quite large. But I would say that if you can go out and measure characters, and you can estimate selection on fluctuating asymmetry, and you get significant selection, it's probably not measurement error you're working on. As has been said repeatedly, fluctuating asymmetry is generally quite small, but there are quite a number of exceptions, and one of them is secondary sexual characters which generally have a very high level of fluctuating asymmetry. Another group of characters that have very large levels of fluctuating asymmetry are different characters in plants. I think it was Livshits who said yesterday that you get quite variable degrees of asymmetry between different characters; when you measure the length and the width of the same character, you get huge variation, and one possibility is that this level of asymmetry is dependent on the functional importance of the trait and perhaps its history of selection. But given that we have very small levels of asymmetry, is it possible to look at selection on these small differences? I think that it's possible to go out and measure these asymmetries and actually look at, for example survivors/nonsurvivors, and compare their degrees of asymmetry, and even controlling for size of these morphological characters. Of course, when you go out and measure asymmetry, particularly in morphological characters, and estimate selection, it's always a problem to know whether this particular character is the target for selection, but I think that can, at least for some characters in some organisms be resolved by doing experiments, and manipulating the level of asymmetry without manipulating other properties of these morphological characters. The problem I want to raise now, and a potential solution, concerns directional selection on size characters in many vertebrates. It has generally been shown that there's strong directional selection for increased size and in many of these examples, it's been shown that these characters have a large heritability. So you expect that given directional selection, there should be a response to selection, but you don't see that, and that's the problem I'll try to address, using fluctuating asymmetry as a tool. One possible explanation for why you don't get this response to selection given this directional selection situation, is that you get genotype/environment interactions, and that could work in a similar way as stabilizing selection. The suggestion is that genotypically large individuals suffer disproportionately under poor environmental conditions. The mechanism behind this could be that individuals that suffer disproportionately under poor environmental conditions develop elevated levels of fluctuating asymmetry, and that levels of fluctuating asymmetry in morphological characters effects the general performance of individuals.

Soulé:
Are you saying that fluctuating asymmetry is the cause or the result?

Moller:
I would say that it potentially could be the cause, even though these asymmetries are quite small. Of course that needs verification, but, at least, for some

of these characters I think it's possible to experimentally verify whether they're the cause. I'll just show you one example. I've been measuring feather characters in birds (barn swallows), and the pattern I've seen in offspring is that, if you divide the offspring into three different groups, individuals of intermediate phenotypes ± 1 standard deviation, and individuals that are smaller or larger, you find this U-shaped relationship between asymmetry and the size of the character. So, the conclusion is that extreme phenotypes are more asymmetric and we've seen similar examples of that earlier. If you have genotype/environment interactions going on, you would expect that these large phenotypes would particularly be effected by poor environmental conditions. And one way in which to analyze this experimentally, is to do poutral cross-fostering experiments and look at the parental phenotypes and subsequently score the offspring phenotypes with respect to the asymmetry. The offspring of genotypically large parents developed larger degrees of asymmetry when they develop under poor environmental conditions. The poor environmental conditions is simply that in these split broods, pairs of broods, I transferred three offspring and two offspring, respectively, from the other nest back to the first and vice versa; so there's one more offspring in one class of bird nests than originally planned by the parents, so the offspring get less food. This effect turns up in these genotypically large individuals, but not in the intermediate or smaller ones. One can, of course, argue if this really matters; does it matter that the wings are slightly longer from one side to the other. I think it does, particularly for such a species like this; they spend something like 70% of daylight hours flying around, and they are long-distance migrants, going from Europe to southern Africa to winter, so it's obvious that such a - character is really important for this long-distance migration, and a slight asymmetry in the wings might effect the ability to move such distances. In fact, if you look at selection on this asymmetry in this particular character, already, after fledging of these offspring, often you see a difference between those that don't survive the first weeks after fledging and those that do. If you look at survival in a longer time perspective, there are those that recruit to the population. The recruits have much less asymmetry than the other individuals that didn't turn up again. So, I think I can conclude from this experiment that this genotype/environment interaction affects fluctuating asymmetry, and perhaps importantly, also this genotype/environment interaction maintains the additive genetic variance in these characters, because these characters are highly heritable. I think that potentially, this could also be used as a method for studying whether evolutionary stasis of fluctuating characters is maintained by stabilizing selection. If that's the case, if you have stasis during long evolutionary time period, you would expect U-shaped phenotypic distributions of asymmetries when plotted against the size of these characters.

Parsons:
Have you looked at any characteristics under domestication? I think it would be quite interesting.

Moller:
Yes, well it's impossible to keep these creatures in captivity.

Parsons:
But you can take chickens or something like that.

Moller:
There's a close relationship between the degree of asymmetry in morphology and the ecology of different bird species. Species that walk around on the ground have generally higher levels of asymmetry than those that fly a lot, so I guess the chickens would have quite higher levels of asymmetry.

Miller:
Have there been explicit experiments to attempt to demonstrate that birds or even *Drosophila* or other insects with asymmetric wing length do, in fact, have difficulty flying for longer distances or flying at all? So far, this appears only correlated.

Moller:
I've done one such experiment where I manipulated the asymmetry of the tail-length. The tail is used for maneuvering, so it's quite important for birds to

have equally long tails on both sides of the body in order to be able to manoeuver. The degree of asymmetry was manipulated, and the birds had to pass through a flight maze, and we counted the number of times they hit barriers they were supposed to fly around. And indeed, if you manipulate asymmetry in this tail character you do get an effect on their maneuverability.

Soulé:
Did you control it by making it symmetrical again after they'd been made asymmetrical to see whether there was a learning element - learning how to fly with an asymmetrical tail?

Moller:
I guess there is such an effect, but I haven't done that.

Polak:
You suggested that there is often an U-shaped relationship between character size and degree of fluctuating asymmetry in characters outside the realm of sexual characters. There is also a negative relationship, often significant, between the size of the sexually selected character and degree of asymmetry. How might you explain the difference in the relationship in the pattern between asymmetry and size in these two different kinds of characters?

Moller:
It's not that simple, because you find quite a number of secondary sexual characters that don't show this nature of relationship between asymmetry and size. Some of them do show this U-shaped distribution. Your question concerns these species that show the relationship between asymmetry and size. I would say that one way to understand this is that phenotypically high quality individuals are able to develop large characters without getting it wrong. An alternative explanation is that individuals with large secondary sexual characters, they have greater problems with asymmetry. But then that's open to experimentation, and in this experiment I just talked about when we manipulate the level of asymmetry and look at maneuverability, it's actually the individuals with the large secondary sexual characters that are performing better within these treatments. So that discards that explanation.

Yuan Yue:
How did you measure the asymmetry? Did you pull out effects of size?

Moller:
I just measured the right and the left wing length.

Yuan Yue:
You didn't consider the size?

Moller:
As I showed, it is the extreme phenotypes that have the large degrees of asymmetry, so these highly asymmetric individuals with generally small or generally large characters.

Yuan Yue:
The measure of asymmetry is just the right side, minus left side?

Moller:
Yes.

Thornhill:
I'm just going to talk about sexual selection and symmetry, and then I'll turn it over to my colleague, Steve Gangested, who will talk about some work we're doing in collaboration. In interaction with Jeff Mitton several years ago, I was introduced to the literature on developmental instability and fluctuating asymmetry. And I've always been interested in sexual selection, and so it seemed that it might be possible that fluctuating asymmetry would reveal phenotypic quality, developmental quality, and thereby be important in sexual selection processes. It seemed that developmental stability correlated with various fitness components, and thus, developmental stability might be important in sexual selection. And also I reasoned that the prob-

lem that behavioral ecologists in general have is they go out and they measure traits, and they see if they correlate with fitness components. And sometimes they do, and sometimes they don't. And there's no *a priori* way to determine what to measure. And what you want to measure, it seems, is a trait for which you know the optima. And with bilaterally represented traits, you know the optima, you know the optima under historical selection; it is perfect symmetry. That's the developmental target. So if you measure deviation from that, presumably, you've got a good fitness handle, a good fitness surrogate. So, with that in mind, I started measuring symmetry in the wings, in the forewings, of an insect that I was working on in Europe at the time, in Germany; it's a German species of scorpion fly. And to my surprise, things worked like a charm. Just take the difference between the left and the right wing of the adults' wing-length, which doesn't change in these insects after they've become adults, and that correlated with the mating success of males; more symmetrical males did better, obtained more mates in competition for mates, and wing-length symmetry correlated with the ability of males and females, as adults, to win competition for a limited resource - food. In addition, I showed that a regression of mean offspring wing asymmetry on father's wing asymmetry was significant, giving a significant heritability, a high heritability, for the conditions of that test. So the next year, I was scheduled to work in Japan; I went to Japan to look at three additional species of scorpion flies. Things, again, worked like a charm. I didn't do the heritability work in Japan, but I did do just the basic studies of the role of fluctuating asymmetry in the sexual selection systems of these Japanese scorpion flies. The most surprising result from this was that, you take two males and give a female a choice between these two males, she can see neither male, but she can smell the pheromones of these two males, and females preferred, had an overwhelming preference - did the experiment twice and got powerful results - the females preferred the more symmetrical males. I have not followed up on this scorpion fly work, it is now sitting on the shelf, but basic results are published. The results I just mentioned are published in papers that have come out in the last year or so.

I've been interested in humans, sexual selection in humans, the role of body symmetry and facial symmetry in our own sexual selection system. And the study I'll mention about humans was done with a colleague in Vienna, Austria, his name is Karl Grammer, and then I'll turn the discussion over in Steve Gangested, who will describe our study of symmetry in human sexual selection in New Mexico. The study in Vienna involved facial symmetry and it's relationship to attractiveness rating of human faces. So, you have a subject come in, and you video the subject's face into the computer, and then you measure facial symmetry. You do that, it turned out, very easily; you take two points at the outside of the eyes, two points at the inside of the eyes, points at the bottom of the nose, points at the edges of the mouth, and then a line at the squareness of the jaw, and then you calculate the distance of those lines and then the midpoint of each of those lines. You subtract the midpoints and that gives you a measure of facial asymmetry for that individual; on a perfectly symmetrical face, all those midpoints will lie on a straight line. In addition we measured the averageness of the individual's facial features too, because a previous study had shown that people with more average faces, in terms of average eye size and so forth, are more attractive. So then we had our faces rated. Fifty men and fifty women rated opposite sex faces in the computer. And lo and behold, facial symmetry is correlated with the attractiveness ratings in this study. The basis of this study was the notion that in human physical attractiveness there is information about phenotypic quality, and that's why we pay so much attention to it. And the phenotypic quality may be from a standpoint of a genetic quality, or it may be from the standpoint of genetic quality and other kinds of benefits that you can get from a potential mate, like, an individual who has great developmental quality may be a better parent, and thus promote one's fitness that way. An individual who has great developmental quality may be healthier overall, which we think is a good candidate for a basic driving of sexual selection in general, that is based on a parasite kind of model of sexual selection. But an individual in better developmental health is less likely to have diseases, so you're better off pairing with such an individual, and so forth. So, from that beginning with facial attractiveness and facial symmetry, there really is apparently a pattern there; more symmetrical faces belong to more attractive people, and I should also add for those of you who are unfamiliar with the human attractiveness literature which is massive - it's been a big subject in social psychology because it matters so much to us - the attractiveness ratings are highly correlated between raters. You can get pictures of Americans and take them to Japan or Africa or vice versa and people rate faces about the same in terms of sexual attractiveness.

McKenzie:
In terms of your attractiveness index, do males just rate females and females just rate males?

Thornhill:
That's the way we did in the computer study and in the study that Steve will describe as well.

McKenzie:
Is there a control? Do you get different answers with within the sex rating?

Thornhill:
Basically, you get about the same answers regardless of who does the ratings, but that has not been done with regard to symmetry per se, for opposite sex and same sex ratings, but the literature indicates you get the same results.

Hedrick:
We have a couple more questions. I think what we'll do; we're almost at break time, so why don't we take these questions and then take a break.

Kieser:
Does it work the other way around as well? Does the male see female quality in the symmetry of effects?

Thornhill:
Yes. Female facial symmetry correlates with female facial attractiveness.

Kieser:
So why do I like Cindy Crawford's face?

Thornhill:
Her face is very symmetrical except for the beauty mark - that's a different story.

Vrijenhoek:
In the scorpion flies, do they have hovering flight in the males?

Thornhill:
They don't have a hovering flight; they're relatively weak fliers.

Vrijenhoek:
Have you actually thought of doing the experiments on the energetics of flight for symmetrical vs. asymmetry individuals and look at the energetics of asymmetry.

Thornhill:
I've thought about it, and I would like to do that kind of work in the future, to follow up on these kinds of things.

Mitton:
Are your more symmetrical flies larger, or are they more average in size?

Thornhill:
The U-shape pattern in general that I described works for wings, when you have big data sets; often you won't find the U-shaped thing because you won't have extreme phenotypes in your sample, and in general it appears that it's the U-shape pattern. So body size will correlate with wing symmetry and body size correlates with sexual success, but symmetry is a better predictor of sexual success than body size. And I think the reason is when you get real big, you fall apart developmentally, so that lowers your sexual success, ruining a good correlation in size and symmetry and success.

Polak:
I have two points to make. As a preliminary comment, though, sexual selection seems to have been somewhat unrepresented at this symposium, largely because any connection between FA and the workings of sexual selection has only recently been made in the literature. As Randy Thornhill has already mentioned, Teri Markow (1987, page 94) originally demonstrated differential male mating success as a function of FA, in *Drosophila melanogaster*. The 1990's have so far been witness to a number of other studies addressing FA and sexual selection. The first point that I'd like to make, then, is that of a potentially confusing aspect related to these studies, that the magnitude of FA in one trait often fails to predict FA in other traits of the same individual. Some advocates of female choice for symmetrical traits, particularly Anders Moller, might argue that females prefer symmetrical males because symmetry reflects phenotypic quality (and presumably genetic quality), including the ability to cope with various environmental stresses encountered during development. However, because measures of FA between traits are typically uncorrelated, low FA in one trait may not reflect general phenotypic quality, if quality is assumed to encompass the many integrated components contributing to individual fitness. Perhaps Anders might later comment on the question of whether FA in one sex character can reveal individual phenotypic quality in a broader sense. My second point deals with a topic related to my own work on a desert-endemic fruit fly and two of its parasites, macrochelid mites and allantonematid nematodes. It was mentioned yesterday that strongly canalized characters may be unsuited as indicators of developmental instability. Support for this exists from my data which shows that wing length FA in flies is unrelated to parasite burden, although bristle number FA was higher among males infected with nematodes than among control groups of flies. The insensitivity of wing length FA to infection may be a product of the extent to which wing length in more canalized (characterized by a smaller coefficient of variation) relative to bristle number. I'd like to emphasize that traits at the other extreme may be equally problematic. A high degree of variation in a character, either between individuals or between sides (eg. abdominal bristles in flies), may overwhelm the effects of variables under study which result in relatively subtle asymmetries. Characters of choice for studies of FA are therefor likely to fall somewhere in between these 2 extremes. Finally, much of the data presented today was largely positive, insofar as they fell out in predicted directions. Our understanding of how FA might fit within the framework of sexual selection is recent in origin, and still very incomplete. It would therefore be useful to see more articles on tests that have failed to support the theory, including those that have uncovered data that fall contrary to prediction.

Markow, T. A., 1987, Genetic and sensory basis of sexual selection in *Drosophila*, pp. 89-95 in Evolutionary Genetics of Invertebrate Behavior, edited by M. D. Huettel. Plenum Publishing.

Moller:
Perhaps it's no surprising if you measure some character, arbitrarily chosen, that you don't find a relationship between mating success and FA. You don't know whether this character is the target of selection. This is an important point to know before you draw any conclusion from these relationships. Your raised this point about sexually selected characters and asymmetries, namely that they aren't correlated with asymmetries in other morphological characters. One reason is perhaps that asymmetries in particular characters, such as secondary sexual characters, relate specific properties that females could use to assess males, such as resistance to parasites. Another possibility is that these asymmetries effect the behavior of males, so there could be direct fitness benefits from choosing symmetrical males in some species where there is paternal care, for example.

Summary

This afternoon's session covered a variety of topics pertaining to the integration of population biology, evolution, and fluctuating asymmetry. These presentations illustrated the diversity of approaches that people have used to examine asymmetry in both evolutionary and environmental contexts and some of the diversity of results that researchers have obtained. Personally, I am skeptical of the generality of the use of asymmetry and at times during this session meeting and I felt like I was an atheist attending a church revival. However, I think my opinion is the majority opinion of evolutionary geneticists outside of this room.

One clear observation from this session is that there exists what could be labelled a 'researcher effect' on the results obtained by different people studying asymmetry. Some researchers always seem to find asymmetry and find it correlated with some important trait while others may or may not find asymmetry and may or may not find an association with a fitness-related character. Paraphrasing the categories suggested by Bob Trivers, there are 'negative' scientists, whose efforts are to falsify some hypothesis, and there are 'positive' scientists, who seek positive results showing the truth of some new phenomenon. I would venture to say that this session is quite overrepresented by the positive scientist category. There may be many negative scientists that have actually measured asymmetry but they are, by and large, not at this meeting.

Is it that some researchers just know how to detect or measure asymmetry better than others or do they simply know how to pick the right organism or the right environmental factor? Reporting bias can be a problem because it is much more exciting to publish positive results, e.g., results that show a correlation of asymmetry and a component of fitness, than to publish negative results. In fact, it is generally much harder to publish negative results, both because a journal may not be as receptive to such results or reviewers that have themselves found positive results have a vested interest in blocking the publication of negative results. I hope that there is not some sort of unconscious or intentional bias on the behalf of the positive asymmetry researchers that leads to their findings and that there are technical explanations for this apparent researcher effect.

There are a number of comments about asymmetry that people have expressed at this session that need to be reiterated. First, there is the general observation that some traits show asymmetry while others do not and that there are some environmental factors that ellicit asymmetry while others do not. It is necessary to understand the factors determining these categories so that it is possible to predict these traits and environmental factors and then verify these predictions experimentally. If we can actually predict associations based upon some underlying phenomena, then the credibility of the positive scientists espousing asymmetry as a fundamental measure would be greatly increased. Further, why isn't there greater correlation in asymmetry over traits if asymmetry is indeed measuring some underlying genetic factors? Maybe this is related to the observation first made by Darwin that deviations from asymmetry are not inherited.

Second, it concerns me that many researchers that measure asymmetry have no burning interest in the underlying basis for asymmetry, either in a genetic sense or in developmental sense. In other words, some researchers appear to be interested only in the superficial and simple measurement of asymmetry for a number of characters and not really concerned why it might be the basis of differential survival or mate selection. For example, why are their so few studies that demonstrate a relationship between known inbreeding coefficients and the level of fluctuating asymmetry? My experience in evolutionary genetics leads me to believe that things are nearly always more complicated than one initially believes. Assuming a simple explanation may be useful in developing a hypothesis to falsify, but it is probably not the real state of affairs.

Finally, I think it behooves all true believers of asymmetry to be as objective and thorough in their research as possible and use good experimental procedures. For example, one must put considerable thought into designing proper controls for asymmetry experiments. Of course, it is assumed in such designs that the traditional statistical techniques of quantitative genetics are appropriate for measuring fluctuating asymmetry and they may not always be. Further, the measurements of various traits must be made with the best precision possible and importance of measurement error by using repeated measurements by the same or different researchers must be documented. Related to this, the extent of variation used in experiments should be in the range encountered in natural populations. In some cases, mutants or other extreme phenotypes may be useful in understanding developmental genetics but the consequences of these variants may not be at all related to selection in natural populations. Finally, good replication, both contemporaneously and in other generations or times and in other populations, should be the standard. As in studies of quantitative genetics, results occurring in one population or one environment may not occur in other situations.

As I said at the beginning, the true believer of asymmetry must realize that most scientists are skeptics and the claim that measuring asymmetry is the greatest thing since sliced bread will be met in most quarters with derision. If the use of asymmetry is to be accepted in the mainstream of evolution and population biology, then a great deal of persuasive work remains and the rabid claims made by some at this meeting must be carefully evaluated so that they will not discredit every asymmetry study or researcher. For example, the statistical approaches suggested by Rich Palmer and the developmental genetics perspective presented by Tom Kaufman should provide the type of rigorous scientific research background that fluctuating asymmetry studies need to become an important scientific player.

T. A. Markow (ed.), Developmental Instability: Its Origins and Evolutionary Implications, 437–440, 1994.

List of Conference Participants

Fred Allendorf
Division of Biological Sciences
University of Montana
Missoula, MT 59812
406-243-5503
FAX: 406-243-4184

Chris Armstrong
Department of Zoology
Arizona State University
Tempe, AZ 85287-1501
602-965-4376
FAX: 602-965-8353

Fred Biddle
Department of Pediatrics
University of Calgary
3330 Hospital Drive N.W.
Calgary, Alberta T2N 4N1
Canada
403-220-7042
FAX: 403-283-4740

Chris Breitmeyer
Department of Zoology
Arizona State University
Tempe, AZ 85287-1501
602-965-4376
FAX: 602-965-8353

Adam Chippindale
Department of Ecology & Evolutionary Biology
University of California at Irvine
Irvine, CA 92717
714-856-6006
FAX: 714-725-2181

Geoff Clarke
CSIRO Division of Entomology
P.O. Box 1700
Canberra City ACT
2601 Australia
61-6-246-4365
FAX: 61-6-246-4000

James P. Collins
Department of Zoology
Arizona State University
Tempe, AZ 85287-1501
602-965-3571
FAX: 602-965-2519

F. Clarke Fraser
McGill Centre for Human Genetics
1205 Ave. Docteur Penfield
Montreal, Quebec H3A 1B1

D. Carl Freeman
Department of Biological Sciences
Wayne State University
Detroit, MI 48202
313-577-2793
FAX: 313-577-6891

Steve Gangestad
Department of Psychology
University of New Mexico
Albuquerque, NM 87131
505-277-3060
FAX: 505-277-1394

Caroline Goldberg
Children's Research Centre
Our Lady's Hospital for Sick Children
Crumlin, Dublin 12
Ireland
353-1-558111
FAX: 353-1-551046

Irving I. Gottesman
Department of Psychology
University of Virginia
Gilmer Hall
Charlottesville, VA 22903
804-924-0661
FAX: 804-982-4766

John Graham
Department of Biology
Berry College
P.O. Box 430
Mt. Berry, GA 30149
706-232-5374 ext. 2617
FAX: 706-236-2286

Philip W. Hedrick
Department of Zoology
Arizona State University
Tempe, AZ 85287-1501
602-965-0799
FAX: 602-965-2519

H. Eugene Hoyme
Department of Pediatrics
University of Arizona
Section of Genetics/Dysmorphology
Tucson, AZ 85724
602-626-5175
FAX: 602-626-3636

Orlay Johnson
Endangered Species Project
National Marina Fisheries Service
2725 Montleke Blvd. E.
Seattle, WA 98112
206-526-6150
206-860-3253

Thomas Kaufman
Department of Biology
Howard Hughes Medical Institute
Indiana University
Jordan Hall
Bloomington, IN 47405
812-855-3033
FAX: 812-855-2577

Jules Kieser
Oral Pathology
Witwatersrand University
Private Bag 3, Wits
Johannesburg, South Africa
2050
407-8624
FAX: 403-1557

Richard Koehn
Vice President for Research
University of Utah
210 Park Bldg.
Salt Lake City, UT 84112
801-581-7236
FAX: 801-585-6212

Larry Leamy
Department of Biology
University of North Carolina at Charlotte
Charlotte, NC 28223
704-547-4040
FAX: 704-547-3128

F. Robb Leary
Division of Biological Sciences
University of Montana
Missoula, MT 59812
406-243-5503
FAX: 406-243-4184

Gregory Livshits
Department of Anatomy & Anthropology
Sackler Faculty of Medicine, Tel Aviv Univer.
Tel Aviv, Israel
03-6409494
FAX: 03-6408287

Therese A. Markow
Department of Zoology
Arizona State University
Tempe, AZ 85287-1501
602-965-4376
FAX: 602-965-8353

John McKenzie
Department of Genetics
University of Melbourne
Royal Parade
Melbourne, Victoria 3052
Australia
613-344-6239/6265 / 613-344-5139

Phil Miller
Department of Zoology
Arizona State University
Tempe, AZ 85287-1501
602-965-4556
FAX: 602-965-2519

Jeff Mitton
Department of Environmental, Population, and Organismic Biology
University of Colorado
Campus Box 334
Boulder, CO 80309
303-492-8956
FAX: 303-492-8699

Anders Moller
Department of Zoology
Uppsala University
Villavagen 9
Uppsala 75227
Sweden
46-18182640
FAX: 46-18559888

Shawn Nordell
Department of Biology
University of New Mexico
167 Castetter Hall
Albuquerque, NM 87131
505-277-9165
FAX: 505-277-0304

Richard Palmer
Department of Zoology
University of Alberta
Edmonton, Alberta T6G 2E9
403-492-3633
FAX: 403-492-9234

Peter Parsons
Waite Institute
University of Adelaide
Glen Osmond
SA 5064
Australia

Scott Pitnick
Department of Zoology
Arizona State University
Tempe, AZ 85287-1501
602-965-4376
FAX: 602-965-8353

Michal Polak
Department of Zoology
Arizona State University
Tempe, AZ 85287-1501
602-965-4376
FAX: 602-965-8353

Andrew Pomiankowski
Department of Genetics & Biometry
University College London
4 Stephenson Way
London NW1 2HE
71-380-7413
FAX: 71-387-3496

Kerri Repa
Department of Psychology
University of New Mexico
Albuquerque, NM 87131
505-277-3060
FAX: 505-277-1394

Patricia Shaw
Department of Psychology
University of New Mexico
Albuquerque, NM 87131
505-277-3060
FAX: 505-277-1394

Ruby Sheffer
Department of Zoology
Arizona State University
Tempe, AZ 85287-1501
602-965-4556
FAX: 602-965-2519

Rebecca Sherrry
Department of Biology & Plant Sciences
University of California
Riverside, CA 92521
909-787-4525
FAX: 909-787-4437

Peter Smouse
Center for Theoretical & Applied Genetics
Cook College, Rutgers University
New Brunswick, NJ 08903-0231
908-932-1120
FAX: 908-932-1126

Rhonda Snook
Department of Zoology
Arizona State University
Tempe, AZ 85287-1501
602-965-4376
FAX: 602-965-8353

Mohammad Sondossi
Department of Microbiology
Weber State University
Ogden, UT 84408-2506
801-626-7509
FAX: 801-626-7445

Michael Soulé
Department of Environmental Studies
University of California Santa Cruz
Santa Cruz, CA 95064
408-429-5473
FAX: 408-459-3518

Curtis Strobeck
Department of Zoology
University of Alberta
Edmonton, Alberta T6G 2E9
Canada
403-492-3633

Robert Thoma
Department of Psychology
University of New Mexico
Albuquerque, NM 87131
505-277-3060
FAX: 505-277-1394

Randy Thornhill
Department of Biology
University of New Mexico
Albuquerque, NM 87131
505-277-2804
FAX: 505-277-0304

Robert Trivers
Department of Biology
University of California at Santa Cruz
Santa Cruz, CA 95064
408-459-2508
FAX: 408-459-4882

Robert Vrijenhoek
CTAG Cook College
Rutgers University
P.O. Box 231
New Brunswick, NJ 08903-0231
908-932-6680 EXT: 368

Jennie Wakefield
Department of Zoology
Arizona State University
Tempe, AZ 85287-1501
602-965-4376
FAX: 602-965-2519

Charles Woolf
Department of Zoology
Arizona State University
Tempe, AZ 85287-1501
602-965-3818
FAX: 602-965-2519

Ronald Yeo
Department of Psychology
University of New Mexico
Albuquerque, NM 87131
505-277-3060
FAX: 505-277-1394

Albert Yuan Yve
Department of Zoology
Arizona State University
Tempe, AZ 85287-1501
602-965-4376
FAX: 602-965-2519

Vladimir Zakharov
N.K. Koltzov Institute for Developmental Biology
Academy of Sciences
26 Vavilov Street
Moscow 117808
Russia
FAX: 011-7-095-135-8012

Eleftherios Zouros
Department of Biology
Dalhousie University
University of Crete
Halifax, N.S. B3H 4J1, CAN
902-494-3570
FAX: 902-494-3736

Contemporary Issues in Genetics and Evolution

1. J.F. McDonald (ed.): *Transposable Elements and Evolution.* 1993 ISBN 0-7923-2338-6
2. Th.A. Markow (ed.): *Developmental Instability: Its Origins and Evolutionary Implications.* Proceedings of the International Conference (Tempe, Arizona, June 1993). 1994 ISBN 0-7923-2678-4

KLUWER ACADEMIC PUBLISHERS – DORDRECHT / BOSTON / LONDON

FSC
www.fsc.org
MIX
Papier | Fördert gute Waldnutzung
FSC® C083411

Kolibri 360 GmbH
Ferdinand-Jühlke-Straße 7
99095 Erfurt, Deutschland
produktsicherheit@kolibri360.de